The Upgrade

Also by Louann Brizendine

The Female Brain
The Male Brain

The Upgrade

*How the Female Brain
Gets Stronger and Better
in Midlife and Beyond*

Dr Louann Brizendine
with Amy Hertz

HAY HOUSE

Carlsbad, California • New York City
London • Sydney • New Delhi

Published and distributed in the United States of America by:
Harmony Books, an imprint of Random House, a division of Penguin Random House LLC, New York. HarmonyBooks.com; RandomHouseBooks.com

Published in the United Kingdom by:
Hay House UK Ltd, The Sixth Floor, Watson House,
54 Baker Street, London W1U 7BU
Tel: +44 (0)20 3927 7290; Fax: +44 (0)20 3927 7291; www.hayhouse.co.uk

Published in Australia by:
Hay House Australia Ltd, 18/36 Ralph St, Alexandria NSW 2015
Tel: (61) 2 9669 4299; Fax: (61) 2 9669 4144; www.hayhouse.com.au

Published in India by:
Hay House Publishers India, Muskaan Complex, Plot No.3, B-2,
Vasant Kunj, New Delhi 110 070
Tel: (91) 11 4176 1620; Fax: (91) 11 4176 1630; www.hayhouse.co.in

Text © Louann Brizendine, MD

Book design and illustrations by Andrea Lau
Jacket design by Anna Bauer Carr

The moral rights of the author have been asserted.

A catalogue record for this book is available from the British Library.

Tradepaper ISBN: 978-1-78817-829-7
E-book ISBN: 978-1-78817-831-0
Audiobook ISBN: 978-1-78817-830-3

To my mother, Louise Ann Stocksdale Brizendine
In memoriam

Never stop trying to solve the puzzle of your life.
Engage until the very end.

———

Contents

Author's Note

I thought I would never write another book again. But as I entered the second half of life myself and started to feel the invisibility reserved uniquely for women of a certain age, something inside me rebelled. I could feel a new power, a new clarity, a laser-like sense of purpose emerging in this phase, and I knew it was time to explore the new science that supported what I was feeling. This was not a slow decline toward the end: I was staring down the most vital, confident, and wise phase of my life. There was a name for the second half of life and we weren't yet using it. I call it the Upgrade. Formerly known as menopause, the Upgrade is the phase of life we emerge into when we exit the hormonal war zone finally able to see and be present to who we are, what we want, and how we want to live. It's a glorious time full of freedom and discovery.

As I put the finishing touches on the manuscript, the first female vice president took the oath of office. The cabinet of the forty-sixth president is full of women, most of them also in the second half of life. I began to feel a hopefulness that I thought had been buried with the defeat of the Equal Rights Amendment in the 1970s and '80s. I felt the door to possibility crack open personally and collectively for embracing a productive second half.

As a neuropsychiatrist specializing in the impact of women's hormones on the brain, i.e., emotions, thoughts, values, priorities, even perception, I think of myself as a medical and psychological scout for the tribe of women. I love going out ahead for reconnaissance, gathering useful information from neuroscience and medical studies to bring back to help others find their own way forward, to understand what is happening to them physically and how that changes everything from mood to sense of identity. This may sound silly, but it encapsulates my motivation to bring you the information about the road ahead in this book.

The Upgrade is about the path to becoming your best deep self—okay, maybe it's a bit grandiose, but if not now, when? In the second half of my life, I want to upgrade my skills and improve my willingness to take responsibility for doing so. I want to develop my enthusiasm, patience, humility, commitment, and determination to make the most of this life transition; recharge my brain and my self-compassion as I search for a new reality, new connections, a new sense of "me." I want a new relationship to myself and others and to return to a larger purpose with renewed focus. Honoring my health span, I want to attend to the care of my body and mind so that the Upgrade becomes possible.

The Upgrade delves into doubts and misinformation in order to help you find your own answers. It is about solving the conflicts you have in your mind about how emotions, hormones, biology, and the brain work together. It is about having the courage to examine all of our qualities, rejoicing in some and improving others. It is about realizing you have a choice about your path in the second half of life.

So much of what feels normal and natural to us, when you look closely, is soft-wired into us by hormones and environment. Some of it is great and helpful, but we receive so many messages about our supposed irrelevance in the second half of life that I've decided it's time to punch through. The more you know about neurochemicals, brain circuits, and the neurons throughout the body, the greater your chances of breaking out of old patterns to create a new life. And the greater the chance we'll be able to reach behind to the next generation and help them too.

I'm acutely aware that my experience and the experience of the women I have encountered in my life will not represent the reality for many women who have struggled under much harsher conditions than I can understand and continue to function while carrying the heavy burden of systemic racism. Many of the women in this book identify as BIPOC (Black, Indigenous, and people of color) and all of them as cisgender, i.e., the female identity they were born with; their Transition and Upgrade stories exemplify experiences that reach across boundaries of race. Out of respect for the individuality of experience, I chose not to write outside my own knowledge and the understanding I have gained from my patients, family, and friends. To that end, I can't adequately address at this time the Upgrade for transgender men and women. There is so much still unknown and under-researched that it might do more harm than good, but where there is something known, it is mentioned. And where there is something known about racial and cultural differences and the Upgrade, it is also mentioned. Structural racism is embedded in the healthcare system; that is a given. I promise to do whatever I can to support the emergence of the Upgrade for all women.

To be seen, heard, and valued is what we all want, and what we need to give to one another and the ones we love. I have laid out the things I know and have experienced, the obstacles and glories in the Upgrade. But as yet there are no real developmental landmarks for women in the second half, and creating them so that we all can find our way forward is a collective effort. I want to learn about your struggles and victories, too. And more than anything I want you to know how much I value your experience and wisdom.

I hope that something here might give you support, knowledge, understanding, and courage to overcome the many obstacles in the second half. We're all in this together. So let's start the conversation that no one seems to want us to have about life after the fertile phase.

Changing the Conversation Starts with Using the Right Words

I am proposing a new vocabulary for menopause and perimenopause, so that we don't have to rely on the M-word and the P-word, which literally refer to the end of fertility.

The Transition: The developmental phase of a woman's life in which the brain and body enter unfamiliar territory as the reproductive-phase circuits are finishing their job. This is the phase formerly known as perimenopause.

Just as the teen years are about so much more than hormones, so is the Transition. The hormonal transition that shuts down fertility is front and center, but it doesn't remotely tell the whole story, nor does it represent who we are. There are also many psychological growth phases within the Transition that include identity and the path toward authenticity. The Transition marks a change in our relationships and societal roles. None of these transitions are explored in any of the literature about the phases of human development.

The Upgrade: The wisdom phase that emerges after spending decades in the hormonal war zone. Emerging into the most powerful identity phase of a woman's life, this is what was formerly known as menopause or postmenopause.

"Perimenopause" and "menopause" are fossil words created by men at pharmaceutical companies. These words arose not because of an interest in helping women reach the Upgrade, which is a whole-person explosion of growth and realization of potential. These words arose as men studied how to maintain elasticity and fullness in the parts of our bodies they, as cisgender men, like to interact with, i.e., breasts and vaginas. I don't believe they encompass the full scope of the Upgrade, and so I decline to use them, other than in this note explaining my reasons. And if you catch me using them anywhere else in the book, snap a picture or screenshot of the page, find me at Dr. Louann Brizendine on Facebook, @louannbrizendine on Instagram, and @DrLouann on Twitter, and take me to task! I can't wait to hear from you all directly.

Hormone replacement therapy (HRT): It is worth noting that when we take synthetic hormones, we aren't replacing, we are adding. This is known as "hormone therapy" or HT in the US.

The Four Phases of the Transition

Pre-Transition: It happens as the number of viable eggs declines and fewer follicles are formed, impacting the amount of sex hormones produced each month. You might notice a bit more anxiety just before your period, a bit more trouble cooling off after an intense aerobic workout, or a little bit of heat or sweating at night. Your sleep might be interrupted around the time of your period. The Pre-Transition begins on average in your late thirties.

Early Transition: This is marked by more noticeable sleep disruption—you may find yourself fully drenched or wake up with the covers thrown off—at least once a week or so. In the Early Transition you might experience occasional irregular bleeding—less volume and fewer days or a bit more blood over a day or two—skipped periods, midmonth breakthrough bleeding or spotting. This happens because of an increase in a common, normal event: anovulation, or an eggless cycle.

Mid-Transition: You have two or three short cycles in a year, going from, say, twenty-eight down to twenty-seven or twenty-six or twenty-five days. This phase is marked by more frequent sleep disruption—several times a week—and feeling hot.

Late Transition: This phase is characterized by nightly sleep disruption, hot flashes, and between three and ten shortened cycles per year.

The Three Stages of the Upgrade

Early Upgrade: Twelve months after your last period marks the beginning of this phase. Some women who still have a uterus choose HRT that mimics their natural cycles; they will continue to bleed monthly. Many others experience a combination of joy over no more periods and shock in the face of a changing body.

Middle Upgrade: This phase involves attempting to make friends with the new reality, exploring new life paths, feeling the pull of needing a break, and desperately wanting to heed the siren's call to disengage.

Full Upgrade: This stage means embodying the fullness of life, embracing the ups and downs, returning to purpose, speaking truth skillfully and effectively, engaging as a mentor and sponsor, being pulled by the desire to leave humanity better than you found it.

I haven't included ages here, because the Transition stages can begin and end at different times. So much depends on the individual. I've seen some reach the Full Upgrade immediately and others remain in Early Upgrade for life. Where you are depends on your attitude and your actions.

Neurochemicals (i.e., Hormones and Neurohormones) You'll Need to Know About

Ovarian hormones:

- Estrogen: Secreted by the follicle that grows and spurts out an egg, estrogen drives growth in the uterine lining and in brain synapses. It controls brain energy and inflammation. It improves mood and bolsters mental acuity, word retrieval, and outgoing, flirty, affectionate behavior.
- Progesterone: The hormone of retaining, it keeps the uterine lin-

ing in place to receive a fertilized egg and puts the brakes on synaptic overgrowth in the brain. It's the comfy, cozy hormone that makes us want to curl up by a fire and eat cake. It's also responsible for the brain fog of pregnancy. "Progestins" is the word for various synthetic progesterones.

- Testosterone: It drives libido, muscle stimulation, and zest for life. I often experience this as the *Out of my way, mother f#$ker* feeling I get when my husband tries to stop to chat as I pass through the kitchen late on my way to the office. Ninety percent of it comes from the ovaries before the transition. After the Upgrade, 90 percent comes from the adrenal glands.

Adrenal hormones:

- Adrenaline (also referred to as epinephrine): Provides the necessary burst of energy to respond physically and mentally in a moment of danger. Causes a feeling of edginess.
- Noradrenaline (also referred to as norepinephrine): The main neurotransmitter coming from the adrenal glands for the sympathetic nerves and the cardiovascular system. It's what causes your heart to pump like crazy so that you can run away from danger.
- Cortisol: The stress hormone. Its impact on the limbic, or emotional brain is ten times higher than its impact on other brain areas. It suppresses the immune system and in excess it can lead to depression, irritability, and cognitive decline.
- Pregnenolone: A parent hormone that creates progesterone, DHEA, testosterone, estrogens, cortisol, and others. It helps us sleep well; it boosts sex drive, mood, memory, and attention; and it's good for the skin, joints, and muscles.
- DHEA (dehydroepiandrosterone): The mother hormone out of which come testosterone and estrogen. It counteracts cortisol, has antidepressant effects, improves sex drive, causes acne, and increases body odor.

Brain and nervous system chemicals
that get used all over the body:

- Follicle-stimulating hormone (FSH): The pituitary produces this chemical to get the ovary to form follicles to get the best eggs ready for launch. The follicles are what produce most of the estrogen, progesterone, and testosterone before the Upgrade.
- Luteinizing hormone (LH): The pituitary produces this chemical to get the best follicle to shoot its egg into the fallopian tube on its way to the uterus. It triggers the release of the egg.
- Oxytocin: This is produced in the hypothalamus and stored in the pituitary. Anytime estrogen kicks up, the hypothalamus will send out a little bucket full of this bonding, cuddling hormone of affection.
- GABA (gamma-aminobutyric acid): Ninety percent of all brain cells have receptors for this chemical, which is the body's natural Valium. It acts like a nice warm bubble bath on the inside.
- ALLO: Allopregnanolone is the molecule that progesterone converts into. It interacts with the calming GABAergic system that runs throughout the brain and entire nervous system.
- ACTH: Adrenocorticotropic hormone is produced and secreted by the pituitary to stimulate the adrenals to make cortisol.
- CRH: Corticotrophin-releasing hormone is produced by the hypothalamus and, when sent to the pituitary, triggers a rise in ACTH. This happens when the amygdala senses danger or threat.
- Acetylcholine: This is the chief neurotransmitter of the parasympathetic nervous system, the calming part of the nervous system, which slows the heart. It's the opposite of epinephrine. It is key to memory consolidation during sleep. Low levels are linked to learning and memory impairments.

Cells You'll Want to Know About

- Neurons: These are garden-variety brain and nerve cells, the ones you read about in high school biology. The surprise is that they are not the most numerous type of cell in the brain. They have a rival!
- Astrocytes: They are as numerous as neurons, and their job is to provide nutrients to the brain cells and keep harmful substances out.
- Microglia: These are scavenger cells that can control the connections between neurons and protect the brain from infection. During the day, the brain sprouts all kinds of connections and also discharges energy and waste from those new connections. During sleep, as the brain shrinks, the channels around the nerve cells can relax and open up, the microglia come in and prune away all the overgrowth.

Concepts You'll Want to Know About

- "Sterile" Inflammation: The chronic low-level immune inflammation unrelated to infections caused by stresses such as environmental conditions like UV radiation, mechanical trauma, cell death, hormones and leaking proteins, lack of sleep, and decreased blood flow. Sex differences in the immune system make it worse with age in males.
- Double X: The typical female has two X chromosomes so we have many brain and immune genes from the second X that may be protective against cognitive decline. This may be the secret to female longevity.

Drawings

The Female Brain

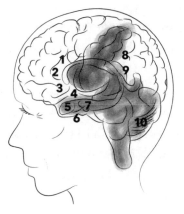

1. Prefrontal Cortex: The queen that rules the emotions and keeps them from going wild. It puts the brakes on the amygdala.

2. Nucleus Accumbens (NAc): The neural interface between desire and action, the bridge between wanting the goody and figuring out a way to get it.

3. Insula: The judge, jury, and executioner especially when it comes to self-image. "Gut feelings" occur here.

4. Hypothalamus: The conductor of the hormonal symphony; kicks the gonads into gear and struggles to manage as they go offline.

5. Amygdala: Snap decisions about life and death threats. Sometimes it perceives life and death when it's only a decision about where to park at the grocery store.

6. Pituitary: Produces hormones of fertility and screams like a junkie in detox when they are cut off.

7. Hippocampus: The elephant, larger and more effective in women, that never forgets a fight, a romantic encounter, or a tender moment, and won't let you forget it either. Though details will begin to fade in the Upgrade.

8/9. TPJ and Precuneus: When these two regions get into lockstep, you will be trapped on the hamster wheel of worry and sucked into the vortex of rumination. It's hard to achieve escape velocity to get out, but with a few tricks and tips, you can!

10. Cerebellum: The part of the brain that coordinates movement and balance for walking and standing. It helps almost every other part of the brain do its job better. It smooths your movements and your thinking and inhibits impulsive decision-making. It helps regulate emotion.

The brain's Default Mode Network (DMN): Connects areas in the front, middle, and back of the brain and turns on when the mind wanders. It rolls around like a big ball of Velcro picking up all kinds of negative thoughts. Meditation and ketamine turn this network off!

Threat/Stress Reaction: The fight-or-flight system, activates behavior to deal with threat; afterward, the calming reset comes from the vagal system.

HPA (hypothalamic pituitary adrenal) Axis, the so-called Stress Axis

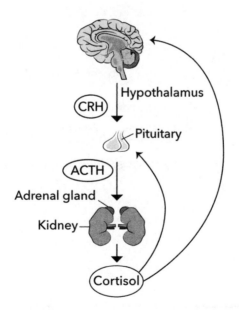

HPA controls release of cortisol (stress hormone) and androgen DHEA

HPO Axis (hypothalamic pituitary ovarian)
Reproductive Axis

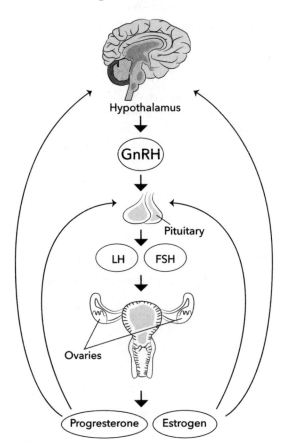

HPO controls release of estrogen, progesterone, and testosterone

Changing the Conversation

Chapter 1

In 1980, while on a research program during medical school at Yale, a mysterious illness landed me in the hospital in London, where I was doing a student clerkship. I was weak, achy, and so low in energy it felt like I was going to die. They did every test possible but never came up with a diagnosis. No lurking cancer, no rare infectious disease, no organ malfunction, no autoimmune issue. Everything came back normal. At one point I realized that some visiting residents were probing my sense of humor to evaluate whether or not I belonged in a psych ward. They concluded I didn't.

After ten days, I was released and returned to the United States for my fourth year of medical school. I was taken off the plane in New York in a wheelchair and paid for a limo back to New Haven. I appealed to the dean for some breathing room, but without a diagnosis, I was given no time off and got very little sympathy from faculty, family, or friends. My social support evaporated as people either didn't believe me or began to see me as difficult.

I did what I had to do to muscle through. I took an apartment across from the hospital in order to be able to go back to sleep after 5:30 A.M. rounds. Somehow I made it through, wrote my dissertation, and graduated.

During surgical and internal medicine rounds, I watched doctors and doctors-in-training stride across hospital rooms at 6:00 A.M., lifting up gowns to begin their examinations without even saying hello to the human beings in the bed. They didn't engage, they didn't ask questions—not even "How are you today?"—and they didn't seem to hear anyone but other doctors. And with the sickest of patients, when they ran out of tools, physicians disappeared, abandoning people who had lives and experiences and families much like their own, without so much as a farewell. They left that to the nurses.

Fresh off my own experience in London, I really knew what that felt like for patients. It's why I decided to go into psychiatry. I refused to be involved in any specialty in which I wouldn't have time to listen.

The brain, in fact the entire human nervous system of which the brain is a part, is hungry for connection. The feedback that we get from conversation, from sincere questions, tells us that we are accepted and that we belong. When we don't get this essential social vitamin, we starve the circuits of the nutrients they need. We get sicker. Our mental acuity withers. Depression creeps into and attacks our system.

Being abandoned, by doctors and by our social support, teaches us to abandon ourselves.

This is what made me start the Women's Mood and Hormone Clinic at UCSF in 1994. I wanted to give women what I didn't get. And when I worked with women as they shifted out of the reproductive phase of life and into uncertain territory, I wanted to know the answers to the questions "How are you feeling?" "What are your days and nights like?" "What scares you?" "What's making you happy now?" and, most important, "Can you tell me your story? I want to hear everything."

For thirty years I have been listening hard to the stories of joy and loss, discovery and fear, freedom and disorientation from women in Transition. And for most of that time, I have been sitting at the vanguard of research at UCSF's medical school, where I've had a front-row seat to groundbreaking, impactful science. I've witnessed the emergence of real actions we can take to come into our fullness, with an increasing number of doctors and organizations focused on exactly this. I wish they were emerging faster, but the more we know, the more we ask for, the more medicine will respond with what we need.

Who Will We Be?

We are asked as children who we want to be when we grow up, and those responses are shaped by the values of the culture that we internalize, or introject. We've lived lives we wanted to live, but also the lives we thought we were supposed to manifest. For many of us it's been marriage or marriages, children or no children, careers outside or work inside the home. So much of it swept us up in a blur; so much of it is still tugging at us to fill specific roles, especially children and grandchildren. But now is the time to ask yourself, Who do you want to be today? Who do you want to be in your transition and beyond? You have decades ahead of you following the reproductive phase of life. Who do you want to be as you enter the fullness of your age? Is it a leader, an artist, a visionary, a mentor and sponsor? Is it a life filled with freedom, purpose, and focus unencumbered by the responsibilities of an earlier time? These are some of the desires I hear from women. I am obsessed with how they answer this question, and how the Transition and freedom from the fertility cycle helps us in forming a new self-image, discovering our authentic selves, and realizing our new opportunity.

I followed the science, and what I discovered is that this phase is an opportunity to grow into the wisdom and strength and resilience we've been primed for across our life span. Just as the female brain is wired for connection and communication at birth, the Upgrade allows those circuits to choose new synaptic partners that have nothing to do with the hormonal drives that dominate decision making during the reproductive phase. Changes in circuits can build a stronger and more confident sense of self, making the wisdom gained from a lifetime of experience a neurological reality.

The Transition can last between two and fourteen years. The unfolding of the Upgrade can take place in a decade and continue over the course of forty or fifty years. It's different for every woman.

From my life, the lives of the women who've come to see me, my friends, and the research, I know that freedom from the reproductive phase is an amazing, extraordinary, and now much longer period of a woman's life than ever. I think the Upgrade can even redefine what it means to be female. Here's what's possible with the Upgrade:

Directness: The massive decrease in estrogen waves changes the way the brain handles anger and disappointment. Whereas the younger female brain diverts the impulse to stand up for herself and for the values she holds dear, biologically trying to impose silence, new hormonal influences unleash those circuits. The force of the impulse to speak out will feel like driving a Maserati for the first time—it takes a little while to get used to the power. But once you do, and you see that the world didn't end, and there is no going back, nor would we want to. We shake up the status quo, redefine the rules of relationship, speak up for what's right.

Focus: The anxiety-provoking stress of multitasking is over. Multitasking and anxiety go hand in hand. We don't know which comes first, but we do know they amplify each other. With the Upgrade, we do one thing at a time. This isn't a deficit. It means you will become more engaged, more thorough, better able to concentrate. It means by being able to keep your eye on the prize, you can be more effective than you've ever been in your life; and you will no longer hesitate to tell someone who is interrupting your flow to come back later.

Validation from within: The great benefit of the cloak of invisibility of age is that we find our authenticity within. No longer driven by fertility hormones to seek external approval, we behave differently and others respond to us differently. Many find that men will listen to us not because of how we look but because of our wisdom and experience.

The return of fearlessness: In the Upgrade, the female brain is no longer stressed by its wiring being hormonally altered by 25 percent every month, and so the freedom to solidify its circuitry allows easier access to feelings of firmness and conviction unlike at any other time in a woman's life. Less concerned than ever with pleasing others, women can use this moment to build a stronger and taller platform from which to see farther and speak out.

Expansiveness: As the window gets shut tightly on the hormone storm of the Transition, the brain circuits have the opportunity to quiet. The hormonal motor that drove the hamster wheel of worry is disconnected at this time, or at least you can now identify an off-ramp. If we learn to tune out everything that interrupts the quiet—judgment, self-flagellation, loops of rumination about situations past, present, and future—a new world of options opens up. The listening circuits for hearing your own mind as well as understanding the thoughts and feelings of others can be recruited. The brain circuits for gratefulness emerge, and knowing how to optimize them determines whether you age with vitality and curiosity or become prone to depression. Compassion for self and others activates joy circuits you never knew you had, and these positive moods create the basis for robust brain health.

Freedom: The urges, obsessions, and delusions that your fertility hormones created around relationships and intimacy have released their grip. The postfertile brain is now free to explore and expand intellectually and emotionally. There is more room in your mind for contemplation, for becoming purpose oriented. Work becomes mission driven; curiosity takes center stage in decision making; a new ease takes over, even in difficult moments. While the younger female brain may pity the older woman's expanding waistline and sagging skin, we know differently. We look back with compassion toward the tortured state of mind of our younger sisters. Sighing a deep *Thank god I don't have to deal with that anymore,* we ask how we can help.

Over the decades that I worked with women at the Mood and Hormone Clinic, I've seen the power of the optimized Upgrade. By the time we've reached this age, we've been through fire. We've likely survived tragedy and begun to thrive again. We are hot molten metal, melted, molded, and burnished by life. We are the unpredictable eruption of a volcano that also builds new landmass. The Upgraded woman glows, radiates, both the sharp sword of wisdom that takes down an unjust culture and the compassion that cherishes her responsibility in creating a future for younger generations, a future she will not live to see.

The Upgrade is bold ownership, like the teen girl who knows she's got it wired and that her mom is an idiot. (Because actually, it's true in her reality.) In the Upgrade, it's the bold authority that women in the second half of life actually do have it wired; and that someday younger women and literally all men will catch up. It's knowing this, saying this, and staring down anyone who hasn't gotten with the program. It's a function of what happens when the female brain is freed from the tortuous hormonal waves that swamp and recede in daily variation during the reproductive phase.

Yes, Virginia, It's Different

When *The Female Brain* first came out, I took a lot of heat. My shocking, newsworthy premise? That female hormones and brains and male hormones and brains are not identical. That idea struck many as politically incorrect, especially other baby-boomer women like me who came of age during the women's movement. We believed that if women were ever to truly gain equality with men in society, we had to avoid emphasizing sex differences; otherwise women would lose out. Feminism itself was built upon a deep sense of inferiority received from the signals in society about who is in power and whom we answered to. And so a mandatory unisex agenda came into existence. Frankly, at the time, I bought into it as much as anyone. But then I went on to study the biology of hormones and followed the evidence—and it led me to see the clear brain differences between men and women. I suspect that future research on the hormonal impact on the brains of transgender men and women will also bolster the case.

Doing a deep dive into the evidence is how I roll, despite the push-back that evidence might prompt. And once I started digging deeper into the reams of newly emerging information about women in the second half of life, I realized there was yet another story about the female brain that I wanted to tell: the riveting, electrifying story of how things get better, things like relationships, self-care, confidence, inner strength, agency, and effectiveness. Given the messages we've received from the youth-obsessed culture about physical peaks coming so early in life, this felt counterintuitive. Because like most of us, I'd internalized what our social structures tell us through persistent pay inequity and our near invisibility in corporate leadership, politics, and research funding—that women over fifty don't matter.

For most of human history, the only way to get ahead seemed to be to agree with those in power. It's just easier to attune to the wavelength of the times. It's the compromise generations of women before us made to keep the peace, to avoid the stress of conflict. A lifetime of being seen as the side dish, the second sex, becomes part of our nervous system. Without being conscious of it, we learn to identify ourselves relative to others, female only relative to male. Taking off those blinders and stepping into our authenticity is a challenge to ourselves: to recognize that "second sex" is not who we are. We are female, as female. And we celebrate our extraordinary strengths.

Taking the Controls

For an Upgrade to happen, you need to have agency. You need to realize how much power you hold to shape the second half of your life. This is the time for creating the map for women in the postfertile, postreproductive phase of human development. The second half of a woman's life is a revelation.

If we don't think the Upgrade is possible, we won't even be able to daydream about it, recruiting the brain's powerful imaginative and planning centers to map out our own future path. We need to be inspired early to envision what we want to do in the years after forty-five. It's harder to dream when you don't have the role models or the landmarks, but if we never begin, if we never try, nothing changes.

Our mothers and grandmothers had few role models for dreaming into what it means to be female in and of itself. Today we have more women to help guide us. And if we make the most of the opportunity, future generations will be able to go even further than we will. Change is possible. Think about how much the world has shifted for the LGBTQ+ community in our lifetimes. Women are 51 percent of the world's population. We can make the Upgrade available for them.

How do we make the world a place that holds hope for older women? As wages for jobs mostly held by women, like waitressing, childcare, cleaning, and healthcare—both in the home and professionally—decline, many will be left without savings or a retirement plan. How many will die early because of lack of access to medical care? Women in the second half of life will take the brunt of the impact of the wealth gap, especially BIPOC women. And the world remains silent as it happens, focusing instead on the loss of jobs for white males.

For the most part, what we have been expected to do in the second half of life is to continue the caretaking role: of older husbands, grandchildren, extremely aged parents and in-laws. Men don't even think about taking on the responsibility of caring for a family member as long as there is a woman around to do it. After decades of raising children, doing the majority of the housework, holding down the second shift at home even if we are the primary breadwinner, again we become unpaid caretakers; our emotional and financial resources are drained, making it impossible to work, to be productive, to soar.

But for anyone who thinks women in the Upgrade are takers, data is proving we are makers. Women over fifty start new businesses at a higher rate than any other group, offering opportunity to others, becoming the nation's economic and caretaking shoulders. To be seen, you must see yourself. And at this moment, when we have survived so many crises both personal and societal, we have the wisdom, stability, courage, and self-knowledge to trust ourselves to make it through. We need to change the conversation in the culture, but more important is the conversation inside our own heads.

———

The world has been in warrior mode for too long, ensnared in one constant zero-sum battle fueled by a testosterone-driven addiction to risk, on the trading floor and on the world stage. Warrior mode, driven by a personality trait of dominance that is specifically male, is not how society progresses. It is not how culture survives. It's time to balance humanity with a different brain style, one that holds as its core value community and preservation of life. The Upgraded female brain has the ability to balance human culture.

My purpose in writing this book is to give you validation for having good ideas, for having a great life, for having a life of your own that you build and choose. And to encourage you to endorse one another as women and endorse ourselves as a result. Amplification has to be put into practice by each and every one of us, each and every day.

In the Upgrade, neurobiology is there to support us in shedding the skin of the fertile-phase identity and in embracing and stepping purposefully into a more authentic, powerful version of ourselves. Ask yourself now, What is your idea of freedom? If money and physical vulnerability were taken out of the equation, what would you do? Who would you be? In the Upgrade, we can begin to dream as if everything were possible. I am here to help you dream.

The Crux of Being Female

Chapter 2

Biological systems love balance. They yearn for, they crave, an easy collaboration, signals communicating like clockwork, pinging gentle reminders to keep everything in working order. They fight for equilibrium. In a biological organism, this drive toward the status quo, otherwise known as health, is called homeostasis. The process turns genes on and off like a massive, multidimensional Rube Goldberg system responding with tiny, rapid countermeasures to any disturbance, from the molecular to a giant external force. Proteins and molecules are generated that help establish balance or demand an action that will right the ship—eat, sleep, snuggle, run—ensuring a smooth operating system that supports health and strength. The pendulum swings perpetually, never finding the perfect middle. This fight to keep things calm and the same is a basic principle for all living things.

So imagine a biological system in which the default drive for balance is buffeted daily by a wild chemical storm, bringing waves of neurohormones determined to evoke an action, a response, a new behavior from the system. As estrogen, progesterone, testosterone, DHEA, cortisol, and many others shift in hourly, daily, weekly, monthly waves, it isn't just menses that is affected and thrown off balance: The brain, the body's entire nervous system, is alternately provoked and soothed, each hormonal molecule demanding an action to further the survival of the system itself. The female brain is pushed by strong hormonal waves to mate, to flirt, to curl up by a fire, to eat chocolate cake, to rage, to hug, to rightly scold, to speak an unvarnished truth no one is ready to hear.

Waves are one of the most powerful and efficient physical forces in existence. Whether ripples in a pond or a twenty-foot storm surge, waves change everything they touch, battering boulders into pebbles, pulverizing shells into sand. Their push builds a coastline, depositing landmass, broken glass, shells, seaweed, ships, sometimes living creatures. Their pull is equally strong, the undertow dragging away what the tide has tossed up, gutting coastlines, digging a steep drop to the ocean floor in deep blue waters.

The Keys are a small chain of islands off the Florida peninsula extending southwest on the limestone, sandy soil, and coral reef upon which most of the state rests. On the west side of the Keys, facing the Florida Bay and the Gulf of Mexico, is calm, perfectly clear cerulean water, sandbars that extend for miles, through which residents wade barefoot all day in the warm, knee-deep sea. The smooth, sandy bottom, marked by soft wave patterns, is clearly visible. The shoreline is regular, inhabited by small creatures at home in shallow saltwater. Dotted with white coral boulders, the gentle tide doesn't wash up much seaweed, nor does it erode the shallows noticeably unless it is hit by an unusually large storm.

The landmass of the islands is narrow, in places less than two miles across. Around the middle of the chain, a ten-minute walk east delivers you to the Atlantic side. There, the ocean is rough, the water murky. Though it is made of the same sand, limestone, warm saltwater, and coral rock, it looks nothing like the western shore. Seaweed carpets the rocky beach like the world's thickest shag. Sea urchins and stingrays hover near the water's edge, hidden in the underwater growth; the shallows are littered with rocks that slice through the most calloused feet. Eroded by the constant pull of a strong undertow, the shore drops off quickly to depths more accommodating to larger, predatory sea life. This is the shore upon which you will find sea glass, its jagged edges rounded, surface softened and turned opaque by the abrasive tides and sand.

Though the elements are the same, the shoreline carved by waves has undergone undeniable structural change. A rock on the calmer gulf side has a structure different from one battered by the relentless push and pull of the open Atlantic Ocean. Dump a truckload of sand on the western shore and it will establish a small beach. Do the same on the Atlantic side and it will be sucked out to sea within a few days.

Waves of Hormones

Waves impact human beings in the form of cycles of behavior-driving hormones. Cortisol, a hormone released by the adrenal glands, flows in a wave that during its push phase wakes us up, keeps our memory sharp, helps us feel alert and eager to learn. At its height in the morning, we feel faster on our feet and excited for the day. During the pull phase, when cortisol bottoms out at 3:00 or 4:00 P.M., we yearn for a nap, chocolate, a caffeinated soda, coffee, anything to recover the day's earlier freshness. If we ride the pull of the cortisol wave out, we relax, slow down, become tired, and if all goes well, another wave of hormones helps us sleep soundly through the night, giving the brain a chance to be cleared of toxins and overgrowth accrued from the stimuli of the day. If anything interrupts this wave—staying up too late, crossing time zones, having a big fight with a loved one, or a deeply traumatic experience—the brain's neurocircuitry is rewired as the cortisol wave becomes irregular, setting off other waves that also behave unpredictably. We might spend half the night awake in agitated, worried rumination. In the case of trauma, if the adrenal hormones get stuck in the *on* position, the brain's circuitry changes so dramatically that it becomes impossible to learn, to relax, or to perceive circumstances and people as anything but threatening.

All terrain, including the neurocircuitry that drives thoughts, emotion, and behavior, is altered by waves. The diurnal cortisol wave is common to both female and male. As long as we stay on a fairly regular schedule of waking, exercising, eating, and sleeping, and we are not facing biological disruption, the waves are less dramatic, and the terrain is less dramatically impacted.

Over the course of a lifetime, the male brain and nervous system will experience the waves common to all human beings, daily ups and downs of chemicals that drive waking, sleeping, circulation, breathing, hunger, thirst, thinking, learning, sex, and emotion. In the male brain at puberty, the pituitary pings the testicles to produce testosterone, and the sex drive ignites in earnest. For boys this feels emotionally dramatic as the urgency of physical intimacy becomes a dominant drive, but the testosterone wave itself is anything but. Testosterone rises and falls daily in a duet with the cortisol wave, making morning erections the strongest. But the male brain's lifelong melody is a single and uncomplicated wave, rising to an early-adult crescendo and falling at a gentle slope into old age. The testosterone wave runs predictably atop the reliable counterpoint of daily hormonal waves controlling other basic drives.

But something different is brewing in the female brain. Between six and thirteen years old, a girl's hypothalamus pings the pituitary to activate several hormonal waves. One is the adrenal wave, which wakes up the little hormone control centers, the adrenal glands, that sit on top of the kidneys, asking for a wave of the androgen DHEA. DHEA, a little bundle of energy, excitement, and zest for life, is often called the mother hormone, because it converts into both estrogen and testosterone. Testosterone, in females as well as males, drives interest in sex.

The tween girl's behavior becomes a bit testy. "Don't touch my hair." "Stop looking at me." "Don't talk to me." "Leave me alone." Other effects of the adrenal wave are the arousal of her nervous system through cortisol, adrenaline, and DHEA, which in the right amounts optimize her ability to both learn new things and remember them and kicks up the vigilance essential to detecting danger and acting successfully in the face of it. It's adrenaline and a bit of cortisol that will put her on alert about going out late at night alone, or heighten her awareness when at some point in her life she walks unaccompanied through a large, deserted parking garage.

This same process of the hypothalamus pinging the pituitary sets off another hormonal interaction, a unique and complex dance of waves crashing and receding on the shores of the brain and the ovaries. Two chemicals from the pituitary, luteinizing hormone (LH) and follicle-stimulating hormone (FSH), will send the ovaries the signal to make the estrogen needed to prepare eggs to be ripened for fertilization and to line the uterus with a soft cushion for a fertilized egg to attach to. For most women, there will be an eightfold increase in estrogen during their cycle. Estrogen pulls behind it a little wagon of oxytocin, the bonding hormone, a strong neurochemical pairing for females. The more estrogen, the more oxytocin. Though males also have oxytocin, the levels are much lower. On the push side of the wave, estrogen and oxytocin help drive flirtatious and warm, cuddly behavior in females.

These waves of LH, FSH, estrogen, oxytocin, and testosterone temporarily reorganize brain networks and stimulate qualities like memory, language, and affection, peaking around ovulation. A few days after ovulation, estrogen crashes, leaving mostly progesterone and a testosterone cattle prod of edginess. The pituitary withdraws the wave of LH and FSH as the ovaries produce the undertow of progesterone that after four days drags away the extroverted, outgoing bubbliness, pulling with it, like strands of seaweed lost in the tide, the brain circuits for interest in social connection and the crisp ability to communicate. While progesterone sends signals to the uterine lining to keep building, it also drives the urge to turn inward, lowers mental acuity, and increases pain sensitivity. For most women, it's an eightfold increase in progesterone at this time. Progesterone, in the form of its metabolite, allopregnanolone (ALLO), dials up the brain's GABAergic activity. GABA is the brain's natural Valium; it is the major calming system in the brain. The system-wide GABAergic activity is what puts the brakes on the growth, connectivity, and stimulation pushed by estrogen. If we didn't have those brakes, we'd be having seizures all the time, or at least be edgy and anxious. The entire nervous system has GABAergic receptors that are influenced by the metabolite of progesterone, ALLO. It sends a calming, antianxiety signal, and is what creates that comfy, cozy time of the month when you mostly want to stay home, watch romantic movies, and eat cake or ice cream. ALLO is what makes you feel incredibly sedated during the first half of pregnancy. But all good things must come to an end, and you know what's next: PMS, bloating, cramps, and blood that will come with the dramatic drop in progesterone, which pulls ALLO down with it. In the last two days of the cycle, when progesterone bottoms out, preparing the uterine lining to shed, our edgy down mood colors everything. We cry during commercials showing adults hugging babies and puppies. An upset over a personal issue can signal the brain to make it feel like the beginning of a global catastrophe.

Negative and Positive

The human brain is notoriously skewed by negativity bias, making it easy to laser-focus in on one little bad thing that happened, while shutting out all the good. When we lived in the wild, this particular cognitive bias helped burn into our memory the location of the watering hole where we saw a lion eating an antelope, so that we would warn others away. Now negativity bias is what makes us hungry for scary alerts or updated news of a disaster, and it explains why good news dissipates quickly. Some neuroscientists describe this as the brain having Velcro for negativity and Teflon for positivity.

Amplified by the steep drop-off of progesterone—around an eightfold decrease—every month, the female brain circuits can get dragged out to a depressive sea. If progesterone is chemically consumed via birth control pill use or given in too high a dose in hormone replacement therapy (HRT), it can dramatically increase depression in a susceptible woman and double her risk for suicide.

As waves of neurochemicals crash and subside in a woman's body, they flip on and off genetic switches and receptors that alter connections in the female brain up to 25 percent every twenty-six to thirty-five or so days. The waves drive radical reality differences week to week—from feeling friendly, smart, and positive to having a mental filter that interprets comments and interactions with others as negative and critical. The pull and tug of these waves has a high-frequency amplitude; over the course of a woman's life, this dramatic churn remakes the female brain's terrain somewhere around five hundred times.

Pregnancy, childbirth, and child-rearing bring a strong, steady wind, inalterably high tides, climate change, and the formation of new landmass, all of which make the female brain think, act, and live for the two, three, four, ten, or however many children she has. Presaged by fogginess during pregnancy, the "self" hard drive in the brain is wiped for recoding to put offspring ahead of all. The lack of mental clarity is a side effect of the major brain resets that happen during puberty, pregnancy, postpartum, Transition, and Upgrade. The wrong words come out, we can't find the keys or remember to calendar the meeting we just arranged—all evidence of the frontal cortex shutting down in order to allow the limbic emotional brain to take over the rewiring that undeniably and irrevocably alters the female biological landscape.

For every action there is an equal and opposite reaction. As the pendulum of the homeostatic process swings hard in the opposite direction of a change, the action will overshoot its mark in its attempt to land in a precise and static equipoise that—in an ever-shifting environment—can never come. In the female brain, the course corrections come rapid fire, every day for a duration of around thirty-five years.

The Transition into the Upgrade can make an already wild fight for homeostasis even more dramatic.

Sometime in a woman's late thirties to early forties, production of adrenal DHEA (which, remember, converts into both estrogen and testosterone) and ovarian estrogen begin to slightly decrease. The hypothalamus and pituitary gland, estrogen junkies since puberty, will send a strong spurt of LH and FSH, demanding relief from withdrawal symptoms. The ovaries comply by spiking a tsunami of estrogen, knocking the brain and body even further off balance. Estrogen has a profound impact on cognition, memory, mental acuity, and mood, and a spike in estrogen can bring about scattered thinking and a crazy amount of testiness. The estrogen spike is followed by a dramatic undertow of progesterone. The outgoing affection caused by estrogen and oxytocin gets interrupted by irritability, sadness, scattered thinking, and pessimism, especially when progesterone drops harder after a more precipitous rise. The woman in transition becomes testy. The behavior provoked by erratic waves of hormones can range from "Don't touch me" to "Get out of my life."

Cycles become irregular when lack of ovulation leads to no estrogen or progesterone surge; it's ovulation that signals for the continuing buildup of the uterine lining. Usually progesterone enters as estrogen declines, keeping the lining plump and in place. If there is no pregnancy, progesterone abruptly declines, signaling the lining to shed. Without progesterone, ALLO doesn't show up to calm the brain and nervous system. And without progesterone's decline, the uterine lining won't shed until it becomes too heavy. At that point, bleeding can become profuse.

Not every woman will experience the shift in the same way, but for the 30 percent who go through a dramatic version of the Transition, the wild push-pull of this neurochemical storm lasts between two and fourteen years, the pituitary's demand for hormones driving extreme swings of mood and behavior. The pulsing GnRH (gonadotropin-releasing hormone) cells in the hypothalamus, which started the whole process in puberty, spike their signals until they get stuck on high, like Maria Callas straining the top of her range, screeching out an endless note to an empty and darkened concert hall.

The biology of the body is tied up in survival. Anytime a system begins shutting down, it creates panic in the organism. That panic, transmitted via the nervous system, becomes an unconscious driver of thought, emotion, and behavior. And does it have an outlet? Can it find relief? The prefrontal cortex, the quintessential human brain area that generates insight, can calm the panicked amygdala and tell us through analysis that this is not life-threatening. We can learn to understand the difference between biological panic and a genuine threat; that the version of us that's projected by neurohormones is not the true us, even though it doesn't feel that way when the waves are crashing in the storm.

There is no male corollary to this process. Though the basic brain material of female and male is mostly the same, waves create measurable structural difference. Just like the two shores of the Florida Keys or the Caribbean and Pacific coasts of Costa Rica, environment, structure, and experience are radically altered by the force and frequency of waves.

Resilience

Waves are the crux of being female, burnishing the essential female power of resilience. Resilience is the ability to quickly return to a zone of well-being when the system, or the person, has been knocked out of phase. Resilience is the single strongest factor that contributes not only to the survival of a system but to the possibility that it will reach its full, maximized potential. Resilience can be built and enhanced only by the buffeting of destabilizing forces, like the waves experienced by the female brain.

The ability to surf those waves in the midst of neurochemical storms is reflected in female psychology as the core of our courage, our strength, our will, our agency, our power. The waves cause just enough suffering to help us connect with our best qualities. The dance of youth, of hormones that trapped us in the arena of nonstop acquisition—marriage, children, career, money—ends when the ovaries go deaf to the brain's demanding signal for estrogen, testosterone, and progesterone.

The crashing of those dramatic waves is wiping the hard drive to prepare it for the Upgrade, a state that can quiet the brain's hypervigilance around survival, kids, and relationships and remove the filters of worry over how we appear to others. When you pass through the door to the Upgrade, your ability to see your most positive and authentic self is no longer clouded. You've walked through fire to come into your true power. You've emerged from the storm into wisdom. What you thought were feminine powers, the minor female powers of sexuality, you realize were just decoration.

Community

As the minds and nervous systems of war veterans are changed and as they are bonded by their experience, so it goes with females. Communities of veterans come together with a shared sense of one another. Women in the Transition and the Upgrade know instinctively that someone who wasn't on the front lines of this particular hormonal battle will never completely understand us.

We are not, in the Upgrade, an old version of ourselves returned. The girl we left behind remains in the past. We have been changed forever by decades lived in a biological war zone, pushed edge to edge every week, chemicals flipping genes connected to fighting disease, stress, sex drive, and mental acuity on and off at a breakneck pace. That war zone has made us who we are, visceral beings, using the complete self to sense and understand people and reality.

In the Upgrade we recognize that our visceral sense makes us stronger. It adds to our intuition, the absorption of information that happens so quickly there is little time to unpack what's coming in. It pushes to the forefront our more available mirror neurons, which allow us to sense another's nervous system, to vibrate with understanding of others. It's not that men don't have similar visceral systems; it's that they don't have the hormonal waves that have sensitized and cracked them open for fully optimized availability.

Unless a Y chromosome comes along, every single embryo will be female. A uterus, ovaries, and clitoris will develop in the absence of a steady production of testosterone. Female is the primordial gender. Waves are nature's primordial force. Male is the female rib, the second sex, unburnished by waves, functioning in a narrower zone of well-being. Female power dances outside that zone, knowing that shaking up normalcy will end in transformation, not destruction.

Out of the wildness that goes with our biology comes an innate ferociousness. Characteristically female, it pushes us to call things out when they go awry, to scream for change when the larger system is in danger, to demand safer conditions when our sisters are burned alive in a clothing factory, to withstand ridicule as we fight to save the planet from disastrous, man-made, *man*-made climate change. That is who we are; this is a great thing, not something to be covered over or stifled or ignored. Our biological extremes are what can make a difference emotionally and socially. We are the change agent. We become capacious in being able to hold so much energy and potential.

Think what happens when the push and pull of all that change settles, how much more resilient and prepared we are for daily life changes, what an opportunity it is for creativity to come to the fore as brain circuits that were engaged nearly full time in managing the waves are now free to be deployed as we see fit, creating a new reality for what could be the best time of our lives. Biology is destiny, unless you know what it's doing to you. The question is, into which destiny would you like to emerge?

Transitioning
into the Upgrade

Chapter 3

My patient Carole is an in-house lawyer with a regional bank in Dallas. At forty-seven, she's been married to Ron for twenty-two years. They have one daughter, and Ron is clearly devoted to both of the women in his life. But lately there seems to have been a lot of stress in their relationship. Carole feels criticized all the time. It's resulting in lots of fights; their old ability to communicate and solve problems seems to have evaporated. Ron feels bewildered as Carole rebuffs his affection one minute and apologizes for being distant the next. "I just want to be able to make her happy again," he tells me when we meet, "but I don't seem to be able to do that right now."

For the first time, things have also been rough at work. Carole has had a meteoric and charmed career, and for years everything she touched turned to gold. Lately, it feels like her hot streak has gone cold. Nothing seems to go right and her boss has become relentlessly impatient. She's been misreading situations and sending flame-o-gram emails for which she finds herself apologizing a day later. She's never been the most laid-back person, but her temper is now hair-trigger, and she's struggling to keep a lid on it.

And then the unthinkable happens: Her boss tells her in their weekly meeting that her contract won't be renewed. She has never lost a job in her entire life, and she is faced with forty-eight hours to pack up a twenty-five-year-career. Where she had always been high energy, ebullient, and optimistic, now Ron holds her every night while she sobs herself to sleep. It's like the bottom dropped out of some innate toughness she had always relied upon in difficult circumstances. It seems like it happened all of a sudden, but, ageism of corporations aside, the hormonal roller coaster caused by the process of wiping her brain's hard drive and getting it ready to be rewired for the Upgrade has been pushing her to this precipice for a few years now. She's had fallings-out with a number of old friends and colleagues over perceived slights. By the time she came to see me, she had partially blown up her life.

Although Carole's situation is complicated and may seem extreme, there were strong biological reasons for what was happening. The physical and identity transition that takes place at this stage in a woman's brain is profound. I want to acknowledge and validate that for you. In this moment when you may feel less like yourself than ever, when your old tricks for calming down, sleeping it off, or losing weight stop working, it's not just stress, and it's not made up. It's real, it's physical, and it's in your brain. You can't just Zen out and think, *Let it go,* as some of Carole's friends suggested, and hope that the symptoms will magically disappear. You can't command your body like you used to, with healthy eating, exercise, and habits, to just figure it out and expect to return to your old normal. While good habits help us turn the Transition into an Upgrade, they won't prevent the symptoms of the Transition for many of us.

What the body is trying to figure out is not that simple. After decades of predictable cycles, suddenly everything just feels off. Cycles may be shorter or longer. You may be bleeding less or more. This isn't just about volume and the calendar: They are hormonal events that can shift dramatically how we feel about ourselves, our lives, our relationships, our careers. And what you are going through or have gone through already may mean you have to get used to being a new and different person, growing into a new and different body, having a new and different sense of who you are. Carole and I were able to find a way forward that worked for her. It's not the same solution for everyone. In this case, hormone replacement therapy (HRT) solved a lot. "It wasn't that I became my old self," she told me when I asked her about how she was feeling. "It's that it helped me emerge into who I was becoming without feeling like I was alternately being dragged away from the shore and tossed back onto the beach. A new job is next!"

For women, there are estrogen and progesterone receptors on every single organ, not just the brain. Not surprisingly, as production of the hormones declines, the impact can feel profound. For many, this time is a fundamental transformation of our physical and emotional identity, almost like a reincarnation taking place within a single lifetime. By going through my own Transition, and by holding the space for so many women as they have gone through it, I have come to respect the process deeply.

As you read this chapter, at some point you are going to think, *Dr. Louann, you've sold me a bill of goods. This book should be called* The Downgrade. And I hear you. Like anything in life, good comes with bad. There are some rough spots during the transition physically, emotionally, and in terms of finding help. None of them are your fault; and the most important message is to learn to become your own advocate. I want to send you in to your doctor armed with the right questions and the best information, so that you can recruit her to your team. Make sure to fill that team with the most supportive and knowledgeable friends and family. If you haven't Transitioned, talk to a friend who has already Upgraded about coming to appointments with you. She might ask questions you may not think of. It's an uncertain time of life; it's so reassuring to have a village in your corner to help you figure out how to make this stage the most productive and rewarding.

Ladies! Start Your Spreadsheets!

It wasn't that long ago, in the nineteenth century, when all it took was a single act of rage, a missed menstrual cycle, or an extended period of grief over the loss of a family member to have a woman sent away. Though we can express ourselves now without being locked in an asylum, the echo of repression is still apparent in how alone and isolated most of us feel during one of the most extraordinary transformations any human being will endure. If you've already gone through the Transition, I hope this map will give you some clarity about the territory through which you've traveled. If you haven't, while it is different for everyone, here are some landmarks you can look out for. If you're still having periods and you're not using your calendar, a spreadsheet, or an app to track your cycles, now is the time to start.

During the Transition, just as in puberty, hormones can cause wild emotional swings. That process alone can cause feelings of isolation, as we wonder if what we are going through is normal, if anyone else has ever had our experience, if we are going crazy or not. If you went through the Transition early, you didn't have your peers to talk to. If you went through it late, your friends who had already completed it might not have wanted to talk about it either. "I asked my friend Ceci," Carole said, "who is about twenty years older than I am, about it. All she had to say was that she remembered being very concerned about her health at that time. But that was it. No details other than she threw some jewelry into the Hudson River when she got mad at her ex-husband." Carole laughed at the memory.

Let's start with a look backward, so that you can orient yourself. Unless you're on hormonal birth control or the pill during the years before the Transition, the pituitary sends out two hormones, luteinizing hormone (LH) and follicle-stimulating hormone (FSH), like clockwork to regulate your cycles. These signals, which are of course blocked by the pill, request that two jobs get done. One prompts the maturation of an estrogen-producing egg follicle, and the other gets the follicle to launch the egg through the fallopian tube to the uterus (ovulation). The estrogen from the follicle nudges the uterus to build a cushy lining in case that egg is fertilized and needs to implant. Whenever estrogen goes up, up with it goes oxytocin, the hormone that triggers an affectionate, cuddly, trusting, connecting state of mind and behavior. After ovulation, that same ovarian follicle turns into an entirely different endocrine organ and produces comfy, cozy, stay-at-home-and-take-care-of-yourself progesterone to keep the thickening uterine lining in place. If the egg isn't fertilized, no implantation happens and progesterone plummets. It is dramatic; we can get pretty cranky as ALLO drops like a rock, and the cushy lining bleeds off.

We are born with about a million eggs, our lifetime supply. By the time we are teenagers, the ovaries' supply of eggs is about half a million. Every month, nine or ten follicles race to maturity. The best, most viable gets picked for ovulation. By our late thirties, the supply is anywhere between ten thousand and fifteen thousand and exponentially declining. Over many decades, as with all aging cells, the DNA of the eggs degrades and we have fewer viable eggs. Genetic mutations in an egg often mean it won't mature or, if it does, those mutations become fatal to an embryo; in less than ten weeks or so, the body's system, sensing that the embryo isn't viable, finds a way to end such a pregnancy via miscarriage. As we run out of mutation-free eggs, none mature. No maturing eggs means no follicular estrogen, which means much less oxytocin and no progesterone. Without those potent neurochemicals signaling the brain the way it's used to being signaled, we already feel a bit destabilized.

You Are Here ↓↓

Looking backward, we can see that there are four stages to the Transition. The first, the Pre-Transition, begins on average at around thirty-seven. It happens as the number of viable eggs declines and fewer follicles are formed, impacting the amount of sex hormones produced each month. You might notice a bit more anxiety just before your period, a bit more trouble cooling off after an intense aerobic workout, or a little bit of heat or sweating at night. Your sleep might be interrupted by this around the time of your period.

The Early Transition, around forty to forty-five, is marked by more noticeable sleep disruption—you may find yourself fully drenched or wake up with the covers thrown off—at least once a week or so. In the Early Transition you might experience occasional irregular bleeding: less volume and fewer days, a bit more blood over a day or two more, skipped bleeding, or midmonth bleeding. This happens because of an increase in a common, normal event: anovulation, or an eggless cycle.

During the fertility phase, we might have one such anovulatory cycle per year, in which there is no viable egg and so the ovary doesn't form an ovulatory follicle. Remember that the follicle becomes its own mini endocrine organ, stimulating the production and withdrawal of hormones that can shift our brain's cognition and mood throughout the month. As we get closer to the Upgrade, we have more such cycles. The sign of anovulation is that your cycle is shorter than usual, maybe by as little as one to three days. If you have two or three short cycles in a year, going from, say, twenty-eight down to twenty-seven or twenty-six or twenty-five days, this can be a sign of the Mid-Transition, and it can happen years before you ever miss a period.

We have all been taught to look for hot flashes and mood swings as the first sign of the Transition, but trust me, it's easy to deny they are happening. It's easy to blame your car's climate control, the air conditioner or bedding in your home, stress, or a tough emotional situation for the first few months or even years of the Transition. But this quantitative marker—the shortening of cycles—is a more reliable sign than testing hormone levels and less subjective than sweating or irritability, though by the Mid- and Late Transition, you may be experiencing sleep interruption on a nightly basis. At the same time, if you're not paying attention, the sign is easy to miss, like crossing a state line at seventy-five miles per hour and whizzing right past a sun-bleached marker. If you haven't tracked your period, now is the time, that is, if you aren't on the pill. Your calendar will tell you everything you need to know about what's coming. If you're on the pill, the only way to find out what's happening with your cycles is to stop taking it.

If you're experiencing between three and ten shortened cycles per year, then welcome to the Late Transition, the gateway to the Upgrade. The average age at the end of this stage is fifty-one, but it's not the same for everyone. The better indicator is a measurable biological event. And for many women, the neurohormonal impact of this time can be profound.

When there's no ripened egg, the very precise hormonal choreography of pituitary, ovaries, and uterus, of LH, estrogen, FSH, and progesterone, becomes a bit loose—like going from the lockstep of the *Swan Lake* corps de ballet to modern improvisation. When there is no maturing egg follicle, the brain doesn't get the estrogen it's used to. So in its demand for more, a strong FSH spike from the brain's pituitary forces a big follicular estrogen spike and an ovulation. Without ovulation, however, there is no progesterone surge, thus no sudden drop that gets the uterine lining to bleed off. Instead, the lining keeps thickening. When it becomes too heavy to hold in place, the law of gravity takes over; the sheer weight and volume of tissue and blood can cause it to shed even without the progesterone trigger.

The amount of blood can be frightening, and a lot of women will think they are miscarrying or dying. Clots the size of golf balls may plop into the toilet. Your bed might look like a crime scene. One friend said it sounded like she was peeing when she took out her tampon over the toilet. If the brain isn't sensing enough estrogen, it can also blast out an FSH spike midcycle, spurring the growth of an entirely new set of follicles before the body has time to deal with the first set. Like trains getting backed up in a station, these LOOP (luteal out-of-phase) or stacked cycles can lead to prolonged bleeding as the uterine lining gets out-of-whack signals and keeps building. The time between cycles continues to shorten, until some women feel like their cycles have flipped—they might bleed for twenty-one days and stop for five. This is not uncommon in the Mid- and Late Transition.

"The best advice I got was to carry a much larger purse," Carole said of her bouts with heavy bleeding. "I had to pack whole boxes of tampons and pads. I needed clean undies and a plastic bag for any that got soaked. A friend told me to give away my white pants, because by the time I would stop bleeding long enough to wear them, they would either be too small or out of style.

"The worst," Carole continued, "was being at a dinner party and feeling a leak about to spring. First, there's the fear of ruining your friend's upholstery. Then the worry of what will happen when you finally stand up. Worse still is that clear white plastic powder room garbage can with no lining. They became my mortal enemy! I started carrying a roll of small opaque garbage bags along with my other supplies."

As many as 25 percent of women will have what we call flooding. Yes, that is the term for it. And no, this isn't due to stress, though travel across time zones and hemispheres might intensify it. At any age, when we do something to throw off the body's circadian rhythm—the wake/sleep/eat hormone cycles—it can destabilize all other hormone signals, including those for ovulation and menstruation. "I used to practice the rhythm method, since I was so regular and was really good at predicting my own fertility," said Liz, fifty-four. "But during my medical internship in my late twenties, I was on call and up all night every fourth night. I couldn't figure out my cycles to save my life during that time, and I had a couple of oopsies with my boyfriend."

"Frequency of overseas travel was one question my doctor asked me when I came in concerned about heavy bleeding," said Patricia, forty-nine, a new patient of mine. Her doctor had been in practice for more than forty years and had seen everything. He noticed that the women he treated in their forties and fifties who had busy travel schedules often complained of heavy bleeding. "Over the course of three weeks around the holidays, I went from winter in New York to summer solstice in Brazil, then winter solstice in Norway. By the third week of January, I was incapacitated." An actress with a touring company, Patricia continued, "I started to pack an extra suitcase and plastic bags for damp clothes that I had washed but didn't have a chance to dry. I never go anywhere now without hydrogen peroxide and Q-tips. It's the best for cleaning blood out of clothes, chairs, couches, sheets, whatever."

I want to reassure you that while much of this is normal, everyone approaches coping with symptoms differently. Some women feel fine riding out the waves until they subside, confident in the body to take its time, even if that happens over the course of years. Others want to beat the body into submission. But the vast majority can find relief in a combination of approaches: medication that may include HRT or an antidepressant, or perhaps an outpatient procedure to reduce the heft of the uterine lining.

One caution: Heavy bleeding can sometimes be a sign of uterine cancer. Since 91 percent of those with uterine cancer have had heavy bleeding, it's a good precaution to get a biopsy of your uterine tissue if your doctor recommends it; but chances are, heavy bleeding just means you're in the early stages of the Transition. Also, if you are bleeding a lot, get your hemoglobin and hematocrit levels checked for anemia. It's possible you will need an iron supplement to get you through this time. If you've suffered from anemia because of heavy bleeding, have your doctor perform an RDW test, which measures red blood cell width, even after you've completed the Transition. That will tell you if your bone marrow, which produces your red blood cells, has caught up in production. Some women need to stay on an iron supplement after the Transition until the Upgrade is complete.

"The hardest part for me," Patricia continued, "was that my old strategies for dealing with cycles started having the opposite effect. Exercise had always helped with cramps and with moodiness. I used to do back-to-back Zumba and cardio-burn classes at the gym on Saturday mornings before heading off to see friends. But once I hit the Transition, those classes would knock me out for two or three hours. And oh, baby, I couldn't believe the cramps."

Exercise can sometimes make Transition cramps worse, especially if you have fibroids. The uterus, which is made of smooth muscle—the type of muscle over which we have no conscious control—also has special little spiral arteries that feed blood to it. If those muscles and arteries didn't contract during menstruation, we would bleed to death. That process of spasm is what makes us feel the pain of cramps and is intensified during exercise. If you think you knew what bad cramps felt like, honey, those of the Transition can be of a whole new order.

"I remember doing a session with a trainer one morning," said Patricia, "and later that night I was curled in a ball on the couch in tears from the pain. Ibuprofen helped a little, but not much. After a lifetime of exercise and activity, my doctors were telling me to sleep, rest, stay quiet, and relax. That was really hard advice to take, but ultimately it worked."

The chemicals of the prostaglandin system can make things contract or relax throughout the body. The prostaglandin system is what causes those smooth muscles and spiral arteries to contract and cut off the blood supply to the uterine lining. Progesterone withdrawal, just prior to monthly bleeding, triggers hyperreactivity of the prostaglandin system, intensifying the spasm of pain. Cramps are obviously not evolutionarily adaptive—anything that puts us flat on our backs is a bad idea living in the wild. But Mother Nature thought it's better to have cramps than to bleed to death. During the Transition, when hormones start to go off balance, pain sensitivity is also heightened.

The only thing that helps is to stay ahead of the pain. By the time you have full-blown cramps, it's too late for ibuprofen or any other medicines that impact the prostaglandin system to have a big impact. This means tuning in to those first twinges and popping the right NSAID (ibuprofen but not aspirin since it causes more bleeding) in order to inhibit the muscle spasm that causes cramping. "That was one of the single most helpful things you told me," said Patricia during a session. "If I was bleeding, I learned to take ibuprofen before exercising, and it worked to keep the cramps in check. It meant I could be a bit more active again."

Et Tu, Estrogen?

Estrogen has for most of your life been your best friend, your driver of language facility, of social connection, sex, and affection. Estrogen is the great preserver of memory in the brain, the great protector of mood and of cognitive functioning in women. Some doctors call it nature's Prozac. But during the Transition, too much of a good thing can create some hairy situations both physically and emotionally.

In addition to building up the uterine lining, estrogen makes fibroids grow like crazy. Most women develop small ones in their thirties, but during the Transition they can become grapefruit-sized or larger, and women of African descent have an increased risk of developing them. During my transition at fifty-three, I had giant fibroids pushing on my bladder that made me pee as often as during the third trimester of pregnancy. Forget about sex. On most days there was a "Don't come near me" sign hanging from my pelvis. Between the cramps, heavier bleeding, and discomfort from fibroids pressing against other organs, the shop was closed.

That's the physical part. Now for the emotions.

Estrogen stimulates growth of brain synapses, intensifies their connections, and reorganizes brain networks. That's why during the early part of normal cycles, we are talkative, we are more outgoing, and our brains feel like they are on high-octane fuel. Growth is great as long as there is a careful gardener, stimulated by progesterone, eventually coming along to cut the weeds, clip the hedges, trim the trees, and clean up the trash. But if the body gets into progesterone deficit, as it does for many women during this time, there is no gardener to do her job in the brain or the uterus. Overgrowth in both areas can result. In the brain, sleep is the best opportunity for neural pruning and taking out the trash; but sleep can be a challenge during the Transition, and without enough of it, focusing and trying to find emotional stability can become a major struggle.

In the Late Transition, the sputtering hormones can trigger mood drops and sadness as the brain tries to adjust to the changes in its neurochemical reality. This is a moment when you may need extra support for dealing with fear or sadness, especially if your Transition is longer than about five years or you've had a major life upheaval.

As cycles continue to shorten, erratic dips in estrogen and progesterone from the ovaries cause neurocircuit withdrawal, setting off alarm bells in the brain. Many doctors recommend testing FSH levels to determine whether or not you've entered the Transition. But in my experience, this test is unreliable until you are already done with the Transition. During this phase, the level can be wildly erratic. It can change by the hour, depending upon whether or not there is an estrogen spike from a follicle that's been forced to squeeze more out. Higher estrogen lowers FSH. If you happen to get tested during an estrogen spike, your FSH will look normal or low, and your doctor will tell you that you are not in the Transition, even though you might be.

Think about what a test result like this might do to your sense of reality if you have been living with wild mood swings, erratic cycles, warm flushes, and just not feeling like yourself anymore. You walk into your doctor's office, hoping for some answers. You get your FSH level tested and it comes back normal. Think about the cognitive dissonance of having to figure out which source is more reliable: your own experience or the data from the lab. If you have an app, calendar, or spreadsheet tracking your cycles, you will have data that is more reliable than your doctor's lab test. The frequency of short cycles will tell a clearer story than the test.

Both Carole and Patricia tracked their cycles after we met, so they were clear about their Transition status. "No matter what my PCP told me about age or getting my hormone levels tested to confirm, I knew I was in the Transition," Carole said in my office. "Having that knowledge helped me remain open to options for getting help to address what was going on in my life." They both opted for HRT, which helped tremendously with one of the biggest issues of the Transition: hot flashes!

Temperature Change and Biological Stress

Besides being the control center that regulates menstrual cycles, the hypothalamus is the brain organ that regulates systems as changes in the environment are detected. That includes heat and cold. The usual range of room temperature tolerance is plus or minus five to ten degrees Fahrenheit, beyond which the biological stress system sends an alert, pushing us, through shivering or sweating, to put on a sweater or to take it off. Temperature variation is such a reliable source of biological stress that for decades researchers have used it in their studies to measure the stress response. They may plunge a subject's hand into near-freezing water. They may shoot burning heat through a wire attached to the sensitive skin on the inside of the wrist. Either way, the sudden shift outside the normal range triggers a biological stress reaction to threat.

Nobody understands what the mechanism is yet, but during the Transition and sometimes well into the Upgrade the range of ambient temperature variation the hypothalamus can tolerate shrinks dramatically. This is called thermoneutral zone narrowing. There is some question as to whether this is due to a reduction in estrogen and progesterone making the hypothalamus hypersensitive to temperature. But here's what this means on a practical level. You are comfortable in your chair by the window in the early morning with your coffee or tea, content with the regularity of a ritual that has begun your day for decades. As the sun comes up, light pours into your room. You smile, anticipating the comfort of the warmth. Instead, irritation arises, followed by a surge of annoyance and a flush of heat from deep inside the core of your upper body that spreads outward over your neck and face, reaching your fingers and toes. If you took your temperature with a thermometer, it wouldn't register as a fever, but you might turn red or sweat or both. Bottom line: While everyone else feels perfectly comfortable, you'll want to swear like a sailor.

More than 80 percent of women experience hot flashes. They can last for a minute, five minutes, even half an hour. You might feel heat, sweating, flushing, chills, irritation, rapid heartbeat, and anxiety. Transgender men on hormone blockers to halt female puberty also experience hot flashes—the hormone blockers mimic menopause. This phase can last four to five years.

"The blow-dryer and humidity from a shower became my sworn enemies," says Carole. "If I had to do hair and makeup, I stood naked and barefoot on a cold floor. Sometimes I had to put cold packs under my feet. Even if the humidity in the bathroom cleared, the sheen of sweat on my face kept me from being able to apply makeup." I know what she means. I gave up everything but a little lipstick and blush. Patricia won't wear a scarf or socks in the winter anymore. "If I can't get my feet cool and my neck free, I will sweat the minute I come in off the street to a heated building."

While on the surface of things, we are talking about garden-variety hot flashes, what science hasn't paid attention to is the biological stress at their foundation. Think about what it means if temperature variation, one of the biggest natural stressors to the body, becomes intolerable to the hypothalamus. Through no fault of her own, constant, normal one- or two-degree room temperature fluctuations can mean a woman in the Transition and even in the Upgrade is living in a perpetual state of biological stress.

Sleep is the body's chance to reset and recover. But heat intolerance means that hot flashes may interrupt sleep during the Transition. Women tend to be too stoic about this. However if you are waking up twice or more during the night, you won't enter REM, and you won't get the rest you need.

Lack of sleep brings serious consequences to mood, ability to concentrate, metabolism, and the health of the heart. For more than 50 percent of women, this kind of distress can last between four and seven years or more after their final period. Though not troubled much by them, Ceci experienced occasional hot flashes into her late seventies. For me, as happens for many women on some form of HRT, they went away when I started wearing the estrogen patch.

Estrogen stimulates the brain's production of the chemicals serotonin, dopamine, and endorphins—the brain's major contributors to well-being and feelings of joy. When those chemicals are lowered, they also bring down norepinephrine, which in turn reduces the ability of the brain's hypothalamus to tolerate heat. It was thought for many decades that estrogen therapy alone was needed to address the lowering of norepinephrine. New research is showing that progesterone might also mitigate the brain's estrogen withdrawal and help reset the temperature tolerance range of the hypothalamus. You just might be able to enjoy that cup of coffee in your sun-filled kitchen again.

Alarm Bells!

There is a little organ in the brain called the insula whose job it is to ping the body's various systems with one question: Are you okay? Before the Transition, on most days, unless you are stressed or sick, the response comes back *yes*. If there is even a small variation in the body's systems, and the response to the are-you-okay ping is *I'm not sure*, then the biological stress system kicks in. Adrenals turn on and cortisol and adrenaline heighten our ability to search for answers, if we are conscious of the stressor. Some of the time, biological stress turns into unconscious anxiety. The biological stress system of brain and adrenals is triggered by the insula's not getting the answer it expected, leaving us with a destabilizing brew of cortisol, adrenaline, and wildly fluctuating signals for follicular estrogen and progesterone. Carole's feelings of instability and her lack of her former resilience was an expression of these neurochemicals.

If you have any worries in your life at this time, like relationship, health of family members, financial concerns, or worry over a child, it's like pouring a geyser of gasoline on an already-burning fire. The biological stress response is in a constant state of alert. It's no wonder that if your husband, your teenager, or a close friend looks at you wrong, it will register as a five-alarm fire. Normal levels of cortisol can heighten our alertness to learning and can bring a little excitement to the day. Even before the Transition, too much cortisol can cause trouble with focus and cognition. The constant flow during the Transition sets the stage for brain fog and confusion. One patient of mine described her wild swings as having scrambled brains. She reported feeling uneasy, moody, and hyperalert yet at the same time having difficulty thinking clearly. "I had all this energy," she said, "but I couldn't harness it. I couldn't stay focused. It was like being a racehorse at the starting gate, ready to run yet unclear about which direction to take off in."

The neurohormones from the ovaries and adrenals are just about the most powerful influencers on the mind the female body has. Waves of wild emotion, out-of-proportion anger, irritability, and intense sadness can be reactions to a drop; overexcitement and agitation can be pushed by a surge of estrogen or progesterone in response to the pituitary's demand for more. In a *New Statesman* article, columnist Suzanne Moore wrote of her own transition, "I don't really have the mood swings that some talk about. I have just the one mood. Rage."

The choreography between the brain and ovarian hormones—estrogen, progesterone, testosterone—and the brain and adrenal hormones—cortisol, adrenaline, and DHEA, which converts into estrogen and testosterone—are treated by doctors as two separate systems. But they are happening in the same body and clearly impact each other. If you are having cortisol and adrenaline surges because of stress—and anybody alive experiences stress—on top of estrogen and progesterone glitches during the Transition, this can be experienced as a powerful emotional meltdown and/or an intensification of hot flashes. Combine an estrogen drop with life stress great or small, and you have confusion, destabilized mood, sleep problems, and memory problems.

It doesn't have to be this way. There is a lot of help out there, and we need to speak up about it more often and more openly. I have helped Carole and Patricia and countless others to stabilize and weather the storm.

Three Stages of the Upgrade

So far we've been focused on the storm of the four stages of the Transition, because identifying the phase you're in helps you better understand what kind of help you'll need and when you'll need it. But when you've closed the window on the tumult, either naturally or surgically, here's what to look for.

Early Upgrade: Twelve months after your last period marks the beginning of this phase. Some women who still have a uterus and choose HRT that mimics their natural cycles will continue to bleed monthly. For many others, it's a combination of joy over no more periods and shock in the face of a changing body. Resistance to the new reality marks this phase for many.

Middle Upgrade: This phase involves attempting to make friends with the new reality, exploring new life paths, feeling the pull of needing a break, and desperately wanting to heed the siren's call to disengage.

Full Upgrade: This stage means embodying the fullness of life, embracing the ups and downs, returning to purpose, speaking truth skillfully and effectively, and engaging as a mentor, pulled by the desire to find ways to leave humanity better than you found it.

I haven't included ages here, because I've seen some reach the Full Upgrade immediately and others remain in Early Upgrade for life. It depends on attitude and action.

———

Almost 90 percent of women turn to a healthcare provider for help during the Transition and the Upgrade. They're a big deal to go through. I want you to have the resources to address symptoms, so that you can think, work, sleep, function, and maintain healthy relationships. Armed with information about medicine, lifestyle, the benefits of staying active, healthy eating, and finally getting consistent sleep, we can teach our doctors how to help. Now that you have a better sense of whether you are in the Pre-, Early, Mid-, or Late Transition or in the Upgrade, you'll learn in the next chapter which actions are best taken to optimize the phase you're in. Knowledge is power; the ability to act upon it means the best is yet to come!

Navigating the Wilderness

Chapter 4

"I'm humming like a well-oiled machine these days," Patricia, fifty-five, told me during a session. I was overjoyed to hear this, as her journey over the previous five years had been a bit of an odyssey. Storms, whirlwinds, rocky shores, numbed-out sadness, drenched sheets, and sleepless nights. It was the kitchen sink of Transition experiences. I knew eventually she'd find her steadiness; the way through is different for everyone, and I wanted to know how she got there. "I'm thrilled, Patricia. Tell me what it's like now."

She sat up and smiled. "For the first time in my life, I can count on my mood. My energy is pretty steady, I'm sleeping better than ever, and I love not having to track cycles or mop up the damage after a hormonal explosion."

Patricia was steadier and more grounded than I'd ever seen her. When she first called me five years earlier, it was a different story. She was tearful, sweating, exhausted, bleeding heavily, and she was profoundly anemic. I was quite worried about her. She had just moved to a new city and was struggling to find a doctor who didn't have a four-month waiting list for an appointment.

"I called seventeen doctors," she reminded me. "I had huge uterine fibroids. I was weak and had heavy bleeding for several years. I had such bad anemia that I could hear the blood pumping in my ears. I couldn't drive myself anywhere, and a doctor friend of mine, when he saw my labs, was stunned that my doctor hadn't been alarmed by how low my red blood cells were. She told me she wanted to continue to treat me with repeated ablations to control the bleeding." Having this procedure meant Patricia would have to go into the doctor's office, be sedated, and have the lining in her uterus burned. I remember her telling me that she would likely have to have this done several times a year. She didn't want to have to keep doing that. She didn't want to keep feeling so weak. She was tired of having to change clothes a few times a day, and tired of ruining sheets and mattresses from the heavy bleeding. She didn't want to have to deal with any more hormonal spikes and dips of the Transition. She'd didn't know how many more years her Transition would last, and she wanted it over with. She sought a complete hysterectomy.

Whether it happens surgically or naturally, the shutting down of the ovaries—making the uterus obsolete—has a direct impact on the brain. It doesn't mean the organs have to be removed. Everyone will feel differently about keeping or losing a body part, and there is no right or wrong decision. It's very, very personal, and surgery for some can feel like a violation of one's wholeness. But Patricia had specific quality of life and medical reasons besides the rough Transition she was going through. "My mother had a history of ovarian cancer at fifty-four," she told me, "and I didn't want the anxiety of quarterly ultrasounds to check my ovaries anymore. I had been doing it, as required by my doctor, since my early thirties. At the same time, I couldn't wrap my head around repeated ablations and embolizations to stop fibroids from causing so much bleeding." She was nearly in tears as she repeated the story, the trauma still fresh.

For women, first transitioning out of childhood and then years later the transition of the Upgrade makes our essential sense of belonging fragile as we wonder if we are alone in what we are feeling and experiencing. Nobody posts the misery of their story on Facebook or Instagram, the golf-ball-sized blood clots that fall into the toilet, or our fears of destroying the upholstery of a friend's couch when Pre-Transition-sized tampons and pads fail. Those who were early or late with their Transition, just as those whose periods started earlier or later than their group, feel left out, with nobody to talk to, no one to validate their experience. Exposing that you are feeling depressed, perpetually exhausted, and in the grip of wild mood swings can feel like you'd be putting relationships, friendships, or career at risk. But the opposite is true. The best thing we can do is talk to others and find the helpers who will hear you. We are here.

Patricia came out of her isolation and asked for help. Eventually her doctor agreed to a hysterectomy and then connected her to a colleague who practiced functional medicine. It took a couple of years, but they got her hormonal cocktail right. She was now on a combination of estrogen, DHEA, and progesterone. She applied a topical estrogen and DHEA in the morning to help with energy, and the progesterone a couple of times a week in the evening to calm and help her sleep. I was witnessing the results in the big warm smile emerging on her face.

To be clear, the Transition and the Upgrade take place almost exclusively in the brain, but what happens to the ovaries and uterus has a direct impact on neurocircuitry. The ovaries make or trigger the production of hormones that push and pull our reality and our mood. In fact, everything we first learned in the 1980s about lack of ovarian estrogen causing mood drops, memory problems, and brain fog came from women who had total hysterectomies without hormone replacement. They were the easiest to study because the complication of hormonal spikes from the ovaries was off the table. To optimize the Upgrade, some women will want or need surgery. But the cultural bias against it can be rough.

My transition was marked by heavy bleeding and some wild mood swings. In 2005 I was finishing my first book, *The Female Brain*, and training a new generation of doctors—who were and still are much more interested in the hormones around pregnancy and the postpartum period—at the Women's Mood and Hormone Clinic at UCSF School of Medicine. I had undergone several D & Cs (scraping of the uterine lining) and embolizations to deal with heavy bleeding from my large fibroids. I was taking the birth control pill to steady my Late Transition fluctuating hormones. I was tired and testy and at fifty-three ready to enter the Upgrade. Knowing what I know about this phase, I felt strongly that a hysterectomy with ovary removal and a post-op estrogen patch, without progesterone, would be the best way forward for me.

I was able to get the surgery, finally, after struggling to find a surgeon who did vaginal hysterectomies along with ovariectomy. Before the surgery I filled out the paperwork indicating I wanted everything removed, including the ovaries so that I wouldn't have to monitor for ovarian cancer. I knew I didn't need a cervix for structural support and keeping it would mean yearly pap smears. So, after being asked repeatedly about my decision, I confirmed, "Yes, take it all out." The other thing I wanted was an estrogen patch to be put on in the recovery room. This is now standard procedure, but it wasn't back then. I knew from the literature and from many of my patients that if my body were surgically deprived of estrogen from my ovaries, that I would hit a wall of brain fog, hot flashes, sleeplessness, and exhaustion that I couldn't afford. I had a book tour to do a few months later, and I needed to be able to function in an Upgrade, not a Downgrade.

It was a difficult time to be asking for hormone replacement because just three years earlier, in spring of 2002, a flawed report came out that, according to colleagues at National Institutes of Health (NIH), set women's health back at least twenty years. The incorrect story this report told made it nearly impossible for me, a doctor who specializes in women's hormones, to get the hormone replacement therapy I wanted.

The Tale of Two Hormones

With estrogen receptors in every organ of the female body, the influence of this neurochemical is profound. In the 1990s, researchers discovered that estrogen is protective of heart health, brain health, bone density, and emotional balance. Studies at that time were clearly showing that estrogen HRT lowers a woman's risk of dementia, heart disease, diabetes, and osteoporosis and so during the 1990s, more women than ever were experiencing the benefits. Doctors, cardiologists, general practitioners, and the newly board-certified functional medicine doctors were prescribing HRT earlier, during the Transitions as opposed to waiting until the Upgrade.

Then the blow came in 2002, with the publication of the confusing and flawed Women's Health Initiative (WHI) report on the risks of using hormones in the Transition and the Upgrade. Warnings of early death, heart disease, stroke, and cancer spread like wildfire from the report's conclusions as the media picked up the story and ran with it. The WHI trial was shut down, and panic around early death spread among doctors and patients. HRT came to a grinding halt.

By 2008, six years after the WHI report, only 5 percent of women in the Transition were taking HRT. And while it focused on risks, the WHI report completely ignored the protective effect estrogen has on brain, bone, metabolism, vaginal, cardiovascular, and emotional health. Those benefits are real and that data predates the WHI report.

Twenty years later, in 2021 and 2022, indications are that some of the data were flat out wrong. For example, the study had left out a crucial piece of information about the participants who were on HRT and had had a heart attack or stroke. These women *already had underlying cardio-vascular health issues* when they started the study. Their diets, smoking habits, alcohol intake, and lifestyles were not factored in. Additionally, at an average of sixty-four years old, they were fourteen years past their last period when they started HRT. We knew then and know now that the benefits of HRT are reaped if it is begun within five years of the Upgrade. We knew then and we know now, that if you wait to start for ten years or more, HRT can be harmful.

The WHI report also indicated a marked increase in the risk for breast cancer for women taking HRT, but again their data were flawed. There were women in the study who had undetected cancer before starting HRT and some were BRCA positive, meaning they carried the gene giving them a 70 percent chance of developing breast cancer. The researchers should have eliminated these women from the data, but they did not. Even so, when the data was reanalyzed, the absolute risk of breast cancer after 5.6 years of combined estrogen and progestin therapy in the study was increased by less than 0.1 percent. But the impact of the report was real. Women stopped taking estrogen, and doctors stopped prescribing it.

With the massive reduction of women taking HRT, if it were really harmful to the health of postmenopausal women, we should have seen an improvement over the two decades following the release of the WHI report. If there were indeed a risk of stroke from HRT, the number of strokes in 2021 should be much lower than the number in 2002, when so many women stopped taking HRT. But the rate hasn't changed. Not even a little bit. It puts a spotlight on the question as to whether or not HRT causes strokes in women at all. And meanwhile, for two decades, women have been suffering symptoms of sleeplessness, sadness, confusion, and brain fog that could have been remedied by HRT.

Increasingly, as doctors recognize the problems with the WHI report, the knowledge around HRT that went dark for twenty years is being retrieved, and many have quietly resumed studying and prescribing hormones again. Disagreement around HRT remains. The North American Menopause Society (NAMS) and the American College of Obstetricians and Gynecologists (ACOG) no longer oppose prescribing it, but the American College of Physicians (ACP) has produced statements claiming the risks of cancer and other diseases are too high. New studies are emerging at a rapid pace, and the bottom line consensus in 2021 has changed again: If women start HRT around the time of the Late Transition and the Early Upgrade, the risk of cardiovascular issues is not only very small, there are cardiovascular and bone protective benefits of HRT, especially from estrogen. Cognitive benefits of estrogen also look promising—it may be a factor in preventing dementia in women.

What always troubled me about the WHI report is that *quality of life* for women was taken off the table, as though getting enough sleep and balancing and improving mood had no contribution to overall health. Patricia felt she was living proof that it saved her mental health and her ability to earn a living. Without at least a low dose of HRT, many women feel deep sadness and a profound lack of energy. Nearly 90 percent of all women have experiences that are troubling enough during the Transition and Upgrade to seek medical help. For some women, sadness can be a bigger health risk by far than that posed by HRT.

If I'm feeling down, whether the cause of the sadness is a life event or low estrogen, I know from research, from decades of work with patients, and from my own experience that self-care will not be at the top of the priorities list. Feeling down never made me want to exercise, even though I am fully aware that moderate cardio four times a week is at least as good as, if not better than, SSRIs (antidepressants) for improving mood. If I am consistently sad, I am less likely to seek social connection, which has a proven impact not only on sadness but on cognition and longevity.

The New Medical Landscape

If you do decide to explore HRT, like many, you'll probably start with the doctor you know best, your gynecologist. The benefit of partnering with this doctor is that likely you've known them for a long time, and the medications they use have FDA approval, which means solid studies, longevity of usage, and a treasure trove of information on how they impact your health. But for your doctor to be of help, they will need to have added to her training like getting certified by the North American Menopause Society (NAMS). It's an entirely different area than what most gynecologists are trained for. They are incredible at their specialty, which is pregnancy, childbirth, cancer detection, and surgical procedures. Their knowledge and skills center on the uterus, which is a muscle, and the ovaries. Yet it's the brain that is center stage during the Transition and the Upgrade. It's entirely the brain. For those out there who still need the low-pitched bellow of a man in order to hear this message and believe it, I give you the former gynecologist and professor at University of Rochester School of Medicine James Woods. He's one of the few earlier gynecologists who decided to specialize in the Transition and the Upgrade because he recognized that he needed a knowledge base other than what he got in his training in obstetrics and gynecology. In his words, "[The M-word/Upgrade] is as different from obstetrics as surgery is from pediatrics." And because of his efforts, maybe we can forgive him for using the M-word.

The aftermath of the WHI report turned HRT into a bit of the Wild Wild West. It led to trying to get estrogens via phytoestrogens in foods like soy, a hairball of an issue. Women with certain cancer risks can't touch it, and its high fat content means packing on the pounds. It doesn't help with hot flashes, vaginal dryness, joint pain, or brain fog either— phytoestrogens don't work on the brain or body in the same way that Premarin, synthetic estrogens, or bioidentical estrogens do. Herbal treatments like dong quai are entirely unregulated, and there are health risks to taking them as well. Dong quai can cause anxiety in some women. Black cohosh has been proven ineffective.

Bottom line is that I don't recommend trying to get your hormones replaced at the health food store. The best bet is seeking out responsible doctors prescribing FDA-approved HRT or customized oral and topical so-called bioidentical hormones like estradiol, progesterone, DHEA, and others. They can track levels through blood and urine tests. Bioidentical hormones are manufactured to be the same structurally as our natural hormones. They are not considered to be better than other prescription forms—in some cases they are not FDA regulated and that can be dicey. But a good practitioner will do regular blood work to check your levels and prescribe hormones via what's called a 503A or 503B FDA-certified compounding pharmacy. You can find a list of qualified pharmacies on the United States Food and Drug Administration website. For some women who are sensitive to dose and sourcing, being able to easily customize is a good option. But it's easier, and probably safer, to try the standard medical approach first. I personally use an FDA-approved twice-weekly estrogen patch. Everyone's different—the best thing you can do is to become an observer in your own science experiment, so keep track of dosages and how you feel each day.

Finding Your New Sweet Spot

Not every woman is going to need to take hormones, but if you're having trouble functioning day-to-day, if your mood feels wildly out of control, or if there is a feeling of lifelessness that is not clinical depression, then HRT might be a good road for you to take. Now that we know it's relatively safe, the risks are low, and it has protective benefits for the brain, heart, vagina, and bones, you can feel confident taking it with the help of your doctor.

You may ask, "Well what about breast cancer?" Taking estrogen plus progesterone for longer than five to ten years may cause a small increase in the risk of breast cancer. It's best to discuss your individual risk and genetics with your doctor. What I can tell you as of now is that if you take HRT make sure you attend regular mammogram screening appointments, keep your weight in the normal range, do moderate exercise, don't smoke, and stop drinking alcohol, since alcohol of any kind, including wine, doubles your risk of breast cancer. I wish I could tell you alcohol in moderation is okay, but the studies so far haven't broken it out that way. They show it as an all-or-nothing variable.

Keep in mind that the advice here is also not the bible on HRT—it would be impossible to write that given the damage done by the WHI report and the fact that so much new research is emerging. The thinking is evolving by the minute. What I'm trying to give you, after studying the evolution of thinking around HRT, is a snapshot of where we are now.

If you're going to start HRT, research is showing that timing matters. If you start HRT in the Late Transition and take it for three to five years after entering the Upgrade, then you will get those health benefits late into life even if you stop taking it. But if you wait too long to start, then HRT can be harmful. Estrogen works well with healthy brain cells, or neurons. But if those neurons have been deprived of estrogen for too long and have started to age too much, then estrogen might even speed that aging process. The same is true of the cells in blood vessels and the heart. One study showed that this was the case in women who were sixty-five and older when they started HRT with estrogen. But don't worry. If you've already missed that window, you'll find a ton of things you can do, especially through food, lifestyle, and exercise, that will protect your brain and heart. (See Chapters 6 and 11.)

If you are in the Early, Mid- or Late Transition, or Early Upgrade, then you can feel confident that there is safe hormonal help out there for you. Tell your doctor about everything you're going through, and that you want to give estrogen supplementation a try. The birth control pill might be right for you in the Pre- or Early Transition, and HRT in the Mid- or Late Transition and Early Upgrade. And if you don't want to have periods or bleeding anymore, tell your doctor so she can adjust your hormone formula accordingly.

If you are in the Pre-Transition—experiencing body temperature regulation issues like having trouble cooling down from a workout but you are not yet having irregular cycles—then it's too early for estrogen supplementation. Raising estrogen alone at this stage can cause too much thickening of the uterine lining. If there isn't a steep enough monthly progesterone drop, that lining will continue to build, and you'll get breakthrough or heavy bleeding. Your doctor will need to use high doses of progesterone or need to scrape the lining out via D & C. Estogen supplementation at this time is just unnecessary.

Progesterone, which you'll need if you still have a uterus, can be calming or depressing, as most women know, because they've experienced PMS. You know what your cycling hormones feel like, and you also know that you can handle anything for a couple of days. But if you've been given progesterone in some form in order to control heavy bleeding and you notice you've got that weepy, cranky, foggy PMS feeling most of the time, it might be the progesterone. Talk to your doctor about changing form, lowering the dose, or other non-progesterone alternatives. If you do still have your uterus, your doctor may need to observe the uterine lining via ultrasound to track thickness, and you might need a quarterly dose of progesterone in order to flush it out.

If you are having hot flashes and are unable to sleep (more on this in Chapter 6), please don't feel like you have to muscle through without help. Small doses of estrogen (17-beta estradiol [oral 1 mg/day or transdermal 0.05 mg/day] or conjugated estrogen 0.625 mg/day) have been effective in eliminating hot flashes in 80 percent of women and can provide relief by reducing them in the other 20 percent.

Testosterone or, for some women, DHEA can help with libido and energy. If your husband is using a topical gel, you'll need much less than he takes, so don't borrow his!

It will be more important than ever to take good care of yourself while you're on HRT because fitness, body fat, diet, and lifestyle can be a factor in whether or not you get the benefits. It's why I've got two chapters on exactly this. I want the neuroscience of health and well-being to be absolutely clear to you. And while studies are showing that obesity, smoking, and alcohol each erase the benefits of HRT, remember that even three to five years of HRT within the window of effectiveness has a lifetime of positive impact on bone and cognitive strength.

The Pill or HRT?

"I had a really scary reaction," Patricia said of her struggle with the pill. "I blew up from bloat, I felt wildly testy, and when it didn't stop the bleeding, my doctor put me on a higher dose. When that didn't work, I started pure progesterone. It not only didn't stop the bleeding, I was catatonic, sobbing, on the couch. I have never been clinically depressed. But those chemicals incapacitated me. I couldn't think. I was completely unable to work, and I was in the middle of trying to start over in New York after my job ended. I had no energy, no zest for life and the consequences were life-altering. I was lucky I had savings to help get me through this time."

If you've started to have the irregular cycles and heavy bleeding of the Mid- or Late Transition, your doctor might have suggested you take the pill. It's the go-to fix every doctor reaches for to stabilize cycles, control bleeding, and provide contraception any time during a woman's life, from a girl's first few years of menstruation to the Transition. The hormones of the pill take the ovarian follicle/pituitary/hypothalamus signaling system offline. If the brain senses enough progesterone and a low steady estrogen, it won't send out FSH or LH to push ovulation or build up the uterine lining. As the pill sends out a consistent signal, the pituitary and hypothalamus can relax. Under control of the pill the cycle becomes regular and the bleeding decreases. On the other hand, there's important information in those non-pill irregular cycles; being able to hear them is one of the Upgrade's greatest lessons in self-care. It can teach us to listen to what the body needs instead of muscling past a call to rest. Opening our ears to those needs becomes crucial to optimizing the Upgrade.

While not knowing when you are going to bleed is inconvenient, having irregular cycles in the Mid- and Late Transition isn't usually a big deal in terms of health. Often, there may be another cause, like biological stress that can result from life stress. Any stress can be intensified by irregular sleep, or inconsistent or excessive exercise habits—at this stage, more is not helpful and overexercising can kick off an explosion of stress hormones. The motto becomes exercise to exhilaration, not exhaustion. Every situation is different, but for many women, supporting the body with good habits and giving it time to find its homeostasis can put you back on track. But if heavy bleeding has been going on for a while, and is becoming incapacitating, that's another story.

There are over fifty variations of the pill. Each one has a different amount and combination of hormones. Doctors will typically try you on several to see what is most effective, starting with their best guess. When you are told you need a stronger pill to regulate bleeding, that usually means you will be getting a higher dose of progesterone. Some women can tolerate that. For many, they just feel rotten—bad mood, brain fog, trouble concentrating, feeling weepy, or flirting with the edge of clinical depression. The symptoms prompt many to quit. For some in the Transition, like Patricia, the progesterone in higher dose birth control is psychiatric poison.

More than 100 million women around the world take birth control pills, and shockingly little is known about the short- or long-term effects. Progesterone can be calming and settling, but if it's just a smidge off or the wrong type, it suppresses enlivening, ebullient connectivity of the brain. The science is becoming clear that progesterone from birth control pills is implicated in a tripling of suicide rates among women, with the highest numbers in teen girls who never had depression or anxiety before.

If there is a dramatic uptick in suicide rates among teen girls on the pill, which is mostly progesterone, to me, it's reasonable to extrapolate serious caution for women in the Transition who are put on progesterone. Worried about this consequence, I started asking the nation's leading experts if anyone was studying how to do better HRT for women who had a variety of reactions to birth control pills or who had reactions to their own hormone fluctuations like premenstrual mood problems. I was told my question was excellent but there aren't any studies with good data. At the same time, between 1999 and 2010, the fastest growing rate of suicide was not white men. It was among women of all ethnicities around the age of sixty. Yet the phenomenon is not being reported nor is it being studied.

Like estrogen, progesterone is a potent neurochemical that can alter our mood or change our reality. If a doctor offers you a strong dose of progesterone in the form of a high dose birth control pill, an implant, a straight progesterone supplement either topical or oral, or a progesterone-infused IUD, make sure you can tolerate the neurochemical first. Before you get a device inserted, ask for an oral or topical (cream) prescription that would be an equivalent dose of progesterone and see what it does to your mood. Track it daily in a journal. Take it for at least three weeks before you decide about an IUD or implant. Please make sure you have a partner or close friend nearby to help you monitor your moods. And note that progesterone might reduce your sex drive.

Even if you had a hard time with a high dose of progesterone in the pill, you might do just fine with the much lower HRT dose of progesterone in combination with estrogen. The only way to find out is to try it and to stay vigilant about mood and outlook. Keep a daily mood journal (see appendix, page 263) and note how pessimistic or joyful you feel for about three weeks. If you find that life sucks and everything is terrible, check your hormone doses with your doctor.

Be prepared to push back when a doctor tells you that you shouldn't be feeling a certain way from a therapy. How you feel may be different from how a sister or a friend felt on the same medication, or after the same procedure. Everyone has a different experience, and to me, the patient is always right. Our job as doctors is to listen to an experience and see if we can translate that experience into medical terms that show us how to help her take the next right step. We have to learn to celebrate the messiness of individual differences. If you've tracked an experience in your mood log that clearly corresponds to the introduction of a new prescription and you're not feeling right, bring it up with your doc and make sure you are heard. You will know better than anyone what's right for you.

Switching from the Pill to HRT

Susan, fifty-one, was taken off the pill for the first time since her late thirties, and her doctor suggested she stay off hormones. "I felt miserable. I begged to go back on." I helped Susan find a new doctor who had been quietly using HRT for certain patients in spite of the WHI and other flawed reports that would emerge in its wake. In an off-the-record conversation, this doctor had told me, "The truth is many women do better with estrogen, and others don't need it. I find that the naturally occurring estrogen, 17 beta estradiol, and progesterone are a fine option for most of my patients." The doctor, age fifty-six, was taking it herself, and she wasn't the only gynecologist I knew who was doing this. One study found that the majority of female gynecologists who are in the Transition or Upgrade use HRT on themselves even if they didn't recommend it to their patients.

How do you find out if going off the pill will be a problem for you? By going off the pill. For 20 percent of women, going off the pill isn't a big deal. But for the vast majority, if they go off the pill and are in or past any of the four phases of the Transition, it might wreak emotional and physical havoc. You can find yourself in horrible withdrawal as the pituitary, hypothalamus, and ovaries struggle to come back online. If you're between forty-eight and fifty-four, Early, Mid- or Late Transition, you can bet the restart is going to be glitchy as the brain, unplugged from any form of hormonal contraception, tries to reboot a faulty system in which there are likely already some power outages. As the system sputters back on, it can feel like you've gone crazy. It can be a WTF moment. Warn your partner first.

Remember that estrogen is an essential joy vitamin for many women, just as testosterone is for men. When estrogen levels crash, you may feel all the color, all the air, even your impulse to smile has suddenly vanished. You can feel bereft for no identifiable cause. The pill is four to eight times stronger than HRT, so a straight switch is like giving up your smartphone for an old flip phone.

"When my new doctor suggested HRT instead of the pill," Susan told me, "I balked. I thought, *that's for old ladies*. But she gave me a prescription for a very low HRT dose. I didn't feel a lot better." Susan and I worked together with her gynecologist to get it right. It's not always a simple switch to HRT, but by increasing her HRT dose, we could support Susan's Transition.

If you are of an age that puts you in Mid- to Late Transition and decide to shift to HRT from hormonal contraception, you have to step the pill dose down gradually, otherwise you fall off a cliff. Lots of docs have learned to start with the higher HRT doses, 1–2 mg of estradiol, which is still one-half to one-fifth of what's in some pills. On the other hand, if you have already started HRT and your doctor wants to raise your estrogen levels to help your mood, libido, and to protect your bones, make sure they do it gradually. There is some evidence suggesting that topical creams and patches, like Vivelle, Climara, Alora, Estraderm, and Menostar, are safer than oral HRT, and may have more flexibility in dosage. For those who have decided to use bioidentical topical creams, make sure you are rubbing it on a part of the body that can both absorb it easily and is far away from the breasts: the forearms, within three inches of the belly button, upper thighs, or for some forms, inserted with a vaginal applicator.

As far as HRT goes, all I can tell you is what the data say and what I have experienced, what patients I've worked with have experienced, and what friends have gone through. For thousands of us, HRT has helped. If you are frightened of it, I totally understand. I have the advantage of being a doctor with knowledge of the risks and benefits identified in research. I was able to make a decision decades ahead of where the thinking was because of the knowledge I had access to.

The most important thing is for you to recognize the impact of what you are taking has on you and to speak up. You and those close to you will know what is working and what isn't, what you can tolerate, and what pushes you over the edge. I encourage you to challenge your doctor, especially if you are sensitive to any substances. If you develop breast soreness or if you are beyond the Transition, are not on cyclic HRT and develop breakthrough bleeding, speak up immediately to have your dose adjusted and your endometrium checked. If you're edgy, your estrogen may be too high. Above all, if you choose HRT, make sure to do what you can to keep your risk factors for other diseases low.

SSRIs and non-HRT Interventions

If you are suffering from hot flashes and cannot take hormones, SSRIs, antidepressants known as serotonin reuptake inhibitors, may be of help. They were first used on men suffering from hot flashes while undergoing hormone blocking therapy for prostate cancer. It turned out that the SSRI paroxetine—most commonly sold in the United States as Paxil—which was being studied in these men on Lupron (the hormone blocking drug) to treat sadness, also helped alleviate their hot flashes. As a result, they decided to test it on women having hot flashes during the Transition. Paroxetine at a dose of 7.5 mg turned out to be effective for some women as well, and so the drug was renamed and rebranded for women as Brisdelle in the United States. It received FDA approval in 2013 for treating hot flashes in women.

For Sarah, a film industry executive who was wary of HRT, antidepressants turned out to be the perfect solution. "At the same time that I was going through the Transition, my boss, who had been a lifelong mentor and sponsor, retired. The woman who replaced him was edgy and explosive, and I needed to keep my balance if my career was going to survive. I had a hard time on hormones, so when my doctor mentioned an SNRI—antidepressants similar to SSRIs—Effexor/venlafaxine, to help me sleep, support my mood, and keep me even I said, 'Give me the pills!'" For her it worked like a charm.

Almost a quarter of women over fifty are on some form of antidepressant and while SSRIs work for some they may not be right for everyone. Tanya, a furniture maker living in Detroit who struggled with intractable melancholy, couldn't tolerate progesterone given to her during the Transition. She said it intensified her sadness. So, her psychiatrist began treating her hot flashes and melancholy with paroxetine. Tanya told me, "I felt like I had been brain damaged," she said. "While traveling in Europe for work, for the first time in my life I had suicidal ideation. I forced myself to go running every morning to keep from thinking about it. With the help of my doctor, I got off it, but it took me several years to climb out of that feeling and get back to myself. By the time I did, the Transition was over, and I was a completely different person. It took me a long time to catch up with myself and find my balance again."

Paroxetine can be effective, but it also has a long list of side effects that can be worse for some than the discomfort they are attempting to alleviate. These include, from the manufacturer's information: "anorgasmia, headache, fatigue, generally feeling unwell (malaise), lethargy, nausea, vomiting, increased dreaming/nightmares, muscle cramps/spasms/twitching, nervousness, anxiety, restless feeling in legs, or trouble sleeping (insomnia)." A new concern from studies is that paroxetine showed an increase in the risk for dementia, the very thing estrogen may protect against. It's not definitive, but there is enough smoke there to make us worry that it's indicating fire.

––––––––––

I'm often asked what happens to the brain during the illness of depression. We know a bit about shifts in neurochemical processing, that the ability to make and utilize the body's natural feel-good neurohormones is interfered with. But what's really important to know is what those neurochemical shifts do to a person's reality. In the depth of their melancholy, they may find it hard to remember ever feeling any differently, even if you've known them to. Their brain circuits will not be able to access that former reality. As the outside world recedes, melancholy draws one ever closer to the siren call of suicidal ideation. It becomes almost a nervous system addiction. As patients recover, the death wish recedes, but for many it remains a persistent whisper. If this happens to you and lasts more than a month, reach out quickly to your medical team since help is available. And with new brain research on the effectiveness of cognitive techniques and medication, those emerging from intractable melancholy can come into a new relationship with their thoughts, to safely face reality.

For close to 60 percent of women, SSRIs can be effective in treating both depression and hot flashes. If you have full blown, clinically diagnosed depression, and your doctor recommends therapy and antidepressants, follow their advice. Estrogen, while for many can boost optimism and zest for life, does not treat severe depression. But the sadness that can accompany the Transition isn't always this more intense state of anhedonia, the lack of joy in clinical depression. It's often a temporary side effect of shifting hormones and a shifting identity; of feeling like we are becoming invisible at work and to the culture—to men and younger women all at the same time. Sadness might be a normal grief reaction. So maybe we need to find a way to better address grief.

The hard truth is that women are more at risk for depression in the Transition and the Upgrade. Somewhere between 45 and 68 percent of women in the Transition report more symptoms of depression compared with around 30 percent of women before the Transition. And if you've had a bout with depression before the Transition, you're two to three times more likely to have one during and/or after. So, protecting your state of mind is important. The evidence is clear that estrogen protects the brain, cognition, and mood. With my patients in the Mid- to Late Transition, and the Upgrade, if they are clinically depressed, I often prescribe a combination of HRT, therapy, and SSRIs. We monitor their sense of well-being closely, together. And while it doesn't treat depression, there is reason to believe that, if taken early enough in the Transition, estrogen, for some women, may stop a long slow slide into clinical depression.

When Sadness or Low Energy Is Anemia or a Thyroid Issue

For the women who experience excessive bleeding during the Transition, it's very common to develop anemia. Patricia did, and I remember her telling me during a session that she was "okay as long as I didn't have to get off the couch!" It's no way to live, and if it goes on long enough, anemia can cause deterioration in cognition as the brain loses oxygen. So, make sure your doctor monitors your hemoglobin and hematocrit levels. And if you've stopped bleeding, have your red blood cell width measured as well. It will tell you whether or not your bone marrow has caught up with cell production. When it has, then you can get off the iron supplements.

The thyroid, a little endocrine gland at the base of the throat, produces hormones (T3 and T4) that have a direct impact on preservation of cognition, on metabolism, and on how energetic we feel. The neurohormone system in the body is deeply interconnected, and for many women, the thyroid can be thrown off by the hormone changes of the Transition. So, if your skin and hair are dry, you are tired all the time, and you can't think straight, ask your doctor for a thyroid checkup since women have ten times more thyroid disease than men.

Tracking Your Moods

"My husband is an engineer," said Patricia, "and he loves spreadsheets. During the Transition, he kept track of my cycles on his calendar. I kept track too, just so I would know when to start loading up my purse with equipment. But that he did it? I was super sensitive about that at first. In fact, I was outraged. It felt so judged because he was doing it to track my mood. But he saved me a bunch of times. I would feel one hundred percent certain that a friend was angry or that a colleague was insulting me, and I would start to gear myself up for a big confrontation. And then he would say, 'Darling, before you burn the house down, would you consider checking your calendar?' The first few times he did this, I shredded him for what felt like a lack of support. But then I saw he was right. I learned to wait it out. And now that I take HRT along with DHEA, he knows when I've been skipping my DHEA since I don't want to be touched down there."

By this point in life, you probably know plenty about the way your mood sinks along with your hormones and vice versa. Starting or stopping hormones can have the same impact. Any time you make a change to the hormones you're taking, let someone close to you know about it. If your behavior or mood shifts, ask them to tell you. Mood shifts can be so powerful that you may not recognize it as a change. It will just feel like your reality. The scary thing about hormones—and for that matter any chemical we take—is that we often can't separate how they're making us feel from what is going on in our lives.

You don't have to make a spreadsheet like Patricia's engineer husband but do consider keeping a daily record when you start new hormones or medications. I like my patients to write it all down every day until we find the correct balance. (See appendix, page 263, for detailed instructions.) When you've charted what's going on for six to eight weeks, you can figure out what is going on and have data to help you adjust the dosage accordingly. With most of my patients, we find their new sweet spot after three to six months. With this kind of information in hand, you have a better chance of making it happen sooner.

No one can feel happy and focused all the time, but you deserve to have more good days than bad during the Transition and in the Upgrade. Your hormones can be an aid to this if you get the balance right. For some women, hormones may improve mood and memory and decrease irritability right away. For others it takes a few tries to get the balance right. The most critical thing about taking hormones is how they make *you* feel.

Balance

Homeostasis, the body's fight for balance, is often visualized as an inverted U-shaped curve, like the perky breast that once upon a time in our teens and twenties stood straight up even when you were lying down. Let's say that homeostasis is at the nipple level and for many years you probably hovered there. You knew what it felt like to be in your body, what the experience of your mind and life force felt like. As hormones shift in the transition, as they pull other neurochemicals with them that regulate sleeping, waking, biological stress, temperature variation tolerance, you get pulled off homeostasis. You end up on the downside of the curve instead of at the top. You feel off, not quite yourself because what you are experiencing in your own skin is not how you remember feeling. It's not how you remember the experience of who you are.

On top of that, you've gone for a checkup and gotten lab work done. After decades of perfect scores in cholesterol, inflammation markers, vitamin and mineral levels, suddenly everything is off. Your cholesterol had been 156 for decades and now it's 210. Your blood glucose was always low and now it teeters on the brink of prediabetes. Your blood pressure has always been 110/70 and now is 132/85. Your vitamin B and D are tanked out. The confluence of the Transition and age conspire to try to make you realize you are not twenty-five anymore. And that is a tough pill to swallow.

I want to reassure you that your out-of-whack numbers will settle, even if it takes a few years. The neurohormonal wildness of the Transition causes unhealthy inflammation in the body and brain, and that will settle with time, maybe HRT, diet, and lifestyle adjustments. You will find your sweet spot again, but it won't be the same one you have been used to. How you experience homeostasis will never be the same. If you can make friends with your new sweet spot you might find it's even better. It requires more patience than you ever thought you would have but I promise, you will find that patience through care for yourself, and it will pay off.

I can't deny that this is a risky and difficult time. Don't make big decisions when you're in the thick of it. Take life planning in chunks. If you are in any of the Transition phases and feel like you want to make a dramatic change, wait one month, then three months, six months, even eighteen months, until you feel more steady and clear. And you will—homeostasis is out there, whether it happens for you naturally or with the help of a doctor. Like a whirling top that's hit a bump in the floor, you might wobble like crazy for a while. But you will recover a new, stable, and effortless spin.

Renewal: Your Brain in Search of a New Reality

Chapter 5

"Honey, I wouldn't look hot in this ever again, even if I could lose the weight," sixty-six-year-old Ceci said to Carole's daughter Dawn. She pulled a tiny, sexy dress out of the closet and gave it to the twenty-two-year-old woman. "It's a dangerous one, so be careful! I remember walking into a room and all the men's eyes were on me. I was wrapped around my first husband's arm when two guys walked over together and tried to pick me up," Ceci said with a laugh. There's not a hint of regret in her voice.

Ceci had first met Carole, Dawn's forty-seven-year-old mother, fifteen years earlier, when the younger woman had come to her for advice. Carole quickly adopted Ceci as a mentor, and before long they became close friends. Dawn had just graduated from college and was about to start work. I was visiting Ceci when Dawn and Carole showed up for a quick celebration and a treasure hunt in Ceci's closet.

Carole knew how important Ceci's looks had been to her and wondered why she didn't seem sadder about how her body and life had changed. Carole herself was dreading the future and for now remained happily in denial that she would ever not be skinny. "You think you will never have baggy arms or watermelons for breasts," Ceci said to Carole, "but say hello to both. Yours will probably be worse. There's nothing to do but embrace being healthy on the inside, and celebrate: NO MORE PE-RIODS!!"

Carole was ogling Ceci's vast collection of anti-aging products laid out on the bed, a pile of high-end skin-care products and makeup received over the years as perks for being in the fashion industry. Ceci noticed. "Take them all, Carole. I don't use them anymore. I don't want one single package in my house that insists I have to anti-age. It pisses me off just looking at them. It took me a long time to remember what I know from the industry—that anti-aging is the marketing response to the fear of young women. It's Dawn and her friends who are the ones freaking out about wrinkles. They are getting their faces sliced up and injected. Not me!"

Ceci is right, though it seems counterintuitive that the vast majority of women undergoing cosmetic procedures are well under fifty. "For a long time I was tortured over losing my looks," Ceci said, continuing her diatribe. "And I remember the torture of trying to hide my age. I'd forget who I told what year I was born or graduated college, and I was always worried I'd be outed. I thought it mattered whether or not my colleagues thought I was ten years younger than I was. But then one thing got super clear: If you get into a fight with reality, who's going to win? I'd been holding on to an idea of who I was and how my looks and my job defined me. No matter how evolved we think we are as women, this stuff is deep in our nervous system. It was a big, big struggle. Don't let anyone kid you about how hard it can be. Once I allowed myself to grieve my old life and my old self, things got much simpler. Now that I am proud of every wrinkle on my face and every year I've lived, I don't have to worry about what I say when someone asks my age. I tell them straight up."

Ceci paused, though, at the memory of the hormonal roller-coaster ride of the Transition. She'd been reluctant to talk about it when Carole had asked. But I was used to drawing women's stories out. After some hesitation, Ceci talked about how it had felt like crazy people were playing a wild tennis match with her brain and moods. With the ups and downs of hormones came the torture of the insula, the part of our brain that scans our body systems to confirm health and check self-image. Beginning with the first transition to the teen-girl brain, the insula compares what we look like to what others look like, influencing how we think about ourselves in relation to others. *Do I look as good? Can I look like that? Would I rather look like someone else? What's my style?* It's the brain area for feeling disgust at the gap between expectation and reality.

For many women the mental and emotional patterns that recruit the insula can set us up for a lifetime of obsession with weight and appearance. But other than making sure she gets healthy fats for her brain and stays away from too many energy-draining carbs, Ceci decided to ignore the insula's prods and stopped dieting after the Transition ended. She discovered what many women figure out after the Transition: that as estrogen drops, your metabolism changes. Eating carbs will just stimulate the brain to want more and to squeeze more insulin out of your pancreas, so carbs make your blood sugar spike and drop much more dramatically. If we keep feeding the brain's craving for carbs, diabetes is more easily triggered; the resulting inflammation can damage joints, arteries, and cognition. To keep her energy even throughout the day, Ceci accommodated her brain's new metabolic needs by feeding it more lean protein and healthy fats. For Ceci, as for many women, it was trial and error until she found the right blend.

"Now that I'm not cycling up and down every month, things are so much more peaceful. I feel comfortable in my own skin in a way I never did in my life. I feel so confident, so strongly rooted in who I am that I wouldn't trade a flat stomach, a wrinkle-free face, or even joint pain for the suffering of those years. My body is what it is. And I get to see happiness in your daughter's face at taking home some great clothes," she told Carole, smiling. "And the best thing about doing business with men is that since they're not trying to figure out how to get me into bed, they finally listen to what I have to say. They are taking me seriously in a way they never did."

Ceci had been through tremendous shifts in her career in the fashion industry, and in her personal life. She had three kids, all grown, the middle one a daughter, Stephanie, who had had a hard time getting her life started. A college dropout, she'd been unable to keep steady work, and

Ceci lived daily with obsessive worry and stress about whether or not Stephanie would find her way and be able to take care of herself. While she loved her craft, her corporate job had always been a source of stress. Ceci had been a textile designer for one of the most prestigious fashion houses in the world, but by the time she hit her late thirties, she had started feeling a pervasive anxiety. "I was having trouble sleeping. Something was making me deeply insecure. It took a long time for me to figure out what it was. And then one day it dawned on me. Every woman I knew who was a bit older than me was getting canned. It started happening in their forties. It didn't matter what industry, and it barely mattered how successful they were. Publishing, fashion, finance, TV, real estate, PR. Some were making it to their midfifties, but most of the women I know who worked for other people were being let go. I suddenly realized that I had a shelf life."

It's not a surprise that Ceci didn't recognize the source of her anxiety right away. Not only do we have an epidemic of it in America, with nearly one-fifth of the population reporting an anxiety disorder, but we now marshal every possible distraction to keep from feeling the discomfort. It's understandable that we'd reach for a smartphone, the TV, a glass of wine, food, anything other than the brain's experience of dread. In times of threat, the body is flooded with stress hormones that can make us go cold, our hearts feel like they are pumping out of our chests. Alertness is intensified. Our hands and feet can go numb as blood rushes to our muscles and brain. If the feeling becomes chronic, if we live with it all the time, then not only does anxiety become a disorder, but we are at higher risk for heart disease, diabetes, cancer, depression, dementia, and more.

But not all anxiety is bad, and research is supporting the truth of Carl Jung's admonition that with any negative state, "the more you resist, the more it persists." If we can be present to the message of anxiety, it can be a warning of real danger, like what happens when we are followed out of a grocery store late at night into a dark parking lot by a person with bad intentions. That is a strong signal, and fear is part of what helps us respond in those moments with potentially lifesaving actions, like returning to the store and asking for an escort or calling someone for help. Signals are there to get the brain to engage the problem-solving abilities of the prefrontal cortex. If we try to circumvent that call for engagement because a feeling is uncomfortable, we could be ignoring important information coming from deep within our own brains.

Dawn and Carole gathered up their bounty, a part of Ceci's former self she was consciously shedding, and took it home. Ceci and I headed out for coffee together. We were new friends—she'd just joined a meditation group I've been part of for twenty-five years. I was curious about her life and her decision to become an entrepreneur, a path many women in the Upgrade take. "When I was in my twenties and thirties," she told me as we sipped our drinks at Cafe Roma, overlooking the San Francisco Bay along Bridgeway in Sausalito, "I remember seeing older women packing their offices to leave. I didn't give it a second thought consciously, other than a passing feeling that they had lost their crispness and it was time to go. But looking back, I started having trouble sleeping after about four or five of them left within a few months of each other."

Ceci paused, remembering how hard it had been to stay alert at work, to keep her energy up when she was sleeping about three hours a night, how whacked her emotions felt. "Every morning I woke up terrified that that day at work would be my last," she continued. "I had no idea what I would do without my job and my identity."

In retrospect she acknowledges that the perceived chronic threat created a distortion field that contributed to the already-high tension at home with her daughter. Threat arising from lack of trust at work makes it difficult to develop real relationships everywhere else in our lives, and she and her husband had several tough years. Our brain is on alert all the time and not calm enough to truly incorporate or build friendships into our circuits. "It felt like a game of *Survivor*," she said. "You could make temporary alliances as long as we held some utility for each other. But as we all got closer to the prize, the path narrowed. You never knew who was going to push you aside in order to leap ahead." Hormones like adrenaline and cortisol are released in response to threat and are great in the short term for enlivening our focus and for inspiring great ideas. But over time, they take their toll and produce the opposite effect. "With all that drama at work," Ceci continued, "I just couldn't be as creative. My ideas weren't good anymore. I was having trouble being fast on my feet. I thought it was just age. What I saw happening to other women was starting to happen to me. I was losing my edge."

"Did it ever occur to you," I asked, "that losing your edge was not necessarily a function of age?"

"Not at first," she replied, "but later, when I had a chance to rest and heal, my mind came roaring back. Once I got away from a threatening environment, my brain got creative again."

It makes sense biologically. Cortisol, the hormone released when we are under pressure, can be friend or foe, depending on the duration and the amount. Women in the Upgrade have a higher cortisol reaction than before the Transition. Cortisol boosts our memory and our interest in learning when it's being released in the right amount. But if we get too much for too long, it kills memory cells.

"So what did you do?" I asked, riveted by her familiar story. I have heard variations of this story from my patients and from my friends. First we face a glass ceiling that keeps us from rising. And then we fall off the glass cliff. We live in an ageist society, and there is no doubt that all older people face discrimination in the workplace. But statistics show that women are aged out of their jobs as much as ten to twenty years earlier than men, depending on the industry.

"I gave up first class for a middle seat," Ceci exclaimed, borrowing a line from Laura Mercier cofounder Janet Gurwitch, who talks about leaving a plum job at Neiman Marcus to start the line of cosmetics that catapulted her to entrepreneurial stardom.

Ceci had an idea for her own company too. She'd been thinking of producing a flattering and more forgiving line of stylish activewear. But starting over in midlife wasn't easy. Where she'd had entrée into any office she wanted while connected to a multinational corporation, doors were no longer opening so easily. With the corporate perks gone, she was traveling around the country on a budget to pitch her new line, using money she'd saved for retirement to fund the new company. "When anxiety about that hit," she said, "I had to use my own brain's cognitive power to remind myself that I was making an investment, that I was betting on myself. If it weren't for the emotional support of my friends, I never would have gotten it off the ground. I nearly gave up a dozen times."

Ceci's journey to entrepreneurship is an increasingly common one among women in the Upgrade, from restaurant workers to executives. When the doors to employment close, those who can are paving their own roads by starting businesses themselves. "A very forthright executive coach was the one to wake me up," Marta, a fifty-four-year-old tax consultant, told me at her appointment. "She flat out said nobody was going to hire me at the level I was used to, and that the sooner I started my own business the better. Building a client base was really hard, and at first I was making a fraction of my former salary. There were so many times I thought I would end up on the streets. But eventually things started working. There is an extraordinary talent pool filled with women who've been aged out of corporations or pushed out because of still having young kids at home, and I am contracting out excess work to them. I can't even believe who I'm getting to partner with on projects, and how much I'm learning from the people I've hired." Her brain had kicked back in in full force.

Marta and Ceci have become part of a significant economic statistic. In the growth of the economy nearing the second decade of the twenty-first century, it's women who are the biggest drivers of job creation. From 2015 to 2016, women started businesses at double the rate of men. As of 2017, companies formed and led by women performed twice as well as those formed and led by men and were responsible for the creation of tens of millions of jobs. Of those female-led companies, the majority of the founders were over forty-five. In a May 2019 article for *Forbes*, Kevin O'Leary of *Shark Tank* talked about his preference for companies run by women. He said that while those led by men meet their targets 65 percent of the time, women have a much better record, meeting financial targets 95 percent of the time. The same article cites research showing that female-led companies are better for employee well-being, especially in the troubling area of engagement at work, where Gallup polls show that 70 percent of employees feel a lack. A lifetime of experience brings the knowledge that fuels success. And a lifetime of facing moments that bring us to our knees means women in an optimized Upgrade can live with uncertainty in ways that the untested among us cannot. We have built a brain that is more resilient under fire.

Early in the growth of her business, Ceci had to sue to collect what she was owed from a large retail chain. "If they didn't pay me," she said, "I was going to go broke. It happens to a lot of people in the clothing business. You spend most of your time as a debt collector. But this one was a whopper that I couldn't absorb. I had to recover the money, or at least a significant portion of it. There was a moment when it didn't look good, where the other side was burying my lawyers in papers and we were starting to miss dates. I was overwhelmed by fear, and for the first time I wasn't sure if the feeling of powerlessness would actually kill me. It was clear that if I remained in this state it was going to seriously damage my health. I remember deciding to stop fighting it and let the feeling take over. I am not religious, but I prayed to all the forces in the universe for help. I wasn't asking that some magical being swoop in and rescue me. I didn't believe that was possible. But I was asking for the ability to be at peace with whatever happened. That if I lost everything, if I was out on the street, that I would be as okay with it as anyone could be." Luckily, Ceci wouldn't end up losing anything. Her lawyers were able to resolve the situation a few days later, but the emotional breakthrough meant that the hard stuff wouldn't ever again shake her up the way it used to. That moment of life bringing you to your knees and coming out with your balance is a sign of the Upgrade, as your brain circuits are primed for resilience.

Centering, feeling at home and at ease in a new reality, is a gift of the Upgrade. For the first time since childhood, many women have the experience of being released from competing agendas—the deep yearning of your authentic self and the reality and behavior driven by the fertility cycle. Hormones are there to shape priorities and to try to force you to act, to make sure you feel intimacy and connection with partners and children. It can feel like it's against your will. The intense estrogen spikes of the Transition can overwhelm the mind with obsession, pushing many women to repeat the habits—like attraction to bad boys—of their teenage years. The intense drop-off at the end of an estrogen spike can leave us so bereft of connection, attraction, and sex drive that we crawl into bed and turn off the phone.

At the same time, I don't want to ignore the fact that the fear of ending up impoverished is the number one threat that haunts older women in America; it is provoked by a signal that is real. While women carry 78 percent of the unpaid caretaking burden, we are, over the course of our careers, making 60 to 70 cents on the dollar compared to men. The wage gap is not the only contributing factor. Unequal pay combined with years of unpaid caretaking, in which women may not have the chance to work outside the home, means we have fewer resources as we age. Because we do not work outside the home for as many years as men do, our contributions to Social Security and retirement funds are lower, making the wealth gap all but uncloseable. Since 1990, divorce rates for couples over the age of fifty have doubled. So if we end up single in the second half of life, and we weren't able to work outside the home for much of our lives, many of us will be trying to survive on a $600-a-month Social Security check. While women like Ceci and Marta were okay—they had savings, and they'd been able to contribute to Social Security during their long working lives—most women after the Transition are not anywhere near financially independent. Knowing this financial reality opens the door to understanding why so many stay in difficult marriages and mimic their husband's attitudes, even those that are clearly misogynistic. Disagreement might mean being cut off from food, clothing, and shelter in old age.

Ceci had financial means, independence, and choices. Her new company not only got off the ground; it exploded. She eventually sold it and used the money to start her own venture capital fund. She focused on funding and mentoring and sponsoring young women, helping them take charge of their financial destiny by planning for the day when they would have to clear their own path. Relying on a corporation to provide employment for the rest of their lives just wasn't a realistic path anymore.

We Are So Much More Than We Have Been Taught to Believe

"It took a long time for me to realize that I actually have something to offer younger women." Ceci had rented a suite of offices in a quasi-open collaborative work space with other entrepreneurs in creative sectors. "They pop in whenever they have a problem," she continued. "I still don't feel like a grown-up myself, so I missed the cues at first, when they started coming around asking questions. Then I remembered having done the same at their age. I had so many older women friends during my twenties and thirties that one mentor called me a 'wisdom junkie.' It's finally dawning on me that they see me as I saw my mentors and sponsors, someone with hard-won wisdom. I keep being surprised by the fact that I often do have real answers to their big questions."

The idea of wisdom is daunting, and not something we connect with easily as women. I know from my own experience. The wildness of the Transition to the Upgrade can make us feel like moody teenagers, constantly regressing and misbehaving, feeling out of control. For most of us, that period of time doesn't feel like the budding of wisdom. It can feel more like all our brain's good qualities are being drained.

By the time you reach your late forties or fifties, you've been through a mountain of experience. You've traversed the turbulent teens, when you were trying to figure out who you were; the unpredictable twenties, as you struggled to find your way in the world; the buckle-down thirties, when you were trying to implement your life plan; and the juggling forties, when you were living the life you'd made for yourself—for better or worse. What's next? This is the big unanswered question.

Women in developed countries are living on average into their mid-eighties and beyond. The Transition can occur anytime in your forties or fifties. Medicine and developmental psychology have lumped what follows for women into one big post-*M*-word category. That's thirty to fifty years unmapped, unexploited, unaccounted for. There are no phases defined, no transitions delineated other than death. It's another reason I refuse to use their word.

I am here to signal to your brain that you are just getting started in becoming who you're meant to be.

Oh, Louann, no. I'm done. I'm tired. I just want to relax! I hear you. Women often feel burned out at this stage. We've been holding up our worlds for a long time—kids, career, caretaking, house maintenance. Taking breaks is fine. Everybody needs them. But I want to ask you to question the feeling of being *done*. What is it you are really done with? Is it something specific, or is it really as global as it feels? After a lifetime of earning less, working harder, being the one responsible for the caretaking, it's normal to feel *done*. But before quitting a job or a long marriage, make sure you have clear answers. Running away before you know what you're running toward can be a big mistake at any age but especially now. Make sure you know what you want. And consider that you might need a sabbatical, some downtime to reflect.

The flip side of this question is our vision for the future. At this moment it is time to ask ourselves the same question we asked in our teens and twenties: What do you want your life to look like now? Is it really one long beach vacation or series of art classes? Is it a new load of caretaking of aging parents, spouse, and grandchildren? Is it being saddled with the family home, which now feels like an albatross? Or is it a new career, a new adventure, a sense of delight that comes with having a beginner's mind? I want you to meet this time with joy, not with regret. I've heard too many of those stories.

Ching was born in San Francisco in the late 1920s. Her story is of another generation, but the expression of not having lived the life she wanted is one I've heard many times in my office. "I always loved America," she said during our third session. "But when my family made the money they planned to earn through trade, they took us all back to China when I was ten. I hated it." Ching had come to see me after having a small brain-stem stroke. We were trying to get her mood and ability to sleep, which had been disrupted by the stroke, back on track. Normally quiet and hard to draw out, she burst into tears and the words came pouring out: "I didn't do anything I wanted to do with my life. I wanted to go to college. I wanted to have a career. I wanted to continue to learn. I was teaching English when World War II hit, and I rode out the war in Macao, cut off from my family in Canton and Hong Kong, where the Japanese had landed. I had always dreamed of returning to America to go to college and go to work. I was a citizen from birth. After the war my parents agreed to let me go and sent me with an aunt to join my older brother, who was already in California. I was just getting settled when the communist revolution came. My family in China lost everything. My brother and I had to go to work to ransom my parents and siblings out one by one. We got my parents and one sister. I worked so hard at my brother's restaurant!"

I paused, waiting for this normally very composed woman to gather her thoughts. "So what about college?" I asked.

"I had to keep working. I finally got a job at a bank and was able to take care of my family. I spent my whole life taking care of everyone. It was hard. I was a woman, an Asian woman on top of that, and I was being underpaid. I knew it because I was part of an audit group. But I couldn't speak up because I would have lost my job if I did."

I tried to hold the space as her grief unfolded. "And now my body and brain don't work well enough for me to pursue any of those dreams. My life is gone and I did nothing I wanted to do with it."

Though Ching's story may sound extreme, many women face a version of this regret, but with some effort, the right help, and vigilance, this kind of deep sadness can be turned around. I worked with Ching to shift her focus to everything she did accomplish at work and with her family, and to remember how proud she was of her granddaughters, who were taking full advantage of so many opportunities by starting careers in molecular biology and environmental science. We talked about this in her sessions, got her sleep and mood back on track, and I gave her journaling homework to help her notice what went right in her life.

I am telling you this story because I want you and your brain to be prepared for the developmental phases of the Transition into a new identity, the early stages of the Upgrade, and the Full Upgrade, when we embrace and acknowledge the depth of our experience and develop a deep concern for feeding it forward, so that future generations have the benefit of our knowledge, taking it even further.

What I mean by a Full Upgrade is when post-Transition women act on the decision to grow into their full female potential. This includes independence of mind, strong compassion for oneself and others, and a view of life that embraces reality as it is, not how we wish it were. It's an act of volition, of personal agency. Passively riding out our later years in a fog of denial about old age and death is not optimized. It's not an Upgrade. It's giving in to the cloak of invisibility, of uselessness, the fossil concept of the M-word. A downgrade.

Growing into Wisdom

In my practice as a psychiatrist, I can track how someone is growing into and embracing their status as a wise old woman. In many cultures around the world, we are comfortable saying "wise old man." But when we swap out "man" for "woman," the phrase gets caught in our throats. When we were younger, we called older women cute, eccentric, little, sweet, loving, dear, but how often did we feel them to be wise? Though I knew instinctively that Grandma was the one who held the real secrets to life and relationships, it was still Dad and Grandpa who were running the show. Though women's wisdom may be revered by families and communities in Africa, in Asia, and among African American, Latino, and a few shamanic cultures, there isn't a single society in the world that has majority female leadership. If you break down leadership by the percentage of women holding the highest political offices and executive positions, and try to find a society that has pay parity over the life of a career, in which women have equal access by most measures to education and capital for success, there isn't a single country that comes close to a fifty-fifty power share. Women in the Upgrade can turn this around.

Men don't have trouble stepping into the role of counselor and sage. But for women it's one we too easily abdicate. "I just don't have anything to contribute," my patient Pauline said to me back in the mid-1990s. She was seventy-two, a well-informed, avid reader. She had worked as an executive assistant in between children. Pauline's sadness was palpable and heavy; I could feel its drag on my own emotions as we talked through her history. In psychiatry, we are trained to watch our own reactions as indicators to help us validate reactions our patients get from others in their daily lives. When I asked her if others felt sad when she told her story to them, she said they did, and she was afraid she would alienate her oldest friends if they really knew how unhappy she was.

We tried an antidepressant and weekly appointments. After a few months, not much had changed, and I was beginning to worry about whether I would really be able to help her. One day she arrived with a bit more energy, and I probed to see if I could figure out the source of the uplift. Pauline mentioned her granddaughter Michelle, who had just graduated from San Francisco State and was moving to Los Angeles for work. Michelle had come to Pauline for style advice, and from what I could see, it was a great choice. Understated and elegant, Pauline had impeccable taste even on a budget. While she and Michelle assembled Michelle's new wardrobe, they talked about family, past wounds, and fears for the future. Michelle was living a life Pauline had wanted for herself. "I wanted to work. I was in love with someone else, but my family pushed me into the marriage with George," Pauline said of her husband of fifty years. "Michelle was full of questions about her parents, about my life with her grandfather, about how she would survive in LA. I don't know why she thinks I have any answers that will help her. What do I know?"

I paused, thinking the interaction with her granddaughter was a good sign and wanting to encourage it. "So do you have plans to spend more time with Michelle?"

"We finished the job. She's ready to go," Pauline said. "I did what I can do. I can show someone how to buy clothes on a budget that will stay in style. I can throw a party. I certainly know how to keep up a strong front. But beyond that, what do I have to give?"

Michelle didn't see her grandmother the way Pauline saw herself. She persisted in the relationship, finding in Pauline a confidant, a safe place to turn when she was afraid about her own brand-new start in the world. "Grandma listens in a different way," Michelle would later tell me in a family session. "She didn't tell me to suck it up the way my mom and dad do. Well, sometimes she laughed at the things I was upset over and told me that in a few years they wouldn't matter, especially the guys I met who didn't call me back. That's a little irritating! But whenever I had trouble with someone at work or with one of my roommates, Grandma seemed to have had the same experience sometime in her life, and she told me what it was like for her. And she often told me the difference between how she handled things at my age and how, looking back, she would handle them given what she knows now."

When we are young, we learn language, concepts, and skills. In the Upgrade, we learn to see the wisdom that grows from the refining experience of success and mistakes, joy and tragedy. All of this comes as the brain settles into a regular pattern, less buffeted by hormonal waves, more able to absorb and integrate the lessons of the memories stored by the hippocampus into the cortex. By the time of the Upgrade, we have a mean, lean, problem-solving machine. Our job is to accept the challenge of embracing this role, to weave the strands of our experience into insight that benefits others. At the beginning of the Transition, the developmental phase plants the seed of wisdom by setting up the brain for resilience in a storm. You have this strength in your brain. One of your jobs is to water these and not be afraid of them, not be afraid of falling short of the wisdom that comes from having survived the storm. At times we will fall short, but it doesn't mean that what we have to give isn't worthy.

The culture does not hand belief-in-self out to women as a birthright the way it does to men. Instead, self-doubt seems to be our birthright. Self-doubt is part of being human, but it's often out of balance in women, affecting our ability to fully embrace the Upgrade. The idea of having wisdom is in competition with self-doubt, and it stops us from passing the baton to others.

I struggle with self-doubt too. Part of the process of writing this book is embracing the reality that I have a lot of knowledge about the female brain and about the life cycle of women. But I also personally have trouble with the idea of growing into a woman who has wisdom that is worthy of transmission. Writing this book means that I am watering that seed in myself and passing along my confidence that you have it too.

Fork in the Road: Upgrade or Downgrade?

At the very moment we have the chance to grow into a new emotional reality, our physical reality changes dramatically; some of those changes may provoke the brain to go into grief, shock, and denial. How we deal with our new realities determines whether we enter an Upgrade or a downgrade.

The first new reality is that your metabolism will slow to a crawl. Testosterone helps drive metabolism in the muscles; that in turn helps us burn more calories and have stronger muscle fibers. And now that all your hormone levels have dropped, including testosterone, it's easy to gain fat and lose muscle, making fat harder to lose. If you have a high school reunion or wedding coming up, it's not a three-week or three-month job to whack off five or ten pounds. It can take six months to a year to really make a difference. For many, it will mean almost no carbs, almost no alcohol, almost no processed food. Just veggies and lean protein, a teeny tiny bit of healthy fats, combined with weight-bearing and aerobic exercise four times a week. Even then, many of us will carry an extra five to ten pounds no matter how hard we try. Personally, I choose slightly looser-fitting clothes over going to war with my body's new metabolism.

Accepting the Challenge

Not stepping into the new reality and the new role of wisdom in the Upgrade might have long-term consequences for our cognitive health. Let me give you two solid reasons for stepping into an optimized Upgrade.

We have long known that when we actively engage in tasks and accomplish them, the reward system of the brain kicks in and releases dopamine. It's the feel-good chemical that floods our system during orgasm, a deep and satisfying conversation, or exercising to exhilaration. But what we didn't know, until research at the University of Washington School of Medicine revealed it, is the role certain cells in the brain play in suppressing dopamine when we give up. When we give in to self-doubt or even a bit of laziness and decide not to accomplish a goal, cells in the brain spring into action and block the reward system from activating. The more we let go of persistence and the more we drop off from follow-through, the stronger these dopamine-suppression circuits become. That change in brain activity toward demotivation presents a huge risk for depression. I am not talking about garden-variety sadness. I'm talking about can't-get-out-of-bed, don't-want-to-take-the-day's-next-step, completely debilitating, shades–closed–until–4:00 P.M. paralysis. Depression poses health risks as it causes us to drop self-care: heart disease, insomnia, dental decay, addiction, inflammation that can cause joint, artery, and cognitive damage. We become isolated as we continually cancel plans, depriving the brain of the essential vitamins it gets from connecting with others. We lose years of longevity as we let meaning and purpose slip away. When we decide not to activate our Upgrade, when we choose not to water our seeds of wisdom, we put everything we've spent a lifetime building at risk. And chances are we could be knocking ten years off our life span.

One way to snap yourself out of the brain's "I give up" circuits is to pull yourself into problem-solving mode. Plan an exercise schedule for a week, like a brief daily walk. Reengage with friends and/or family. In Pauline's case, I asked her to find ways to keep the dialogue open with her granddaughter, perhaps going to Los Angeles to help her get settled in her new apartment. Think about travel, learning something new, going to church, reading and finishing a book, calling one person every day.

The second reason for stepping into the Upgrade, for not giving up on our own wisdom, our own bodies, on our own health, is for others. Regardless of our self-doubt, it looks like the world could use our help. Not that we have to solve every major global problem—that would be impossible. But what we do in our daily interactions makes a difference. How we relate to others contributes either to peace or to conflict. It's up to us how we want to participate.

This is the opposite of "senioritis," those final days of high school or college when the grades are already in and your destiny is set, when you don't feel like going to class and you do the bare minimum to get by. Instead, ask yourself what else you want to do. Pull back to the thirty-thousand-foot view and notice what you might not have seen while still in the weeds of caretaking or the roles we are trapped in by cultural bias. It's time for us to go on the road, at least internally, with the same excitement to explore that Kerouac and the Beat poets had as they shattered convention and sought new meaning and purpose. We can, as they did, see ourselves and the world through new eyes, even if we need reading glasses to bring the details into focus.

Talking with Ceci, it was clear that her life had completely shifted. "The truth is," she said, "I'm feeling better than I ever have. I'm taking barre classes three times a week and a restorative yoga class once a week. I try to take some long walks in between. The best thing about that is I can sleep through the night and I can cough or sneeze without peeing!" I was really interested in Ceci's news. Many women struggle with some form of incontinence, especially after childbirth or hysterectomy, but may not be aware that pelvic-floor therapy or exercise like Pilates and barre can strengthen the muscles enough to alleviate the issue.

I could see that Ceci was really taking the Upgrade seriously. It was a joy to hear. She couldn't stop talking about what she was noticing. "I'm really seeing now how much food impacts my body—joint pain, energy, everything," she told me. "I've pretty much stopped drinking wine. It wakes me up with hot flashes and my hands hurt the next day. It just isn't worth it anymore. And it isn't that my problems have all evaporated. At my age, there's a greater likelihood that I will lose more friends, get sick, have more really big, really bad things happen. There's nothing I can do about it. There's no way to control anything in life. "

There is a lie about every decade that hides the truth about life's suffering. We tell those in their midtwenties that it's the time of their life, that they are hitting their stride, when those same young women are probably suffering their quarter-life crises and feel like they've been fourteen for ten years. We tell them that they'll calm down in their thirties, but then anxiety over marriage and kids—whether or not we have them—takes over our lives. We say that life begins at forty, but that's exactly when the Transition starts and can mean we take a wrecking ball to our relationships and career. The zero-Fs-given fifties are fabulous, and just when you feel great, you reach that phase when back-to-back catastrophes of losing friends and family and career become expected events. We don't know when catastrophe is going to hit, but we know it *is* going to hit. As long as we are alive, we will experience joy and suffering. But with every obstacle, we find the courage to keep going. We engage the problem solving of the prefrontal cortex, we keep moving physically to help the hippocampus find the energy to store new memories, and we spring to excitement as cortisol, particularly for women in the Upgrade, boosts our interest in learning new things.

The biggest benefit to having gone through those moments that bring us to our knees is learning the lesson of setting aside pride. We know what we don't know, and when you can't figure things out, as Mr. Rogers said, we find the helpers.

The Upgrade Circle

When you surround yourself with those who do what you want to be doing and are who you want to be, then that's what you preoccupy yourself with and that's what you become. Neurons that fire together wire together. If you are problem solving and seeing ahead, you are engaging circuits that get lost without use.

What you spend time pondering, that's what you become. The thought or emotion takes over your mind, and soon it impacts your behavior. Too many women get stuck in a phase I call Transition-mind. It's a necessary developmental stage that comes with the Transition. It is dominated by grief over what's been lost and confusion about what's to come. Many women cope by clinging to their before-the-Transition identity, dressing like teenagers and in some cases even dying on the operating table during a cosmetic procedure.

As little girls, teenagers, and young women, it was easier to shift out of transitions—there were clear cultural models for what we might look like in the next phase of life. But the Upgrade isn't yet visible. We've seen the cat-lady downgrade; that's not a place we are longing to go and it's not our only option.

I invite you to consider that even if you are done working, you are not done growing. Take time to really think about how you want to live this part of your life. Take a postretirement sabbatical if you need to, as I did. Listen to your body instead of overriding it. Spend some time becoming familiar with who you are at this point. Your brain may be hungry to learn new things. There are options other than volunteering or taking a job for much less money than a retired man would make. Choose what reignites your passion. I am challenging you to press forward in the cone of growth and fearlessly make yourself, your femaleness, visible. Repeat: You're not done yet. Choose the way you want to move through these years, and build the life of which you have just begun to dream.

The Neuroscience
of Self-Care

Chapter 6

Before the Transition, my massage therapist used to call me her Maserati; like the finicky sports car, I was always in the shop. I pushed myself to the limit and called on her when I blew past it, which was often. I muscled through everything, regardless of my body's signals of fatigue, illness, or pain. When I felt sick, I reached for something to suppress the aches caused by the storm of proteins called cytokines released by the immune system to attack an invading virus or bacteria. I learned to take paracetamol (acetaminophen) for those aches instead of ibuprofen; ibuprofen kills antibodies and cytokines (and your stomach), which you need so that the immune system can do its job of killing hostile invaders like viruses and bacteria. You can bet I wrote myself a prescription for Tamiflu at the first sign of a virus, so that I wouldn't miss a beat at work.

For a while after I quit my job to help my husband recover from a surgery, I didn't change my habits. But I woke up one day in my early sixties with that familiar achy, fluish feeling and found myself doing something different. Instead of reaching for my prescription pad, I decided to rest and to trust my body, to respect the healing process and embrace feeling ill. I decided for the first time in my life to get out of the way and let my body use its natural powers. Respecting the wisdom of the body is huge for a doctor, especially when we can pretty much prescribe anything for ourselves except opioids or drugs like Valium. And for the first time, I didn't need to override symptoms and prop myself up to work.

Think about what it was like to be a newborn. Cycles of eating, sleeping, and getting rid of waste controlled not only your day but that of your caretaker. Every thought, plan, or movement was governed by the very strict schedule your tiny body needed. Meals, naps, and baths were all kept on a strict schedule, because neurohormones are stimulated by a regular light/dark cycle and a reliable routine lays the best foundation for brain development and overall health.

So let me ask you a question. When does the body stop needing a schedule? I didn't ask when we stop adhering to the body's schedule. That happens the minute we get woken up for school, still tired, not ready to be alert. Later we do it to ourselves, overriding the body's needs in order to get to an early meeting after being out or up too late the night before with kids, sick parents, friends, or a spouse.

The body we inhabit and the brain we count on never stopped being the baby that needs a schedule. Yes, your body had tons of resilience in your teens and twenties, but that was a temporary illusion, the last of it evaporating with the Transition. The stress buildup of being part of the sandwich generation, working and taking care of kids and aging parents, chips away at our physical health—sleeplessness, daytime agitation and anxiety, more colds and flus. As this wave gathers speed and size, you can bet it crashes right over the health of neurons, the brain cells we need for strong cognition.

When I was studying neurobiology at Berkeley, I was lucky enough to have neuroscientist Marian Diamond as my adviser. She was one of the main researchers to prove that as we age, brain-cell death is not inevitable, that we can indeed grow new neurons well into our eighties and nineties. She also let me in on another cool observation: that unlike what most of us thought, the brain's most important and populous cells are not neurons.

More than half of the brain is made of cells that clean up synaptic trash and bring nutrition to neurons, cells that can cause or reduce inflammation in the brain. Called microglia and astrocytes (or astroglia), until the late 1990s they were simply thought of as the mortar that held the brain together: "glia" is Greek for "glue." But they keep the neurons nourished and function as the gardeners and the Environmental Protection Agency of the brain. Sleep is the state in which they operate best.

Getting to sleep and staying asleep requires the balancing of a complex system of neurochemical waves rising and falling, hormones released and withdrawn that drive behavior. At least fifteen brain regions are involved in modulating the sleep/wake cycle, and neuroscientists still don't know how they all interact. But we do know something about what influences a proper cascade: sunlight, movement, food, beverages, and feelings of safety. Sleep is key to the unfolding of the Upgrade. If we don't get enough, we won't feel like exercising. If we don't exercise, the ability to think clearly diminishes, and it impacts our decision making on everything from food to relationships. We can initiate a virtuous cognitive cycle or a vicious cognitive cycle, depending on our actions and priorities.

Throughout the day, the brain is busy solving problems, creating new memories and new connections. The waste by-products of those synaptic firings are discarded proteins and molecules that need to be carted off before they rot or clump together in tangles that can spread through the brain, causing all kinds of cognitive trouble. The lymphatic system in the brain, the glymphatic system, is a system of rivers and streams of cerebrospinal fluid that run around neurons and synapses. When you're awake, the cells puff into that place, slowing down the flow of fluid. When you sleep, they retract, so that the rivers can swell and flush all the wasted proteins out, sending them off to the body's lymphatic system for filtering to the liver, which converts waste into forms the kidneys and bowel can expel. If that flushing in the brain doesn't happen, a buildup of sticky, toxic proteins gets in the way of synapses firing. If you've ever experienced brain glitches after a night of no sleep, sticky, toxic proteins left behind could be the cause.

With at least six uninterrupted hours of healthy, natural sleep, the brain's immune cells, microglia, have the chance to emerge like careful nighttime gardeners to trim away the overgrowth and carry out trash like extra tau and amyloid protein, while astrocytes restore and nourish neurons, rejuvenating their ability to communicate with one another. Astrocytes also form the brain's protective barrier, keeping toxins out, sorting for proper brain nutrients, and delivering those nutrients to neurons, like a mama bird finding food to bring to her babies. Keeping the brain clean and well fed, astrocytes and microglia are crucial to brain and cognitive health. Without restful sleep, they can't do their job of balancing debris and inflammation, and that can lead to the emergence of zombie cells.

The Dementors in the Harry Potter series have a special quality in their ability to drain happiness and hope from anyone to whom they come near. They are given one job in the books: to guard the prison of Azkaban. They are effective because proximity dampens the will of the prisoners to be free and the vitality needed to plan escape. In a state of inflammation, whether from lack of sleep or a proinflammatory lifestyle, the body has the capacity to create cells that act just like Dementors; cells that have stopped dividing and exist in a half-dead-half-alive state, like zombies, spewing toxic substances. Zombie cells are the primary drivers of toxic inflammation in the brain, especially the hypothalamus. When chronic inflammation happens in the brain, microglia and astrocytes are especially prone to going into zombie mode. Once that happens, those cells cease to clean out brain garbage. Instead, they generate new toxins that create even more inflammation, setting the stage for brain degeneration, faulty connections, and finally severe cognitive decline.

There may be hope in dealing with zombie cells. Studies in mice are showing that even when things in the brain look bad because of disease, destroying zombie cells helps restore the ability to retain newly formed memories. In the experiment, once the zombie cells were destroyed, signs of toxic inflammation were eliminated, there was no brain shrinkage, and there were no physical signs of cognitive impairment in the brain. Newer experiments in mice are indicating that flipping a genetic switch on astrocytes might turn them into neurons, i.e., make new brain cells. Drugs that mop up wasted brain proteins and zombie cells are being tried in humans now. For all of us and especially anyone suffering from brain degeneration, this line of inquiry could be promising.

Permission to Sleep

"I had never been a good sleeper," Chittra told me over her cup of herbal tea, "but at least I used to be able to get six straight hours." She looked exhausted and desperate; for three years, she had been unable to sleep more than a few hours. "I fall asleep easily. By ten thirty or eleven P.M. I am out cold. But I wake up at two A.M., and if I'm lucky I'll get another hour between six and seven A.M. I feel like I'm losing my mind. I now have a hair-trigger temper, I'm incredibly emotional, and my window for being able to work efficiently and think clearly during the day is very small."

Chittra had tried sleeping pills but concluded that they didn't help. While they may make you sleep, they don't actually improve sleep quality. Drugs interrupt our natural cycles of light sleep, deep sleep, and rapid eye movement (REM) sleep, during which we dream about what we've learned, giving the nervous system and brain a chance to rehearse and integrate new knowledge and information. The side effects of poor-quality sleep—irritability, moodiness, foggy thinking—are pretty much the same as those of insomnia, with the added obstacle of potential drug addiction. Long-term insomnia and use of sleeping pills are both associated with cognitive decline.

"So what are you doing now? Did you find any other sleep solutions?" I was fishing to see if Chittra had gone to common over-the-counter sleep remedies.

"Yes, I did find something that helped. A friend used Benadryl on her kids to get them to sleep on an overnight flight." She paused when she saw the look of horror on my face. "I know it's a last resort and it shouldn't be done," she interjected, "but I figured it wouldn't hurt me as an adult."

That's where Chittra was dead wrong. I wasn't all that upset about the impact of Benadryl on a child's brain. I was worried about Chittra's Upgrade.

Acetylcholine is a neurotransmitter with many roles in the nervous system and brain. It is a chemical messenger that activates muscles, makes us alert and able to learn new things, and plays a strong part in the formation and retention of memory. Healthy rising and falling of acetyl-choline during waking and sleep cycles is essential to memory function. During waking hours, when acetylcholine is high, what we experience and learn during the day goes through a kind of first-draft wiring into the brain. During sleep, those memories undergo a reorganization process called memory consolidation, so that when we wake up in the morning, we remember not only what happened the day before but also what we need to do in the present. Consolidation requires us to practice in order to improve something we are working on learning and helps us remember what happened thirty years ago, in case any of it is relevant now. REM sleep, crucial to memory consolidation, changes in the second half of life. As women age, their nighttime sleep can be less steady because of drops in estrogen and hot flashes, among other things. Microawakenings become more common due to destabilized levels of melatonin, the neuro-hormone that helps us fall asleep and stay asleep. Sleep apnea becomes more common. All of this awakening contributes to problems with short-term memory.

Benadryl (diphenhydramine), like paracetamol, Nytol, Unisom, drugs to help with incontinence, tricyclic antidepressants like nortriptyline, and many other medications in this class, also share an additional property, what we call anticholinergic—the chemicals in these medications block acetylcholine, destroying the brain's ability to lay down memory path-ways. Over time the suppression of acetylcholine can cause memory glitches and can block those muscles that release your poo and pee, mak-ing it harder to defecate or urinate. And acetylcholine blockage can also cause cognitive difficulties. The cognitive outcomes reported include mild cognitive impairment, confusion, forgetfulness, dizziness, falls, de-lirium, and decreased psychomotor speed and executive function. I've had patients come in looking like they are in the throes of dementia, when many times they've been taking too many anticholinergic drugs. If I catch it in time, getting patients off the medicine usually restores mem-ory and mental clarity, not to mention bowel and bladder function. It turned out Chittra had sleep apnea.

Cortisol, Sleep, and Memory

Memory consolidation in sleep also requires low cortisol levels, especially during those first couple of hours of sleep. Cortisol is that hormone produced by the adrenals in a regular daily wave: high in the morning and low at night but also when the brain and nervous system get ready to respond to threat. So if you've had a fight or watched the news or read about violence right before sleeping, your brain will have a harder time with memory function from the cortisol alone. Chronic stress means chronic elevation of cortisol, making it nearly impossible to form new accurate memories during periods of extended grief and trauma.

A bit of neuroscience trivia dropped in my lap by Professor Diamond stuck with me over the years. When Albert Einstein passed away, he donated his brain to be studied. Its size and shape were unremarkable. But his astrocyte count was off the charts. He was famous for sleeping ten hours a night and taking a nap almost every day. A short nap, that is. If you sleep too long during the day, it will interrupt the nighttime cycle.

Einstein's high astrocyte count meant he had a clean and well-nourished brain. His habit of ten hours of sleep and regular, brief naps offers a lesson for the Upgrade. I know what it feels like when I've slept well: I am alert and energetic, the little things don't bother me as much, and I suddenly feel like my brain got a massive broadband upgrade. But making the time to get enough sleep may feel impossible or even counter-intuitive. If you've had children and you work, you've spent most of your life sleep-deprived. Add to that the caretaking of aging parents, and it seems like all bets are off. As clarity fades from the morning jolt of caffeine and a stressed-out, foggy brain takes over, it makes getting through the day feel like a Herculean effort. Over time, it just feels true that life is exhausting.

Research is showing that as we head into our sixties, we need between seven and nine hours a night for the Upgrade to unfold. Less than six hours is associated with sterile inflammation, the kind of inflammation not caused by an infection. Yet insomnia is just about the most common complaint in adults sixty-five and older. But sleep problems are not inevitable, and usually they are not permanent. We can take simple actions during the day to prevent them. Here's Louann's Daily Sleep Plan, step one in activating the neuroscience of self-care:

- **Get direct sunlight.** Spend at least ten minutes in the sun on a sunny day, forty minutes on a cloudy day or use a light box. A thirty-minute walk first thing in the morning usually takes care of it.
- **Exercise to the edge of feeling tired.** Do something vigorous for thirty minutes before 3:00 P.M. If you do it later than that, chances are you will still have cortisol levels at bedtime that will interfere with sleep.
- **Limit caffeine.** Drink no more than one cup of something caffeinated, and only in the morning; for the best results, cut out caffeine altogether if you can.
- **Consume no stimulating substances in the afternoon.** Skip foods and drinks like dark chocolate and caffeinated sodas (which you should skip forever anyway).
- **Drink no alcohol in the evening or with dinner.** Alcohol may make you fall asleep faster, but it will likely cause you to wake up in the middle of the night.
- **Focus on protein and nonstarchy vegetables in the evening.** Carbs produce sugar, which creates short-term bursts of energy that raise cortisol.
- **Try to finish dinner by 6:00 p.m.**
- **Go to bed at the same time every night.** Schedule your winddown and bedtime routine.
- **Get up one hour earlier if you've been feeling down.** This act alone cuts depression by a double-digit percentage. It means you have to go to sleep one hour earlier as well.
- **Shut off screens and devices at least thirty minutes before bedtime.**

- **Emphasize foods with L-tryptophan in the evening.** This is nature's sleep amino acid. You can find it in turkey, milk, cottage cheese, chicken, eggs, fish, pumpkin and sesame seeds, tofu, even bananas, but I'd eat those sparingly because of the high sugar load.
- **Nap sparingly, so as not to interrupt your nighttime sleep cycle.** Limit naps to ten to twenty minutes.
- **Invest in blackout shades and make sure the bedroom is quiet.** Or at least use an eyeshade and earplugs. Yes, you can get used to them. (I use custom-molded earplugs, since we live on a busy street.)
- **Make it cold.** Air temperature during sleep should be between sixty-six and sixty-eight degrees Fahrenheit. That doesn't mean you can't keep warm with blankets, and if your spouse complains, tell him to man up and get a heating pad. The super cold room temperature is good for him too.
- **If you get less than six straight hours of sleep, take a timed twenty-minute nap before 4:00 P.M.** Some who suffer chronic insomnia have been able to use three timed ten-minute naps per day to begin to downregulate the nervous system, aiding nighttime sleep. Long naps will absolutely interrupt nighttime sleep, so make your naps snack-length.

The Transition and Sleep

For me the Transition was a debilitating time. I woke up four or five times every night, the sheets soaked. I was going through two or three sets of pajamas nightly, and my days were miserable. Following the plan, I did start to sleep. My colleague Lynn, who was a few years younger and clearly going through the Transition, was showing signs of the irritability and memory glitches that come from insomnia. When I asked, she said she was doing just fine with sleep. Knowing women tend to be stoic, I probed a bit more.

"Are you really sleeping straight through the night? No little episodes of waking? No getting up to pee?" I asked.

"Ever since I did pelvic-floor therapy and have kept up with barre classes, nope, not waking up to pee anymore," Lynn said.

"No hot flashes? No warmth at night?" I pushed a little more.

"Well, I don't sweat or anything like you do, but three or four times a night I wake up hot and throw the blankets off. My husband usually covers me back up when he sees I'm getting cold," she admitted.

Waking up a little warm, rolling over, and falling back to sleep is fine if that happens once a night. But more than that and your restorative sleep is wrecked. Lynn's battle with the blankets was enough to keep her brain from getting the benefits of restorative sleep. I had a hunch a little adjustment to her HRT would make the difference. I asked her to show me her compounding pharmacy prescription. As I suspected, her combination of estrogen, DHEA, and progesterone were bundled into one topical dose at night. For some people this works fine, but for many, the behaviors that these three hormones elicit from the brain can come into conflict. Estrogen and DHEA are a bit activating. They are great for getting the engine started and firing up the brain's cognitive powers. Perfectly suited to morning. But progesterone is the comfy, cozy hormone of self-care. Remember, progesterone increases the activity of the calming, Valium-like GABA system in the brain, relaxing us and making us sleepy. It's a robust behavior provocation that will demand actions like cuddling in a warm blanket near a fire with the best hot chocolate in the world during a cold winter rain. Self-nurturing is just what the body needs to settle in before sleep.

Estrogen stimulates the growth of new connections in the brain; it's like ultrafertilizer causing branches to grow everywhere. Before the Transition, in the second half of a cycle, progesterone comes in and downregulates that overgrowth. Estrogen spikes during the Transition will cause much more overgrowth. On a microgram scale from ten to four hundred, estrogen can plummet to ten and spike to four hundred in the same day. The overgrowth can feel scrambling and disorienting and cause a lot of unfocused energy. Progesterone can be good for downregulating for better sleep and pruning the brain's overgrowth. But if you hit your brain with progesterone in the morning, it may make getting the day going a struggle.

Lynn was used to trying out new things, and she switched to using her progesterone at night. "I am sleeping like a baby," Lynn told me after three weeks. "I sleep straight through, no hot flashes, no waking up. I do wake up a little groggy some days and need to exercise and use the estrogen and DHEA right away."

"Great. Did you have to play with the dose at all?" I asked, because Lynn got her topical cream in syringes so that she could adjust the dose within a prescribed range. She's pretty sensitive and her doctor knows that. "I did," she said. "I ended up cutting everything in half. My doctor was upping estrogen to take care of hot flashes. It turned out that the higher dose wasn't making the hot flashes better, and I was getting really edgy. Once I found the right dose for me, I started to settle into a really comfortable sleep rhythm."

Estrogen plus DHEA can push us to the razor's edge of vigilance. If you have that wound-up feeling that doesn't go away no matter how much yoga you do, consider taking progesterone at bedtime. If you do, add it in tiny doses, upping it slowly over time, giving your brain and nervous system the time they will need to adjust and consider stopping DHEA. The same should be done with any hormone. Add it slowly, increasing in tiny increments if you are sensitive like Lynn. Most doctors won't know to do this. Remember, their training is for handling high-risk pregnancies and cancer surgery. Neurologists won't know either. They are trained for treating MS, Parkinson's, strokes, Alzheimer's, and others. None of them are trained in how our hormones impact the brain. Endocrinologists are trained but generally don't practice in this area. If you can find one with experience outside thyroid, diabetes, and rare endocrine issues like the disease of the adrenals, Addison's, and Cushing's, you're lucky. Although as a psychiatrist I was trained to understand the hormone as a neuro-chemical, I hadn't been trained in the specific issues of women until I got interested in them myself. There is no medical subspecialty that particularly trains in HRT for women, so don't expect your doctor to know. Gynecologists, functional medicine doctors, and nurse practitioners who prescribe hormones will know the most. Whoever you work with, you'll likely have to teach them how to treat you. That idea doesn't go over well with many doctors, so keep searching until you find a medical *partner*.

The thing to remember about hormones is that even if they are making you feel good, more is not always better. There is a sweet spot that will work best for you. It will be different from someone else's, and it may be different from a so-called normal range reported by a lab. You'll know what's normal for you because you will sleep, you will not have hot flashes, and you will feel better.

Reducing Stress

Chronic sterile—not caused by an infection—inflammation from the body's stress and threat system might be the single biggest long-term threat to healthy cognition in the Upgrade. By now you know the cascade of problems from chronic cortisol release, especially the risk of intractable melancholy. And it doesn't take a brain scientist to tell you that stress can make it hard to sleep. But the body's nervous system offers plenty of inflection points for teaching it to stand down from red alert.

Even before her Transition, Diane, now sixty, admits to having had a fairly explosive personality, triggered easily by insecurity and fear. "My sense of self got tugged at daily during what you call the fertile phase," she said in a conversation about her journey to self-care. "It was hard to stay strong when the men around me seemed to be working overtime to prove I wasn't good enough, or to shut me out of opportunities." I knew what she meant. It happened to me all through med school and in several early jobs. "It got harder to handle as soon as I hit the Transition. I kept getting knocked over by things that hadn't bothered me in a long time. Sleep was harder. Internally I felt like I was completely out of control.

The confidence and certainty that had sustained me earlier in my life just vanished."

I asked Diane how she started to find her balance. "When I was a kid," she said, "I was a pretty serious musician. I played the oboe; I competed and performed a lot. If you get too nervous, your hands get cold and you lose the fine control you need over your fingers. Anxiety makes it harder to control the tiny muscles in the face and around the mouth that you need to recruit in order to produce a steady sound. And if you can't calm down, the stomach tightens, keeping you from getting a deep enough breath to play through a phrase." Everything Diane was telling me was an indication of the sympathetic stress reaction takeover.

"So what helped?" I asked.

"They taught us alternate-nostril breathing. When my mind started feeling scrambled during the Transition, somehow I remembered this technique. I did alternate-nostril breathing while waiting for my toast or the Nespresso machine to finish, any opportunity, until it became a habit during stressful moments. It led me to yoga and meditation, which have given me more targeted and sophisticated tactics to keep the threat system in my brain from overwhelming my body."

Diane had something there; alternate-nostril breathing is an ancient technique that can trick the vagus nerve into activating the body's calming circuits. Try it sometime. In a moment of activation, plug your right nostril and take a fine, slow breath through your left nostril. Then plug your left nostril and breathe out a fine, slow breath through the right. Try to keep the breath from being audible, even to yourself. It can help to imagine breathing in and out through a pinhole. Do that three times— plug the right on the inhale, plug the left on the exhale—and then switch, plugging the left on the inhale, the right on the exhale for three rounds. The final three breaths should be fine, slow inhales and exhales through both nostrils. (For more on this technique, see appendix, page 258.)

There are nine breaths in one round of this practice, and here's how it works on the brain. Inside your nostrils are tiny little hairs that have nerves at their roots. Their job is to communicate changes within the sinus passages to the brain. Those little hairs expect that when you breathe in, the nerves in both nostrils will be stimulated. When we disrupt that expectation by closing one nostril, it alerts the brain to pay attention to the breath. The cerebellum registers that balance is off and recruits a keen, settled attentiveness. Following a predictable pattern engages the cortex, the thinking organ of the brain, to connect with that little bit of alertness that has been provoked. I use this technique throughout the day to settle triggered brain circuits. It allows the nervous system to get out of threat-response mode, and anything I can do to get myself out of distress supports cognitive function both short term and long term.

In the Upgrade, we can consciously activate healthy circuits for social connection, safety, and self-nurturing. Visualization is one of the most powerful ways for this to occur. At Emory University's Compassion Center, they have studied the effects of meditative engagement with what they call a *nurturing moment*. As participants are asked to recall or imagine a moment of protection, safety, peace, or nurturing, whether spiritual, among other people, or in nature, what surprises people the most is that all the circuitry that has experienced caring, whether giving or receiving, can be activated easily. It takes just a bit of concentration on the details of an experience or imagined moment of safety or protection, calm or refuge, to reawaken it as though it were happening in the present moment. Whether we've had good or bad parenting, good or bad marriages, the fact that we are alive today means that at some point someone cared unconditionally for us, even if it was just for one moment. I know I have a nurturing-moment memory of sitting on my father's lap while we were reading a big animal book and my mother was beaming while taking a photo of us. Rehearsing these moments to the point that we feel the presence of their impact means that in tough moments we are more likely to find ways to reconnect, to activate the bonds of the Upgrade. I use mine every morning as part of my prayer and meditation practice (see appendix, page 253). It's the first step in a compassion training protocol. And just so you know, the entire protocol has been proven to have a greater impact on helping the brain and body recover more quickly from stress and reducing inflammation than simple mindfulness of the present moment or even support groups. I now teach this to my patients.

The Gut Brain

What and when we eat has almost as big an impact on mood and cognition as do sleep and stress. And it's all because every single body has two brains. And no, it's not what you think. I'm talking about the one that stretches from your esophagus to your anus: the gastrointestinal tract.

Terri had been waking up every morning at 5:00 A.M. in a cold sweat, heart pounding, her insides in turmoil, feeling terrified. She called me because her GP had diagnosed these episodes as panic disorder and wanted her to start on antianxiety meds. Terri was concerned about taking them; her sister had had a bad reaction to a similar drug, and her experience of getting off it had taken her to a psychologically frightening brink. Having known Terri for so many years, and having treated so many patients with panic disorder, my instincts told me that this wasn't panic. So I asked her to try an experiment. "Get your doctor to write a prescription for five tablets of five milligrams of Valium," I suggested. "And then do the following: When you have one of these episodes, break one of the pills into quarters. Put one quarter under your tongue and let it dissolve." When Valium is dissolved under the tongue, it gets the chemicals into the bloodstream and to the brain more quickly; Terri would feel the effect of it within a few minutes, rather than waiting the twenty or more it could take to absorb into the bloodstream through her digestive tract.

An overactive amygdala is a key brain organ that triggers that heart-racing, out-of-body, sweaty feeling of impending doom. If Terri was having a panic attack, then Valium, which quiets the amygdala's fear circuits, should make both the symptoms and the panic lessen or even disappear. But if it wasn't a panic attack, then while the fear might subside, the symptoms would remain. And that would mean something else was causing Terri's attacks.

In Terri's case, it was not panic. "Louann, I did what you suggested for a week," she told me, "and every single time, the fear went away but the intestinal turmoil, cold sweat, and heart pounding didn't stop." My hunch was right. Terri's emotional overwhelm was coming from another part of her body, a part whose language is vague yet powerful in the control it exerts over the brain and nervous system.

Further testing showed that Terri had severe, undiagnosed food allergies. I can already hear the question bubbling up, because I once asked the same one: How does an allergy that's not life-threatening, an allergy that's not causing your throat to close (anaphylaxis), cause a feeling of panic?

An allergy happens when ingestion of a normally benign substance, like pollen, peanuts, gluten, or soy, provokes an immune response from the body, which attacks the substance as though it were a dangerous invader that needed to be annihilated. Allergens leave traces along the various filtration layers of the digestive tract. The immune system, a huge proportion of which is in the GI tract, attacks whatever tissue these traces attach to, causing inflammatory reactions—just as a wound turns red and swells.

The entire gastrointestinal tract, from your esophagus to your anus, is wrapped in a web of interconnected neurons burrowed within the bowel wall that regulates motility and secretions. It has 50 to 100 billion neurons, the same number of neurons as in the entire spinal cord. It's the GI tract's very own nervous system, which regulates the key functions of the bowel: movement of food and nutrients, secretion and absorption of fluid, repair of the bowel lining, and blood flow through it. Called the ENS (enteric nervous system), it's such a huge number of neurons that it's called the body's second brain. The ENS can signal the brain via the vagus nerve about what's going on in the GI tract, especially if there is distress, as in an allergic response, and try to get the conscious mind to form an idea for corrective action. That's the usual mechanism for the nerve/brain dance. It works great when signals take the form of a clear message to the conscious mind: *Ouch, the stove is hot. Move your hand.* Or if you just ate something rotten: *Throw up now.* But when the signal is less clear and travels up a long nerve like the vagus that attaches to and communicates with the parts of the brain that control unconscious functions, those signals can be more subtle and harder to register.

The Vague Signals of the Vagus Nerve

Widely distributed through the voice box, chest, and gut, the vagus nerve—"vagus" comes from the same Latin root as "vagabond"—is a long and wandering nerve. It attaches at the brain stem—the part of the brain that is in charge of functions like breathing, heart rate, blood pressure, body temperature, wake and sleep cycles, emotional well-being, digestion, sneezing, coughing, vomiting, and swallowing. It is partly responsible for balancing both our threat response (sympathetic nervous system) and our calming response (parasympathetic nervous system). Without the potential calming effect, sensations like tightness or uneasiness in the stomach and throat that can be a precursor to fear; a burning in the throat, chest, or upper belly that can come just before an explosion of anger; or a lump in the throat that is a harbinger of sadness can make the difference between being able to wait out a feeling and having an emotional explosion. It's often called the sixth sense because of how sensitive it is, and it also connects the brain to the gut.

The inside of the intestines is also populated by a vast jungle of single-celled creatures that help us process what enters the body and bloodstream, extracting nutrients and expelling waste. The usually healthy bacteria that take up residence in our intestines help us grow, develop, and live a healthy life. The flora and fauna in our intestines—collectively known as the microbiome—are essential to helping our immune system do its job of maintaining balance between inflammation—when confronted with an invader—and anti-inflammation (regulation when the threat is gone). We've known that for decades. But fascinating new research is indicating that the microbiome has several ways to impact the brain; it may strongly influence mood, clarity of thought, even emotions.

Through the ENS–vagus–brain stem circuit, the microbiome sends up *We're okay, so you're okay* or *We're not okay, so you're not okay* signals that are influenced by the environmental shifts caused by food, sleep, travel, and exercise habits. It makes you feel tired, moody, or even anxious when you switch up a diet or keep an irregular schedule. Think about the last time you spent a day or two not being able to eat properly. No energy? A little irritated? Cytokines, chemicals released from immune cells, can be responsible for that feeling of sickness—so lethargic you can't get out of bed, foggy-headed, fearful, down, or even depressed. As the insula scans to see if everything is okay, via the ENS–vagus–brain stem circuit, the microbiome might be complaining loudly if there are inflammatory cytokines because of an autoimmune response: *No! Not okay! A lot of inflammation in here. It's too hot and too acidy and we can't survive!*

At the same time, the language of the vagus—the nerve responsible for transmitting the microbiome's signals to the brain—is, well, vague. It has two extremes that issue clear commands: *Total well-being, so keep doing what you're doing* or *This is poison, so vomit now.* Everything in between is pretty unclear. When the microbiome is unhappy, the signal might be to eat more, sleep more, flop on the couch, cry, or panic.

Scientists in Canada proved the microbiome/anxiety connection in 2011 by transplanting poop from a group of mice genetically bred to be anxious into the intestines of a group that was not. After a short period of time, the nonanxious mice behaved like the anxious mice just from the change in their microbiome caused by the poop from the anxious mice. The opposite turned out also to be true: Transplanting poop from a healthy donor made the anxious mice calmer.

When the immune system goes into overdrive, as it did in Terri's case, and kicks up inflammation in the gut in response to allergens, the ENS will detect a threat and will set off any alarm it can. The lives of the microbiome are on the line, and they don't have a great translator in the vagus. As Terri's allergens caused a massive destabilization to the community of roughly 100 trillion bacteria communicating with her ENS, the poor little creatures were screaming for help, but because of the vagus, the message was arriving in an unrecognizable language.

Upgrading the Gut-Brain Team

It wasn't a surprise that Terri's symptoms were worsening during her Transition. Microbiome composition changes with age, influenced by the shift in hormone levels. Progesterone helps bolster the microbiome's population of lactobacillus, which is protective against depression and anxiety. Estrogen helps regulate the microbiome and maintain the diversity needed to keep it in balance. Lower estrogen means the body has a harder time adjusting to inflammation that can disturb the balance of bacteria. When that happens, it can kick off a hostile environment, allowing the growth of hostile bacteria that can eventually impact mood and cognition. When unfriendly bacteria start to outpace the supportive bifidobacterium and lactobacillus, particularly *Lactobacillus reuteri*, it not only impacts mood but also can interfere with memory. *L. reuteri*, which we receive from our mothers at birth, provides protection from unfriendly species. Its presence in human microbiomes decreases both with age and with overuse of antibiotics in childhood and beyond—antibiotics kill not only harmful bacteria but your friendly ones as well. Artificial sweeteners, including stevia, also cause microbiome imbalance and changes in the gut/brain signaling. Combine this with the effects of dwindling estrogen and progesterone, and the imbalance of bacteria not only exacerbates the brain fog and mood changes of the Transition but also can make Swiss cheese out of the digestive tract's lining. When that happens, toxins and bacteria can leak through tiny gaps into your body instead of being flushed out. When these foreign substances hang around where they aren't supposed to be, it triggers chronic inflammation, which in turn can start a cascade of issues that lead to cognitive decline. This is how inflammation in one part of the body can trigger inflammation in the brain; inflammation in the brain is a cause of cognitive decline.

New studies are showing that the microbiome can be a contributing factor in brain diseases like MS, stroke, Parkinson's, and Alzheimer's. In 2019 researchers at Johns Hopkins did an experiment with mice to explore the connection between the gut microbiome and brain health. In the experiment, the mice were given bacteria that would harm their gut and cause them to develop Parkinson's. Parkinson's is a disease that causes nerve cell damage and cell death which in turn causes the decrease in dopamine in the brain. Without dopamine, movement gets harder, everything from trying to put a spoon to one's mouth to swallowing. In the experiment, they separated the mice into two groups. In one, they simply observed the progression of the disease. In the other group, they cut the vagus nerve, interrupting the gut's messages to the brain. The group with the intact vagus nerve developed Parkinson's. The group with the severed nerve did not. They concluded that some injurious gut proteins created by the bacteria were signaling the brain via the vagus nerve and causing Parkinson's.

The same study also showed that we don't have to sever the vagus nerve to prevent the microbiome from sending unhealthy messages to the brain. We can take action through diet and exercise, both of which have a positive impact on microbiome health. The healthier the microbiome, the healthier the signals sent to the brain, and the better chance we have for optimizing cognition in the Upgrade. Everything that causes weight gain is also bad for the microbiome. (See appendix, page 268.) And what's bad for the microbiome is bad for cognition. This doesn't necessarily mean you should go out and start gobbling down a bunch of probiotics. What it does mean: Focus on foods high in fiber, which promotes regeneration that may help protect the brain from both Parkinson's and Alzheimer's diseases. And to grow and support the right kinds of helpful bacteria, follow the Mediterranean diet, with lean proteins, lots of leafy greens, and a small amount of healthy fats (see appendix, page 268). And timing is everything: Make sure you leave twelve to sixteen hours between dinner and breakfast. It stresses the metabolism in a good way, just like building muscles in the gym. Giving the GI tract a rest during these hours helps the microbiome flourish. And even without changing your diet, moderate aerobic exercise, thirty to sixty minutes three times a week, has been shown to increase the diversity of microbes in your gut and reduce intestinal inflammation in just six weeks, although scientists don't yet understand why.

Unless you have other medical conditions, a probiotic supplement can be helpful, especially after you've been given a course of antibiotics, which destroy the good, brain-supporting bacteria. The microbiome needs to be seeded and given good fertilizer to grow back healthy. If it's recommended that you take some, get a good concentrated probiotic supplement with 200 billion cells, and make sure it contains *L. reuteri*.

When the microbiome is populated by friendly flora, those bacteria trigger balance and well-being. You can eat and exercise your way to health, but you'll have to do it without the usual high-sugar and high-fat comfort foods or artificial sweeteners, which are the fastest route to a downgrade. There is a beautiful garden in our guts working to nourish our body, cognition, and mood, growing what we need to fuel the Upgrade. But it needs our support in order to do its job.

Food, Alcohol, and Cognition

I have been involved in Weight Watchers for twenty-five years. I am pretty compliant and I have the data to prove it, collected over the course of time. I don't drink much these days because it interrupts my sleep and I don't feel well afterward. But Sam and I were out at a friend's one night for dinner, and I decided I was going to join in on everyone's fun. The couple we were visiting are serious gourmets and have amazing knowledge of wine, so I had a little red and a second piece of the flourless chocolate torte they had made. It was delicious. All that sugar gets the brain very excited—glucose is its favorite fuel, easy to access and to use. So I was in a bubbly mood. We went home, I fell asleep quickly, and I was happy with my FOMO-driven splurge.

At 3:00 A.M., I woke up, heart pounding, drenched in sweat. I was scared and spent two hours thinking I was having a heart attack. At around five thirty I realized it was reflux, so I took an antacid and fell back to sleep around six. When I got out of bed at nine, I was groggy and my hands hurt. My fingers and knuckles were red and every joint ached.

Inflammation in the body is, at its best, a choreography of healing processes. We cut ourselves, and proinflammatory cytokines rush to the site to start building a bridge for healing. Similarly, we can trigger a wound response inside the body through the wrong food and drink, lack of sleep, or lack of exercise. Just a tad too much sugar can stimulate the release of those proinflammatory cytokines that cause redness and swelling, but this time inside the body. It happens more easily as we age; the body becomes more susceptible to inflammation.

In the brain, proinflammatory cytokines reduce the signaling in areas that stimulate muscles, causing that feeling of not wanting to move, suppressing motivation. Brain inflammation shuts down energy production in neurons, slowing cognitive processes including memory. In a state of inflammation it's harder to read, work, or concentrate for any length of time. When cytokines remain in an overactivated state, as they are in chronic inflammation, they can have significant impact on the female brain by disrupting sleep and decreasing the release of feel-good neurochemicals such as serotonin and dopamine. Instead of making the feel-good neurochemicals free-floating and available to the right circuits so that we will feel and sleep better, cytokines encourage the brain to suck them all back up. When that happens, you won't feel their benefit, the experience of well-being.

In one fell swoop, proinflammatory cytokines can cause moodiness, low libido, anxiety, insomnia, intractable melancholy, and muscle weakness. In a healthy homeostasis, the inflammatory process is a beautifully choreographed dance performed by the immune system, in which proinflammatory and anti-inflammatory cytokines are in balance. But things get tricky when cytokines linger too long, past their usefulness in repairing tissues. When we ingest a consistent flow of sugary food and drink, cytokines overstay their welcome and inflammation becomes chronic; the result is progressive tissue *damage* rather than repair.

The decrease in estrogen after the Transition makes it harder for the body to be resilient in the face of inflammation, and with weight gain around the middle, toxic belly fat also releases proinflammatory cytokines, which in turn sparks chronic low-grade inflammation. A 2013 study in Iowa even showed an association between inflammation and a shortened life span.

In amounts greater than one glass of wine, alcohol becomes an inflammatory brain toxin, damaging the ends of neurons and impairing the growth of new cells, especially in the hippocampus, the brain's organ of memory. Several brain imaging studies and tests for memory and learning skills show that excessive drinking increases the risk of cognitive decline and dementia, and some are pretty alarming. For instance, a 2013 study in Australia found that 78 percent of people diagnosed with alcohol-use disorder displayed some form of dementia or brain pathology. The effects of alcohol on the brain are so powerful that there are even specific types of dementia, distinct from Alzheimer's disease or vascular dementia, that are called "alcohol-related." Heavy alcohol use can literally restructure your brain, causing permanent damage, especially in the frontal cortex, which is your planning, judgment, and cognition center. Structural changes can be permanent.

The kicker: The bad effects of alcohol on an aging brain are more damaging in women than in men. The toxic effects can make the symptoms of the Transition worse, slowing metabolism and increasing the risk of osteoporosis and insomnia.

And if that weren't bad enough, women in their sixties today are drinking heavily more than ever, even more than men of the same age. Studies at the National Institutes of Health show that while men drink out of social pressure, women do so in response to emotional pain. Research on women and alcohol in the last twenty years has shown that damage to the brain, liver, and metabolism happens faster for us, even for those of similar muscle mass and weight to a man. Light alcohol use—no more than one drink per day for women—may actually be neuroprotective against dementia. But even one glass can exacerbate hot flashes and sleeplessness.

I know. It's not the best news, but we work with what we've got.

The combination of sugar and alcohol in the female brain can turn the Upgrade into a downgrade faster than you can uncork a bottle of champagne. In many post-Transition women, sugar and alcohol can trigger some concerning inflammatory consequences. Increases in inflammatory markers, which can be measured by your doctor, are a direct threat to the female brain and manifest themselves in both cognitive decline and intractable melancholy. A 2019 study showed that high levels of CRP (C-reactive protein, a common marker of inflammation) in older adults predicted worsening mood symptoms just twelve months after elevated levels were detected. And over time, that lack of positive mood also predicted increasing inflammation—a vicious cycle becomes established. In a French study of eighty-to-eighty-five-year-olds, those with the highest levels of IL-6, another cytokine marker of inflammation, showed the greatest shrinkage in their brains, which is a sign of predementia. Inflammation may be a powerful factor in brain shrinkage—and thus cognitive decline—as we age. And it's a powerful factor in chronic melancholy and insomnia.

On the flip side, food, sleep, and exercise can be medicine. Eating anti-inflammatory foods at the right times, keeping belly fat under control, and making sure you're moving enough can turn inflammation around, making a huge difference in mental sharpness now and for the future. We can eat, sleep, and exercise our way to an Upgrade.

———

I don't remember where this came from, but when I heard it, it stuck in my mind: "'Eat, pray, love' takes on another meaning at this time: I eat less, I pray more that it doesn't make me fat, and I try and fail every day at loving the body I have." Like most women in our culture, I've spent a lifetime denying myself food that I wanted in order to look a certain way. Now I've tried to flip that formula on its head and think about what's important to me in the Upgrade. Yes, I want to be healthy and look great, but my priority is having a strong, sharp, clear, and quick brain. That's the only way to have a second half that's even better than the first.

Keeping Your Reserve Tank Full

During the fertility era, what we prioritized was pretty much dictated by biology and fear. If you ask most women what they want more time for in the Upgrade, they say self-care, whether it's exercise, rest, or transformative spiritual practice. It's a wake-up call in the Transition. It announces itself with a feeling that you are burning too deeply into your reserves, that if you don't do something soon you'll be in serious trouble. I hear this from women as they enter the Transition all the time. We tend to ignore those signals until we break.

"I've been saying for years that I needed a break," said my colleague Lynn, "that I needed a vacation. And when I used to blow past something like that, I might get a cold or a flu that would put me in bed for a few days or a week. But this time I got a cold, the flu, and pneumonia, and I was out for two months. And recovery? They say six months before I can feel normal, two years before my lungs stop being triggered by irritants."

What I urge women to hear at this moment is one message: Trust your body, trust your brain, hear the cries of the nervous system. If you are alive in the second half, you came from people who genetically were tough. They were survivors. You made it to the second half because you have better-than-average genes. So even if they don't work the way they used to, recognize that you can trust your body and brain. But the approach has to be different.

You have to give the body the right conditions to heal. Rest, nutrition, sleep, exercise, daylight, managing your threat response, being silent to listen for what the body needs. Sometimes that will be cardio, sometimes restorative yoga. Listen to and trust what it says, and do your best not to override it with fear or anxiety over how long it's taking to get better. The time frame of healing changes in the Upgrade. It's messy. Things we thought we got over come and go and take their own time.

The Upgrade, if we listen, teaches us that if we want to keep our cognition, we have no option but to hew closely to what the brain and body need. And it's often not an option. When the burdens are too big, learn to unbite your tongue. Find a way to say, "It's your turn. My body needs to fall apart. Stop asking me for things. I need ten minutes. I need a day. Please figure it out yourself." We can begin to shift the environment around us so that it becomes more healing and supportive. We have the choice to recognize our responsibility for our decisions. We have the freedom now to change everything.

Your Brain in Search of Connection

Chapter 7

"I never imagined how in love I would fall," Sherry told me on our weekly call. Widowed for five years, she had sold her café in Connecticut and moved to Arizona to be near her son David's family. She has three grandchildren and a great relationship with her daughter-in-law, Elise. She's a big part of the children's lives, participating in their school activities and staying with them when Mom and Dad need a getaway. She's become a coconspirator to her sixteen-year-old granddaughter's fashion rebellions: If her mother won't let her have it, Grandma will get it for her—within reason, of course. All done with a wink and a nod to Elise. "That part of my life with the grandkids is great," the seventy-two-year-old tells me. "But don't kid yourself. It's hard and lonely. I am nowhere near my friends on the East Coast, and there's only so much David wants me around. With that move I went through a really big depression a couple of years ago. As you know, I didn't think I would ever climb out of it."

The part of the brain that registers social pain from isolation, the anterior cingulate cortex (ACC), is also part of our basic primate empathy circuit. They are the same circuits that help us figure out what to do when a baby cries or another's face expresses sadness. The ACC is part of what enables us to feel and react to another's emotions and pain. High levels of cortisol from feeling isolated can make that wiring go askew over time. We misread the cues of language and faces; we stop responding to sadness. If we remain isolated long enough, reconnecting won't feel like an option. The attempts of others to reach out to us will feel hostile. We are more likely to lash out and push family, friends, and qualified mental health practitioners away.

The phrase "I could die of loneliness" is not just a melodramatic cliché. It's both a deep evolutionary memory and a physical reality. Wired to live in the wilderness, human beings, like most other mammals, have the best chance of thriving in groups. When there are others to help share the load of gathering or cultivating food, of caring for one another, of fending off attackers, survival comes from collaboration. For women, this instinct is strong. For us, smaller and often with young children in tow, safety has always come in numbers, especially when faced with a violent male.

One of the jobs of the brain and its nervous system is to signal us to engage in acts that lead to survival. When we are thirsty, we feel dry, headachy, tired, and foggy. We hydrate and the nervous system stops sending hormones and chemicals to the brain that push us to drink. In time, symptoms go away. When we are hungry, the stomach tightens and growls, sometimes painfully; we feel exhausted, stressed, and moody. When we eat, the symptoms dissipate and the nervous system tells the brain that it is safe to close the refrigerator door and leave the kitchen.

Connecting with others—the need to belong—is as deep a survival instinct as eating, drinking, and sleeping. Social connection activates the brain's physical reward circuitry far more than researchers imagined. At the University of Chicago, they found that the brain's strong pleasure pathways related to social connection have evolved over hundreds of millions of years in response to the success of collaboration in survival. The joy of belonging brings an accompanying feeling of lightness and ease in the body. At the same time, they found that social pain piggybacks on the body's physical pain circuits. So when we are hurt or heartbroken or alone, that feeling can become actual pain. There is no louder signal to most nervous systems than a blow or cut to the body. It stops everything and lets us know that we are in danger. Researchers hypothesize that it's probably evolution—the deadly consequences of being alone in the wild or left unprotected as a helpless newborn—that made it possible for social pain to hijack physical pain circuits.

Belonging circuitry is among the strongest in any mammal. But recognizing the symptoms of lacking connection is much harder than recognizing the symptoms of hunger, thirst, or sleepiness. We can attribute lethargy to hunger, lack of sleep, or exertion. We can attribute body aches to the onset of flu, an allergy, arthritis, lack of sleep, or a life stress. But we rarely hear the nervous system's calls to solve isolation, drowned out as they are by a collective cultural boom box playing a rugged, individual-worshipping "survival of the fittest" anthem on full blast.

Let's just bust that lone-wolf lie here and now. No one does anything alone. It's neither theoretically nor logically possible. Without others you don't have life. Without others you have nothing to eat or drink, no home to live in, no roads to travel, no work, no source of income. Whether they intend to or not, the actions of others help us survive every moment we are alive.

Columbia University psychologist Scott Barry Kaufman analyzed the results of several studies on gender and personality conducted across various academic institutions. Using personality traits, he was able to determine whether a brain was female or male with 85 percent accuracy. So when we say that dominance, aggression, and rigid hierarchies are male personality traits, we can say, according to Kaufman's work, that the statement is true. The research shows just as reliably that women are more likely to be agreeable and men more likely to be disagreeable. How this plays out is that women tend to value collaboration, transparency, and fluid organizational structures, while men admire the lone wolf and value fixed hierarchies that stay in place for long periods of time.

So when we buy into the worship of self-sufficiency, we are mostly operating from an introjected male value. When we are sad and roll up like a potato bug, refusing to reach out for fear of being seen as weak, we are overriding the life-giving survival mechanism of connection. And we are endangering our health. Not just as women but as human beings.

The Tent of "Me"

It wasn't just the mere act of moving to a new city that was disorienting to Sherry. It was incredibly hard as her brain, her mirror neurons—those cells in the brain responsible for reflecting the nervous system of another, helping to create a bridge of empathy—and nervous system went through the adjustment of losing the former affirmations of her visceral sense of herself. It wasn't just the loss of the familiar friends; it was also the dry cleaner, the woman who did her hair, the clerks she saw at the grocery store, who in regular interactions shaped her sense of self. It would take years for the nervous system to adjust to the shift in feedback: both the loss of familiar human beings and the new input of those who were beginning to enter into her life.

Awareness of how the brain and nervous system incorporate others into our sense of self means we can collaborate with it instead of being at its mercy. We can Upgrade that sense of self into who we have always wanted to be.

When the brain constructs a sense of self, it must do so via the experience transmitted by the entire nervous system. An embodied sense of self is generated and contained as a unique combination of patterns across multiple neurons in each circuit that contributes to the construction. The feeling of "Me" is stored in patterns of nerve impulses across widely distributed systems, which interact not only with many complex circuits of the brain but with all the large and teeny tiny nerves in your skin, your organs, your blood vessels. Experience, encounters with others, and the imprints they leave on the nervous system are major components of our sense of what it feels like to be "Me" in my own skin. With every connection and bond, the nervous system matches and incorporates the expressed thoughts and unexpressed behavior patterns of those with whom we become familiar. The more frequently our nervous system comes into contact with another's nervous system, the more influenced our own becomes by theirs, the more we incorporate the feeling of being around them into the feeling of "Me."

While the brain's motivation and reward system, via the neurochemicals dopamine and oxytocin, is triggered intermittently by new people and new discoveries, it is turned on full blast and kept purring along through consistent contact with those to whom we feel closeness and love.

Sitting or walking or talking together, our heartbeats attune, our blood pressure syncs, and we begin to breathe in unison. Our nervous system triggers muscle movements in response to unnoticed cues; like musicians playing in an ensemble, we sense more than see what others need from us and find ourselves responding to subtle signals without conscious awareness. This attunement, when reciprocated, grows into attachment: the very deep sense that a partner, child, parent, or sibling is an integral part of who we are.

When they are not there, not only do we miss them, but we physically miss who we are when we are in their presence. Nerve signals that we rely on to tell us everything is okay with the "Me" that feels familiar become scrambled when those most familiar to us are no longer in our lives, just as they are scrambled when a body part is removed. If there is a dramatic or abrupt change in a relationship with someone to whom we are close, to whom we are used to being around, the brain and nervous system crave their proximity intensely. The nervous system goes into acute withdrawal with the death of or rupture with a spouse, child, parent, close family member, or old friend. Signals indicating lack of familiarity start setting off alarms that something isn't right. Grief can become complicated by the nervous system's distress, overwhelming us with anxiety, fear, and sleep interruption, as well as dramatic changes in appetite, food preference, and libido. It can feel extreme, like our world is coming to a catastrophic end.

Anticipation of the lack of social connection, when you feel you've been treated unfairly or you're afraid someone will be pissed over what you did or didn't do or said or didn't say, is among the most painful physical and emotional experiences we have. Women who divorce around the time of the Transition have told me this feeling is particularly strong as their married friends begin to drop them, in some cases because they had socialized as couples but in others because some married women are afraid a divorced woman will go after their partners. The pain circuits triggered by the feeling of being ostracized are very physical. It feels apocalyptic.

During the struggle of the Transition, as she focused on rebuilding her work life, my patient Terri had lost touch with an ecumenical contemplative group she'd participated in for twenty years. Out of the blue, Sandra, a colleague who had become a very close friend when they met at the beginning of their careers, called and suggested she join her in attending a candlelight vigil for peace during the upcoming holidays. Terri declined. It was across town in Atlanta, and was to start at 6:00 P.M.; traffic would be at a standstill. Terri's attempt to build her own graphic design firm outside of the corporate umbrella she was used to working within was taking longer than she'd hoped, and a low-grade, pervasive anxiety over future income meant she was working more hours than she should. The stress was chipping away at her sharpness and creativity; she was forgetting the lessons of self-care she had once known to restore her powers. She couldn't remember the last time she had taken a real vacation.

During the days after the call from Sandra, a persistent sensation dogged Terri that something had been missing in her life. It felt physical as memories of Sandra's and her search for meaning, something beyond what they'd valued in their early lives, began to activate her brain circuits of connection. At the last minute, Terri realized she desperately wanted to attend the vigil. She'd been working hard and mostly alone for so many years, and she was once again hungry for what the group had to offer. She decided to go. Looking up at the clock, she hesitated; there was no way she would make it on time. But a wave of determination—she didn't know where it came from—pushed her to override the urge to give up, sit back down, and turn her attention to work. She stood up, packed her things, and left the office.

The decisions Terri seemed to be making consciously were happening in concert with an activated brain and nervous system. By the time she reached for her car's door handle, she was receiving strong, contradictory signals simultaneously urging her onward and making navigation difficult: eager to get there, stressed over arriving late. It's the same push-pull we get when we are really hungry for breakfast but running late and need to get out the door quickly. Hijacked by the stress of haste, the brain is less able to focus on the small actions needed to make breakfast. Hence opening the box of oatmeal while rushing to grab the blueberries, both spilling, a lamp knocked over while stumbling to reach for the vacuum, errant feet smashing blueberries into the kitchen floor. The mishaps pile up, causing more stress and further delay.

In response to Terri's feeling both lonely and anxious about being late, her hypothalamus and pituitary fired up the adrenal glands to send out stress hormones to stimulate vigilance in the face of danger. The amygdala, the brain's fear center, came online too strongly, clouding her judgment and overpowering the vagus nerve's ability to keep heart rate and blood pressure in the sweet spot of well-being. Terri hated the idea of missing something and hated the idea of disturbing others, especially when she hadn't been around the group for the last four or five years.

In the right amount, stress hormones can make us alert and eager to learn; but combined with low blood sugar, financial worry, and a nervous system deprived of a sense of belonging, levels of cortisol and adrenaline were just high enough to make Terri feel jittery and confused. With the emotional amygdala cutting off pathways to the calmer frontal lobe—the seat of higher reason—she lost focus and swung a left a block too early. The GPS app showed her arrival would be delayed another ten minutes.

By the time Terri pulled into the poorly lit parking lot, her stress system was on fire, activating inflammation, the body's response to infections and wounds, as well as an overdose of cortisol and adrenaline, which is what the hypothalamus and adrenals throw at us when our sense of well-being and belonging are shaken. If inflammation goes unnoticed and unchecked, it lays the foundation for heart disease, cancer, diabetes, loss of memory, and decreased cognitive functioning.

When Terri finally opened the door to the meditation hall, the flames from the candles held by some of the participants were creating a soft glow. As she took in the low light, the spiritual icons, the Gregorian chant rising from hidden speakers around the comfortably full room, her body absorbed the signal of safety, which calmed the adrenals and brought the brain's vagus nerve—the body's largest—online to begin settling her heart rate, blood pressure, and hormones of threat. Inflammation would now have a chance to calm.

As hands reached out in greeting while she searched for an open spot, the brain's reward center activated. The hormones of bonding, such as oxytocin, and feel-good chemicals like the dopamine we get in surges from orgasm and deep conversation, streamed into her body as she felt enfolded by familiar and unfamiliar hugs. Oxytocin was modulating her brain's stress response, signaling the hypothalamus and pituitary to stand down in their call for energy from the adrenals. She was beginning the recovery process. Endorphins, the same neurochemicals that trigger a runner's high, were released as the peace and good-hearted intentions of those in the room cooled her system. Through connection, Terri's brain was beginning to connect the neural pathways to contentment.

Her nervous system was beginning to allow her to be present, to think clearly, to shift her attention from worry about her work and the logistics of the day to connecting with a community and its larger purpose. Her memory of so many years of feeling like she belonged in this room stimulated the hypothalamus to release oxytocin to be received by specialized receptors in the brain's nucleus accumbens, which in turn sent signals to other pathways to release a flood of serotonin, the neurochemical of well-being. The deep enjoyment of connection began to flood Terri's being once again. She was feeling more like herself with every passing moment, more like the "Me" she had been accustomed to experiencing.

The feeling that had dogged her earlier in the week, that something was missing, was a craving for the feeling of being present with a group, the feeling of her mirror neurons helping her to reflect the peace and well-being of others in a known community. Her body was seeking the stimulation of the neurochemicals of familiarity. That Terri would arrive home feeling calmer, more peaceful, more alert, more friendly, more hopeful and creative, is not a psychological or spiritual mystery. It's a brain and nervous system response to the cascade of neurohormones essential to the well-being of the social animal, *Homo sapiens*. It's a response to the familiar feeling of the tent of "Me."

Think for a minute about where the label of "Me" falls when you say it out loud. When we are young, a simple declaration of "I am" seems to do the trick. As we take on different responsibilities, our identity may feel powered by various roles—wife, employee, doctor, mother, friend, activist, caretaker—where we fit into our communities, with which friends and family members we feel we belong. Depending on circumstances, life may force us to question whether any of these things are "Me." Do I feel like "Me" without my kids at home, without a spouse? Do I feel like "Me" living outside the city and home in which I have lived for thirty years? Do I feel like "Me" without my job of more than three decades, without the co-workers and friends who have defined my social life? If not, then what happened to the "Me" that felt so certain? It may feel, in the second half, as though your tent is emptying. Trying to refill it may not always prompt the best choices.

Tricks and Traps of Refilling the Tent

Loneliness, defined by having fewer relationships in which you feel heard and understood than you need to maintain a basic sense of well-being, has become a big enough epidemic that more than sixty million Americans say it's a major source of unhappiness in their lives. The culprit, it turns out, may not be loneliness but the lack of a sense of belonging. This is what happens when the sense of "Me" contracts as a result of kids moving out, friends and loved ones moving away or passing away, leaving the home and community in which we raised a family, careers ending and roles changing; it's as though the tent of "Me" becomes empty of belonging. These chronic feelings of not belonging can drive a cascade of biological events that accelerate the aging process, even increasing the risk of dementia. It shows up in the stress hormones, immune function, and cardiovascular function, and some research is suggesting that it can cut your life span by as much as fifteen years.

Knowing this now, Sherry is working hard to shift internally. "I can't afford to move back east at this point, though if I could, I'd do it in a heartbeat. But I learned I have to fight the urge to curl up at home. Whenever I gave in, I started drinking too much," she confessed. Alcohol had not been a big problem in the past. "But it got to a point that it was threatening my relationship with David. And I can understand. He was protecting his family."

Sherry used the brain's wiring for social connection to turn things around and repopulate her tent of "Me." "I've started to take exercise classes," she told me, scrunching up her face, "which I used to hate, but they give me a chance to meet new people. I've made a few new walking buddies that way. And I'm volunteering to take meals to people who are housebound. Seeing the struggles of so many not much older than I am puts things in perspective. I have no choice but to stop feeling sorry for myself."

Connection in real time, even virtually, with longtime friends, family, and colleagues, reminds the nervous system that *Yes, I am still "me."* The brain's insula, in its scan to ascertain the body's well-being, will engage in helping to confirm that *I still feel the same patterns; my mirror neurons are stimulated to react in the same way; my systems are receiving feedback that I still matter.* When the insula gets a response it isn't used to, the alarm bells ring. It can drive compensatory behaviors through a cascade of neurohormones to create a craving for familiarity. It can take the form of an impulse to drink or overeat; or after the end of a long relationship, it can reawaken all those adolescent brain circuits for pursuing bad boys. It depends on what kind of familiarity the nervous system is craving.

Toxic Familiarity

At first, after her divorce at fifty-seven, Rena had a hard time finding someone to connect with. "I don't understand, Louann," she said to me, "because it's not for lack of availability. There is not a shortage of men to date, and many of them at this age want to get serious. They all want to remarry. But somehow, there's just no spark for me with any of them. And if there is, it keeps hitting a dead end."

Knowing Rena's history, this made sense. Her father had been an alcoholic. Dennis, her husband of twenty years, had gambled in secret, and if I had to diagnose him from afar, I'd say he was also a narcissist. For twenty-seven years, Rena did a dance responding to his incessant demands that she be responsible for making his world perfect. No matter what she said or did, it was wrong. She was constantly accused of being selfish, of not considering his feelings, of not wanting to make him happy. "It got to the point that when he asked me my opinion, I would ask him, 'What would you like me to say?' And of course that would send him into a rage," Rena said. "And I started to get scared for my safety. His temper was becoming extreme, and he would chase me around the house yelling. I used to go to the guest room and lock the door." Years later I could see how sad she still was about how things ended. "A part of me still feels like it's my fault."

Over the next two years, Rena found herself dating again. She fell for a couple of men just like Dennis. She felt instant chemistry with them, lots of excitement at the beginning. When friends set her up with really nice guys who weren't narcissists, she felt like there was no magic and didn't call them back.

Through her childhood and long marriage, Rena's nervous system had been trained to crave an unreasonable and demanding partner. The familiarity made her feel alive. Lack of familiar triggering will make a safe setting feel boring or uninteresting. When we are used to being continually activated by a rage-aholic, the state of fear becomes equivalent to feeling alive and like "me." When we don't feel that, we feel bored and adrift. The nervous system signals the brain to search for meaning by connecting with familiarity; this is one of the reasons that even after a spouse has stopped drinking and is sober, things don't *feel* right. Rena had her work cut out for her, and it would take some time to help her rebalance.

Toxic familiarity is poison to the brain's sense of self, deepening familiar patterns that send self-destructive signals to the brain. In the reproductive phase, we can muscle through a lot of misery. But in the second half, the stress of things going badly is more apparent, and the opportunity to wake up to the symptoms is more available than at any other time in our lives. The social stress circuits become more sensitive as women get older because the HPA (hypothalamic, pituitary, adrenal) axis responds more quickly to social stress but takes more time and attention to calm and reset. While we recognize more readily through experience that we can't demand perfection from ourselves or our friends—that would leave us quite alone and isolated—we can also recognize that if the stress of a specific relationship is paralyzing, it's time to reconsider the connection. Terri was finally, miraculously, ready to do just that. In the Upgrade our lives no longer have to include familiar experiences that lock us into toxic patterns.

Resetting the Nervous System

One patient of mine, Jean, who lost her husband to cancer when she was seventy-four, didn't call anyone for months. Somehow, in her mind-numbing isolation, she became convinced that because her husband died of cancer, she was an unlovable failure whom no one would want to see. The thought was entirely irrational and had nothing to do with reality. But the incorporation of her husband into the tent of "Me" meant that when that piece of her was gone, she interpreted its absence as inadequacy. "It's like every voice that ever told me I wasn't good enough, that made me feel incompetent, came roaring back to fill the space that Jack left. I know now, after talking it through, that it was crazy, but that's how it felt," she told me.

Through nervous system harmonization, we physically incorporate the people we are close to. You don't realize it's like singing in harmony until your choirmate is gone. What we feel in their absence is the nervous system and brain patterns being stuck on a search for the familiar.

Luckily friends were persistent in their efforts to help Jean get back into the world. It took years for her nervous system to recalibrate in the healing space of belonging. "I'll never get over it," she said, and I encouraged her to allow that visceral sense of loss. It's impossible to replace the people we lose. "But I can see a way forward," she concluded.

Acts of social withdrawal are a razor's edge. When we need to calm agitation, rest, and refresh the nervous system, a little alone time can be good. But there is a tipping point of solitude that increases agitation. Belonging is one of the best ways to calm the nervous system in the long run.

Even a pleasant exchange with a stranger provides the nervous system with the essential affirmation that we matter. These exchanges can help balance us cognitively.

This is why solitary confinement is so cruel. We mirror others, so when we are off, a balanced nervous system can reset us, just by being in proximity. We can reset one another or we can destabilize one another. We all know that friend we should not call when we are fragile, or that person we should call when we are fragile, or that fragile friend we know we can help reset. It's not mystical or weird; it's the neuroscience of mirror neurons and oxytocin, which attune us to one another through body language, vocal inflection, and a million tiny, unconscious cues.

Think about what it feels like when you are out to dinner or at someone's home with a bunch of friends. Remember the rapid-fire conversations, the group laughter, the food, the one-on-one connections. If you close your eyes and allow yourself to pull up a strong memory, chances are it's making you happy. That feeling is good for your nervous system and good for your brain. If you visualize yourself at a party from your youth, turning on the music you heard then and dancing as if the room were filled with your oldest and best friends, there is evidence that it has a positive impact on memory, physical strength, vision, and hearing.

When we lack social stimulation, whether it's actual or imagined, parts of us wither. Without its counterbalance, we lose those essential brain vitamins of social connection and with them our resolve and resilience. In isolation after her retirement as an executive assistant, Pauline started reexperiencing hurt she thought she had set aside decades earlier. Old pain from her marriage, even her childhood, began to resurface. It wasn't just because of the lack of distraction through a schedule that was no longer full; it was the lack of nourishment via oxytocin and serotonin for the brain circuits that help us remain in the zone of well-being. In the second half of life, that zone narrows, its borders become more porous, and the signals that we are out of it are stronger; we can become suddenly overwhelmed in ways that feel paralyzing. It was helping her granddaughter move to Los Angeles that made her feel connected and useful again.

The experience of reciprocity that happens when we are a part of a circle of others feels good to the brain and nervous system. Mirror neurons and oxytocin play a huge role in how we regulate one another. In a harmonious group, mirror neurons ping signals of happiness and belonging among the members. The bond is reinforced and rewarded by oxytocin sent from the hypothalamus to the nucleus accumbens bringing one of the brain's calming and pain relieving systems online with a cocktail of serotonin, dopamine, and other endorphins that are like a balm of instant well-being to the brain and body's nerves.

The presence of trust and a calmer nervous system means heightened cognitive powers are retained longer. It also means a longer life, as the immune system is allowed to function properly instead of reacting to signals of threat stimulated by an isolation-activated stress system. Trust and hugs become like rainy-day funds to use in moments of stress. They create our reserve for meeting life's ups and downs with flexibility and strength, the resilience that helps us return to our optimal emotional and physical zone of well-being in the Upgrade.

The correlation between social connection and long-term benefits for brain, heart, and systemic health is strong. By lowering stress hormones, social connection reduces inflammation, improving cognition and immune function. Through stimulation of feel-good neurochemicals, we become motivated to engage in healthier habits. The feeling of care coming from others helps us engage in self-care.

Belonging is like jet fuel for any habit that supports good health and well-being; it's as essential to improving cognition and decreasing risk of diabetes as getting enough sleep. New research is showing that isolation is likely to be as damaging to health as smoking, lack of exercise, or a poor diet. While it shortens a woman's life span by as much as fifteen years, those who feel supported by their spouses to go out in groups of three or more girlfriends twice a week live years longer.

Trust, one of the key ingredients of belonging, feels good to the brain and nervous system. Because of changing hormones in the Transition and Upgrade, we again have a unique opportunity to break free of misplaced trust. Before the Transition, a hug even from a partner you distrusted would trigger a release of oxytocin, making you forget every suspicion.

I used to warn a woman before the Transition not to let her partner hug her if she suspected him or her of lying. A hug would set off a flood of oxytocin, boosted by spiking levels of estrogen, that would make her believe whatever she was told.

In the Upgrade, we get the best of both worlds. We can still lead with love and trust, but with constant lower estrogen, our thinking is no longer overwhelmed by big surges of oxytocin. Trust becomes a choice, rather than a neurohormonal destiny. We have the ability to notice when the nervous system is tugged by a familiar yet unhealthy nervous-system pattern of another.

Giving the brain the essential vitamins of connection and validation is key to the Upgrade. For just a moment I'm going to ask you to put aside your pride in independence and your denial of aging. Ladies, if you are leaking, if you're accusing everyone of mumbling, or if you've got back pain, chances are you are going to start saying no to invitations. Stop it. Just stop it. If you can't walk well, get the damn scooter you've been resisting. Sixty percent of women over sixty have a mobility problem, so I know I am talking to some of you. If you think everyone around you is suddenly speaking too softly, go to the doctor and get your hearing tested. Stubbornness born of pride is one of the biggest obstacles to optimizing the Upgrade. Do everything you can to stay connected.

It's not just longevity we are after in the Upgrade; it's joy, emotional strength, and sharpness. Who we are arises in a context, not in isolation. Who we are is a composite formed out of connection; our interaction with others is the ground on which we plant the tent pole that supports our sense of self. Engaging in healthy connection isn't an optional part of life, something that we do when we have extra time; it's intrinsic to who we are as humans. Finding belonging is not a frivolous, secondary activity. It determines whether we enter a downgrade or an Upgrade.

As we move into the Upgrade, whom we surround ourselves with becomes more important in the tent of "Me 2.0." Find those who help you discover the best in yourself. Immerse yourself among those who are on the same quest in the renovation of their tent of "Me." As part of its evolution, the human brain became primed for living in groups, and the bonding hormone, oxytocin, cemented the need. All these millennia later, the evidence for this primal need is confirmed by studies that show girl-friends improve our health. So head out or video-chat with a group of four; for full health and stress-reduction benefits, drop any trace of guilt. And to amplify those benefits, set aside all your responsibilities and take a trip with the ladies. In times of stress, the female brain, even after the Upgrade, releases oxytocin for bonding, not just the stress hormone cortisol, which makes us flee or fight. In the Upgrade we seek out bonds with others who give us permission to be totally ourselves.

Upgrading the Mommy Brain

Chapter 8

"Seriously, she got a headline alert that there was a blizzard coming, and my mom found it necessary to stop what she was doing at work in Dallas, pick up the phone, and call me in Chicago to remind me to wear my gloves and hat," says thirty-five-year-old communications executive Latanya. "Does she really think I'm still ten?"

We may have helped push them out of the house to live independent lives, but whether or not we ever push them out of our tent of "Me" is a separate question. But it may be a necessary step if you want to keep them in your lives. As my son will attest, too many of these kinds of interferences will mean our adult children begin to shut us out. And those moments or days when they don't return our texts or calls can feel like part of us is left hanging in the breeze, slowly being ripped away by the wind. I have found it utterly defeating how tied to the moods of my son I still am.

When the mommy-brain circuits turn on, it changes who you are. The passage of the baby's head through the cervix and vagina triggers a surge of oxytocin, as does nipple stimulation during nursing; but skin-to-skin contact and the smell of the top of the baby's head alone initiate a powerful cascade of bonding hormones. Through motherhood we are re-wired to detect the smell of their bodies and the smell of their poop, hear the different sounds of their cries as their very being gets incorporated into our tent of "Me." Whether you gave birth to the child or not doesn't matter: Skin-to-skin contact changes everything.

No matter how quickly they've grown up, no matter how old we are, the mommy-brain circuits are very hard to turn off. My brain is hijacked by those mommy circuits frequently, even though my son is now an adult. When Whitney first moved out of the house, I used to go into his room to smell his smell. Lately I have to practice hard to cultivate loving detach-ment, because at a certain point, our kids really wish our mommy brains would leave them alone to live their lives.

Even if our kids are doing well, we will still worry. Our frontal lobes are always on for problem solving. Though everything may be okay, our overthinking and overconcern will create problems where they don't exist, pushing us to interfere and rescue when there is no crisis to be rescued from. For that, our kids have the most powerful weapons: distance, si-lence, and separation. Regardless of age, the female brain's circuitry is still wired for connection. Their silence is DEFCON 1 to the female brain and its nervous system. Jane Isay writes honestly and deftly of her experience in her book *Walking on Eggshells*:

> There were some rocky years when I thought I'd lost him and he didn't want to hear anything I had to say. Very painful. He had to teach me how to separate from him. I had to learn to pay more attention to myself and less to him. I had to become accustomed to living on the periphery of his life. I had no power and had to learn (still learning) to give up expectations. It turns out he is a lot of fun and is witty, knowledgeable and smart. It has taken me years to adjust to the new normal. And challenges still come up all the time. We keep on trying to stay close to each other without driving each other crazy. Accepting my own mistakes and failings— which he and I laugh at together now. I know he loves me and occasionally he will even tell me.

When Whitney pushed back against how often I was stopping by his apartment, I said, "I want to see you a lot because I think you're lonely." He responded, "Yes, I am lonely, but seeing you isn't going to do anything about it." I thought my presence was giving him someone to hang out with. If anything, the opposite was true. My anxious presence was keeping him from exploring how to address his own loneliness. Showing me that let me off the hook. I was no longer responsible for addressing his wounds.

For the relationship with our adult children to survive, we have to expand the mommy circuits to a broader target. If we keep hovering, their experience is that we don't respect them or validate their adulthood. And if we don't find a way to validate their adulthood, they will erect a wall. They will block us on social media, screen our calls, or simply be dishonest with us. They will find their own ways to push our relationship with them to a place they can tolerate, even if it hurts us. Trust me.

For the majority of women who have become mothers, the mommy brain has dominated our adult reality, motivated by the urgent need to keep them alive until they are eighteen. Our social lives and work lives revolved around getting them to the right place at the right time, connecting with the school and other moms, engaging in a daily game of high-level calendar chess. But that's about as far ahead as we've thought. In the sweep of history, female survival beyond the fertility phase is new. The completely absorbing balancing act of job, social life, and children's activities that hijacks the brain and nervous system is over. There is no road map to transitioning into being a mother of adult children. In the new era of longevity, this territory has to be mapped, this developmental phase has yet to be named or explored.

You've probably heard the story of the old man walking through the woods who comes across a butterfly struggling to get out of its cocoon. He thinks the butterfly is stuck, and so he opens the cocoon. The butterfly has shriveled wings and a fat body and is unable to fly, leaving it more vulnerable to birds looking for a meal. The struggle to break the cocoon is what makes the wings strong and the body smaller. The old man's instinct to rescue deprived the butterfly of its birthright: strength to fly.

I don't know where the story comes from, but I've heard it told many times by various people. Learning when not to open the cocoon for our kids is as important to them now as when we would reassure them they were okay after a fall on the playground. We can give our adult children their own dignity by treating them as we would other adults.

One sign of a mommy brain in a downgrade is thinking we know our kids. We don't. We can honor them by giving them space, and that is how we will get to know them as adults. In the Upgrade, we approach them as if we were getting to know a new friend, listening rather than interrogating: Did you pay the bill? Where is the form you were supposed to fill out? Did you call the dentist? Leave them alone. They will figure it out, or grow from the consequences if they don't.

The transition into the second half of life is a long shift into letting go, practicing loving detachment, remaining silent to listen to adult kids, spouses, and siblings. We learn the hard way that things grow best when they are tended lightly and learn to hold the space for our loved ones as opposed to marking and guarding the perimeter of their sandbox. It's not easy to stop being at the ready to lay down our lives for them. It's not easy to keep from violating the space we formerly felt entitled to enter through interrogation. We would never cross that line with other adults we know; the questions would be disrespectful. And rescue-mommy-brain mode only makes them feel disempowered.

Extreme Mommy-Brain Challenges

It takes love and humility to accept our own weaknesses and to come to terms with the failings of our children.

—*Jane Isay,* Walking on Eggshells

If you are the mother of an adult child with a self-destructive habit like alcohol or drugs, or one with a mental health issue for which they have opted not to follow treatment, how you deal with the incorporation of your child into your tent of "Me" can have a dramatic impact on your own life.

"I was at the end of my financial rope," Wendy, sixty-four, told me after a talk I gave in Denver. "I come from rural Nebraska and I grew up very poor. But I was lucky enough to go to college and be very successful in the financial industry. My husband and I made a lot of money together, and so when Chris, our oldest of three kids, became ill with his addiction at fifteen, we started writing six-figure checks to rehab facilities. He was in five different in-house programs by the time he was twenty-two." I could see and feel her pain. As a psychiatrist, I know that the recovery rate for addicts and alcoholics is barely 30 percent. The vast majority never get out from under their disease. And as a doctor familiar with these programs, I also know that Wendy was telegraphing that they had spent more than a million dollars on various facilities.

"After my husband died five years ago," Wendy continued, "I just didn't know what to do. I had retired and needed to make sure the money we'd saved was going to last. I also have another son and daughter and wanted to be able to help them when they needed it. But Chris kept getting into trouble, and as a mother, what do you say to your child when he's in pain and he's begging you for help?"

When we have incorporated another into our tent of "Me," whether a spouse or child or close friend, the nervous system sends us powerful alerts when something is wrong in their lives or they are trying to create distance. It's this unconscious intertwining of our nervous system with theirs that can keep us hooked to people who can cause us damage. We become codependent, literally addicted to them. The nervous system seeks familiarity and repetition. It seems weird, but if we are used to someone making us miserable, we will oddly crave the familiar misery unless we find the strength, the friends, and the support to help us break the cycle. We will need help. (See appendix, page 274.) Because while it feels bad to keep toxic loved ones in our tent of "Me," it can feel worse to push them out. The dilemma can send us into isolation as we push away those who challenge our addiction to their problems. We don't have a road map for dealing with this. If we have even a tiny ounce of the instinct for self-preservation, it will kick our social circuits back into gear; almost against our own desire, it will push us out of the house to at least browse in a department store, talk to a salesperson, call a therapist, take up an invitation to a dinner with a group of people we might not normally seek out. If we can't hear or don't honor those impulses, the nervous system will pull us down the rabbit hole of obsession into intractable melancholy.

"By the time I looked up from my relationship with Chris," Wendy said, "I was completely alone. Other parents would talk about how great their kids were doing. I didn't know how to relate to people without talking about his problems, and I didn't know how to be present with others without checking for his texts or making sure he was still alive. I would feel jealous their kids were doing great. It made me testy and unpleasant to be around. My nervous system was so wired to his well-being that I no longer had a life or even a sense of myself. I was on my knees."

Different people are drawn to different tools for addressing isolation and repairing the tent of "Me." "I started with Al-Anon," Wendy told me about her decision to make a change. "I worked the program as hard as I could to learn how to make that separation with my son and start to care for myself. Two years ago I did the hardest thing I've ever done in my life. I changed my phone number and moved. I didn't tell Chris how to find me. I had to finally come to grips with the reality that I am powerless to help him. I've let him go at thirty-five. Keeping him close was keeping me from living, it was damaging my relationship with my other two kids, and it was keeping him from finding help. I had to get out of his way and let him walk his own path. My other kids were worried about what Chris's illness was doing to me, especially the times he physically threatened me. By the time I made the decision, I had lost touch with most of my friends. Anyone outside of Al-Anon had no way of understanding the hell I was living."

The nervous system is trained to respond and behave habitually by our social connections. So if our nervous system has been engaged with the toxic familiarity of a loved one who is addicted, mentally ill, or is a rage-aholic, breaking those connections within the tent of "Me" can feel apocalyptic. In the first weeks and months after that kind of separation, anxious boredom often sets in. "I didn't know what to do with myself after I moved and got settled in," Wendy told me. "I was so used to organizing my entire being around Chris's needs that disconnecting felt like an identity crisis. I didn't know who I was without the chaos." Wendy's nervous system was craving familiarity, and it was hard for her not to fill that void by becoming a codependent crisis machine, looking for or creating drama where there wasn't any.

When the tent of "Me" is unconsciously ensnared by the profound disorder of a child, it short-circuits the Upgrade. But as we grow more conscious of how the nervous system and brain can become a source of feedback in forming the tent of "Me," we can take more control over whom we let in and how much influence they get. Wendy realized that not only had she given Chris a place of prominence within her tent, but "I let him take over the tent pole. It was like he had control over how I judged myself and how I set my daily priorities. I was always reacting to something, a phone call or visit from him or a crisis, so I couldn't plan and was always canceling things. Yet at the time, it felt like it was the right way to be. Now I know I need to focus on self-care."

The Upgrade is our chance to radiate strength and goodness powered by the wisdom and courage to hold the space for whatever is emerging for the adult child. It takes patience. It will hurt when they lash out from pain. But once they hold their proper place in our tent of "Me," their unhappiness with us is no longer an identity-threatening pain. We can finally hold the space for ourselves and for them, in healthy, intimate separation.

Granny Brain

"The biggest hit of euphoria and the biggest speed bump in my family relationships showed up as the grandkids appeared on the scene," Ceci told me over our regular coffee.

I had wanted to know what had most challenged her stability in the Upgrade. "That's interesting," I said, feeling a little puzzled. I don't have grandchildren yet, so I have to rely on what I know from neuroscience and the experiences of others. "You hear all the time about the joy. A friend from Brazil always jokes that children are a necessary evil for getting grandchildren. But tell me about the challenges."

"When I held those babies," Ceci continued, "I thought my heart would burst. I never knew how big that love could be. It broke me open. The pull to be around my grandkids is so strong; our conversations as they grow up are so intimate, so filled with unconditional acceptance, so much mutual respect, so much deep listening. I'm in love. It's the only way to describe it."

The echoes of mommy-brain circuits are powerful, and even with lower levels of bonding hormones, they are easily reactivated. The blessing of the biological distance makes room for a more unconditional love, free of the fears, responsibility, and judgments we had with our own kids. The brain's nervous system doesn't incorporate them into the tent of "Me" with the same sense of daily urgency as it does our own children, so it's easier to approach their suffering with an open heart, ready to listen instead of going into control mode, fixing, and interfering. Still, the bond is shatteringly powerful.

This is the reactivation of the caretaking circuits pounded into the female brain and nervous system from the first years of life, the flood of oxytocin, the comfy, cozy progesterone, the nervous system's memory of the pleasurable part of bonding. There can be peace in those echoes, and clearly Ceci was experiencing that. "This sounds so great, Ceci. It's hard to imagine the downside."

"It's there," she replied. "You learn about it pretty suddenly: the first time you say no to your kids when they ask you to take a grandchild for a day or overnight. I said yes a lot after my first one was born. But then I started to feel like I was losing access to the goals I had set out for this period of my life. I'm still working pretty hard, and it's exhausting taking care of active grandkids. I want time for walks, for thinking, for rest, for reevaluating life, priorities, choices. If I am growing consciously into a better person, I can be better for everyone. If I can find peace and happiness, that will make a difference to others around me. It took a long time to understand, to really deeply get that my happiness is not selfish—it's a contribution to the happiness of others. I don't mean that small thing of being satisfied. I mean the big happiness that comes when you drop your worries about looks, clothes, money, possessions. When you can embrace life's joys and its scary uncertainties. That's the kind of centering and strengthening of my tent pole I am engaging in. This is my time for that. I'm worth it. And I have the right to claim it."

I could sense her now-or-never urgency, a feeling of breaking out of prison. And that might be just the right analogy.

By the time most girls are a year old, we are handed our first dolly and told to feed it with a fake bottle. The culture is already grooming us to become martyr moms, irrevocably and primarily responsible for the survival of our families. We are being prepared for what can at times feel like a prison of expectation and obligation, one we happily accept out of love but also in order to avoid the social price and guilt we pay for not complying. But in the second half, in the Upgrade, we can break that chain: We are no longer primarily responsible for the survival of small children, though society and the echoes of old wiring will try to tell us otherwise.

Martyr-mom wiring is very hard to undo. The shift to being centered comes when we are able to be alert to and resist the tug of the imprints of old waves, when we can shift our goals for the tent of "Me" from outward approval to our own well-being, making sure that we are meeting our best interests in a way that actually makes us better to and for everyone around us.

"So when I claim that time," Ceci said, "or I am busy with work and I say no, you'd think by my son or daughter's reaction that I had just said I hate them all and never want to see them again. You can't imagine the nasty things they say. 'Why am I not surprised. You were never there for us as kids. Why would I think you'd be any different as a grandmother?' They don't say these things to their father when he says no. And they don't ask him as often as they ask me."

"Wow, that's rough," I said. "There are a lot of emotions that are bound to come up, but how does that feel, viscerally, that tug of grandchildren and the anger of your kids?"

"For a long time it was like the first gun going off in the biggest internal battle I've ever fought." Ceci paused, sadness coming over her face. "Those caretaking impulses are real. The guilt over not acting on them is powerful." Seeing as women end up with 78 percent of all caretaking duties, data backs up her feeling. "And the things my kids say cut deep. But I go back to that tent metaphor you use. I have really been looking at who is controlling my tent pole. It got clear to me during my morning walks, which I started taking alone a couple times a week instead of with my group. As I upped my activity and felt stronger, I started to see that I wasn't the one always in control of my tent pole. I started to see who I had given access to. And it was a big shock. My kids, my friends, even my dead mother was still tugging on it. So I walked, exercised, did yoga, some writing and breathing exercises. All of that helped me see what was going on."

Because I'm in Charge!

I've known Diane through the medical community for years. She's an internist who is winding down her practice. She's been married for thirty-seven years—she met her partner in med school—has two kids in their twenties, and like many, cares for nearby aging parents. She's always been busy taking care of everybody's needs, and it hasn't always been easy getting a spot on her schedule. But she's been more available lately and I wondered what had changed. "It's easy to feel pulled in a million directions," she said. "It's easy to fall into the caretaking trap and just not have a life. . . ." Her voice trailed off as she looked at a message on her Apple Watch. "Hang on. Josh just texted me to say he lost his passport."

I knew her twenty-six-year-old son was traveling in Europe, and I was already moving toward my car keys to get out of her way. She had always been intensely involved in her kids' lives and I assumed our visit was over. "Oh no!" I said. "Do you have to go call him?"

"Absolutely not," she said, to my surprise. "I'll be right back." Diane disappeared into the kitchen and emerged with her phone. She appeared unmoved. "So what happened?" I asked.

As she handed me a cup of tea, she said, "I texted back and said, 'You're an adult. I'm sure you can find people at the airport to help you figure it out.'"

I was impressed. I am not confident I would be able to show the same restraint; my son pulls hard at my tent pole, and if I don't respond to what I perceive as his crises, then something doesn't feel quite right.

"Two years ago," she said, seeing the question arising in my face, "I would have said goodbye to you and run to his rescue. But that kind of behavior was crippling him and exhausting me. Same for my mom, who calls me obsessively. I answered every call no matter what I was in the middle of. So I found some things to keep her occupied and I stopped picking up the phone when she calls, which is literally every hour and a half. When I didn't answer, she started texting more, and I could see what was urgent and what wasn't. Not answering felt at first like a betrayal, and then I understood I was actually breaking an addiction. Kinda like codependency. As for my husband, I have learned to ignore any question emanating from the kitchen that begins with 'Where's the . . .' or 'How do I . . .' Somehow, he manages to find whatever he's looking for all by himself. I just have to sit on my hands until it passes."

Before the Transition you may have felt that it was just fine to let a whole bunch of people control your tent pole and that many things were your responsibility to fix: people, situations, the lives and relationships of others. That fixing impulse was a big part of the tent of "Me." It gave us confidence in our competence and our agency, until the day our plan didn't work, or it backfired. In the face of something going wrong comes a deflation of our sense of power and control, bringing on anxiety, fear, agitation, sadness; a collapse, whole or partial, of the tent of "Me." By pursuing what we can't fix, we get constant failure feedback in the tent of "Me."

In the Upgrade, we recognize that our tent pole doesn't belong to anyone else, and we can see there is little we can do to fix the minds of others. There is only one mind and one tent pole for which we must care: our own. After decades of trying to control others, acceptance of the truth of powerlessness over most everything frees us to finally know ourselves. I'm hoping that by the time I go, my son will agree that my epitaph will be "Finally learned to mind her own business." I'm still working on that.

The Relationship Brain

Chapter 9

*The biggest predictor of how healthy you will be at eighty
is your satisfaction with your relationships at fifty.*

From birth, the female brain is a mean, lean, observing machine. The ability to read faces and hear vocal inflection can make women seem like mind readers, all because of an awesomely fine-tuned nervous system and the right mix of neurochemicals to support this particular talent. By now, you've probably spent a lifetime perfecting your ability to anticipate the needs and moods of others. You can feel their exhaustion, hunger, or sadness before they do, and before you know it, you've set about trying to fix the problem. Part of that ability is societal—experiments have shown that anyone in a subordinate position develops this skill—and part of it is hardwired into us as women.

It can be powerful, having this ability, and it can be exhausting.

I was thinking about Sylvia and Robert, the couple we followed in *The Female Brain*. Their long marriage had broken up, but I was curious to know if while married she had felt what so many of the women I knew also experienced during long partnerships. It's the feeling of absorbing your partner's nervous system to the point that it seems to drown out your own nervous system's signals—a kind of extreme empathy. "I always had a hard time not physically feeling my husband's emotions," Sylvia, now seventy-seven, told me. "I would walk into a room, and if he were down and feeling angry, I got a knot in my stomach. I could feel the gripping in my chest and sweat in my palms when he was anxious or angry. I could also feel his contentment and peace, so that was a plus. But for the most part I kept feeling like I'd lost my center, that I couldn't feel my own emotions, my own discomfort. I had a split in my soul. It felt like I didn't have a self that wasn't absorbed in his reality."

"That must have been really disconcerting," I said, having known the same feeling. Emotional contagion is a large field of study, and from the nervous-system side, our oxytocin and mirror neurons act as amplifiers for "catching" someone else's feelings. When you add the closeness of an intimate relationship, it's much harder to resist resonating with the emotional state of a loved one. "It was enraging, actually. After forty years I looked up and realized I knew him inside and out; but I had become a stranger to myself. It's part of why I left. I needed to find and stand on my own ground again."

In the beginning of an intimate relationship, the brain's love-attachment system, driven by neurohormones—more oxytocin in females, more vasopressin in males—is primed for us to take the habits, tastes, and preferences of the person to whom we are becoming attached as our own. As the tent of "Me" expands to include this new person and their feelings—their likes and dislikes, their experiences—how we feel in our skin is altered. When we or a spouse are away for a long business trip, that feeling of being out of sorts is a signal from the nervous system that some kind of essential feedback is missing. That first hug upon return so often sets things right, as it activates the calming circuits of belonging.

There is a flip side to this deep connectedness. Many times I've found myself holding attitudes or behaving in ways that don't feel native to me, attitudes and behaviors I might not be proud of. Neuroscience shows how many of our attitudes and habits and how much of our physiology come from what we absorb from those closest to us; what we think and say and do may not always be the real us, though it may feel like it in the moment.

This deep, unconscious absorption of another's nervous-system patterns is amplified during the fertility phase by estrogen: It makes orgasm more readily available and pumps much more oxytocin into our systems. The calming of those bonding and caretaking circuits during the Upgrade can give us more bandwidth for looking inward, for analyzing and focusing on what's necessary to do for ourselves. It can give us the strength to resist being pushed by the echoes of old caretaking waves that would have driven us to race to the rescue of everyone else. Especially when we begin to understand the neuroscience of empathy.

Empathy, feeling the feelings of another, is a gift at the heart of compassion. But according to studies at the Max Planck Institute in Germany and the University of Wisconsin at Madison, it has a potential dark side. That we feel what someone else feels, especially their suffering, sounds noble and virtuous. But think about what happens to us when we are faced with the suffering of those close to us, how easy it is to feel debilitating sadness over what they are going through, how easy it is to drown in their pain. We don't need to see the results of fMRI (functional MRI) studies to tell us that feeling someone else's suffering lights up the pain circuits in our own brains. This kind of passive resonance can easily slide into empathetic distress, a powerful nervous-system force that can release a tsunami of stress neurochemicals and overwhelm our ability to think clearly, function normally, or engage in helpful problem solving. Sylvia struggled with feeling porous to Robert's moods. So did Ceci with her ex-husband, and pretty much every woman I've ever spoken to who has been in a long-term partnership. I know that feeling myself with both my husband and son. Even in a healthy relationship, if we are swamped by a partner's depression, we won't be any good to ourselves or to them. The stress of another can wake up our fear-triggering amygdala; when that little almond-sized brain organ takes over, it cuts off access to clear thinking. We fly, hair on fire, into rescue-survival mode.

Finding nervous-system independence was becoming a crucial self-care issue for Sylvia. When we last left off with their story, Sylvia had found freedom in the Upgrade; she had started setting new boundaries and was determined to start a career in her midfifties. The changes in Sylvia frightened Robert, and their forty-year marriage began to unravel. She moved out and they divorced. I caught up with them again to see if anything had changed.

It turns out that after about six months, things cooled off enough that they could begin to rediscover their friendship. They dated other people, but when they had trouble, they still turned to each other. When Robert got a bad flu, Sylvia came over every day for two weeks to help him. When Sylvia's mother needed legal work, Robert continued to care for his former mother-in-law. His attempts at new relationships were unsuccessful. "I had a woman leave in the middle of dinner," Robert told me, "because she realized I was still in love with Sylvia. It was clear I was never going to be able to commit to anyone else. So I made a commitment to Sylvia to go back to therapy and see if we could work things out."

After seven years apart, Sylvia moved back in, and they joked with me about living in sin. They traveled the world and really enjoyed themselves. Sylvia continued her career, starting and running a mental health nonprofit and later handing it off to the next generation but still conducting workshops for families and educators. Robert kept practicing law but reduced the firm's capacity and moved his practice into the den.

Robert is fourteen years older than Sylvia, and his cognitive decline had a profound impact on her life. Sylvia was still traveling for work, but Robert steadily became more easily frightened and needier. He became obsessed with worry that Sylvia would leave again, especially for a younger man, so she traveled only with him. But as he became more frail, they stayed home.

Over time, as his memory worsened, he became more like the old Robert, griping and complaining all the time, not well enough to be fully independent but stubborn enough to insist he could take care of himself. His routine made him happy: breakfast in the kitchen, a couple of hours at the computer in his home office, lunch in the kitchen, news in the library the rest of the day. He could no longer hear the TV, so he read the ticker at the bottom. At four thirty he had a martini, fueling evening irritability and insomnia.

Sylvia tried everything to keep him engaged with the world, to see friends and family. As his memory slipped, she became more isolated, not only because she was trapped at home but because having a normal conversation was no longer an option. "I was losing my friend," she told me. "I couldn't talk to him about daily events or discuss politics. He couldn't retain memory for anything except the law. On that front, he was still sharp as a tack."

Sylvia was becoming depressed and angry, and though she herself is a mental health expert, and has recommended therapy for other women in similar situations, she believed she was handling things just fine. She couldn't see that she needed someone to talk to, someone with expertise in elders who are in the process of dementing, who could help her with strategies for retaining some of her own hard-won independence. That's when her daughter reached out to me again. Age was taking down Robert's cognition. The isolation and loneliness were threatening Sylvia's nervous system as well, turning her Upgrade into a downgrade.

I knew that in order to disconnect the nervous system from contagion with a partner's requires mind/body techniques for closing off circuitry and containing the tent of "Me." I asked Sylvia how she managed to separate, to remain connected to Robert but not be overwhelmed by his moods. "I was working with a yoga teacher who helped me recognize when my husband's emotions were too prominent in my tent of 'Me,'" Sylvia told me. When she felt pulled by Robert's emotions, Sylvia was taught to first consciously slow her breath and bring her attention to her belly, a point just below and behind the navel. Once she was able to feel that connection, she rested her mind lightly on that spot. By bringing her attention to the center of balance in her body, it helped her focus on remaining centered. And while her attention was occupied, it meant she was able to disconnect from the habit of searching for signals of distress coming from Robert; it meant she had the option not to be so reactive to them.

When Sylvia did her breathing and body-awareness meditation, her tent flap could gently close—not lock, just create enough of a seal to contain the nervous system's circuit relaying the visceral sense of self. "This is what I found keeps the integrity of me," Sylvia told me as she described how much more powerful she felt in this practice and how it made her better able to respond not only to her husband but also to the ups and downs of life circumstances.

I like to think of this process as remaining in my own Hula-Hoop. I use this image when I feel too pulled by my husband's emotions. So many of us have spent a lot of our energy trying to step inside other people's Hula-Hoops, fixing and managing what we find there. But doing that also takes away the dignity of their feelings and their autonomy. So staying within my circle—like the ones they painted on the ground in parks during the COVID-19 pandemic to set boundaries for physical distancing—means I'm supporting the integrity of myself and everyone else around me.

Change and the Fear Circuits

Sylvia's practice to disconnect from Robert's mood felt wrong to her at first. Any change, good or bad, can throw a system's homeostasis off. In the early stages of shifting to the Upgrade, the nervous system needs to get used to the new shape of the tent of "Me" and the new, firmer placement of the tent pole. Like learning any new skill, or inhabiting any new state, it takes practice. The new habits almost always feel wrong at first.

The brain on default evaluates change through a lens of fear. We are wired to scan for threat to normalcy, for an indication of the slightest thing being a little off, to be constantly alert to change, ready to interpret it as a possible threat to our lives. The process by which bad news burns so deeply into us is called negativity bias. It can be a good thing, triggering healthy fear to avoid real danger, and life-threatening if we ignore the fear signals. If we didn't have negativity bias, we would be like those who don't have pain receptors and burn themselves on a hot skillet. Yet it's important to remember every day that the brain is like Velcro for negativity and Teflon for positivity. When the brain detects change, the lack of familiarity breaks through as a 9-1-1 call. It's that same bias that can mistake a healthy change, as in perhaps dating someone who is different but better for us, for danger.

Before the Transition, with full-bloom hormone levels that drive us to seize the world, our discovery circuits, the dopamine and adrenaline rush of the new, overwhelm the fear of negativity bias. We muscle through, fueled by the high of the new and the next. But the Transition and Upgrade are different. The discovery circuits are quieter without the strong pulse of androgens like DHEA and testosterone. Perception of change can manifest as deep anxiety during the Transition. What feels dangerous now might not have felt dangerous in our late thirties and early forties.

Caution and hypervigilance are core parts of the wiring that develops in women through acculturation, but they aren't our natural state; they're the wiring that emerges in anyone in a subordinate position. We become hyperalert to the needs of the "boss," able to anticipate and ameliorate quickly to avoid what our cavewoman wiring perceives as a threat to our source of food, clothing, and shelter. We hold being always ready, willing, and able as the peak of a highly functioning, thriving female, but it's a time-limited coping skill whose wiring comes undone in the Transition. The opportunity is freedom from the feeling of being on call 24-7 to respond to the needs of others. Explaining this to Sylvia made her ready for this kind of freedom.

The Brain Benefits of Groups

The pressure we put on a one-on-one long partnership to serve so many of our needs may actually be damaging to cognition in the Upgrade. In a study on aging female mice at Ohio State University, researchers put postretirement-aged rodents into two different social situations. One group lived in a cage that held only two mice—the old-stay-at-home-couple model—while the other group lived in a cage with six mice. After three months, the mice were tested for memory and cognition. The six group-housed mice won hands down. Their brains were working like the brains of healthy young mice. The couple-housed mice not only had decreased memory but also had increased brain inflammation, highly correlated with eroded cognitive health. The brains of the group-housed mice had fewer signs of inflammation, so their brains literally functioned and looked "younger" than the less interactive mice. The part of the brain responsible for memory was physically more robust.

"I've dreamed of group living since my forties," Ceci told me. "It would be fun and practical at the same time. I've always wanted to get a group together to buy a small building in Santa Fe, where I have a strong community. About ten or twelve units, an elevator, obviously. Keep an apartment for younger people to live there and help out with maintenance. We could share resources and caretaking, read, and talk about books and issues to keep each other cognitively in the game. It would be easier for us to spell each other when someone has a partner who's having trouble. It's a built-in support system without being in a nursing home!"

Preservation of cognition is key to the Upgrade, and the people we are bonded with have the greatest impact on long-term cognition than most anything else in our lives. Even the mice are telling us that being stuck at home alone or with only one person can be less stimulating and more damaging to brain function. Inflammation, so present in the brains of the coupled mice, can be responsible for cognitive decline. Keeping our social circles as wide as possible will not only preserve the Upgrade but also help take the pressure off at home.

For better or worse, we master what we repeat. Just as the improving dexterity of a violinist's left hand manifests as increased connectivity in the brain's right motor cortex, women with larger social networks are better at making new social connections. The knock-on effect, according to one study, is that they are less likely to develop dementia than those with smaller networks. Women who had daily contact with friends and family cut their risk of dementia by nearly half. Connection and belonging trigger a soothing cocktail of neurotransmitters that regulate mood and provide a buffer in times of stress and anxiety. Getting together with friends nourishes the brain; girlfriends become the secret weapon of the Upgrade. These interactions may even reset the aging clock.

In a study done in New Hampshire in 1979, psychologist Ellen Langer created a setting for men in their late seventies and early eighties that would bring them back to 1959. The men spent a week in an environment that had Nat King Cole on the radio, newspaper headlines about Castro coming to power in Cuba, brand-new books by Ian Fleming and Leon Uris, and Jackie Gleason on TV. Compared to both a control group that was not immersed in 1959 and to their own results in tests done before they entered the experiment, the men exhibited clear improvements in memory, vision, hearing, and strength. Before and after photos showed that the men looked objectively younger after spending a week together as though it were 1959 instead of 1979. Anything that turns back the aging clock and reduces inflammation is great for the brain. So watch your old childhood favorites on television, listen to and dance to the music of your teen years, and head off to that reunion you've been avoiding. Those old memories are great for resetting the aging clock.

We can reproduce some of the effects of the experiment by consciously activating healthy brain and nervous-system circuits. Visualization is one of the most powerful ways for this to occur, and one of the fastest ways to make change. Try sitting down and taking a few deep breaths, letting your body relax. See if you can remember a moment in your childhood, teen years, or early adulthood when you felt happy, full of fun, at the top of your game. Maybe it was a party, a sleepover, or your first big career boost. Maybe it was the loving excitement your parents showed when you learned a new skill as a child. Spend some time conjuring the details as though you were there right now: the sounds, smells, sights, people, environment, and feelings, both emotions and sensations. Hear the music and feel the laughter. Sense the joy of youthful movement. Let in the deep care for old friends. Take a few minutes and rest in that memory, and notice if there are any changes in your body. It may take a few tries to trick the body into releasing the cascade of feel-good hormones, but it's possible. And if you do this often enough when you're calm, you might have more resilience during tough times, as studies done at Emory University are indicating. I now practice this mental exercise daily and recommend it to my patients.

———

The echoes of old tent-of-"Me" relationship circuits are strong, and not all of us find our way to upgrading them. But we have a choice in the new space that opens in the Upgrade. We can go down the path of despair and worry that comes with the old patterns, with powerlessness over aging, with helplessness over the direction of our children's lives. Or we can stand and make friends with the realities of life. For many of us who value our ability to make things happen, discovering in the Upgrade that there is so much over which we are not in control threatens our sense of competence. But there's so much in store in the Upgrade. We can engage new circuits, shifting our roles in relationships to ones that give us the most freedom to be ourselves. We can feel our own feelings, find our own desires, separate from what's happening for those closest to us. Centering in the Upgrade gives us the space to come to terms with things as they are, to confidently ride the ups and downs instead of fighting them.

Centering

Chapter 10

"So how do you feel about seeing everyone?" I asked my friend Diane, fifty-eight, as she was getting ready for her fortieth high school reunion. Personally, these kinds of gatherings had been my most feared moments of the Transition. I imagined the anxiety Diane must be feeling as the insula (the area of the brain that began torturing us in puberty as we compared ourselves to others and found ourselves lacking) activated memories of social hierarchies, feeling left out if she hadn't been invited to the right party, or not feeling as pretty as the popular girls. I was calling Diane to give a little moral support, making sure she wasn't going through what so many of us do when we are headed into a gathering of people who likely remember us only when our looks were the best they were ever going to be.

When we are teenagers, that high school image of "Me" imprints on the insula as a template for who we are and what we should strive to look like. Its wiring stands at the ready to signal self-disgust when someone shows us a recent photo taken in bright light, when we see ourselves on a video call, when we see the clothes we can no longer wear, when we meet up with a high school friend whom we haven't seen in forty years and think they look younger than we do. Encounters like this can be a perfect nervous-system storm for the female brain, and I was worried that those old tracks in Diane's brain would be prompting a lot of fear and anxiety.

But Diane had an unexpected reaction. "Hey, babe, I'm glad you called, but I am fine," she said. I could hear her rifling through her closet as we spoke, and then the line was silent.

"Diane, honey, you there?" I asked.

"What's this still doing here?" she said quietly. "I thought I tossed it."

"What did you find?"

"It's an old cocktail dress that doesn't fit anymore. I was traumatized by this one," she said. "Remember that phase when you still thought you could beat your changing body into submission?" I remembered all too well. There were a lot of tears involved.

"Back then, I tried to squeeze myself into this dress for my niece's wedding," Diane continued. "My husband hurt his hand trying to zip this one. He knew how badly I wanted to wear it, so he didn't give up. He went out to the garage and got a pair of pliers. He used the pliers to pull the zipper up." Diane started to giggle. "First the little tab on the zipper broke off. Then he managed to get a grip on the zipper itself. He got it zipped, all right; and the whole dress ripped!"

By the time she finished the story, Diane and I were in full-on guffaw. I knew she had been making some changes in her life, but I didn't know she had also been hard at work shifting the neurological and neurochemical tracks in her brain. In particular, she had already laid down new tracks in the insula, tracks that allowed for a sense of "Me" that aligned with her current reality instead of fighting with her "Me" of the past. The insula maps out what is considered beautiful in a culture and then engages in constant comparison. Diane had upgraded from the teenaged Insula 18.0 to Insula 58.0. She was staying healthy, eating well, and exercising, and she wasn't setting unrealistic goals, like trying to fight her way down to the weight she was at eighteen. She had come to terms with a larger midsection, crow's-feet, lip lines, a wrinkled neck, and a little bit of jowl. She had decided life was too short to feel bad about her neck.

"It took me a while to get to this place," she said when I asked about how she was managing the Upgrade. "That scene wasn't funny at the time. I tried on two more dresses, and none of them fit. I was sobbing and I had black lines streaming down my face. An hour of work on hair and makeup was down the drain. I love my niece and we are very close, but by the time I squeezed into a third dress, I didn't want to go to the wedding. My husband was amazing. I was falling apart, and he was comforting and persistent. He knew I'd regret it for the rest of my life if I didn't get myself there. I finally found something loose and black—I know, you're not supposed to wear black to a wedding, but it was that or yoga pants. We arrived an hour late; I missed the ceremony altogether, and I didn't know what to tell my sister. I got as close to the truth as I could and told them that I had been sick. I guess after they read this they'll know what really happened!"

"Oh, honey," I said, "I've been there. I remember those battles." I had many in my fifties, in the Early Upgrade. Listening to Diane, I could feel the creeping sense of disgust at myself, echoes of my own Insula 18.0. The battle with the body and the shift to embracing a new reality is a developmental phase in the Early and Middle Upgrade, one that we don't talk about.

Like Ceci, Diane had decided to give away everything that didn't fit and look for a style that made her feel comfortable now. "I had to get rid of the reminders," she said, "for the sake of my own mental health, and my husband's poor fingers."

Diane's action made sense from an Upgrade perspective. If she was opening a closet twice a day to a rack full of clothes that didn't fit, she was activating the insula in its task of comparison, which all too easily escalates into self-blame and self-flagellation. Those attitudes signal the pituitary to send out a hormone prompting the adrenals to open a fire hose of distress hormones. Those hormones trigger the brain and nervous system to cause jitters, anxiety, and testiness. If we allow those old neurological tracks to take over, hopelessness, even intractable melancholy, can be the outcome. In those with diagnosed depression, that pituitary signal to the adrenals is already stuck on high, and stress hormones pour out nonstop. Not everyone can cope without medication, but many of us, like Diane, are not helpless. Developing strategies for shifting the tracks of Insula

18.0 was an important step in upgrading the insula to align with Diane's sense of the tent of "Me" at fifty-eight.

"I thought my closet was cleaned out," she said, "but this one must have escaped. It's going to Goodwill first thing tomorrow!"

So many patients and friends have reported that during the Transition and in their Early Upgrade, issues they thought they'd dealt with re-emerge, hijacking the nervous system with unexpected force. The energy to muscle through is gone.

It can be triggered by something as small as someone disagreeing with our opinion or leaving us out; it can be a time when the world no longer responds to who we believe we are. It can happen when we encounter a reminder of an old trauma. Every day, and more so as we enter the Transition and the Upgrade, life circumstances, old wounds, and people's reactions to us challenge our tent of "Me." It happens even to the strongest of us, those who went through years of therapy and lots of hard work to address issues large and small. It happens as we become fed up with old caretaking roles, only to find ourselves sucked back into them. And as I can attest, it happens a decade after we negotiated peace with our body, when we think we long ago ended the lifetime of war. It happens when we find ourselves taking on extra work for free because someone managed to find and push our "good girl" button and we jumped to volunteer. It happens when we realize those occupying the space closest to our tent pole may have wrested control right out of our hands (or we voluntarily handed the control over once again), throwing our sense of priorities into disarray.

Waves leave imprints. Wading through calm waters, we feel their ridges in the sand beneath our feet, evidence of their passage. Though our hormonal waves may have calmed in the Upgrade, their echoes are still imprinted in the tent of "Me." In certain situations, echoes can become amplified, activating tracks that were laid in another time, reminders that push our neurohormonal (and emotional) buttons, pulling us off balance, sending us spiraling out of healthy self-care into sadness, fear, isolation—"I didn't want to go to the wedding"—and sometimes self-loathing.

Becoming centered is a gift of the Upgrade; yet unrecognized echoes of the old waves threaten our ability to manifest that gift. Those old patterns of who we thought we were as a woman in her fertility phase—how we should look, feel, socialize, exercise, eat, engage in care of others and ourselves—can pull us out of our zone of well-being faster than you can say "hot flash." To emerge into an optimized Upgrade means using our attention and our intelligence to take full control of our own tent pole, learning how to keep it from being yanked out of the ground or snapped in half by echoes of waves that no longer pound the shores. It means being able to remain still even as the echoes of old hurts, old wounds, try to pull us off center. Hitting an obstacle doesn't throw you off your horse anymore. When something is emotionally upsetting, you realize that if you put one foot in front of the other, within forty-eight to seventy-two hours the sting reduces on its own. We develop the strength of holding the space to see what happens, resisting the impulse to act. As those waves weaken, we can watch them dissolve into a warm, gentle foam as we wade more easily along the shore where we once sat and brooded.

Tina Sloan is an actress who played the role of nurse Lillian Raines on *Guiding Light* for twenty-six years. In her book *Changing Shoes: Getting Older—NOT OLD—with Style, Humor, and Grace,* she talks about the day she experienced the *coup de vieux,* the "blow of age." On a break from shooting a party scene, she and her on-screen daughter headed out for coffee in ball gowns. As a soap opera star who had been counseled as a teenager by the cream of French society on how to command a room and the attention of men, she was used to wielding that power. But on this day, in this café in Midtown Manhattan, at nearly fifty years old, she could feel something was off. It was unsettling, and she wasn't sure what it was. And then she realized that the attention of the men in the café was riveted by her costar, not her. She felt disoriented as the tent of "Me" she was accustomed to experiencing felt like it was being blown away.

It was the insula noticing something was off-kilter as her nervous system waited for the familiar stares of admiration. When those signals did not come, her memory circuits set off the alarm: *We are not in balance.* In parallel, the vagal system—the body and brain's largest nerve, responsible for both activating and calming the body's threat system—picked up the alert. Acting as a superhighway among the intestines, the heart, and the brain, the vagus triggered a literal gut instinct that something was wrong. The nerves were begging her brain, *Do something to get us back to our cushion of familiarity at the center of the tent of "Me."*

The push to *do something* comes in the form of brain signals to produce neurohormones that drive behavior. There are so many responses that can satiate the hunger for a familiar response, and the responses we choose can be determined by whether or not we've been able to upgrade the insula. If we are slaving under the oppression of Insula 18.0, then the signals can be torture. Accustomed to being recognized as the hot one, when a hotter one comes on the scene, the insula sends out a red alert that you're not going to get the good sperm. It flashes *insufficient* to your sense of self, sending a hot poker to your nervous system to make your prefrontal cortex start doing its job of problem solving. Even though those old fertility hormones might be gone, the echoes of them can create a flurry of activity. I've seen those urges manifest in friends, patients, and myself when Insula 18.0 rips at the tent pole and stability of our sense of "Me."

Resisting the impulse to react to the insula's commands barked at the prefrontal cortex is hard, but finding a way to take our foot off the gas is a quality-of-life decision. When Jacqueline de La Chaume moved from Paris to the United States with her husband, Yul Brynner, she was warned by her old friend, Karl Lagerfeld, not to give in to cosmetic-procedure pressure in Hollywood. And she said that Yul was against it as well, telling her that it would erase the map of her life as it appeared on her face. The pressure can feel real. I've made and canceled appointments for procedures dozens of times myself.

"It's hard to watch what's going on among the women here," Sarah said after moving to Houston. "You see women in their late eighties not eating, wobbling around in high heels, wigs, and heavy makeup. Their cheeks and lips are stuffed full of fillers. Many in my generation—in our fifties—just aren't doing this anymore. There's something tragic about seeing women being pulled away from themselves like this, unable to express who they really are." Sarah is right. Statistics are showing that those getting cosmetic procedures are skewing younger, to the before-the-Transition years. It's too soon to tell if women in the Early Upgrade aren't getting the procedures because they already did them in their thirties and forties or if this is a real trend away from cosmetic procedures.

When we become centered, able to remain grounded in order to honor the woman who's emerging, the neurological mindset shift of the Upgrade frees us from the struggle to prolong the appearance of the fertility phase. Although these decisions are very personal and I don't feel it's right to judge, I do admire Tina Sloan's choice as an actress never to have her face altered by cosmetic procedures. It tells me that she made a decision to take control of her own tent pole, her visceral sense of who she is.

When it comes to appearance in the Upgrade, what I'm talking about goes beyond acceptance to loving engagement and unconditional embrace. A begrudging stance toward ourselves is still destructive, like continuing to sneak nourishment to Insula 18.0 even as we are trying to shape the tent of "Me" 2.0. As our peer group makes different decisions—some opting for an Upgrade, others opting to go to war against nature, friendships shift.

What We Are Up Against

Female fertility is at the core of our social fabric. As children we played house and perhaps fought over who got to play the mommy. We bonded secretly over the beginning of our periods and our first sexual encounters. Many of us built our own families; around our kids another social life formed as we got to know the parents of their friends at school. Those of us without children formed bonds at work and through volunteering. Disruption first comes if you are divorced, and possibly shut out by married girlfriends, or are aged out of a job and career. And when the kids grow up and leave, the social bonds formed around them and the roles we've played are also gone.

As older women we are marginalized, dehumanized, labeled as cute, called "young lady" or "a little holiday angel of cheer," as two young women in Atlanta addressed fifty-four-year-old Terri one time while she was dropping off her dry cleaning. If we have allowed the nervous system to absorb that narrative, then it amplifies the signal that we are not relevant. It indicates that we should stop caring for ourselves and paves the path to social isolation. This doesn't become garden-variety loneliness; the nervous system begins to lose the ability to engage the circuitry that keeps us reaching out to connect, to keep those essential brain vitamins we receive through community that nurture the desire to keep putting one foot in front of the other every day. This is likely an underlying contributor to the alarming rise in suicide among women over sixty. Twenty-five percent of women will have a major clinical depression sometime in their life. If you've had a major melancholic episode before the transition, it means that you are two to three times more likely to have another one post-Transition. If we absorb the message that we don't matter, melancholy is often a consequence. How we feel about how the world sees us can make the Upgrade challenging.

I was listening to NPR one Friday morning in early March of 2020 when I heard about a United Nations Development Programme (UNDP) report about attitudes toward women. The study was conducted across seventy-five countries, and the report concluded that nearly 90 percent of all human beings are biased against women at work, in government, for their opinions, in every area of life. I had two reactions. First I thought, *Did I hear that right? Ninety percent?* I called a friend, who found the *Guardian* story on the report. Indeed, I had heard it right.

Deep down I wasn't surprised by this news. The bigger surprise was that this massive obstacle to feeling free to claim our common human dignity, to stand firmly in the experience that our rights and needs are equal to others', was finally being reported on.

And then the harder truth hit me square in the face. The report didn't find that 90 percent of men were biased against women; the finding indicated that on average, nearly 90 percent of all human beings are biased against women, and that 86 percent of women were biased against women. It made me question who was in the 14 percent; I suddenly had to consider that if women were biased against their own gender, then there were likely men in that tiny group of allies who weren't. I also had to face that I might be part of that 86 percent of women who engage in hostile sexism toward my own gender.

In my psychiatric training I learned about the concept of introjection, when an individual absorbs the perspective of another or their culture and makes it part of themselves. It is particularly common in anyone in a weaker position who feels the need to agree with the stronger one in order to survive. I knew that at various points, especially in my twenties, I had introjected some form of hostile sexism. It happens unconsciously during the epoch of competition for the best sperm. No matter how feminist we are, the hormones coursing through our bodies during the fertility phase ping the brain continuously to drive actions and behaviors dedicated to optimizing procreation. As young women dominated by neurohormones, we often turned away from anything we perceived as being distasteful to men. If younger men derided sexism and feminism, we often did too. I remember sometimes feeling that women's groups were stupid; I was convinced older women were just jealous of us because they had lost their hotness. That I feel the opposite now is not just about education, experience, and the fact that I am now an older woman. It has a lot to do with the easing of neurohormones that drove those earlier attitudes. In the Upgrade, this is another bias that can drop away.

New Source of Validation

Many of us talk about having been much more disciplined in our younger lives about taking care of ourselves, and that post-Transition our efforts feel intermittent, scattered, uneven. The signals we got to take care of ourselves when we were younger were, no matter how enlightened we thought we were, driven by fertility hormones and cultural messages about showing off our health and ability to procreate. It was important to the survival of the species that we be healthy when young; cultural obsession with youth and beauty is really just a symptom. It's not a coincidence that in the second half we have a harder time focusing and maintaining our efforts—we don't feel society's deep care about our health. It can feel harder to remember which tools to use or which people to call who have helped us emerge from problems in the past. The signals we get after the Transition about our importance to the culture at large are devastating: that once we lose our eggs, we don't matter; that we should give up, especially since nearly everybody we encounter in a single day is biased against us as women. Though the UNDP report didn't break out bias by age, it is reasonable to conjecture that for women over the age of forty-five or fifty, the experience of bias could easily climb higher.

Every decision takes bandwidth. We have the opportunity to choose how much bandwidth we want to expend on diet plans, hair color, dermatologists, and makeup. Some of those decisions didn't feel optional during the fertility phase. But now they might be. "I am wearing the same stretchy shirts and pants," Ceci said. "And in my world, being chic was very important. I paid a lot of attention to it, shopped in Italy, and treated my clothes like an investment portfolio. But at a certain point I realized it was such a big waste of time and money," she continued. "Now I have the same freedom I had when I wore a uniform in high school. It wasn't restrictive. It was freeing. I didn't have to decide what to wear." In her Upgrade, Katharine Hepburn wore black pants and a white collared shirt every day for the last thirty years of her life. Jacqueline de La Chaume, who in the 1960s worked for designer André Courrèges and for *Vogue Paris,* began making her own simple clothes in her sixties. In my sixties, I choose athleisure. I can free up my nervous system's bandwidth by not stressing over a decision that no longer feels important to me.

Many women in the Upgrade found this freedom in the lockdown during the COVID-19 pandemic, when we suddenly saw how much energy was dedicated to purely fabricated standards around appearance. Now we know that without contact with others, we wear whatever we want. The deep relief many felt is a window on what it would be like to make that choice under normal circumstances.

When we are no longer trying to have it all, when we recognize that there are real limits to what can be accomplished and we can accept them, we can be free of the torture of unrealistic goals remaining constantly out of reach. We can see freedom in the Upgrade instead of lack.

We can shift from mourning the loss of fertility to enjoying the freedom of not having to freak out about pregnancy, mood swings, blood, or cramps. We can enjoy the freedom of walking down the street without the burden of the voracious stares of men, the whistling, the catcalls. We can enjoy that drugstore shopping time is cut because there is an entire aisle we never have to walk down again—unless our daughter-in-law or granddaughter is visiting.

Remember, biology is destiny if you don't know what it's doing to you. By recognizing the biological and neurochemical principles in the female brain, we can see which circuits we are reinforcing through our physical, mental, and emotional habits. Becoming mindful and alert means we have a unique opportunity to reshape the circuits during this time. The Upgrade supports us in this endeavor. And if we are lucky, the internal voice of the centering circuits can become so strong that they are no longer suggestible to old habits. The tent pole becomes unshakable. And what you decide to do with your tent matters to me.

Body Hacks for the Mind

Chapter 11

Let's say I went to sleep angry about something that happened with my husband, my son, or a friend. I'm in a fitful state most of the night, not really resting. I wake up thinking about the argument, replaying it in my head, rehearsing the words I wish I'd been quick enough to think of, willing a different outcome with all my might, priming myself for more anger and probably another argument.

The mind serves up all kinds of thoughts during the night in the form of dreams, and more when we wake up. This is completely normal. Sometimes a whole scene from the past, something that is very upsetting, will arise full blown and hijack my thinking circuits. Trying to block or suppress the thoughts or memories that make us uncomfortable doesn't work—they tend to come back in another form and sometimes even stronger. While we are busy trying to keep old anger from surfacing, we might take it out on someone nearby.

The biggest biological impact on emotion is hormones. Hormones drive behavior. Ghrelin makes us want to eat. Testosterone makes us want to have sex. Oxytocin makes us want to repair a relationship. Progesterone makes us want to curl up under a blanket. Cortisol and adrenaline, the stress hormones set off by frustration, danger, fear, or anger, can make us want to lash out to protect ourselves.

A burst of cortisol, say from a big argument, can continue affecting the body and brain for many hours—assuming there isn't another stressor to set off another round of it. Then it can take up to five days for the brain to reset to the new normal and for access to circuits of higher reason and judgment to become fully accessible again. And in the Upgrade, the effects of cortisol can be prolonged, especially if you have a stressed partner. So that old adage about cooling off for twenty-four hours before responding to a highly charged situation? It isn't nearly long enough. A week and a day is more like it. If we try to respond or to problem-solve while we are at the mercy of our hormones, we are more likely to develop what psychologists call emotional incontinence. We just can't hold it in; we can't contain the extremes of emotion, so we cross a line with words or the slam of a door. Escalation increases stress and the cortisol cycle starts all over again.

Cortisol, the classic stress hormone, naturally rises early in the morning to help us wake up, raise our blood pressure, and achieve alertness, getting us ready to learn new things and be at the top of our mental game. It also rises when we perceive a threat of any kind, from an irritating news story to a narrowly avoided car accident. Levels can rise with aging and be higher in older females than in older males. But chronic rumination or intractable melancholy can intensify cortisol, making levels high enough to disable clear thinking and problem-solving altogether. It can also set off a chain reaction that causes sterile inflammation of tissues and blood vessels, even in the brain. Add to that lower estrogen, which in our earlier days protected cognition against stress, and we can be set up for a more rapid mental decline. Even if we are taking HRT, chronic stress and rumination can run roughshod over estrogen's protective benefits. Stress can be one of the most dangerous enemies to cognition in the second half.

As a doctor who specializes in the brain, of course I prioritize its health. It's a natural bias when you love something as much as I love the female brain. But there's also deep justification. Strong and healthy cognition is key to being able to manifest the best qualities of the female brain in the Upgrade. When we make choices that hinder the brain's capacity to remain strong, vibrant, and sharp, we can all too easily slip into smaller and sadder worlds that can spiral into isolation and melancholy. In my practice over the decades, I've seen this happen too many times, watched too many women slip into a preventable downgrade. But simple changes—even changes as simple as moving your body—play a huge role in preserving cognition, especially in the face of stress.

Movement Is Cognition; Cognition Is Movement

My patient Rhonda, sixty-eight, is a social worker. Super comfortable with her age, she let her hair go white in her early sixties. It took ten years off her face and helped revive her part-time acting career: She was able to land a few small parts in some big films. In that regard, she shifted her tent of "Me" to incorporate her changing physical identity. She also decided never to have any cosmetic procedures on her face.

Rhonda takes her Upgrade seriously. She is a gregarious and outgoing woman, a friendly and generous person—until a younger woman or man dares to "young lady" or "dear" her. "It started happening in my fifties," she told me. "I won't answer when someone asks, 'And how are we today, young lady?' But if they persist, I have my response cued up: 'I hope you aren't addressing me. I haven't been called *young lady* since I was five and in trouble with my mother. I am a grown woman. I am proud of every single year and every single wrinkle. You can call me *ma'am*. No, not *dear*. Ma'am.'"

Rhonda had been working out regularly, taking yoga, strength training, and going on long, fast walks for cardio conditioning. When she feels physically strong, her confidence rises and she feels able to stay centered in the face of slights and difficulties. But life has many blows to deliver.

After a long period of stability, Rhonda's thirty-four-year-old daughter Seana, who suffers from bipolar disease, went off her meds and off the rails. Seana walked out on her husband and four-year-old daughter and took up residence on the streets of LA, trading sex for drugs, as she had done in her teen years. After a few weeks Seana disappeared from the neighborhood and Rhonda had no way of finding her daughter. Rhonda became despondent, listless, unable to think or express herself clearly. Her tent of "Me" as a mother was in collapse. As a social worker, she knew intellectually that her daughter's disease was not her fault and was not within her control to cure. But as a mother, she turned every drop of guilt she could find into a psychological bat with which to beat herself. It didn't change her outward circumstances or help her find her daughter, but it intensified her stress and sorrow.

In times of profound chronic stress, the cortisol-releasing hormone CRH can get stuck on high, eventually depleting the system and its circuitry of the fuels that might otherwise motivate us to get up and move. It's like running an engine at its highest RPM, burning out the motor to the point that it seizes up. Now scientists believe that when people get profound melancholic depression, and in particular melancholic depression post-Transition, it affects their movement system. In these extreme cases patients report feeling stuck curled in the fetal position. We used to think the will to move was all that was missing. Now we know that the motor system also goes into a clinical depression along with motivation and mood. This is why, as a psychiatrist, after I've tried everything else, I now recommend taking an integrated mind-body approach to profound sadness.

Rhonda was an old patient of mine for whom antidepressants had terrible side effects. When her husband, Mike, called concerned about Rhonda's state of mind, I knew drugs wouldn't be an option. But new information about a brain organ called the cerebellum was opening the door to a different way to help.

Many times, when babies start to crawl, they will pull themselves along the floor using only their arms, dragging the lower part of the body behind them. Getting their knees underneath them, coordinating their alternating movements—right arm, left leg, left arm, right leg—is the job of the headband-shaped motor cortex, which wraps around the top of the brain and communicates down to the coordinator of motion, the cerebellum. But different systems of the brain turn on at different times during the first year of life, and sometimes nature needs a little help from Mom and Dad. This is often a simple fix: With your baby lying on her back, make a game of holding her feet and bicycling her legs. Then help her stand and hold her hands while she stomps her feet. Feeling and watching the movement sends a signal backward from the eyes and muscles via the nerves to the cerebellum: *Hey up there! Start your engine for this part of the body. I need to move!* It's not usually very long before the wiring develops a strong and steady signal that kicks the cerebellum into gear to organize the signals sent by the motor cortex. Baby is off and running, and you are getting more exercise than you planned chasing after her.

When Mike got Rhonda to agree to a phone consultation, I decided to test a thought that we could do something similar for her, using movement to trigger emotional stability. "Rhonda, can you put me on video and set the phone down?" I asked her during our first call. Knowing that she had practiced yoga since her twenties, I took a chance on her muscle memory. She placed the phone on a table and I asked her to stand nearby. "Now, feel your feet on the ground. Feel the sensation of contact on the bottoms of your feet. Close your eyes and let your attention rest there for a bit."

"Oh my goodness, my legs are shaking!"

"I know," I responded. This can happen as attention signals nerves to fire, as in Rhonda's case, to make her legs feel stronger than she felt emotionally. "Don't worry about it. Just let them shake. Feel your feet."

"Okay," she said after a silence. "They stopped. I'm ready. What's next?"

"Make sure your feet are hip distance apart. Now settle there, and when you are comfortable, slowly shift your weight to one side."

"That feels really unsteady. I feel like I might fall."

"Okay, open your eyes a little. Don't focus on anything. Let your attention drop to the foot you have your weight on. Tell me when you feel comfortable. Don't rush."

"Okay. I'm okay."

"Now lift the other foot. Stay there for a while. Get your balance. Focus on a spot on the floor and breathe for a few counts. Feel the sole of the standing foot. Then gently put the lifted foot down again; feel the sensation of contact with the floor. When you are ready, shift to the other side and lift the other foot." We did this a few times until Rhonda could maintain her balance on one foot with some reliability. When her physical balance began to stabilize, I started to hear a tiny bit of light in Rhonda's voice. When she lifted the phone again, I could see her face had begun to relax.

"Okay, this is great, Rhonda," I said. "Now let's try a little more movement. Is it nice outside?"

"It's cloudy," she said, "but it's okay to go outside."

"Can you plug in your headphones and put the phone in your pocket?" I asked.

"Yes, sure."

"Okay, let's go, then. I just want you to walk slowly, deliberately, paying attention to each step to the end of the block and back."

"That's it?"

"Yes, that's it." I remained silent on the phone. Rhonda reported feeling unsteady and off balance. I kept suggesting she return her attention to the soles of her feet, feeling the movement of her legs, finding balance by noticing that spot in the center of the body behind the navel, the center of gravity for the female body.

"Boy, this is starting to feel good," she said after a few minutes. "But my legs feel a little shaky at this slow pace. I can't balance when the anxiety starts to come up again."

"Yes, that's normal. So breathe slowly, walk deliberately," I reminded her. "And let's try it again."

As Rhonda moved, her pronunciation became clearer and her words came more easily. Her voice relaxed and sounded less constricted. Movement began to impact Rhonda's mood and her cognition.

I asked her to practice alternate-nostril breathing (see appendix, page 258) and a balancing exercise every day, followed by a very slow brief walk. As she became steadier, I suggested she try walking mindfully along a seam in her wood floor as though it were a tightrope. Over time, Rhonda's clarity began to return. Mike confirmed that Rhonda had become much calmer and was beginning to return to her old talkative self again.

Old Wounds, New Approach

Deep in the back, at the bottom of the brain, hanging near the base of the skull like two big ovaries, is the cerebellum. For more than one hundred years, neurologists and neuroscientists have known that the cerebellum is a command center for movement, balance, and fine motor skills. When combined with visual and inner-ear input, it is responsible for how we learn new dance steps and drills in soccer, how we know where to put our feet as we watch where we are walking or running. With the addition of a mirror, the cerebellum helps us perfect positions in yoga or ballet. It's one of the oldest parts of the brain; it's present in our reptile and aquatic cousins, helping fish orient their movement in the undulations of the sea.

The cerebellum is super sensitive to alcohol, and it's why, after a few glasses of wine, we might drive erratically and have a tough time maintaining balance on that straight line on which the officer asks us to walk. If we are nervous or very upset, the cerebellum is inhibited by the body's threat response; right after we receive very bad news, the flood of cortisol and other adrenal neurohormones makes it more likely we will bump into things, break things, or crash our cars. If the cerebellum is injured, we will have a hard time buttoning our clothing or guiding a spoonful of oatmeal to our mouths.

Most fMRIs that study the brain's functioning and connectivity easily image the brain areas above the cerebellum, like the frontal cortex, the visual cortex, the parietal cortex, the amygdala, and other brain organs. We are, as a result, more familiar with the blood flow and connectedness of these areas of problem-solving, interpreting visual stimuli, processing sensory information, and initiating muscle movement, and the brain's center of fear. Because of its location so deep and below the brain's big hemispheres, it's been tougher to get a clear picture of the cerebellum and study its connectivity to other parts of the brain. For the better part of the last century, neuroscientists felt the role and responsibility of the cerebellum for coordinating motion and balance was settled, so it remained ignored in the basement of the brain.

But in 2018 we got a big surprise. One of the very few labs in the world that has the capacity to image the cerebellum via a special functional MRI was able to see that its connectivity to other parts of the brain was dramatically more complex than anyone had imagined. Scientists at Washington University in St. Louis discovered that motor coordination was only one-fifth of the cerebellum's job. By observing connectivity to other parts of the brain, they learned that 80 percent of the cerebellum's functioning was related to areas that deal with judgment, problem-solving and the ability to think in the abstract, emotion and mood, memory, and language. The cerebellum, the hub of movement, balance, and coordination, is involved in editing almost everything we ask our brain to do. It fine-tunes emotional and cognitive function just the way it fine-tunes muscle movement. It makes learning and social engagement possible. It's a key to motivation, that feeling of being psyched up. It's only 10 percent of the brain's volume, but it holds more than 50 percent of the brain's neurons (brain/nerve cells).

Just as estrogen stimulates the prefrontal cortex (PFC), boosting our mental acuity and access to the problem-solving functions of the brain, progesterone helps boost functionality in the cerebellum. When it functions optimally, the cerebellum is a kind of filter, checking thoughts, emotions, and sensory information for spam, phishing, or scams before letting you act on or express them. As the original researchers point out, the cerebellum doesn't think, balance, or judge; it corrects errors and helps the parts of the brain that are responsible for those activities do their job more efficiently. A healthy cerebellum will check impulses as they emerge to make sure acting on them is good for your ability to thrive. If it's not a good idea, the cerebellum will delete it. If it is a good idea, the cerebellum will help recruit brain and nervous system resources to express it and act upon it. If it's impaired by alcohol, it won't be able to stop us from speaking or acting impulsively. Drunk driving and drunk texting both emanate from the same source.

Psychologists and researchers have long known that when it comes to sadness and clinical depression, aerobic exercise is at least as effective as antidepressants in making a productive shift. The new information on the cerebellum tells us much more about how this might work. Because it is connected to the emotion, reward, and judgment centers of the brain, physical and emotional balance can be understood as part of the same process, part of the same drive for overall homeostasis.

Homeostasis is the dynamic balance that every living system strives for in its mission to optimize survival. It's not a static state; it's a process of unending microadjustments made to approximate some mythical permanent state of well-being. Like the ballast correcting a sailboat's angle in a strong wind, the cerebellum helps us push back against what throws us off, both physically and emotionally. When physical balance is upset, the cerebellum triggers movement-stabilizing circuits, fine-tuning them so that we stand or sit firmly upright again. When emotional balance is upset, we can enlist muscle movement to help bring the cerebellum's capacity back online—practicing physical balance can help with melancholy, for example. And in some cases we might be able to jump-start the engine of stability by combining attention and movement, as Rhonda did.

This new way of understanding the cerebellum has led me to include it in my evaluations of patients, especially those who are falling more often. If Rhonda, for example, had not been able to regain her physical balance, and if she had been falling more than normal, after getting a brain MRI, I might have started to look for decline in cognition, for lapses in judgment. If the cerebellum is not able to coordinate movement and balance, it is likely having trouble coordinating language, emotion regulation, and stress response.

These days I don't just prescribe meds for profound sadness. For those who need the support, I definitely still do write prescriptions. But I also prescribe a gentle ten-to-twenty-minute walk once a day, a walking meditation, a gratefulness or prayer walk, extending the time gradually to an hour if it's possible. It doesn't cure everything, but it helps bring feelings of sadness into a manageable range, so that we can apply other strategies to regain control of our tent pole. Being able to restart our habit of self-care determines our resilience in the face of the difficulties we know are bound to come.

It took about two months for Rhonda to restart her fitness routine. "I'm still utterly anxious and worried over Seana," she told me. She'd since gotten news that her daughter had been found and had agreed to go into a treatment facility. It was a relief, but Rhonda had been here many times before, and she was still quite worried over whether or not this round would stick for Seana. It's normal to be in a fearful state when times are tough, either personally or globally, to have anxiety levels that remain at a persistent five out of ten. But we have a chance to change things once the hormonal spikes of the Transition have evened out. We can use that new steadiness to introduce a few tricks for nudging the brain's system for clarity and calm.

Cultivating Joy

"I was running five miles three times a week, and on the weekend I often did back-to-back classes at the gym," Sarah, forty-eight, told me. "I'd been doing this for years, and it was magical. I was tiny, eating healthy mostly but knowing if I cheated I could just burn it off. Regular, intense exercise every morning before work made me sleep better at night for a long time. But when I hit the Transition, I went through an insomnia nightmare. I kept waking up at three and couldn't go back to sleep. It felt like I couldn't handle the stress of work anymore. I lost the certainty I had when I was in my thirties and started second-guessing myself all the time. I couldn't absorb the normal ups and downs of life. I couldn't sleep if I watched the news past eight P.M. I couldn't sleep if my husband mentioned money at ten P.M. Working out had been my go-to stress reducer and my most reliable weight-control plan. And when I hit the Transition and it got, well, a bit spectacular, with bleeding and irregularity, I got devastating advice from my doctor: Rest. Do yoga. Stop running. Meditate. Take gentle walks. Sit on a park bench and look at the river. I had never sat still!"

I knew just what she meant and why that prescription felt devastating. Sarah continued, "I had no choice but to follow directions, because I didn't have the energy to keep up my old pace anymore. I was terrified I would start to feel sad because I couldn't kick up the endorphin high of running. I was terrified of becoming depressed if my weight ballooned. And it did. I gained twenty-five pounds in one year."

Sarah went into a tailspin, canceling social events and refusing to travel to her close friend Kim's fiftieth birthday party, putting the relationship at risk. "My moods were so unpredictable, I couldn't fit into my wardrobe, and I couldn't bear for anyone to see me like this. I tried to explain to Kim that I didn't want to ruin her big day by being so depressed at such a happy occasion, but she said I was ruining it by not coming. She didn't talk to me for a year."

Lack of sleep, feelings of uncertainty, and torture by self-image constantly trigger our threat systems, causing cortisol spikes that make us lash out and pack on the pounds. Grief over the loss of everything that comes with fertility—energy, appearance, attention, vibrance—sets in, and the sense of who we are begins to destabilize. Insula 18.0 had struck again.

The Transition can overwhelm the mind with deep insecurity as the sense of who we are seems to unravel. The longer we are subject to excess chronic cortisol and adrenaline, the more they alter our perception of the motives of others. Seeing threats and insults everywhere, we retreat from connection, and in isolation, the stress cocktail kicks off negative rumination. On the other hand, playful acceptance stimulates joy and optimism; optimism is another key to cognitive health and can be protective against dying from cardiovascular issues.

"We are not going to talk it out," Sarah's husband, Dan, said one night. She was spiraling into hypercriticism over everything and everyone at their book club. Her perception of others' negativity was interfering with their ability to socialize as a couple and sparking huge fights. Dan was determined to break the cycle, and the pressure made him creative. He had passed through the living room one day while she was watching *Grey's Anatomy*. The two main characters had been talking about their problems, and they started dancing instead of continuing the conversation. "Get up," he said. "We are not going to talk this out. We are going to dance it out." Sarah felt completely foolish, and at first she couldn't get into it. Part of it was the age difference; Dan was thirteen years older, and the music he was playing didn't resonate with her junior high school dance years. Once he switched from the Beatles to Earth, Wind & Fire, Sarah couldn't help herself. She had to move. "I couldn't stop smiling," she said. "I forgot why I was so upset."

Movement is connected to our first success at survival as babies. We reach for the breast and get it. We grab a piece of food from the table and guide it toward our mouths. The reward system for movement is enormous because survival depends on it. We reach, grab, practice, master. *I got that sucker. Yay! I will live.* The basic neuromuscular circuitry of survival rewards us with feelings of success.

The expression of joy is hardwired to muscle movement. Throw your arms up and out to the sides and it begins to stimulate the process, or strike a power pose, the superhero stance that is proven to stimulate confidence and lift the mood through the neuromuscular feedback loop. Add music to movement and the entire nervous system is recruited, bypassing cognition, igniting the heart, circulation, and breathing. I like to blast "Start Me Up" by the Rolling Stones and dance like crazy on my deck, throwing my arms up with almost every beat.

Play is commingled with joy in the nervous system, and before you tell me you've outgrown the need, I'm going to let science set you straight: It's one of our oldest evolutionary instincts, and its impact on cognition is profound. Play is a key nutrient for cognition and the Upgrade. The neurobiology of play impacts key circuitry essential to social skills, creativity, adaptability, problem-solving, and raw intelligence. Scientists have found that the inability to play is a sign of lack of healthy social interactions in rats, and of psychiatric disorder in humans.

Age can determine how much play we need, but not the way you might think it does. We need it most during tough times or after periods of isolation—think post-COVID-lockdown-years—and that need can be more frequent in adulthood, as there is often more to be stressed about. In difficult times, whether personal or societal, we need to play more than ever; it's the very means by which we prepare for the unexpected, search out new solutions, and remain optimistic. It's key circuitry for the brain's hopefulness.

Joy feeds back hopefulness into the nervous system, and hope is pure heroin to the brain. Triggering dopamine to course through the system, it's a massive stimulus to cognition. It ignites creativity and problem-solving, and it becomes contagious in our relationships. The familiar banter among girlfriends that can make you laugh so hard you cry or pee a little, collaborative efforts like working a puzzle with family, or trivia games with friends, stimulate the release of all kinds of feel-good neuro-chemicals.

We can bring a playful attitude to all physical activity. Instead of barking orders at the body, try asking permission, making sure you have consensus from body parts, muscles, and the nervous system. At the beginning of my swim class, I smile at all the parts that make me move and say, "Come on girls, let's jump in!" That lets me start the class with joy. It helps me benefit even more from the exercise and deepen my connection to myself. I become more attuned and aligned. If you can deepen that connection, you not only find new energy, focus, and efficiency but also have a wider bandwidth for others. You can start having fun with yourself and even smile at everyone's foibles.

Drink a shot of joy every day by moving to music, walking in nature, jumping into a pool. Do something playful and something challenging or new every day. Take up space. When you go to the gym, occupy the weight room. Like the warrior stance, it tells the brain that you are powerful. Just don't let exuberance override your better judgment about how much you can lift.

When the Joy Circuits Need a Jump Start

Movement and play don't work for everyone, and they might not work for those with clinically diagnosed issues. Intractable melancholy is a risk for women in the second half of life; if you've had an episode before the Transition, your risk is increased of having another one after.

In the face of threat, the brain's amygdala kicks the hypothalamus to send out corticotrophin-releasing hormone (CRH) to the pituitary, and that signals it to release ACTH (adrenocorticotropic hormone), which gets the puppy-dog adrenals to release cortisol and adrenaline. It's been well known for over thirty years that CRH gets stuck on high in clinical depression. Instead of the usual waves of adrenal cortisol that wake you in the morning, and the drop at 3:00 P.M. that helps you wind down for the day, CRH turns the adrenals into a nonstop cortisol firehose. The body's adrenaline keeps you running like you're taking diet pills or speed. You become hyperfocused, and negative rumination spins out of control day and night. You don't want to eat and you can't sleep. It's a recipe for intractable melancholy. If the rumination spiral becomes fueled by pessimism, it can impact cognition. When you throw in the estrogen spikes and plunges of the Transition, it scrambles the brain completely.

For many women in the Transition, adding estrogen and progesterone can soften the wildness of the waves by calming the brain's spiky demands for a dwindling supply in the ovaries. But since the Women's Health Initiative (WHI) report, doctors have started to increase the amount of SSRIs (selective serotonin reuptake inhibitors) they are prescribing for mood and hot flashes instead of hormones. (See appendix, page 257.) SSRIs work predictably for mood and hot flashes in only about 30 to 60 percent of women. I have had some women respond nicely, but there is a chance they won't work well for you. In addition, no one has ever done a proper long-term study on how these drugs interact with the Transition.

Over the course of my thirty years at the UCSF clinic, by the time a woman in the Transition would come to see me, she would be at the end of her rope emotionally. She had usually already been through three or four docs and was feeling hopeless, as each had told her that her despondency was "just stress" and "I can't find anything wrong with you." I could see what was happening—it was a familiar story—and for many, SSRIs were part of the solution, along with hormones. These were cases where low mood and lack of joy were the standout symptoms. Yet most women still fear the weight gain associated with antidepressants. So I always chose combinations of hormones and antidepressants that would have the least impact on appetite. I want women to find relief, and weight isn't just an emotional issue. If too much weight is going to hinder a woman's ability to exercise, that's also going to impact the quality of her Upgrade.

The key is individuality of dose and timing. Prozac, or "Vitamin P" as I like to call it, has the lowest impact on weight gain, and I found it to be the most flexible of the SSRIs. Average doses of Prozac for clinical depression are between 20 and 80 milligrams. But that's often way too high for women in the Transition. I liked to use the liquid form so I could help my patients microdose and find the best amount for them individually. I started many of my patients on doses from 2 to 5 milligrams. That's ten times less than the dose for clinical depression.

Whatever combination of HRT and SSRI I prescribed, I started everyone on a tiny dose. We did test-drives of several formulas, starting small and increasing or decreasing slowly. We charted the dose and the timing, and patients recorded their feelings throughout the day, bringing those journals and charts to every visit. We even changed the time of the dose depending on those feelings. I wanted the women to have the best chance of the antidepressant cooperating with the body's natural rhythms and hormones. Those rhythms are in constant flux. Catching the right wave is crucial, and it takes time to figure out what works for you. I usually tell women it will take up to six months to get it right.

In the meantime, a new understanding of the neurochemical influence on mood is just beginning to emerge. For decades, the focus has been on SSRIs, drugs like Prozac that make serotonin more available. The thinking was that serotonin was what was needed to shift mood. New scientific techniques have allowed researchers to see that serotonin and dopamine are the two key players in mood, decision making, the reward system, depression, movement disorders, and thought disorders. It has thrown the field of treating mood and melancholy wide open. With the research being done into new compounds and approaches, I'm guessing we are going to see more effective treatments in the years to come.

Whether you opt for medical help or not, if you can calm the amygdala's and hypothalamus's alarm that pushes CRH, kicking off the body's threat-response system, it can help right the brain's ship. If we can find ways to increase the feelings of safety, nurturing, joy, it goes a long way to decreasing levels of cortisol and adrenaline. And tools for handling runaway negative rumination may save us from going down the road of intractable melancholy.

Think about Sarah's doctor and the recommendation to rest, be peaceful, and get some sleep. If she's feeling crappy because Insula 18.0 is torturing her over the fact that she can't run anymore, then she is being activated by threat, not safety. "I felt like a failure," Sarah said, "because I couldn't rest, and it seemed like I woke up progressively sadder every day. And then I tried to meditate with a friend. My mind wouldn't sit still. All I could think about was that this tiredness and bleeding would never end and that I would never get my life back. I felt like a complete failure."

Our female biology predisposes us to rumination because our stress system takes longer to downregulate; and Sarah got stuck in her sadness and self-flagellation. Remember, the brain has Velcro for negativity and Teflon for positivity. Any potential danger is perceived to be larger, and anything positive is dwarfed by it. If I wake up with a sequence of negative thoughts and scenes running through my head, they can take on a life of their own if I don't take action to put the brakes on rumination. It can ruin my morning, my day, even my mental health.

For me sometimes life events are so big that it's hard to focus, hard to settle down. So I make sure I find ways to get some distance from what's bothering me by taking a walk, exercising, or calling a friend to keep from crashing and burning on the altar of negativity. Deliberate breathing can also activate the cerebellum and trick the vagus nerve into signaling the parts of the nervous system that help us calm down. You can try alternate-nostril breathing or the box breathing that everybody is talking about. (See appendix, page 258.)

Tiny Muscles and Essential First Thoughts (EFTs)

When it comes to making change, I like to start small.

I have a new plan in the morning. I wake up, I wiggle my toes, and I smile, even if I don't feel like it, especially on the days I feel rotten, the hamster wheel of negative thoughts or worry already spinning off its axis by the time I open my eyes. Sometimes a scene from the past is there full blown on the viewing screen. If I stay there too long, then I will be sucked down the gravitational vortex of rumination about things that I can't change. Now I deploy my antinegativity missiles, like EFTs (essential first thoughts), better known as tools for coping. After that initial big, genuine smile, in which I make sure to engage my eyes, I imagine whatever force for goodness may be supporting our existence saying to me, "Good morning, Louann. I love you. I will be working for you today, and I won't be needing your help. Have a nice day. God." Emotionally, this helps me adjust expectations around what I am capable of controlling just for one day. Feeling cared for helps me make room for more kindness toward others. And much of that mood shift has to do with the neurobiological intervention of muscle movement.

It seems trivial to say that one way to improve your day is to wiggle your toes and smile first thing when you wake up, but each aspect of this new wake-up routine kicks off a positive neurological feedback loop that potentially reduces stress and inflammation. Anything that reduces inflammation helps the brain: Lower inflammation means preservation of cognition. The healthy effects get amplified as the seeds of friendliness impact relationships, which in turn helps me maintain the social connections that provide essential nutrition to my brain. It begins with the intentional firing of teeny tiny muscles like the levator anguli oris and zygomaticus, which raise the corners of the mouth, and the orbicularis oculi, which crinkle the eyes, the real signal to the brain and to others that the smile is not false but a genuine expression of warmth and friendship.

The theory that facial muscles can cause a shift in mood has been around since Darwin first wrote about it in 1872. Psychological studies over the course of the past fifty years continue to support the idea, and scientists have been trying to find the neurological trigger for just as long. Theories abound as to which facial nerves send the signal to the brain to begin its chemical cascade: facial muscle movement, nerves in facial skin, contraction or expansion of veins and arteries changing airflow in the nasal passages so that the brain detects a temperature change. Getting proof of how each mechanism works is still extremely hard, but based on what we know about the nervous system, it's impossible that there is no neurophysiological feedback loop operating here. Smiling, and making sure you involve the eyes, can elevate mood for a moment; even this can be enough of a window to feel a temporary release from difficult emotional states like uncertainty, melancholy, or even rage. Once that happens, we have opportunities to deploy more coping mechanisms, like balancing on one foot, which started a trajectory toward exercise and out of sadness for Rhonda.

Wiggling your toes activates the sciatic, the longest nerve in your body, sending a signal from the big toe to the sacrum and lumbar spine that gets the brain ready to wake up the legs and to move. The mound of the big toe is a powerful source of balance and stability as we stand and walk. It's the power behind every step, whether running or walking. It's the first place we bring our attention in any standing yoga posture to stabilize the pose and engage the legs. And wiggling your toes may activate circuits of fun, if you have memories of people playing with your toes and tickling your feet as a child. There are keys to cognition in moving our feet.

Flora's standard answer to "How are you?" was "Still kicking!" She meant that literally. Always a dancer, at age ninety-eight, she was in great shape. In her mideighties she started an elders tap-dancing troupe that visited senior-living communities to inspire others with their moves. "Dr. B.," she told me when I asked her how she kept up with all her activities, "as long as you keep moving, the buzzards won't get you." She was right, especially when it comes to supporting mental clarity. The neuromuscular feedback loops of movement are key to keeping the buzzards of declining cognition away.

Flora's "moving" and "kicking" are a key factor in her mental sharpness. A study of 120 older adults in Pittsburgh showed that consistent exercise, both aerobic and strength-training, changes the brain's anatomy by creating new cells and increasing the size of the hippocampus and the prefrontal cortex. That's a direct impact on the brain organs for memory, clearheaded thinking, and decision making. They are also the two brain areas most vulnerable to decline and aging.

Step ball change. Cross, cross. Skip and slide, jump and clap. Pump your arms, punch and curl, spin around. Balance and coordination challenge the cerebellum, which keeps signaling balance in emotion and judgment and the unleashing of creativity. Conquering a complex pattern stimulates the reward circuits, refreshing the cognitive centers in the brain. The intensity variation of starting and stopping also plays the adrenals to exhilaration, releasing just the right amount of the stress hormones to set off a cascade of neurochemical rewards, giving us that feeling that we can conquer the world. Intervals, even gentle ones, are key.

Before the Transition, going for a long run might have been one way of goosing endorphins to lift mood and brighten the mind. "I know I missed the high," said Sarah, lamenting that between exhaustion and arthritic hips, five-mile runs weren't an option anymore. "But my trainer gave me a twenty-minute interval program that was magical. [This can be modified to involve only the arms if you're having trouble walking.] There is a two-minute warm-up walk and then a sequence of four one-minute intervals: a fast walk, a slow jog, a faster jog or slow run, and a run at a pace I can handle. Repeat that four-minute cycle four times, ending with a two-minute walking cooldown. It seemed like nothing when she first described it, but I couldn't believe how quickly I was sweating! By the time I got to the middle of the second cycle of four minutes, I was feeling the kind of brain high that used to kick in around my second or third mile."

Here is what's happening to Sarah's brain. The adrenals pump a little cortisol, adrenaline, and norepinephrine, waking up just a little bit of vigilance as she completes the one-minute intervals. As soon as levels get just to the edge of stress, the adrenals are given a break; they calm down and stand down from kicking into high alert for fight or flight. The highs and lows of the cortisol waves mean that Sarah's brain doesn't get flooded with stress and inflammation. On the contrary, it's just enough to ask for the dopamine and endorphin neurochemical release that makes exercise so joyful—a runner's high without the exhausting run. You can apply intervals like this to any aerobic exercise: cycling, walking outside, rowing, arm cycles (especially important for anyone wheelchair bound), swimming. Dance has intervals in its DNA.

I knew that Sarah had had a bout with pneumonia that left her in bed for six weeks, so I asked her what she did to get back to exercise. "Did you use the same routine?"

"In a way I did," she said. "But I modified it with help. When I was younger, I would have tried to come back with the same intensity on day one, and probably would have gotten away with it. But since the Transition I have learned to start more slowly. So instead of basing the intervals on pace, we used incline. I kept the walking pace slow, used a flat incline for the warm-up and cooldown, and steadily increased the incline through the four minutes, starting over each time at the lower level. It took a month to get a little energy back, and two months before I was ready to jog a little. Now I'm switching it up even more. I'm focusing on things like cardio dance and adding barre classes with light weights and Pilates. Combining them, I find I burn more calories and I'm more exhilarated by exercise instead of being drained."

After just one exercise session, your brain responds by putting out more neurotransmitters like endorphin, dopamine, noradrenaline, and serotonin, all of which raise your mood as soon as you towel off, and this becomes protective against persistent melancholy. A single workout is capable of amping up our ability to focus for at least two hours afterward and improves reaction time—it's a great weapon against reflexes slowing with age. And even if you haven't been very physically active before, exercise will bring benefits at any age. A study of 8,206 women from the Women's Health Initiative showed that women who started working out at ages seventy to seventy-nine benefited after just twelve weeks.

Movement reminds the cerebellum that it has to keep the automatic functions strong; otherwise it will begin to close the shop. When movement is limited, the cerebellum isn't pinged as much to stimulate other functions, such as emotion regulation and problem-solving. The nervous system's activation and motivation for basic survival of the body declines. The heart and lungs and cardiovascular system don't get the signal that they need to function optimally in order to support life. Muscles aren't stressed in a way that builds them; the brain, in turn, gets a weaker signal to maintain heart, lungs, and circulation to support basic nervous-system needs for survival. It doesn't take long for the downward spiral to begin.

Without movement, muscles begin to waste. Called sarcopenia, this wasting can become the enemy of balance, and it's often an indicator that severe cognitive decline is not far behind. Muscle loss, like bone loss, actually starts soon after age thirty but can become a rapid, progressive, debilitating condition after age sixty unless muscle-strengthening exercises are done. Within twenty-four hours of being bedridden, muscle fibers start to lose strength. If you're in bed for several weeks, you lose muscle strength at a rate of about 12 percent a week. After three to five weeks of bed rest, almost half of the normal strength of a muscle is gone. I remember that feeling after my hip replacement. It comes back fast, but you have to take small steps; if you jump back in too fast, injury and pain put you back to square one.

A study in Sweden found that in people aged sixty to eighty, the more fit someone was, as measured by a session on a stationary bike, the more words they were able to mentally access. In animal trials, we're finding that exercise significantly increases the number of new neurons being made in a few critical parts of the brain, meaning: neurogenesis! It is possible to make new neurons at any age, and exercise helps bulk up the parts of the brain that can protect us from memory loss and cognitive decline. The female brain is begging us to work out in the Upgrade. It's time to answer the call.

If you've been sitting too much, the core muscles of the body begin to atrophy, and the glutes—our butt muscles, which power forward motion—get a kind of amnesia. Crucial to the movement behind getting up out of a chair, they forget to do their job, putting our balance and the cerebellum at risk. If you've been putting off a knee or hip replacement, it's time to get it done for the sake of your brain. If you've been too sedentary, like sitting in a chair for hours at the computer like me, squeeze your butt and change your brain. Twenty chair squats a day, and then a hundred butt squeezes (squeeze for three counts, relax for three counts). Break them into batches of twenty-five if you have to. I do mine while brushing my teeth. You can do them sitting, standing, or lying down. But doing them will fuel your centers of balance and judgment.

So move those toes, stand up, and when you get to the bathroom, stop at the mirror. Smile at yourself. Get the muscles around your eyes into the act. Using the mirror engages the visual cortex, the largest part of the brain, and starts a neurological feedback loop that elicits joy. It only takes ten to fifteen seconds of your own eyes twinkling back at you to make an impact, so forget about the crow's-feet. In fact, wrinkle them as much as possible. When, during the day, the brain serves up random uncomfortable thoughts that have nothing to do with reality, reigniting the memory of the feeling from the morning is usually enough to keep them from ruining your day.

Your body knows that movement will signal your brain, *I am alive and well*, and that well-being follows it. It craves movement snacks. They can be five or ten minutes if you're too busy to do a sustained workout or don't have access. Get up and walk around the house during commercials if you're watching TV. Park a good distance from where you need to go, if you can. Get off public transportation a stop early, or walk to the next one to pick it up. Do ten chair squats whenever you can. Some of us need help remembering to take a movement snack, especially if we are stuck inside or at a desk. Get a reminder app, or if you have a smart watch, turn on the reminders to stand and move, and try to be compliant. Your brain will reward you with sharpness, alertness, and protection of cognition that will last longer. If you want to boost your sharpness, the effects of a moderate burst of exercise are strongest for the first two hours afterward. If you are not able to take HRT, we now know that movement provides a massive boost to brain health. And if you are on HRT, studies are indicating that fitness not only boosts the effects of estrogen on the brain but offsets risks associated with HRT as well.

If you haven't moved in a while, or if you're recovering from surgery or an illness, microdose at the beginning. Walk around your dining room table. Then move it up to hallways, until you have the stamina to walk around the block. Do it every single day. After you build strength, ramp up your distance slowly, and your energy will rise exponentially. If you aren't mobile enough to walk, exercise your arms. Raise them up and move them for a few minutes. Build up to five, then ten. Turn on a symphony and conduct it. Feel the joy of play again. Your body and brain are begging you for it!

The Return of Purpose

Chapter 12

Around the dining room table overlooking Manhattan's East River sat several women, among them two former media executives, a (current) venture capitalist, a former White House counsel, a former curator at one of the world's most important art museums, and a designer who had done iconic work that made her famous. Among them ages ranged between near fifty and early seventies. The collective body blows around the table included loss of children, spouses, careers, money, social status, health, looks, and community support. Each woman would tell you she had been torn down to the bone, pride stripped, life shattered. Everything each once held dear had been released, including old strategies—drinking, travel, shopping, excessive exercise, and obsessive self-improvement—to acquire peace. Everything that they had once thought essential to survival and success no longer seemed important.

Though their lives and origins looked very different—hailing from European upper-class to Hollywood, from mid-Atlantic working class to middle-class Manhattan native—all of them agreed on one point. The blows life had delivered provided the best opportunity to step into who they really are. "I wouldn't trade most of these experiences for anything in the world," said Jean, who had lost her husband. "Hardship has made me who I am. Not a drop of it was easy or fun, and if you'd have told me any of this at the time, I probably would have decked you."

"Yes, there's nothing more infuriating than someone coming at you with spiritual pablum like 'You'll see this will be a blessing,'" Marian interjected. "Especially when you're midtrauma. People can be really unskillful, even if what they are saying might turn out to be true. It takes time to heal, time for the trauma of it all to unwind."

Nancy, now in her late sixties, was a legendary beauty. She was interviewed by a national newspaper along with two actresses who were complaining about losing their looks. But she didn't see aging as a drawback; in fact she was thrilled. "For the first time in my life, I can get a man to listen to and accept my ideas and advice," she said. No longer pulling male attention meant Nancy could begin to get things done even with men in the room.

As for the others, many found new careers, found new communities, and started their own businesses. They allowed tough times to bring out their best, relishing their ability to meet a struggle, even when it was hard. They found purpose again, and not a bit of it was driven by anyone or anything outside their own internal engine. In no longer having anything to lose, they found in the Full Upgrade the freedom to be and do what was most authentically aligned with who they are.

The attention we paid over our lifetimes—no matter our level of achievement in the world—to hair, makeup, clothes, was part of nature's brainwashing that these rituals were what would attract a mate and get our genes reproduced. All that primping is part of nature's strategy around propagation of the species.

The wiring of fertility's survival imperative keeps us restless, always seeking satisfaction and approval from the outside. It locks us in an excruciating tension—*It worked! He's checking me out!* and *He only likes me for my breasts.* The internal scream *This isn't all of me!* is nearly impossible to stifle. It was certainly hard for me. Even though intellectually we may know appearance isn't everything, the feeling that if we are not attractive we will die is real: Nature makes sure you will believe a delusion if it means you'll find someone to reproduce with. That includes picking the wrong person. So right now you can forgive yourself many of the errors in judgment during the fertility years. We've all made mistakes. The misperceptions about relationships, the misreadings of character, the lopsided priorities, were designed to a degree by natural selection. Getting stuck in those old stories can be an off-ramp into a downgrade.

By the time you've reached the Transition, the selfish gene-survival imperative to reproduce itself is over. In the Upgrade, that game is over. And with it comes a new beginning, with new rules and a survival imperative that drives a new sense of purpose. It's different for everyone.

"I was shocked by how my thinking changed," said seventy-year-old Alina. "I've done corporate interiors most of my life, but everything changed when I had grandchildren. It's like my heart exploded and my sense of protectiveness extended to all children. It was completely spontaneous. I couldn't look at another child without thinking, *This is my grandchild, too.* And right after that came the thought *Shit, we have wrecked the planet! How will they survive? I need to do something to protect the grandchildren now!*"

Alina was still running her firm but started spending much more time engaged with environmental and political activism. "I am a fierce recycler, and my local and state representatives are sick of hearing from me about enforcing the laws. But I don't care. This is too important."

Alina had had insomnia during the Transition, so I wondered how she was faring these days. "How are you sleeping with all this on your plate?" I asked.

"Actually, I'm happier, more energetic, and more focused than I've felt in years," she said. "Staying engaged with something that matters so much has given my life new meaning."

Alina's engaged, purpose-driven mindset was also activating protective circuitry in the brain that blocks depression. Whether or not those circuits are activated can be shaped by the decision made when faced with the fork in the Transition road that some women face: heed the siren call to exit into the comfortable Transition, or build your Upgrade.

Life Extension

Jane, at fifty-four, realized she was only ever going to have one grandchild. It was a tremendous sadness at first because she loves children. Instead of grieving, she went back to school for a degree in social work and, finding new purpose, became an infant mental health specialist and a play therapist. "It meant I could go into preschools and play with kids all day long," she said, beaming. "Meanwhile, I was helping to keep at-risk kids from becoming at-risk teens and at-risk adults. If you intervene early with cognitive, behavioral, and emotional issues, you can change the course of a person's life."

The stimulation of the joy circuits through regular play was magical fuel. Over time, she built a foundation that focused on helping families, toddlers, and preschool teachers create the best learning environment possible in underserved neighborhoods. At seventy-five she's handed the reins to a younger director, but she is still engaged with educating educators and adults in foster-grandparenting programs. "I just can't seem to get myself to stop. I love it."

It's long been known that finding purpose and having a direction for your life lowers mortality risk beyond other factors that impact longevity, like a healthy diet and lifestyle. With diet and lifestyle, the sooner you begin, the greater the impact on your protection from disease. Researchers in Canada assumed the same would be true for purpose and embarked on a study to confirm their assumption. They were sure that the earlier you discover purpose, the bigger the impact on your health. But the biggest surprise in the fourteen-year follow-up period was that finding your purpose had the same benefits to longevity regardless of the age of discovery. Looking at six thousand participants, it was clear: Age didn't matter when it comes to the return of purpose, and purpose might even be more health protective in older adults than it is in younger ones. It's never too late to start asking questions around what you feel you were meant to do. In the Upgrade you'll finally have the bandwidth to do it, even though you might have trouble remembering things.

Engaging in purpose to fuel the Upgrade is a choice that's not always so clear. In 2016, when I handed over my role at the UCSF Women's Mood and Hormone Clinic, a part of me felt a huge relief. For the first time in decades, the weight of taking care of students and patients, training new staff, and raising money for my program and the institution was off my shoulders. I felt like I could breathe. And I did. As I stared out over San Francisco Bay, I started taking photos. Liking what I saw, I got interested in learning real technique and using a real camera. So I took classes and got a lot of validation. I even had a couple of exhibits. I went into a state of joy for several years, getting into a creative flow, learning a new skill set, and hanging out with artists instead of doctors. I learned oil painting and textile design, things I loved when I was young. It felt like I was picking up a dropped stitch in the knitting of my life. But I didn't realize that the comfortable Transition was a bit of a dead end.

I got to a point where I realized that the only way forward with photography would be to study a lot of technical stuff in Photoshop, and that just wasn't what I wanted to do with my life. I was missing the outward orientation of my profession, the interactional interdependence and intimate engagement of caring for patients. I had felt euphoric doing art, but—boom!—just like that I was depressed and alone, bereft of the colleagues, life, and career I had spent decades building. Just like wrinkles and watermelon boobs, I had always thought, *This is not going to happen to me,* and then—surprise!—isolation and uselessness busted down the door. Like the transitional men I had dated between marriages, even though the relationships felt real and all-consuming at first, I had to recognize that art was a phase, the comfortable Transition. There were more meaningful things I was meant to take care of, like helping other women access the Upgrade.

I like to think in pie charts. When I think about the brain bandwidth that the survival imperative—*stay healthy and attractive in order to mate and reproduce*—took up in the fertility era, it was almost the entire pie. Neurohormones were screaming at us to put most of our attention on everything that supports fertility—relationships, looks, fitness. But if I look at the pie chart of the survival imperative for the Upgrade, it's blue sky for new priorities. It's blue sky for purpose, for turning the arrow of concern outward. This is a major ingredient for joy in the brain.

If my survival-imperative pie chart is taken up with a narrow, fearful version of the question, like a constant refrain of *What's going to happen to me?*, then I might be setting myself up for depression. Yes, I have to be practical about taking care of myself, but if my time is spent worrying about what I'm losing or how I'm going to struggle to hang on to what's past, then I may be activating the neurocircuitry for depression. When the arrows of concern only point inward toward ourselves, we can hit bottom pretty fast. In extreme cases, too much self-concern can activate debilitating states that can keep us from functioning, like profound melancholy and even obsessive-compulsive disorder. But flip the arrow outward, make the survival-imperative pie chart focus on engaging with things that matter to others, and we become connected and energetic. Purpose reemerges.

This opposite of self-cherishing, focusing on others, is for me as good as it gets, and the neurochemicals and circuitry that get turned on explain why that is. When advanced meditators were studied in fMRIs, they were asked to cultivate compassion, the empathetic mental state of wanting to be of help to others. During their meditations, the minute this compassion arose from their hearts, the joy circuits in the brain's region of higher thinking exploded. This isn't the same thing as being a martyr or a doormat. Those feelings don't bring joy in the brain. They bring frustration and probably some negative self-talk. You can't obliterate yourself and be helpful at the same time. You have to bring your whole calm, clearheaded being to the table. You have to have no agenda other than being helpful.

Finding the right fit, the best way forward in returning to purpose after leaving a work environment, can be incredibly challenging, not only to the tent of "Me" but in terms of options presented by the culture.

When men leave big positions, they are generally offered paid board seats and consulting gigs that aren't available to women. "You get asked to volunteer," said Alina. "You get asked to give a lot of free advice. But a paying gig? I know a lot of men who got them, but even my most powerful female friends have struggled with the fact that they are expected to be corporate caretakers and work for free."

It doesn't matter where you worked, what your career was, or if you worked. You will be asked to volunteer, while men of a similar age will be paid. There are plenty of resources for connecting older people to volunteer opportunities, but what about the woman with no means? A woman whose care for elderly parents drained her resources after her husband left her and then she lost her job? "I was aged out of my job at the bank a year ago," said Carole. "All I wanted was another job, but I was chased by every major advocacy organization to sit on their boards for free. The ladies who lunch," she continued, "expected me to contribute my knowledge to their causes for free, to do all their finances for free, but I was a lady who still needed to make money to pay for that lunch. The feeling of isolation during that time was profound."

The isolation of female professionals in the Upgrade is real. During our careers, colleagues can feel closer than blood relatives, but many find that when we are downsized or retire, that social world evaporates. Men have experienced this, but this emotional gut punch is new in history for women; it's a new developmental stage that, if left unaddressed, can keep us stuck in Transition-mind or Early Upgrade.

The minute you're out of the loop, all the old junior high school circuits of feeling left out by the popular kids are reactivated. No matter how high your social status, no matter how elevated your success, it happens. "Even in my activism," said Alina, "if I discover I've been left out of a meeting, it puts me right back to the twelve-year-old girl who wasn't invited to the cool kids' party."

During her Transition and Upgrade, remember that Sherry surprised herself by taking a bunch of exercise classes to meet new people after moving from the East Coast to Arizona to be with her son David and his family. But she also found new purpose. "One of the ways I pulled myself out of sadness after moving to Arizona to be with David and the grandkids was to become a consultant to café and restaurant owners," she recalled. "I charged a small fee for helping them create structure and organization around the business side. I was mostly working from home at the time. Many of my meetings were by video chat. When I realized I was going to classes because I missed having colleagues, I found an office to rent in a fantastic co-working space for entrepreneurs who had been running a business for three or more years. Most of the tenants were women and minority business owners, and that was magic. Through new connections and hallway conversations, I rediscovered what I loved about my work and started my first growth spurt after having left the East Coast. That growth was driven by my hiring subcontractors, women out there struggling on their own who needed to reignite their confidence, their skills, and even their careers. Seeing how many women I was starting to help, well, that made me pop out of bed every day."

Clearing the Obstacle of Worry

The pre-Upgrade female brain is wired to see danger everywhere, to be on high alert to protect the helpless around her. High waves of estrogen and the stress hormone cortisol keep this instinct front and center, inhibiting access to questions that seem less like life-and-death, and more like *What should I do with my life after the Transition?* Not only is this not an issue for the male brain, but testosterone makes men much more tolerant of risk. The higher the testosterone, the less they perceive risk. If you've partnered with a man, or been in a car with a couple that's been together for a long time, you have heard the difference of perception of risk manifest in an old, familiar argument. "For god's sake, slow down! You're going to kill us all! The kids are in the car!" "Honey, what are you talking about? I'm driving just fine. When's the last time I had an accident? Never. That's right. Now stop criticizing me."

The difference in the perception of risk is estrogen keeping the female brain hypervigilant to danger in its drive to protect the vulnerable ones. When hormonal waves calm and estrogen lowers, it dials down that constant alertness.

With the quieting of the hormonal siren call to worry, there is more silence in which to hear our own voice. There is more room in our mind for the return of purpose. The answers to who we are and how we want to live become clearer.

It will seem unfamiliar at first, the urge to explore the world alone, to start a business—*What am I doing at my age?*—to climb mountains and fly solo. But once you start taking new adventures, you become addicted. "Grandma, can I come with you to Japan?" While at times you may say yes, at other times you might feel, *Are you kidding? My worst nightmare! I just want to do this by myself.* You say, "Oh, honey, not this time. You would be bored with this old lady. Maybe the next trip."

Being on purpose brings all the parts of yourself into focus. Work becomes mission driven; curiosity takes center stage in decision making; a new ease, even in difficult moments, takes over. It doesn't mean that the worry circuits never turn on again—I have certainly struggled with that in regard to my son, and that may never change. But there is enough release to start exploring new ways to engage with proactively undoing a lifetime of brainwashing about who we are and what we can contribute.

Keeping Purpose on Track

When I was a child, things pretty much went my way. Of course I had my traumas, but I was the smartest kid in class and successful at most things I tried. However, then I got my period and reality changed. What struck me to my core was the deep unfairness of my own biology, that I was going to have this for what seemed like the rest of my life and boys would never have to experience it. When my biology announced itself, I was unprepared for it, even though I was educated about it. I didn't know how dramatically cramps and moodiness would alter my world for days each month. I had a similar experience during pregnancy, when morning sickness came with a nausea so powerful it shut out the rest of existence. And then the neurochemicals of motherhood rewired my brain for overprotectiveness and a deep dread of relationship rupture with my child and my spouse.

Evolutionary biologist Richard Dawkins wrote a seminal book in the

1970s called *The Selfish Gene*. In it he argues that the only thing Mother Nature cares about is getting sperm into the next generation, ergo that's all men care about. The book that needs to be written about women's biology is *The Unselfish Gene*. Think about it. Since a mother and baby don't have matching DNA, the female immune system has to be suppressed so that the baby doesn't die. The Upgrade is the first time in a woman's life she doesn't have to biologically or psychologically suppress something so that another may survive. But the social expectation that she will continue to do this, and to suppress her own purpose, remains. Thus the burden of caretaking, which she expects of herself as well. It falls on women as inevitably as puberty.

"I remember being so overwhelmed," said Mariana, fifty-two. "My mother-in-law had a stroke and had been taken to rehab. I was the primary breadwinner; my husband's accounting practice had hit the rocks while my career was soaring. But the nurses and social workers wouldn't talk to him or his brother. They waited for me to visit and they asked me, not them, 'So what's the plan for how you will take care of her after she is discharged?' I had no idea what to do. And I had no idea that my now-ex-husband would actually be useless. So was his brother. I was earning all the money, and I became responsible for things like picking up her laundry from rehab, which was an hour away, taking it home, washing it, and bringing it back. I ended up being in charge of managing caretakers when she moved in with us. It was another full-time job. I was so busy that I had to turn down a bunch of opportunities for career growth. I ended up blowing a promotion."

I hear this all the time. Chantal, fifty, a former news anchor, roared in frustration when her brother refused to help her care for their mother when she was diagnosed with Alzheimer's. Chantal's mother lived with her until it became unmanageable. She sobbed the day she signed an agreement with a facility. It was a combination of despair over feeling she hadn't done enough and the loneliness of the process. Chantal had been building her own media business while cleaning up after her mother's incontinence, responding to 3:00 A.M. demands for dinner, and putting her child through college. All of her friends saw that she was squeezing out her last ounce of love, strength, and care. It was outrageous to her friends to think she was awash in guilt and remorse over failing her mother, yet those feelings were so real and so solid. That's an indication of a mommy-brain-circuit reality hijack, one that brings shame and social isolation with it. The gift of hearing the hard truth from a friend—that there was nothing more anyone else could have done and that her brother was a class-A jerk—is part of what got her back out into the world.

For some women, caretaking is purpose, it's a career, it's life; that's not what I'm talking about here. I'm talking about how much of it women get saddled with without choosing it consciously. Most of us talk about being cared for at home in our final days, but if there isn't a woman around to make sure it happens, who will do it? If you are a woman with dementia who was abandoned by her partner and you have only brothers, who will care for you? How will it be paid for? Just as 78 percent of childcare is done by women—even if they have full-time jobs—the same thing has to be true of the other end of life. They don't keep statistics on it yet, and that is an indication of just how pervasive this unspoken expectation is.

It's a blinding reflex for some of us. My patient Natalie, sixty-three, can have a chockablock-full calendar, but if there is a caretaking need, it's as though the clouds part and her schedule becomes clear blue sky. She's let herself take sole responsibility for her elderly parents, her mother-in-law, and her sister-in-law who had a stroke in her fifties. There are siblings and spouses, but she does it all. It's hard for her to resist what I call the helper circuits.

From the time we are teeny tiny, "Help Mommy with this, please" is a siren call to feel useful. Many in our lives count on those helper-brain circuits being activated when they mindlessly call out, "Where's the . . ." or "How do I . . . ," to suck us into helping them with minutiae that pull us off task and out of our own zone of purpose, as though we were the family's Google, on call and available 24-7.

Before getting married, Natalie had a career with an international management consulting firm, one she didn't want to give up, but felt she had to in order to help her husband grow his burgeoning and extremely successful fitness franchise. She designed events for big meetings and top franchisees while raising the kids and caring for the older generation. "I see my daughter with her kids, a career, and a partner who fully participates in the caretaking, and I think about what I gave up that she didn't have to," Natalie tells me. "It's hard. Motherfucker, it's what I wanted. I take those feelings out, look at them, and then stuff them back inside, because really, what else can I do?" The fear that she'll be criticized if she doesn't do it all is overwhelming. But there is real evidence that caretaking stress not only is an obstacle to the return of purpose but may shorten a woman's life.

Part of the way we keep living is that cells replicate. At the end of the strands of DNA that make up chromosomes are telomeres, a kind of protective cap that keeps the ends of chromosomes from fraying and making them sticky. The common image used is the plastic on the end of a shoelace. Intact telomeres are essential to healthy genetic material. Science now has a way to measure telomeres; the longer they are, the lower the impact of aging on all cells, including brain and nervous system cells. A 2014 study done on unpaid female family caregivers in Wisconsin showed significant shortening of telomeres. Conclusion: Caretaker stress accelerates aging.

From the outside, Natalie looks like one of the most badass women you'll ever meet. But scratch the surface and you find she is exhausted and often bedridden from caregiving fatigue. It doesn't come without reward. There is gratification in being in the space of compassion, because of the positive feedback women get socially and culturally for overly developed empathy and guilt circuits. We get to be the rescuer, everyone approves, we get a dopamine reward for success, and so we become chronic, daily rescuers. Purpose gets buried once again, unless caretaking is the chosen path for it.

The expectations of the culture will fall on you as inevitably as your biology. I had no choice about my period, but there is a choice about caretaking. If we know to question it before it happens, to recruit others and make a plan ahead of time, we have a chance to live the second half the way we envision it, not hijacked by and drowning in biological and sociological guilt. Of course, in the thick of it you will have to change some of your plans. It's inevitable that your care for an ill partner or close family member will change the dynamics. But it's good to have a plan and some backup help, so that caretaking can become episodic rather than chronic.

Awakening Purpose by Speaking Up

In my childhood, there was no fiercer truth teller than my grandmother. She'd spill the family secrets at Thanksgiving and in front of everyone ask a cousin why he snuck that cookie when he'd been told not to take one before dinner—*yes, of course she saw him.* Everyone attributed her lack of inhibition to eccentricity or a possibly failing mind. But this wasn't the case for my grandmother. It was simply the lifting of the hormonal shroud of accommodation, allowing her to find a new comfort in her own skin.

As young girls, say four to six years old, chances are we were scolded more than once for using our outdoor voice to tell an older relative they had bad breath, or they were fat, or they had a mole that looked gross. We spoke out not only because we weren't yet properly socialized; we spoke out because the brain circuits for judgment and especially conflict avoidance hadn't yet come online. We don't yet know how or why it happens, but during puberty the female brain changes dramatically. The amygdala, a center for fear and anger response, develops more in puberty and is triggered easily by estrogen and testosterone. Since men have more testosterone, their anger and physical aggression are triggered more readily together. But for women, with higher estrogen and much less testosterone, the signals to leap into battle are less direct. When the fertility cycle begins in earnest, the effect is that an outspoken little girl suddenly becomes a confused, frustrated, and sad preteen, especially when you ask her to explain what's wrong.

Nature selected for this characteristic: In the wild, a lower aggression flash point probably meant better chances of survival for the smaller gender. Starting fistfights with much larger primitive males probably wasn't a great bet for success, so evolution wired a lock on the dam of female aggression. As estrogen lowers in the Upgrade, this lock on aggression can open. The part of the brain that used to hijack anger and strangle our voice is now neurologically released. Given the female acuity with language, we are more likely to throw words instead of fists. No more mulling over what to do or say. A gusher of well-timed zingers can blow, and so Grandma starts telling you the truth. She reads situations and acts.

In trying to categorize the Upgrade's new freedom, I often hear women talking about recovering the girl we left behind, her attitudes, and her behaviors. But that implies nothing has happened since you were eight. A lifetime of hormonal waves has changed your brain circuits. A lifetime of experience has brought wisdom. There may be dreams and hopes and wishes that ache to be pulled forward, but it doesn't mean you are going back. It means that you are finally able to incorporate that girl into who you are now, instead of leaving her abandoned by the side of the road. It can mean integrating her drive to fuel the return of purpose.

The voice in our head that used to censor us, always telling us to be quiet so that we didn't offend or upset anyone, becomes less loud in the Upgrade. We are less pulled by our peripheral vigilance over the needs of others, and we can discover a power pack we didn't know we had. All that power is difficult to control at first, so don't be surprised by sudden explosions. It can be like driving on a racetrack with a Maserati after decades of struggling to get a bald-tired VW Beetle up an icy hill. This force is necessary. If a child is walking into traffic, there is nothing gentle about the bellowed warning or the force needed to push that child out of harm's way. If we need to carve out the space for the return of purpose, to be our authentic selves, it's essential. It's powerful. It's fiercely female.

"No" Is a Complete Sentence

"I'm sorry," said Lisa, sixty-five, after speaking her mind. I was staying in her large home, and I was one of three houseguests. As the chef/owner of a restaurant, she was doing all the cooking for us. She had asked that, after four days of nonstop meal preparation, we figure out dinner on our own. She was tired, and we all had cars and the means to go out. Making that request felt like anger to her, but to the rest of us, not only did it sound reasonable, but we felt a little guilty that we hadn't offered to bring in food for everyone sooner.

Lisa and I sat at the kitchen table overlooking the woods behind her house. Over a cup of tea, I asked her about that apology for setting what I perceived to be a healthy boundary. "It felt super aggressive," she said. "I felt angry. It just burst out of me, and I thought I hurt everyone's feelings."

I could understand where she was coming from, but none of us experienced her as being angry. It was clear she was overloaded, and I wondered why she had agreed to cook meals for so many people for so many days. Looking down at her cup of tea, she said, "It just never occurred to me that I could say no."

One of the most important lessons I've learned in the Upgrade, specifically in keeping purpose on track, is that "No" is a complete sentence. But it's one we might not have spoken with confidence since we were toddlers.

"No" is for many children their first word. We don't have to learn this assertion of will; we are born knowing how to signal distress, whether verbally or physically, by squirming, using our arms and legs to escape something uncomfortable or threatening. From two years old on, we were taught to stuff our "no," opening the door to shattered boundaries and choices that might not be right for us. Is it any surprise that by the time we reach the Upgrade we almost don't know where to find it? Or that it bursts out when you least expect it, catches you by surprise? It can feel like you have no idea where it came from, like it was somewhere deep inside, stored in the nervous system's lockbox. When it emerges in the Upgrade, it comes through the tempered steel of age, not out of a childhood stubbornness. This is not a reversion. It's an integration, a reweaving of the part of ourselves our culture and biology told us to pack away.

In the fertility phase I felt compelled to say yes to everything, squeezing too much into the corset of each day. At some point "yes" breaks you and every promise you've made to yourself about how you want to live.

You didn't need to learn to say no as a toddler. But who taught you to say no as a woman? If you can't remember, that is a problem that we need to fix now, in this uncharted, post-Transition developmental stage. I had to relearn this lesson myself if I was going to be healthy enough to stick to my new plan of getting back to books and back to helping women. I was in horrible pain from a pinched nerve in my neck when I got on a plane to visit a friend in New Mexico. This pain wasn't new—it had been lingering for almost a month, and I was incapacitated by it. I sat through therapy sessions pinching my other side to distract myself from the pain so that I could be present to others. Knowing my friend was looking forward to the time together on a big birthday, I didn't want to disappoint her, so I said yes and made the trip when I should have simply stated, "This isn't good for me right now." I was trying to work on this book, and it turned out to be a massive disruption.

It was a wake-up call that with every request I have to remember to run these questions through my brain and body: *Is this good for me? Is this good for living on purpose?* We usually override that doubt about adding another responsibility with *I'm being selfish; I can handle extra pain; this is more important than my own pain.* But at a certain point, we will run through our reserves; muscling through now comes with serious consequences. Sincerely asking ourselves the question *Is this good for me?* can be scary, because sometimes a truthful answer will mean a fight or rupture if others aren't used to hearing "no." But in the Upgrade, we no longer put our health, well-being, and living on purpose second.

I had to coach myself to say no by watching women who do it well. I started by writing down phrases that got me more comfortable with not saying yes. Practicing got my brain and nervous system used to a new response. When I didn't feel comfortable saying no right off the bat, I responded by saying, "Let me sleep on that and I'll get back to you." It gave me time to gather the courage. And if I wasn't able to get that phrase out and immediately regretted agreeing to something I didn't want to do, then I would say something like "On second thought, I need to sleep on that" or "On second thought, let me check my calendar" (or talk with my family, etc.).

Setting boundaries that help us protect our purpose might be biologically harder for women to do. Everything about the wiring of the female brain is about connecting, and that includes our threat response. It's been more than twenty years since researchers at UCLA discovered that women respond differently from men to stress. We don't just get a burst of cortisol and adrenaline triggering the fight-or-flight response and firing up our muscles; the bonding hormone oxytocin is also part of the female stress response. When faced with threat, we will do everything we can to maintain relationships. This female stress response is called tend and befriend.

If you've ever gotten into a fight with a child, spouse, or close friend and found yourself compulsively insistent on talking things out until you felt okay with the relationship, it's likely oxytocin driving that urge to make peace immediately. It's the imprints from waves of fertility hormones and an echo of life on the savanna, when a ruptured relationship could mean death in exile. When the waves calm in the Upgrade and that intense drive to bond no longer overwhelms our brain, we can find some freedom for purpose if we can overcome decades of neurochemically driven habits.

The quieting of the tend-and-befriend neurohabit helps create respectful space in relationships. When we don't rush in to fix, resolve, and control, all parties have a chance to let their stress circuits calm and the decision-making process reignite. A side effect is that we honor the other person's process. Just ask your spouse and kids if that tend-and-befriend, we-need-to-talk thing feels invasive. They will tell you it does. It takes a kind of fierce confidence to let go and trust that with time and distance things will work themselves out.

There is a learning curve on pushback, and one of the things we discover is that when handled straightforwardly, it brings respect, not the dislike we fear. Think about those you know who stand their ground well: Aren't they women you respect? If you've thought of yourself as a pleaser your whole life, then this will be a struggle. But the liberation is amazing. Just think back over all the times you wished you'd said no, the times of feeling buried under the mountain of "yeses," the mountain of regrets over not saying no. It's time to push back on demands. It can change the way we live the second half of our lives.

Making Friends with Boredom: The Discovery of Purpose

Remember, hormones are molecules that drive behavior, so when estrogen and testosterone drop, that sense of persistent striving releases its grip as well. When the hormonal waves quieted for me in the Upgrade, everything felt odd. I had just left the clinic, and after a lifetime of constantly reacting and responding to my role, I couldn't shut that habit off. I had ideaphoria. Millions of ideas were popping up, and I chased each one like a squirrel collecting acorns. I would report each new one to my husband, Sam, along with the progress I'd made in pursuing it, and he would say, "Louann's going fast nowhere." It took me a while to realize that my brain and nervous system resisted adjusting to the lack of schedule and role. That new lack of stimulus registered as boredom. And boredom really frightened me at first.

The first step in the return of purpose is to learn that boredom won't kill you, though the shift will make you feel like something is wrong—remember, the brain and nervous system love the familiar. With boredom comes not death but stillness. Being able to sit with uncertainty about how my life would unfold became crucial to the unfolding of new purpose. Facing the essential boredom of figuring out what you will do with the second half will be necessary whether you've worked inside or outside the home.

There are secrets in boredom. All kinds of things can come up, old memories, old triggers, long-buried dreams, long-buried regrets. It's like opening that neglected storage closet that's full of everything you threw in there that you didn't want to deal with at the time. You don't have to organize and clean that closet all at once. You may only need to look once and do nothing. Or you may take it a little at a time. Bite off what you can chew. There is no need for heroics in the Upgrade.

To keep myself from spinning out over every new idea, a friend helped me to just celebrate that my brain could generate so many. It helped me to watch each idea arise and smile as it floated away. If it came back, I considered it again, but mostly they were just a product of my nervous system churning in response to the unfamiliar stillness.

The answer to the question of your purpose may not be immediately clear. So many of us didn't know that asking what we want at this stage would be an option. It may be belonging, love, friendship, quiet. It may be activism, a new career, a new role in the family. Don't be frightened if you don't know right away. The answers will come in silence.

New Focus

Chapter 13

My friend Diane, a successful internist at a large private health system, had come of age after the first wave of feminism. She entered adulthood with the expectation of having it all—a fulfilling career and family. But the details of exactly how to juggle everything had never been ironed out, and she had found herself *doing* it all, never thinking she could ask for help. Between raising her kids, managing the household, and her work as a doctor, she had little time left to focus on her marriage or cultivate close friendships. She hit the Transition feeling lost and alone, in a fog of anxiety and difficulty in concentrating at work.

But the Upgrade was a different story. She felt stronger and more focused than perhaps ever in her life. Her kids had launched their own lives, she started reconnecting with old friends, and she tabled her efforts to "fix" her marriage. *Who knows,* she thought, *maybe I'll decide to leave. Maybe things will change.* She could finally stop juggling.

She threw herself into her career with a vigor she hadn't felt since medical school. Every case fascinated her; every patient interaction felt new. She found herself drawing on her wealth of experience to support her views and decisions; she dropped the second-guessing of the Transition for the careful analysis of the Upgrade. For the first time, she saw the younger women around her not as a threat to her position but as a new generation to be mentored and sponsored. She found a new focus in her Upgraded circuitry.

Waves of fertility hormones can increase stress hormones. The more stress hormones we have, the more threatening things seem. It's easier, during the fertile phase and the Transition, to tip into overwhelm. And the part of the brain that controls attention shifting—i.e., multitasking—gets overloaded and attention to detail begins to erode. The feeling of too much to do might have meant we'd like to shut off the phone, curl up on the couch, and put everything off for another day. After the Transition, Diane noticed that she wasn't feeling as overwhelmed anymore. She found herself ready to face challenges on the spot: starting a new project, having a tough conversation, speaking up or nipping a problem in the bud. "Yeah, that Scarlett O'Hara 'I'll think about it tomorrow' thing vanished after the Transition," Diane said.

"One thing that's happening," she told me, "is that I'm finding myself flustered in conversations with younger female colleagues. If one of them comes in with a rapid-fire stream of ideas, problems, and insights and wants my attention before I've finished what I'm working on, I lose it." Diane looked uncomfortable. "I don't want to make them feel bad. I want to support them. But it's hard to get them to take a breath, let me finish thinking through what they've said and not lose my train of thought on my own work."

Before the Transition, Diane might have done the same thing. Before the Transition, like most women, Diane would have been a rapid-fire multitasker.

Multitasking and anxiety go hand in hand. We don't know which comes first, but we do know they amplify each other. Doing too many things at once demands rapid switches of attention; that constant gear shifting might have given us a jolt of excitement over our own competence in our twenties and thirties, but now, during the Late Transition and the Upgrade, it makes the mind fuzzy, distracted, and unfocused. The key now is to calm the anxiety circuits that are triggered by multitasking so they don't set off another round of multitasking and anxiety. It can become a vicious cycle if we don't consciously intervene.

Many women who do learn to adjust to the Upgrade's new brain circuitry can experience a career boost, becoming better at each thing they do and more effective than they've ever been. Without the distractions of multitasking and the need to put things off, many reexperience a career momentum not felt since their twenties. That process of reenergizing is good for you on so many levels: Studies show that those with high career momentum in the Upgrade score better on measures of self-acceptance, independence, and effective functioning in our fifties and sixties, including stronger physical health.

Diane and I had a chance to talk about her frustration over losing the ability to multitask. "I feel like I'm losing my edge," she said, "and I don't want to stop working." I understood the feeling, but there's a reframe here that I have found very powerful in my own life. "What if," I responded, "you thought of it a bit differently? In the old days, doing a million things at once might have felt good at some level. But you and I both know from the science that we did everything less well as a result. Now your brain is demanding that you stay good and focused on one thing. You have a better chance of getting it right."

"So it might be a gift?"

"Yes, and not only for you, but if you're asking others to slow down as well, it can help them stay more focused and be more careful in their thinking. Your focus can act like a grounding wire for them."

Diane was pensive. Finally she said, "I'm starting to see how engaging my own brain's wiring can help me make better decisions. By going with the natural flow, I can allow the space for taking a good hard look at things, in a way I don't think I've ever given myself permission to do. This is true for my marriage, family, work, everything." By riding in the direction the horse was going, Diane was beginning to optimize her Upgrade.

Memory and Distractibility

Although the medical profession views the Transition and the Upgrade primarily as a reproductive transition, the impact is almost entirely neurological. Hot flashes, brain fog, anxiety, and sleep disruption indicate glitches in the pervasive estrogen-regulated nervous systems. In all human brains, including male, estrogen is a master regulator that works to ensure the brain gets its energy supply.

One of the neurological impacts of the decline of estrogen is change in memory function. By the time we have gone through the Transition, we have less estrogen in our brains than men—their testicles produce testosterone throughout their lives, and that converts to estrogen in the brain. Very low estrogen is hypothesized to be one reason women are more susceptible to dementia. And it's another infuriating inflection point for me: HRT could likely have brain-protective benefits, and it looks like those benefits outweigh the risks pointed out in previous studies. It's been staring us in the face, but the doctors who shaped the research and the remedies missed it, focused as they have been on the body parts that interest the reproductive culture most. Hint: It's not our brains.

On the flip side, the impact of steady, lower estrogen on memory may have some psychological benefit. UCSF neurology professor Adam Gazzaley has done seminal work on memory and distractibility in the over-sixty female brain. What he discovered is that before the Transition, we could hold two things in mind, one in each hemisphere of the brain, and switch between the two channels by momentarily suppressing one and then switching back to the other. If you tried to hold a third thing, you would have to drop one of the other two. In the Upgrade, Gazzaley found, even a second thought cannot be suppressed and will knock the first one into the black hole of lost words and forgotten ideas. The two ideas collide in a bottleneck and one has to go. Just as the loss of collagen changes the structure of your skin, this is a normal structural change in how the brain works.

At first, not being able to hold more than one thing in the mind by switching back and forth can feel disconcerting. "I think I'm developing dementia," said Diane when she came to visit me over worries about memory. "My partner interrupted me and changed the subject. I got annoyed with him for doing it because I couldn't remember what I was about to say and I knew it was important. It took hours for the thought to come back." She looked visibly distressed.

I told her, and I'm telling you: That's not dementia. It's just normal. If you have one thing in your mind, a second idea will push the first one out and you won't be able to retrace to what the first one was. You just need a few work-arounds.

The hippocampus, prefrontal cortex, amygdala, neocortex, and cerebellum work in a beautiful choreography of laying down, while awake and asleep, short-term, long-term, emotional, and movement-coordination memory. A fight, a threat, or intense joy—any strong feeling—means the amygdala helps burn the details of an emotional memory into our minds forever. That's a memory we will retain. But if we were holding one thought when another enters, we will lose the original thought.

Dr. Gazzaley's work shows how the multitasking ability gets disrupted in the Upgrade. It's like this. You've got an email from a new friend with her phone number; you want to call her and have to input the number manually. You see the number, repeat it to yourself as you enter the digits. An alert pops up for an email that you click on. You might not only forget the phone number; you might forget that you wanted to call someone at all.

The brain's memory circuits in the Upgrade are forever changed, forcing us to newfound heights of focus. Because a new thought or interruption will knock the thing you were just about to do or say out of your short-term memory banks, you will learn to say, "Wait a minute. Let me just finish what I'm doing first." You must hold your ground in a new, forceful way so as not to lose your train of thought or momentum in your action.

Since I have an older husband, I have watched what he does. Here are some tips I've gleaned from watching him. The first is to schedule tasks according to the time of day when you have the best focus. The second is that as soon as you know you have to do something, jump on it right away and complete it. If you have to wait to do it, write it down in a place where you'll remember to read it. Sometimes I forget to read my reminder list, but that's another story. If you try to chase down a lost train of thought, you won't find it. You have to let it come back around on its own. It may be too late by the time it does, but that's just the new reality. You learn to let it go and giggle about it with friends.

Once Diane accepted that her brain network for memory had changed, she made a few adjustments. The first was that she relied on the emptier memory brain organs of younger colleagues. She started to warn younger colleagues about catching her in the hallway without her phone or her organizer. "They would stop me and tell me something on my way to the bathroom. By the time I got back to my desk, my mind was on something else," she said, "and I would have completely forgotten. I carried my phone and to-do list around more frequently, but if I didn't have it with me, I asked them to email me what they had just said, so that it would be at the top of my list when I got back to my desk. That worked."

My friend Kathryn turned distractibility into a communications su-perpower. Standing up to speak in front of a group used to scare her so badly in her thirties and forties that she was constantly battling her intes-tines for a long enough break from needing to go to the bathroom so that she could deliver her speech. Now, at sixty-five, she strides out onto the stage, faces 1,200 salespeople gathered for a conference on television ad sales, and barely looks at the ten words on the scrap of paper she's got tucked into her right hand. "I forget what I'm saying in the middle of a talk all the time. I lose my place, and instead of panicking, I ask the audi-ence to help me. 'What was I saying?' or 'Now why am I talking about that?' I find they are not only willing to help, but they are excited that I've asked them to engage. And by the number of people who shout out re-minders, I know whether or not I'm reaching them. If I'm not doing my job, they won't respond, or they'll give me the wrong answer. I spent de-cades in the bathroom freaking out about things like this when I was younger. If I made a mistake during a talk, I would pretend like nothing was wrong and white-knuckle it through the rest of the speech, hoping nothing else would go awry."

"When did everything change?" I asked.

"I was exhausted during a presentation. It was a really rough period—my career and first marriage were teetering on the brink—and I was rattled. I stopped to take a sip of water and spotted a friend in the audience whom I hadn't noticed earlier. We made eye contact and smiled. It was such a relief to see her face that I completely forgot where I was. I don't know what made me do it, but I blurted out what had just happened. All of it. That I had been nervous, super stressed about my life falling apart, and that seeing my friend made me feel so relieved I forgot what I was talking about. I was shocked when so many in the audience smiled and started helping me find my bearings. All those years I had thought they were waiting for me to fail. And now I could see they wanted me to succeed."

Memory and the Hamster Wheel of Worry

It seemed so benign, that coffee after breakfast. As she sat alone on a sunny day, looking out the kitchen window and just letting the mind wander, the memory of our conversation floated up in Kathryn's mind, she later told me. She smiled at her success and newfound ease. But that day, that month, that year, had been really bad, she thought. Then a flood of other memories: The couple who owned the business she worked for were fighting, and it was clear they were headed for a split. Kathryn's life was a daily hell of being put into positions of choosing sides. In the middle of that, she'd discovered some mysterious hotel credit card charges and had confronted her now-ex-husband. *How could you do this to us?* she remembered screaming at the top of her lungs as he confessed his affair. The entire fight was now in full bloom in her mind: the tired evening hours, the door slamming, the retreat to the guest bedroom, the sobbing, the lack of sleep. She thought of what she should have said instead of what was said, what she hoped to say if she saw him again. And those stepkids! She had put all her energy into them when their mother had checked out of their lives, and as soon as they heard their father was splitting with her, they blocked her. It was all as though it were happening in that moment, at breakfast, while sipping coffee. Living in these past and imagined future scenarios, she couldn't find her way to the present. She couldn't break the cycle. That's when she called me.

Kathryn was caught in rumination. Locked into the hamster wheel of worry, she'd lost her ability to be present, and that is a recipe for an unhappy brain state.

More than two thousand people signed up to participate in research on happiness conducted at Harvard in 2010 by two psychologists, Matthew Killingsworth and Daniel Gilbert. During the study, participants received text messages randomly throughout the day asking three simple questions: What are you doing? What are you thinking about? How happy are you feeling? Nearly half the time, people reported that their mind was wandering. And nearly all of the mind wanderers reported feeling much less happy than those who were focused on the task at hand. Those who responded that they felt connected to what they were doing in the moment reported being pretty happy.

Wild and unfettered, we let the mind and imagination go where it will. After school and work and raising families, this kind of freedom sounds like just the ticket to joy and happiness. Well, it turns out the opposite can be true. Just as children and teenagers crave structure or they become anxious and afraid, an unsupervised wandering adult mind quickly becomes unhappy. It doesn't mean you should spring into action and keep busy. Your brain will need the quiet. But unless you're a very experienced meditator, you'll need some tricks and tips for managing that quiet in order to make the new calm delicious.

Science has only recently discovered why, when the mind is left to daydream without a goal or direction, things tend to go negative pretty quickly. The key turns out to be the brain's default mode network (DMN), which was discovered in 1979. Part of the function of this system is to allow us to do familiar tasks like making the bed, fixing breakfast, and driving safely almost on autopilot. We repeat some things so often that the DMN incorporates them into its circuitry and we no longer have to think about what we are doing. Because the DMN is wired into the memory network, it uses what we know already, so that a thing we do or think about over and over again becomes effortless over time. It becomes a skill.

One of those skills is being alert to danger. It was super useful when humans lived in the wild: That alertness made sure we would see and remember places that could put the lives of ourselves and our tribe on the line.

Now that we are not living in the wild, you'd think that the danger vigilance would quiet a bit. Guess again. Even after millennia of living in increasingly safe times, the brain has turned this instinct into a full-on cognitive distortion called negativity bias. It acts as a filter and makes it feel realistic to be pessimistic, to look for problems where there are none.

As a result, if we leave the mind to its own devices, unsupervised daydreaming turns negative quickly. Our dreams of a happy and free mind turn unpleasant fast. When the mind is wandering, i.e., in default mode, the brain turns into that big ball of Velcro that rolls around picking up negative thoughts. Since the DMN is also connected to the long-term memory network, those negative thoughts will tend to be autobiographical memories: Your worst, your scariest, your most depressing experiences jump to the front of the line, screaming like an eager, smart-ass kid in school, *Me! Me! Pick me! I know the answer!* And you will pick that eager negative thought almost every single time. The mind will get pulled, as if by a magnet, to worry, bad relationships, regrets about the past, big fears about the future. When memories and scenarios arise in full bloom, our wandering mind leads us not to joy; it gets sucked into the vortices of unhappiness, shame, guilt, anger. Everything will feel and look so much worse than it is. But at the time, we won't know that.

As Kathryn was discovering over her morning coffee, when the unsupervised mind is stuck in negative thoughts, memories, and scenarios, parts of the prefrontal cortex become locked into a pattern we experience as rumination. It's the hamster wheel of worry that sucks us into the downward spiral of agitation, fear, and sadness. Once the thought pattern starts, it feels as though you can't get off. And in 2020, researchers discovered why it's so hard to break out of rumination once it starts.

After priming research participants for moments of rumination by having them remember times of rejection, researchers observed brain activity in an fMRI. What they saw was as riveting as rumination itself: Two areas of the cortex, the TPJ (temporoparietal junction) and the precuneus, lit up in a robust, exclusive partnership. These two brain areas are part of the DMN. Their functions, having to do with memory, integration of information from the body and the environment, and perception of the environment, can hijack cognition into a feedback loop that becomes something akin to an unbreakable magnetic lock. When this brain-organ partnership is in control, we can start believing that all our thoughts are accurate and real, regardless of how crazy they might seem when faced with the truth. The mag-lock state of mind prevents new insights from entering; it disables creativity and innovation; it blocks us from being able to find new solutions to old problems. Remember, the cascade that led to the trap of rumination began with a wandering, daydreaming mind. When we start to believe the thoughts of failure, doom, and gloom, it can snap the tent pole in half.

Breaking the Rumination Mag Lock

The good news is that in the same 2020 study, researchers documented the effectiveness of cognitive techniques for breaking the magnetic lock of the TPJ and precuneus that are stuck in rumination. The techniques had to do with how we talk to ourselves, what we believe to be true about ourselves, and how we can put a setback or even a flaw into perspective.

On that sunny morning, Kathryn was stuck reliving the moment of her ex-husband's betrayal. She told me that along with rumination came labels: *I'm alone. I'm a failure at marriage. I must be unlovable if my husband would cheat on me. I'm a nasty, rageful person because I was ruthless in slicing his ego to the bone when I confronted him.* Soon these labels formed into solid beliefs that became difficult to shake; they became the bait that sucked her back into replaying and reimagining scenes again and again. To break the lock, I asked Kathryn if she was the only one who was ever betrayed, if she was the only one who wanted revenge when they were hurt, if no one ever recovered after an event like this, or if there might be a positive side to her rage, like not being intimidated easily and being able to speak up for herself. I asked her if there was any way to accept that it's okay to have negative emotions in difficult times and that having tough times is part of being human. And finally I asked her to consider that negative emotions and setbacks can be overcome. I asked her to give herself the time to think clearly about tiny steps she could take that, over days, weeks, and months, could form solutions. She agreed to spend ten minutes daily journaling about these possibilities.

This is called reframing. When participants' brains were scanned in the fMRI using the same kinds of reframing and reappraisal techniques I was suggesting to Kathryn, they were able to track the progress they were making in decoupling the mag lock of the TPJ and the precuneus. Forty percent of participants were successful in physically and emotionally emerging from the brain lock of rumination.

New interventions for the rumination that comes with depression will be found as a result of this work, and I believe it's going to show us what monks and meditators, and Gilbert and Killingsworth's 2010 Harvard study on focus, have already shown: that being present, focused on what we are doing, makes us happy, and that with the right tools and practice, we can make a difference in our brains, in the happiness in our minds, and make room for the joy in our lives. After I talked all of this through with Kathryn, she added to her morning ritual one minute of focusing on her breathing before beginning her reframing exercise in her journal.

When I first started digging into this subject, I was full of questions. Reappraisal and reframing have long been known to be effective interventions in cognitive behavioral therapy (CBT). They free us to think and see differently by helping us uncover and challenge distorted thoughts and feelings about who we are. Seeing the neuroscience behind how we can get stuck in ruminative distortions, going around and around about something in the past or an imagined future conflict, and learning how we can use that science to break out of the cycle is an illuminating first step.

Why do I say *first* step? Because smashing a ruminative cycle is a momentary intervention. This pull to negative thoughts is an ancient habit, deeply wired into the oldest parts of the human brain. Like a heavy smoker mindlessly picking up the next cigarette, we will go there again and again. In the Upgrade, I personally want my days, nights, thoughts, and emotions to be about comfort in my skin, quality connection with others, the ability to pursue my real passions. I want the clear mental and emotional bandwidth to imagine into the Upgrade, to use creativity to innovate in the second half of life. I want to see clearly the path toward manifesting purpose. If I'm trapped in rumination, I will be stuck repeating old patterns. And I didn't gain all of this wisdom and experience just to have it locked up by the magnetic pull of rumination.

I want to become skilled at prevention, skilled at working with attention, focus, and developing meta-awareness—alertness to patterns of thoughts, feelings, and emotions. It's an awareness that gives us the chance to catch the mind in the act of wandering, to stop it before it goes down a rabbit hole. It gives us the chance to change direction when imagination steers itself toward that magnetic dark place. Kathryn and I have been practicing these skills and talking about our experiences. It has helped us both stay on track.

Focus: The Steering Wheel of the Mind

There are a lot of potholes around San Francisco, and if I'm not careful when I drive there, my wheels get knocked out of alignment. When that happens, if I let go of the wheel for just a moment, the car will veer out of its lane. When we let go of the steering wheel of the mind, we let go of meta-awareness, awareness of what we are thinking and feeling right now. If we don't keep our hands on the wheel, we will veer out of the lane of well-being and straight into rumination.

The amount of worry, the type of worry, differs between the genders. Women worry more about safety, of others especially. Mothers constantly suppress ruminations about their children getting into accidents. That doesn't change as we get older. At ninety you worry about your seventy-year-old kid getting cancer. The echoes of mommy-brain wiring will produce these scenarios, these unsupervised daydreams. These negative scenarios lead to fixation, but that isn't focus; it's just the mag lock of rumination. Fixation is part and parcel of the hamster wheel of negative thought. It's the rigidity of obsession over a person, a substance, an object, or status. It's the earworm you can't get out of your mind that replays a song in your head for hours.

Focus is flexible, allowing you to notice when you're headed off into fixation, stop, and bring the mind back to the present. It's the opposite of unsupervised daydreaming, the wandering default-mode mind. It sounds awkward and self-conscious to stay alert and aware, like having to constantly chase a rambunctious toddler. But once upon a time, putting a spoon to your mouth, brushing your teeth, riding a bike, driving a car, all were awkward, self-conscious affairs. Over time, those skills were transferred into the DMN, the same network responsible for the wandering, ruminating, perseverating mind. As a new habit was practiced, as coordination and muscles became stronger, those activities became automatic. A lightly focused, relaxed, ready, clear, alert, and aware mind can also become a habit. As the activity of lightly steering the mind becomes habit, it can be transferred to the DMN. Over time (and I don't mean days, weeks, or months; we are talking years) it too becomes effortless. Without the waves of monthly hormone cycles pushing and pulling at the brain's circuitry, focus gets easier.

Like a squirmy toddler told to sit still, attention will wander off the minute you try to gently rest it on anything. It's what minds do. We can either get frustrated or enjoy its energy and creativity. The act of bringing the mind back from wandering, or fuzziness if you get sleepy, is what makes us stronger. Like repeated bicep curls, it's how muscles develop.

We can use new muscles of awareness to keep from going down the usual what-if rabbit holes of worry, negative comparisons to others, unrealistic expectations, and belief that our fears about the future are true. In the Upgrade it is our job to find better alternatives to the mind being triggered into negative wandering: shutting off social media, going for a walk, using the Breathe app on our iPhone or Apple Watch, saying no to a call or lunch with a toxic person.

For those who have an overactive mind or a mind that won't sit still, sometimes movement or stretching first helps. For Kathryn, images of balance helped. "I have always had a hard time settling down," she said, "and it got more intense during the Transition. Concentration was really hard. The only time I could focus was while doing balancing poses in yoga." It makes sense: There's a lot at stake, maybe even broken bones if you fall, so the cerebellum activates the network necessary for alertness to balance. You can leverage the same neurocircuitry by conjuring the feeling of trying to balance on a stability disk, a beam, whatever works, to recruit the alertness of balancing the mind on the present moment or an object or subject—like a nurturing moment (see appendix, page 258), safety, relaxation, or compassion—of meditation.

The brain in meditation kicks up the best of the nervous system and neurochemicals. Oxytocin and dopamine, bonding, and reward chemicals are released. Anytime you focus on the breath, the vagal system is activated. Steadying your attention on the breath signals the vagus nerve to activate the relaxing parasympathetic nervous system, counteracting the sympathetic nervous system—the threat-response system that is fired up by the adrenals. A nurturing moment signals safety and prepares the vagus nerve to stimulate the parasympathetic system.

Alternate-nostril breathing, watching the breath, body scan, present-moment awareness, prayer, the nurturing moment, compassion training, twelve-step programs (see appendix, pages 251 and 257). There are so many methods, and you've likely heard or seen them or been guided through one during a yoga class. If you take nothing else away from this, I want you to remember that focus holds the key to the Upgrade. If you can learn and practice a method for very brief periods daily and eventually without being guided, you will start to be able to take the wheel of your own mind and steer it happily in the direction you'd like to go. You can start with just thirty seconds. That's the Dalai Lama's recommendation.

The opposite of the perseverating mind is one that is motivated by something that brings it lasting joy: Diane's rediscovered focus at work;

Alina's new passion for the environment; Kathryn's harnessing of her mind; my excitement about the Upgrade. Imagining, manifesting, and stabilizing our own Upgrade becomes a choice. The best news is that the quieting of hormonal waves puts the focus we need more within our reach. It gets easier to choose.

Life Span or Health Span

Chapter 14

My mother's friend Belle divorced in the 1940s and never remarried. An avid golfer and outdoor enthusiast, she went on safari alone in the fifties. She had lovers in Australia and Italy. She never used a golf cart; she loved walking the course. When she reached her eighties, she opted for playing nine holes over getting a lift.

She was sharp as a tack and never turned the volume down on her New York voice. Grandchildren giggled as they held up the handset of the old touch-tone phone for others to hear across the room. She read avidly, played a weekly bridge game, and indulged in a sip of sherry after dinner until the day she died at eighty-four. Though she wasn't ill, she was climbing the stairs to her doctor's office for a checkup when she got tired and sat down on the steps, and her heart gave out.

My great-aunt Harriett, who was married to a difficult man with a sometimes violent temper, lived until she was ninety-five. She didn't exercise most of her life, including after her husband died, when she was in her late sixties. She was a heavy martini drinker and didn't much worry about what she was putting into her body, other than the concern over what might make her gain weight. Her cognitive decline began in her early seventies—it was clear to her family she had been struggling with her hearing for a decade, but she refused to do anything about it. Hearing loss and cognitive decline can come together. No one is sure if hearing aids stop its progress, but we are sure that they decrease the chance of depression from isolation; depression does impact our ability to think clearly and regulate emotions. The family had good reason to worry about the fact that Harriett was too proud for hearing aids.

Harriett's knees gave out and she stopped being able to travel by the time she turned eighty, remaining homebound and mostly alone other than frequent trips to the doctor to keep adjusting her increasing list of medications for blood pressure, fear and sadness, joint pain, and insomnia. She spent the last seven years of her life with full-time care, her mind ravaged by Parkinson's. Unable to recognize her family members, her emotions were in a constant state of turmoil. She suffered until the end.

Harriett's life span was eleven years longer than Belle's. But her health span, that period of time in which she was able to enjoy a quality of life that included both physical and mental energy, was perhaps fourteen years shorter. Everyone would like to exercise their own preference for how they want to live; in the Upgrade, personally, I choose health span over life span. But how available is that choice?

Measuring Medical Success

Think for a minute about how success in medicine is measured. There is one marker: whether or not a patient dies. Success is not measured by quality of health or life of the person who doesn't die. Success could mean someone is bedridden or in a lifelong coma. When avoiding death is the only marker for success, what does that mean for quality of life?

For more than one hundred years, medicine's single-pointed focus on increasing life span has resulted in world-changing innovations powered by the young men, and a few women, who set out on the noble mission of saving lives. We have penicillin; vaccines for smallpox, polio, pneumonia, and COVID-19; heart medications; chemotherapy for cancer; organ transplants; cholesterol-lowering medications; artificial joints. All of these medicines and protocols support life; and in focusing on cancer and cardiovascular health, we have bent the curve on early death. That's a great thing. Yet at the same time it has come at a huge cost.

America now has the biggest and most expensive medical system in the world. Residents spend more on "healthcare" than any other nation in the world, and so it would seem that those who live in the United States would have a longer life span and better health outcomes than everyone else. But it isn't true. When you compare the life span and health of Americans to those of residents of all other developed nations, the United States has the worst outcomes in nearly every category, including longevity, heart disease, cancer, and especially maternal and infant mortality.

Women in the second half are bound to fare a bit worse in terms of health span, though we don't have any way of knowing directly, since the data have never been collected. But it's not hard to extrapolate; women's health, other than reproductive health, has barely been part of the equation in medical education. It's been ignored that women over sixty are the fastest-growing demographic in suicides, even though the numbers are right there staring us all in the face. Yet with all of these unfavorable outcomes, young doctors who see defeating death as the central goal of research and practice continue to set the agenda not just for medicine but for the quality of our lives as we age.

Knowledge May or May Not Be Power

We can't control every factor that impacts our health span. There are a vast number of factors and circumstances that can cause illness. It's impossible to account or control for everything. Poverty can increase our risk for heart and respiratory disease; so can our mother's smoking while pregnant with us. Trauma has an impact on life span as well as disease susceptibility, and we can be unknowingly poisoned by what's in the air, water, or soil around us. Those who grew up in a certain area of New Jersey have a higher risk for bladder cancer because of toxins in the water.

Those who grew up around the Gulf Coast's Golden Triangle area, where oil is refined, have a higher risk for asthma and blood cancers. The death of a child or spouse can lead to broken-heart syndrome, impacting our health. None of these things is our fault. We are not losers, nor is anyone else, because of being affected by circumstances beyond our control, because of the suffering that is inevitably part of the human condition. That life itself always ends in death is not a mistake; it is reality.

Genetics do play a role that is also beyond our control. Nancy Wexler is a renowned geneticist who devoted her career to studying the gene for Huntington's disease, a debilitating and fatal hereditary condition that runs in her family. If you have one parent who carries the gene, there is a fifty-fifty chance you will get it. It is what we call a single-gene dominant inheritance disorder: If the one gene is present, the chance of developing Huntington's is 100 percent. Dr. Wexler's mother died of the disease; it took the lives of her uncles and her grandfather as well. And she inherited it too.

There are many conditions for which we can detect genetic causes. The BRCA mutation for breast cancer; the APOE4 variant for Alzheimer's; the sickle cell gene for sickle cell disease. Not every mutation means that a gene gets expressed. With Huntington's, we know that it does, and the same is true for those who carry two copies of sickle cell mutation. One hundred percent of those who carry two sickle cell gene mutations will get sick. But with others, just because you have the gene doesn't mean you will get the disease. With the BRCA gene, up to 80 percent of the women with this mutation will develop breast cancer, but not everyone. Even two APOE4 variants don't inexorably lead to Alzheimer's disease in all cases.

When you have a family history of a disease, there is a higher likelihood that you are carrying some of the relevant genes, and there are some disorders, such as breast cancer, in which we already have DNA tests that can inform you about your relative risk. Whether or not you want to get that information is a deeply personal decision. Sometimes knowing can mean you have the chance to engage in preventive activities, like more vigilance with diagnostic technology or even prophylactic mastectomy and ovariectomy if you have the BRCA mutation. You can also learn if a certain kind of chemotherapy is likely to work on a particular cancer based on your own genetic makeup. So genetic information can in some cases guide a course of action. In many cases, genetic information has no practical actionable value. And don't expect it to tell you what you should be eating. The testing just isn't that sophisticated yet.

What we learn from doctors and genetics can feel like destiny and cause us to make life-altering decisions. Rachel Naomi Remen, the author of *Kitchen Table Wisdom,* was told at a young age that her severe Crohn's disease would shorten her life. The doctors informed her that she wouldn't live past forty. As a result, she chose not to marry or have children, because she didn't want to inflict the suffering of losing a wife or mother on a family. But she didn't die at forty. She's in her eighties as of this writing.

How Remen's life would have been different if she had not been informed of a likely early death is impossible to know. But there is some evidence that could make an argument for not knowing. There is a category of older people whom medicine has labeled super-agers. These are people in their seventies, eighties, and beyond whose mental acuity and physical activity reflect the age of someone decades younger. As part of a study in Southern California, the brains of some super-agers were scanned after their deaths. In the cases of several women in their nineties who had been active physically up until the end, engaged in things like a weekly bridge game, researchers were shocked by their MRI scans: Their brains showed the same kind of moth-eaten appearance as the brains of some Alzheimer's patients. And when their brains were dissected after death, they were full of the tangles and plaques of typical Alzheimer's patients—yet their minds were cognitively intact. No one ever told them that they had scans that should reflect extreme cognitive decline. The question I have is how the knowledge of these brain scans might have impacted the way these super-agers lived. Would they have remained super-agers, or would they have given up if they had been told by a doctor that their scans were showing strong signs of dementia? Would they have been given medicines whose side effects might have decreased their quality of life or set off more cognitive decline?

There is in medicine something we call "incidental findings." In my midforties I kept spraining my ankle. Orthopedists did repeated X-rays, and not finding anything, they put me in orthotics. When that didn't work, I was sent for an MRI. This time they found a disease: osteochondritis dissecans, which means "a hole in the ankle bone." The doctors decided that this hole in my ankle was the cause of all my trouble. But when I learned that fixing that hole could end up creating more trouble, I decided to work with a physical therapist instead. While doing some massage, the therapist released some tendon and muscle contractions that were pulling my ankle in a funny direction. Within six months the problem was fixed. Twenty years later, I haven't had another sprained ankle.

The hole in my bone had nothing to do with the problem in my ankle. It was an "incidental finding." Sometimes a finding is a red herring, and doctors will end up telling a story about it that isn't really the story. Not every finding in a scan is the direct cause of a complaint. It might or might not be.

Nancy Wexler opted not to have the genetic test for Huntington's. If she discovered too young that she carried the gene, she felt it would disempower her from living a full life. So she decided to live as healthy and vigorous a life as possible. By her early seventies, she had become fairly debilitated by the symptoms of Huntington's, but she told me she didn't regret her decision for a minute.

This is not to say that ignorance is always bliss. But you can make a discerning choice about how information may impact your decision making. Doctors have noted that people "die on time"; when told they are expected to have a certain period of time to live, they tend to live exactly that amount of time. Because it happens so frequently, many medical professionals have begun to think this is not just because they are such good prognosticators. The power of suggestion is huge. If you are going down the road of genetic testing, find a doctor and genetic counselor who will take the time to really explore options. Get multiple opinions and perspectives. This is not a one-stop internet shopping moment.

Since my Transition, I am clear about how I want to age and that health span is my marker for success. I want my mind to be sharp and my body to be mobile. I know I don't have total control over all the factors that go into how I age, but there are things available to me that I can do to make sure I do my best to remain cognitively sharp, fearless, active, and as energetic as I can be. There's no guarantee that if I do everything right I will end up with the perfect outcome. Anything can happen. But how I emerge into the later decades of the Upgrade can be shaped by what I do now.

Our Real Risks: Cognitive Decline and Alzheimer's

Cancer is so mysterious. In its early forms it's painless, and you might only feel a lump or find it on a scan. As a medical student, while reading about the disease, I kept feeling my abdomen and breasts a couple of times a day, looking for signs of painless lumps. The more I read, the more I checked. In those days I didn't know what fibrocystic or lumpy breasts were, but I had them; I thought every bump was breast cancer and that I was a goner. It wasn't true. Reading about a disease makes us suggestible, and it's a common phenomenon among medical students that we become sure we have what we are reading about.

Women have some real risks for Alzheimer's, but it doesn't mean you will get it. I don't want you to have to go through what I went through as a med student, thinking you have a disease you are learning about. So while absorbing the following, remember the super-agers, who were in full control of cognition into their nineties even though their brain scans looked like they should have had dementia. Remember Belle walking the golf course and playing bridge. Remember normal structural changes to the brain that keep us from multitasking in the Upgrade. Keep all of this in mind as you read the next section, and maybe it will help keep you from getting stuck on the hamster wheel of worry.

While women on average are more resilient to cognitive decline and memory loss that comes with age, one-fifth of all women will be diagnosed with Alzheimer's disease (AD). As with cancer and heart disease in the 1950s, we don't have good treatment options. The most common form of dementia is AD, and the risk for AD is higher in women than in men. The biggest reason appears to be that women live longer. Both sexes show a similar risk of developing AD until advanced ages—in the eighties, when women show increased risk, one in five women versus one in ten men. But the theory is that by the time people reach their eighties, the less hardy men have already died off. If more of them had survived, researchers assume that they'd be at equal risk.

With all the brain benefits of estrogen, there's been a lot of speculation about a possible role in prevention of AD. But multiple studies have shown that starting estrogen after the age of sixty doesn't help. And in 2003 a large study on women who carried at least one APOE4 variant showed that starting estrogen plus progestin after the age of sixty-four was associated with an increase in dementia. While we know that estrogen is good for memory and learning, starting it at the wrong time can be more harmful than helpful. So if you're more than ten years past the Upgrade and haven't started HRT, it's probably better not to. But if you are between forty-five and fifty-five, studies suggest that taking HRT long-term may reduce your risk of getting any neurodegenerative disease, including Alzheimer's. Remember that all the other strategies—self-care, movement, sleep, diet, stress modulation, care for the microbiome—will have a positive impact on preservation of cognition.

And while estrogen on its own hasn't been shown to treat Alzheimer's, estrogen does have proven brain benefits. HRT sets the brain up for helping you take action against inflammation that might speed cognitive decline. The growth in the brain stimulated by estrogen brings a more positive outlook, an enthusiasm for life, and energy for motivation to engage in self-care. All of this helps preserve cognition. Studies have shown that women who carried the APOE4 variant but engaged in moderate cardio for five to six hours per week and kept cholesterol low with a better diet and statins preserved cognition for a longer period.

If you are among the 2 to 3 percent of the population that carries two APOE4 variants, you may have up to a 90 percent chance of developing Alzheimer's. And for the 25 percent of women who carry one APOE4 the risk is higher too. But keep in mind that only 37 percent of those who have AD have the APOE4 variant. And for many of us the risk is less. So whatever your risks, there are things you can do to mitigate the odds. As soon as you realize you are in the Mid- or Late Transition, exercise, stop smoking and drinking alcohol, and keep your weight and cholesterol under control. Discuss taking HRT with your doctor. The impact on inflammation in the brain and on protecting cognition is well documented for each of these strategies. Do them all together if you can.

Puppies. Kittens. Babies. Hug one or find a video to get your oxytocin going. Then breathe to stop the spiral into ruminative certainty that you have Alzheimer's. There's a very good chance you don't. But since women have a higher risk for it, I want you to know the reality.

Here's the case for HRT and brain health. Estrogen is essential to a cascade of neurochemical processes in the brain. It helps neurons connect. It helps the brain use glucose, sugars, for energy. It helps the immune system protect neurons from harm, strengthening the brain's ability to bounce back from inevitable assaults like inflammation and to find its balance again. It kicks off a bunch of feel-good reactions, including production of GABA, the antianxiety neurochemical, and the release of endorphins, like those we get from a runner's high, which soothe pain and depression. Estrogen is beneficial to circulation, and in the brain that means a good strong flow of blood, oxygen, and nutrients to keep it healthy. Estrogen helps regulate inflammation in the brain, which means that those astrocytes and microglia have a chance to perform their function of cleaning out the trash and feeding the brain cells, more so in female brains than male. Regulation of inflammation means a lower chance of microglia and astrocytes going into zombie mode, spewing toxins and further endangering cognition. This may be one reason that women in their forties, who still have estrogen, are less likely to get early onset Alzheimer's than men of the same age.

The Early- to Mid-Transition is a shock to the brain and the circuits in the female brain change quite a lot compared with those of men of the same age. The bottoming out of estrogen causes sudden crashes in brain energy. This inhibits the brain's ability to access its usual source of energy and use that energy efficiently. Estrogen is the master regulator of metabolism in the female brain and promotes glucose uptake—glucose is the brain's favorite fuel. Estrogen dials up the energy of mitochondria, the powerhouse of all cells in the body. It protects cells from the damage caused by toxins emitted by aging cells; it prevents cell death; and it helps balance calcium in the body and builds strong bones. It helps keep the blood-brain barrier healthy, preventing damaging substances circulating in the blood from getting into the brain. The role of estrogen in the brain is the same for men, but they never experience the Transition, this sudden hormone drop, the accompanying loss of gray matter and white matter, or wholesale remodeling of brain circuits. Testosterone converts into estrogen, and so they have a fairly continuous lifetime supply. Some scientists are now investigating if the loss of estrogen for women in the Transition is a tipping point for cognitive decline.

The drop in estrogen means the mechanism for transporting glucose through the blood-brain barrier is dramatically diminished, triggering a cascade of metabolic effects as the brain tries to adjust to lower levels of glucose. It means that proinflammatory cytokines in the brain are less well regulated. Lower estrogen is why we have hot flashes, why we become more likely to develop diabetes, and why inflammation becomes less easily managed. "I was surprised," Terri told me, "that after lifelong perfect blood work, my inflammatory markers skyrocketed during the Transition. I couldn't absorb why that was happening, and nobody told me it was because of my hormone changes."

The brain fog that comes with the Transition and Early Upgrade is not dementia, and it's not presaging Alzheimer's disease. Its cause and cure is estrogen. The brain fog is a result of lower estrogen, and it can be fixed through HRT. I repeat: It's not dementia. It's not the onset of dementia. It's temporary and fixable.

It's the reason I had my surgeon slap an estrogen patch on my butt in the recovery room after my total hysterectomy in 2005. Even while conventional medical wisdom was terrified by that 2002 WHI report warning of the dangers of HRT, I knew the report was flawed and decided my brain health outweighed the risks.

This was my personal choice. Yours may be different.

If you decide not to take HRT, or if it's contraindicated, do everything you can to fight inflammation: move, eat right, sleep well, destress wherever you can. You'll find some plans in the appendix.

Many women, knowing they have a family history of Alzheimer's or knowing they have one or two APOE4 variants, are getting baseline brain scans to track what they might feel is an inevitable slide into dementia. For me personally, knowing what I know about super-agers, I question whether or not it's information I would want to have. If the information indicated a course of action, then, yes, I would want it. But with APOE4 status? The only thing I could do to reduce my risk would be to engage in habits that reduced inflammation and take a statin so that the brain didn't decay quite as quickly. It turns out that statins reduce risk of dementia by almost 30 percent. And that's why I'm doing everything I can—focus, meditation, food, controlling cholesterol, movement, sleep, social connection, calm—to keep my brain in the Upgrade.

I'm afraid that if I knew I had the APOE4 variant, I'd be a wreck, constantly reading the tea leaves for signs that my disease had begun. But that's not true for everyone. I'd become obsessive about it. For Paula Spencer Scott, author of *Surviving Alzheimer's*, knowing her APOE4 status gave her power. Instead of causing her stress and depression, as some researchers feared, it motivated her to get a complete workup. When she saw that some of her inflammation markers were high and learned that these could cause a more rapid decline in brain health, it motivated her to act. She writes:

There's nothing like proof of weakness to motivate behavior change. For the first time in my life, I joined a gym and hired a trainer and began brain-beneficial high-intensity interval training.

(Just walking—my old exercise—wasn't enough, I learned.) I became even more vigilant about my Mediterranean diet, took prescribed supplements, learned "brain breaks" like mindfulness and more.

Two years on, my improved cognitive scores and lower cholesterol keep me going. As does finding myself in stronger shape at 60 than I was at 40. (True, I gained several pounds . . . but of muscle.)

I now run every choice I make through the lens of my brain health: *Is this helping or harming?*

For someone else, testing might raise urgency about lousy sleep, hypertension, poor glucose control, or stress. Swiss researchers have suggested that having motivational reserve—motivation to improve one's cognitive health—may even have a protective effect on the course of mild cognitive impairment.

The Hidden Brain Dangers of OTC and Prescription Drugs

Carole was beside herself. "Louann, my mom has gone downhill so quickly, and I just don't understand it," she said, on the edge of tears. "She's been in great shape—exercising, super active, and super sharp. I don't know what's going on. In less than two weeks she went from totally independent to needing full-time care. She's completely disoriented."

I had met Carole's mother, Helen. She was in her early eighties and didn't have genetic risks for Alzheimer's. Carole's grandmother had lived into her nineties and had her cognitive powers until the end. I asked if we could all get together and if Carole could collect any over-the-counter and prescription drugs her mother was taking.

We sat down in the living room, and on the coffee table was the answer. Her bottles of pills were all laid out, and there in full view was a bottle of nortriptyline. It is a tricyclic antidepressant being used in many pain clinics, and it had been prescribed two weeks earlier for chronic arthritic pain. It can cause dry mouth, dry eyes, and difficulty urinating. If drying up and trouble controlling functions is happening in one part of your body, it's happening also in your brain. For some people, this effect of anticholinergics can deliver a wallop to focus, memory, concentration, the entire memory system, as an anticholinergic. Remember, choline and acetylcholine are neurochemicals that govern the wake/sleep cycle and are part of the neurochemical process involved in memory consolidation. paracetamol and Benadryl also have anticholinergic effects.

Not everyone has side effects from nortriptyline, but as we age, side effects become more likely. When you hit seventy-five, due to a decrease in metabolism, it takes longer for the body to clear a drug. At age fifty, you might have been able to metabolize Valium in twenty-four hours, but at seventy-five, it could still be in your body three to ten days later. When patients continue to take it every day, the buildup gets too high and they can exhibit signs of dementia.

Once-a-day dosing of drugs is based on the average metabolism of twenty-to-fifty-year-old males. All pediatricians know you have to adjust doses for children to levels that are based on their body weight and metabolism. But it's not always done for small women and people over the age of fifty. Not even for those over seventy-five, whose body weight and muscle mass have gone down, taking their metabolism down too. With decreased metabolism, the half-lives of drugs, the time it takes for the body to clear them, becomes longer. This applies to supplements too. It's not inconceivable that effective doses of certain drugs could mean taking them once a week.

If a patient is in good communication with the doc, the doc will lower the dose. But with memory problems, confusion, or emotional issues, usually we will think, *It's just me. I'm getting old, and this is what happens.* But if family, friends, spouses, and doctors are smart about this and can get a patient off the meds, things can turn around immediately. Getting Helen off the nortriptyline cleared her cognitive problems within a few days. Anticholinergic drugs are a well-known *reversible* cause of memory problems.

The decision to take something preventively has to be driven by knowing that it will be more beneficial than harmful. More women are harmed by bleeding stomach ulcers caused by daily baby aspirin than are prevented from having strokes. Not only that, but daily baby aspirin can cause the very effect we might take it to prevent. If you already have plaques and you're taking aspirin or a blood thinner, you can develop little hemorrhagic bleeds in the arterial plaques. Those plaques then come loose and can cause a stroke—the very thing you're taking aspirin or a blood thinner to prevent. That's why the recommendation was finally dropped, especially for women. If you are in a high-risk group for heart disease, then taking something for prevention or taking an action to reverse the risk is the right thing to do. If you are having a heart attack, call 9-1-1, and immediately chew a full-sized regular aspirin. If you are at low risk, prevention may not only be useless but might be harmful. If you're in a moderate-risk zone, it's a question mark. Discuss with your doctor.

There have been a lot of heart attacks in my family. I had slightly elevated LDL (bad cholesterol), and my blood pressure was borderline high at 130/80, and so in my early sixties I wanted to see if I was also at risk of a heart attack. I had a sophisticated scan to see if I had calcium buildup indicating the possibility of plaque formation. I was in the 56 percent calcium range, and so my doctor put me on a statin called Crestor. Within a few days I developed blurred vision and brain fog. I couldn't function on the drug. So I asked, "Am I at high risk of heart attack?" The answer was "no." I probed further about her reasons for prescribing the drug. She began her sentence with "Studies show . . ." I stopped her there. Studies show lots of things, and the science constantly changes. I interrupted and said, "Okay, I've got to get off this statin." She said, "Fine. Let's follow up with more detailed tests and plan to check in every six months and monitor the situation." Since my LDL stayed high, I have followed the course of action and tried another statin. For now, we are trying to find one I can take. At the same time, I switched my estrogen from oral to topical. With oral estrogen, there is an increased risk of stroke. But that risk goes away with topical application, so I started using the patch again.

There is nothing that can replace the advice of a doctor you know and trust, so please don't mistake this for my telling you to get off your medications or to start HRT. Your GP, *in partnership with* **you**, will know better. But do ask questions. Review all your medications and supplement dosages every year, and drop what you don't need. And take an advocate with you if your doctor is under fifty. The doctor may not hear your voice, and you'll need backup.

———

We need new markers for success to help shift the core values of medicine. We need to add quality-of-life markers, but until we can talk about death openly, as a reality, not a mistake, this conversation won't happen. Facing death, talking about it, may go against the nervous system's survival instinct, but not facing it means that we are all dealing with a persistent underlying anxiety caused by denial.

The best way to establish these markers comes from knowing what you want the biggest end point of all to look like for you. Communicating that verbally, and in writing, to your loved ones can open so many doors.

So Many Transitions

Chapter 15

I'm not Catholic, but my friend Diane convinced me to attend a ceremony in Washington, DC, for a cohort of novitiates who, after eleven years of training, would take their full nun's vows. Seated near the back of the cathedral, we watched as, one by one, dozens of bishops and cardinals processed toward the altar in towering scarlet hats making them look eight feet tall. Censers swinging, organ playing, each took a velvet-covered seat on the platform in the nave. I had never been to anything like this—I grew up in a politically progressive Protestant church—and I found it deeply uncomfortable. I felt the oppression of thousands of years of male power in that moment.

The procession of bishops ended and then things got very quiet, very still. It was similar to that moment of anticipation at a wedding, when the atmosphere and music shift, cueing participants to turn for the bride's entrance. I did not expect to experience what came next. A tiny woman, barely five feet tall, stood in the doorway, illuminating the entire church with her power, miniaturizing the posturing and pomp that had come before. Mother Teresa walked slowly, smiling in her simple blue-and-white sari, toward the altar and took her seat at the front left corner next to several cardinals. She radiated peace. It was palpable. Her presence transformed the cathedral atmosphere from rigidly ceremonious into uncontrived love and communion. I had never felt that kind of power emanate from a human being before, much less a woman.

Looking back, I see that Mother Teresa was manifesting her Upgrade. I don't have any illusions of becoming a saint, nor would I want to be. I didn't often agree with Mother Teresa's politics, but her love and service were undeniable. She was indefatigable. She never stopped caring or giving. She didn't retreat to a slower, cozier, or more comfortable life. She never retired, seemingly impervious to society's unspoken axiom that at a certain age women are repulsive and have nothing left to offer.

Everything I've learned about the female brain and neurobiology has shown me that we women have been leaving our gifts at the door of the second half of life. It took me a long time to realize that giving in to that feeling, *I'm done; it's time to be cozy,* while comfortable, might not be an Upgrade. Instead, it might be a trap that leads to suffering as feelings of uselessness grow. It might cause suffering for generations of women who aren't being benefited by what we know. Escape into that kind of ease leaves future generations in the lurch to struggle for what we've failed to gain.

By the time women are over fifty, we've figured out a lot of things and we've accumulated wounds we've gone numb to (until our adult kids remind us of them). It's time to have a heart to heart with ourselves about being heard. "Before I got help with HRT," said Terri, "I felt like I'd hit a wall. I kept waiting for the clarity and energy to come back so that I could take on complex, intense design projects again. I started to feel like I wasn't living a useful life anymore." I had a similar experience when I came up against the start-up energy of the very young. At their pace, I couldn't figure out what I had to offer. They had hired me for my thirty years of experience but didn't seem to slow down long enough to benefit from it. Our society doesn't honor wisdom, and so we tend not to honor the wisdom in ourselves. It took me a while to get comfortable with the difference in pace and to find my place in helping them see the big picture.

Younger people convince themselves, and you, that they are the ones who have it together, and they are shocked when you see around corners they don't even know exist. It can take them a while to recognize what you have to offer, and they won't until we recognize that our time on the planet is a gift to them. In the same way we trusted our robust engagement in our twenties, we must continue to trust ourselves that we will find the place that lets us stand in the center of our authenticity. If we don't trust it now, when will we trust it?

Avoiding the cozy post-Transition trap and stepping into our own authority is a developmental phase of the second half of life, a phase that isn't mapped or charted or spoken about. It's one in which we refuse to agree that seeing women over fifty in an underwear commercial is disgusting. It's a developmental phase in which we don't quit because others are talking over us; we simply outlast them.

Genius is evenly distributed throughout society, regardless of race, gender, zip code, or gray hair dyed purple. We have too much to contribute to step off the stage now, when the world needs our wisdom, the wisdom of those who have put so much energy into solving problems for the preservation of life.

In psychology and psychiatry, we map developmental stages because it helps us contextualize someone's mindset so that we can better be of help. Common sense tells us that we can't use the same words or the same psychology for a fifteen-year-old boy as for a forty-eight-year-old woman. Each phase brings different challenges and has different goals. The burgeoning drive for independence in puberty can be a source of conflict in the family. As we grow into early adulthood, dreams of the future take hold and the prefrontal cortex is absorbed in running scenarios for building family, career, achievement, accumulation. Life during the fertility phase is lived in the future; psychologists have well mapped the developmental stage of so-called adulthood. But what are the stages like when there is no more runway for the future? Where does the imagination go and what are the new dreams? What are the problems to be solved between the end of the fertility phase and the end of life?

The Upgrade is a complete expression of the developmental stages of the second half of life. And to make the Upgrade possible, we need to know what the developmental stages are.

Defeating Transition Mind, Stage 1: Winning the Battle with the Demon of Appearance

It is sudden and disorienting when it happens in childhood, the moodiness and awkwardness that come with awakening sexuality. Over the years we learn how to inhabit this new identity. If we don't, the dissonance is enormous: The personality and outlook of a nine-year-old girl in a twenty-five-year-old body might get that woman sent straight to my office for help integrating and growing up, if it's not the result of a disorder about which nothing can be done.

There is a similar shift in sexual identity in the Transition and Early Upgrade. For the first time since puberty, our identity as sexual beings might feel as though it is beginning to crack. It's a shock to the system when we start to see that we no longer inhabit men's fantasies. It's hard to resist the pull of the insula wanting us to feel the way we are used to feeling in our own skin. It's hard to resist the demands of the insula to maintain that old identity through any means possible, including the brutally surgical. Yet just as it's disturbing to see a sexualized child or a childlike woman, there is a similar dissonance in seeing women try to inhabit an old identity in a new body and mind. It is exhausting to try to keep up an old appearance, holding reality at bay. And the people around you will feel uncomfortable. They won't tell you what they are feeling. They will instead exclaim wholeheartedly that the steps you've taken to prop up the old identity look great. The strong modulation in their tone of voice is often a compensation for shock, because the falseness can be alarming. They may take on a similar tone when we let our hair go gray and our skin wrinkle. At this stage of life, women can't win for losing.

As the natural preoccupation of our reproductive years, thinking about hair, makeup, clothes, body, and weight took up a lot of our day-dreaming bandwidth. When I was younger, I remember that a new haircut or hair color often signaled—to both myself and others—that a new version of me was emerging. In the Upgrade, maybe that looks different: Instead of coloring my hair, maybe it's my natural color emerging. Instead of clothes that told the story of the sexy striver, maybe a new wardrobe and aesthetic tell more of the story of what's happening on the inside, one that reflects the quieting of those hormones. I find it deeply personal and quite scary to expose this internal process of Upgrading. I fear the reactions of my husband, family, friends, and enemies (especially!), that if I turn away from all that surgery, hair color, and cosmetics have to offer, they will find me gross. I fear being ostracized and humiliated, no longer invited to speak, to attend high-level conferences, to be in the center of the action, where I've spent most of my adult life. That reaction is normal. We are social beings as humans and particularly as females, so the idea of no longer belonging is, as you know by now, kryptonite to the female brain, even in the Upgrade. The fear of not belonging is one of the biggest obstacles to finding the courage to experiment with being myself at this age.

I'm still exploring a lot of this, but I've gotten this far: If my inside is kinder, friendlier, and less jealous, then I want my outside to more naturally reflect that I am no longer competing with other women for men or jobs. I want my outside to reflect competence, wisdom, and openness as opposed to sexiness and striving.

Knowing the power of the insula, I've made a conscious effort to combat its directives. I've had to think long and hard about what I want my appearance to say *now* about the inside of *me*, about how I want them to match up. It hasn't been a straightforward process. As I let go of the power of appearance, up pops the seemingly urgent feeling that if I want to stay in the game, I've got to look younger. I ask myself if I would still be taken seriously if I let my hair go white. And if the answer is "no," then do I adhere to the status quo or experiment with acts of rebellion that might move the needle for others? At the same time, I have yet to find a woman who has gray hair whom I think any less of. And so I ask myself to find the beauty in gray and to be honest about what game I'm trying to stay in. Regardless, it's a very personal decision.

It's a big adjustment in the early Upgrade to imagine and step into our new appearance. "My grandmother," Terri said, "was white haired and elegant. She wore simple, tailored clothes and very little makeup. Her bearing was regal. It was so natural to her that she didn't even know that's how people saw her. And that's what I want," she concluded, "though maybe a little updated with more athleisure and more comfortable shoes!" She's lucky to have had a role model.

Stage 2: Sex Becomes an Afterthought

How we feel about sex in the Upgrade is as much related to the demon of appearance as it is to the hormone shift that changes the reality of drive. It's another aspect of the same developmental shift that isn't mapped. "My body changed and I didn't feel sexy anymore," said Penny, fifty-three. "I didn't look good in sexy clothes, and it was hard to look at myself in the mirror." She had to spend time helping her brain and nervous system catch up to her new outer reality. But the inner reality was even more powerful.

After the Transition, when estrogen, oxytocin, and testosterone all go quiet, for many women, sexual reality shifts completely. It's as though the drive just evaporates. "When the symphony of intense bleeding finally came to a grand finale," said Penny, "I thought, *That's it? I don't have a sex drive?* It wasn't that I couldn't get my motor started. It's that I didn't even have a motor anymore. I had read enough to know that testosterone was the issue, and I begged my doctor to give me some. I was desperate not to lose my sex life. Until I got it, I had to trick myself into wanting sex," she said.

There was a moment in writing this book when I realized I had almost forgotten to write anything about sex. It's not that it hadn't been an important part of my life and relationship; of course it was. But one look at my testosterone levels, and those of many women in the Upgrade, tells the story of a reality shift, of a hunger that's gone away.

My friends Janet, fifty-five, and Sergio, sixty, had moved from New York to Argentina to Chicago and to three different homes in Dallas. Both had full-time jobs, but Janet ended up being project manager for every single move, even handling the Argentinian move in her broken Spanish. The evening after they made the second move within Dallas, she climbed into bed and told Sergio she was utterly exhausted, that she felt on the verge of a collapse. Without missing a beat, Sergio turned and asked her to help figure out the final plans for the electronics in the house. "My mind exploded," Janet said. "I looked at him and sobbed. Through the tears I said, 'I just told you I'm broken because of the move, and your response is to ask me to do something else for the house? To take on another project?'" Janet paused and, with a twinge of disbelief in her voice, she continued: "And then he capped it off by saying the best thing for me in that moment would be to have sex. Right then and there. As I was weak, and sobbing, and telling him I was broken. For the first time in more than a decade, I wanted to sleep in another room. I couldn't believe what I was hearing. Sex was the last thing I wanted, and it felt like another project I had to take on because it's what he wanted."

By the time we hit the Upgrade, unless we are taking supplemental testosterone, ours is up to three times lower than it was during our fertile years. It's not that we won't ever want sex again. It's more like not being hungry, yet still opening the refrigerator door to get something to eat just because it's dinnertime. The incredibly rich flourless chocolate torte you once craved with your whole being just doesn't seem as appealing. Okay, that's never happened to me, but you know what I mean.

Of all the women I've talked to, many find they don't care if they regain their desire. But if you are in a committed relationship with a man, he will care that you don't seem to want him. He will be 100 percent certain that if you don't want to have much sex with him, it means you don't love him anymore. He will be 100 percent certain that your lack of desire means you are having sex with someone else. Because this is his hormonal brain reality. If he loves you, he wants to have sex with you. If he doesn't want to have sex with you, he's probably getting it somewhere else, unless there is a health issue that is preventing erection. But no matter what you tell him about your hormones, he won't believe that the lights are off because the power is out. As I speak with more women in the Upgrade, they tell me that this very difference in hormonal state of mind has been a source of tremendous stress and sadness in their relationship.

With the exception of the hormonal storm of new love, male and female sex drives are almost always unmatched. Just as ours tanks, though, he might be taking Viagra or testosterone, making the gap even wider. My patient Michelle is a sixty-three-year-old teacher still working full time. Her husband, sixty-nine, is still running his packaging and shipping business. He is taking testosterone, and there is a brewing resentment about how his needs shape her life. "If I don't organize my day so that we have sex every morning, our relationship becomes a living hell," she said. "He's awful to be around. Once I get into it, it's okay, and it's nice to have an orgasm, but I don't crave it as much anymore. I'd rather just get out of bed in the morning and get on with my day."

For many women in the Upgrade, a vibrator does just fine for having an orgasm. "You lose the need to have a face in front of you when you have an orgasm," said Doris, eighty-one. "And the desire for one also becomes less urgent." Another woman, when asked if she minded if her husband slept with other women, said, "Not anymore. It's one less time I have to have sex."

The dynamic of the conversation about the gap can be dehumanizing for both partners. He can feel as though we are reducing him to a brutish animal, derided for having a drive that we no longer have. And his insistence on needing to *put it in* can make us feel like a blowup doll.

There is no place for women to talk about the standoff going on in so many homes. "I developed a repertoire of excuses," says Janet. "It's too late, and you know I have so much trouble sleeping. . . . I'll have to get up and wash because I keep getting bladder infections, and I'm just too tired. . . . I don't feel good about my body. . . . I'm exhausted. . . . I have a headache. . . . My back/hip/knee/neck hurts. Sergio started snapping back sarcastically with things like 'Oh, wait, there is a speck of dust on the floorboards. I have to clean it, so I can't have sex.'"

We need permission to open the dialogue so that we can understand each other. Without sex, he will feel crabby, unwanted, panicky, and unloved, just as we do when he shuts down and shuts us out. That same feeling of destabilization and fear that the relationship is ending when he won't talk is exactly what he feels when we don't want to have sex. It's how his brain is wired. And by fifty-five, his testosterone has been dropping for fifteen years, and the lower level drags his confidence down with it. He'll be looking for more reassurance from you because of it, and that emotional reassurance comes from the physical intimacy of making love.

"Sergio and I learned to talk about his need as an emotional yearning for connection," said Janet. "We were able to see that we suffered similarly through lack of intimacy of different kinds. When he shuts down and won't talk, I feel panicked about whether or not he loves me. I finally got it that when we don't make love, he feels panicked about whether or not I love him. So when he asks for connection, for feeling loved again, for feeling at peace, then I can respond to that." Since our sex drive isn't motivated by hormones anymore, it has to be activated by other means. Emotion and connection can do it for women. The men in the second half who learn this will get laid more often by using language she can hear. "I need sex [for my health or for my prostate]" will get the door slammed in men's faces.

Talking, connection through intimate conversation, is what sparks joy in the female brain, even during the Upgrade. Janet started calling it "girl sex." "Sergio and I had to learn to balance both girl sex and boy sex, and that means we've had to make sure we have quality time together every day," she continued. "If we each get our different needs for connection met, things are good between us. If we lose one aspect, we both get crabby."

Before the quiet of the Upgrade, the storm of hormone spikes in the Transition can kick up libido in wild bursts. For Beth, forty-seven was a crazy year. "I split with my husband, and I was like a teenager. I found my inner Samantha. She took over and I couldn't get enough. I got into bed with my boyfriend on a Sunday afternoon, and by the time I looked at the clock again, six hours had gone by." She was experiencing a double hit of hormones: Transition waves and the novelty that comes with a new partner. "It was short-lived," she said with some sadness in her voice. "By the time heavy bleeding started, I could feel the drive draining out of me. After the Transition was over, so was my libido."

Regardless of phase, Transition or Upgrade, and regardless of age, hormones can still kick up in your seventies and eighties with new love. "I'm not going to stick with the schlub I'm seeing," said Jean, "but for now, we are going at it like teenagers." Another patient of mine, a widow at seventy-two, was dating the husband of her late best friend. After a few months she broke it off, saying, "I'm just not that sexually attracted to him. It wouldn't be fair to continue, because I want him to have that joy in his life. I'm just not that into him sexually. I'm not into mercy screws." She wanted chemistry and wanted that for him too.

I opted not to keep taking testosterone when mine dropped. I tried it and didn't like how irritable it made me feel. It felt counter to the Upgrade in how it pushed the brain to hold on to the striving of the fertility phase. Just think what it would be like if, in the second half, you had the sex drive of a nineteen-year-old. How would it change the priorities of each day? We'd think more about sexy clothes, makeup, and seeking male attention, all the things we did to find partners during the fertility phase. I realized that wasn't how I wanted to live in my sixties and beyond. It was one thing to be consumed by these drives when I had my whole life ahead of me. It's another when I can count the average remaining years in front of me on my fingers and toes. I can remember my mother saying to me and my friend Janet over lunch when I was in my forties and she was in her midsixties, "Girls, frankly, I'm not so interested anymore, and I've had enough good sex to last me the rest of my life." In the Upgrade we have the chance to escape the control of hormone-driven thought and behavior. Like everything else, it's a personal decision, different for everyone.

Stage 3: Nothing Left to Lose

"My granddaughter turned to me one day," said Sylvia, "and asked the most surprising question. She wanted to know if my proximity to death was changing the way I thought about living."

"And what did you say?" I wondered how she'd address this concern of an eleven-year-old.

"I told her of course it changes everything," Sylvia responded. "How could it not?"

She saw the surprise on my face. "I had to be straight with her," Sylvia continued. "The big lesson for me was watching Robert die of cancer. He was so frightened, so completely freaked out. I had tried to talk to him about dying for many years, but he just kept slamming the door on the conversation. It was torture seeing him go through it with so much distress. But it's understandable. We don't talk about it in our culture. We don't examine it. We spend our lives pretending it won't happen. And I didn't want my granddaughter to grow up feeling the same way."

Facing change, loss, and death is another developmental phase that is not mapped, whether it occurs in the second half or, tragically out of time and place, early in life. Grief over dramatic change, including the loss of our own lives, is inevitable. Psychology doesn't treat any of this as a developmental stage but as episodic, as though it were optional, something surprising that might or might not happen across what is assumed to be the ideal of a steady flow of growth. Change and loss are seen as hiccups instead of what they really are: the actual condition of life itself. Change and loss happen with every heartbeat and breath, and without them, there would be no life, no nervous system and brain to power us. Without change and loss, the generation of neurochemical processes that influence our feelings, thoughts and actions, body and mind, would be dead, static, frozen. The opposite of alive. The opposite of engaged.

Change and loss manifest in every stage of life, and in the second half of life, all the discomforts of disease, illness, aging, and limitations are on steroids. Lost roles, lost partners, lost children, lost communities, lost friends, and confrontation with the reality of losing your own life. All of it can feel like an endless series of tornadoes crashing through our emotional lives. Relationships, homes, and the cities we live in change. The older we get, the more dramatic change we experience, in ourselves and our loved ones. "I lost eight people in one year," Natalie told me, reacting as though this were inconceivable. It's not. It's normal. It's life; it's death.

But to the nervous system, dramatic change is inconceivable. The nervous system craves certainty and stability in order to do the one job it has: keeping us alive. Yet the craving of the nervous system is in direct opposition to reality, and that is a major source of suffering. The more we accommodate the nervous system's delusion, the harder it is to manage the inevitable when it arrives, the harder it is to allow the tent of "Me" to adjust to rapidly shifting circumstances. Psychology may point out this issue, but it doesn't provide effective tools for shifting our experience of permanence to match the reality of impermanence.

There is no sugarcoating that life ends in death. Since the job of the brain and nervous system is survival, there will be nothing viscerally comfortable in this statement or in its exploration. Just the mention fires up the adrenals, squeezing out a massive flood of threat hormones, turning the survival instinct on full blast in the brain. The natural response is to try to close the door on the source of discomfort. But turning away can spiral us out into a debilitating death anxiety. When fear goes underground, it has a nasty tendency to be transformed into chronic stress. Persistent denial shuts off our problem-solving capacity. Instead of recruiting our creativity for addressing reality, our ingeniousness ends up scaffolding more delusion, like *I will not die, somehow.* Distress builds and reality breaks through, no matter how hard we try to block it out.

I get it. It's hard to face death. We would rather live in denial and take the consequences. But to complete the Upgrade, there is a quality-of-life and quality-of-death decision to be made. There is growth and opportunity right up until the very end. But first we have to look the fear right in the face. If you're already experiencing some strong feelings and need to take a break, go ahead. If you're ready to ask the hard questions, let's go.

Stage 4: Facing the Monster Under the Bed

Terri remembered asking her mother at age three if everybody dies. "My mother said yes. Then I asked her if I was going to die. She responded, 'Yes, but not for a very long time, so don't worry about it.'" What that fairly typical conversation does is teach avoidance as coping: Just don't think about it. "And I remember," Terri said, "that I became much more anxious even in that moment. That fear resurfaced as full-blown anxiety when I faced big health issues in my twenties."

After a lifetime of the worry echoing in our brain circuits, death anxiety can take control of our lives, wrecking the time we have left by keeping us from being present to what's right in front of us as fear becomes a filter for everything. It's not fun for us, and it's not a picnic for those who end up being our fear's punching bags. The only way to deal with that anxiety is to open the door and face it.

"My friend Rosa had breast cancer that had metastasized to the bone, and she fought dying with all her might," said Terri. "She was miserable. She was in constant pain from useless procedures she demanded in the vain hope of prolonging her life. She screamed at everyone, alienating caretakers, friends, and family. It was awful to witness. Even at the end, she wouldn't let anyone use the word 'cancer' in her presence. She died terrified, clutching at the bed." If we don't look death in the face, the stress and threat systems remain in low-grade, chronic activation. Attuning to reality is part of how humans heal, especially when we can't be cured.

"I was in my early twenties and I was single in New York, just out of the hospital from a second major surgery, when the AIDS crisis was in full bloom," said Terri. "We lived in a constant state of terror. It was before we really knew how it was transmitted, and we were all afraid we'd picked it up somehow."

I remember those days in San Francisco. So many young people were suffering and dying. For those of us who stood on the sidelines, our personal fear collided with a global health crisis. For some it can be a recipe for collapse. But Terri chose another way.

"I decided to run headlong in the direction of my fear," she said. "I read everything I could. I learned about the newly emerging hospice movement. I sat at bedsides and held hands with friends I was losing. I did everything I could to be present for them as fearlessly as possible. I started to understand a bit more about the process, seeing how sensitive their nervous systems became, how much peace and solitude they needed. So I ran interference with difficult family members and made sure rowdy friends who might trigger painful memories had short visits. I left false optimism, false hope, and the imposition of my own grief at the door. It's the last thing a dying person needs. It might make us feel good, but it stresses them out."

The psychological intervention cognitive behavioral therapy (CBT) is particularly effective in addressing anxiety and full-blown phobias. One of the tactics is called exposure therapy. In her twenties Terri, like many of us, hadn't experienced the loss of many people in her life. To address her fear, she instinctively sought out the experience in her own form of exposure therapy. Familiarity gave her the courage to begin to open the door to her own mortality.

When I got sick in London during medical school and no one knew what was wrong with me, I was so depleted I felt I might die. But being gravely ill at a developmental stage when your peers are getting revved up for life can be isolating. It was for me, and it was for Terri.

The courage to face death anxiety is different in the Upgrade. We have a database of experience: loss of friends and family, our own brush with death, or a frightening series of tests. We've spent years wondering what's going to take us out, and at this stage we know it's going to be something sooner or later. We've participated in plenty of frightening dress rehearsals. In the aftermath, the adrenals power down, the parasympathetic nervous system turns on, we relax, we breathe, we smile, and we sleep . . . until reality sets in once again. It may feel counterintuitive to walk right into the heart of anxiety when everything in your brain and nervous system may tell you that facing death will bring it on faster. But just as with the monster under the bed, we have to look at it in order to quell our fear.

I was thinking back to a friend's grandmother, who didn't have any spiritual beliefs but read Sherwin Nuland's bestseller *How We Die* and said it gave her peace. Nuland, a physician who was also my professor at

Yale, wrote of the various ways the body shuts down as it approaches death. Knowing what could take place physically helped my friend's grandmother attune to what was actually happening instead of clinging to what she wished were happening. Her internal adjustment sparked a shift from the crankiness of loss to the joy of the present. Her family noticed the difference in her personality. She was more fun than ever to be around. She wrote out her medical directives and had her DNR (do not resuscitate) order taped to the wall above her bed, so that if a caregiver called an ambulance, she'd have a better chance of having her wishes respected. When she realized she was near death, and her son needed to go abroad for work, she told him to go. "I might not be here when you get back," she said, "but I'll see you another time."

Stage 5: Facing the Dying Brain and Body

When my mother was dying, my brother was told to take her home so that she could be in familiar surroundings and receive hospice care. For days all four siblings talked with her, held her hand, stroked her head, and tried to make her feel loved. My sister and I helped her into the bathroom one morning so that she could have a bowel movement. We brought her back to bed, and after some time the hospice nurse had us leave the room. Shortly after we complied, our mother passed. In our desire to comfort her and make her feel loved, our touch had caused too much peripheral stimulation, which had to be quieted in order for her to leave. Many years of practice and lots of anecdotal evidence make it clear that people need solitude and they need their loved ones to give them space to focus on their own process. It's hard for loved ones of the gregarious. It's a reversal that may feel like a rejection, but their need for you not to be there is a reflection of how deep their attachment is.

I watched how my mother changed, how totally peaceful and loving she became, how open and accepting of everyone she came into contact with. The kind of emotional and spiritual growth she went through made it clear that facing death is a developmental stage. If we don't align with the reality of it, we miss a huge opportunity to enlarge our playing field as wisdom fills our tent of "Me." We can be very brave and compassionate when we are fully, joyfully, tragically present.

When we don't understand the developmental stage, we will say and do the wrong things for others at every turn. "My friend Rosa," said Terri, "when she was dying of breast cancer, limited visitors to just a few of us. It was too hard for her to be around those who were out of tune with her process, who couldn't be present to it." Feelings were hurt because of lack of understanding.

The outer signs of dying—the sunken cheeks, eyes, and temples, the translucent pallor of the skin, the glassy eyes and lack of focus—are not hard to spot. Having to pee a lot, a feeling of heaviness in the limbs, inability to swallow easily, are all signs that the body is shutting down. The nervous system becomes very sensitive to others' moods and energy, harsh sounds, bright lights. The late stages bring a slowing of the breath, some difficulty breathing, and the need for lots of sleep. There can be a last-minute surge of energy and clarity—many report long, deep, meaningful conversations hours before a loved one departs.

Researchers are beginning to collect stories of the less tangible indicators of the proximity of death. Buffalo, New York, hospice physician Christopher Kerr has compiled more than a thousand cases for a peer-reviewed study indicating the commonality of experiences at the end of life. Of particular interest to him has been waking, conscious experiences of interactions with loved ones who have already passed. These experiences are real to those who have them, and they occur while wide awake. He learned during his experience with the AIDS crisis that no matter how a patient seemed to be doing physically, as soon as they had one of these experiences, it meant that death was near, almost always within days or weeks.

As doctors and researchers collect more information and first-person accounts of the dying, new questions are being asked about the relationship between the brain and consciousness. For more than a century, medicine has labeled consciousness as an epiphenomenon of the brain, meaning it is assumed that consciousness arises out of the brain, but there is no explanation for why this happens. The other assumption is that the brain must be in a state of neurochemical and structural organization and awake for consciousness to function. But the advent of resuscitation medicine in the 1970s has begun to shift the thinking.

Near-death experiences (NDEs) have been recorded across time and culture, but the number of people having them grew exponentially in the last third of the twentieth century. More and more people were emerging from surgery or resuscitation with detailed knowledge of conversations and instruments used, as well as reports of spiritual experiences that were practically the same. In fact, one British researcher, Sam Parnia, who is a professor at New York University's Langone Medical Center, worked closely with ICU doctors in New York to study the connection between consciousness and wakefulness. They came into rooms in which patients were not awake and hung paintings for a period of time. They removed the paintings before the patients awoke, and a surprising percentage asked their caregivers what had happened to the paintings. Consciousness was working without wakefulness, and Parnia will be setting up more studies to confirm the findings.

In death, the heart stops but the brain doesn't at first. Connections between synapses happen through electricity generated by neurochemicals. Brain cells suck up the last of the blood's oxygen and hang on as long as they can. There is a surge of serotonin, which may be responsible for a tremendous feeling of peace reported by many. When the oxygen runs out, a tsunami of reactions halt the neurochemical electricity in the cells. The lights, so to speak, are turned off pretty much all at once.

It was always assumed that when the brain becomes disorganized in this process, consciousness shouldn't be able to function. But it does in NDEs. Neuroscience is proving that you can be awake and not be conscious. No one is saying that people have died and come back, but we are able to confirm that consciousness is occurring at a moment when not only is wakefulness impossible but life itself might not be possible. And in those experiences, people commonly report exquisite clarity, expansive and immersive feelings of love and compassion, and a visceral sense of the continuation and vastness of consciousness itself.

Maybe it's strange that a neuroscientist is talking about these experiences. It feels a bit odd for me to be exploring how the brain and biology are intertwined with spiritual experience. For someone who has spent most of my life finding a material cause for feeling states, this is new. It is the brain and nervous system that mediate our experience of reality, our visceral sense of what is real, right, and true, so it makes sense that there might be spiritual answers too in neurochemistry and neurobiology. The question of consciousness and the brain has opened a whole new field of contemplative neuroscience and exploration into why NDEs and even the use of psilocybin (the psychotropic chemical in psychedelic mushrooms) can wipe away the fear of death for many.

I haven't made up my mind about a lot of it, but as a scientist I think it's important to remain open to mystery.

Stage 6: Planning for a Good Death

Developmentally we spend our late childhood through early adult years expanding and practicing the use of the prefrontal cortex (PFC) to solve, plan, decide, anticipate risks and rewards, all in the process of solving the puzzle of what makes up a good life. We do it so that we get our priorities straight and so that our days and relationships are full of quality. Our health is absolutely part of the equation, but improving our health span adds only about two and a half years to our life span. I think it's equally important to quality of life to ask what makes a good death, to calm the fear circuits and begin to recruit the PFC to the task.

If family and friends were to say to each other, "She had a good death," what would that mean? In my family, it meant that all of the siblings were able to be peaceful around our mother, to show our love for one another during the last two weeks of her life. My mother told us what she wanted, and we worked on her memorial together, her at the kitchen table, me on the floor, patting her knees. She said she couldn't believe we were planning her final transition together, and she made it clear she didn't want anyone to wear black or to mourn. She wanted a joyful, loving, God-filled celebration of her life, so I went out and bought a hot-pink minidress with rhinestones. My sister wore a bright woven jacket my mother had brought back from Guatemala. No one behaved badly, and that had a lot to do with the tone of unconditional love set by my mother. It allowed all of us to share a space of reverence, respect, and dignity, to be present enough to feel the enormity of her passing.

We can each harness the power of daydreaming, imagination, and innovation that we have used to build a good life and visualize a snapshot of the deathbed scene we want, the feelings we want our loved ones to come away with. What would that look and feel like for us?

Maybe we can begin with what we don't want it to feel like, the burdens we might have been left with that we'd like not to place on our loved ones because of unfinished business. Birth and death, our two big transitions, often happen in an overly medicalized hospital setting. Maybe that's what we want, to fight medically until the last. Or maybe being surrounded by strangers; bright lights; whirring, beeping machines; constant procedures; and separation from loved ones might cause too much distress. There is no such thing as a perfect death. But what makes a good death for you needs to be defined and talked about. It's as important as a good life.

There are so many practice opportunities if we stop to think. A mammogram, waiting for clarifying results of an abnormal test, even a checkup can be a source of much more anxiety at this stage. We've been through it with others, holding their hands, seeing the different outcomes. To be clear: not the same as going through it yourself. Those were dress rehearsals. Your own is not. Facing it down yourself is another matter.

There are lots of resources on how to manage the phase when the doctors give up on you, to grieve the loss of your own life, and to map out palliative care. But how the visceral brain-and-nervous system aspect is influenced by emotion and thoughts is important to respect. I know I want to find my way to calm, peace, and clarity. In his book *Good Life, Good Death*, Tibetan lama Gehlek Rimpoche wrote that a good death begins with a good life, one of practicing "patience, love, and compassion in your daily life." And regardless of our habit, he strongly recommends we plan and think about it well ahead of time, because when we are actually dying, it will be too late to say what we need to say or to create the circumstances for the transition we want to have. There are communities, death cafés, death doulas, and more who can shed light on the process, guides who can help us learn to focus on creating the most supportive internal and external environments. By connecting with the helpers who can address the terrors in the night, we are triggering calming neurochemicals of bonding; mirror neurons come into play as we begin to match the emotional state of an experienced guide. The brain and nervous system can support the Upgrade right up until the very end. Working with this is a rest-of-your-life process.

I WILL UPGRADE!

Chapter 16

Women are born with pain built in. It's our physical destiny—period pains,
sore boobs, childbirth. We carry it within ourselves throughout our
lives. . . . We have pain on a cycle for years and years and years, and then
just when you feel you are making peace with it all, what happens? The
menopause comes. The fucking menopause comes and it is the most
wonderful fucking thing in the world. Yes, your entire pelvic floor crumbles
and you get fucking hot and no one cares, but then you're free. . . . You're
just a person.

—*Belinda,* Fleabag, *season 1*

There is a reason this soliloquy has been repeated and shared relentlessly
since the moment it aired: It's true. For me, hearing it was a turning point
in getting this book finished. The Upgrade is an exquisite portal into the
best years of our lives. I wrote it to start, not end, the conversation about
it, and I hope we will all contribute to creating a new map of the second
half together.

I've learned through my journey that the Upgrade is not about reaching some mythical plane of awareness. It's a deeper engagement with and

understanding of this life, this world, this body, this mind. It's remembering to listen to the part of myself that knows the truth of who I am and how things work, the truth of my relationships, successes, and failures, all that I deafened my ears to during the reproductive years in order to keep the peace, the status quo. Blocking out the truth made me at times feel like I was about to lose my mind, and I never understood why that feeling dogged me. But with the earplugs of the fertile phase removed, I can learn to hear what my own will is asking for again, lifting the lid of internal strife created by the competing priorities of what my biology was dictating and who I am as a woman.

The pacemaker of your nervous system and brain changes in the Upgrade, providing a beat that supports emerging out of who we thought we were in the fertile phase into who we actually are. I find I have become deeply interested in improving the tent of "Me," facing, accepting, and adjusting faults; finding ways to bring out the best I have to give, in order to be effective in offering it up every day.

For me, human potential, personal growth, has become real in the Upgrade. I finally have the willingness to see how some of my old emotional and behavioral patterns bring suffering to others and to myself; that realization is compelling me to find effective ways to change. In wanting to be the best version of myself, I seek out a reliable plan, a program with clear goals, clear results, and support from others who have succeeded in doing what I am attempting to do.

There are people to learn from in a way that is different from how we have learned in the past. I find I have to be willing to be shaken to my core, to have a fearless look at my own tent, at what doesn't serve anymore, to weed out the parts of my character that are harmful and consciously fertilize the good parts, helping them grow. The result is more peace, more joy. I have earned the right to pursue this in the Upgrade.

I most often see patients for the first time during a crisis in their thirties or forties, and almost always a huge part of solving their issue is around hormones and/or antidepressants. When they are stable enough, they leave therapy. Often, some fifteen years later they will circle back, having been brought to their knees again by life's events. They are past the fertile phase of hormonal waves, and so I no longer start with medications or hormones for most. I listen carefully to hear what remains broken from the past. I also look for where they have already stepped onto the path of the Upgrade and find ways to turn their attention to that path in order to continue. I want to know what they are doing to care for themselves. We consider meditative and spiritual resources to help them fertilize the best in them and to weed out what's choking their garden. I do my best to help them disable the hamster wheel of worry. I begin showing them how to wire feelings of safety and self-compassion through meditative techniques instead of looking to medication to temporarily create those pathways. By using cognitive practices, they more easily find the strength to fix the broken parts and to continue the Upgrade.

Scientists have now shown that it's possible to dial up or down just about any cognitive or behavioral trait—aggression, compulsion, sociability, learning, memory, compassion, gratitude—in the brain's soft wiring by means of various interventions, whether chemical or meditative. Prozac can make the shy bold and the solemn cheerful. Compassion meditation can melt the fearful into courageous openheartedness. If some may see this as manipulative of our own genetic or epigenetic destiny, well, so be it. The tent of "Me" is more under our control than most of us realize. And I celebrate this widening of possibility.

There is no road map for making changes to our nervous system and brain; I'm building my own and sharing it with you to encourage you to build your own as you go. What are the qualities we value, and what neural support would we like to give ourselves? This brain is OUR brain, this story is OUR story; it's up to us to ask ourselves who we want to be in the Upgrade and to learn what is possible to achieve.

Like dieting, attending to shifting what's in the tent of "Me" requires sticking to a plan and accepting that change doesn't happen overnight. But if we spend as much time on the Upgrade as we did on weight control and appearance in the fertility phase, our tent of "Me" will be awesome.

Upgrading takes courage, but without it we walk backward into old age, obsessed with the past. Going down memory lane, reminiscing in retirement, is an old model in which we abandon growth and shrink into senility. I want to walk wholeheartedly into the future, with my eyes on the way ahead.

Fear kills microglia, the cells in the brain we need to keep it nourished and clean. When remodeling the tent of "Me," it's important to consider the impact of fear and look for the helpers who can support us in unearthing the courage we need. There will be people who won't want you to remodel your tent. They will fight you. You may have to move them out of the inner circle, away from the tent pole. What I've found is that those who support us in the way we need often show up rather magically.

Pregnant with Compassion

When the huge responsibility for family and little kids is unplugged, it unleashes decision-making bandwidth as never before. When the near ones don't factor so heavily into daily life choices, the love brain has a chance to expand. The neurocircuitry of the love brain and tent of "Me" can now enlarge. What's included will feel new, weird, and maybe even selfish: What's best for "Me" takes priority in a new way; and the impulse to act on what's best for others outside the family can fill us with a new motivation to act.

The spirit of altruism seeps into the second half, sometimes for the first time. You see suffering outside the family and want to help. The grandmothering impulse, the nurturing maternal instinct, may creep in even for women who haven't had their own children. I've seen it drive women in the Upgrade on a mission to become a peer role model, to share what we've learned with other women in this phase of life. It's a key imperative, to walk this road with enough people so that it is no longer a semisecret trail in the woods but a superhighway with very clear signage. In order to be ready to embrace this role, we have to think about our health. We give primacy to longevity and health because we have work to do. We are needed to mother the world.

The first part of life was primarily about becoming self-conscious, and wanting to come out ahead in comparison to others. The Upgrade is about unwinding self-consciousness into a generosity toward ourselves and others, shifting the focus to becoming who we wished we'd met when we were younger. Becoming, for others, the person we needed to help us through life.

In the optimized Upgrade, women fight for women. We get daughters out of bad marriages; we help our granddaughters rebel, our nieces succeed. I know with my nieces I am often a sounding board, once in a while the source of some much-needed extra cash in an emergency, and a neutral shoulder to cry on in the wake of some bad decisions. Being real rather than nice becomes a gift to others.

In the Upgrade we learn to refuse to accept unacceptable behavior from others—and from ourselves. The old rhythms of the fertility phase, the traces of those habits, will always remain, ready to be pricked in the circuits of my heart, gut, and brain. So I am vigilant to make sure I counterbalance those pulls by enhancing the circuits of the Upgrade. I've learned to meet my own needs more and stop expecting that they be met by others. I've worked on building the confidence to tell the truth and being willing to pay the price for speaking my truth. I've learned to tell my story without flinching, and I've become a boundary-setting ninja. I've learned to say what I mean and mean what I say, and try my best to do it without saying it mean. It's better for me and it's better for everyone around me.

Reaching out to the younger generation becomes an imperative. I think of Elizabeth Warren's tireless selfies and pinky promises with little girls as she asked them during her 2020 presidential campaign to step into power when they grow up.

Becoming the Woman We Needed to See

The Talmud states, "Do not be daunted by the enormity of the world's grief. Do justly now, love mercy now, walk humbly now. You are not obligated to complete the work, but neither are you free to abandon it."

So often I've felt in the Upgrade that we come into a full appreciation of being female just as society writes us off. But there is a vast historical library of female vocal authority in Eleanor Roosevelt, Shirley Chisholm,

Dolores Huerta, Carol Moseley Braun, Ann Richards, Toni Morrison, Jane Goodall, Maya Angelou, Angela Merkel, Ruth Bader Ginsburg, Sonia Sotomayor, Marie Yovanovitch, Fiona Hill, Stacey Abrams, Greta Thunberg, and Kamala Harris, America's first female vice president. Even *we* forget when we hear those voices.

It's not easy to step into our power. When we don't introject secondary status as expected, we are brutalized from all sides. If you don't believe me, name one woman who stepped out of her gender role who hasn't been attacked outrageously. What's different in the Upgrade is how much less the familiar onslaught gets under our skin.

With the quieting of the ovaries' caretaking hormones we have the opportunity not to let any of it cut us deeply, as it did during the fertile phase, when oxytocin would spark an apocalyptic fire in the nervous system, telling us the world was coming to an end if one single person disliked us. In the Upgrade, we can speak our minds, knowing we will not be destroyed by the fire we draw.

In another era we would have been called a battle ax. Now we are badasses. We are becoming leaders in the Upgrade. What do some of our most formidable, fearless leaders have in common? They're women who came to the fullness of their powers on the other side of menopause, and now they're running Congress (Nancy Pelosi), running for president (Elizabeth Warren, Kamala Harris, Amy Klobuchar), sitting on the Supreme Court (RBG, rest in power, Sonia Sotomayor, Elena Kagan). Think of Christine Lagarde, Patti Smith, Ruth E. Carter, Sister Helen Prejean— all forces to be reckoned with as younger women, but none of them as deeply visionary, as thoroughly glorious, as when they got to the other side.

No longer the crazy old crone in the attic, we are emerging into a world in which we are taking more control than ever.

I WILL UPGRADE

I'm tired of having yet another bar to rise to, I found myself thinking about the Upgrade. But what I've learned is that seeing that I'm worth more than I ever imagined allows the Upgrade to flow naturally. I just need to get out of the way of its unfolding. I want to give myself the best chance for a bright future, and I hope you give yourself and the women around you that chance too.

Like everyone, I drag my previous selves behind me. Although I occasionally look back, I try not to stare. I've learned to love myself in my own eyes, not in everyone else's. I feel gratitude for having this once-in-a-lifetime opportunity to make the transition into the Upgrade.

My friend Janet went through many crises in her early years: disastrous relationships, dire illness, dislocation of work taking her to a country she never wanted to live in. In the Upgrade she sat alone in a restaurant in Paris having a last indulgent meal before heading into a retreat. "And I told my younger self," she recalled, "who was so freaked out by all the change, not to worry, that everything would be just fine. It's not that my outer circumstances had changed. It's that inside I was at peace with whatever happened."

Age wrinkles your skin; loss of enthusiasm wrinkles the soul. I will not let my being shrivel. I hope you won't let yours either. I was never entirely ready for anything in my life, and I'm not always ready for the Upgrade either. But to get myself psyched up, I thrust my arms up to the sky and shout, often, to no one and everyone, "I WILL UPGRADE!" It helps trigger the brain and nervous system to generate enthusiasm for becoming who I want to be: less jealous, more open, a warmer heart, a wiser mind.

Visualize your future, your Upgrade. Start by thinking about how you want to feel and who you want to be an hour from now. Work your way up to a day, a week, a month, a year. Visualize and fake it till you make it. It will materialize if you can see it. It will materialize if you remain willing to engage in solving the puzzle of your life until the very end. I look forward to meeting you in the Upgrade.

Appendix

I wanted to create a few shortcuts so that you'd easily be able to find the actionables that emerge from the book. I hope this might help you create your own plan for optimizing your cognitive and emotional superpowers in the Upgrade.

Daily: Before getting out of bed, I wiggle my toes and smile, practice EFTs, essential first thoughts: say good morning to God, meditate, and pray that I might be of use to others by being kind, easygoing, and calm. I ask for the courage to be direct, honest, and patient; and to keep my mouth shut when I shouldn't speak, especially when people need to learn things on their own without my interference. I stretch, squeeze my butt cheeks (to strengthen glutes), and get bright morning light for fifteen to twenty minutes (in winter I use a full spectrum lamp); I do moderate exercise in pool aerobics, reclining bike, dance on my deck, and eat a Mediterranean diet with fifteen hours of intermittent fasting (no food after 6:00 P.M.). I supplement with vitamin D_3, magnesium, CoQ10, vitamin K, and a fiber supplement. I take my medications: rosuvastatin (a statin drug), levothyroxine (thyroid hormone), and an estradiol patch. I aim to go to sleep at 10:30 P.M. and wake at 7:00 A.M.

Upgrade Principles to Honor

- Circadian rhythms: Sleep is key to cognition and control of inflammation. If you can manage to make a habit of getting to bed by 11:00 P.M. at the latest—in bed with screens off at 10:30 P.M.—you will set yourself up for success the following day. You'll be more alert and more energetic, and you'll want to engage in the movement your brain needs to power cognition.

- Waking up one hour earlier than usual, and sleeping earlier, gives you a double-digit improvement in mood and protection from depression.

- Move to stay alive: It will improve your mood, preserve cognition, battle melancholy, reduce inflammation, and remind your brain and body that you still need them to keep pumping because you've got a lot left to do in the world. If you can't take HRT, movement, even if it's just your arms, is your cognitive safety net.

- Eat for the Upgrade by defeating inflammation and feeding your muscles: Limit carbs, grains, and fruits; lower animal fat as much as possible; rely mainly on protein, healthy fats, and veggies for optimal functioning.

- Stimulate your memory by staying engaged and by exercising. Squeeze your glutes 100 times a day. Engaging and strengthening this key core muscle for standing, balancing, and walking will help you reap maximum cerebellum benefits.

- I believe in snacks! Movement and meditation snacks, that is. Don't make change insurmountable by thinking you have to run a marathon or start a monthlong silent retreat tomorrow. Start with "snacks." Movement snacks of ten minutes—stretching, a quick walk, some quick work with weights—can be the foundation for building a new habit. The same goes for meditation: If you haven't done it before or very much, the mind's attention muscles won't be strong, so expecting to remain calm and focused for more than a minute or two is unrealistic. Even the Dalai Lama recommends starting with thirty seconds a few times a day. A friend of mind uses the Breathe app on her Apple Watch for daily reminders.

- Calm the nervous system: It's going to be key to clearheaded decision making. Sleep and exercise are huge; so are breathing and meditation. Recommendations for practices are in "Supportive Modalities for the Upgrade" below.

Set Yourself Up for Sleep

- Go to bed at the same time every night.
- Get at least fifteen minutes of bright morning light every sunny day, forty minutes on cloudy days (use a full spectrum light lamp in winter).
- Do something physical that makes you a little tired before 2:00 P.M., cardio and strength. You need to stay active so that your brain registers that your nerves and muscles are still in strong need of its services. Otherwise both your brain and muscles start to weaken. There's a strong correlation between leg strength and mental acuity over age eighty: The stronger your legs, the sharper your mind.
- Focus on eating as many nonstarchy vegetables as you can manage in a day, and have lean protein with every meal; consume under 100 grams of carbs per day. You can use a carb counting app to track. Learn to eat to best feed the body, and see if you can't take a hard look at emotional eating—I know when I reach for ice cream and peanut butter, I'm not actually hungry, so I try to distract myself to keep from setting off harmful inflammation. If you keep the body happy and healthy, you don't want to wreck it.
- Eat eggs, turkey, or cottage cheese a couple of times per week; choline from egg yolks helps with sleep and memory, as does tryptophan in turkey and cottage cheese.
- Consume omega-3 fatty acids that cross the blood-brain barrier. I eat salmon several times a week because most omega supplements

don't cross the blood-brain barrier as well as food does. You can get omegas in chia and flaxseeds too.

- Minimize caffeine. My one cup of coffee in the morning is two-thirds decaf, one-third caffeinated, and that's it for the day, unless I'm determined not to sleep!
- Alcohol doesn't suit. It causes hot flashes, makes your hands red and swollen, and makes you sleepy at the wrong time. Two hours after your last drink the brain becomes alert. Mostly, you'll fall asleep quickly, then wake in the middle of the night and have a helluva time getting back to sleep. If you really want to drink, lunch on a weekend is best. To keep people from thinking I'm antisocial at dinners, I let the host pour, and I may take a sip or two. If pressed, I'll say something like, "I'll feel better if I don't."
- Do intermittent fasting five days a week: Finish dinner by six or seven, exercise in the morning on an empty stomach, and eat breakfast around ten or eleven.
- Sleep seven to eight hours per night.
- Sleep with an eye mask and custom earplugs, and keep the room between sixty-five and sixty-eight degrees. It helps prevent wake-fulness from ambient light and sounds in the room, and a low temperature helps you sleep.

Apnea

Be on the lookout: If you are sleeping through the night but waking up exhausted, if someone tells you that you snore, if you fall asleep within minutes of sitting down during the day, if you have a hard time staying awake through meetings or movies, if inflammation markers and blood pressure are rising, and if you're carrying more weight than is healthy, consider getting a sleep study to see if you might have undiagnosed sleep apnea. It's an incredibly dangerous condition in 10 to 20 percent of women that can cause heart attacks, brain fog, and strokes and has been linked to diabetes and cancer. Don't take it lightly. If your doctor prescribes a CPAP, wear it. It can take time to find one that's comfortable. Don't give up.

SSRI: Use for Hot Flashes and Depression

Estrogen is the gold standard for treating hot flashes, but for those who cannot take it and have severe hot flashes, an SSRI or SNRI is worth trying and in some women can reduce hot flashes by as much as 65 percent. They can also help with sleep. The SSRIs identified in studies as helpful are paroxetine (Paxil or Brisdelle), paroxetine extended release (Paxil CR), citalopram (Celexa), and escitalopram (Lexapro). Venlafaxine (Effexor XR) was identified as a potential first-line SNRI. Paroxetine extended release was the most effective at doses of both 12.5 milligrams per day and 25 milligrams per day. Venlafaxine worked more quickly on symptom relief than the SSRIs did, but in study participants there were more side effects like nausea and constipation. Fluoxetine (Prozac) and sertraline (Zoloft) work too. SNRIs may increase blood pressure, so if you have an issue with blood pressure, use extreme caution. For depression without hot flashes, buproprion (Wellbutrin), psychotherapy, and CBT (cognitive behavioral therapy) are well established treatments. The NAMS (North American Menopause Society) recommendations include the following SSRIs and SNRIs: paroxetine salt (Brisdelle) 7.5 milligrams per day; paroxetine or paroxetine extended release 10–25 milligrams per day; escitalopram 10–20 milligrams per day; citalopram 10–20 milligrams per day; desvenlafaxine 50–150 milligrams per day; and venlafaxine (Effexor XR) 37.5–150 milligrams per day. I agree with their recommendation to start at the lowest available dose and raise it in tiny increments if needed. I like working with the SSRIs you can get in liquid form, which allows me to individualize the dose more easily. All of them now come in liquid except Brisdelle, which comes only in 7.5 milligram doses. It is the only non-estrogen drug that is FDA approved for hot flashes. If you do start an SSRI or SNRI, remember to keep a regular and detailed mood and side-effect log. Some people have had dangerous emotional reactions. If you find yourself having suicidal ideation, talk to someone immediately, and get back to your doctor for help getting off or changing the drug. Don't attempt to stop it on your own. If no one is around, call the Samaritans: 116 123 (from any phone) or visit samaritans.org.

Alternate-Nostril Breathing

Known as nine-round breathing in some Buddhist circles, maybe you've been guided through this at a yoga class as well. It's a pretty widespread technique used in contemplative traditions for getting the mind to settle before meditation, prayer, or even before needing to focus at work.

Round 1: Sit up straight but relaxed. If you can sit on a cushion or the edge of a chair without support, it's great, but if you need support, please use it. Start by using your right hand to block your right nostril. Breathe in as slowly and quietly as possible through the left nostril. At the top of the breath, use the right hand to block the left nostril, and breathe out through the right nostril as slowly and quietly as possible. Breathe in through the left and out through the right three times.

Round 2: Without taking a break (unless you feel dizzy, in which case, stop, of course), switch the process, using your left hand to cover your left nostril. Breathe in through the right nostril as slowly and quietly as possible. At the top of the breath, use the left hand to block the right nostril, and breathe out through the left nostril as slowly and quietly as possible. Breathe in through the right and out through the left three times.

Round 3: Again without taking a break unless you are straining or are dizzy, breathe in and out through both nostrils three times at the same slow and quiet pace.

You can add a visualization of breathing in good, healing energy and breathing out a dark, smoky cloud of what no longer serves. You can do the nine rounds once or repeat them three, seven, or even twenty-one times.

Nurturing Moment Meditation

The following is an excerpt, reprinted with permission, from "The Compassion Shift," part of a twenty-one-day compassion challenge hosted by the Compassion Center at Emory University. It's an introduction to the foundational practice of Cognitively Based Compassion Training (CBCT). You can learn more at compassion.emory.edu. You can also find a guide to brief daily meditations there.

When we feel safe and cared for it opens the door to possibility. Connecting with a personal experience of receiving kindness calms the body and mind, helping us to extend care and kindness to others.

Description

Born to a fifteen-year-old addict partnered with an abusive drug dealer, LaTonya Goffney's childhood was full of violence and uncertainty until she was nine. She found refuge at school, under the care of teachers who recognized her brilliance. She flourished after moving in with her grandmother, and went on to get a degree in education. As a superintendent, Goffney has embraced the challenge of turning around some of the most troubled school systems in Texas. The care and safety she received allowed her to thrive and to extend care and safety to other children, many times over.

Adversity can knock us out of what psychologists call our Zone of Wellbeing. The Zone of Wellbeing is an optimal physical and psychological balance. When our biological threat systems are calmed, the release of neurochemicals makes us feel strong and at peace, able to weather life's ups and downs. The Zone of Wellbeing is where we experience resilience, the ability to bounce back from adversity. We might experience it after a good night's sleep, a relaxing vacation, praise from a colleague, the embrace of a loved one, a joyful welcome from a beloved pet, a peaceful moment by a river, lake, or sea, a calming walk in nature, or an inspiring spiritual setting.

Having ready access to feelings of safety and emotional warmth bolsters resilience—our ability to bounce back from upsets and challenges, large or small. When we are able to recall or imagine feelings of refuge, nurturance, peace, or freedom, and when we remind ourselves of the benefits of these feelings, we can begin to have some influence over reentering and remaining more reliably in our Zone of Wellbeing. Research has shown that when our nervous system is balanced, our health improves, along with our ability to remain calm and think clearly even in difficult times.

Awakening to the value of kindness, of being safe and cared for, we become ready to venture beyond ourselves. If we can abide for a moment in a sense of security, it gives us the foundation to imagine how wonderful it would be if others could also feel safe and secure. It can motivate us to commit more wholeheartedly to compassion.

Before we begin the meditation, **Connecting to a Moment of Nurturance,** let's engage in a three- to five-minute writing exercise. Take a moment to find something to write with. Choose, or imagine, a moment of feeling secure, comforted, or nurtured. Don't worry about the story that came before or after that instant. Just see if you can isolate in your mind a soothing moment sitting by a river, receiving kindness from a stranger, support from a friend. Maybe it's a time in the arms of a loved one, or a family member. If you are alive today, you had a moment like this, whether you remember or not. Someone took care of you. Someone fed and nurtured you. So imagine what that might have felt like to be held and protected, safe, and comforted. Or remember how you felt holding a small child, or a time you were able to offer welcome comfort to someone in distress. Perhaps it is like when you dance, holding your partner close. Perhaps it is the experience of unconditional love from a pet who welcomes you home with joy. If you grew up near the sea or a lake, think of the peace of floating on calm pristine waters. If that moment of safety and security comes from a spiritual source, evoke that presence or figure, the place of worship, or the community itself.

Once you have chosen your moment of nurturance, write down as many details as you can remember or imagine, especially the sights, smells, sensations, textures, the feeling of the presence of others if they are involved.

With these in mind, let's get ready to meditate.

Meditation

Please take a moment to find a comfortable posture and connect with your body and current feelings. If you notice tension in any part of the body, feel free to stretch or move gently to help relax. You may close your eyes or keep them slightly open. Feel your body in your seat, and allow yourself to settle into the present experience.

When ready, take a few deep breaths if it is comfortable to do so. Gently inhale, and if you like, have the sense that nourishing air, rich with oxygen, is infusing your entire being. As you breathe out, see if you can release tensions and worries to some degree to allow the body and mind to settle into an unfolding sense of calm or ease.

Let's now take a moment to bring to mind a nurturing moment: something that makes us feel better, safer, or happier. When have we experienced that? In nature? While being cared for—by a friend or loved one, or a mentor, or a figure from a faith tradition? If such a resource does not come to mind, see if we can just imagine a person, or an environment, that would support feeling safer or better.

Choosing one, let's take a moment to immerse ourselves in this experience by bringing it to mind as vividly as we can. Where is this scene happening? What do we see; which colors, textures; what's the light like? What about the surroundings? Are there sounds? Sensations? Are there scents in the air?

If this is a moment of shared kindness with others, do we recall facial expressions or body language, or do we hear a comforting tone of voice?

Let's continue immersing ourselves in this nourishing moment for a minute or so.

Having connected with this nurturing resource, let's now bring attention to our body in the present moment, noticing sensations and feelings. Has anything shifted? If we find pleasant or neutral sensations, such as a warmth in the chest, relaxation of the shoulders, a smile on the face, we may rest with those. If instead we notice areas of discomfort, we may take a few breaths to settle the body and mind, or direct our attention back to the nurturing resource for a few moments, or shift our attention to a different part of the body that feels better.

Finally, let's reflect: How important are such moments of comfort and safety for our well-being?

And how important are acts of kindness and compassion to create a safe and secure world where our fellow human beings can thrive?

Seeing the value of kindness and compassion, how might that shift our perspective on our own life and our relationships with others?

Let's dedicate our practice today to those we know to be in need of health and well-being and, as we are able, expand this dedication to include a widening circle of beings on this earth.

And let's conclude by setting an intention to extend the skills and insights from this practice into everyday life.

Melancholy, Medicine, and the Transition

It's often hard to know if what you're experiencing requires an antidepressant, HRT, a thyroid supplement, or vitamin and mineral supplements. Intractable sadness, irritability, sleeplessness, and more can be signs of any of these things. If you've had depression before you can be more vulnerable now. In a moment when you won't feel ready, you'll have to take charge of getting the right tests and tracking symptoms when you take anything new. You'll need to be proactive about working with your doctor to make adjustments to doses of antidepressants and hormones. During the Transition and in the Upgrade, ask for these tests:

- Thyroid: TSH, free-T4-to-T3 conversion
- Vitamin and mineral levels: vitamin B_{12}, vitamin D
- Post-Transition: free estrogen, free testosterone, and DHEA-S

Keep a log of how you feel when you start taking something new. Often something that is supposed to help can make you feel worse.

If you do start an antidepressant or HRT, ask your doctor for forms that can allow you to track your dose to find your own sweet spot.

Remember that the simplest way to know if you've hit the Transition is by seeing that your cycle shortens by one or more days two or more times in a year.

Tracking Your Moods: How to Make a Log

Make a few columns for time of day, ideally morning, afternoon, and evening, to track mood on a 1–10 scale: Am I feeling happy (10) or sad (1)? Optimistic (10) or pessimistic (1)? Also three times a day, note energy or a feeling of zest for life on a scale of 1 to 10, 10 being totally engaged and energetic. Track mental clarity an hour before and two hours after taking any progestin or SSRI, 10 being super clear, 1 a total fog. Also track libido daily (raring to go being 10; "Are you kidding?" being 1), note how frequently you feel hot or sweaty, and in the morning, write about your quality of sleep the night before. How many times did you wake? How long did it take to go back to sleep? How many times were you hot or sweaty? Note the time and dose of all medications and supplements so you can see if they are interacting or if any of them might be at the root of some overwhelming good or bad feelings. Having your journal as hard evidence of your mood prior to taking hormones or medication can be a lifesaver.

After the first few visits, I tracked with my patients the top three issues for them weekly, noting how they characterized them. For example, brain fog, lack of joy, irritability, anxiety, insomnia, tearfulness, anger, depression, and so forth. Because I was able to see what they kept track of it was much easier for me to adjust medication each time in a way that was more targeted to their actual individual symptoms. Remember: NSAIDs can block SSRIs's effectiveness. When you've charted these things for six to eight weeks, you can figure out what is going on and have data to help you adjust the dosage accordingly. With most of my patients, we'd find their new sweet spot after about three to six months. With this kind of information in hand, you have a better chance of making it happen sooner.

The Tricky World of Supplements

Check manufacturers of supplements for independent testing. Supplements are not regulated, and they don't always contain what's advertised. And beware the hype, since no FDA approval is required to make sure they do what they claim to do.

Vitamins and supplements can interact with one another, with your prescription drugs, and with your OTC drugs. And you may have sensitivities that are unexpected. Many women tell me they take a quarter of what's recommended and that suits them just fine. You'll need help from a qualified professional to figure out what works and what doesn't. This area is still new and under-researched, so be ready to sort through conflicting advice.

As with all medicines, keep a log of symptoms for everything new that you take. Some things that are meant to make you feel better can make you feel worse. For example, some probiotics can cause brain fog. Don't be quick to blame symptoms on stress or on the way the stars are aligned. It might be that new substance causing the problem.

My approach in general is that less is more, so I stay away from taking too many things preventively. *Harvard Women's Health Watch* warns for example that calcium supplements may make some women as much as seven times more likely to develop dementia, and we are already at higher risk in later years. If it ain't broke, don't fix it. Keep the pill popping to a minimum. (With all the hype I know that's hard.)

Taking Charge of Your Prescriptions: Checking the Beers Criteria Anticholinergic Scale

It happens like this: You visit a doctor to resolve reflux and are given medication for it. Then you hurt your arm and start taking an NSAID. Your stomach gets upset and you get medicine for that as well. Side effects pile up and symptom whack-a-mole snowballs with more medication. You get foggy and disoriented, and another doctor decides it's dementia because in tests your cognitive levels have clearly declined. This problem, polypharmacy, has to stop. And it may rest on your shoulders to stop it.

Every visit to the doctor has to include a review of all medication and supplements. Take a complete list of all your pills, patches, creams, and supplements. You can start your own review by researching a few scales, including the Anticholinergic Cognitive Burden, the Drug Burden Index–sedative component, and the Drug Burden Index–anticholinergic component, as well as the Beers criteria for inappropriate medications.

Remember, choline is needed for healthy brain function. Many drugs are anticholinergic, including those given for incontinence. Incontinence in many cases can be resolved through physical activity and physical therapy. See "The Vagina Gym" below.

When Common Conditions Are Mistaken for Dementia

If you or a loved one seems to be sliding quickly into strong cognitive decline, check these common conditions before jumping to conclusions about dementia. These are conditions in which disorientation and memory loss can be corrected.

- Urinary tract infections: As we age, we don't feel the symptoms as strongly, the burning and discomfort from peeing. The infection doesn't get caught as quickly, and in the body's effort to fight it, there will be inflammation throughout the body and brain that can cause cognitive issues. An antibiotic can clear this up within days.
- Urinary retention: If you just started a new medication and you can't pee or you find it a struggle, this could be an anticholinergic effect, and cognitive impact may be a simultaneous symptom. This can even be caused by taking high doses of paracetamol recommended by a doctor for pain after a procedure or surgery.
- Depression: Symptoms of disorientation may seem like dementia, but they may instead be signs of early clinical depression. An SSRI or SNRI may clear the confusion caused by depression.
- Drug or alcohol side effects, particularly the anticholinergic side effects of many drugs discussed above.
- Subdural hematoma: Did you hit your head and now find yourself losing your balance easily or sitting a bit lopsided? It could be a brain bruise, which is easier to develop as we age. Get to the doctor and have it checked. This is nothing to mess around with. It can be reversed if caught soon enough.
- Normal pressure hydrocephalus (NPH): This happens when extra spinal fluid seeps into the crevices of the brain but doesn't raise intracranial pressure. The excess fluid can interrupt brain function. It can be treated by placing a simple shunt to drain the excess fluid.

Eat Your Way to an Upgrade

There are a lot of crazy, extreme diets out there, but over time, two principles for optimal nutrition have held true: the Mediterranean diet and meal timing.

Intermittent Fasting: It's All in the Timing

Cleo, fifty-one, had been running for years. She ran ten miles daily—if she couldn't schedule a single long run, she ran to and from work, five miles each way. She was always a healthy eater, but somehow she kept gaining weight, and the mental fog just wouldn't lift. She'd been thin all her life and she was now thirty-five pounds overweight. She was getting sadder by the minute. She didn't know if it was body image or something else that was making her feel so down.

Humans did not evolve eating three big meals and two snacks a day around a fire. They were kinetic, always on the move, they were grazers, and they probably weren't overweight. Emerging evidence shows that eating all your daily calories within an eight-hour window, making sure there are at least twelve to sixteen hours between dinner and breakfast, is great for metabolism and in promoting brain health. The reason for extending daily fasting is to mobilize your body fat to be metabolized. When you don't eat, your insulin goes down. The liver's glycogen reserves are available as fuel for the brain for only twelve hours after your last meal. When that runs out, the brain turns to using stored fat in other parts of the body for fuel. The brain doesn't have its own fat, so it prompts the body to break down fat and transport it through the blood supply.

On the advice of a nutritionist, Cleo began intermittent fasting (IF) combined with a Mediterranean diet. She started eating two big meals a day—and one snack: waiting until noon for her first and finishing dinner by 7:30 P.M.—and if she was headed out for an extra-long morning run, she would have a spoonful of nut butter and a small cappuccino beforehand. Not only did she lose the thirty-five pounds, but her brain fog lifted. "The weight melted off," she said, "and I feel sharp and alert again. Like my old self." Researchers at MIT found that this approach stimulates neural stem cells, which are key for replacing old, worn-out cells—keeping your brain resilient and flexible. And it has been shown to help you lose more fat and retain the muscle that triggers the brain to keep cognition, emotion, and judgment at peak levels.

Leaving twelve to sixteen hours between dinner and breakfast stresses the metabolism in the way that weights and exercise stress your heart and muscles in order to strengthen them. A little bit of metabolic stress from fasting can end up making your cells and tissues stronger and more resistant to disease. And anything that reduces chronic inflammation makes the gut-brain team function at its best. That in turn improves a range of health issues from arthritic pain to asthma, while lowering the risk for cancers by clearing toxins and damaged cells

Pack On the Protein

One last piece of the puzzle is protein. You know by now that movement and cognition are linked, and muscle strength is key to remaining active. Protein is the nutritional key to maintaining muscle, which in turn helps us keep our clarity and memory and lifts our mood.

As we age, we lose muscle strength faster, so you actually need to eat *more* protein than when you were younger. It is estimated that 41 percent of adult women have dietary protein intakes below the recommended daily allowance (RDA). In a 2018 study that followed more than 2,900 people over sixty-five for more than twenty-three years, researchers found that those who ate the most protein were 30 percent less likely to become functionally impaired than those who ate the lowest amount. Even if you have to drink it, get your protein intake up to between 60 and 90 grams per day, depending on your size. It will take planning to get that much into your diet. You'll need to save room and calories for that second piece of fish at lunch or dinner.

Healthy Microbiome: Honoring the Gut Brain

Keeping inflammation in check means a healthier microbiome. Too much sugar/carbs causes inflammation, and researchers around the world are learning that artificial sweeteners might be even worse. Most of the sweeteners—including natural forms like stevia and birch xylitol—cause alterations in the microbiome, creating an environment for the growth of unhealthy bacteria that can invade the bloodstream. Sugar substitutes can also send signals to activate the brain's dopamine circuits. This is the intense pleasure/reward system; overstimulation is the mechanism at the heart of addiction. Satisfying the cravings in turn can trigger glucose intolerance, making us crave more sugar, setting up the condition that leads to diabetes. Diabetes, or prediabetes, equals chronic inflammation. Chronic inflammation equals brain shrinkage. Brain shrinkage equals cognitive decline: a downgrade, not an Upgrade.

Eating too much; eating high-salt fast foods, low-fiber foods, fried foods, trans fats, sugars, or chemical additives in processed foods; nicotine; drinking alcohol; poor sleep quality; and being sedentary can all destroy the healthy microbiome. If we are having carb cravings, it could mean we don't have friendly bacteria signaling contentment to the brain, and so the quickest way to remedy that is to goose the brain into boosting the feel-good chemicals, dopamine and serotonin, which are most easily triggered by sugars. The high doesn't last long, and pretty soon we are either exhausted on the couch or reaching for caffeine or another candy bar. It's a vicious cycle that disables the all-important gut brain.

Add to the list antibiotics and laxatives; even the chlorine and fluoride found in some tap water can be dangerous to the microbiome. Chlorine evaporates after a while at room temperature, but you can boil your water to remove it more quickly. Treasure your microbiome. Feed and cultivate it carefully. Work with your specialist to find the right fermented foods and/or probiotic for you if your microbiome needs repopulating.

Actionable Numbers on Inflammation

Your doctor can help you gather the information you need to understand how much work you might have to do to battle inflammation and preserve health and cognition. Here are some markers to track via blood work and genetic testing that can help you make good decisions on diet and supplements:

- Heart: A calcium scan can help you understand how much animal fat you can eat and if you need a statin.

- Brain: Check for the APOE4 gene variation to make an informed decision about HRT. If you have even one mutation and have a family history of dementia, it might influence you to add estrogen during the Transition, since we know it can protect cognition.
- Hemoglobin A1C, C-reactive protein, triglycerides, LDL, VLDL (very low-density lipids): These tests can indicate diabetes, heart and vascular issues, and general "zombie-cell" inflammation. If any of these numbers is off, cut the sugars/carbs/unhealthy fats and emphasize lean protein and nonstarchy vegetables. Consider a statin if LDL is over 100.
- Hemoglobin, hematocrit, and red blood cell width: If you have heavy bleeding during the Transition, it's very likely you're anemic. Most doctors check hemoglobin and hematocrit levels to determine whether or not you need an iron supplement. But there is another test, red blood cell width, that indicates whether or not the bone marrow has caught up in red blood cell production. That measurement will tell you if your anemia is actually resolved.

Remember, the body's inflammation goes higher naturally with age, so it's more important than ever, if you want an Upgraded brain, to keep inflammation under control. The keys to doing this are:

- Getting enough sleep
- Exercising to exhilaration, not exhaustion
- Relying on nonstarchy vegetables and lean protein for most of your nutrition
- Avoiding alcohol, sugar, too much salt, fried foods, and overeating and overexercising

Be alert to new discoveries. They used to tell you not to eat eggs. Now eggs are considered a source of healthy omegas and their yolks a particularly good source of choline (great for the brain). That gets confusing. On Monday you hear A; on Friday you hear not A. Both could be true, because A will work for one woman while it will harm another. What has panned out over time is what I have emphasized here, though keep watching for new and helpful markers that are beginning to come into sight.

The Vagina Gym

In other news of adjustments down there during the Transition, if you've had babies and/or pelvic surgery, you're probably familiar with the panic that comes just before a cough or a sneeze, or the moment you have to run to catch a green light or a bus. *How much pee is going to come out? A little drop, or a noticeable squirt?* "I just can't bring myself to buy Depends or Poise pads," says Christina, sixty-five, who still runs to try to keep the post-Transition weight off. After three kids she is having trouble holding her pee. "It makes me feel too old to buy anything in the incontinence aisle, so I use maxi pads." Christina, however, is on her third round of antibiotics for a stubborn urinary tract infection. It's not a surprise. Pads for your period are designed to hold a lot less liquid. If you use menstrual pads, the urine will overwhelm the pad's capacity and the liquid will stay close to your skin. Can you say "middle-aged diaper rash"? Christina's recurring UTIs are most likely being made worse by not using the right equipment.

Christina doesn't want to change her habits, but targeted exercises like those in a barre class will help tone the pelvic floor and help her hold her pee. Pelvic floor physical therapy—and yes, sometimes it means they manipulate muscles from inside the vagina—can cure the problem for some women. When I broach the subject with her, she wants neither. "I hate barre classes. Running keeps me thinner. And I'm not letting anyone poke around down there unless I have to let them."

Insistence on running while not addressing pelvic floor issues or not remaining sexually active also can be a recipe for prolapse of the vagina, the rectum, the bladder, the uterus. This happens more often than you think, and women sometimes wait years before telling their doctor they are having a problem. Meanwhile, these issues can destroy a woman's confidence and intimate life, putting her at risk of depression. She doesn't want surgery, which entails pinning organs back into place, even though the results—putting you back into social settings you've been afraid to enter—are life-changing. So get it fixed if other things haven't worked!

You have choices in how you meet changes like this. You can fight them, resist them, deny what's happening, hide them from everyone, and take the well-trodden path to a downgrade. Or you can turn to face the road less traveled, work with what you have, and create the conditions for making this time of your life an Upgrade. We have agency. Peeing when you sneeze is not inevitable. How our destiny unfolds is up to us.

Estrogen builds thickness and lubrication in mucous membranes, so as estrogen declines, so will vaginal lubrication. Even with HRT it might be time to break out the lube, like Slippery Stuff, and you may need vaginal estrogen cream. And wash the minute you are done. The skin is more sensitive to small tears, and as men age and their semen becomes more acidic, your body will like it less. You'll be much more prone to UTIs. Yes, this will all take away some spontaneity, but it's better than nothing. Find a way to make a game of it.

If you're having trouble with vaginal dryness or painful intercourse, estrogen inserted up to twice weekly can be super helpful. You'll find it in the form of a cream like Estrace, Estring, Vagifem, and others, or in vaginal suppositories or rings. Ask your doctor for a prescription. The upside of vaginal insertion is that these forms of HRT have been shown to be safe even for women with a breast cancer risk.

Supportive Modalities for the Upgrade

We know you'll find your own way to support movement, nutrition, and nervous-system maintenance for the Upgrade, but people I trust have found a few things that have been helpful. I can't guarantee results, and you'll have to monitor your own risks, but maybe here's a start:

- **Nutrition and cognitive change:** Noom, a weight-loss app, has been helpful in shifting attitudes toward food and health and sustaining healthy diet, exercise, and emotional habits.

  If you just want a weight, nutrition, and calorie counter, Lose It! is a great app for that.

- **Exercise for pelvic floor strength, joint lubrication, and core strength:** Xtend Barre classes on the Openfit app. Many of the other programs on Openfit may be a bit too intense, but Xtend Barre, XB Pilates, and XB Stretch have great basics for fitness and for strengthening our weakest parts.

  Since many of us are more sedentary than we think, getting a Fitbit or an Apple Watch and using the activity app will give great feedback on how much you're moving and how many calories you're actually burning through exercise, as opposed to what you think you're burning. Yes, most of us overestimate, eat more than we burn, and then decide that exercise doesn't help us lose weight.

- **Nervous-system maintenance and conscious Upgrading:**
  Body scans: These relaxation techniques are great for destressing, resetting, and sleeping. There are many to be found on YouTube, Netflix, Spotify, and Apple Podcasts. Headspace or Calm is a good place to begin if you've never done anything like this before.

  Learning to focus: Concentration meditation is different from relaxation. We're not looking to just calm; we are also looking to sharpen, train, and steady the mind. Once we accomplish that, we can start to focus on change and developing the qualities we've always wanted to have as the woman we've always wanted to be. The compassion protocol in the next paragraph also has great training for steadying the mind.

  Cultivating compassion: Emory University has developed a protocol that does it all: calms the nervous system, steadies the mind, helps you find compassion for yourself, and teaches you to develop compassion that energizes, keeping you from burning out from the world's grief.

For help dealing with long-haul Covid brain fog issues: Visit https://www.verywellhealth.com/dealing-with-covid-brain-fog-5209460.

For help dealing with teen or adult children who struggle with addiction or mental health issues, Al-anon has an excellent twelve step program for parents: Visit www.al-anon.org and search for "parents programs".

Acknowledgments

This book had its beginnings when several fans of *The Female Brain* reached out to me and wondered if I would please do a book on the next phase of a woman's life, after fertility and into the years of empowerment and wisdom. At first I was skeptical, since all the medical science focused only on what goes wrong in the second half, but I realized that no one had yet mapped out these stages of development for women. So here it is, *The Upgrade*, and I have so many to thank for what I have learned, what I am becoming every day. To anyone I missed, you know who you are: thank you.

I want to thank my UCSF colleagues, supporters, and mentors: Cori Bargmann, Lynne Krillich Benioff, Marc Benioff, Liz Blackburn, Jennifer Cummings, Mary Dallman, Alison Doupe, Dena Dubal, Elissa Epel, Laura Esserman, Adam Gazzaley, Anna Glezer, Mindy Goldman, Lyn Gracie, Mel Grumbach, Steve Hauser, Dixie Honig, Holly Ingraham, Cynthia Kenyon, Joel Kramer, Rob Malenka, Sindy Mellon, Michael Merzenich, Bruce Miller, Nancy Milliken, Tammy Neuhouse, Thomas C. Neylan, Kim Norman, Faina Novosolov, Aoife O'Donovan, Christina Pham, Ricki Pollycove, Sandy Robertson, John Rubenstein, Alla Spivak, Brandon Staglin, Garen Staglin, Shari Staglin, Matt State, Marc Tessier-Lavigne, Owen Wolkowitz, and Kristine Yaffe.

I want to thank my UC Berkeley teachers, mentors, and colleagues, especially Frank Beach, Marian Diamond, Peter Hornick, Daniel Mazia, Clyde Willson, and Fred Wilt.

I want to thank my Yale teachers, mentors, and colleagues, especially Marilyn Farquar, Florence Hazeltine, Stanley Jackson, Eric Nestler, Sherwin Nuland and Philip Sarrel.

I want to thank my UC London and Wellcome Institute teachers, mentors, and colleagues, especially Bill Bynum, Charlotte MacKenzie, Roy Porter, Janet Thompson, and Richard Wollheim.

I want to thank my Harvard teachers, mentors, and colleagues, especially Mary Anne Badaraco, Myron Belfer, Beth Blessing, Herb Goldings, Kathy Kelly, William Meissner, George Valliant, Bessel van der Kolk, and Peggy Wingard.

I want to thank my other colleagues in the women's neuroscience field, especially Roberta Brinton, John Cacioppo, Larry Cahill, Lee Cohen, Neill Epperson, Stephani Faubion, Jill Goldstein, Jennifer Gordon, Victor Henderson, Melissa Hines, Sarah Hrdy, Hadine Joffe, Claudia Kawas, Eleanor Maccoby, Martha McClintock, Bruce McEwen, Pauline Maki, Lisa Mosconi, Barbara Parry, Jennifer Payne, Natalie Rasgon, David Rubinow, Peter Schmidt, Barbara Sherwin, and Deborah Tannen and Myrna Weissman.

I want to thank my friends who keep me going in my own Upgrade: my Saturday girlfriend lunch gang: Jennifer Curley, Mickey Berg Kordelos, Lin Repola, Glenda Seidel, Maria Carini Sneed, and Marty Wolf; my women's dinner group gang: Kimberely Cameron Brody, Maggie Cox, Juannie Eng, Katherine Honig, Kerry King, Sandy Kleiman, Susan Lopes, Kary Schulman, and Brenda Way; my Mill Valley photography group; my go-to writers: Julie Margaret Hogben, Jena Pincot, and Michelle Stacey; my Mill Valley meditation group; my Jewel Heart meditation teachers and group; my aquatics class teachers and teammates, Patricia Chytrowski, Peggy Conrad, Kathy King, Suzanne Paynovich, and Marianne Wilman; my girlfriends who always pick up the phone when I call: Diane Cirrincione, Alice Corning, Jean Hietpas, Lisa Klairmont, Adrienne Larkin, Sharon Agopian Melodia, Sue Rosen, Wendy Shearn, Mary Sherman,

Nancy Todes-Taylor, and Jody Yeary; my mother's dearest girlfriend from childhood, Caroline "Maury" Knight, who has shared some juicy stories; and my dear friend the late Janet Durant.

I want to thank my literary team at Harmony; my wonderful editor Donna Loffredo, whose skills have made this book what it is; my agent, Elizabeth Kaplan, whose dedication, availability, and advice have been everything I could hope for and more.

And a big shout-out and thank-you to all the fans of *The Female Brain* for your support, letters, and questions over the years. With every letter and question you have challenged me to hear different voices, to learn more so I can become more helpful along our Upgrading path.

I would like to thank my family, especially my sister Diana Brizendine, who supports me every day with her love and prayers, my loving son John Whitney Brizendine, whose support and lightning-fast computer skills helped me finish this book. And to my brother Buzz Brizendine, in memory of my sister Paula Brizendine Kuntz, my stepdaughters Jessica and Elizabeth Barondes, uncles Tom and John Brizendine and aunts Kath and Shirl Brizendine, and nieces Jessica Johns, Morgan Brizendine, Rachel Llanes, Nicole Kuntz, and Louisa Llanes, and nephews Ryan Brizendine and Derek Kuntz who cheered me on. The memory of my mother, Louise Ann Brizendine, who is with me every day and knew what was important in life.

I would like to thank my husband, Samuel Barondes, whose love, wisdom, and support keep me going. He is my gem.

It's hard to put into words my appreciation and gratitude to my writing partner, editor, and friend Amy Hertz, whose professional skills, talent, and breadth as a person have challenged and guided me through the process of writing not only *The Upgrade* but also *The Female Brain*. Without her neither of these books would have come to fruition.

I especially want to thank all those patients whose stories I told but who, for confidentiality reasons, must remain anonymous. Their struggles and joys in Upgrading are gifted to you to help guide you in your own Upgrade.

Selected References

Author's Note

Davis, Emily J., Iryna Lobach, and Dena B. Dubal. 2018. "Female XX sex chromosomes increase survival and extend lifespan in aging mice." *Aging Cell* 18 (1). doi: 10.1111/acel.12871.

Santoro, Nanette, and John F. Randolph. 2011. "Reproductive hormones and the menopause transition." *Obstetrics and Gynecology Clinics of North America* 38 (3):455–66. doi: 10.1016/j.ogc.2011.05.004.

Sripada, Rebecca K., Christine E. Marx, Anthony P. King, Nirmala Rajaram, Sarah N. Garfinkel, James L. Abelson, and Israel Liberzon. 2013. "DHEA enhances emotion regulation neurocircuits and modulates memory for emotional stimuli." *Neuropsychopharmacology* 38 (9):1798–1807. doi: 10.1038/npp.2013.79.

Taylor, Caitlin M., Laura Pritschet, and Emily G. Jacobs. 2021. "The scientific body of knowledge—whose body does it serve? A spotlight on oral contraceptives and women's health factors in neuroimaging." *Frontiers in Neuroendocrinology* 60 (6). doi: 10.1016/j.yfrne.2020.100874.

Chapter 1: Changing the Conversation

Archer, John. 2019. "The reality and evolutionary significance of human psychological sex differences." *Biological Reviews* 94 (4):1381–415. doi: 10.1111/brv.12507.

Luine, Victoria, and Maya Frankfurt. 2020. "Estrogenic regulation of memory: The first 50 years." *Hormones and Behavior* 121. doi: 10.1016/j.yhbeh.2020.104711.

Monteleone, Patrizia, Giulia Mascagni, Andrea Giannini, Andrea R. Genazzani, and Tommaso Simoncini. 2018. "Symptoms of menopause—global prevalence, physiology and implications." *Nature Reviews Endocrinology* 14 (4):199–215. doi: 10.1038/nrendo.2017.180.

Protsenko, Ekaterina, Ruoting Yang, Brent Nier, Victor Reus, Rasha Hammamieh, Ryan Rampersaud, et al. 2021. "'GrimAge,' an epigenetic predictor of mortality, is accelerated in major depressive disorder." *Translational Psychiatry* 11 (1). doi: 10.1038/s41398-021-01302-0.

Sartori, Andrea C., David E. Vance, Larry Z. Slater, and Michael Crowe. 2012. "The impact of inflammation on cognitive function in older adults." *Journal of Neuroscience Nursing* 44 (4):206–17. doi: 10.1097/JNN.0b013e3182527690.

Torréns, Javier I., Kim Sutton-Tyrrell, Xinhua Zhao, Karen Matthews, Sarah Brockwell, MaryFran Sowers, and Nanette Santoro. 2009. "Relative androgen excess during the menopausal transition predicts incident metabolic syndrome in midlife women." *Menopause* 16 (2):257–64. doi: 10.1097/gme.0b013e318185e249.

Wang, Yiwei, Aarti Mishra, and Roberta Diaz Brinton. 2020. "Transitions in metabolic and immune systems from pre-menopause to post-menopause: Implications for age-associated neurodegenerative diseases." *F1000Research* 9. doi: 10.12688/f1000research.21599.1.

Xin, Jiang, Yaoxue Zhang, Yan Tang, and Yuan Yang. 2019. "Brain differences between men and women: Evidence from deep learning." *Frontiers in Neuroscience* 13. doi: 10.3389/fnins.2019.00185.

Chapter 2: The Crux of Being Female

Hill, Sarah. *This Is Your Brain on Birth Control: The Surprising Science of Sex, Women, Hormones, and the Law of Unintended Consequences.* New York: Penguin, 2019.

Mueller, Joshua M., Laura Pritschet, Tyler Santander, Caitlin M. Taylor, Scott T. Grafton, Emily Goard Jacobs, and Jean M. Carlson. 2021. "Dynamic community detection reveals transient reorganization of functional brain networks across a female menstrual cycle." *Network Neuroscience* 5 (1):125–44. doi: 10.1162/netn_a_00169.

Paul, Steven M., Graziano Pinna, and Alessandro Guidotti. 2020. "Allopregnanolone: From molecular pathophysiology to therapeutics. A historical perspective." *Neurobiology of Stress* 12. doi: 10.1016/j.ynstr.2020.100215.

Pritschet, Laura, Tyler Santander, Caitlin M. Taylor, Evan Layher, Shuying Yu, Michael B. Miller, et al. 2020. "Functional reorganization of brain networks across the human menstrual cycle." *NeuroImage* 220 (4). doi: 10.1016/j. neuroimage.2020.117091.

Syan, Sabrina K., Luciano Minuzzi, Dustin Costescu, Mara Smith, Olivia R. Allega, Marg Coote, et al. 2017. "Influence of endogenous estradiol, progesterone, allopregnanolone, and dehydroepiandrosterone sulfate on brain resting state functional connectivity across the menstrual cycle." *Fertility and Sterility* 107 (5):1246–55.e4. doi: 10.1016/j.fertnstert.2017.03.021.

Weber, Miriam T., Mark Mapstone, Jennifer Staskiewicz, and Pauline M. Maki. 2012. "Reconciling subjective memory complaints with objective memory performance in the menopausal transition." *Menopause* 19 (7):735–41. doi: 10.1097/gme.0b013e318241fd22.

Witchel, Selma Feldman, Bianca Pinto, Anne Claire Burghard, and Sharon E. Oberfield. 2020. "Update on adrenarche." *Current Opinion in Pediatrics* 32 (4):574–81. doi: 10.1097/mop.0000000000000928.

Chapter 3: Transitioning into the Upgrade

Bluming, Avrum Z. 2021. "Progesterone and breast cancer pathogenesis." *Journal of Molecular Endocrinology* 66 (1):C1–C2. doi: 10.1530/jme-20-0262.

Boyle, Christina P., Cyrus A. Raji, Kirk I. Erickson, Oscar L. Lopez, James T. Becker, H. Michael Gach, et al. 2020. "Estrogen, brain structure, and cognition in postmenopausal women." *Human Brain Mapping* 42 (1):24–35. doi: 10.1002/hbm.25200.

Craig, A. D., K. Chen, D. Bandy, and E. M. Reiman. 2000. "Thermosensory activation of insular cortex." *Nature Neuroscience* 3 (2):184–90. doi: 10.1038/72131.

El Khoudary, Samar R., Gail Greendale, Sybil L. Crawford, Nancy E. Avis, Maria M. Brooks, Rebecca C. Thurston, et al. 2019. "The menopause transition and women's health at midlife." *Menopause* 26 (10):1213–27. doi: 10.1097/gme.0000000000001424.

Engel, Sinha, Hannah Klusmann, Beate Ditzen, Christine Knaevelsrud, and Sarah Schumacher. 2019. "Menstrual cycle-related fluctuations in oxytocin concentrations: A systematic review and meta-analysis." *Frontiers in Neuroendocrinology* 52 (6):144–55. doi: 10.1016/j.yfrne.2018.11.002.

Fan, Yubo, Ruiyi Tang, Jerilynn C. Prior, and Rong Chen. 2020. "Paradigm shift in pathophysiology of vasomotor symptoms: Effects of estradiol withdrawal and progesterone therapy." *Drug Discovery Today: Disease Models* 32 (1):59–69. doi: 10.1016/j.ddmod.2020.11.004.

Freedman, Robert R. 2014. "Menopausal hot flashes: Mechanisms, endocrinology, treatment." *Journal of Steroid Biochemistry and Molecular Biology* 142:115–20. doi: 10.1016/j.jsbmb.2013.08.010.

Goldstein, S. R., and M. A. Lumsden. 2017. "Abnormal uterine bleeding in perimenopause." *Climacteric* 20 (5):414–20. doi: 10.1080/13697137.2017.1358921.

Hanstede, Miriam M. F., Martijn J. Burger, Anne Timmermans, and Matthé P. M Burger. 2012. "Regional and temporal variation in hysterectomy rates and surgical routes for benign diseases in the Netherlands." *Acta Obstetricia et Gynecologica Scandinavica* 91 (2):220–25. doi: 10.1111/j.1600-0412.2011.01309.x.

Henderson, V. W., J. R. Guthrie, E. C. Dudley, H. G. Burger, and L. Dennerstein. 2003. "Estrogen exposures and memory at midlife: A population-based study of women." *Neurology* 60 (8):1369–71. doi: 10.1212/01.Wnl.0000059413.75888.Be.

Hodis, H. N., and P. M. Sarrel. 2018. "Menopausal hormone therapy and breast cancer: what is the evidence from randomized trials?" *Climacteric* 21 (6):521–28. doi: 10.1080/13697137.2018.1514008.

Nanba, Aya T., Juilee Rege, Jianwei Ren, Richard J. Auchus, William E. Rainey, and Adina F. Turcu. 2019. "11-Oxygenated C19 steroids do not decline with age in women." *Journal of Clinical Endocrinology & Metabolism* 104 (7):2615–22. doi: 10.1210/jc.2018-02527.

Pletzer, Belinda, Ti-Anni Harris, Andrea Scheuringer, and Esmeralda Hidalgo-Lopez. 2019. "The cycling brain: Menstrual cycle related fluctuations in hippocampal and fronto-striatal activation and connectivity during cognitive tasks." *Neuropsychopharmacology* 44 (11):1867–75. doi: 10.1038/s41386-019-0435-3.

Pouba, Katherine, and Ashley Tianen. 2006. "Lunacy in the 19th Century: Women's Admission to Asylums in the United States of America." *Oshkosh Scholar* 1:95–103.

Santoro, Nanette, and John F. Randolph. 2011. "Reproductive hormones and the menopause transition." *Obstetrics and Gynecology Clinics of North America* 38 (3):455–66. doi: 10.1016/j.ogc.2011.05.004.

Süss, Hannah, Jasmine Willi, Jessica Grub, and Ulrike Ehlert. 2021. "Estradiol and progesterone as resilience markers?—Findings from the Swiss Perimenopause Study." *Psychoneuroendocrinology* 127. doi: 10.1016/j.psyneuen.2021.105177.

Taylor, Caitlin M., Laura Pritschet, and Emily G. Jacobs. 2021. "The scientific body of knowledge—whose body does it serve? A spotlight on oral contraceptives and women's health factors in neuroimaging." *Frontiers in Neuroendocrinology* 60. doi: 10.1016/j.yfrne.2020.100874.

Weber, M. T., L. H. Rubin, R. Schroeder, T. Steffenella, and P. M. Maki. 2021. "Cognitive profiles in perimenopause: Hormonal and menopausal symptom correlates." *Climacteric* 24 (4):1–7. doi: 10.1080/13697137.2021.1892626.

Zorumski, Charles F., Steven M. Paul, Douglas F. Covey, and Steven Mennerick. 2019. "Neurosteroids as novel antidepressants and anxiolytics: GABA-A receptors and beyond." *Neurobiology of Stress* 11. doi: 10.1016/j.ynstr.2019.100196.

Chapter 4: Navigating the Wilderness

Allais, Gianni, Giulia Chiarle, Silvia Sinigaglia, Gisella Airola, Paola Schiapparelli, and Chiara Benedetto. 2018. "Estrogen, migraine, and vascular risk." *Neurological Sciences* 39 (S1):11–20. doi: 10.1007/s10072-018-3333-2.

Beral, Valerie, Richard Peto, Kirstin Pirie, and Gillian Reeves. 2019. "Type and timing of menopausal hormone therapy and breast cancer risk: individual participant meta-analysis of the worldwide epidemiological evidence." *The Lancet* 394 (10204):1159–68. doi: 10.1016/s0140-6736(19)31709-x.

Brinton, Roberta Diaz, Richard F. Thompson, Michael R. Foy, Michel Baudry, JunMing Wang, Caleb E. Finch, et al. 2008. "Progesterone receptors: Form and function in brain." *Frontiers in Neuroendocrinology* 29 (2):313–39. doi: 10.1016/j.yfrne.2008.02.001.

Bromberger, Joyce T., and Cynthia Neill Epperson. 2018. "Depression during and after the perimenopause." *Obstetrics and Gynecology Clinics of North America* 45 (4):663–78. doi: 10.1016/j.ogc.2018.07.007.

Cagnacci, Angelo, and Martina Venier. 2019. "The controversial history of hormone replacement therapy." *Medicina* 55 (9). doi: 10.3390/medicina55090602.

Cummings, Jennifer A., and Louann Brizendine. 2002. "Comparison of physical and emotional side effects of progesterone or medroxyprogesterone in early postmenopausal women." *Menopause* 9 (4):253–63. doi: 10.1097/00042192-20020 7000-00006.

"Dietary Supplements: What You Need to Know." National Institutes of Health Office of Dietary Supplements, September 3, 2020.

Edwards, Alexis C., Sara Larsson Lönn, Casey Crump, Eve K. Mościcki, Jan Sundquist, Kenneth S. Kendler, and Kristina Sundquist. 2020. "Oral contraceptive use and risk of suicidal behavior among young women." *Psychological Medicine*:1–8. doi: 10.1017/s0033291720003475.

Genazzani, Andrea R., Patrizia Monteleone, Andrea Giannini, and Tommaso Simoncini. 2021. "Hormone therapy in the postmenopausal years: considering benefits and risks in clinical practice." *Human Reproduction Update* doi: 10.1093/humupd/dmab026.

Gibson, Carolyn J., Yixia Li, Guneet K. Jasuja, Kyle J. Self, Karen H. Seal, and Amy L. Byers. 2021. "Menopausal hormone therapy and suicide in a national sample of midlife and older women veterans." *Medical Care* 59:S70–S76. doi: 10.1097/mlr.0000000000001433.

Gordon, Jennifer L., Tory A. Eisenlohr-Moul, David R. Rubinow, Leah Schrubbe, and Susan S. Girdler. 2016. "Naturally occurring changes in estradiol concentrations in the menopause transition predict morning cortisol and negative mood in perimenopausal depression." *Clinical Psychological Science* 4 (5):919–35. doi: 10.1177/2167702616647924.

Heath, Laura, Shelly L. Gray, Denise M. Boudreau, Ken Thummel, Karen L. Edwards, Stephanie M. Fullerton, et al. 2018. "Cumulative antidepressant use and risk of dementia in a prospective cohort study." *Journal of the American Geriatrics Society* 66 (10):1948–55. doi: 10.1111/jgs.15508.

Hill, Sarah. *This is Your Brain on Birth Control: The Surprising Science of Sex, Women, Hormones, and the law of Unintended Consequences.* New York: Penguin, 2019.

Kolata, Gina. "Rate of hysterectomies puzzles experts." *New York Times*, September 20, 1988.

Lobo, Roger A. 2016. "Hormone-replacement therapy: Current thinking." *Nature Reviews Endocrinology* 13 (4):220–31. doi: 10.1038/nrendo.2016.164.

Loprinzi, Charles L., Debra L. Barton, Lisa A. Carpenter, Jeff A. Sloan, Paul J. Novotny, Matthew T. Gettman, and Bradley J. Cristensen. 2004. "Pilot evaluation of paroxetine for treating hot flashes in men." *Mayo Clinic Proceedings* 79 (10):1247–51. doi: 10.4065/79.10.1247.

Maki, Pauline M., Susan G. Kornstein, Hadine Joffe, Joyce T. Bromberger, Ellen W. Freeman, Geena Athappilly, et al. 2019. "Guidelines for the evaluation and treatment of perimenopausal depression: Summary and recommendations." *Journal of Women's Health* 28 (2):117–34. doi: 10.1089/jwh.2018.27099.mensocrec.

Mazer, Norman A. 2004. "Interaction of estrogen therapy and thyroid hormone replacement in postmenopausal women." *Thyroid* 14 (S1):27–34. doi: 10.1089/105072504323024561.

Pokras, R., and V. G. Hufnagel. 1988. "Hysterectomy in the United States, 1965–84." *American Journal of Public Health* 78 (7):852–53. doi: 10.2105/ajph.78.7.852.

Rasgon, Natalie L., Jennifer Dunkin, Lynn Fairbanks, Lori L. Altshuler, Co Troung, Shana Elman, et al. 2007. "Estrogen and response to sertraline in postmenopausal women with major depressive disorder: A pilot study." *Journal of Psychiatric Research* 41 (3–4):338–43. doi: 10.1016/j.jpsychires.2006.03.009.

Russell, Jason K., Carrie K. Jones, and Paul A. Newhouse. 2019. "The role of estrogen in brain and cognitive aging." *Neurotherapeutics* 16 (3):649–65. doi: 10.1007/s13311-019-00766-9.

Singh, Meharvan, Chang Su, and Selena Ng. 2013. "Non-genomic mechanisms of progesterone action in the brain." *Frontiers in Neuroscience* 7. doi: 10.3389/fnins.2013.00159.

Skovlund, Charlotte Wessel, Lina Steinrud Mørch, Lars Vedel Kessing, Theis Lange, and Øjvind Lidegaard. 2018. "Association of hormonal contraception with suicide attempts and suicides." *American Journal of Psychiatry* 175 (4):336–42. doi: 10.1176/appi.ajp.2017.17060616.

Willi, Jasmine, Hannah Süss, Jessica Grub, and Ulrike Ehlert. 2020. "Prior depression affects the experience of the perimenopause—findings from the Swiss Perimenopause Study." *Journal of Affective Disorders* 277:603–11. doi: 10.1016/j.jad.2020.08.062.

Writing Group for the Women's Health Initiative, Investigators. 2002. "Risks and benefits of estrogen plus progestin in healthy postmenopausal women: principal results from the women's health initiative randomized controlled trial." *JAMA: The Journal of the American Medical Association* 288 (3):321–33. doi: 10.1001/jama.288.3.321.

Chapter 5: Renewal: Your Brain in Search of a New Reality

Freeman, Ellen W., Mary D. Sammel, David W. Boorman, and Rongmei Zhang. 2014. "Longitudinal pattern of depressive symptoms around natural menopause." *JAMA Psychiatry* 71 (1). doi: 10.1001/jamapsychiatry.2013.2819.

Gordon, Jennifer L., Alexis Peltier, Julia A. Grummisch, and Laurie Sykes Tottenham. 2019. "Estradiol fluctuation, sensitivity to stress, and depressive symptoms in the menopause transition: A pilot study." *Frontiers in Psychology* 10. doi: 10.3389/fpsyg.2019.01319.

Handa, Robert J., and Michael J. Weiser. 2014. "Gonadal steroid hormones and the hypothalamo–pituitary–adrenal axis." *Frontiers in Neuroendocrinology* 35 (2):197–220. doi: 10.1016/j.yfrne.2013.11.001.

Herrera, Alexandra Ycaza, Howard N. Hodis, Wendy J. Mack, and Mara Mather. 2017. "Estradiol therapy after menopause mitigates effects of stress on cortisol and working memory." *Journal of Clinical Endocrinology & Metabolism* 102 (12):4457–66. doi: 10.1210/jc.2017-00825.

Koch, Patricia Barthalow, Phyllis Kernoff Mansfield, Debra Thurau, and Molly Carey. 2005. "'Feeling frumpy': The relationships between body image and sexual response changes in midlife women." *Journal of Sex Research* 42 (3):215–23. doi: 10.1080/00224490509552276.

Mosconi, Lisa, Valentina Berti, Jonathan Dyke, Eva Schelbaum, Steven Jett, Lacey Loughlin, et al. 2021. "Menopause impacts human brain structure, connectivity, energy metabolism, and amyloid-beta deposition." *Scientific Reports* 11 (1). doi: 10.1038/s41598-021-90084-y.

Ochsner, Kevin N., Jennifer A. Silvers, and Jason T. Buhle. 2012. "Functional imaging studies of emotion regulation: a synthetic review and evolving model of the cognitive control of emotion." *Annals of the New York Academy of Sciences* 1251 (1):E1–E24. doi: 10.1111/j.1749-6632.2012.06751.x.

Pariante, Carmine M., and Stafford L. Lightman. 2008. "The HPA axis in major depression: Classical theories and new developments." *Trends in Neurosciences* 31 (9):464–68. doi: 10.1016/j.tins.2008.06.006.

Parker, Kyle E., Christian E. Pedersen, Adrian M. Gomez, Skylar M. Spangler, Marie C. Walicki, Shelley Y. Feng, et al. 2019. "A paranigral VTA nociceptin circuit that constrains motivation for reward." *Cell* 178 (3):653–71.e19. doi: 10.1016/j.cell.2019.06.034.

Protsenko, Ekaterina, Ruoting Yang, Brent Nier, Victor Reus, Rasha Hammamieh, Ryan Rampersaud, et al. 2021. "'GrimAge,' an epigenetic predictor of mortality, is accelerated in major depressive disorder." *Translational Psychiatry* 11 (1). doi: 10.1038/s41398-021-01302-0.

Sapolsky, Robert M. 2000. "Glucocorticoids and hippocampal atrophy in neuropsychiatric disorders." *Archives of General Psychiatry* 57 (10). doi: 10.1001/archpsyc.57.10.925.

Sherwin, Barbara B. 2012. "Estrogen and cognitive functioning in women: Lessons we have learned." *Behavioral Neuroscience* 126 (1):123–27. doi: 10.1037/a0025539.

Woods, Nancy F., Molly C. Carr, Eunice Y. Tao, Heather J. Taylor, and Ellen S. Mitchell. 2006. "Increased urinary cortisol levels during the menopause transition." *Menopause* 13 (2):212–21. doi: 10.1097/01.gme.0000198490.57242.2e.

Zaki, Jamil, Joshua Ian Davis, and Kevin N. Ochsner. 2012. "Overlapping activity in anterior insula during interoception and emotional experience." *NeuroImage* 62 (1):493–99. doi: 10.1016/j.neuroimage.2012.05.012.

Chapter 6: The Neuroscience of Self-Care

Allen, Andrew P., Timothy G. Dinan, Gerard Clarke, and John F. Cryan. 2017. "A psychology of the human brain-gut-microbiome axis." *Social and Personality Psychology Compass* 11 (4). doi: 10.1111/spc3.12309.

Baer, R. A., J. Carmody, and M. Hunsinger. 2012. "Weekly change in mindfulness and perceived stress in a mindfulness-based stress reduction program." *Journal of Clinical Psychology* 68 (7):755–65. doi: 10.1002/jclp.21865.

Bancos, Simona, Matthew P. Bernard, David J. Topham, and Richard P. Phipps. 2009. "Ibuprofen and other widely used non-steroidal anti-inflammatory drugs inhibit antibody production in human cells." *Cellular Immunology* 258 (1):18–28. doi: 10.1016/j.cellimm.2009.03.007.

Besedovsky, Luciana, Tanja Lange, and Monika Haack. 2019. "The sleep-immune crosstalk in health and disease." *Physiological Reviews* 99 (3):1325–80. doi: 10.1152/physrev.00010.2018.

Boehme, Marcus, Marcel van de Wouw, Thomaz F. S. Bastiaanssen, Loreto Olavarría-Ramírez, Katriona Lyons, Fiona Fouhy, et al. 2019. "Mid-life microbiota crises: Middle age is associated with pervasive neuroimmune alterations that are reversed by targeting the gut microbiome." *Molecular Psychiatry* 25 (10):2567–83. doi: 10.1038/s41380-019-0425-1.

Bonaz, Bruno, Thomas Bazin, and Sonia Pellissier. 2018. "The vagus nerve at the interface of the microbiota-gut-brain axis." *Frontiers in Neuroscience* 12. doi: 10.3389/fnins.2018.00049.

Braun, Theodore P., Xinxia Zhu, Marek Szumowski, Gregory D. Scott, Aaron J. Grossberg, Peter R. Levasseur, et al. 2011. "Central nervous system inflammation induces muscle atrophy via activation of the hypothalamic-pituitary-adrenal axis." *Journal of Experimental Medicine* 208 (12):2449–63. doi: 10.1084/jem.20111020.

Breit, Sigrid, Aleksandra Kupferberg, Gerhard Rogler, and Gregor Hasler. 2018. "Vagus nerve as modulator of the brain-gut axis in psychiatric and inflammatory disorders." *Frontiers in Psychiatry* 9. doi: 10.3389/fpsyt.2018.00044.

Brunt, V. E., R. A. Gioscia-Ryan, J. J. Richey, M. C. Zigler, L. M. Cuevas, A. Gonzalez, et al. 2019. "Suppression of the gut microbiome ameliorates age-related arterial dysfunction and oxidative stress in mice." *Journal of Physiology* 597 (9):2361–78. doi: 10.1113/JP277336.

Burokas, Aurelijus, Silvia Arboleya, Rachel D. Moloney, Veronica L. Peterson, Kiera Murphy, Gerard Clarke, et al. 2017. "Targeting the microbiota-gut-brain axis: Prebiotics have anxiolytic and antidepressant-like effects and reverse the impact of chronic stress in mice." *Biological Psychiatry* 82 (7):472–87. doi: 10.1016/j.biopsych.2016.12.031.

Bussian, T. J., A. Aziz, C. F. Meyer, B. L. Swenson, J. M. van Deursen, and D. J. Baker. 2018. "Clearance of senescent glial cells prevents tau-dependent pathology and cognitive decline." *Nature* 562 (7728):578–82. doi: 10.1038/s41586-018-0543-y.

Cai, Dongsheng, and Sinan Kohr. 2019. "'Hypothalamic microinflammation' paradigm in aging and metabolic diseases." *Cell Metabolism* 30 (1):19–35. doi:10.1016/j.cmet.2019.05.021.

Casaletto, Kaitlin B., Fanny M. Elahi, Adam M. Staffaroni, Samantha Walters, Wilfredo Rivera Contreras, Amy Wolf, et al. 2019. "Cognitive aging is not created equally: Differentiating unique cognitive phenotypes in 'normal' adults." *Neurobiology of Aging* 77:13–19. doi: 10.1016/j.neurobiolaging.2019.01.007.

Collins, F. L., N. D. Rios-Arce, S. Atkinson, H. Bierhalter, D. Schoenherr, J. N. Bazil, et al. 2017. "Temporal and regional intestinal changes in permeability, tight junction, and cytokine gene expression following ovariectomy-induced estrogen deficiency." *Physiological Reports* 5 (9). doi: 10.14814/phy2.13263.

Cryan, John F., Kenneth J. O'Riordan, Kiran Sandhu, Veronica Peterson, and Timothy G. Dinan. 2020. "The gut microbiome in neurological disorders." *Lancet Neurology* 19 (2):179–94. doi: 10.1016/s1474-4422(19)30356-4.

Daghlas, Iyas, Jacqueline M. Lane, Richa Saxena, and Céline Vetter. 2021. "Genetically proxied diurnal preference, sleep timing, and risk of major depressive disorder." *JAMA Psychiatry* 78 (8):903–10. doi: 10.1001/jamapsychiatry.2021.0959.

Dantzer, Robert. 2018. "Neuroimmune interactions: From the brain to the immune system and vice versa." *Physiological Reviews* 98 (1):477–504. doi: 10.1152/physrev.00039.2016.

D'Mello, C., N. Ronaghan, R. Zaheer, M. Dicay, T. Le, W. K. MacNaughton, M. G. Surrette, and M. G. Swain. 2015. "Probiotics improve inflammation-associated sickness behavior by altering communication between the peripheral immune system and the brain." *Journal of Neuroscience* 35 (30):10821–30. doi: 10.1523/jneurosci.0575-15.2015.

Erickson, Michelle A., William A. Banks, and Robert Dantzer. 2018. "Neuroimmune axes of the blood-brain barriers and blood-brain interfaces: Bases for physiological regulation, disease states, and pharmacological interventions." *Pharmacological Reviews* 70 (2):278–314. doi: 10.1124/pr.117.014647.

Evrensel, A., B. Onen Unsalver, and M. E. Ceylan. 2019. "Therapeutic potential of the microbiome in the treatment of neuropsychiatric disorders." *Medical Sciences (Basel)* 7 (2):21. doi: 10.3390/medsci7020021.

Felger, Jennifer C. 2018. "Imaging the role of inflammation in mood and anxiety-related disorders." *Current Neuropharmacology* 16 (5):533–58. doi: 10.2174/1570159x15666171123201142.

Felger, Jennifer C., and Michael T. Treadway. 2016. "Inflammation effects on motivation and motor activity: role of dopamine." *Neuropsychopharmacology* 42 (1):216–41. doi: 10.1038/npp.2016.143.

Fuhrman, B. J., H. S. Feigelson, R. Flores, M. H. Gail, X. Xu, J. Ravel, and James J. Goedert. 2014. "Associations of the fecal microbiome with urinary estrogens and estrogen metabolites in postmenopausal women." *Journal of Clinical Endocrinology and Metabolism* 99 (12):4632–40. doi: 10.1210/jc.2014-2222.

Fung, T. C., H. E. Vuong, C. D. G. Luna, G. N. Pronovost, A. A. Aleksandrova, N. G. Riley, et al. 2019. "Intestinal serotonin and fluoxetine exposure modulate bacterial colonization in the gut." *Nature Microbiology* 4 (12):2064–73. doi: 10.1038/s41564-019-0540-4.

Gibson, Glenn R., Robert Hutkins, Mary Ellen Sanders, Susan L. Prescott, Raylene A. Reimer, Seppo J. Salminen, et al. 2017. "Expert consensus document: The International Scientific Association for Probiotics and Prebiotics (ISAPP) consensus statement on the definition and scope of prebiotics." *Nature Reviews Gastroenterology & Hepatology* 14 (8):491–502. doi: 10.1038/nrgastro.2017.75.

Goverse, Gera, Michelle Stakenborg, and Gianluca Matteoli. 2016. "The intestinal cholinergic anti-inflammatory pathway." *Journal of Physiology* 594 (20):5771–80. doi: 10.1113/jp271537.

Greenfield, Shelly F., Sudie E. Back, Katie Lawson, and Kathleen T. Brady. 2010. "Substance abuse in women." *Psychiatric Clinics of North America* 33 (2):339–55. doi: 10.1016/j.psc.2010.01.004.

Griswold, Max G., Nancy Fullman, Caitlin Hawley, Nicholas Arian, Stephanie R. M. Zimsen, Hayley D. Tymeson, et al. 2018. "Alcohol use and burden for 195 countries and territories, 1990–2016: A systematic analysis for the Global Burden of Disease Study 2016." *Lancet* 392 (10152):1015–35. doi: 10.1016/s0140-6736(18)31310-2.

Harand, Caroline, Françoise Bertran, Franck Doidy, Fabian Guénolé, Béatrice Desgranges, Francis Eustache, and Géraldine Rauchs. 2012. "How aging affects sleep-dependent memory consolidation?" *Frontiers in Neurology* 3. doi: 10.3389/fneur.2012.00008.

Hardeland, Rüdiger. 2019. "Aging, melatonin, and the pro- and anti-inflammatory networks." *International Journal of Molecular Sciences* 20 (5). doi: 10.3390/ijms20051223.

Harper, C. 2009. "The neuropathology of alcohol-related brain damage." *Alcohol and Alcoholism* 44 (2):136–40. doi: 10.1093/alcalc/agn102.

Hilderbrand, Elisa R., and Amy W. Lasek. 2018. "Estradiol enhances ethanol reward in female mice through activation of ERα and ERβ." *Hormones and Behavior* 98:159–64. doi: 10.1016/j.yhbeh.2018.01.001.

Kim, Jee Wook, Dong Young Lee, Boung Chul Lee, Myung Hun Jung, Hano Kim, Yong Sung Choi, and Ihn-Geun Choi. 2012. "Alcohol and cognition in the elderly: A review." *Psychiatry Investigation* 9 (1). doi: 10.4306/pi.2012.9.1.8.

Kim, Sangjune, Seung-Hwan Kwon, Tae-In Kam, Nikhil Panicker, Senthilkumar S. Karuppagounder, Saebom Lee, et al. 2019. "Transneuronal propagation of pathologic α-synuclein from the gut to the brain models Parkinson's disease." *Neuron* 103 (4):627–41.e7. doi: 10.1016/j.neuron.2019.05.035.

Kowalski, K., and A. Mulak. 2019. "Brain-gut-microbiota axis in Alzheimer's disease." *Journal of Neurogastroenterology and Motility* 25 (1):48–60. doi: 10.5056/jnm18087.

Küffer, Andreas, Laura D. Straus, Aric A. Prather, Sabra S. Inslicht, Anne Richards, Judy K. Shigenaga, et al. 2019. "Altered overnight levels of pro-inflammatory cytokines in men and women with posttraumatic stress disorder." *Psychoneuroendocrinology* 102:114–20. doi: 10.1016/j.psyneuen.2018.12.002.

Lindbergh, Cutter A., Kaitlin B. Casaletto, Adam M. Staffaroni, Fanny Elahi, Samantha M. Walters, Michelle You, et al. 2020. "Systemic tumor necrosis factor-alpha trajectories relate to brain health in typically aging older adults." *Journals of Gerontology: Series A* 75 (8):1558–65. doi: 10.1093/gerona/glz209.

Lionnet, Arthur, Laurène Leclair-Visonneau, Michel Neunlist, Shigeo Murayama, Masaki Takao, Charles H. Adler, et al. 2017. "Does Parkinson's disease start in the gut?" *Acta Neuropathologica* 135 (1):1–12. doi: 10.1007/s00401-017-1777-8.

Liu, Jing, Fei Xu, Zhiyan Nie, and Lei Shao. 2020. "Gut microbiota approach—a new strategy to treat Parkinson's disease." *Frontiers in Cellular and Infection Microbiology* 10. doi: 10.3389/fcimb.2020.570658.

Lobionda, Stefani, Panida Sittipo, Hyog Young Kwon, and Yun Kyung Lee. 2019. "The role of gut microbiota in intestinal inflammation with respect to diet and extrinsic stressors." *Microorganisms* 7 (8). doi: 10.3390/microorganisms7080271.

Martin, Dominique E., Blake L. Torrance, Laura Haynes, and Jenna M. Bartley. 2021. "Targeting aging: Lessons learned from immunometabolism and cellular senescence." *Frontiers in Immunology* 12. doi: 10.3389/fimmu.2021.714742.

Michaud, Martin, Laurent Balardy, Guillaume Moulis, Clement Gaudin, Caroline Peyrot, Bruno Vellas, et al. 2013. "Proinflammatory cytokines, aging, and age-related diseases." *Journal of the American Medical Directors Association* 14 (12):877–82. doi: 10.1016/j.jamda.2013.05.009.

Miller, Andrew H., Ebrahim Haroon, Charles L. Raison, and Jennifer C. Felger. 2013. "Cytokine targets in the brain: Impact on neurotransmitters and neurocircuits." *Depression and Anxiety* 30 (4):297–306. doi: 10.1002/da.22084.

Milton, David C., Joey Ward, Emilia Ward, Donald M. Lyall, Rona J. Strawbridge, Daniel J. Smith, and Breda Cullen. 2021. "The association between C-reactive protein, mood disorder, and cognitive function in UK Biobank." *European Psychiatry* 64 (1). doi: 10.1192/j.eurpsy.2021.6.

Moieni, Mona, Kevin M. Tan, Tristen K. Inagaki, Keely A. Muscatell, Janine M. Dutcher, Ivana Jevtic, et al. 2019. "Sex differences in the relationship between inflammation and reward sensitivity: A randomized controlled trial of endotoxin." *Biological Psychiatry: Cognitive Neuroscience and Neuroimaging* 4 (7):619–26. doi: 10.1016/j.bpsc.2019.03.010.

Mu, Qinghui, Vincent J. Tavella, and Xin M. Luo. 2018. "Role of *Lactobacillus reuteri* in human health and diseases." *Frontiers in Microbiology* 9. doi: 10.3389/fmicb.2018.00757.

Mursu, Jaakko, Lyn M. Steffen, Katie A. Meyer, Daniel Duprez, and David R. Jacobs. 2013. "Diet quality indexes and mortality in postmenopausal women: The Iowa Women's Health Study." *American Journal of Clinical Nutrition* 98 (2):444–53. doi: 10.3945/ajcn.112.055681.

Niles, Andrea N., Mariya Smirnova, Joy Lin, and Aoife O'Donovan. 2018. "Gender differences in longitudinal relationships between depression and anxiety symptoms and inflammation in the health and retirement study." *Psychoneuroendocrinology* 95:149–57. doi: 10.1016/j.psyneuen.2018.05.035.

Oslin, David W., and Mark S. Cary. 2003. "Alcohol-related dementia: Validation of diagnostic criteria." *American Journal of Geriatric Psychiatry* 11 (4):441–47. doi: 10. 1097/00019442-200307000-00007.

Pace, Thaddeus W. W., Lobsang Tenzin Negi, Daniel D. Adame, Steven P. Cole, Teresa I. Sivilli, Timothy D. Brown, et al. 2009. "Effect of compassion meditation on neuroendocrine, innate immune and behavioral responses to psychosocial stress." *Psychoneuroendocrinology* 34 (1):87–98. doi: 10.1016/j. psyneuen.2008.08.011.

Peterson, Christine Tara. 2020. "Dysfunction of the microbiota-gut-brain axis in neurodegenerative disease: The promise of therapeutic modulation with prebiotics, medicinal herbs, probiotics, and synbiotics." *Journal of Evidence-Based Integrative Medicine* 25 (1). doi: 10.1177/2515690x20957225.

Polloni, Laura, and Antonella Muraro. 2020. "Anxiety and food allergy: A review of the last two decades." *Clinical & Experimental Allergy* 50 (4):420–41. doi: 10.1111/ cea.13548.

Rehm, Jürgen, Omer S. M. Hasan, Sandra E. Black, Kevin D. Shield, and Michaël Schwarzinger. 2019. "Alcohol use and dementia: A systematic scoping review." *Alzheimer's Research & Therapy* 11 (1). doi: 10.1186/s13195-018-0453-0.

Ridley, Nicole J., Brian Draper, and Adrienne Withall. 2013. "Alcohol-related dementia: An update of the evidence." *Alzheimer's Research & Therapy* 5 (1). doi: 10.1186/alzrt157.

Sayed, Nazish, Yingxiang Huang, Khiem Nguyen, Zuzana Krejciova-Rajaniemi, Anissa P. Grawe, Tianxiang Gao, et al. 2021. "An inflammatory aging clock (iAge) based on deep learning tracks multimorbidity, immunosenescence, frailty and cardiovascular aging." *Nature Aging* 1 (7):598–615. doi: 10.1038/s43587-021-00082-y.

Schiffrin, E. J., D. R. Thomas, V. B. Kumar, C. Brown, C. Hager, M. A. Van't Hof, et al. 2007. "Systemic inflammatory markers in older persons: the effect of oral nutritional supplementation with prebiotics." *Journal of Nutrition, Health & Aging* 11 (6):475–79.

Sovijit, Watcharin N., Watcharee E. Sovijit, Shaoxia Pu, Kento Usuda, Ryo Inoue, Gen Watanabe, et al. 2021. "Ovarian progesterone suppresses depression and anxiety-like behaviors by increasing the Lactobacillus population of gut microbiota in ovariectomized mice." *Neuroscience Research* 168:76–82. doi: 10.1016/j. neures.2019.04.005.

Sohn, Emily. 2021. "Why autoimmunity is most common in women." *Nature* 595 (7867):S51–S53. doi: 10.1038/d41586-021-01836-9.

Ticinesi, Andrea, Claudio Tana, Antonio Nouvenne, Beatrice Prati, Fulvio Lauretani, and Tiziana Meschi. 2018. "Gut microbiota, cognitive frailty and dementia in older individuals: A systematic review." *Clinical Interventions in Aging* 13:1497–511. doi: 10.2147/cia.S139163.

Van Houten, J. M., R. J. Wessells, H. L. Lujan, and S. E. DiCarlo. 2015. "My gut feeling says rest: Increased intestinal permeability contributes to chronic diseases in high-intensity exercisers." *Medical Hypotheses* 85 (6):882–86. doi: 10.1016/j. mehy.2015.09.018.

Vieira, Angélica T., Paula M. Castelo, Daniel A. Ribeiro, and Caroline M. Ferreira. 2017. "Influence of oral and gut microbiota in the health of menopausal women." *Frontiers in Microbiology* 8. doi: 10.3389/fmicb.2017.01884.

Vulevic, Jelena, Alexandra Drakoularakou, Parveen Yaqoob, George Tzortzis, and Glenn R. Gibson. 2008. "Modulation of the fecal microflora profile and immune function by a novel trans-galactooligosaccharide mixture (B-GOS) in healthy elderly volunteers." *American Journal of Clinical Nutrition* 88 (5):1438–46. doi: 10.3945/ ajcn.2008.26242.

Walker, Keenan A., Ron C. Hoogeveen, Aaron R. Folsom, Christie M. Ballantyne, David S. Knopman, B. Gwen Windham, et al. 2017. "Midlife systemic inflammatory markers are associated with late-life brain volume." *Neurology* 89 (22):2262–70. doi: 10.1212/wnl.0000000000004688.

Xie, Ruining, Pei Jiang, Li Lin, Jian Jiang, Bin Yu, Jingjing Rao, et al. 2020. "Oral treatment with *Lactobacillus reuteri* attenuates depressive-like behaviors and serotonin metabolism alterations induced by chronic social defeat stress." *Journal of Psychiatric Research* 122:70–78. doi: 10.1016/j.jpsychires.2019.12.013.

Xu, Ming, Tamar Pirtskhalava, Joshua N. Farr, Bettina M. Weigand, Allyson K. Palmer, Megan M. Weivoda, et al. 2018. "Senolytics improve physical function and increase lifespan in old age." *Nature Medicine* 24 (8):1246–56. doi: 10.1038/ s41591-018-0092-9.

Yoon, Kichul, and Nayoung Kim. 2021. "Roles of sex hormones and gender in the gut microbiota." *Journal of Neurogastroenterology and Motility* 27 (3):314–25. doi: 10.5056/jnm20208.

Zahr, N. M., and A. Pfefferbaum. 2017. "Alcohol's effects on the brain: Neuroimaging results in humans and animal models." *Alcohol Research* 38 (2):183–206.

Chapter 7: Your Brain in Search of Connection

Cacioppo, John T., and William Patrick. *Loneliness: Human Nature and the Need for Social Connection.* New York: W. W. Norton, 2009.

Dölen, Gül, and Robert C. Malenka. 2014. "The emerging role of nucleus accumbens oxytocin in social cognition." *Biological Psychiatry* 76 (5):354–55. doi: 10.1016/j.biopsych.2014.06.009.

Eisenberger, Naomi I., and Matthew D. Lieberman. 2004. "Why rejection hurts: A common neural alarm system for physical and social pain." *Trends in Cognitive Sciences* 8 (7):294–300. doi: 10.1016/j.tics.2004.05.010.

Holt-Lunstad, J., T. B. Smith, M. Baker, T. Harris, and D. Stephenson. 2015. "Loneliness and social isolation as risk factors for mortality: A meta-analytic review." *Perspectives on Psychological Science* 10 (2):227–37. doi: 10.1177/1745691614568352.

Jiang, Luo-Luo, Tamas David-Barrett, Anna Rotkirch, James Carney, Isabel Behncke Izquierdo, Jaimie A. Krems, et al. 2015. "Women favour dyadic relationships, but men prefer clubs: Cross-cultural evidence from social networking." *PLOS ONE* 10 (3). doi: 10.1371/journal.pone.0118329.

Kaufman, Scott B. 2019. "Taking Sex Differences in Personality Seriously." *Scientific American.*

Kudwa, Andrea E., Robert F. McGivern, and Robert J. Handa. 2014. "Estrogen receptor β and oxytocin interact to modulate anxiety-like behavior and neuroendocrine stress reactivity in adult male and female rats." *Physiology & Behavior* 129:287–96. doi: 10.1016/j.physbeh.2014.03.004.

Laakasuo, Michael, Anna Rotkirch, Max van Duijn, Venla Berg, Markus Jokela, Tamas David-Barrett, et al. 2020. "Homophily in personality enhances group success among real-life friends." *Frontiers in Psychology* 11. doi: 10.3389/fpsyg.2020.00710.

Luo, Ye, Louise C. Hawkley, Linda J. Waite, and John T. Cacioppo. 2012. "Loneliness, health, and mortality in old age: A national longitudinal study." *Social Science & Medicine* 74 (6):907–14. doi: 10.1016/j.socscimed.2011.11.028.

Chapter 8: Upgrading the Mommy Brain

Isay, Jane. *Walking on Eggshells: Navigating the Delicate Relationship Between Adult Children and Their Parents.* New York: Broadway Books/Flying Dolphin Press, 2008.

Li, Tong, Ping Wang, Stephani C. Wang, and Yu-Feng Wang. 2017. "Approaches mediating oxytocin regulation of the immune system." *Frontiers in Immunology* 7. doi: 10.3389/fimmu.2016.00693.

Wirth, Michelle M. 2014. "Hormones, stress, and cognition: The effects of glucocorticoids and oxytocin on memory." *Adaptive Human Behavior and Physiology* 1 (2):177–201. doi: 10.1007/s40750-014-0010-4.

Chapter 9: The Relationship Brain

Caldwell, Heather K., and H. Elliott Albers. 2016. "Oxytocin, vasopressin, and the motivational forces that drive social behaviors." *Current Topics in Behavioral Neurosciences* 27:51–103. doi: 10.1007/7854_2015_390.

Dumais, Kelly M., and Alexa H. Veenema. 2016. "Vasopressin and oxytocin receptor systems in the brain: Sex differences and sex-specific regulation of social behavior." *Frontiers in Neuroendocrinology* 40:1–23. doi: 10.1016/j.yfrne.2015.04.003.

Huang, M., S. Su, J. Goldberg, A. H. Miller, O. M. Levantsevych, L. Shallenberger, et al. 2019. "Longitudinal association of inflammation with depressive symptoms: A 7-year cross-lagged twin difference study." *Brain, Behavior, and Immunity* 75:200–207. doi: 10.1016/j.bbi.2018.10.007.

Langer, Ellen J. *Mindfulness*. Massachusetts: Addison-Wesley, 1989.

Maffei, L., E. Picano, M. G. Andreassi, A. Angelucci, F Baldacci, L. Baroncelli, et al. 2017. "Randomized trial on the effects of a combined physical/cognitive training in aged MCI subjects: The Train the Brain study." *Scientific Reports* 7 (1). doi: 10.1038/srep39471.

Niles, A. N., M. Smirnova, J. Lin, and A. O'Donovan. 2018. "Gender differences in longitudinal relationships between depression and anxiety symptoms and inflammation in the health and retirement study." *Psychoneuroendocrinology* 95:149–57. doi: 10.1016/j.psyneuen.2018.05.035.

Sharma, Animesh N., Paul Aoun, Jean R. Wigham, Suanne M. Weist, and Johannes D. Veldhuis. 2014. "Estradiol, but not testosterone, heightens cortisol-mediated negative feedback on pulsatile ACTH secretion and ACTH approximate entropy in unstressed older men and women." *American Journal of Physiology-Regulatory, Integrative and Comparative Physiology* 306 (9):R627–35. doi: 10.1152/ajpregu.00551.2013.

Shrout, M. Rosie, Randal D. Brown, Terri L. Orbuch, and Daniel J. Weigel. 2019. "A multidimensional examination of marital conflict and subjective health over 16 years." *Personal Relationships* 26 (3):490–506. doi: 10.1111/pere.12292.

Smith, B. M., X. Yao, K. S. Chen, and E. D. Kirby. 2018. "A larger social network enhances novel object location memory and reduces hippocampal microgliosis in aged mice." *Frontiers in Aging Neuroscience* 10:142. doi: 10.3389/fnagi.2018.00142.

Chapter 10: Centering

Boyle, Christina P., Cyrus A. Raji, Kirk I. Erickson, Oscar L. Lopez, James T. Becker, H. Michael Gach, et al. 2020. "Estrogen, brain structure, and cognition in postmenopausal women." *Human Brain Mapping* 42 (1):24–35. doi: 10.1002/hbm.25200.

Uddin, Lucina Q. 2014. "Salience processing and insular cortical function and dysfunction." *Nature Reviews Neuroscience* 16 (1):55–61. doi: 10.1038/nrn3857.

Chapter 11: Body Hacks for the Mind

Adamaszek, M., F. D'Agata, R. Ferrucci, C. Habas, S. Keulen, K. C. Kirkby, et al. 2016. "Consensus paper: Cerebellum and emotion." *Cerebellum* 16 (2):552–76. doi: 10.1007/s12311-016-0815-8.

Bloss, E. B., W. G. Janssen, B. S. McEwen, and J. H. Morrison. 2010. "Interactive effects of stress and aging on structural plasticity in the prefrontal cortex." *Journal of Neuroscience* 30 (19):6726–31. doi: 10.1523/jneurosci.0759-10.2010.

Borhan, A. S. M., Patricia Hewston, Dafna Merom, Courtney Kennedy, George Ioannidis, Nancy Santesso, et al. 2018. "Effects of dance on cognitive function among older adults: A protocol for systematic review and meta-analysis." *Systematic Reviews* 7 (1). doi: 10.1186/s13643-018-0689-6.

Brown, Stuart L., and Christopher C. Vaughan. *Play: How It Shapes the Brain, Opens the Imagination, and Invigorates the Soul.* New York: Avery, 2010.

Casaletto, Kaitlin B., Adam M. Staffaroni, Fanny Elahi, Emily Fox, Persephone A. Crittenden, Michelle You, et al. 2018. "Perceived stress is associated with accelerated Monocyte/Macrophage aging trajectories in clinically normal adults." *American Journal of Geriatric Psychiatry* 26 (9):952–63. doi: 10.1016/j.jagp.2018.05.004.

D'Angelo, Egidio. 2019. "The cerebellum gets social." *Science* 363 (6424):229. doi: 10.1126/science.aaw2571.

de Kloet, E. Ron, Marian Joëls, and Florian Holsboer. 2005. "Stress and the brain: From adaptation to disease." *Nature Reviews Neuroscience* 6 (6):463–75. doi: 10.1038/nrn1683.

de Oliveira Matos, Felipe, Amanda Vido, William Fernando Garcia, Wendell Arthur Lopes, and Antonio Pereira. 2020. "A neurovisceral integrative study on cognition, heart rate variability, and fitness in the elderly." *Frontiers in Aging Neuroscience* 12. doi: 10.3389/fnagi.2020.00051.

Devita, Maria, Francesco Alberti, Michela Fagnani, Fabio Masina, Enrica Ara, Giuseppe Sergi, et al. 2021. "Novel insights into the relationship between cerebellum and dementia: A narrative review as a toolkit for clinicians." *Ageing Research Reviews* 70. doi: 10.1016/j.arr.2021.101389.

Ding, Kan, Takashi Tarumi, David C. Zhu, Benjamin Y. Tseng, Binu P. Thomas, Marcel Turner, et al. 2017. "Cardiorespiratory fitness and white matter neuronal fiber integrity in mild cognitive impairment." *Journal of Alzheimer's Disease* 61 (2):729–39. doi: 10.3233/jad-170415.

Erickson, K. I., M. W. Voss, R. S. Prakash, C. Basak, A. Szabo, L. Chaddock, et al. 2011. "Exercise training increases size of hippocampus and improves memory." *Proceedings of the National Academy of Sciences* 108 (7):3017–22. doi: 10.1073/pnas.1015950108.

Killingsworth, M. A., Gilbert, D. T., 2010 "A wandering mind is an unhappy mind." *Science,* 330 (6006):932. doi: 10.1126/science.1192439.

Lavretsky, Helen, and Paul A. Newhouse. 2012. "Stress, inflammation, and aging." *American Journal of Geriatric Psychiatry* 20 (9):729–33. doi: 10.1097/JGP.0b013e31826573cf.

Leggio, Maria, and Giusy Olivito. 2018. "Topography of the cerebellum in relation to social brain regions and emotions." *Handbook of Clinical Neurology* 154:71–84. doi: 10.1016/B978-0-444-63956-1.

Marek, Scott, Joshua S. Siegel, Evan M. Gordon, Ryan V. Raut, Caterina Gratton, Dillan J. Newbold, et al. 2018. "Spatial and temporal organization of the individual human cerebellum." *Neuron* 100 (4):977–93.e7. doi: 10.1016/j.neuron.2018.10.010.

Matyi, Joshua M., Gail B. Rattinger, Sarah Schwartz, Mona Buhusi, and JoAnn T. Tschanz. 2019. "Lifetime estrogen exposure and cognition in late life: The Cache County Study." *Menopause* 26 (12):1366–74. doi: 10.1097/gme.0000000000001405.

McEwen, Bruce S. 2019. "What is the confusion with cortisol?" *Chronic Stress* 3. doi: 10.1177/2470547019833647.

Mishra, Aarti, Yuan Shang, Yiwei Wang, Eliza R. Bacon, Fei Yin, and Roberta D. Brinton. 2020. "Dynamic neuroimmune profile during mid-life aging in the female brain and implications for Alzheimer risk." *iScience* 23 (12). doi: 10.1016/j.isci.2020.101829.

Murri, Martino Belvederi, Federico Triolo, Alice Coni, Carlo Tacconi, Erika Nerozzi, Andrea Escelsior, et al. 2020. "Instrumental assessment of balance and gait in depression: A systematic review." *Psychiatry Research* 284. doi: 10.1016/j.psychres.2019.112687.

Ochs-Balcom, Heather M., Leah Preus, Jing Nie, Jean Wactawski-Wende, Linda Agyemang, Marian L. Neuhouser, et al. 2018. "Physical activity modifies genetic susceptibility to obesity in postmenopausal women." *Menopause* 25 (10):1131–37. doi: 10.1097/gme.0000000000001134.

Schmahmann, Jeremy D. 2019. "The cerebellum and cognition." *Neuroscience Letters* 688:62–75. doi: 10.1016/j.neulet.2018.07.005.

Shrout, M. Rosie, Megan E. Renna, Annelise A. Madison, Lisa M. Jaremka, Christopher P. Fagundes, William B. Malarkey, and Janice Kiecolt-Glaser. 2020. "Cortisol slopes and conflict: A spouse's perceived stress matters." *Psychoneuroendocrinology* 121. doi: 10.1016/j.psyneuen.2020.104839.

Siviy, Stephen M. 2016. "A brain motivated to play: Insights into the neurobiology of playfulness." *Behaviour* 153 (6–7):819–44. doi: 10.1163/1568539x-00003349.

Söderkvist, Sven, Kajsa Ohlén, and Ulf Dimberg. 2017. "How the experience of emotion is modulated by facial feedback." *Journal of Nonverbal Behavior* 42 (1):129–51. doi: 10.1007/s10919-017-0264-1.

Strata, Piergiorgio. 2015. "The emotional cerebellum." *Cerebellum* 14 (5):570–77. doi: 10.1007/s12311-015-0649-9.

Topp, Robert, Marcia Ditmyer, Karen King, Kristen Doherty, and Joseph Hornyak. 2002. "The effect of bed rest and potential of prehabilitation on patients in the intensive care unit." *AACN Clinical Issues: Advanced Practice in Acute and Critical Care* 13 (2):263–76. doi: 10.1097/00044067-200205000-00011.

Traustadóttir, Tinna, Pamela R. Bosch, and Kathleen S. Matt. 2005. "The HPA axis response to stress in women: Effects of aging and fitness." *Psychoneuroendocrinology* 30 (4):392–402. doi: 10.1016/j.psyneuen.2004.11.002.

Vecchio, Laura M., Ying Meng, Kristiana Xhima, Nir Lipsman, Clement Hamani, Isabelle Aubert, et al. 2018. "The neuroprotective effects of exercise: Maintaining a healthy brain throughout aging." *Brain Plasticity* 4 (1):17–52. doi: 10.3233/bpl-180069.

Weitz, Gunther, Mikael Elam, Jan Born, Horst L. Fehm, and Christoph Dodt. 2001. "Postmenopausal estrogen administration suppresses muscle sympathetic nerve activity." *Journal of Clinical Endocrinology & Metabolism* 86 (1):344–48. doi: 10.1210/jcem.86.1.7138.

Chapter 12: The Return of Purpose

Baune, Bernhard, Antoine Lutz, Julie Brefczynski-Lewis, Tom Johnstone, and Richard J. Davidson. 2008. "Regulation of the neural circuitry of emotion by compassion meditation: Effects of meditative expertise." *PLOS ONE* 3 (3). doi: 10.1371/journal.pone.0001897.

Hedegaard, Holly, Sally C. Curtin, and Margaret Warner. "Suicide mortality in the United States, 1999–2017." NCHS Data Brief No. 330, November 2018.

Litzelman, Kristin, Whitney P. Witt, Ronald E. Gangnon, F. Javier Nieto, Corinne D. Engelman, Marsha R. Mailick, and Halcyon Gerald Skinner. 2014. "Association between informal caregiving and cellular aging in the Survey of the Health of Wisconsin: The role of caregiving characteristics, stress, and strain." *American Journal of Epidemiology* 179 (11):1340–52. doi: 10.1093/aje/kwu066.

López-Otín, Carlos, Maria A. Blasco, Linda Partridge, Manuel Serrano, and Guido Kroemer. 2013. "The hallmarks of aging." *Cell* 153 (6):1194–217. doi: 10.1016/j.cell.2013.05.039.

Chapter 13: New Focus

Al-Hashimi, Omar, Theodore P. Zanto, and Adam Gazzaley. 2015. "Neural sources of performance decline during continuous multitasking." *Cortex* 71:49–57. doi: 10.1016/j.cortex.2015.06.001.

Anguera, J. A., J. Boccanfuso, J. L. Rintoul, O. Al-Hashimi, F. Faraji, J. Janowich, et al. 2013. "Video game training enhances cognitive control in older adults." *Nature* 501 (7465):97–101. doi: 10.1038/nature12486.

Anguera, Joaquin A., Jessica N. Schachtner, Alexander J. Simon, Joshua Volponi, Samirah Javed, Courtney L. Gallen, and Adam Gazzaley. 2021. "Long-term maintenance of multitasking abilities following video game training in older adults." *Neurobiology of Aging* 103:22–30. doi: 10.1016/j.neurobiolaging.2021.02.023.

Bratman, Gregory N., J. Paul Hamilton, Kevin S. Hahn, Gretchen C. Daily, and James J. Gross. 2015. "Nature experience reduces rumination and subgenual prefrontal cortex activation." *Proceedings of the National Academy of Sciences* 112 (28):8567–72. doi: 10.1073/pnas.1510459112.

Clapp, W. C., M. T. Rubens, J. Sabharwal, and A. Gazzaley. 2011. "Deficit in switching between functional brain networks underlies the impact of multitasking on working memory in older adults." *Proceedings of the National Academy of Sciences* 108 (17):7212–17. doi: 10.1073/pnas.1015297108.

Frago, Laura M., Sandra Canelles, Alejandra Freire-Regatillo, Pilar Argente-Arizón, Vicente Barrios, Jesús Argente, et al. 2017. "Estradiol uses different mechanisms in astrocytes from the hippocampus of male and female rats to protect against damage induced by palmitic acid." *Frontiers in Molecular Neuroscience* 10. doi: 10.3389/fnmol.2017.00330.

Kim, Yu Jin, Maira Soto, Gregory L. Branigan, Kathleen Rodgers, and Roberta Diaz Brinton. 2021. "Association between menopausal hormone therapy and risk of neurodegenerative diseases: Implications for precision hormone therapy." *Alzheimer's & Dementia: Translational Research & Clinical Interventions* 7 (1). doi: 10.1002/trc2.12174.

Reuter-Lorenz, Patricia A., and Denise C. Park. 2014. "How does it STAC up? Revisiting the scaffolding theory of aging and cognition." *Neuropsychology Review* 24 (3):355–70. doi: 10.1007/s11065-014-9270-9.

Sherwin, Barbara B. 2008. "Hormones, the brain, and me." *Canadian Psychology/ Psychologie canadienne* 49 (1):42–48. doi: 10.1037/0708-5591.49.1.42.

Tsuchiyagaito, Aki, Masaya Misaki, Obada Al Zoubi, Martin Paulus, and Jerzy Bodurka. 2020. "Prevent breaking bad: A proof of concept study of rebalancing the brain's rumination circuit with real-time fMRI functional connectivity neurofeedback." *Human Brain Mapping* 42 (4):922–40. doi: 10.1002/hbm.25268.

Chapter 14: Life Span or Health Span?

Brinton, Roberta Diaz. 2008. "Estrogen regulation of glucose metabolism and mitochondrial function: Therapeutic implications for prevention of Alzheimer's disease." *Advanced Drug Delivery Reviews* 60 (13–14):1504–11. doi: 10.1016/j. addr.2008.06.003.

Chowen, Julie A., and Luis M. Garcia-Segura. 2021. "Role of glial cells in the generation of sex differences in neurodegenerative diseases and brain aging." *Mechanisms of Ageing and Development* 196. doi: 10.1016/j.mad.2021.111473.

Dubal, Dena B. 2020. "Sex difference in Alzheimer's disease: An updated, balanced and emerging perspective on differing vulnerabilities." *Handbook of Clinical Neurology* 175:261–73. doi: 10.1016/B978-0-444-64123-6.00018-7.

Dumas, Julie, Catherine Hancur-Bucci, Magdalena Naylor, Cynthia Sites, and Paul Newhouse. 2008. "Estradiol interacts with the cholinergic system to affect verbal memory in postmenopausal women: Evidence for the critical period hypothesis." *Hormones and Behavior* 53 (1):159–69. doi: 10.1016/j.yhbeh.2007.09.011.

Etnier, Jennifer L., Eric S. Drollette, and Alexis B. Slutsky. 2019. "Physical activity and cognition: A narrative review of the evidence for older adults." *Psychology of Sport and Exercise* 42:156–66. doi: 10.1016/j.psychsport.2018.12.006.

Fassier, Philippine, Jae Hee Kang, I. Min Lee, Francine Grodstein, and Marie-Noël Vercambre. 2021. "Vigorous physical activity and cognitive trajectory later in life: Prospective association and interaction by apolipoprotein E e4 in the Nurses' Health Study." *Journals of Gerontology: Series A*. doi: 10.1093/gerona/glab169.

Fisher, Daniel W., David A. Bennett, and Hongxin Dong. 2018. "Sexual dimorphism in predisposition to Alzheimer's disease." *Neurobiology of Aging* 70:308–24. doi: 10.1016/j.neurobiolaging.2018.04.004.

Grodstein, Francine, Jennifer Chen, Daniel A. Pollen, Marilyn S. Albert, Robert S. Wilson, Marshal F. Folstein, et al. 2000. "Postmenopausal hormone therapy and cognitive function in healthy older women." *Journal of the American Geriatrics Society* 48 (7):746–52. doi: 10.1111/j.1532-5415.2000.tb04748.x.

Jiménez-Balado, Joan, and Teal S. Eich. 2021. "GABAergic dysfunction, neural network hyperactivity and memory impairments in human aging and Alzheimer's disease." *Seminars in Cell & Developmental Biology* 116:146–59. doi: 10.1016/j. semcdb.2021.01.005.

Klosinski, Lauren P., Jia Yao, Fei Yin, Alfred N. Fonteh, Michael G. Harrington, Trace A. Christensen, et al. 2015. "White matter lipids as a ketogenic fuel supply in aging female brain: Implications for Alzheimer's disease." *EBioMedicine* 2 (12):1888–904. doi: 10.1016/j.ebiom.2015.11.002.

Leng, Yue, and Kristine Yaffe. 2020. "Sleep duration and cognitive aging—beyond a U-shaped association." *JAMA Network Open* 3 (9). doi: 10.1001/ jamanetworkopen.2020.14008.

Marin, Raquel, and Mario Diaz. 2018. "Estrogen interactions with lipid rafts related to neuroprotection. Impact of brain ageing and menopause." *Frontiers in Neuroscience* 12. doi: 10.3389/fnins.2018.00128.

Mehta, Jaya, Juliana M. Kling, and JoAnn E. Manson. 2021. "Risks, benefits, and treatment modalities of menopausal hormone therapy: Current concepts." *Frontiers in Endocrinology* 12. doi: 10.3389/fendo.2021.564781.

Mulnard, Ruth A., Carl W. Cotman, Claudia Kawas, Christopher H. van Dyck, Mary Sano, Rachelle Doody, et al. 2000. "Estrogen replacement therapy for treatment of mild to moderate Alzheimer disease." *JAMA* 283 (8):1107–15. doi: 10.1001/jama.283.8.1007.

Paganini-Hill, Annlia, Claudia H. Kawas, and Maria M. Corrada. 2016. "Lifestyle factors and dementia in the oldest-old: The 90+ Study." *Alzheimer Disease & Associated Disorders* 30 (1):21–26. doi: 10.1097/wad.0000000000000087.

Shin, Jean, Stephanie Pelletier, Louis Richer, G. Bruce Pike, Daniel Gaudet, Tomas Paus, and Zdenka Pausova. 2020. "Adiposity-related insulin resistance and thickness of the cerebral cortex in middle-aged adults." *Journal of Neuroendocrinology* 32 (12). doi: 10.1111/jne.12921.

Shumaker, Sally A., Claudine Legault, Stephen R. Rapp, Leon Thal, Robert B. Wallace, Judith K. Ockene, et al. 2003. "Estrogen plus progestin and the incidence of dementia and mild cognitive impairment in postmenopausal women." *JAMA* 289 (20): 2651–62. doi: 10.1001/jama.289.20.2651.

Subramaniapillai, Sivaniya, Anne Almey, M. Natasha Rajah, and Gillian Einstein. 2021. "Sex and gender differences in cognitive and brain reserve: Implications for Alzheimer's disease in women." *Frontiers in Neuroendocrinology* 60. doi: 10.1016/j. yfrne.2020.100879.

Thurston, Rebecca C., Howard J. Aizenstein, Carol A. Derby, Ervin Sejdić, and Pauline M. Maki. 2016. "Menopausal hot flashes and white matter hyperintensities." *Menopause* 23 (1):27–32. doi: 10.1097/gme.0000000000000481.

Wang, Dan, Xuan Wang, Meng-Ting Luo, Hui Wang, and Yue-Hua Li. 2019. "Gamma-aminobutyric acid levels in the anterior cingulate cortex of perimenopausal women with depression: A magnetic resonance spectroscopy study." *Frontiers in Neuroscience* 13. doi: 10.3389/fnins.2019.00785.

Watermeyer, Tamlyn, Catherine Robb, Sarah Gregory, and Chinedu Udeh-Momoh. 2021. "Therapeutic implications of hypothalamic-pituitaryadrenal-axis modulation in Alzheimer's disease: A narrative review of pharmacological and lifestyle interventions." *Frontiers in Neuroendocrinology* 60. doi: 10.1016/j.yfrne.2020.100877.

Wingfield, Arthur, and Jonathan E. Peelle. 2012. "How does hearing loss affect the brain?" *Aging Health* 8 (2):107–9. doi: 10.2217/ahe.12.5.

Yaffe, Kristine, Warren Browner, Jane Cauley, Lenore Launer, and Tamara Harris. 1999. "Association between bone mineral density and cognitive decline in older women." *Journal of the American Geriatrics Society* 47 (10):1176–82. doi: 10.1111/j.1532-5415.1999.tb05196.x.

Yaffe, Kristine, Cherie Falvey, Nathan Hamilton, Ann V. Schwartz, Eleanor M. Simonsick, Suzanne Satterfield, et al. 2012. "Diabetes, glucose control, and 9-year cognitive decline among older adults without dementia." *Archives of Neurology* 69 (9):1170–754. doi: 10.1001/archneurol.2012.1117.

Chapter 15: So Many Transitions

Gehlek, Nawang, Gini Alhadeff, and Mark Magill. *Good Life, Good Death: Tibetan Wisdom.* New York: Riverhead Books, 2001.

Kondziella, Daniel. 2020. "The neurology of death and the dying brain: A pictorial essay." *Frontiers in Neurology* 11. doi: 10.3389/fneur.2020.00736.

Parnia, Sam, Tara Keshavarz, Meghan McMullin, and Tori Williams. 2019. "Abstract 387: Awareness and cognitive activity during cardiac arrest." *Circulation* 140 (Suppl_2). doi: 10.1161/circ.140.suppl_2.387.

Rady, Mohamed Y., and Joseph L. Verheijde. 2016. "Neuroscience and awareness in the dying human brain: Implications for organ donation practices." *Journal of Critical Care* 34:121–23. doi: 10.1016/j.jcrc.2016.04.016.

Appendix

American Geriatrics Society 2019. "American Geriatrics Society 2019 updated AGS Beers Criteria® for potentially inappropriate medication use in older adults." *Journal of the American Geriatrics Society* 67 (4):674–94. doi: 10.1111/jgs.15767.

Baum, Jamie, Il-Young Kim, and Robert Wolfe. 2016. "Protein consumption and the elderly: what is the optimal level of intake?" *Nutrients* 8 (6). doi: 10.3390/nu8060359.

Baune, Bernhard, Antoine Lutz, Julie Brefczynski-Lewis, Tom Johnstone, and Richard J. Davidson. 2008. "Regulation of the neural circuitry of emotion by compassion meditation: Effects of meditative expertise." *PLOS ONE* 3 (3). doi: 10.1371/journal.pone.0001897.

Blumenthal, James A., Patrick J. Smith, and Benson M. Hoffman. 2012. "Is exercise a viable treatment for depression?" *ACSM'S Health & Fitness Journal* 16 (4):14–21. doi: 10.1249/01.FIT.0000416000.09526.eb.

Borghesan, M., W. M. H. Hoogaars, M. Varela-Eirin, N. Talma, and M. Demaria. 2020. "A senescence-centric view of aging: Implications for longevity and disease." *Trends in Cell Biology* 30 (10):777–91. doi: 10.1016/j.tcb.2020.07.002.

Buchowski, Maciej S., Kathrin Rehfeld, Angie Lüders, Anita Hökelmann, Volkmar Lessmann, Joern Kaufmann, et al. 2018. "Dance training is superior to repetitive physical exercise in inducing brain plasticity in the elderly." *PLOS ONE* 13 (7). doi: 10.1371/journal.pone.0196636.

Cai, Dongsheng, and Sinan Kohr. 2019. "'Hypothalamic microinflammation' paradigm in aging and metabolic diseases." *Cell Metabolism* 30 (1):19–35. doi: 10.1016/j.cmet.2019.05.021.

Collins, Nicholas, Natalia Ledo Husby Phillips, Lauren Reich, Katrina Milbocker, and Tania L. Roth. 2020. "Epigenetic consequences of adversity and intervention throughout the lifespan: Implications for public policy and healthcare." *Adversity and Resilience Science* 1 (3):205–16. doi: 10.1007/s42844-020-00015-5.

de Cabo, Rafael, Dan L. Longo, and Mark P. Mattson. 2019. "Effects of intermittent fasting on health, aging, and disease." *New England Journal of Medicine* 381 (26):2541–51. doi: 10.1056/NEJMra1905136.

Gaspard, Ulysse, Mélanie Taziaux, Marie Mawet, Maud Jost, Valérie Gordenne, Herjan J. T. Coelingh Bennink, et al. 2020. "A multicenter, randomized study to select the minimum effective dose of estetrol (E4) in postmenopausal women (E4Relief): part 1. Vasomotor symptoms and overall safety." *Menopause* 27 (8):848–57. doi: 10.1097/gme.0000000000001561.

Monteiro-Junior, Renato Sobral, Paulo de Tarso Maciel-Pinheiro, Eduardo da Matta Mello Portugal, Luiz Felipe da Silva Figueiredo, Rodrigo Terra, et al. 2018. "Effect of exercise on inflammatory profile of older persons: Systematic review and meta-analyses." *Journal of Physical Activity and Health* 15 (1):64–71. doi: 10.1123/jpah.2016-0735.

Robinson, M. M., S. Dasari, A. R. Konopka, M. L. Johnson, S. Manjunatha, R. R. Esponda, et al. 2017. "Enhanced protein translation underlies improved metabolic and physical adaptations to different exercise training modes in young and old humans." *Cell Metabolism* 25 (3):581–92. doi: 10.1016/j.cmet.2017.02.009.

Sparkman, Nathan L., and Rodney W. Johnson. 2008. "Neuroinflammation associated with aging sensitizes the brain to the effects of infection or stress." *Neuroimmunomodulation* 15 (4–6):323–30. doi: 10.1159/000156474.

Uchoa, Mariana F., V. Alexandra Moser, and Christian J. Pike. 2016. "Interactions between inflammation, sex steroids, and Alzheimer's disease risk factors." *Frontiers in Neuroendocrinology* 43:60–82. doi: 10.1016/j.yfrne.2016.09.001.

Vamvakopoulos, N. C., and G. P. Chrousos. 1993. "Evidence of direct estrogenic regulation of human corticotropin-releasing hormone gene expression. Potential implications for the sexual dimorphism of the stress response and immune/inflammatory reaction." *Journal of Clinical Investigation* 92 (4):1896–902. doi: 10.1172/jci116782.

Index

About the Author

Dr Louann Brizendine is the Lynne and Marc Benioff Endowed Chair in Clinical Psychiatry at the University of California, San Francisco, founder of UCSF's Women's Mood and Hormone Clinic and the *New York Times* bestselling author of *The Female Brain* and *The Male Brain*. She lives in Sausalito, California, with her husband, Samuel Barondes.

Also by bestselling author
DR LOUANN BRIZENDINE

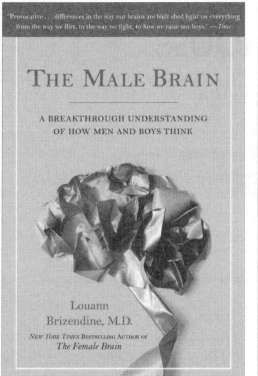

"Provocative . . . differences in the way our brains are built shed light on everything from the way we flirt, to the way we fight, to how we raise our boys." —*Time*

THE MALE BRAIN

A BREAKTHROUGH UNDERSTANDING
OF HOW MEN AND BOYS THINK

Louann
Brizendine, M.D.

NEW YORK TIMES BESTSELLING AUTHOR OF
The Female Brain

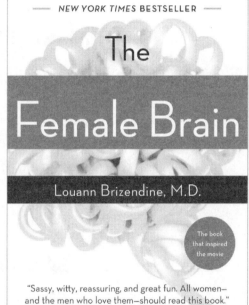

NEW YORK TIMES BESTSELLER

The

Female Brain

Louann Brizendine, M.D.

The book
that inspired
the movie

"Sassy, witty, reassuring, and great fun. All women—
and the men who love them—should read this book."
–Christiane Northrup, M.D., author of *The Wisdom of Menopause*

10TH ANNIVERSARY EDITION

HAY HOUSE

Look within

Join the conversation about latest products, events, exclusive offers and more.

 Hay House

 @HayHouseUK

 @hayhouseuk

We'd love to hear from you!

Contents

10 Unjust Dismissal 249

11 Marketing, Advertising, and Product Safety 273

12 Occupational Health and Safety 311

13 Ethics in Finance 339

Preface

The field of business ethics has grown in recent years into an interdisciplinary area of study that has found a secure niche in both the liberal arts and business education. Credit for this development belongs to many individuals—both philosophers and business scholars—who have succeeded in relating ethical theory to the various problems of ethics that arise in business. They have shown not only that business is a fruitful subject for philosophical exploration, but also that practicing managers in the world of business can benefit from the results.

Ethics and the Conduct of Business, fourth edition, is a comprehensive and up-to-date discussion of the most prominent issues in the field of business ethics and the major positions and arguments on these issues. It is intended to be used as a text in a philosophical business ethics course or one taught in a school of business, on either the undergraduate or M.B.A. level. The substantial number of cases included provides ample opportunity for a case study approach or a combined lecture-discussion format. There has been no attempt to develop a distinctive ethical system or to argue for specific conclusions. The field of business ethics is marked by reasonable disagreement that should be reflected in any good textbook.

The focus of *Ethics and the Conduct of Business* is primarily on ethical issues that corporate decision makers face in developing policies about employees, customers, and the general public. The positions on these issues and the arguments for them are taken from a wide variety of sources, including economics and the law. The study of ethical issues in business is not confined to a single academic discipline or to the academic world. The issues selected for discussion are widely debated by legislators, judges, government regulators, business leaders, journalists, and, indeed, virtually everyone with an interest in business.

An underlying assumption of this book is that ethical theory is essential for a full understanding of the positions and arguments offered on the main issues in business ethics. Fortunately, the amount of theory needed is relatively small, and much of the discussion of these issues can be understood apart from the theoretical foundation provided here. The book also contains a substantial amount of legal material, not only because the law addresses many ethical issues, but also because management decision making must take account of the relevant law. Many examples are used throughout the book in order to explain points and show the relevance of the discussion to real-life business practice.

Preparing the fourth edition of *Ethics and the Conduct of Business* provides an opportunity to incorporate new developments and increase its value in the classroom. The most significant changes are the addition of a new chapter on ethics in finance and an overhaul of the chapter on international business ethics. Although finance involves many substantial ethical issues, the field of ethics in largely unformed. As a result the topic has been neglected in business ethics textbooks. The material in this chapter is adopted from my own work *Ethics in Finance* (Blackwell Publishers, 1999), which is the first textbook in the field. Except for the section on insider trading, which has been retained from previous editions, the chapter on ethics in finance contains new material covering ethical issues in financial markets, financial services, and hostile takeovers, with cases on the takeover of Pacific Lumber, check-kiting at E.F. Hutton, and the bond trading scandal at Salomon Brothers.

International business ethics is perhaps the fastest growing area in business ethics, and so a revised treatment is appropriate with a new edition. Because the problem of international "sweatshops" has received great attention in recent years, this chapter begins with a case on Nike and includes a section on the topic of foreign contractors.

In addition to the new cases in the chapters on ethics in finance and international business ethics, the fourth edition contains four others: the case "Beech-Nut's Bogus Apple Juice" in Chapter 2; a case on AIDS drugs in South Africa in Chapter 4; a case on an industrial espionage dispute between Procter & Gamble and Unilever in Chapter 6; and the case "The Nun and the CEO" in Chapter 14. Other additions include sections on managing conflict of interest in Chapter 6 and privacy and the Internet in Chapter 7.

The task of revising *Ethics and the Conduct of Business* for a fourth edition has been aided by the advice of many instructors who have used it. To these contributors, whose names are too numerous to mention, I express my thanks. I also wish to thank the reviewers for this edition: Elwin Myers, Texas A&M University and Robin Rathke, University of Texas, San Antonio. I am grateful for the support of Loyola University Chicago, especially the School of Business Administration and Dean Henry Venta. I have benefited from the resources of the Raymond C. Baumhart, S.J., chair in business ethics, which was created to honor a former president of Loyola University Chicago, who was also a pioneer in the field of business ethics. To Ray Baumhart I owe a special debt of gratitude. I am thankful for the secretarial support of Kathleen King and the research assistance of Pragnesh Hariawala and Mary Koenig. Finally, my deepest expression of appreciation goes to my wife Claudia whose affection, patience, and support have been essential for the preparation of the fourth edition, as they were for the ones previous.

John R. Boatright

Acknowledgments

I would like to express my gratitude for permission to use material from the following sources:

A. Carl Kotchian, "The Payoff: Lockheed's 70-day Mission to Tokyo," *Saturday Review Magazine*, 9 July 1977. Copyright ©1977, reprinted by permission of *Saturday Review Magazine*.

Thomas Donaldson and Lee E. Preston, "The Stakeholder Theory of the Corporation: Concepts, Evidence, and Implications," *Academy of Management Review*, 20 (1995). Reprinted by permission.

Johnson & Johnson, the statement *Our Credo*, reprinted by permission.

John R. Boatright, *Ethics in Finance* (Malden, MA: Blackwell Publishers, 1999). Copyright ©1999 by John R. Boatright, reprinted by permission of Blackwell Publishers.

The letter from T. J. Rodgers to Sister Doris Gormley, dated May 23, 1996. Copyright © 1996 by T. J. Rodgers, reprinted by permission.

1

Ethics in the World of Business

CASE 1.1 Johnson & Johnson: The Tylenol Crisis

On September 30, 1982, James Burke, the CEO of Johnson & Johnson, received word that several deaths in the Chicago area might be linked to one of the company's products.[1] Reports were sketchy at first, but authorities eventually determined that seven people died from taking cyanide-laced capsules of Extra-Strength Tylenol. The news riveted the nation's attention, and Burke faced the challenge of his career.

Tylenol is one of Johnson & Johnson's most successful products. The nonaspirin pain reliever was developed in the mid-1950s by McNeil Laboratories and sold initially as a prescription drug, primarily for hospital use. After Johnson & Johnson acquired McNeil Laboratories in 1959, the company recognized the drug's potential and gained approval to sell it as an over-the-counter medication. Sales increased slowly but steadily, and by 1982 Tylenol had captured over 35 percent of the $1 billion analgesic market—over three times the market share of its nearest competitor. The product provided 7 percent of Johnson & Johnson sales and a whopping 17 percent of the company's profits.

Tylenol's success was achieved by heavy advertising and price reductions. Because the only active ingredient, acetaminophen, can easily be manufactured, the drug was open prey to competitors, and when Bristol-Myers introduced Datril in 1975, Johnson & Johnson had to move quickly. The company protected its market share by slashing prices by 30 percent and boosting advertising to $4 million in 1976 (from $142,000 the year before). The figure rose to $40 million in 1982, and during the seven-year period from 1976 to 1982, Johnson & Johnson spent over $155 million to promote the Tylenol brand. Developing a best-selling brand had been expensive, but by 1982 the company was reaping the rewards of its investment.

The cyanide that caused the deaths was placed in capsules of Extra-Strength Tylenol. Advertised as "the most potent pain reliever you can buy without a prescription," the extra-strength product—which contains 500 milligrams of acetaminophen as compared with 350 milligrams in regular Tylenol—was sold at the time in both capsule and tablet form. Extra-Strength Tylenol was instantly popular, and by 1979 it accounted for 70 percent of all Tylenol sales. The capsule form appealed to consumers because it was easier to swallow and also because of an association of capsules with strength. Capsules are susceptible to tampering, however, because they can be pulled apart and refilled, and the fact that all of the deaths occurred in one area—and that only a few capsules in each of the bottles contained cyanide—suggested that someone (who has never been apprehended) took advantage of the ease with which capsules can be used to kill.

James Burke and his staff quickly concluded that the tampering had not occurred at a McNeil production facility, and so the company was not responsible for the cyanide contamination itself. Still, the poisonings were associated with the Tylenol name, and sales of the brand were dropping rapidly. The public reaction was fueled by confusion, which was furthered by false reports of Tylenol-related deaths in several other states, and the company feared a rash of copycat incidents, especially as Halloween approached.

Johnson & Johnson faced two problems. One was how to respond to public concerns—and to the possibility that more cyanide-laced capsules might still be found. What information should the company release to the media and to worried customers? What changes, if any, should be made in the advertising for Tylenol, or should all advertising be suspended? Should Johnson & Johnson recall capsules of Extra-Strength Tylenol only in the Chicago area, or should the company pull all capsules from store shelves nationwide? A nationwide recall would involve an estimated 31 million bottles with a retail value of $100 million. Removal of the product would also cede valuable shelf space to Tylenol's competitors and risk a permanent loss of market share.

The problem, moreover, was tampering, which could occur with any medication in capsule form, and indeed with any over-the-counter drug. The pharmaceutical industry, and not merely Johnson & Johnson, was threatened by the Tylenol poisonings. The cost of a recall outside of the Chicago area was out of proportion to the risk, and consumers faced far greater risk from adverse reactions to many pharmaceutical products that were regarded as safe. On the other hand, capsules of Extra-Strength Tylenol on store shelves (which no longer belonged to Johnson & Johnson) was so much "dead stock" that might never be sold, and the loss would be borne by distributors and retailers.

The second problem was whether—and if so, how—to save the Tylenol brand name. Many marketing experts were convinced that the brand was doomed—that the public would forever associate Tylenol with death—and that a new identity should be sought for an acetaminophen product before the company's competitors could capture the market. James Burke was convinced that the brand name could be saved if Johnson & Johnson restored public confidence quickly, but doing so would require not only reassuring words but steps to make tampering less likely. An aggressive campaign could backfire, however, if another poisoning occurred, because the company could be accused of being more concerned with profits than consumer safety.

In addressing these problems, James Burke and others at Johnson & Johnson had the benefit of goodwill from a public that recognized that the company was also a victim of a senseless crime. The company also had an invaluable resource in the relation of trust that had been built through decades of adherence to the Johnson & Johnson credo. This declaration of the company's responsibility to customers, employees, and the community—as well as stockholders—had enabled Johnson & Johnson to prosper so far. Could it also be a guide in the current crisis?

INTRODUCTION

The Tylenol crisis created an unusual, and fortunately rare, test for the managers of Johnson & Johnson. The ethical dilemmas of management are generally less dramatic. Still, this case exhibits some typical features of ethical decision making in business. Johnson & Johnson managers faced ethical issues that were inextricably bound to practical business concerns. Their job was not merely to do the right thing but also to make a sound business decision. The future of the company was at stake. They had to act, moreover, in a highly competitive market environment under severe time pressure without adequate information. In the Tylenol crisis, the managers of Johnson & Johnson had to consider issues that were outside their expertise and not commonly a part of corporate decision making. They benefited, however, from the values expressed in the Johnson & Johnson credo.

This book is about the ethical issues that arise for managers—and, indeed, for all people, including employees, consumers, and members of the public. Corporate activities affect us all, and so the conduct of business is a matter of concern for everyone with a stake in ethical management. The ethical issues we will be examining are those considered by managers in the ordinary course of their work, but they are also issues that are discussed in the pages of the business press, debated in the halls of Congress, and scrutinized by the courts. This is because ethical issues in business are closely tied to important matters of public policy and to the legislative and judicial processes of government. They are often only part of a complex set of issues.

Case 1.2 Four Business Decisions

The Sales Rep

A sales representative for a struggling computer supply firm has a chance to close a multimillion-dollar deal for an office system to be installed over a two-year period. The machines for the first delivery are in the company's warehouse, but the remainder would have to be ordered from the manufacturer. Because the manufacturer is having difficulty meeting the heavy demand for the popular model, the sales representative is not sure that subsequent deliveries can be made on time. Any delay in converting to the new system would be costly to the customer; however, the blame could be placed on the manufacturer. Should the sales representative close the deal without advising the customer of the problem?

The Research Director

The director of research in a large aerospace firm recently promoted a woman to head an engineering team charged with designing a critical component for a new plane. She was tapped for the job because of her superior knowledge of the engineering aspects of the project, but the men under her direction have been expressing resentment at working for a woman by subtly sabotaging the work of the team. The director believes that it is unfair to deprive the woman of advancement merely because of the prejudice of her male colleagues, but quick completion of the designs and the building of a prototype are vital to the success of the company. Should he remove the woman as head of the engineering team?

The Marketing VP

The vice president of marketing for a major brewing company is aware that college students account for a large proportion of beer sales and that people in this age group form lifelong loyalties to particular brands of beer. The executive is personally uncomfortable with the tasteless gimmicks used by her competitors in the industry to encourage drinking on campuses, including beach parties and beer-drinking contests. She worries about the company's contribution to underage drinking and alcohol abuse among college students. Should she go along with the competition?

The CEO

The CEO of a midsize producer of a popular line of kitchen appliances is approached about merging with a larger company. The terms offered by the suitor are very advantageous to the CEO, who would receive a large severance package. The shareholders of the firm would also benefit, because the offer for their stock is substantially above the current market price. The CEO learns, however, that plans call for closing a plant that is the major employer in a small town. The firm has always taken its social responsibility seriously, but the CEO is now unsure of how to balance the welfare of the employees who would be thrown out of work and the community where the plant is located against the interests of the shareholders. He is also not sure how much to take his own interests into account. Should he bail out in order to enrich himself?

BUSINESS DECISION MAKING

These four hypothetical examples give some idea of the ethical issues that arise at all levels of business. The individuals in these cases are faced with questions about ethics in their relations with customers, employees, and members of the larger society. Frequently the ethically correct course of action is clear, and people in business act accordingly. Exceptions occur, however, when there is uncertainty about ethical obligations in particular situations or when considerations of ethics come into conflict with the practical demands of business. The sales representative might not be sure, for example, about the extent to which he is obligated to provide information about possible delays in delivery. And the director of research, although convinced that discrimination is wrong, might still feel that he has no choice but to remove the woman as head of the team in order to get the job done.

In deciding on an ethical course of action, we can rely to some extent on the rules of right conduct that we employ in everyday life. Deception is wrong, for example, whether we deceive a friend or a customer. And corporations no less than persons have an obligation not to discriminate or cause harm. Unfortunately, business activity also has some features that limit the applicability of our ordinary ethical views. What we ought to do depends to some extent on our situation and on the particular roles we occupy, and slightly different rules or codes of ethics are needed to guide us in the different departments of our lives. The CEO, by virtue of his position, has responsibilities to several different constituencies, and his problem in part is to find the proper balance.

One of the features that distinguishes business activity is its *economic* character. In the world of business, we interact with each other not as family members, friends, or neighbors but as buyers and sellers, employers and employees, and the like. Trading, for example, is often accompanied by hard bargaining, in which both sides conceal their full hand and perhaps engage in some bluffing. And a skilled salesperson is well versed in the art of arousing a customer's attention (sometimes by a bit of puffery) to clinch the sale. Still, there is an "ethics of trading" that prohibits the use of false or deceptive claims and tricks such as "bait-and-switch" advertising.

Employment is also recognized as a special relation with its own standards of right and wrong. Employers are generally entitled to hire and promote whomever they wish and to lay off or terminate workers without regard for the consequences. (This right is being increasingly challenged, however, by those who hold that employers ought to fire only for cause and to give employees an opportunity to defend themselves.) Employees also have some protections, such as a right not to be discriminated against or to be exposed to workplace hazards. There are many controversies in the workplace, such as the rights of employers and employees with regard to drug testing.

The ethics of business, then, is at least in part the ethics of economic relations—such as those involving buyers and sellers and employers and employees. So we need to ask, what are the rules that ought to govern these kinds of relations? And how do these rules differ from those that apply in other spheres of life?

A second distinguishing feature of business activity is that it typically takes place in large, impersonal *organizations*. An organization, according to organizational theory, is a hierarchical system of functionally defined positions designed to achieve some goal or set of goals. Consequently, the members of a business organization, in assuming a particular position—such as sales representative or vice president for marketing or CEO—take on new obligations to pursue the goals of the firm. Thus, the marketing executive is not free to act solely on her own standards of good taste and social responsibility at the expense of sales for the brewing company.

Nor can the CEO rightfully ignore the interests of shareholders and consider only the impact of the merger on one group of employees and their community any more than he can consider only his self-interest.

Levels of Decision Making

Decision making occurs on several distinct levels: the level of the *individual,* the level of the *organization,* and the level of the *business system.* Situations that confront individuals in the workplace and require them to make a decision about their own response are on the level of individual decision making. An employee with an unreasonably demanding boss, for example, or a boss who is discovered padding his expense account faces the question: What do I do? Whether to live with the difficult boss or to blow the whistle on the padding are questions to be answered by the individual and acted on accordingly.

Many ethical problems occur at the level of the organization in the sense that the individual decision maker is acting on behalf of the organization in bringing about some organizational change. Sexual harassment, for example, is an individual matter for the person suffering the abuse (and for the harasser), but a manager in an office where sexual harassment is happening must take steps not only to rectify the situation but also to ensure that it does not occur again. The decision in this case may be a disciplinary action, which involves a manager acting within his or her organizational role. The manager may also institute training to prevent sexual harassment and possibly develop a sexual harassment policy, which not only prohibits certain behavior but also creates procedures for handling complaints. Developing a policy with regard to sexual harassment, as opposed to dealing with harassment of one's self, involves decisions on the organizational level rather than the level of the individual.

Problems that result from accepted business practices or from features of the economic system cannot effectively be addressed by any single organization, much less a lone individual. Sales practices within an industry, for example, are difficult for one company to change singlehandedly, because the company is constrained by competition with possibly less ethical competitors. The most effective solution is likely to be an industrywide code of ethics, agreed to by all. Similarly, the lower pay of women's work, which is discussed in Chapter 9, results from structural features of the labor market, which no one company or even industry can alter. A single employer can adopt a policy of comparable worth as one possible solution because the problem is systemic, and consequently any substantial change must be on the level of the system. Systemic problems are best solved by some form of regulation or economic reform.

Identifying the appropriate level for a decision is important, because an ethical problem may have no solution on the level at which it is approached. The beer marketer described in Case 1.2 may have little choice but to follow the competition in using tasteless gimmicks, because the problem has no real solution on the individual or organizational level. An effective response requires that she place the problem on the systemic level and seek a solution appropriate to that level. Richard T. DeGeorge has described such a move as "ethical displacement," which consists of addressing a problem on a level other than the one on which the problem appears.[2] The fact that some problems can be solved only by displacing them to a higher level is a source of great distress for individuals in difficult situations, because they still must find some less-than-perfect response on a lower level.

Three Points of View

Decision making in business involves many factors, of which ethics is only one. In order to gain an understanding of the relevance of ethics for the conduct of business, it will be useful to begin with a description of three points of view from which decisions in business can be made: the economic, the legal, and the moral. Then we can see how these points of view may be integrated to form an approach to business decision making that can aid people facing difficult ethical situations.

Closing the deal for the sale of an office computer system is good from a strictly economic or business point of view, as long as there are no repercussions. The quick completion of the design for the airplane component and increasing beer sales among college students are good for the same reason. The individuals involved might also evaluate different courses of action from the point of view of their own careers and ask: What is best for me? What should I do from the point of view of self-interest? Or the individuals involved might consider the law and take a legal point of view. Would removing the woman from her job as head of the engineering team be considered illegal sexual discrimination? Would it be legal to oppose the merger merely to avoid closing the plant? Finally, the individuals involved might consider the moral point of view when making their decisions. Morally speaking, what is the best thing to do?

The Moral Point of View. In order to understand what it means to decide something from the moral point of view, let us consider the case of the sales representative. In deciding whether to disclose the possible delays in delivery, he might ask: What is accepted business practice? What would my boss expect me to do? What would other sales representatives in my company or the industry do? What kind of conduct is generally regarded as legally permissible? To proceed in these ways is to seek guidance from what is conventionally thought by one's peers or society at large to be right or wrong. Unable to find an answer, or perhaps wanting to make sure that he had found the correct one, the sales representative might push further and ask for the reasons that he ought to act in one way rather than another. Three reasons readily suggest themselves.

First, informing the customer about the possibility of a delay might result in losing the sale and a handsome commission. Judged purely by considerations of benefit to himself, he ought to close the deal, unless, of course, he would suffer greater harm if the company is unable to fulfill the contract. If benefit and harm to himself are reasons for acting in some way, why should he not consider the benefit and harm for the customer as well? The fact that the customer might suffer substantial losses would seem to be a morally relevant reason for revealing the possible delivery problems. But the sales representative might think, "I have no obligation to look out for this customer's welfare. If he suffers a loss, that's his problem, not mine; I am not going to pass up a sale to protect him."

On further reflection, however, he might realize that trust is essential in his line of work. If he and the company acquire a reputation for dishonesty, doing business will be more difficult in the future. In addition, if trust is lost in business as a whole so that buyers and sellers can no longer rely on each other's word in their dealings, then everyone suffers. Full disclosure in trade is of value because, on the whole, it helps everyone. In terms of benefit and harm for all concerned, therefore, it is a good policy to inform a customer of matters such as the possibility of delays.

Second, insofar as not revealing the information is misrepresentation, it is a form of lying, and we have been taught since childhood that lying is wrong. Misrepresentation does not require that something false actually be said. If the sales representative assures the customer that there will be no problem with the deliveries, then he is lying. But a person can lie by remaining silent or even saying something true. Consider a person selling a used car who says that the transmission was checked by a mechanic only last week but fails to add that the mechanic found serious problems. If the seller's words would lead a hearer to conclude that the transmission is in sound condition, misrepresentation has occurred.

If we wish to push the matter further, we can ask, what is morally wrong with lying or misrepresentation? If we appeal to the harm done, so that the rule "Do not lie" is itself based on benefit and harm, then this second reason is no different from the first. However, a different line of reasoning can be sketched as follows. To intentionally bring about a false belief so that another person cannot make a rational decision about some matter of importance is to manipulate that person. Manipulating or using another person is morally objectionable because it involves treating people as "things" for satisfying our desires and preventing them from acting to satisfy their own desires. In short, manipulation shows a lack of respect for the essential humanity of others. And the idea of respect for persons is an important moral consideration—different from benefit and harm—that supports the commonly accepted view that lying is wrong.

Third, the sales representative might ask: How would my action appear to the customer were he to know the full facts? Or how would I view it if I were in the customer's place? Would I want to be treated in the same way? These questions suggest a line of reasoning that is commonly expressed by the Golden Rule: Do unto others as you would have them do unto you. Part of the force of this rule is its insistence on equality, which is an important element of fairness or justice. To treat others in ways that we would not like to be treated is to make exceptions for ourselves and hence to depart from strict equality. The point is also expressed in the familiar slogan "What's fair for one is fair for all." Assuming that the sales representative would not want to have important information withheld from him, then it would be wrong for him to withhold the information from the customer.

Two Features. The moral point of view has two important features.[3] First is a willingness to seek out and act on *reasons.* The best action, according to one writer, is "the course of action which is supported by the best reasons."[4] This does not get us very far without some account of what are the best reasons, but it indicates a commitment to use reason in deliberating about what to do and to construct moral arguments that are persuasive to ourselves and others. Moral rules should not be accepted merely because they are a part of the prevailing morality. Rather, we should attempt to justify the rules we act on by means of the most general and comprehensive kind of reasons available.

Second, the moral point of view requires us to be *impartial.* We must regard the interests of everyone, including ourselves, as equally worthy of consideration and give all interests equal weight in deciding what to do. The moral point of view is the opposite of being purely self-interested. The idea of a personal morality—that is, a morality to be followed only by ourselves—is absurd. Morality by its very nature is public, in the sense that it involves a shared set of rules that can be observed by everyone.[5] A good test of the moral point of view is whether we would feel comfortable if our colleagues, friends, and family were to know about a decision we had made. Would we be willing to have an article on it appear on the front page of the local newspaper, for example? A decision made from the moral point of view can withstand and even invites this kind of openness and scrutiny.

An Integrated Approach

The approach advocated in this book is that decision making in business should involve an integration of all three points of view: the economic, the legal, and the moral. Business ethics is, in part, the attempt to think clearly and deeply about ethical issues in business and to arrive at conclusions that are supported by the strongest possible arguments. An integrated approach requires that we give proper weight to the economic and legal aspects of a problem, but to think that sound business decisions could be made solely from a perspective that excludes ethics is just as wrongheaded as it is to think that they could be made on the basis of ethical reasoning alone.

Integrating different points of view is nothing new; we do it all the time. Managers must juggle financial, production, marketing, personnel, and a host of other factors in taking just the economic point of view. Inevitably, there is tension between the three points of view, but the ideal resolution is not a trade-off between ethics and other considerations. The outcome, instead, should be a decision that is ethically defensible while at the same time satisfying the legitimate demands of economic performance and a company's legal obligations.

An example of an integrated approach is provided by Johnson & Johnson's response to the Tylenol crisis (see Case 1.1). Under the leadership of CEO James

Burke, the company quickly cleared all bottles of Extra-Strength Tylenol capsules off store shelves in the Chicago area and recalled all bottles from the two batches that had been identified in the poisonings. A decision was made to be completely candid with the medical community, the media, and the public. Accordingly, the company issued warnings to physicians and hospitals around the world, briefed the press fully, and provided a toll-free telephone number for answering consumer inquiries. In addition, all advertising for Tylenol was temporarily suspended. On October 6, 1982, one week after the first Tylenol-related deaths, Johnson & Johnson instituted a nationwide recall of the 31 million bottles of Regular- and Extra-Strength Tylenol capsules that remained unsold.

James Burke was convinced that the Tylenol brand name could be saved with a well-designed strategy. The key was to protect consumers by developing a tamper-resistant package that would both discourage tampering and make tampering more evident. The Tylenol crisis had alerted the pharmaceutical industry to the need for tamper-resistant packaging, and on November 4, 1982, the Food and Drug Administration (FDA) made such protection mandatory. By acting quickly to develop the technology, Johnson & Johnson could establish an industry leadership position. The company decided on a three-point protection system consisting of a tape seal on the box, a wrapper on the cap, and a seal over the mouth of the bottle. By late November, Extra-Strength Tylenol in capsule form was once again on store shelves. Advertising was resumed in December, and more than 80 million coupons were distributed for the product in its new tamper-resistant packaging. The direct cost for this recovery effort has been estimated at $150 million.

The response of Johnson & Johnson to the Tylenol crisis can be explained as sound business decision making. The effort to restore a valuable brand name was costly, but the high-stakes gamble paid off. The decision for a nationwide recall of all capsule products may appear obvious in retrospect because of the crucial role it played in assuring consumers of the company's commitment to safety. Strong arguments were made at the time against a recall, however, including the tremendous expense and the success of other companies in dismissing product tampering as isolated incidents.

An important factor in deciding on a recall was the Johnson & Johnson credo (see Exhibit 1.1). Developed in the 1940s as an expression of the company's way of doing business, the credo guided several generations of Johnson & Johnson employees. One of James Burke's first acts upon becoming CEO was to hold a series of credo "challenges," to determine whether the document was still relevant to management decision making, and the outcome was a gratifying reaffirmation of the credo's continuing force. Although economic and legal considerations played a role in Johnson & Johnson's response to the Tylenol crisis, the credo served too, and the clear statements of responsibility that it contained enabled the company to make some really tough, but ultimately right, decisions.

In retrospect, however, the company might be faulted for one decision: to continue marketing Extra-Strength Tylenol in capsule form. Consumer surveys found that a segment of the market strongly preferred medication in a capsule. But

---

Exhibit 1
Johnson & Johnson: Our Credo

We believe our first responsibility is to the doctors, nurses and patients, to mothers and fathers and all others who use our products and services. In meeting their needs everything we do must be of high quality. We must constantly strive to reduce our costs in order to maintain reasonable prices. Customers' orders must be serviced promptly and accurately. Our suppliers and distributors must have an opportunity to make a fair profit.

We are responsible to our employees, the men and women who work with us throughout the world. Everyone must be considered as an individual. We must respect their dignity and recognize their merit. They must have a sense of security in their jobs. Compensation must be fair and adequate, and working conditions clean, orderly and safe. We must be mindful of ways to help our employees fulfill their family responsibilities. Employees must feel free to make suggestions and complaints. There must be equal opportunity for employment, development, and advancement for those qualified. We must provide competent management, and their actions must be just and ethical.

We are responsible to the communities in which we live and work and to the world community as well. We must be good citizens—support good works and charities and bear our fair share of taxes. We must encourage civic improvements and better health and education. We must maintain in good order the property we are privileged to use, protecting the environment and natural resources.

Our final responsibility is to our stockholders. Business must make a sound profit. We must experiment with new ideas. Research must be carried on, innovative programs developed and mistakes paid for. New equipment must be purchased, new facilities provided and new products launched. Reserves must be created to provide for adverse times. When we operate according to these principles, the stockholders should realize a fair return.

Johnson & Johnson, the statement *Our Credo*, reprinted by permission.

---

a tampering incident about three years later, in February 1986, took the life of a 23-year-old New York State woman and finally convinced Burke that the safety of capsules could not be assured, even with tamper-resistant packaging. Production of Tylenol in capsule form was terminated, and a new product, Tylenol caplets—tablets shaped like capsules—was introduced. At the time, approximately one-third of all Tylenol sold was in capsule form, and the change was estimated to reduce Tylenol sales by about 6 percent annually. Perhaps Johnson & Johnson should have taken this step three years earlier.

Case 1.3   The Ethics of Hardball

Toys "R" Us: Fair or Foul?

Hardball tactics are often applauded in business, but when Child World was the victim, the toy retailer cried foul.[6] Its complaint was directed against a major competitor, Toys "R" Us, whose employees allegedly bought Child World inventory off the shelves during a promotion in which customers received $25 gift certificates for buying merchandise worth $100. The employees of Toys "R" Us were accused of selecting products that Child World sells close to cost, such as diapers, baby food, and infant formula. These items could be resold by Toys "R" Us at a profit, because the purchase price at Child World was barely above what a wholesaler would charge, and then Toys "R" Us could redeem the certificates for additional free merchandise, which could be resold at an even higher profit. Child World claims that its competitor bought up to $1.5 million worth of merchandise in this undercover manner and received as much as $375,000 worth of gift certificates. The practice is apparently legal, although Child World stated that the promotion excluded dealers, wholesalers, and retailers. Executives at Toys "R" Us do not deny the accusation and contend that the practice is common in the industry. Child World may have left itself open to such a hardball tactic by slashing prices and offering the certificates in an effort to increase market share against its larger rival.

Home Depot: Good Ethics or Shrewd Business?

When weather forecasters predicted that Hurricane Andrew would strike the Miami area with full force, customers rushed to stock up on plywood and other building materials.[7] That weekend the 19 Home Depot stores in southern Florida sold more 4-foot-by-8-foot sheets of exterior plywood than they usually sell in two weeks. On August 24, 1992, the hurricane struck, destroying or damaging more than 75,000 homes, and in the wake of the devastation, individual price gougers were able to sell basics like water and food as well as building materials at wildly inflated prices. But not Home Depot. The chain's stores initially kept prices on plywood at prehurricane levels, and when wholesale prices rose on average 28 percent, the company announced that it would sell plywood, roofing materials, and plastic sheeting at cost and take no profit on the sales. It did limit quantities, however, to prevent price gougers from reselling the goods at higher prices. In addition, Home Depot successfully negotiated with its suppliers of plywood, including Georgia-Pacific, the nation's largest plywood producer, to roll back prices to prehurricane levels. Georgia-Pacific, like Home Depot, has a large presence in Florida; the company runs 16 mills and distribution centers in the state and owns 500,000 acres of timberland. Although prices increased early in anticipation of Hurricane Andrew, Home Depot was still able, with the cooperation of suppliers, to sell half-inch plywood sheets for $10.15 after the hurricane, compared with a price of $8.65 before, thereby limiting the increase to less than 18 percent. Home Depot executives explained their decision as an act of good ethics by not profiting from human misery. Others contend, however, that the company made a shrewd business decision.

ETHICS, ECONOMICS, AND LAW

Businesses are economic organizations that operate within a framework of law. They are organized primarily to provide goods and services as well as jobs, and their success depends on efficient operation. In a capitalist system, firms must compete effectively in an open market and make a profit. American business has often been described as a game, in which the aim is to make as much profit as possible while staying within the rules of the game, which are set mainly by government.[8] On this view, it may be helpful and even essential to observe certain ethical standards, but doing so is merely a means to the end of profit making.

Both economics and law are critical to business decision making, but the view that they are the only relevant considerations and that ethics does not apply is plainly false. Even hard-fought games like football have a code of sportsmanship in addition to a rule book, and business, too, is governed by more than the legal rules. In addition, a competitive business system, in which everyone pursues his or her self-interest, depends for its existence on ethical behavior and is justified on ethical grounds. However, the relation of business ethics to economics and the law is very complicated and not easily summarized. The following discussion is intended to clarify these relations.

The Relation of Ethics and Economics

According to economic theory, firms in a free market utilize scarce resources or factors of production (labor, raw materials, and capital) in order to produce an output (goods and services). The demand for this output is determined by the preferences of individual consumers who select from among the available goods and services so as to maximize the satisfaction of their preferences, which is called utility. Firms also seek to maximize their preferences or utility by increasing their output up to the point where the amount received from the sale of goods and services equals the amount spent for labor, raw materials, and capital—that is, where marginal revenues equal marginal costs. Under fully competitive conditions, the result is economic efficiency, which means the production of the maximum output for the least amount of input.

Economics thus provides an explanatory account of the choices of economic actors, whether they be individuals or firms. On this account, the sole reason for any choice is to maximize utility. However, ethics considers many other kinds of reasons, including rights and justice and noneconomic values. To make a choice on the basis of ethics—that is, to use ethical reasons in making a decision—appears at first glance to be incompatible with economic choice. To make decisions on economic grounds and on ethical grounds is to employ two different kinds of reasoning. This apparent incompatibility dissolves on closer inspection. If the economists' account of economic reasoning is intended to be merely an explanation, then it tells us how we do reason in making economic choices but not how we *ought* to reason. Economics as a science need do no more than offer explanations, but economists generally hold that economic reasoning is also justified. That is, economic actors ought to make utility maximizing choices.

The Justification of the Market System. The argument for this position is the classical defense of the market system, which is discussed in Chapter 4. In *The Wealth of Nations*, Adam Smith, the "father" of modern economics, justified the pursuit of self-interest in exchange on the grounds that by making trades for our own advantage, we promote the interests of others. The justification for a free market capitalist system is, in part, that by pursuing profit, business firms promote the welfare of the whole society. Commentators on Adam Smith have observed that this argument assumes a well-ordered civil society with a high level of honesty and trust and an abundance of other moral virtues. Smith's argument would not apply well to a chaotic society marked by pervasive corruption and mistrust. Furthermore, in his defense of the free market in *The Wealth of Nations*, Smith was speaking about *exchange*, whereas economics also includes *production* and *distribution*.[9] The distribution of goods, for example, is heavily influenced by different initial endowments, access to natural resources, and the vagaries of fortune, among other factors. Whether the vast disparities in wealth in the world are justified is a question of distribution, not exchange, and is not addressed by Smith's argument.

Moreover, certain conditions must be satisfied in order for business activity to benefit society. These include the observance of minimal moral restraints to prevent theft, fraud, and the like. Markets must be fully competitive, with easy entry and exit, and everyone must possess all relevant information. In addition, all costs of production should be reflected in the prices that firms and consumers pay. For example, unintended consequences of business activity, such as job-related accidents, injuries from defective products, and pollution, are costs of production that are often not covered or internalized by the manufacturer but passed to others as spillover effects or *externalities*. Many business ethics problems arise when these conditions for the operation of a free market are not satisfied.

Some Conditions for Free Markets. A common view is that ensuring the conditions for free markets and correcting for their absence is a job for government. It is government's role, in other words, to create the rules of the game that allow managers to make decisions solely on economic grounds. However, the task of maintaining the marketplace cannot be handled by government alone, and the failure of government to do its job may create an obligation for business to help. Although government does enact and enforce laws against theft and fraud, including such specialized forms as the theft of trade secrets and fraud in securities transactions, there are many gray areas in which self-regulation and restraint should be exercised. Hardball tactics like those allegedly employed by Toys "R" Us (Case 1.3) are apparently legal, but many companies would consider such deliberate sabotage of a competitor to be an unacceptable business practice that is incompatible with the market system.

Recent work in economics has revealed the influence of ethics on people's economic behavior. Economists have shown how a reputation for honesty and trustworthiness, for example, attracts customers and potential business partners, thus creating economic opportunities that would not be available otherwise. Similarly, people and firms with an unsavory reputation are punished in the market. People are also motivated in their market behavior by consideration of fairness. This is illustrated by the "ultimatum bargaining game," in which two people are given a certain amount of money (say $10) on the condition that one person proposes how the money is to be divided (for example, $5 to each) and the second person accepts or rejects the proposed division. The first person can make only one proposal, and if the proposal is rejected by the second person, the money is taken away and each person receives nothing. Economic theory suggests that the second person would accept any proposal, no matter how small the share, if the alternative is no money at all. Hence, the first person could offer to share as little as $1 or less. But many people who play the game will refuse a proposal in which they receive a share that is considered too small and hence unfair.[10]

Economists explain the behavior of companies like Home Depot (Case 1.3) by the fact that considerations of fairness force firms to limit profit-seeking behavior. Consumers remember price gouging and other practices that they consider unfair and will punish the wrongdoers by ceasing to do business with them or even engaging in boycotts. One study found that people do not believe that scarcity is an acceptable reason for raising prices (despite what economists teach about supply and demand),[11] and so Home Depot and Georgia-Pacific, which are there for the long haul, have more to lose than gain by taking advantage of a natural disaster. Evidence also indicates that people in a natural disaster feel that everyone ought to make some sacrifice, so that profit seeking by a few is perceived as shirking a fair share of the burden.[12]

Finally, when economics is used in practice to support matters of public policy, it must be guided by noneconomic values. Economic analysis can be applied to the market for cocaine as easily as the soybean market, but it cannot tell us whether we should allow both markets. That is a decision for public policy makers on the basis of other considerations. A tax system, for example, depends on sound economic analysis, but the U.S. tax code attempts to achieve many aims simultaneously and to be accepted as fair. A demonstration that a particular system is the most efficient from a purely economic perspective would not necessarily be persuasive to a legislator in drafting a new tax code.

The Relation of Ethics and the Law

Business activity takes place within an extensive framework of law, and some people hold that law is the only set of rules that applies to business activity. Law, not ethics, is the only relevant guide. The reasons that lead people to hold this view are varied, but two predominate.[13]

Two Schools of Thought. One school of thought is that law and ethics govern two different realms. Law prevails in public life, whereas ethics is a private matter. The law is a clearly defined set of enforceable rules that applies to everyone, whereas ethics is a matter of personal opinion that reflects how we choose to lead our own lives. Consequently, it would be a mistake to apply ethical rules in business, just as it would be a mistake to apply the rules of poker to tennis. A variant of this position is that the law represents a minimal level of expected conduct that everyone should observe. Ethics, on the other hand, is a higher, optional level. It's "nice" to be ethical, but our conduct *has* to be legal.

The other school of thought is that the law embodies the ethics of business. There are ethical rules that apply to business, according to this position, and they have been enacted by legislators into laws, which are enforceable by judges in a court. As a form of social control, law has many advantages over ethics. Law provides more precise and detailed rules than ethics, and the courts not only enforce these rules with state power but also are available to interpret them when the wording is unclear. A common set of rules known to all also provides a level playing field. Imagine the chaos if competing teams each decided for themselves what the rules of a game ought to be. For these reasons, some people hold that it is morally sufficient in business merely to observe the law. Their motto is "If it's legal, then it's morally okay."[14]

Why the Law Is Not Enough. Despite their differences, these two schools of thought have the same practical implication: Managers need consider only the law in making decisions. This implication is not only false but also highly dangerous. Regardless of the view that a practicing manager takes on the relation of law and ethics, reliance on the law alone is prescription for disaster, as many individuals and firms have discovered. Approval from a company's legal department does not always assure a successful legal resolution, and companies have prevailed in court only to suffer adverse consequences in the marketplace. As a practical matter, then, managers need to consider both the ethical and legal aspects of a situation in making a decision for many reasons, including the following.

First, the law is inappropriate for regulating certain aspects of business activity. Not everything that is immoral is illegal. Some ethical issues in business concern interpersonal relations at work or relations between competitors, which would be difficult to regulate by law. Taking credit for someone else's work, making unreasonable demands on subordinates, and unjustly reprimanding an employee are all ethically objectionable practices, but they are best left outside the law. Some hardball tactics against competitors may also be legal but ethically objectionable. Whether the effort of Toys "R" Us to sabotage a promotion by its competitor is acceptable behavior (see Case 1.3) is open to dispute, but not every legal competitive maneuver is ethical. Generally, legislatures and the courts are reluctant to intervene in ordinary business decisions unless significant rights and interests are at stake. They rightly feel that outsiders should not second-guess the business judgment of people closer to a problem and impose broad rules for problems that require a more flexible approach. Companies also prefer to handle many problems without outside interference. Still, just because it is not illegal to do certain things does not mean that it is morally okay.

Second, the law is often slow to develop in new areas of concern. Christopher D. Stone points out that the law is primarily reactive, responding to problems that people in the business world can anticipate and deal with long before they come to public attention.[15] The legislative and judicial processes themselves take a long time, and meanwhile much damage can be done. This is true not only for newly emergent problems but also for long-recognized problems where the law has lagged behind public awareness. For example, racial and sexual discrimination was legal—and widely practiced in business—before the passage of the Civil Rights Act of 1964. It should not take a major piece of legislation to make corporate managers aware that discrimination is wrong. They should have recognized this and changed their discriminatory practices long before Congress finally got around to passing a law. At the present time, legal protection for employees who blow the whistle and those who are unjustly dismissed is just beginning to develop. Employers should not wait until they are forced by law to act on such matters of growing concern.

Third, the law itself often employs moral concepts that are not precisely defined, so it is impossible in some instances to understand the law without considering matters of morality. The requirement of *good faith*, for example, is ubiquitous in law. The National Labor Relations Act requires employers and the representatives of employees to bargain "in good faith." One defense against a charge of price discrimination is that a lower price was offered in a good-faith attempt to meet the price of a competitor. Yet the notion of good faith is not precisely defined in either instance. Abiding by the law, therefore, requires decision makers to have an understanding of this key moral concept.

The *fiduciary duty* of a person, such as a trustee, to act in the best interests of a beneficiary is another example. The classic statement of this duty was given by Justice Benjamin Cardozo in the case *Meinhard* v. *Salmon*:

> Many forms of conduct permissible in a workaday world for those acting at arm's length, are forbidden to those bound by fiduciary ties. A trustee is held to something stricter than the morals of the market place. Not honesty alone, but the punctilio of an honor the most sensitive, is then the standard of behavior.[16]

A person in a fiduciary relation, then, must act with reference to a very high standard that is properly a part of morality.

A fourth argument, closely related to the preceding one, is that the law itself is often unsettled, so that whether some course of action is legal must be decided by the courts. And in making a decision, the courts are often guided by moral considerations. Many people have thought that their actions, although perhaps immoral, were still legal, only to discover otherwise. The courts often refuse to interpret the law literally when doing so gives legal sanction to blatant immorality. Judges have some leeway or discretion in making decisions. In exercising this discretion, judges are not necessarily substituting morality for law but rather expressing a morality that is embodied in the law. Instead of the motto "If it's legal, it's morally okay," another motto is perhaps more accurate: "If it's morally wrong, it's probably also illegal." Where there is doubt about what the law is, morality is a good predictor of how the courts will decide.

Fifth, a pragmatic argument is that the law is a rather inefficient instrument, and an exclusive reliance on law alone invites legislation and litigation where it is not necessary. Many landmark enactments, such as the Civil Rights Act of 1964, the

National Environment Policy Act of 1969, the Occupational Safety and Health Act of 1970, and the Consumer Protection Act of 1972, were passed by Congress in response to public outrage over the well-documented failure of American businesses to act responsibly. Although business leaders lament the explosion of product liability suits by consumers injured by defective products, for example, consumers are left with little choice but to use the legal system when manufacturers themselves hide behind "If it's legal, it's morally okay." Adopting this motto, then, is often shortsighted, and businesses may often advance their self-interest more effectively by considering ethics in making decisions.

ETHICS AND MANAGEMENT

Most managers think of themselves as ethical persons, but some still question whether ethics is relevant to their role as a manager. It is important for people in business to be ethical, they might say, but being ethical in business is no different than being ethical in private life. The implication is that a manager need only be an ethical person. There is no need, in other words, to have specialized knowledge or skills in ethics.

Nothing could be further from the truth. Although there is no separate ethics of business, situations arise in business that are not easily addressed by ordinary ethical rules. We have already observed that the obligation to tell the truth is difficult to apply to the dilemma faced by the sales rep in Case 1.2. In addition, the manager of sales reps might face the task of determining the rules of acceptable sales practices for the whole organization and ensuring that the rules are followed. More broadly, high-level managers have a responsibility for creating and maintaining an ethical corporate climate that protects the organization against unethical and illegal conduct by its members. Furthermore, a well-defined value system serves to guide organizations in uncertain situations and to guard against the pursuit of unwise short-term gains.

Ethical Management and the Management of Ethics

A useful distinction can be made between *ethical management* and the *management of ethics*. Business ethics is often conceived as acting ethically as a manager by doing the right thing. This is *ethical management*. Acting ethically is important, both for individual success and organizational effectiveness. Ethical misconduct has ended more than a few promising careers, and some business firms have been severely harmed and even destroyed by the actions of a few individuals. Major scandals in the news attract our attention, but people in business face less momentous ethical dilemmas in the ordinary course of their work. These dilemmas sometimes result from misconduct by others, as when a subordinate is ordered to commit an unethical or illegal act, but they are also inherent in typical business situations.

The *management of ethics* is acting effectively in situations that have an ethical aspect. These situations occur in both the internal and the external environment of a business firm. Internally, organizations bind members together through myriad rules, procedures, policies, and values that must be carefully managed. Some of these, such as a policy on conflict of interest or the values expressed by the Johnson & Johnson credo, explicitly involve ethics. Effective organizational functioning also depends on gaining the acceptance of the rules, policies, and other guides, and this acceptance requires a perception of fairness and commitment. For example, an organization that does not "walk the talk" when it professes to value diversity is unlikely to gain the full cooperation of its employees. With respect to the external environment, corporations must successfully manage the demands for ethical conduct from groups concerned with racial justice, human rights, the environment, and other matters.

In order to practice both ethical management and the management of ethics, it is necessary for managers to possess some specialized *knowledge*. Many ethical issues have a factual background that must be understood. In dealing with a whistle-blower or developing a whistle-blowing policy, for example, the managers of a company should be aware of the motivation of whistle-blowers, the measures that other companies have found effective, and, not least, the relevant law. Some of this background is provided in Chapter 5 on whistle-blowing. In addition, many ethical issues involve competing theoretical perspectives that need to be understood by a manager. Whether it is ethical to use confidential information about a competitor or personal information about an employee depends on theories about intellectual property rights and the right to privacy that are debated by philosophers and legal theorists. Although a manager need not be equipped to participate in these debates, some familiarity with the theoretical considerations is helpful in dealing with practical situations.

To make sound ethical decisions and to implement them in a corporate environment are *skills* that come with experience and training. Some managers make mistakes because they fail to see the ethical dimensions of a situation. Other managers are unable to give proper weight to competing ethical factors or to see other people's perspectives. Thus, a manager may settle a controversial question to his or her satisfaction, only to discover that others still disagree. Moral imagination is often needed to arrive at creative solutions to problems. Finally, the resolution of a problem usually involves persuading others of the rightness of a position, and so the ability to explain one's reasoning is a valuable skill.

The need for specialized knowledge and skills is especially acute when business is conducted abroad.[17] In global business, there is a lack of consensus on acceptable standards of conduct, and practices that work well at home may fare badly elsewhere. This is especially true in less developed countries with lower standards and weak institutions. How should a manager proceed, for example, in a country with exploitive labor conditions, lax environmental regulation, and pervasive corruption? Even the most ethical manager must rethink his or her beliefs about how business ought to be conducted in other parts of the world.

Ethics and the Role of Managers

Every person in business occupies a role. A role is a structured set of relationships with accompanying rights and obligations. Thus, to be a purchasing agent or a personnel director or an internal auditor is to occupy a role. In occupying a role, a person assumes certain rights that are not held by everyone as well as certain role-specific obligations. Thus, a purchasing agent is empowered to make purchases on behalf of an organization and has a responsibility to make purchasing decisions that are best for the organization. To be a "good" purchasing agent is to do the job of a purchasing agent well.

The obligations of a particular role are sometimes added to those of ordinary morality. That is, a person who occupies a role generally assumes obligations over and above those of everyday life. Sometimes, however, role obligations come into conflict with our other obligations. In selecting people for promotion, a personnel director, for example, is obligated to set aside any considerations of friendship and to be wholly impartial. A person in this position may also be forced to terminate an employee for the good of the organization, without regard for the impact on the employee's life. A personnel director may even be required to implement a decision that he or she believes to be morally wrong, such as terminating an employee for inadequate cause. In such situations, the obligations of a role appear to be in conflict with the obligations of ordinary morality.

The idea of a role morality—that is, a morality that is specific to a particular role—is especially applicable to the situation of *professionals*, such as physicians, lawyers, engineers, and accountants. For example, accountants, including internal auditors and public accountants, have a stringent obligation to ensure the accuracy and completeness of financial information. This responsibility requires that they not only observe high standards of objectivity and integrity but also follow a prescribed course of action in reporting any financial irregularities. Internal auditors are required by their professional code of ethics to disclose their findings to top management and to resign if appropriate action is not taken. But they are prohibited from releasing any information to the public, even if doing so would protect some groups such as investors from great harm. The assumption is that the auditor's role is merely to inform top managers of any financial irregularities and that it is the role of those managers to take appropriate action.

Various justifications have been offered for role obligations. One justification is simply that people in certain positions have responsibilities to many different groups and hence must consider a wide range of interests. The decisions of a personnel director have an impact on everyone connected with a business organization, and so denying a friend a promotion or terminating an employee may be the right thing to do, all things considered. A more sophisticated justification is that roles are created in order to serve society better as a whole. A well-designed system of roles, with accompanying rights and obligations, enables a society to achieve more and thereby benefits everyone. A system of roles thus constitutes a kind of division of labor. As in Adam Smith's pin factory, in which workers who perform specific operations can be more productive than individuals working alone, so, too, a business organization with a multiplicity of roles can be more productive and better serve society. Of course, this justification requires that any system of roles be well designed, and there can be disagreement over the design of any given system. Some have argued, for example, that internal auditors ought to report financial irregularities to outside authorities if management does not take appropriate action.

What Is the Role of Managers?

We cannot understand the rights and obligations of managers without knowing their specific role. Managers serve at all levels of an organization and fulfill a variety of roles. Usually, these are defined by a job description, such as the role of a purchasing agent or a personnel director. Uncertainty arises mainly when we ask about the role of high-level corporate executives who make key decisions about policy and strategy. These questions often take the form: To whom are top managers responsible? Whose interests should they serve? What goals should they strive to achieve? To these questions, three main answers have been proposed.

Managers as Economic Actors. According to one widely accepted view, the manager's role is to make sound economic decisions that enable a firm to succeed in a competitive market. As economic actors, managers are expected to consider primarily economic factors in making decisions, and the main measure of success is profitability. The position is commonly expressed by saying that managers are agents of the shareholders, with an obligation to operate a corporation in the shareholders' interests. Legally, managers are not agents of the shareholders, but the law does impose an obligation on managers to seek a maximum return on all investments. This is the goal of managers who serve as economic actors even if they operate a sole proprietorship, a partnership, or any other kind of business enterprise.

Managers as Trustees. As leaders of business organizations, managers are entrusted with enormous assets and given a charge to manage these assets prudently. Employees, suppliers, customers, investors, and other so-called stakeholders have a stake in the success of a firm, and managers are expected to meet all of their legitimate expectations and to balance any conflicting interests. Generally, trustees have a fiduciary duty to act in all matters in the interests of the designated beneficiaries. A critical question, therefore, is whether managers are trustees or fiduciaries for shareholders alone or for all corporate constituencies.

Managers as Quasi-Public Servants. Managers exert enormous power both inside and outside their organizations. Although they are not elected in a democratic process, they nevertheless have many attributes of government officials, such as the power to make major investment decisions for society. Moreover, managers exercise their power by participating in the political process and cooperating with political bodies, including regulatory agencies. In any political system, power must be legitimized by showing how it serves some generally accepted societal goals, and managerial power is no exception. So managers are expected to demonstrate corporate leadership.

The debate over the role of managers at the highest level of business organizations cannot be settled here. Many of the issues of corporate ethics are discussed in Chapter 14. However, it is important to note that the relation of ethics and management cannot fully be explained without developing a full account of the manager's role, with its attendant rights and obligations.

MORALITY, ETHICS, AND ETHICAL THEORY

Before proceeding further, we need to clarify the meaning of the key terms *morality* and *ethics* and the cognates *moral* and *ethical* and *morally* and *ethically*. Generally, *morality* and *ethics*, *moral* and *ethical*, and so on are interchangeable. The presence of two words in the English language with the same meaning is due to the fact that they derive from different roots: *morality*, from the Latin word *moralitas*, and *ethics*, from the Greek *ethikos*. There is no difference, therefore, between describing discrimination as a moral issue and as an ethical issue, or between saying that discrimination is morally wrong and that it is ethically wrong. There are some subtle differences, however, between *morality* and *ethics*.

Morality is generally used to describe a sociological phenomenon, namely, the existence in a society of rules and standards of conduct. Every society has a morality, because this constitutes the basis for mutually beneficial interaction. Without such fundamental rules as "Do not kill" and "Do not steal," for example, stable communities would be impossible. Not all rules and standards are part of morality, of course. Eating peas with a knife, for example, is a breach of etiquette but not a moral wrong, and the rule "Look both ways before crossing the street" is a rule of prudence, not morality. Etiquette and prudence, therefore, constitute sets of nonmoral rules and standards. Morality also has many complex ties to the law, as we have already observed.

Moralities are also specific to societies and exist at certain times and places. Thus, we can speak of the morality of the Trobriand Islanders or the colonial settlers. The morality of Americans in the 1990s is different from that in the 1950s or the 1850s. We can even speak, as Karl Marx did, of the morality of different classes in society. In a highly developed society such as our own, morality also includes a complex vocabulary and patterns of reasoning that permit the members of the society to engage in moral discourse for the purpose of evaluating the actions of individuals and the practices and institutions of the society. *Ethics* is roughly a synonym for *morality*, but it is often restricted to the rules and norms of specific kinds of conduct or the codes of conduct for specialized groups. Thus, we talk about the ethics of stockbrokers or the code of ethics for the accounting profession but usually not about the morality of these groups.

The term *ethics* also has another, quite different, use which is to denote the field of *moral philosophy*. Ethics, along with logic, epistemology, and metaphysics, is a traditional area of philosophical inquiry that dates back to the time of the ancient Greeks. Ethics as a philosophical endeavor is the study of morality. Such a study is either *descriptive* or *normative*. Descriptive ethics may involve an empirical inquiry into the actual rules or standards of a particular group, or it could also consist of understanding the ethical reasoning process. A sociological study of the values of American business managers would be an example of the former, and the work of psychologists on moral development would illustrate the latter. Normative ethics is concerned largely with the possibility of justification. It takes morality as its subject matter and asks such questions as: Are there any means for showing that the rules and standards of our morality are the right ones? Are there any ultimate moral principles that can be used to resolve inconsistencies or conflicts? Normative ethics is concerned not with what people *believe* we ought to do but with what we *really* ought to do and is determined by reasoning or moral argument. Philosophical ethics is not a substitute for morality; rather, it seeks to organize our ordinary moral beliefs in a precise and consistent manner and to discover whatever justification they have.

Conclusion

In the next few chapters, three major theories of ethics are presented as the basis for our beliefs about moral obligations, rights, and justice, and they are applied in discussions of a number of issues concerning the rights and obligations of employees in a firm, employee relations, the protection of employees and consumers, and the responsibility of corporations to the public at large. We will discover a large measure of disagreement among these ethical theories in their content and in the results of applying them to specific cases. The differences between theories should not lead us to despair of resolving ethical issues or to conclude that one resolution is as good as another. Nor should we be discouraged by the fact that agreement on complex ethical issues is seldom achieved. The best we can do is to analyze the issues as fully as possible, which means getting the facts straight and achieving definitional clarity, and then develop the strongest and most complete arguments we can for what we consider to be the correct conclusions.

Case 1.4   A Sticky Situation

Kent Graham is still on the telephone, receiving the good news that he has just secured his largest order as an account manager for Dura-Stick Label Products.[18] His joy is tinged with uncertainty, however.

Dura-Stick is a leader in label converting for the durable-products marketplace. Label converting consists of converting log rolls of various substrates (paper, polyester, vinyl) into die-cut, printed labels. The company specializes in high-performance labels for the automotive, lawn and garden, and appliance industries. Dura-Stick has a well-deserved reputation for quality, technical knowledge, and service that enables the company to command a premium price for its products in a very competitive market.

Kent Graham has been with Dura-Stick for two years. Because he came to the company with ten years in the label industry, he was able to negotiate a very good salary and compensation plan, but his accomplishments since joining Dura-Stick have been mediocre at best. Kent fears that his time with Dura-Stick might be limited unless he starts closing some big accounts. Furthermore, with a wife and two children to support, losing his job would be disastrous. Kent was on a mission to land a big account.

Kent called on Jack Olson at Spray-On Inc., a manufacturer of industrial spraying systems for the automotive painting industry. Dura-Stick has been providing Spray-On with various warning and instructional labels for about twenty years. Jack has been very pleased with Dura-Stick's performance, especially the quality of its manufacturing department under the direction of Tim Davis. After giving Kent another excellent vendor evaluation report, Jack began to describe a new project at Spray-On, a paint sprayer for household consumer use that needs a seven-color label with very precise graphics. This label is different from the industrial two-color labels that Dura-Stick currently supplies to Spray-On.

Jack explained that this was the biggest project that Spray-On has undertaken in recent years and that it would generate a very large order for some label company. Jack then asked Kent, "Does Dura-Stick produce these multicolor, consumer-type labels?" Kent thought for a moment. He knew that a "yes" would give him a better shot at the business, and Dura-Stick might be able to handle the job, even though the company's experience to date was only with two-color labels. Almost without thinking, he replied, "Sure we can handle it, Jack, that's right up our alley!" "That's great news," Jack shot back. "Now take this sample and give me your proposal by Monday. Oh, and by the way, I hope your proposal looks good, because I would really feel confident if this important project were in the hands of your production people!"

Kent gave the sample to Marty Klein, who is responsible for coordinating the costs and price quotes for new opportunities. Marty took one look at the sample and said emphatically, "We'll have to farm this one out." Kent's heart sank down to his shoes. He knew that Jack would want to work with Dura-Stick only if the labels were produced at Dura-Stick's facility. Yet, he still allowed Marty to put the numbers together for the proposal. Kent presented the proposal to Jack at Spray-On. "Gee, Kent, these prices are pretty high, about 20 percent higher than your competition. That's pretty hard to swallow."

Kent knew that the price would be high because it included the cost of another company producing the labels plus Dura-Stick's usual profit margin, but he countered cheerily, "You know the quality that we provide and how important this project is to your company. Isn't it worth the extra 20 percent for the peace of mind that you will have?"

"Let me think about it," Jack replied.

The next day, Kent got a phone call from Jack. "Congratulations, Kent, Dura-Stick has been awarded the business. It was a tough sell to my people, but I convinced them that the extra money would be well spent because of the excellent production department that you have. If it wasn't for the fact that Tim Davis will personally oversee production, you guys probably would not have gotten this business."

Kent had to bite his tongue. He knew that Tim would not be involved because the labels would be produced in Kansas City by Labeltec, which would then send the finished labels to Dura-Stick for shipment to Spray-On's facility. Kent also knew that Jack would be completely satisfied with the quality of the labels. Besides, this order was crucial to his job security, not to mention the well-being of his company.

While Jack continued to explain Spray-On's decision, Kent pondered how he should close this conversation.

Case 1.5   Argus Incorporated: A Leasing Triangle

Susan Solomon walked slowly down the hall to Craig Dunston's office.[19] She was pondering a telephone call she had just received from an irate Mr. Hayes, who was demanding his missing monthly lease payments. Susan, a computer operations manager for Argus Incorporated, was disturbed because the lease in question was with TekUSA, not Mr. Hayes, and the lease had been terminated four months ago. In fact, the early termination was such an achievement that the Argus CEO had personally commended Susan and Craig. Now their triumph might be turning into a disaster.

As the computer operations manager for Argus Incorporated, a commercial real estate and property management company, Susan was responsible for the lease, purchase, maintenance, and disposition of all computer equipment and services for Argus. Craig hired her as a senior project manager four years ago because of her hands-on technical background and flair for explaining technology to employees and clients. During this time, Susan worked on a wide variety of projects under Craig's direction and showed her range of talents and skills. When the previous computer operations manager left, eight months ago, Craig immediately promoted Susan to this position.

As director of shared services, Craig was Susan's immediate superior, with responsibility for computer services in all of Argus's offices. Although Craig had strong managerial abilities and a thorough knowledge of the company's business, he lacked Susan's technical expertise. As a result, Argus had lagged behind the competition in offering sophisticated services. Craig had hoped that he and Susan could combine their strengths to make Argus the leader in their market.

Susan shared Craig's vision, and they became a powerful team. Their latest success had been to terminate the TekUSA lease eight months early. The agreement required Argus to make a lump-sum payment of $380,000 and return the old equipment to TekUSA. The savings in ongoing costs and the strategic opportunities provided by the new leased equipment more than compensated Argus for the cost of terminating the old lease. A critical factor in securing the early lease termination and negotiating a new lease was Susan's relationship with TekUSA. Although TekUSA had been one of Argus's most unreliable and expensive suppliers, Susan had confidence in its potential. The problem, she concluded, was that TekUSA did not understand Argus's business requirements. However, after a series of meetings that clarified each company's expectations, TekUSA was awarded the bulk of Argus's business for leased technology products.

The Master Lease Agreement allowed TekUSA to sell its interest in the leased equipment to a third party. Mr. Hayes claimed that he had bought the lease rights from TekUSA more than a year ago. If so, then Argus should have sent the final $380,000 payment and the old equipment to Mr. Hayes rather than TekUSA, not to mention the lease payments. TekUSA was required by the lease agreement to notify Argus if any portion of its rights were assigned to a third party. If a Notice of Assignment had been sent by TekUSA, then Argus had made a mistake that would be difficult, and costly, to rectify. If, on the other hand, a Notice of Assignment had not been sent, then the fault lay with TekUSA, which would be responsible for dealing with Mr. Hayes. Because all lease documents and invoices were sent to Craig, Susan made a beeline for his office.

"Well, Susan," Craig sighed as he pulled out the TekUSA lease contract file, "I can't imagine that we could've let something like this fall through the cracks, can you?" Shaking her head, Susan scanned the contract and quickly found the section that allowed TekUSA to assign its rights to a third party, provided notice was given to Argus. When Craig pulled out the invoice folder, they discovered that the invoices were generated by TekUSA with their logo on top, but Mr. Hayes's name and address were in the "remit to" box. Susan's heart sank. Now they had to determine whether Argus had ever received a Notice of Assignment.

Craig reached nervously into another folder and pulled out the Notice of Assignment from TekUSA. It was addressed to Craig and dated just over one year ago, four months before Susan's promotion. The Notice clearly stated that TekUSA was transferring the rights to the lease payments and the equipment to Mr. Hayes. Stunned, Susan realized that Argus had negotiated early termination with the wrong party. Argus was now liable for a second $380,000 and a year's lease payments to Mr. Hayes. Worse, Susan and Craig knew that TekUSA had destroyed as junk the equipment that rightly belonged to Mr. Hayes.

In spite of TekUSA's profitable relationship with Argus, the company was struggling with a serious slump in sales, and it had just been forced to lay off 300 people, a fifth of its work force. The odds of getting the payments back from TekUSA to send to Mr. Hayes were bleak. Worse, because the equipment had been junked, Mr. Hayes might ask for compensation based on its fair market value, which could run as high as $1 million. In its current state, TekUSA could not meet these demands without severe damage to its business. In addition, Argus could not permit such a blow to fall on TekUSA. As a critical supplier, TekUSA had to be kept afloat.

Craig spoke first. "Look, Susan, this Notice of Assignment has no Argus countersignature on it, nor was it sent certified mail with return receipt requested. No one knows we have this document, and no one can prove that we ever got it. Let me shred it and you just tell this Hayes guy that Argus never received a Notice of Assignment. Then his problem is with TekUSA, not with us."

Susan had never known Craig to suggest anything unethical, so they must be in serious trouble. TekUSA was seriously remiss in accepting the payments and negotiating the termination of a lease that it had assigned to someone else. Mr. Hayes had also contributed to the problem by not complaining months earlier. But Argus, and especially Craig, were at fault for overlooking the Notice of Assignment. Should Susan go along with Craig's suggestion in order to protect him and Argus?

Susan had a reputation for resolving tough situations. Her usual strategy was to encourage all parties to cooperate in developing a "win-win" solution. Perhaps she could persuade TekUSA to repay the misdirected funds by reducing Argus's monthly lease payments. Perhaps Mr. Hayes would be satisfied to receive similar equipment in place of the original which had already been destroyed. This would be a hard sell with no guarantee of success, and there were a lot of details to be worked out. Claiming that Argus had never received the notice would be a lot easier, and attempting to work out a solution would put Susan's reputation, and even her career, on the line. She wondered whether the effort was worth it.

NOTES

1. Material on this case is taken from "James Burke: A Career in American Business (A) & (B)," Harvard Business School, 1989; "The 1982 Tylenol Poisoning Episode," in Ronald M. Green, *The Ethical Manager: A New Method for Business Ethics* (New York: Macmillan, 1994), 208–19; and Robert F. Hartley, *Business Ethics: Violations of Public Trust* (New York: Wiley, 1993), 295–309.
2. Richard T. DeGeorge, *Competing with Integrity in International Business* (New York: Oxford University Press, 1993), 97–99.
3. The concept of the moral point of view is developed in Kurt Baier, *The Moral Point of View: A Rational Basis of Ethics* (Ithaca, NY: Cornell University Press, 1958). A similar account is offered in James Rachels, *The Elements of Moral Philosophy*, 3d ed. (New York: Random House, 1998), chap. 1.
4. Baier, *Moral Point of View*, 88.
5. Ibid., 195–96.
6. This case is adapted from Suzanne Alexander, "Child World Says Rival Cheats; Toys 'R' Us Answers: 'Grow Up,'" *Wall Street Journal*, 19 September 1991, B1, B10.
7. This case is adapted from Steve Lohr, "Lessons from a Hurricane: It Pays Not to Gouge," *New York Times*, 22 September 1992, D1, D2.
8. Albert Z. Carr, "Is Business Bluffing Ethical?" *Harvard Business Review*, 46 (January–February 1968), 148.
9. Amartya Sen, "Does Business Ethics Make Economic Sense?" *Business Ethics Quarterly*, 3 (1993), 45-54.
10. The results of experiments with the ultimatum bargaining game are presented in Robert H. Frank, *Passions within Reason: The Strategic Role of the Emotions* (New York: W. W. Norton, 1988), 170-74. For a discussion of the implications for business ethics, see Norman E. Bowie, "Challenging the Egoistic Paradigm," *Business Ethics Quarterly*, 1 (1991), 1-21.
11. Daniel Kahneman, Jack L. Knetch, and Richard Thaler, "Fairness as a Constraint of Profit-Seeking: Entitlements in the Market," *American Economic Review*, 76 (1986), 728–41.

12. Douglas C. Dacy and Howard Kunreuther, *The Economics of Natural Disasters* (New York: Free Press, 1969), 115–16.

13. Lynn Sharp Paine, "Law, Ethics, and Managerial Judgment," *Journal of Legal Studies Education,* 12 (1994), 153-69.

14. This phrase is taken from Norman E. Bowie, "Fair Markets," *Journal of Business Ethics,* 7 (1988), 89–98.

15. Christopher D. Stone, *Where the Law Ends: The Social Control of Corporate Behavior* (New York: Harper & Row, 1975), 94.

16. *Meinhard* v. *Salmon,* 164 N.E. 545 (1928).

17. See Thomas Donaldson, "Values in Tension: Ethics Away from Home," *Harvard Business Review,* 4 (September–October 1996), 48-62.

18. This case was prepared by Kerry Winans under the supervision of Professor John R. Boatright. Copyright 1995 by John R. Boatright.

19. This case was prepared by Kate Abele under the supervision of Professor John R. Boatright. Copyright 1998 by John R. Boatright.

2

Utilitarianism

Case 2.1   Lockheed in Japan

When Carl Kotchian, president of Lockheed Aircraft Corporation, made a trip to Japan in August 1972, the company he headed was in a very precarious financial situation. Lockheed had failed to get contracts with several major European carriers. Cost overruns on the C5A Galaxie transport plane and performance problems with the Cheyenne helicopter had caused the Defense Department to cancel its orders for these aircraft. Lockheed had avoided bankruptcy in 1971 only with a $250 million loan guarantee from the federal government. The survival of Lockheed as a company was riding on the effort to sell the new L-1011 TriStar passenger jet to All Nippon Airways.

Shortly after landing in Tokyo, Kotchian asked a representative of the Marubeni Corporation, a trading company that Lockheed had engaged to aid in negotiations with All Nippon Airways (ANA), to arrange a meeting with Kakuei Tanaka, the prime minister of Japan.[1] Kotchian knew that Tanaka would be meeting with President Richard Nixon in Hawaii in a few days and that Nixon would ask him to improve the U.S. balance of payments by buying more American products. He felt that it was important for the prime minister to be informed beforehand about the merits of the TriStar.

The representative of Marubeni, Toshiharu Okubo, informed Kotchian that a "pledge" of 500 million yen (about $1.6 million) would be required to set up such a meeting. Without specifically being told the destination of the money, Kotchian assumed that it was intended for the prime minister's office. Kotchian was hesitant about making an irregular payment of this size to the highest official in the Japanese government, but he knew that refusing to do so would hamper Lockheed's efforts and that the blame for any failure would rest squarely on his shoulders. So he agreed to pledge the amount requested, and a meeting was held at 7:30 the next morning. At the meeting, which Kotchian did not attend, the president of Marubeni allegedly secured Tanaka's help on behalf of Lockheed with an offer of 500 million yen.

After more than two months of complex negotiations, executives of ANA were on the verge of placing an order for six planes with an option to buy eight more. Late in the evening of Sunday, October 29, Carl Kotchian received a telephone call from Okubo informing him that the sale was assured if he would do three things. Two of them were minor, but the third was a bombshell. Kotchian was asked to have $400,000 in Japanese yen ready the next morning. Of this amount, $300,000, or 90 million yen, was to be paid to Tokuji Wakasa, the president of All Nippon Airways. This figure was based on $50,000 for each of the six planes ordered. The remaining $100,000, or 30 million yen, was to be divided among six Japanese politicians. When Kotchian protested that it would be impossible to raise that much cash so quickly, he was told the 30 million yen for the politicians was essential; the rest could wait.

By 10:00 the next morning, 30 million yen in cash was delivered to Okubo, and the 90-million-yen payment to the president of ANA was made a week later. Kotchian returned to the company's headquarters in Burbank, California, amid general celebration and apparently forgot about the pledge of 500 million yen for Prime Minister Tanaka. Eight months later, though, Okubo called Kotchian to say that now was the time to follow through. Kotchian asked whether the payment was necessary because the deal had been concluded such a long time ago. Okubo assured him that if he did not honor the pledge, Lockheed would never be able to do business in Japan again, and he hinted darkly that the president of Marubeni, who had made the offer to Tanaka, would have to leave the country.

INTRODUCTION

The case of Lockheed in Japan presented Carl Kotchian with a complex ethical dilemma. He had to act quickly in a situation where he would be called upon to justify whatever decision he made. Kotchian needed to think through the ethical implications of the alternatives and select a course of action that could be defended ethically. Fortunately, Carl Kotchian has provided us with an account of the reasoning behind his decisions, which we can use not only to understand what it means to reason ethically but also to judge for ourselves the defense he offered. In an account of his experiences, Kotchian wrote:

> After hanging up the telephone, I went home and thought about the matter overnight. I decided on the basis of what Okubo had told me that we could not possibly risk any retaliation against Lockheed or against Marubeni. If we did not make the payment on this matter, Hiyama [the president of Marubeni] would be forced into exile, Lockheed might not be able to sell anything in Japan again, and our relations with Marubeni might be completely disrupted. Consequently, the more I thought about it, the more I was convinced that there was no alternative but to make the payment. In the end, after talking it over with other Lockheed executives, I called Okubo and told him we would honor the pledge.[2]

Later, when All Nippon exercised the option to buy eight more TriStar planes, Okubo requested $400,000, based on $50,000 per plane. Kotchian again felt that he had no choice but to comply and ordered that the payment be made. In all, Lockheed paid about $12.5 million in bribes and commissions to sell 21 TriStars in Japan.

The arguments used by Carl Kotchian appeal primarily to *consequences*. Tens of thousands of jobs were saved, thereby benefiting Lockheed workers, their families, and the communities in which they lived. The cost to the company was negligible given the size of the deal, and stockholders were saved from the loss that would have resulted from the collapse of the company. Consequences aside, some people hold that bribery is wrong because of the violation of *duty* it involves. Officials both in the Japanese government and at All Nippon Airways hold positions of trust, in which they are pledged to serve the interests of the people of Japan.

Who is correct in this case? What is the best way to reason about these kinds of ethical issues? Are there any theories of ethics that can aid us in handling cases such as those faced by Carl Kotchian? This chapter examines one major ethical theory known as utilitarianism which bases justification on the consequences of actions. In subsequent chapters we will look at other important theories, including some that base right action on the notion of duty. It will also be useful in this chapter to clarify the distinction already made between two types of ethical theories; namely, those based on consequences and those based on duty—or teleological and deontological theories, to use their technical names.

TWO TYPES OF ETHICAL THEORIES

It is customary to divide ethical theories into two groups, usually called teleological and deontological. The most prominent historical examples of a teleological and a deontological theory are utilitarianism and the ethical theory of Immanuel Kant, respectively. A third kind of ethical theory is one based on the concept of virtue, which is discussed in the next chapter. Aristotle's ethics is the best example of a theory of this kind.

Teleological Theories

Teleological theories hold that the rightness of actions is determined solely by the amount of good consequences they produce. The word *teleological* is derived from the Greek word *telos*, which refers to an end. Actions are justified on teleological theories by virtue of the end they achieve, rather than some feature of the actions themselves. Thus, the concept of goodness is fundamental in teleological theories, and the concepts of rightness and obligation, or duty, are defined in terms of goodness. According to utilitarianism, our obligation, or duty, in any situation is to perform the action that will result in the greatest possible balance of good over evil.

Obviously, a great deal depends on what is regarded as good and as evil. In classical utilitarianism, pleasure is taken to be ultimately the only good, and evil is the opposite of pleasure, or pain. But utilitarianism can be understood more broadly, so that goodness is human well-being. Whatever makes human beings generally better off or provides some benefit is good, and whatever makes them worse off or harms them is evil. Differences of opinion exist, of course, on what constitutes benefits and harms, or being better or worse off. Generally, utilitarianism does not attempt to resolve these differences but accepts each person's own conception of what being better off means for him or her.

Strengths and Weaknesses. Teleological theories have many strengths. One is that they are in accord with much of our ordinary moral reasoning. The fact that an action would provide some benefit or inflict some harm is generally a morally relevant reason for or against performing it. So utilitarianism is able to explain why such actions as lying, breaking a promise, stealing, and assault are wrong and their opposites—truth telling, promise keeping, respect for property, and the like—are right. Lying, for example, creates false beliefs that often lead people to make disadvantageous choices, and if lying were to become common, then trust would be eroded, with a resulting decline in welfare for everyone. Having true beliefs, by contrast, is generally beneficial, and a society that values truth telling will have a higher level of welfare. Utilitarianism can also explain why lying in some circumstances is the right thing to do. It would be wrong to tell a murderer the location of an intended victim, for example, because the harm that would be done in this instance outweighs any benefit from telling the truth.

Second, teleological theories provide a relatively precise and objective method for moral decision making. Assuming that the goodness of consequences can easily be measured and compared, a teleological decision maker need only determine the possible courses of action and calculate the consequences of each one. For this reason, utilitarianism is attractive not only for matters of individual choice but also for decisions on issues of public policy. Utilitarian reasoning has also found favor among economists, who use the assumption that individuals seek to maximize their own utility or welfare to explain and predict a wide range of economic phenomena, such as prices and the allocation of resources. The ethical theory underlying classical economic theory is broadly utilitarian.

The weaknesses of teleological theories derive from the same features that constitute their strengths. Although much of our ordinary moral reasoning is teleological, some of it is decidedly nonteleological in character. Generally, we have an obligation to keep our promises, even when more good might be achieved by breaking them. If we promise another person to store some food that belongs to them, for example, it would be wrong to give the food away to hungry beggars merely because doing so would have better consequences, although there might be stronger obligations that override the keeping of a promise, such as an obligation to save a life.

Role Obligations, Rights, and Justice. Role obligations, which occupy a prominent place in business, often seem to be nonteleological. Parents have obligations to their children, for example, that are created by the special relationship of parenthood. There is nothing wrong with parents providing for their children—even when the money could be better spent on orphan relief. Indeed, parents who contribute to the good of more children by their donations to orphan relief while neglecting their own children would be properly regarded as failing to fulfill their duty as parents.

The concepts of rights and justice pose an especially difficult challenge for teleological theories. The right of free speech, for example, generally entitles us to speak freely—even when restricting this right might produce better consequences. Similarly, we believe that justice ought to be done regardless of consequences. Even if it could be shown that discrimination against women or racial minorities, on balance, produces better consequences, discrimination is a violation of a basic principle of justice. Both rights and justice appear to be nonteleological in character.

Deontological Theories

Those who argue that the actions of Lockheed in Japan were wrong not because of their *consequences* but because we have a *duty* not to bribe are taking a deontological rather than a teleological approach to ethical reasoning. Deontological theories, in contrast to teleological theories such as utilitarianism, deny that consequences are relevant to determining what we ought to do. Deontologists typically hold that certain actions are right not because of some benefit to ourselves or others but because of the nature of these actions or the rules from which they follow. Thus, bribery is wrong, some say, by its very nature, regardless of the consequences. Other examples of nonconsequentialist reasoning in ethics include arguments based on principles such as the Golden Rule (Do unto others as you would have them do unto you) and those that appeal to basic notions of human dignity and respect for other persons. Obligation, or duty, is the fundamental moral category in deontological theories, and goodness and other concepts are to be defined in terms of obligation, or duty. (The word *deontological* derives, in fact, from *deon*, the Greek word for duty.)

An example of a deontological theory consisting of a set of absolute moral rules is that presented by the twentieth-century British philosopher W. D. Ross. The seven rules in Ross's system are the following:

1. Duties of fidelity—to keep promises, both explicit and implicit, and to tell the truth.
2. Duties of reparation—to compensate people for injury that we have wrongfully inflicted on them.
3. Duties of gratitude—to return favors that others do for us.
4. Duties of justice—to ensure that goods are distributed according to people's merit or deserts.
5. Duties of beneficence—to do whatever we can to improve the condition of others.
6. Duties of self-improvement—to improve our own condition with respect to virtue and intelligence.
7. Duties of nonmaleficence—to avoid injury to others.[3]

Strengths and Weaknesses. One strength of deontological theories such as Ross's is that they make sense of cases in which consequences seem to be irrelevant. Especially in justifying the obligations that arise from *relations,* such as contracts and roles, it is more plausible to appeal to the relations themselves than to the consequences. Thus, a manufacturer has an obligation to honor a warranty on a defective product even if the cost of doing so exceeds the benefit of satisfying a consumer. And an employee has an obligation to an employer to be loyal and to do his or her job. Another strength of deontological theories is the way they account for the role of *motives* in evaluating actions. Two people who give large amounts to charity—one out of genuine concern to alleviate suffering and the other to impress friends and associates—produce the same amount of good, yet we evaluate the two actions differently. Deontologists generally hold that the rightness of actions depends wholly or in part on the motives from which they are performed and not on consequences.

The main weakness of deontological theories lies in the failure to provide a plausible account of how we can know our moral obligations and resolve problems of moral conflict. Although the rules in Ross's theory are plausible, no reason is offered for accepting these rules and not others. Ross's rules are also open to the charge of ethnocentrism, that is, of erroneously accepting the rules of our own society as though they were universal. People at different times and in different places might reject Ross's rules and regard others as being equally worthy. Worse, he suggests no order of priority among the rules, so that we have no guidance in cases where they conflict. For example, should we keep a promise or tell the truth when doing so will harm someone? Or should gratitude for a past favor alter in some way a distribution of goods according to merit or desert?

CLASSICAL UTILITARIANISM

Different parts of the utilitarian doctrine were advanced by philosophers as far back as the ancient Greeks, but it remained for two English reformers in the nineteenth century to fashion them into a single coherent whole.[4] The creators of classical utilitarianism were Jeremy Bentham (1748–1832) and John Stuart Mill (1806–1873). In their hands, utilitarianism was not an ivory tower philosophy but a powerful instrument for social, political, economic, and legal change. Bentham and Mill used the principle of utility as a practical guide in the English reform movement.

The Principle of Utility

Jeremy Bentham's version of utilitarianism is set forth in the following passage:

> By the principle of utility is meant that principle which approves or disapproves of every action whatsoever, according to the tendency which it appears to have to augment or diminish the happiness of the party whose interest is in question: or, what is the same thing in other words, to promote or to oppose that happiness.

So stated, the principle requires that consequences be measured in some way so that the pleasure and pain of different individuals can be added together and the results of different courses of action compared. Bentham assumed that a precise quantitative measurement of pleasure and pain was possible, and he outlined a procedure that he called the *hedonistic calculus* (*hedonistic* is derived from the Greek word for pleasure). The procedure is to begin with any one individual whose interest is affected:

> Sum up all the values of all the *pleasures* on the one side, and those of all the pains on the other. The balance, if it be on the side of pleasure, will give the *good* tendency of the act upon the whole, with respect to the interests of that *individual* person; if on the side of pain, the *bad* tendency of it upon the whole.

If this process is repeated for all other individuals whose interests are affected, the resulting sums will show the good or bad tendency of an action for the whole community.

Bentham's theory is open to some rather obvious objections. Among them is the long-standing opposition of many philosophers to the thesis of *hedonism*: Pleasure and pleasure alone is good. Critics at the time complained that pleasure is too low to constitute the good for human beings and pointed out that even pigs are capable of pleasure, which led to the charge that utilitarianism is a "pig philosophy" fit only for swine. One absurd consequence of Bentham's view, according to critics, is that it would be better to live the life of a satisfied pig than that of a dissatisfied human being such as Socrates. Utilitarianism does not require the thesis of hedonism, however, and many things besides pleasure have been regarded as good by utilitarian theorists, including friendship and aesthetic enjoyment.

Mill's Version

John Stuart Mill was aware of the objections to Bentham's theory, and in his major work on ethics, *Utilitarianism* (1863), he attempted to develop a more defensible version of the utilitarian position. His initial statement of the principle of utility is

> The creed which accepts as the foundation of morals, Utility, or the Greatest Happiness Principle, holds that actions are right in proportion as they tend to promote happiness, wrong as they tend to produce the reverse of happiness. By happiness is intended pleasure, and the absence of pain; by unhappiness, pain, and the privation of pleasure.

Mill departed from Bentham's strict quantitative treatment of pleasure by introducing the idea that pleasures also differ in their *quality*. The charge that utilitarianism is a pig philosophy can be met, Mill claimed, by holding that human beings are capable of enjoying higher pleasures than those experienced by swine. Because human beings, but not pigs, can enjoy the arts and intellectual pursuits, Mill concluded:

> It is better to be a human being dissatisfied than a pig satisfied; better to be Socrates dissatisfied than a fool satisfied. And if the fool, or the pig, are of a different opinion, it is because they know only their side of the question.

Even if some pleasures are better than others, this insight does not succeed in saving the thesis of hedonism or the utilitarian principle that we ought to produce the greatest possible amount of pleasure. This is so because the higher pleasures enjoyed by a few people with elevated tastes are unlikely, in any actual society, to outweigh the total sum of the low pleasures enjoyed by the bulk of the population. As long as the number of people who prefer trashy television shows, for example, greatly exceeds the number who appreciate fine drama on television, a utilitarian decision maker would be forced, most likely, to give preference to soap operas and situation comedies over the plays of Shakespeare. In any event, Mill's writings give us no guidance for comparing the quality with the quantity of pleasure.

THE FORMS OF UTILITARIANISM

Classical utilitarianism can be stated formally as follows:

> AN ACTION IS RIGHT IF AND ONLY IF IT PRODUCES THE GREATEST
> BALANCE OF PLEASURE OVER PAIN FOR EVERYONE.

So stated, the utilitarian principle involves four distinct theses:

1. *Consequentialism.* The principle holds that the rightness of actions is determined solely by their consequences. It is by virtue of this thesis that utilitarianism is a teleological theory.
2. *Hedonism.* Utility in this statement of the theory is identified with pleasure and the absence of pain. Hedonism is the thesis that pleasure and only pleasure is ultimately good.
3. *Maximalism.* A right action is one that has not merely some good consequences, but also the greatest amount of good consequences possible when the bad consequences are also taken into consideration.
4. *Universalism.* The consequences to be considered are those of everyone.

The first two of these have already been explained, but the last two call for some comment.

Virtually every action produces both pleasure and pain, and the principle of utility does not require that only pleasure and no pain result from a right action. An action may produce a great amount of pain and still be right on the utilitarian view as long as the amount of pleasure produced is, on balance, greater than the amount of pleasure produced by any other action. Both Bentham and Mill assumed that the amount of pain produced by an action can be subtracted from the amount of pleasure to yield the net amount of pleasure—in the same way that an accountant subtracts debts from assets to determine net worth.

The thesis of universalism requires us to consider the pleasure and pain of *everyone alike.* Thus, we are not following the principle of utility by considering the consequences only for ourselves, for our family and friends, or for an organization of which we are a part. Utilitarianism does not require us to ignore our own interest, but we are permitted to place it no higher and no lower than the interest of anyone else. The utilitarian principle does not insist that the interest of everyone be *promoted*, though. In deciding whether to close a polluting plant, for example, we need to consider the citizens of the community who suffer from the pollution, the workers who will lose their jobs if the plant is shut down, the owners of the company that operates the plant, and, possibly, consumers as well. No matter what decision is made, the interests of some people will be harmed. Utilitarian reasoning obligates us to include only the interests of everyone in our calculations, not to act in a way that advances every individual interest.

Act- and Rule-Utilitarianism

In classical utilitarianism, an action is judged to be right by virtue of the consequences of performing *that action*. As a result, telling a lie or breaking a promise is right if it has better consequences than any alternative course of action. Utilitarian morality thus seems to place no value on observing rules, such as "Tell the truth" or "Keep your promises," except perhaps as "rules of thumb," that is, as distillations of past experience about the tendencies of actions that eliminate the need to calculate consequences in every case. This result can be avoided if we consider the consequences of performing not just particular actions, but also *actions of a certain kind.* Although some instances of lying have consequences that are better than telling the truth, lying in general does not. As a kind of action, then, truth telling is right by virtue of the consequences of performing actions of that kind, and any instance of truth telling is right because actions of that kind are right.

This suggestion leads to a distinction between two versions of utilitarianism, one in which we calculate the consequences of each act and another in which we consider the consequences of following the relevant rule. These two versions are called act-utilitarianism (AU) and rule-utilitarianism (RU), respectively. They may now be expressed formally in the following way:

(AU)　An action is right if and only if it produces the greatest balance of pleasure over pain for everyone.

(RU)　An action is right if and only if it conforms to a set of rules the general acceptance of which would produce the greatest balance of pleasure over pain for everyone.[5]

Act- and rule-utilitarianism each has its merits, and there is no consensus among philosophers about which is correct.[6] Act-utilitarianism is a simpler theory and provides an easily understood decision procedure. Rule-utilitarianism seems to give firmer ground, however, to the rules of morality and to role obligations, which are problems for all teleological theories. A further advantage of rule-utilitarianism, according to its proponents, is that it eliminates the difficult task of calculating the consequences of each individual act.

Problems with Calculating Utility

Classical utilitarianism requires that we be able to determine (1) the amount of utility (that is, the balance of pleasure over pain) for each individual affected by an action and (2) the amount of utility for a whole society. Whether we can measure utility in the way that utilitarianism requires has been a concern not only of philosophers but also of economists, because utility has been generally accepted as a basis for economic theory. Because of the need in economics for precise calculations of utility—and misgivings about Bentham's simplistic hedonistic calculus—economists have introduced a number of important refinements into the theory of utilitarianism.

There is little difficulty in calculating that some actions produce more pleasure for us than others. A decision to spend an evening at a concert is usually the result of a judgment that listening to music will give us more pleasure at that time than any available alternative. Confronted with a range of alternatives, we can usually rank them in order from the most pleasant to the least pleasant. A problem arises, however, when we attempt to determine exactly how *much* pleasure each course of action will produce, because pleasure cannot be measured precisely in terms of quantity, much less quality. Utilitarianism also requires that we calculate the pleasure of alternative courses of action not only for ourselves but also for everyone.

Some critics contend that this requirement imposes an information burden on utilitarian decision makers that is difficult to meet. In order to buy a gift for a friend that will produce the greatest amount of pleasure, for example, we need to know something about that person's desires and tastes. Consider, for example, the task faced by a utilitarian legislator who must decide whether to permit logging in a public park. This person must identify all the people affected, determine the amount of pleasure and pain for each one, and then compare the pleasure that hiking brings to nature lovers versus the pain that would be caused to loggers if they lost their jobs. The abilities of ordinary human beings are inadequate, critics complain, to acquire and process the vast amount of relevant information in such a case. The response of utilitarians to these problems is that we manage in practice to make educated guesses by relying on past experience and limiting our attention to a few aspects of a situation.

Comparing the pleasure and pain of different people raises a further problem about the *interpersonal comparison of utility*. Imagine two people who each insist after attending a concert that he or she enjoyed it more. There seems to be no way in principle to settle this dispute. Some philosophers and economists consider this problem to be insoluble and a reason for rejecting utilitarianism both as an ethical theory and as a basis for economics.[7] Others argue for the possibility of interpersonal comparisons on the basis that regardless of whether we can make such comparisons precisely, we do make them in everyday life with reasonable confidence.[8] We may give away an extra ticket to a concert, for example, to the friend we believe will enjoy it the most. The problem of the interpersonal comparison of utility is not insuperable, therefore, as long as rough comparisons are sufficient for utilitarian calculations.

Applying the Two Principles

Although bribery is morally wrong in most situations, Carl Kotchian contended that it is an accepted practice in many parts of the world and a necessity in a competitive climate where other companies do the same. If Lockheed had refused to make the requested payments, the company would quite possibly have lost the Japanese market to less scrupulous competitors and gained nothing in return. The expenditure was worthwhile to the company inasmuch as the $12 million payment amounted to less than 3 percent of expected revenues from the sale of the 21 TriStars. Kotchian argued further that his actions did not constitute bribery but rather that Lockheed was an innocent victim of extortion demands made by the Japanese. Lockheed's motive in making the payments, moreover, was not to make a sale by corrupting Japanese officials; the company was simply trying to do business in an atmosphere tainted by the corruption of others. Kotchian insisted that Lockheed violated no American law.[9] Finally, the payments were actually offered by the middlemen of the Marubeni Corporation, who were apparently well versed in the ways of decision making, Japanese style.[10]

Carl Kotchian was apparently reasoning as an act-utilitarian when he authorized the payments in Japan. One may question, however, whether he considered all of the consequences of his action. Lockheed was competing for sales with two other American companies: Boeing with its 747 and McDonnell Douglas with its DC 10. So the jobs that were saved in Burbank, California, were lost in Seattle, St. Louis, and elsewhere, and the workers for these two companies, their families, subcontractors, and the stockholders of Boeing and McDonnell Douglas were all harmed as a result. In the aftermath of the Lockheed affair, many careers were destroyed. Carl Kotchian resigned from Lockheed. Prime Minister Tanaka was forced from office and prosecuted along with other Japanese officials. Many Japanese feel that the country itself was hurt because of the distrust of political institutions that the revelations created in Japan. When the principle of act-utilitarianism is correctly applied in this case, Carl Kotchian's decision would not appear to be justified.

The principle of rule-utilitarianism would have us consider the consequences of a general practice of bribery. Judged by this standard, Carl Kotchian's decision is even less justified. Bribery usually results in higher prices and reduced quality, because people are led by personal advantage to make decisions on considerations other than the value of the goods and services being offered. A widespread system of bribery would lead to less competitive conditions and a resulting decline in efficiency. Lockheed alone cannot eliminate bribery in the sale of planes abroad, of course, but by engaging in the practice so readily, it perpetuated the practice and encouraged others to do the same. Moreover, Lockheed could have worked with others to remove corruption wherever possible. On rule-utilitarian grounds, therefore, bribery is properly regarded as morally wrong.

COST-BENEFIT ANALYSIS

Bentham's ideal of a precise quantitative method for decision making is most fully realized in *cost-benefit analysis*. This method differs from Bentham's hedonistic calculus primarily in the use of monetary units to express the benefits and drawbacks of various alternatives. Any project in which the dollar amount of the benefits exceeds the dollar amount of the damages is worth pursuing, according to cost-benefit analysis, and from among different projects, the one that promises the greatest net benefit, as measured in dollars, ought to be chosen.[11]

From an economic point of view, cost-benefit analysis is simply a means for achieving an efficient allocation of resources. Business decisions that evaluate investment opportunities in terms of their return are thus instances of cost-benefit analysis. However, there is an important difference between business decision making and what is generally described as cost-benefit analysis. Companies usually calculate the anticipated costs and benefits only for *themselves*, whereas legislators, social planners, regulators, and other users of cost-benefit analysis generally ask what the costs and benefits are for *everyone* who is affected.

The chief advantage of cost-benefit analysis is that the price of many goods is set by the market, so that the need to have knowledge of people's pleasures or preference rankings is largely eliminated. The value of different goods is easily totaled to produce a figure that reflects the costs and benefits of different courses of action for all those concerned. Money also provides a common denominator for allocating resources among projects that cannot easily be compared otherwise. Would scarce resources be better spent on a Head Start program, for example, or on the development of new sources of energy? In cost-benefit analysis, decision makers have an analytic framework that enables them to decide among such disparate projects in a rational, objective manner.

Because of the narrow focus on economic efficiency in the allocation of resources, cost-benefit analysis is not commonly used as a basis for personal morality but as a means for making major investment decisions and decisions on broad matters of public policy. It is not a complete ethical theory, therefore, but a utilitarian form of reasoning with a limited but important range of application. Cost-benefit analysis is also not necessarily intended to be the *only* means for making decisions. The goal of protecting the environment, for example, sometimes overrides considerations of efficiency and leads to the adoption of regulations in which the costs exceed the immediate benefits. Still, determining the appropriate level of environmental protection requires an awareness of the costs and benefits of different regulatory schemes and of the trade-offs being made.

A distinction is commonly made, moreover, between cost-benefit analysis and cost-effectiveness analysis. *Cost-effectiveness* analysis assumes that we already have some agreed-upon end, such as reducing injuries from defective products or protecting the environment, so that the only remaining question is, what is the most efficient means for achieving this end? *Cost-benefit analysis*, by contrast, is used to select both the means to ends and the ends themselves.[12] Nonutilitarians who reject the use of cost-benefit analysis to settle questions about ends, such as the rights of consumers and the value of environmental protection, should have no quarrel with a requirement that we pursue morally justified ends in the most efficient manner possible.

The Problems of Assigning Monetary Values

Cost-benefit analysis is criticized on many different grounds. First, not all costs and benefits have an easily determined monetary value. The value of the jobs that are provided by logging on public land can be expressed precisely in dollars, as can the value of the lumber produced. But because the opportunity for hikers to enjoy unspoiled vistas and fresh-smelling air is not something that is commonly bought and sold, it has no fixed market price. There is also no market to determine the price of peace and quiet, the enjoyment of the company of family and friends, and freedom from the risk of physical injury and death. The same is true of many public goods, such as police protection, roads and bridges, and public health programs.

In addition, the market for some goods is distorted by various factors so that the price of these goods does not reflect their "true" value. In economic terms, the market price of goods does not always correspond to their opportunity cost, as determined by consumers' marginal rates of substitution, which economists regard as the proper measure of value. The fact that a yacht costs more than a college education, for example, does not mean that consumers value yachts more highly than education. It would be a mistake, therefore, to use cost-benefit analysis to support policies that foster the ownership of yachts by the rich while making it more difficult for the poor to attend college. Critics complain, however, that noneconomic goods and mispriced goods tend to be left out of cost-benefit analyses entirely or else are assigned arbitrary values.

Some applications of cost-benefit analysis require that a value be placed on human life. Although this may seem cold and heartless, it is necessary if cost-benefit analysis is to be used to determine how much to spend on prenatal care to improve the rate of infant mortality, for example, or on reducing the amount of cancer-causing emissions from factories. Reducing infant mortality or the death rate from cancer justifies the expenditure of some funds, but how much? Would further investment be justified if it reduced the amount available for education or medical care for the elderly? No matter where the line is drawn, some trade-off must be made between saving lives and securing other goods.

Addressing the Problems. Experts in cost-benefit analysis attempt to overcome the problem of assigning a dollar figure to noneconomic goods with a technique known as *shadow pricing*. This consists of determining the value reflected by people's market and nonmarket behavior. For example, by comparing the prices of houses near airports, busy highways, and the like with the prices of similar houses in less noisy areas, it is possible to infer the value that people place on peace and quiet. The value of life and limb can be similarly estimated by considering the amount of extra pay that is needed to get workers to accept risky jobs. Several methods exist, in fact, for calculating the value of human life for purposes of cost-benefit analysis.[13] Among these are the discounted value of a person's future earnings over a normal lifetime, the value that existing social and political arrangements place on the life of individuals, and the value that is revealed by the amount that individuals are willing to pay to avoid the risk of injury and death. When people choose through their elected representatives or by their own consumer behavior not to spend additional amounts to improve automobile safety, for example, they implicitly indicate the value of the lives that would otherwise be saved. Using such indicators, economists calculate that middle-income Americans value their lives between $3 million to $5 million.[14]

The purpose of assigning a monetary value to life in a cost-benefit analysis is not to indicate how much a life is actually worth but to enable us to compare alternatives where life is at stake. If we know that spending $5 million on automobile safety will save one life whereas the same amount spent on childhood vaccinations will save two, then it is rational to spend the money on the latter, if a choice must be made. When we are comparing alternatives that can save lives, the dollar amount that we place on a life is irrelevant; all that counts is the number of lives saved. But we must also compare $5 million spent on childhood vaccinations to save two lives with the same amount spent on schooling that will create the opportunity for prosperous, rewarding lives for 200 children. Now we are forced to compare the value of saving lives with other things that we value, and it would not be rational to spend all of our money on medical care and none on education. Beyond a certain dollar amount, the cost of saving a life is not worthwhile because of the other goods we could buy with those resources. Experts in risk assessment calculate that the "break-even" point where the amount expended to save a life is worth the cost is about $10 million.[15]

There are some pitfalls in using the technique of shadow pricing, especially when human life is involved. Many people buy houses in noisy areas or accept risky jobs because they are unable to afford decent housing anywhere else or to secure safer employment. Some home buyers and job seekers may not fully consider or appreciate the risks they face, especially when the hazards are unseen or speculative. Also, the people who buy homes near airports or accept work as steeplejacks are possibly less concerned with noise or danger than is the general population. We certainly do not want to assume, however, that workplace safety is of little value simply because a few people are so heedless of danger that they accept jobs that more cautious people avoid.[16]

Finally, people's individual and collective decisions are not always rational. People who drive without seat belts are probably aware of their benefit *for other people* but are convinced that nothing will happen *to them* because they are such good drivers.[17] As a result, they (irrationally) expose themselves to risks that do not accurately reflect the value they place on their own life. Mark Sagoff observes that the choices we make as consumers do not always correspond to those we make as citizens. He cites as examples the fact that he buys beverages in disposable containers but urges his state legislators to require returnable bottles and that he has a car with an Ecology Now sticker that leaks oil everywhere it is parked.[18]

Should All Things Be Assigned a Monetary Value?

A second criticism of cost-benefit analysis is that even if all the technical problems of shadow pricing could be solved, there are still good reasons for not assigning a monetary value to some things. Steven Kelman argues that placing a dollar value on some goods reduces their perceived value, because they are valued precisely because they cannot be bought and sold in a market. Friendship and love are obvious examples. "Imagine the reaction," Kelman observes, "if a practitioner of cost-benefit analysis computed the benefits of sex based on the price of prostitute service."[19] In *The Gift*

Relationship: From Human Blood to Social Policy, Richard M. Titmuss very perceptively compares the American system of blood collection with that of the British. In the United States, about half of all blood is purchased from donors and sold to people who need transfusions.[20] The British system, by contrast, is purely voluntary. No one is paid for donating blood, and it is provided without charge to anyone in need. As a result, the giving of blood and blood itself have an entirely different significance. If blood has a price, then giving blood merely saves someone else an expense, but blood that cannot be bought and sold becomes a special gift that we make to others.[21]

Although some things are cheapened in people's eyes if they are made into commodities and traded in a market, this does not happen if goods are assigned a value merely for purposes of comparison. It is the actual buying and selling of blood that changes its perceived value, not performing a cost-benefit analysis. Moreover, Titmuss himself argues in favor of the British system on the grounds that the system in the United States is (1) highly wasteful of blood, resulting in chronic acute shortages; (2) administratively inefficient because of the large bureaucracy that it requires; (3) more expensive (the price of blood is five to fifteen times higher); and (4) more dangerous, because there is a greater risk of disease and death from contaminated blood.[22] In short, a cost-benefit analysis shows that it is better not to have a market for blood.

Other Values in Cost-Benefit Analysis

Cost-benefit analysis is offered as a method that is itself value-free and applies only the values that people express in the market. Critics charge, however, that the method is heavily value-laden because the values of the analyst cannot be excluded entirely.[23] First, the analyst must choose the range of alternatives to be considered, and some may be deliberately excluded or inadvertently overlooked because of the analyst's values. Cost-benefit analysis requires the analyst to determine what constitutes a cost and a benefit. Loud outdoor music at rock concerts, for example, is a benefit to fans of that kind of music, but it is a cost to others who are forced to endure the noise. Whose values will determine whether the music is a cost or a benefit?

Analysts must further decide what is to count as a *consequence* of a particular course of action. Should the costs and benefits of various packaging materials be calculated only up to the time that the product reaches consumers, for example, or should the calculations continue to include the costs of disposing of the materials? If better packaging encourages people to eat more fast food to the detriment of their health, should this also be included as a consequence in the analysis? In economic terms, the question is, to what extent should "spillover effects" or externalities be considered? And the answer to this question will be decided to some extent by the analyst's values. Finally, how far into the future should consequences be calculated? Applying cost-benefit analysis presupposes some commitment to a time frame for calculating the consequences of alternatives for future generations.

Defenders of cost-benefit analysis reply that any theory is "value-laden," and an advantage of cost-benefit analysis is that it makes its value commitments explicit, so they can be "flagged" and properly taken into account. The validity of any particular cost-benefit analysis depends, moreover, on the objectivity of the people who conduct it. And we must also have an understanding of the proper role of cost-benefit analysis in public decision making. It is not intended to be the sole means for arriving at the choices we make as a society. Efficiency in the allocation of resources is not our only value. The key question is: What matters should be turned over to the analysts? This question cannot itself be decided by a cost-benefit analysis.

Conclusion

Utilitarianism is a powerful and widely accepted ethical theory that has special relevance to problems in business. Not only does it enable us to justify many of the obligations of individuals and corporations, but the principle of utility, as we will see in subsequent chapters, provides a strong foundation for rights and justice. Utilitarianism fits easily, moreover, with the concept of value in economics and the use of cost-benefit analysis in business. We will find in this book that utilitarianism is relevant to most of the topics considered. Utility arguments occur in just about every chapter, and they provide a relatively firm and coherent basis for business ethics.

Case 2.2   Exporting Pollution

As an assistant to the vice president of environmental affairs at Americhem, Rebecca Wright relishes the opportunity to apply her training in public policy analysis to the complex and emotion-laden issues that her company faces.[24] Rebecca is convinced that cost-benefit analysis, her specialty, provides a rational decision-making tool that cuts through personal feelings and lays bare the hard economic realities. Still, she was startled by the draft of a memo that her boss, Jim Donnelly, shared with her. The logic of Jim's argument seems impeccable, but the conclusions are troubling—and Rebecca is sure that the document would create a furor if it were ever made public. Jim is preparing the memo for an upcoming decision on the location for a new chemical plant. The main problem is that atmospheric pollutants from the plant, although mostly harmless, would produce a persistent haze; and one of the particles that would be released into the atmosphere is also known to cause liver cancer in a very small portion of the people exposed. Sitting down at her desk to write a response, Rebecca reads again the section of the memo that she had circled with her pen.

> From an environmental point of view, the case for locating the new plant in a Third World country is overwhelming. These reasons are especially compelling in my estimation:
>
> 1. The harm of pollution, and hence its cost, increases in proportion to the amount of already existing pollution. Adding pollutants to a highly polluted environment does more harm than the same amount added to a relatively unpolluted environ-

ment. For this reason, much of the Third World is not efficiently utilized as a depository of industrial wastes, and only the high cost of transporting wastes prevents a more efficient utilization of this resource.

2. The cost of health-impairing pollution is a function of the forgone earnings of those who are disabled or who die as a result. The cost of pollution will be least, therefore, in the country with the lowest wages. Any transfer of pollution from a high-wage, First World country to a low-wage, Third World country will produce a net benefit.

3. The risk of liver cancer from this plant's emissions has been estimated at one-in-a-million in the United States, and the resulting cancer deaths would occur mostly among the elderly. The risk posed by the new plant will obviously be much less in a country where people die young from other causes and where few will live long enough to incur liver cancer from any source. Overall, the people of any Third World country might prefer the jobs that our plant will provide if the only drawback is a form of cancer that they are very unlikely to incur.

4. The cost of visibility-impairing pollution will be greater in a country where people are willing to spend more for good visibility. The demand for clear skies—which affects the aesthetics of the environment and not people's health—has very high income elasticity, and so the wealthy will pay more than the poor to live away from factory smoke, for example. Because the cost of anything is determined by how much people are willing to pay in a market, the cost of visibility-impairing pollution in a First World country will be higher than the same amount of pollution in a Third World country. Thus, people in the United States might prefer clear skies over the benefits of our plant, but people elsewhere might choose differently.

Case 2.3 Beech-Nut's Bogus Apple Juice

When Lars Hoyvald joined Beech-Nut in 1981, the company was in financial trouble.[25] In the competitive baby food industry, the company was a distant second behind Gerber, with 15 percent of the market. After faltering under a succession of owners, Beech-Nut was bought in 1979 by Nestlé, the Swiss food giant, which hoped to restore the luster of the brand name. Although he was new to Beech-Nut, Hoyvald had wide experience in the food industry, and his aim, as stated on his resumé, was "aggressively marketing top quality products."

In June 1982, Hoyvald was faced with strong evidence that Beech-Nut apple juice for babies was made from concentrate that included no apples. Since 1977, the company had been purchasing low-cost apple concentrate from a Bronx-based supplier, Universal Juice Company. The price alone should have raised questions, and John Lavery, the vice president in charge of operations, brushed aside tests that showed the presence of corn syrup. Two employees who investigated Universal's "blending facility" found merely a warehouse. Their report was also dismissed by Lavery. A turning point occurred when a private investigator working for the

Processed Apple Institute discovered that the Universal plant was producing only sugared water. After following a truck to the Beech-Nut facility, the investigator informed Lavery and other executives of his findings and invited Beech-Nut to join a suit against Universal.

Although some executives urged Hoyveld to switch suppliers and recall all apple juice on the market, the president was hesitant. Even if the juice was bogus, there was no evidence that it was harmful. It tasted like apple juice, and it surely provided some nutrition. Besides, he had promised his Nestlé superior that he would return a profit of $7 million for the year. Switching suppliers would mean paying about $750,000 more each year for juice and admitting that the company had sold an adulturated product. A recall would cost about $3.5 million. Asked later why he had not acted more decisively, Hoyvald said, "I could have called up Switzerland and told them I had just closed the company down. Because that is what would have been the result of it."

Fearful that state and federal investigators might seize stocks of Beech-Nut apple juice, Hoyvald launched an aggressive foreign sales campaign. On September 1, the company unloaded thousands of cases on its distributors in Puerto Rico. Another 23,000 cases were shipped to the Dominican Republic to be sold at half price. By the time that state and federal authorities had forced a recall, the plan was largely complete. In November, Hoyvald reported to his superior at Nestlé, "The recall has now been completed, and due to our many delays, we were only faced with having to destroy approximately 20,000 cases." Beech-Nut continued to sell bogus apple juice until March 1983.

In 1988, Hoyvald and Lavery were tried and convicted on charges of consumer fraud, and each received a sentence of one year and one day and fined $100,000. Previously, Beech-Nut had settled charges by paying a $2 million fine. The company also settled a class-action suit brought by consumers for $7.5 million. The parent company, Nestlé, kept Hoyvald and Lavery on the payroll and paid their legal expenses, which amounted to several more million dollars. In issuing the sentences, the judge rejected a proposal from Hoyvald's lawyer that the former president of Beech-Nut be placed on probation and required to give lectures to business students so that they would not make the same mistakes.[26]

NOTES

1. The most complete source of information on this case is Robert Shaplen, "Annals of Crime: The Lockheed Incident," *The New Yorker*, 23 January 1978, 48–74; and 30 January 1978, 74–91. Kotchian's own reflections are contained in A. Carl Kotchian, "The Payoff: Lockheed's 70-Day Mission to Tokyo," *Saturday Review*, 9 July 1977, 7–12. This article is adapted from a memoir, *Lockheed Sales Mission: Seventy Days in Tokyo*, published in Japan.

2. Kotchian, "Payoff," 11.

3. W. D. Ross, *The Right and the Good* (Oxford: Oxford University Press, 1930), 21.

4. For a brief account of the historical origins of utilitarianism, see Anthony Quinton, *Utilitarian Ethics* (New York: St. Martin's Press, 1973).

5. This formulation of RU follows that given by David Lyons for what he calls ideal rule-utilitarianism. David Lyons, *Forms and Limits of Utilitarianism* (Oxford: Oxford University Press, 1965), 140. There is considerable controversy over the correct formulation of RU and the relation between AU, RU, and another principle usually called utilitarian generalization. In addition to the discussion in Lyons, *Forms and Limits of Utilitarianism*, see Marcus G. Singer, *Generalization in Ethics* (New York: Alfred A. Knopf, 1961), chap. 7; Jan Narveson, *Morality and Utility* (Baltimore, MD: Johns Hopkins University Press, 1967), 129–40; and R. B. Brandt, *A Theory of the Good and the Right* (Oxford: Oxford University Press, 1979), 278–85. Early discussions of the distinction between AU and RU are R. F. Harrod, "Utilitarianism Revised," *Mind*, 45 (1936), 137–56; Jonathan Harrison, "Utilitarianism, Universalization, and Our Duty to Be Just," *Proceedings of the Aristotelian Society*, 53 (1952–1953), 105–34; J. J. C. Smart, "Extreme and Restricted Utilitarianism," *The Philosophical Quarterly*, 6 (1956), 344–54. Perhaps the most ambitious attempt to formulate an adequate statement of RU is R. B. Brandt, "Toward a Credible Form of Utilitarianism," in *Morality and the Language of Conduct*, ed. Hector-Neri Castaneda and George Nakhnikian (Detroit: Wayne State University Press, 1963), 107–43. Criticism of Brandt is contained in Alan Donagan, "Is There a Credible Form of Utilitarianism?" in *Contemporary Utilitarianism*, ed. Michael Bayles (Garden City, NY: Anchor Books, 1968), 187–202. Brandt's fullest statement of a roughly rule-utilitarian position, which he calls "a pluralistic welfare-maximizing moral system," is in *A Theory of the Good and the Right*.

6. Some philosophers hold that there is no difference between the two formulations. See Lyons, *Forms and Limits of Utilitarianism*, chap. 3. Also, R. M. Hare, *Freedom and Reason* (Oxford: Oxford University Press, 1963), 130–36; and Alan F. Gibbard, "Rule-Utilitarianism: Merely an Illusory Alternative?" *Australasian Journal of Philosophy*, 43 (1965), 211–20.

7. See, for example, Lionel Robbins, *An Essay on the Nature and Significance of Economic Science* (London: Macmillan, 1932), 140; and Kenneth Arrow, *Social Choice and Individual Values*, 2d ed. (New York: John Wiley, 1963), 9.

8. A defense of interpersonal comparisons of utility by a prominent economist is I. M. D. Little, *A Critique of Welfare Economics*, 2d ed. (Oxford: Oxford University Press, 1957), chap. 4.

9. Although the payments were legal under U.S. law at the time, Lockheed was found guilty in 1979 of four counts of fraud and four counts of making misleading statements for deducting payments from its taxes as "marketing cost" in violation of Section 162C of the Internal Revenue Code.

10. For the distinctions between bribery, kickbacks, and extortion, see Michael Phillips, "Bribery," *Ethics*, 94 (1984), 621–36; Thomas L. Carson, "Bribery, Extortion, and the 'Foreign Corrupt Practices Act,'" *Philosophy and Public Affairs*, 14 (1985), 66–90; John Danley, "Towards a Theory of Bribery," *Business and Professional Ethics Journal*, 2 (1983), 19–39; and Kendall D'Andrade, Jr., "Bribery," *Journal of Business Ethics*, 4 (1985), 239–48. A comprehensive study is Neil Jacoby, Peter Nehemkis, and Richard Eells, *Bribery and Extortion in World Business: A Study of Corporate Political Payments Abroad* (New York: Macmillan, 1977).

11. For an authoritative exposition, see E. J. Mishan, *Cost-Benefit Analysis* (New York: Praeger, 1976).

12. Michael S. Baram, "Cost-Benefit Analysis: An Inadequate Basis for Health, Safety, and Environmental Regulatory Decision Making," *Ecological Law Quarterly*, 8 (1980), 473.

13. See M. W. Jones-Lee, *The Value of Life: An Economic Analysis* (Chicago: University of Chicago Press, 1976). Also, Michael D. Bayles, "The Price of Life," *Ethics*, 89 (1978), 20–34. For trenchant criticism of these methods, see Steven E. Rhoads, "How Much Should We Spend to Save a Life?" in *Valuing Life: Public Policy Dilemmas*, ed. Steven Rhoads (Boulder, CO: Westview Press, 1980), 285–311.

14. Peter Passell, "How Much for a Life? Try $3 to $5 Million," *New York Times*, 29 January 1995, D3.

15. Ibid.

16. These points are made by Steven Kelman, "Cost-Benefit Analysis: An Ethical Critique," *Regulation*, (January–February 1981), 33–40.

17. Rosemary Tong, *Ethics in Public Policy Analysis* (Upper Saddle River, NJ: Prentice Hall, 1986), 20.

18. Mark Sagoff, "At the Shrine of Our Lady of Fatima, or Why Political Questions Are Not All Economic," *Arizona Law Review*, 23 (1981), 1283–98.

19. Kelman, "Cost-Benefit Analysis: An Ethical Critique," 39.

20. Richard M. Titmuss, *The Gift Relationship: From Human Blood to Social Policy* (London: Allen & Unwin, 1971).

21. Peter Singer, "Rights and the Market," in *Justice and Economic Distribution*, ed. John Arthur and William H. Shaw (Upper Saddle River, NJ: Prentice Hall, 1978), 213.

22. Titmuss, *Gift Relationship*, 246.

23. See Alasdair MacIntyre, "Utilitarianism and Cost-Benefit Analysis: An Essay on the Relevance of Moral Philosophy to Bureaucratic Theory," in *Values in the Electric Power Industry*, ed. Kenneth Sayre (Notre Dame, IN: University of Notre Dame Press, 1977), 217–37.

24. This case is based on a memo written by Lawrence Summers, then chief economist at the World Bank. See "Let Them Eat Pollution," *The Economist*, 8 February 1992, 66.

25. Much of the information for this case is taken from James Traub, "Into the Mouth of Babes," *New York Times Magazine*, 24 July 1988.

26. Leonard Bruder, "Jail Terms for 2 in Beech-Nut Case," *New York Times*, 17 June 1988, sec. 3, p. 1.

3

Kantian Ethics, Rights, and Virtue

Case 3.1   Big Brother at Procter & Gamble

In early August 1991, a former employee of Procter & Gamble telephoned *Wall Street Journal* reporter Alecia Swasy at her Pittsburgh office to report some disturbing news.[1] "The cops want to know what I told you about P&G," he said. This 20-year veteran of the company had just been grilled for an hour by an investigator for the Cincinnati fraud squad. The investigator, Gary Armstrong, who also happened to work part time as a security officer for P&G, had records of the ex-manager's recent long-distance calls, including some to Swasy.

Alecia Swasy had apparently angered CEO Edward Artz with two news stories about troubles at P&G that the company was not ready to reveal. An article in the *Wall Street Journal* on Monday, June 10, 1991, reported that B. Jurgen Hintz, executive vice president and heir apparent as CEO, had been forced to resign over difficulties in the food and beverage division. The next day, on Tuesday, June 11, a long article on the division's woes quoted "current and former P&G managers" as saying that the company might sell certain product lines, including Citrus Hill orange juice, Crisco shortening, and Fisher nuts. Swasy believes that Artz had deliberately lied to her when she tried to confirm the story of Hintz's departure in a telephone conversation on Saturday, and that he tried to sabotage the *Journal* by allowing the news to be released to the rival *New York Times* and the Cincinnati newspapers in time for the Sunday editions while the public relations department continued to deny the story to Swasy.

Immediately after the two articles appeared in the *Wall Street Journal*, Artz ordered a search of P&G's own phone records to determine the source of the leaks to the press. When this investigation failed to uncover any culprits, the company filed a complaint with the Hamilton County prosecutor's office, which promptly opened a grand jury investigation. The grand jury then issued several subpoenas calling for Cincinnati Bell to search its records for all calls in the 513 and 606 area codes, which cover southern Ohio and northern Kentucky, and to identify all telephone calls to Alecia Swasy's home or office and all fax transmissions to the newspaper's Pittsburgh office between March 1 and June 15. The search combed the records of 803,849 home and business telephone lines from which users had placed more than 40 million long-distance calls.

P&G contended that it filed the complaint because of "significant and ongoing leaks of our confidential business data, plans and strategies," which included not only leaks to the news media but also leaks to competitors as well. The legal basis for the grand jury probe was provided by a 1967 Ohio law that makes it a crime to give away "articles representing trade secrets" and a 1974 Ohio law that prohibits employees from disclosing "confidential information" without the permission of the employer. However, reporters are generally protected by the First Amendment right of freedom of the press, and Ohio, Pennsylvania, and 24 other states have so-called "shield laws" that protect the identities of reporters' confidential sources.

Information about an executive's forced departure is scarcely a trade secret on a par with the formula for Crest toothpaste, and the use of the phrase "articles representing trade secrets" has been interpreted in the Ohio courts to mean documents such as photographs and blueprints, not word-of-mouth news. Any law that limits First Amendment rights must define the kind of speech prohibited and demonstrate a compelling need, but the 1974 law does not specify what constitutes confidential information or the conditions under which it is protected. Thus, some legal experts doubt the law's constitutionality. P&G denied that any reporter's First Amendment rights were being violated: "No news media outlet is being asked to turn over any names or any information. The investigation is focused on individuals who may be violating the law."

The response to P&G's role in the investigation was quick and angry. The Cincinnati chapter of the Society of Professional Journalists wrote in a letter to CEO Artz: "The misguided action Procter & Gamble is taking threatens to trample the First Amendment and obviously reflects more concern in identifying a possible leak within the company rather than protecting any trade secrets.... Your complaint has prompted a prosecutorial and police fishing expedition that amounts to censorship before the fact and could lead to further abuse of the First Amendment by other companies also disgruntled by news media coverage." An editorial in the *Wall Street Journal* asked "What possessed P&G?" and questioned the legality by saying, "We understand that P&G swings a big stick in Cincinnati, of course, and maybe the local law can, like Pampers, be stretched to cover the leak. It is not funny, though, to the folks being hassled by the cops."

The sharpest criticism came from William Safire, the *New York Times* columnist, who objected to Edward Artz's contention that P&G's mistakes are not "an issue of ethics." Safire concluded a column entitled "At P&G: It Sinks" with the words: "It's not enough to say, 'our leak hunt backfired, so excuse us'; the maker of Tide and Ivory can only come clean by showing its publics, and tomorrow's business leaders, that it understands that abuse of power and invasion of privacy are no mere errors of judgment—regrettably inappropriate—but are unethical, bad, improper, wrong."

In the end, no charges were filed against any individual, and the company continued to deny any wrongdoing. A spokesperson for P&G asserted that the press "has the right to pursue information, but we have the right to protect proprietary information." Fraud squad investigator Gary Armstrong later went to work for Procter & Gamble full time.

INTRODUCTION

The critics of Procter & Gamble's heavy-handed investigation did not cite any harmful consequences beyond the chilling effect it might have had on employees and members of the press. They complained instead about the abuse of power and invasion of privacy. In particular, P&G was charged with violating certain *rights*—the right of reporters to search out newsworthy information and the right of ordinary citizens not to have their telephone records searched. Violations of rights can have very serious consequences, and so we should be alarmed by any occurrences. Consequences aside, however, there is something objectionable about a company snooping on its own employees and using law enforcement officials for company purposes.

    This chapter examines two approaches to ethics that do not appeal to consequences. One of these is the most prominent deontological theory, namely Kantian ethics; the other approach is virtue ethics, which is most commonly associated with the name of Aristotle. Rights also figure prominently in our thinking about ethical issues, but rights do not themselves constitute a theory of ethics. A right is instead a moral concept like "ought," so judgments about it need to be justified by some ethical theory. Although utilitarians have offered a foundation for rights, the concept has been thought to pose a stumbling block for any teleological theory, because rights seem to be independent of consequences. So deontological theories, such as Kantian ethics, have been considered more promising candidates for justifying the rights that we have. Before we can examine the justification of rights, though, we need to understand Kant's theory of ethics.

KANT'S ETHICAL THEORY

Immanuel Kant (1724–1804) wrote his famous ethical treatise *Foundations of the Metaphysics of Morals* (1785) before the rise of English utilitarianism, but he was well acquainted with the idea of founding morality on the feelings of pleasure and pain, rather than on reason. Accordingly, Kant set out to restore reason to what he regarded as its rightful place in our moral life. Specifically, he attempted to show that there are some things that we ought to do and others that we ought not to do merely by virtue of being rational. Moral obligation thus has nothing to do with consequences, in Kant's view, but arises solely from a moral law that is binding on all rational beings.

The Main Features of Kant's Theory

The main features of the ethical theory presented in the *Foundations* can be illustrated by considering one of Kant's own examples:

> ... [A] man finds himself forced by need to borrow money. He well knows that he will not be able to repay it, but he also sees that nothing will be loaned him if he does not firmly promise to repay it at a certain time. He desires to make such a promise, but he has enough conscience to ask himself whether it is not improper and opposed to duty to relieve his distress in such a way.

What (morally) ought this man to do? A teleological theory would have us answer this question by determining the consequences of each alternative course of action. Making a promise that he knows he cannot keep might enable the man to extricate himself from his immediate troubles, but more likely the long-term consequences of being in debt and losing the trust of others would outweigh any possible gain. No rule that would make such an action morally right could possibly be justified, moreover, by the consequences of everyone's acting in that way. According to both act- and rule-utilitarianism, therefore, the man probably ought to refrain from making the untruthful promise.

Kant regarded all such appeals to consequences as morally irrelevant. As a deontologist, he held that the duty to tell the truth when making promises arises from a rule that ought to be followed without regard for consequences. Even if the man could do more good by borrowing money under false pretenses—by using it to pay for an operation that would save a person's life, for example—the action would still be wrong. Kant denied, furthermore, that any consequence, such as pleasure, could be good. In a deontological theory, duty rather than good is the fundamental moral category. As a result, the only thing that can be good without qualification, according to Kant, is what he called a *good will*, performing an action solely because it is our duty.

Hypothetical and Categorical Imperatives. The main difficulty of deontological theories that include a list of absolute moral rules is the lack of any convincing answer to the question, How do we know that these are the rules we should follow? Kant attempted to answer this question by seizing on a difference between the moral and nonmoral senses of the word *ought*. Consider the following examples:

1. If you want to improve your serve, then you ought to take lessons from a tennis pro.
2. If you want to lower your cholesterol level, then you ought to eat less red meat.

Kant called these *hypothetical imperatives*, because they tell us to do something only on the condition that we have the relevant desire. They have the form "If you want _____, then do _____." If we do not care about improving our serve, though, or lowering our cholesterol level, then there is nothing that we *ought* to do. In neither case does the use of the word *ought* express a moral obligation.

Kant characterized moral rules as imperatives that express what we ought to do *categorically* rather than hypothetically. That is, they are uses of the word *ought* that tell us what to do regardless of our desires. Imperatives that command categorically are of the form "Do _____ (period)." Thus, we cannot evade the force of the moral rule "Tell the truth" merely by saying, for example, "But I don't care about being trusted." The question of how we can know which rules to follow was posed by Kant, then, as follows: How is it possible for there to be imperatives that command categorically, that is, that command us to perform actions no matter what desires we happen to have? His answer is that they follow from a principle, called the categorical imperative, which he believed every rational being must accept.

The Categorical Imperative. Kant's own turgid statement of the *categorical imperative* is

ACT ONLY ACCORDING TO THAT MAXIM BY WHICH YOU CAN AT THE SAME TIME WILL THAT IT SHOULD BECOME A UNIVERSAL LAW.

Rendered into more comprehensible English, Kant's principle is, act only on rules (or maxims) that you would be willing to see everyone follow.[2] The categorical imperative suggests a rather remarkable "thought experiment" to be performed whenever we deliberate about what to do. Suppose, for example, that every time we accept a rule for our own conduct, that very same rule would be imposed, by some miracle, on everybody. Under such conditions, are there some rules that we, as rational beings, simply could not accept (that is, will to become universal law)?

Let us see how this would apply to Kant's example. If the man were to obtain the loan under false pretenses, the rule on which he would be acting would be something like this: Whenever you need a loan, make a promise to repay the money, even if you know that you cannot do so. Although such a rule could easily be acted on by one person, the effect of its being made a rule for everyone—that is, of becoming a universal law—would be, in Kant's view, self-defeating. No one would believe anyone else, and the result would be that the phrase "I promise to do such-and-such" would lose its meaning.

Kant has sometimes been accused of being a "closet utilitarian" by slipping in an appeal to consequences. To Kant's own way of thinking, however, the objection to the rule just stated is not that everyone's following it would lead to undesirable consequences but that everyone's following it describes an impossible state of affairs. Willing that everyone act on this rule is analogous to a person making plans to vacation in two places, say Acapulco and Aspen, *at the same time*. Both are desirable vacation spots, but willing the metaphysical impossibility of being in two places at once is not something that can be done by a rational person.

The Principle of Universalizability

Regardless of whether Kant is successful in his attempt to show how it is possible for some imperatives to command categorically and not merely hypothetically, many philosophers still find a kernel of truth in Kant's principle of the categorical imperative which they express as the claim that all moral judgments must be *universalizable.* That is, if we say that an act is right for one person, then we are committed to saying that it is right for all other relevantly similar persons in relevantly similar circumstances. By the same token, if an act is wrong for other people, then it is wrong for any one person unless there is some difference that justifies making an exception. This *principle of universalizability* expresses the simple point that, as a matter of logic, we must be consistent in the judgments we make.

The principle of universalizability has immense implications for moral reasoning. First, it counters the natural temptation to make exceptions for ourselves or to apply a double standard. Consider a job applicant who exaggerates a bit on a résumé but is incensed to discover, after being hired, that the company misrepresented the opportunity for advancement. The person is being inconsistent to hold that it is all right for him to lie to others but wrong for anyone else to lie to him, or that there is nothing wrong with lying if it is necessary to get a job, but companies ought to be completely truthful in their dealings with potential employees. An effective move in a moral argument is to challenge people who hold such positions to cite some morally relevant difference. What is so special about you specifically or job applicants in general? If they can give no acceptable answer, then they are forced by the laws of logic to give up one or the other of the inconsistent judgments.

Second, the universalizability principle can be viewed as underlying the common question "What if everyone did that?"[3] The consequences of a few people cheating on their taxes, for example, are negligible. If everyone were to cheat, however, the results would be disastrous. The force of "What if everyone did that?" is to get people to see that because it would be undesirable for everyone to cheat, no one ought to do so. This pattern of ethical reasoning involves an appeal to consequences, but it differs from standard forms of utilitarianism in that the consequences are hypothetical rather than actual. That is, whether anyone else actually cheats is irrelevant to the question "What if everyone did that?" The fact that the results would be disastrous *if everyone did* is sufficient to entail the conclusion that cheating is wrong.[4]

Although this kernel of truth in Kant's categorical imperative is of great importance, it is too limited to serve as the basis for a complete ethical theory.[5] First, people accused of wrongdoing can insist that significant differences exist between their actions and those of others. Of course it would be wrong if everyone were to cheat on their taxes, but my case is different, a tax evader might say; it would be okay for everyone to do exactly what I am doing. Second, the principle of universalizability is incapable of refuting fanatics who would be content for everyone to act as they do. R. M. Hare[6] asks us to consider the case of a Nazi who is so convinced of the rightness of the ideal of a pure Aryan race that even evidence proving him to be a Jew does not change his mind. Most people are unlikely to accept such repugnant consequences. But as long as there are fanatics who are willing for everyone else to act in the same way, their positions—no matter how immoral—cannot be refuted by appealing merely to the principle of universalizability.

Respect for Persons

The *Foundations* contains a second formulation of the categorical imperative, which Kant expressed as follows:

> ACT SO THAT YOU TREAT HUMANITY, WHETHER IN YOUR OWN PERSON OR IN THAT OF ANOTHER, ALWAYS AS AN END AND NEVER AS A MEANS ONLY.

These words are usually interpreted to mean that we should respect other people (and ourselves!) as human beings. The kind of respect that Kant had in mind is compatible with achieving our ends by enlisting the aid of other people. We use shop clerks and taxi drivers, for example, as a means for achieving our ends, and the owners of a business use employees as a means for achieving their ends. What is ruled out by Kant's principle, however, is treating people *only* as a means, so that they are no different, in our view, from mere "things."

The moral importance of human beings is not unique to Kant's theory, of course. Virtually all systems of ethics require that we respect other persons. The distinctiveness of Kant's contribution lies in his view of what it means to be a human being and what we are respecting when we have respect for persons. For utilitarians, human beings are creatures capable of enjoying pleasure, and so we are morally obligated to produce as much pleasure as possible, taking into consideration the pleasure of everyone alike. Kant holds, by contrast, that the morally significant feature of human beings is not their capacity for enjoying pleasure but their *rationality*. Lower animals are capable of enjoying pleasure, too; what distinguishes human beings, for Kant, is their capacity for using reason. To respect persons, therefore, is to respect them as rational creatures.

Kant's Conception of Rationality. Rationality is what gives persons a greater moral value than anything else in creation, including pleasure. Indeed, rational beings are the only things that ultimately have value, according to Kant; other things have value only because we place a value on them as means to our ends. Books, music, fine food, travel, and the company of friends, for example, have value as a result of our enjoyment of them. Apart from rational beings who want them and act to obtain them, they have no value at all. Kant expressed this point by saying that their value is *conditional*. The only thing that has *unconditional* value and is an end in itself is that which gives value to other things, namely, human beings. In Kant's own words:

> Now, I say, man and, in general, every rational being exists as an end in himself and not merely as a means to be arbitrarily used by this or that will. In all his actions, whether they are directed to himself or to other rational beings, he must always be regarded at the same time as an end.

From this follows the second formulation of the categorical imperative.

Our account of the second formulation is not complete, however, without an explanation of how Kant understood reason. Reason, in Kant's view, is what enables human beings to have a free will, which is possible because we are able to create the rules that govern our conduct. This idea of acting on rules of our own devising is also conveyed by the term *autonomy*, which is derived from two Greek words meaning "self" and "law." To be autonomous is quite literally to be a lawgiver to oneself, or self-governing. A rational being, therefore, is a being who is autonomous. To respect other people is to respect their capacity for acting freely, that is, their autonomy.

Strengths and Weaknesses. The principle of respect for persons has some significant strengths and weaknesses. The main weakness is that it does not lend itself to a precise method for decision making. Certainly, slavery and gross exploitation of labor are forbidden by the principle, but utilitarianism judges these practices to be immoral as well. Respect for persons is likely to yield different results, though, in cases where utilitarians would sacrifice the interests of a few individuals to increase the overall welfare of society. Kant's principle lays a greater stress on the welfare of every person, thereby providing greater protection for the claims of individuals over those of society at large. However, it does not tell us where to draw the line. For example, any company that claims to treat its employees with respect must assure them of reasonable job security, but the principle of respect for persons provides little help in managing the inevitable trade-offs.

Respect for persons is also likely to yield different results in cases where the welfare of individuals comes into conflict with their freedom to choose. In matters of occupational health and safety, for example, utilitarians tend to favor government regulations to protect workers. An alternative that is more compatible with Kant's principle is to inform workers of the dangers and allow *them* to decide whether to assume the risk. Regulation, according to some critics, is *paternalistic;* that is, it treats workers as a parent treats a child. Utilitarians argue in reply that a lack of regulation leaves workers vulnerable to exploitation by employers, so that outlawing unsafe working conditions shows greater respect for workers as persons than leaving the choice to them. Regulation is justified, they say, because workers cannot effectively exercise individual choice in this case.

Despite shortcomings, Kant's ethical theory yields at least two important results. The principles of universalizability and respect for persons, although not sufficient for deciding all questions of ethics, are still important avenues of ethical reasoning that serve as valuable correctives to the utilitarian approach. Kantianism also provides a strong foundation for rights, which is another part of morality.

THE CONCEPT OF A RIGHT

Rights play an important role in business ethics—and, indeed, in virtually all moral issues. Both employers and employees are commonly regarded as having certain rights. Employers have the right to conduct business as they see fit, to make decisions about hiring and promotion, and to be protected against unfair forms of competition. Employees have the right to organize and engage in collective bargaining and to be protected against discrimination and hazardous working conditions. Consumers and the general public also have rights in such matters as marketing and advertising, product safety, and the protection of the environment. The language of rights is commonly employed in debates over the treatment of workers in less developed countries by transnational corporations. Some American manufacturers have been accused of violating the rights of workers by offering low wages and imposing harsh working conditions.

Beyond business, the debate over abortion, the use of life-support systems, access to medical care, and discrimination in housing and education all involve the rights of different parties. Each side in the abortion debate, for example, appeals to a supposed right: the right to life of the fetus or the right of a woman to choose whether to bear a child. There is growing support for the recognition of the right of the terminally ill to choose to die, the right of people to a minimal level of health care, and so on. Many of our constitutional protections in the Bill of Rights arouse controversy. Does the First Amendment right of free speech, for example, protect obscenity or flag burning? In 1948, the United Nations adopted the Universal Declaration of Human Rights which set forth basic human rights for all peoples.

The introduction of rights into the discussion of ethical issues is often confusing. First, the term *rights* is used in many different ways, so that the concept of a right and the various kinds of rights must be carefully distinguished. Second, many rights come into conflict. The right of an employee to leave his or her employer and join a competitor conflicts with the legitimate right of employers to protect trade secrets, for example, so that some balancing is required. Third, because of the moral significance that we attach to rights, there is a tendency to stretch the concept in ways that dilute its meaning. For example, the rights to receive adequate food, clothing, and medical care, mentioned in the Universal Declaration of Human Rights, are perhaps better described as political goals rather than rights. Finally, there can be disagreement over the very existence of a right. Whether employees have a right to due process in discharge decisions, for example, is a subject of dispute. For all these reasons, the claim of a right is frequently the beginning of an ethical debate rather than the end.

The Nature and Value of Rights

The concept of a right can be explained by imagining a company that treats employees fairly but does not recognize due process as a right.[7] In this company, employees are dismissed only for good reasons after a thorough and impartial hearing, but there is no contract, statute, or other provision establishing a right of due process for all employees. Something is still missing, because the fair treatment that the employees enjoy results solely from the company's voluntary acceptance of certain personnel policies. If the company were ever to change these policies, then employees dismissed without due process would have no recourse. Contrast this with a company in which due process is established as a right. Employees in this company have something that was lacking in the previous company. They have an independent basis for challenging a decision by the company to dismiss them. They have something to stand on, namely, their rights.

Rights can be understood, therefore, as *entitlements*.[8] To have rights is to be entitled to act on our own or to be treated by others in certain ways without asking permission of anyone or being dependent on other people's goodwill. Rights entitle us to make claims on other people either to refrain from interfering in what we do or to contribute actively to our well-being—not as beggars, who can only entreat others to be generous, but as creditors, who can demand what is owed to them. This explanation of rights in terms of entitlements runs the risk of circularity (after all, what is an entitlement but something we have a right to?), but it is sufficiently illuminating to serve as a beginning of our examination.

Kinds of Rights. Several different kinds of rights have been distinguished.

1. *Legal and Moral Rights.* *Legal rights* are rights that are recognized and enforced as part of a legal system. In the United States, these consist primarily of the rights set forth in the Constitution, including the Bill of Rights, and those created by acts of Congress and state legislatures. *Moral rights*, by contrast, are rights that do not depend on the existence of a legal system. They are rights that we (morally) *ought* to have, regardless of whether they are explicitly recognized by law. Moral rights derive their force not from being part of a legal system but from more general ethical rules and principles.

2. *Specific and General Rights.* Some rights are *specific* in that they involve identifiable individuals. A major source of specific rights is contracts, because these ubiquitous instruments create a set of mutual rights as well as duties for the individuals who are parties to them. Other rights are *general* rights, because they involve claims against everyone, or humanity in general. Thus, the right of free speech belongs to everyone, and the obligation to enforce this right rests with the whole community.

3. *Negative and Positive Rights.* Generally, *negative* rights are correlated with obligations on the part of others to refrain from acting in certain ways that interfere with our own freedom of action. *Positive* rights, by contrast, impose obligations on other people to provide us with some good or service and thereby to act positively on our behalf.[9] The right to property, for example, is largely a negative right, because

no one else is obligated to provide us with property, but everyone has an obligation not to use or take our property without permission. The right to adequate health care, for example, is a positive right insofar as its implementation requires others to provide the necessary resources.

With this examination of the concept of a right now complete, let us turn to the more difficult and controversial matter of the foundation of rights. Are rights fundamental moral categories with their own source of support? Or are rights part of some more general ethical theory, such as utilitarianism or Kantian ethics? We begin by considering the natural rights tradition which offers a historically influential account of rights separate from these other possible grounds.

Natural Rights Theory

One prominent foundation for rights focuses on what are called *natural* rights or, more recently, *human* rights. Rights of this kind, which are prominent in historical documents, are rights that belong to all persons purely by virtue of their being human.[10] They are characterized by two main features: *universality* and *unconditionality*. Universality means that they are possessed by all persons, without regard for race, sex, nationality, or any specific circumstances of birth or present condition. Unconditionality means that natural or human rights do not depend on any particular practices or institutions in society. The unconditionality of rights also means that there is nothing we can do to relinquish them or to deprive ourselves or others of them. This feature of natural, or human, rights is what is usually meant by the term *inalienable*.[11]

The idea of natural rights has a long and distinguished history going back to the ancient Greeks, who held that there is a "higher" law that applies to all persons everywhere and serves as a standard for evaluating the laws of states.[12] Both Roman law and the medieval church adopted this idea and developed it into a comprehensive legal theory. Perhaps the most influential natural rights theory, though, is that presented by John Locke (1633–1704) in his famous *Second Treatise of Government* (1690).[13] Locke began with the supposition of a state of nature, which is the condition of human beings in the absence of any government. The idea is to imagine what life would be like if there were no government and then to justify the establishment of a political state to remedy the defects of the state of nature. Locke held that human beings have rights, even in the state of nature, and that the justification for uniting into a state is to protect these rights. The most important natural right for Locke is the right to property. Although the bounty of the earth is provided by God for the benefit of all, no one can make use of it without taking some portion as one's own. This is done by means of labor, which is also a form of property. "Every man has property in his own person," according to Locke, and so "[t]he labor of his body and the work of his hands… are properly his."

Locke's theory of natural rights represents a significant advance over the traditional natural law theory in at least two respects. First, rights, in Locke's view, are protections against the encroachment of the state in certain spheres of our lives, so that individuals have moral standing as persons independent of their role as citizens. Second, the particular rights listed by Locke, especially the right to property, are precisely those required for the operation of a free market. Thus, Locke's theory is important in the rise of modern capitalism. However, natural rights theory, in its traditional form or the version offered by Locke, fails to provide a wholly satisfactory basis for the full range of rights in our modern society. A number of contemporary philosophers have attempted to resurrect the notion of natural rights with only limited success.[14]

Utility and Rights

Any attempt to found rights on utility might seem to be doomed from the start, because a standard objection against utilitarianism is that the theory is incapable of accounting for rights. The major stumbling block for the utilitarian theory is that rights often serve to protect individual interests against claims based on general welfare. For example, the right of free speech protects unpopular and dangerous opinions that society might be better off without. A tough-minded utilitarian such as Bentham can respond by simply holding that we have a right to free speech only insofar as respecting this right has good consequences. At first glance, it is difficult to understand how a utilitarian could justify unbridled liberty for inflammatory racist diatribes or sordid pornographic works. So the principle of utility would seem to support some restrictions on the expression of socially undesirable views.

However, Mill contends, in his famous defense of free speech in Chapter 2 of *On Liberty*, that by denying a right to free speech we run the risk of suppressing the truth along with error, and to suppose that we can distinguish between the two is to make an assumption of infallibility. History is full of examples of the persecution of people holding unpopular opinions, including Socrates, Jesus, Galileo, and Gandhi. Even the expression of false beliefs ought to be allowed, because by public discussion their falsity is all the more clearly revealed. Also, the absence of a right of free speech creates an atmosphere in which thinkers are afraid to challenge prevailing orthodoxies and to entertain new and unfamiliar ideas. "The greatest harm done," Mill said, "is to those who are not heretics and whose whole mental development is cramped, and their reason cowed, by the fear of heresy."

The justification for a right to free speech, then, is that once we start restricting people's expression, we lose much of the benefit of the open exchange of ideas. The utilitarian optimum would be to permit only beneficial speech, while suppressing speech that is socially undesirable, but this ideal is unattainable. Because the utility of suppressing unpopular views is often immediate and evident, whereas the utility of permitting them distant and speculative, there is a great temptation to deny freedom of speech. Giving free speech the status of a right, therefore, serves to keep us from being shortsighted and acting precipitously. Mill thus offers a plausible utilitarian foundation for rights that considers their overall benefit to society.

A Kantian Foundation for Rights

Kant himself did not give a great deal of attention to rights, but subsequent philosophers in the Kantian tradition have developed a number of justifications based on the concepts of rational agency and respect for persons. Rights are founded by Kant on his conception of humans as rational agents, that is, as beings who are capable of acting autonomously. In order to be agents of this kind, it is necessary for us to be free from limitations imposed by the will of others. Complete freedom is impossible, however, because one person can be free only if others are constrained in some way. We are free to speak, for example, only if everyone else is prevented from interfering. Being rational, autonomous agents requires, therefore, that we be justified in placing restrictions on what others can do, which is to say that we have rights.

Our rights must be in accord with universal law, however. That is, we have rights that justify our limiting the freedom of others only to the extent that others have these same rights to limit us. (This is the requirement of universality contained in the first formulation of the categorical imperative.) It follows that there is only one fundamental innate right, according to Kant, and that is the right to be free from the constraint of the will of others insofar as this is compatible with a similar freedom for all. From this fundamental right follow several subsidiary rights, including the right to equal treatment, equality of opportunity, and the ownership of property.

One criticism of the Kantian foundation for rights is that the fundamental right it supports—namely, the right to freedom—is excessively narrow.[15] In particular, Kant argues primarily for negative rights to noninterference and not for positive rights to welfare. A further difficulty is that Kantian theory provides few resources for determining what rights we do have. Without a doubt, slavery and torture are wrong because they violate the minimal conditions for rational action or dignity and respect. But the minimal conditions in other instances are less clear. Exactly how much freedom and well-being must we have to be rational, or autonomous, agents?

VIRTUE ETHICS

Despite their differences, utilitarian and Kantian ethics both address the question, What actions are right? Virtue ethics asks instead, What kind of person should we be? Moral character rather than right action is fundamental in this ethical tradition which originated with the ancient Greeks and received its fullest expression in Aristotle's *Nicomachean Ethics*. The role of ethics according to Aristotle is to enable us to lead successful, rewarding lives—the kind of lives that we would call "the good life." The good life in Aristotle's sense is possible only for virtuous persons—that is, persons who develop the traits of character that we call the virtues. Aristotle not only made the case for the necessity of virtue for good living but also described particular virtues in illuminating detail.

A complete theory of virtue ethics must do three things. First, it must define the concept of a virtue. Second, it must offer some list of the virtues—and a list of their corresponding vices. Finally, the theory must offer some justification of that list and explain how we decide what are virtues and vices. Honesty, for example, is likely to be on any list of the virtues. But what is it about honesty that makes it a virtue? How could we defend honesty as a virtue to someone who disagreed?

What Is a Virtue?

Defining virtue has proven to be difficult, and philosophers are by no means in agreement.[16] Aristotle described virtue as a character trait that manifests itself in habitual action. Honesty, for example, cannot consist in telling the truth once; it is rather the trait of a person who tells the truth as a general practice. Only after observing people over a period of time can we determine whether they are honest. Mere feelings, like hunger, are not virtues, according to Aristotle, in part because virtues are *acquired* traits. A person must become honest through proper upbringing. A virtue is also something that we actually *practice*. Honesty is not simply a matter of knowing how to tell the truth but involves habitually telling the truth. For these reasons, Aristotle classified virtue as a *state* of character, which is different from a feeling or a skill. Finally, a virtue is something that we *admire* in a person; a virtue is an excellence of some kind that is worth having for its own sake. A skill like carpentry is useful for building a house, for example, but not everyone need be a carpenter. Honesty, by contrast, is a trait that everyone needs for a good life.

A complete definition of virtue must be even more encompassing, because a compassionate person, for example, must have certain kinds of feelings at the distress of others and also the capacity for sound, reasoned judgments in coming to their aid. Virtue, for Aristotle, is integrally related to what he calls *practical wisdom* which may be described roughly as the whole of what a person needs in order to live well. Being wise about how to live involves more than having certain character traits, but being practically wise and being of good moral character are ultimately inseparable. Although the problems of defining virtue are important in a complete theory of virtue ethics, the idea of virtue as a trait of character that is essential for leading a successful life is sufficient for our purposes.

Most lists of the virtues contain few surprises. Such traits as benevolence, compassion, courage, courtesy, dependability, friendliness, honesty, loyalty, moderation, self-control, and toleration are most often mentioned. Aristotle also considered both pride and shame to be virtues on the grounds that we should be proud of our genuine accomplishments (but not arrogant) and properly shamed by our failings. More significantly, Aristotle lists justice among the virtues. A virtuous person not only has a sense of fair treatment but can also determine what constitutes fairness. The conception of justice as a virtue led Aristotle to develop a useful classification of kinds of justice and a complex theory of justice that are discussed in the next chapter.

Defending the Virtues

Defending any list of the virtues requires consideration of the contribution that each character trait makes to a good life. In particular, the virtues are those traits that everyone needs for the good life no matter his or her specific situation. Thus, courage is a good thing for anyone to have, because perseverance in the face of dangers will improve our chances of getting whatever it is we want. Similarly, Aristotle's defense of moderation as a virtue hinges on the insight that a person given to excess will be incapable of effective action toward any end. Honesty, too, is a trait that serves everyone well because it creates trust, without which we could not work cooperatively with others.

In defending a life of virtues, we cannot consider merely their contribution to some end, however; we must also inquire into the end itself. If our conception of a successful life is amassing great power and wealth, for example, then would not ruthlessness be a virtue? A successful life of crime or lechery requires the character of a Fagin or a Don Juan, but we scarcely consider their traits to be virtues—or Fagin and Don Juan to be virtuous characters. The end of life—that at which we all aim, according to Aristotle—is *happiness*, and Aristotle would claim that no despot or criminal or lecher can be happy, no matter how successful such a person may be in these pursuits. Defending any list of virtues requires, therefore, that some content be given to the idea of a good life. What is the good life, the end for which the virtues are needed?

The virtues, moreover, are not merely means to happiness but are themselves constituents of it. That is, happiness does not consist solely of what we get in life but also includes who we are. A mother or a father, for example, cannot get the joy that comes from being a parent without actually having the traits that make one a good parent. Similarly, Aristotle would agree with Plato that anyone who became the kind of person who could be a successful despot would thereby become incapable of being happy because that person's personality would be disordered in the process.

To summarize, defending a list of the virtues requires both that we determine the character traits that are essential to a good life and that we give some content to the idea of a good life itself. Virtue ethics necessarily presupposes a view about human nature and the purpose of life. It needs some context, in other words. This point is worth stressing because the possibility of applying virtue ethics to business depends on a context that includes some conception of the nature and purpose of business.

Virtue Ethics in Business

Virtue ethics could be applied to business directly by holding that the virtues of a good businessperson are the same as those of a good person (period). Insofar as business is a part of life, why should the virtues of successful living not apply to this realm as well? However, businesspeople face situations that are peculiar to business, and so they may need certain business-related character traits. Some virtues of everyday life, moreover, are not wholly applicable to business. Any manager should be caring, for example, but a concern for employee welfare can go only so far when a layoff is unavoidable. Honesty, too, is a virtue in business, but a certain amount of bluffing or concealment is accepted in negotiations. Regardless of whether the ethics of business is different from that of everyday life, we need to show that virtue ethics is relevant to business by determining the character traits that make for a good businessperson.

Applying virtue ethics to business would require us, first, to determine the end at which business activity aims. If the purpose of business is merely to create as much wealth as possible, then we get one set of virtues. Robert C. Solomon, who has developed a virtue ethics-based view in his book *Ethics and Excellence,* argues that mere wealth creation is not the purpose of business. According to Solomon,

> The bottom line of the Aristotelean approach to business ethics is that we have to get away from "bottom line" thinking and conceive of business as an essential part of the good life, living well, getting along with others, having a sense of self-respect, and being part of something one can be proud of.[17]

Solomon contends that individuals are embedded in communities and that business is essentially a communal activity in which people work together for a common good. For individuals this means achieving a good life that includes rewarding, fulfilling work; and excellence for a corporation consists of making the good life possible for everyone in society.

Solomon finds that the character traits that lead to success in life also enable us to work together in economic production. Thus, he argues that

> ... such notions as "honest advertising" and "truth in lending" are not simply legal impositions upon business life nor are they saintly ideals that are unrealistic for people in business. They are rather the preconditions of business and, as such, the essential virtues for any business dealing.[18]

Honesty in business is not necessarily the same as honesty in other spheres of life, however. The ethics of negotiation, for example, permits some concealment that would be unacceptable between family members or friends, but the differences may be accounted for by the purposes of the different activities. Whether any given character trait is a virtue in business, then, is to be determined by the purpose of business and by the extent to which that trait contributes to that purpose.

Evaluating Virtue Ethics

The main strengths of virtue ethics derive from a better fit with our own moral experience. In particular, an ethical theory that emphasizes character over the rules of right conduct is better suited to explain the following.

How We Actually Think about Decisions. Although utilitarianism and Kantian ethics provide universal moral principles that can be applied to specific cases, the proponents of a virtue ethics approach respond that people generally do not reason that way. The response of most people to a complex ethical dilemma is to ask what they feel comfortable with or what a person they admire would do. Codes of professional ethics generally offer abstract principles and sometimes specific rules, but they also stress that a professional should be a person of integrity, and this conception of the character of a professional may be a more effective guide than principles and rules. This concern with shaping character reflects the belief that a disposition to do the right thing is more important than merely knowing what is right. Ideally, morality should be something that we do not think about at all—but merely do out of habit.

The Importance of Relations in Morality. Utilitarianism and Kantian ethics would have us treat the interests of everyone impartially, but, in fact, we consider the interests of family members, friends, and members of a local community to be of greater moral importance. Insofar as virtue ethics views individuals as embedded in a community and holds that a web of close relationships is essential for a good life, it is better able, so its proponents claim, to give an account of the importance of relations in morality. A concern for the virtues also echoes some themes in feminist ethics which models ethics on family relations in which love and caring are central.[19] Because business activity consists so much of roles and relationships, then perhaps an ethics of virtue is more relevant to the experience of people in the workplace.

Virtue ethics also has some well-known weaknesses that afflict both Aristotle's theory and modern attempts to restore the Aristotelean tradition.

Incompleteness. For all its importance, a virtuous character can take us only so far in dealing with genuine ethical dilemmas. Some dilemmas involve the limits of rules—such as when not revealing information becomes a lie—or conflicts between rules. Merely relying on the virtues of honesty and compassion does not give us much guidance in such situations, although some writers offer lengthy and subtle descriptions of particular virtues. There are also some difficult ethical situations to which the virtues do not readily apply. Decisions about whether to bribe a foreign official or to require drug testing of employees or to give preference to minorities in hiring involve complex considerations of people's rights and the demands of justice, which are not easily reduced to the character traits that we call virtues. Some virtue ethics theorists respond that the importance of dilemmas or quandaries in ethics has been overstated and that questions of rights and justice should not be the central concerns of ethics.[20] Although these are important matters for some people, ethical theory should deal instead with the more mundane problems of everyday life that affect us all.

Conflicting Interests. Because he assumes that we can all achieve happiness through a life of virtue, Aristotle is accused of overlooking the fact that our interests often conflict. Insofar as our goals in life include possessing goods that are in limited supply, then not everyone can be successful. Consequently, a major task of traditional ethical theory is to manage conflict and to enable us to live in peace with one another. This view of morality is also congenial to economic theory because the market provides a mechanism for moderating conflict and enabling self-interested individuals to interact in ways that benefit all. One response of virtue ethics theorists is to play down the problem of conflicting interests and to insist that morality is more a matter of living cooperatively in communities than of moderating conflict. It is true that business is about cooperation, but it is also about power and curbing the abuse of power, and virtue ethics has little to say about that.

The idea of virtue in business is not hopelessly out of place. There are character traits that not only lead to success in business but also elevate the tone of business, and the world of business contains leaders and ordinary workers with exemplary character. No ethical theory neglects the virtues entirely, however; both utilitarianism and Kantian ethics consider character to be important in leading us to perform right actions. The distinguishing feature of virtue ethics is its insistence that being of certain character and not performing right actions is central to morality. Although this view fits some features of our moral experience, it fares less well with others. Ultimately, the choice lies between two different conceptions of what we expect of an ethical theory. If we expect an ethical theory to help us solve the really hard and complex problems of life, then an ethics of right action may be more helpful. If, on the other hand, we are more concerned with living our daily life in a community with others, then perhaps an ethics of character is more appropriate.

Conclusion

This concludes our discussion of the three major theories of ethics considered in this book, along with the foundation they provide for rights. As theories about the ultimate justification of our moral judgments—that is, about what makes right actions right—they are very different. To the extent that they suggest patterns of moral reasoning, however, they are not wholly incompatible. For example, a utilitarian can also accept the principle of universalizability and even combine it with the principle of utility to form a hybrid ethical system.[21] Utilitarians no less than Kantians consider the principle of respect for persons and the concepts of dignity and autonomy as essential elements of morality, although they offer different reasons for their importance. And all three theories agree that certain virtues are important to have, although only virtue ethics makes being virtuous the essential element in leading a moral life. Our treatment of ethical theory would not be complete, however, without a consideration of one remaining concept—namely, justice—which is the subject of the next chapter.

Case 3.2   Clean Hands in a Dirty Business

Even with her newly minted M.B.A., Janet Moore was having no luck in finding that dream job in the marketing department of a spirited, on-the-move company. Now, almost any job looked attractive, but so far no one had called her back for a second interview. Employers were all looking for people with experience, but that requires getting a job first. Just as she began to lose hope, Janet bumped into Karen, who had been two years ahead of her in college. Karen, too, was looking for a job, but in the meantime she was employed by a firm that was planning to add another marketing specialist. Janet was familiar with Karen's employer from a case study that she had researched for an M.B.A. marketing course, but what she had learned appalled her.

The company, Union Tobacco, Inc., is the major U.S. manufacturer of snuff, and her case study examined how this once staid company had managed to attract new customers to a product that had long ago saturated its traditional market.[22] Before 1970, almost all users of snuff—a form of tobacco that is sucked rather than chewed—were older men. The usual form of snuff is unattractive to nonusers because of the rough tobacco taste, the unpleasant feel of loose tobacco particles in the mouth, and the high nicotine content, which makes many first-time users ill. Snuff, to put it mildly, is a hard sell.

The company devised a product development and marketing campaign that a federal government report labeled a "graduation strategy." Two new lines were developed, a low-nicotine snuff in a tea-bag-like pouch with a mint flavor that has proved to be popular with young boys and a step-up product with slightly more nicotine, a cherry flavor, and a coarse cut that avoids the unpleasantness of tobacco floating in the mouth. Both products are advertised heavily in youth-oriented magazines with the slogan "Easy to use, anywhere, anytime," and free samples are liberally distributed at fairs, rodeos, and car races around the country.

The strategy has worked to perfection. Youngsters who start on the low-nicotine mint- and cherry-flavored products soon graduate to the company's two stronger, best-selling brands. Within two decades, the market for snuff tripled to about seven million users, of which one million to two million are between the ages of 12 and 17. The average age of first use is now estimated to be 9-1/2 years old. Janet also reported in her case study that snuff users are more than four times more likely to develop cancers of the mouth generally and fifty times more likely to develop specific cancers of the gum and inner-cheek lining. Several suits had been filed by the parents of teenagers who had developed mouth cancers, and tooth loss and gum lesions have also been widely reported, even in relatively new users.

Karen admitted that she was aware of all this but encouraged Janet to join her anyway. "You wouldn't believe some of the truly awful marketing ploys that I have been able to scuttle," she said. "Unless people like you and me get involved, these products will be marketed by people who don't care one bit about the little kids who are getting hooked on snuff. Believe me, it's disgusting work. I don't like to tell people what I do, and I sometimes look at myself in the mirror and ask what has become of the idealism I had starting out. But there will always be someone to do this job, and I feel that I have made a difference. If you join me, the two of us together can slow things down and avoid the worst excesses, and maybe we'll even save a few lives. Plus, you can get some experience and be in a better position to move on."

Janet admitted to herself that Karen had a strong argument. Maybe she was being too squeamish and self-centered, just trying to keep her own hands clean. Maybe she could do others some good and help herself at the same time by taking the job. But then again....

Case 3.3   An Auditor's Dilemma

Sorting through a stack of invoices, Alison Lloyd's attention was drawn to one from Ace Glass Company. Her responsibility as the new internal auditor for Gem Packing was to verify all expenditures, and she knew that Ace had already been paid for the June delivery of the jars that are used for Gem's jams and jellies. On closer inspection, she noticed that the invoice was for deliveries in July and August that had not yet been made. Today was only June 10. Alison recalled approving several other invoices lately that seemed to be misdated, but the amounts were small compared with the $130,000 that Gem spends each month for glass jars. I had better check this out with purchasing, she thought.

Over lunch, Greg Berg, the head of purchasing, explains the system to her. The jam and jelly division operates under an incentive plan whereby the division manager and the heads of the four main units—sales, production, distribution, and purchasing—receive substantial bonuses for meeting their quota in pretax profits for the fiscal year, which ends on June 30. The bonuses are about one-half of annual salary and constitute one-third of the managers' total compensation. In addition, meeting quota is weighted heavily in evaluations, and missing even once is considered to be a death blow to the career of an aspiring executive at Gem. So the pressure on these managers is intense. On the other hand, there is nothing to be gained from exceeding a quota. An exceptionally good year is likely to be rewarded with an even higher quota the next year, because quotas are generally set at corporate headquarters by adding 5 percent to the previous year's results.

Greg continues to explain that several years ago, after the quota had been safely met, the jam and jelly division began prepaying as many expenses as possible—not only for glass jars but for advertising costs, trucking charges, and some commodities, such as sugar. The practice has continued to grow, and sales also helps out by delaying orders until the next fiscal year or by falsifying delivery dates when a shipment has already gone out. "Regular suppliers like Ace Glass know how we work," Greg says, "and they sent the invoices for July and August at my request." He predicts that Alison will begin seeing more irregular invoices as the fiscal year winds down. "Making quota gets easier each year," Greg observes, "because the division gets an ever-increasing head start, but the problem of finding ways to avoid going too far over quota has become a real nightmare." Greg is not sure, but he thinks that other divisions are doing the same thing. "I don't think corporate has caught on yet," he says, "but they created the system, and they've been happy with the results so far. If they're too dumb to figure out how we're achieving them, that's their problem."

Alison recalls that upon becoming a member of the Institute of Internal Auditors, she agreed to abide by the IIA code of ethics. This code requires members to exercise "honesty, objectivity, and diligence" in the performance of their duties but also to be loyal to the employer. However, loyalty does not include being a party to any "illegal or improper activity." As an internal auditor, she is also responsible for evaluating the adequacy and effectiveness of the company's system of financial control. But what is the harm of shuffling a little paper around? she thinks. Nobody is getting hurt, and it all works out in the end.

NOTES

1. The information in this case is taken from Alicia Swasy, *Soap Opera: The Inside Story of Procter & Gamble* (New York: Touchstone, 1994); "Procter & Gamble Calls in the Law to Track News Leak," *Wall Street Journal*, 12 August 1991, A1, A4; "P&G Says Inquiry on Leak to Journal Was Done Properly," *Wall Street Journal*, 13 August 1991, A3; "What Possessed P&G," *Wall Street Journal*, 13 August 1991, A16; "No Charges Are Expected in P&G Affair," *Wall Street Journal*, 14 August 1991, A3; Mark Fitzgerald, "Cops Investigate News Leak," *Editor & Publisher*, 17 August 1991, 9, 39; "P&G Won't Bring Criminal Charges as a Result of Probe," *Wall Street Journal*, 19 August 1991, A3, A5; William Safire, "At P&G: It Sinks," *New York Times*, 5 September 1991, A25; "P&G Looks for a News Leak," *ABA Journal*, November 1991, 32; "Biggest Business Goofs of 1991," *Fortune*, 13 January 1992, 80–83.

2. A helpful commentary on the *Foundations* is Robert Paul Wolff, *The Autonomy of Reason* (New York: Harper & Row, 1973). For a full-length study of the categorical imperative, see Onora Nell, *Acting on Principle: An Essay on Kantian Ethics* (New York: Columbia University Press, 1975).

3. For a discussion of the logical force of this question, see Colin Strang, "What if Everyone Did That?" *Durham University Journal*, 53 (1960), 5–10.

4. For Marcus G. Singer, this is the generalization argument, which utilizes the generalization principle but also requires the support of a utilitarian principle that he calls the principle of consequences. See Marcus G. Singer, *Generalization in Ethics* (New York: Alfred A. Knopf, 1961), chap. 4.

5. An ambitious attempt to develop the principle of universalizability into a complete ethical theory is Alan Gewirth, *Reason and Morality* (Chicago: University of Chicago Press, 1978).

6. R. M. Hare, *Freedom and Reason* (Oxford: Oxford University Press, 1965), 160.

7. This method of explaining rights is derived from Joel Feinberg, "The Nature and Value of Rights," *The Journal of Value Inquiry*, 4 (1970), 243–57. Reprinted in Joel Feinberg, *Rights, Justice, and the Bounds of Liberty* (Princeton, NJ: Princeton University Press, 1980), 143–55; and in David Lyons, ed., *Rights* (Belmont, CA: Wadsworth, 1979), 78–91.

8. This term is used in H. J. McCloskey, "Rights," *The Philosophical Quarterly*, 15 (1965), 115–27; and also in Richard A. Wasserstrom, "Rights, Human Rights, and Racial Discrimination," *The Journal of Philosophy*, 61 (1964), 628–41. This latter article is reprinted in Lyons, *Rights*, 46–57. For a useful survey of different accounts, see Rex Martin and James W. Nickel, "Recent Work on the Concept of Rights," *American Philosophical Quarterly*, 17 (1980), 165–80.

9. Closely related is a distinction between negative and positive liberty. A poor man is free (in a negative sense) to buy a loaf of bread, for example, as long as no one stands in his way, but he is not free (in a positive sense) unless he has the means to buy the bread. Compare these two senses in the case of "fee for service" medical care, which secures free *choice*, and socialized medicine, which provides free *access*. Which system of medical care is more "free"? The classic discussion of this distinction is Isaiah Berlin, "Two Concepts of Liberty," in *Four Essays on Liberty* (Oxford: Oxford University Press, 1969), 118–72.

10. Among the many works describing natural or human rights are Maurice Cranston, *What Are Human Rights?* (New York: Basic Books, 1963); B. Mayo, "What Are Human Rights?" in *Politi-*

cal Theory and the Rights of Man, ed. D. D. Raphael (Bloomington: Indiana University Press, 1967), 68–80; D. D. Raphael, "Human Rights, Old and New," in *Political Theory and the Rights of Man*, 54–67; Maurice Cranston, "Human Rights, Real and Supposed," in Raphael, *Political Theory and the Rights of Man*, 43–53; and A. I. Melden, *Rights and Persons* (Berkeley and Los Angeles: University of California Press, 1977), 166–69.

11. See Stuart M. Brown, Jr., "Inalienable Rights," *The Philosophical Review*, 64 (1955), 192–211; and B. A. Richards, "Inalienable Rights: Recent Criticism and Old Doctrine," *Philosophy and Phenomenological Research*, 29 (1969), 391–404.

12. A good survey of the early history of natural rights is Richard Tuck, *Natural Rights Theories: Their Origin and Development* (Cambridge: Cambridge University Press, 1979).

13. Locke's theory of natural rights is the subject of great controversy. Two articles that provide a good introduction are W. Von Leyden, "John Locke and Natural Law," *Philosophy*, 21 (1956), 23–35; and William J. Wainwright, "Natural Rights," *American Philosophical Quarterly*, 4 (1967), 79–84. The thesis that Locke was not a natural rights theorist is developed in Leo Strauss, *Natural Right and History* (Chicago: University of Chicago Press, 1953). One response is Charles H. Monson, Jr., "Locke and His Interpreters," *Political Studies*, 6 (1958), 120–35.

14. For a survey, see Tibor R. Machan, "Some Recent Work in Human Rights Theory," *American Philosophical Quarterly*, 17 (1980), 103–15. Two important articles are Margaret MacDonald, "Natural Rights," *Proceedings of the Aristotelian Society*, 47 (1947–1948), reprinted in *Human Rights*, ed. A. I. Melden (Belmont, CA: Wadsworth, 1970), 40–60; and H. L. A. Hart, "Are There Any Natural Rights?" *The Philosophical Review*, 64 (1955), 175–91, also reprinted in *Human Rights*, 61–75.

15. An examination of Kant's account of rights, on which the following discussion is largely based, is Bruce Aune, *Kant's Theory of Morals* (Princeton, NJ: Princeton University Press, 1979), 141–52.

16. For analyses of the virtues, see Philippa Foot, *Virtues and Vices and Other Essays in Moral Philosophy* (Berkeley: University of California Press, 1978); Peter T. Geach, *The Virtues* (Cambridge: Cambridge University Press, 1977); Gregory E. Pence, "Recent Work on the Virtues," *American Philosophical Quarterly*, 21 (1984), 281–97; James D. Wallace, *Virtues and Vices* (Ithaca, NY: Cornell University Press, 1978); G. H. Von Wright, *The Varieties of Goodness* (London: Routledge & Kegan Paul, 1963).

17. Robert C. Solomon, *Ethics and Excellence: Cooperation and Integrity in Business* (New York: Oxford University Press, 1992), 104.

18. Ibid.

19. Two books influential in defining feminist ethics are Carol Gilligan, *In a Different Voice: Psychological Theory and Women's Development* (Cambridge, MA: Harvard University Press, 1982); and Nel Noddings, *Caring: A Feminine Approach to Ethics and Moral Education* (Berkeley: University of California Press, 1984). A good survey is Rosemary Tong, *Feminine and Feminist Ethics* (Belmont, CA: Wadsworth, 1993).

20. This point is made in Edmund L. Pincoffs, *Quandaries and Virtues: Against Reductionism in Ethics* (Lawrence: University of Kansas Press, 1986).

21. This is proposed by R. M. Hare in *Moral Thinking: Its Levels, Method and Point* (Oxford: Oxford University Press, 1981).

22. Material for this case is taken from Alix M. Freedman, "How a Tobacco Giant Doctors Snuff Brands to Boost Their 'Kick,'" *Wall Street Journal*, 26 October 1994, A1, A6; and Philip J. Hilts, "Snuff Makers Are Accused of a Scheme to Lure Young," *New York Times*, 27 April 1995, A13.

4

Justice and the Market System

Case 4.1   Green Giant Runs for the Border

To generations of consumers, the image of the Jolly Green Giant with his trusty side-kick Little Sprout has symbolized the convenience and dependability of frozen vegetables from America's agricultural heartlands.[1] However, the "Ho, Ho, Ho" of this genial figure is now being heard increasingly from the fields of Mexico, where Green Giant's production has been shifted in recent years. And Pillsbury, the owner of Green Giant, was itself the victim of a 1989 hostile takeover by Grand Metropolitan, PLC, a British food, liquor, and retailing concern, whose other properties include Burger King, Pearl Vision Centers, and Häagen-Dazs ice cream. The Jolly Green Giant, it turns out, is not an American but a citizen of the world with no fixed home.

The idea that companies such as Green Giant are part of the American economy and will stay put, providing jobs for local residents and benefits for communities, is deeply ingrained. The citizens of Watsonville, California, were shaken, then, when Green Giant, which once employed more than one thousand people in plants that packaged frozen broccoli and cauliflower, began laying off workers and hiring new employees at a facility in Irapuato, Mexico. Today, Green Giant employs fewer than 150 people in Watsonville and over 850 in Irapuato, and in California as a whole, roughly one-half of all frozen-food processing firms have gone south, taking over 16,000 jobs with them. An estimated 40 percent of all frozen vegetables sold in the United States are now imported from Mexico.

The economic benefits for Green Giant's move to Mexico are immense. The wages of plant workers in Watsonville averaged $7.50 per hour or about $15,600 annually. Their counterparts in Irapuato receive $4.50 a day for an annual income of around $1,400. Even after other costs are counted, Green Giant is estimated to save $13,224 per worker by moving jobs to Mexico. With fringe benefits that raise the pay by almost 60 percent, Green Giant's wages are slightly above the local rate, and unsuccessful union organizers have had to admit that the company's workers are quite content. Still, their income allows them only a very minimal standard of living and falls short of the guarantee in the Mexican constitution of a minimum wage sufficient to support a wife and children. Many of the unemployed in California are legal immigrants who came in search of economic opportunity and now find that American companies are going to Mexico for the same reason. The irony of the situation is understood by one worker who remarked: "The company is going away to get rich. Just like we did."

Forces other than the immediate gain are behind Green Giant's move to Mexico. Pillsbury planned the phaseout of production in California as part of a futile effort to avoid the takeover by Grand Met. The company realized that its customers in the East could be served more cheaply by growing and packaging vegetables in Mexico and the Midwest and that a failure to capitalize on this savings would only make it more vulnerable to a hostile raider. Remaining independent requires that managers take advantage of every opportunity for increasing profit, because other managers are always ready to step in and reap the benefit themselves. Furthermore, in an age of worldwide competition, other countries are forming their own trading blocs and allowing their companies to move production into low-wage areas. As Robert Reich observes: "Today, corporate decisions about production and location are driven by the dictates of global competition, not by national allegiance.... The global manager who fails to take advantage of global opportunities will lose profits and market share to global managers who do."[2]

Green Giant's move imposed great hardships on those workers who have few job skills. The average laid-off worker has been described as a 45-year-old Hispanic woman with no high school diploma and 14 years' experience at the single task of trimming broccoli. An expanding economy, sparked in part by an increased demand for American products in Mexico, will create new jobs for other workers, but few of these jobs will be filled by the workers laid off by Green Giant. The transfer of low-wage jobs to Mexico might increase the job prospects of American workers as a whole, but the gain of some workers will come at a painful cost to others. American consumers also gain as cheaper vegetables imported from Mexico stretch everyone's pocketbook. Spending a dollar less for broccoli is as good as a dollar raise in salary, but that is little consolation for the newly unemployed.

The job gains in Mexico benefit the new hires at Green Giant, which is known locally as *Gigante Verde*, and the Mexican economy as a whole receives a needed boost, not only from the foreign investment but also from the exposure to more advanced technology. The move of Green Giant and countless other American companies to Mexico increases the competitiveness of both countries in the global economy, but the benefit to Mexican citizens also comes at a price. The conversion of vast acreage to vegetables for the American table increases the price of beans and corn, which are the staples of the local diet. The Green Giant facility in Irapuato pumps more than half a million gallons of well water each day, and the water table, which used to be between 30 and 60 feet, is now over 450 feet. Few local residents can afford to drill wells that deep, and so most are forced to rely on the heavily polluted Guanajuato River. The use of well water to process vegetables protects American consumers from contaminants in the river but increases the health hazards for the citizens of Irapuato.

American companies are not required to observe U.S. standards for worker health and safety and for protection of the environment, and the applicable Mexican standards are generally much lower and weakly enforced. Although Green Giant claims to operate by the same standards on both sides of the border, the record of many other companies is not as clean. In *maquiladora* plants, which receive special tax treatment, sweatshop conditions are know to prevail, and toxic wastes—which, by law, should be shipped back to the United States for proper disposal—now pollute the Rio Grande.

Green Giant did not merely walk away from the problems in Watsonville. The company gave some severance pay to the displaced workers and provided them with resources for retraining and job relocation. The federal government also contributed half a million dollars to the effort, which bore some fruit, although in 1991 the unemployment rate in Watsonville approached 15 percent. For 25 years, Green Giant was a good corporate citizen, paying taxes and supporting a variety of community services. So what does the company owe the community now that it is leaving? After summing up the gains and losses to all the different groups affected by Green Giant's move, the plant manager in Irapuato concluded: "Is it right or wrong? I can't tell you. But it's the way the world is going."

INTRODUCTION

Justice, like rights, is an important moral concept with a wide range of applications. We use it to evaluate not only the actions of individuals but also social, legal, political, and economic practices and institutions. Although the word *just* is sometimes used interchangeably with *right* and *good*, it generally has a more restricted meaning that is closer to *fair*. Questions of justice often arise when there is something to distribute. If there is a shortage of organ donors, for example, we ask, what is a just, or fair, way of deciding who gets a transplant? If there is a burden, such as taxes, we want to make sure that everyone bears a fair share. Justice is also concerned with the righting of wrongs. It requires, for example, that a criminal be punished for a crime and that the punishment fit the crime by being neither too lenient nor too severe. To treat people justly is to give them what they deserve, which is sometimes the opposite of generosity and compassion. Indeed, we often speak of tempering justice with mercy.

The concept of justice is relevant to business ethics primarily in the distribution of benefits and burdens in situations like the move of Green Giant to Mexico. Economic transformations often involve an overall improvement of welfare that is unevenly distributed, so that some groups pay a price while others reap the rewards. Is the resulting distribution just, and if not, is there anything that is owed to the losers? Justice also requires that something be done to compensate the victims of discrimination or defective products or industrial accidents. Because justice is also an important concept in evaluating various forms of social organization, we can also ask about the justice of the economic system in which business activity takes place.

In succeeding sections, four theories of justice are examined, namely, Aristotle's principle of proportionate equality, the utilitarian theory, and the prominent contemporary theories of John Rawls and Robert Nozick. One important kind of justice is economic justice, and a system of free markets is both criticized and defended on grounds of justice. This chapter includes, therefore, a discussion of the justification of the market system.

ARISTOTLE'S ANALYSIS OF JUSTICE

The first thorough analysis of the concept of justice is still the best. Aristotle observed in Book V of the *Nicomachean Ethics* that the word *justice* has a double meaning. In one sense it applies to the whole of virtue. A just or morally upright person is one who always does what is morally right and obeys the law. Justice in this sense is called *universal justice* by Aristotle. The other sense of justice, which he called *particular justice*, is concerned with virtue in specific situations. More precisely, particular justice consists of taking only a proper share of some good. An unjust person is thus a grasping person who takes too much wealth, honor, or other benefit that society offers, or a shirker who refuses to bear a fair share of some burden. Aristotle divided particular justice into

1. Distributive justice, which deals with the distribution of benefits and burdens.
2. Compensatory justice, which is a matter of compensating persons for wrongs done to them.
3. Retributive justice, which involves the punishment of wrongdoers.

Both compensatory and retributive justice are concerned with correcting wrongs. Generally, compensating the victims is the just way of correcting wrongs in private dealings, such as losses resulting from accidents and the failure to fulfill contracts, whereas retribution—that is, punishment—is the just response to criminal acts, such as assault or theft.[3]

Questions about distributive justice arise mostly in the evaluation of our social, political, and economic institutions, where the benefits and burdens of engaging in cooperative activities must be spread over a group. In some instances, a just distribution is one in which each person shares equally, but in others, unequal sharing is just if the inequality is in accord with some principle of distribution. Thus, in a graduated income tax system, ability to pay and not equal shares is the principle for distributing the burden. Generally, distributive justice is *comparative*, in that it considers not the absolute amount of benefits and burdens of each person but each person's amount relative to that of others.[4] Whether income is justly distributed, for example, cannot be determined by looking only at the income of one person but requires us, in addition, to compare the income of all people in a society.

The rationale of compensatory justice is that an accident caused by negligence, for example, upsets an initial moral equilibrium by making a person worse off in some way. By paying compensation, however, the condition of the victim can be returned to what it was before the accident, thereby restoring the moral equilibrium. Similarly, a person who commits a crime upsets a moral equilibrium by making someone else worse off. The restoration of the moral equilibrium in cases of this kind is achieved by a punishment that "fits the crime." Both compensatory and retributive justice are *noncomparative*. The amount of compensation owed to the victim of an accident or the punishment due a criminal is determined by the features of each case and not by a comparison with other cases.

A useful distinction not discussed by Aristotle is that between just *procedures* and just *outcomes*.[5] In cases of distributive justice, we can distinguish between the procedures used to distribute goods and the outcome of those procedures, that is, the actual distribution achieved. A similar distinction can be made between the procedures for conducting trials, for example, and the outcomes of trials. If we know what outcomes are just in certain kinds of situations, then just procedures are those that produce or are likely to produce just outcomes. Thus, an effective method for dividing a cake among a group consists of allowing one person to cut it into the appropriate number of slices with the stipulation that that person take the last piece. Assuming that an equal division of the cake is just, a just distribution will be achieved, because cutting the cake into equal slices is the only way the person with the knife is assured of getting at least as much cake as anyone else. Similarly, just outcomes in criminal trials are those in which the guilty are convicted and the innocent are set free. The complex procedures for trials are those that generally serve to produce those results.

Aristotle on Distributive Justice

Aristotle observed in the *Politics* that "all men think justice to be a sort of equality," but what sort of equality justice is remains a source of controversy to the present day.[6] The extreme egalitarian position that everyone should be treated exactly alike has found few advocates, and most who call themselves egalitarians are concerned only to deny that certain differences ought to be taken into account.[7] A more moderate egalitarianism contends that we ought to treat like cases alike. That is, any difference in the treatment of like cases requires a moral justification.

Aristotle expressed the idea of treating like cases alike in an arithmetical equation that represents justice as an equality of ratios.[8] Let us suppose that two people, A and B, each receive some share of a good, P. Any difference in their relative shares must be justified by some relevant difference, Q. Thus, a difference in pay, P, is justified if there is a difference in some other factor, Q, that justifies the difference in P—such as the fact that one person worked more hours or was more productive. Aristotle added the further condition that the difference in each person's share of the good must be *proportional* to the difference in his or her share of the relevant difference. If one person worked twice as many hours as another, then the pay should be exactly twice as much—no more and no less. Aristotle's principle of distributive justice can be stated in the following manner.[9]

$$\frac{\text{A'S SHARE OF P}}{\text{B'S SHARE OF P}} = \frac{\text{A'S SHARE OF Q}}{\text{B'S SHARE OF Q}}$$

This account of Aristotle's principle of distributive justice is obviously not complete until the content of both P and Q are fully specified. What are the goods in question? What features justify different shares of these goods? Among the goods distributed in any society are material goods—such as food, clothing, and housing—and income and wealth, which enable the people to purchase material goods. There are many nonmaterial goods, including economic power, participation in the political process, and access to the courts, which are also distributed in some manner. Finally, Aristotle counted honor as a good, thereby recognizing that society distributes status and other intangibles. This multiplicity of goods has led philosophers to speak of many different kinds of justice, such as economic justice, legal justice, political justice, and social justice. It is unclear whether Aristotle's principle is intended to be an account of all these or only some.

Among the many different justifying features that have been proposed are ability, effort, accomplishment, contribution, and need.[10] In setting wages, for example, an employer might award higher pay to workers who have greater training and experience or greater talent (*ability*); to workers who apply themselves more diligently, perhaps overcoming obstacles or making great sacrifices (*effort*); to workers who have produced more or performed notable feats (*accomplishment*) or who provide more valued services (*contribution*); or, finally, to workers who have large families to support or who, for other reasons, have greater *need*.

Criticism of Aristotle's Account.   Each of these justifying features has some merits and some defects. Ability alone is not a suitable criterion because it rewards workers with great ability who exert little effort and consequently do not accomplish much. The criterion of effort rewards people with less ability who must try harder to make the same contribution. We also use different justifying features in different contexts. Thus, ability is an appropriate criterion for selecting budding tennis players for a training program, whereas invitations to a tournament ought to be based on accomplishment. The socialist slogan, "From each according to his abilities, to each according to his needs," holds that different criteria are relevant to determining what a person ought to contribute to society and what a person ought to receive in return. Finally, more than one criterion may be relevant to a single decision. College scholarships, for example, are often awarded on the basis of a combination of ability, effort, achievement, contribution, and need. Disagreements are possible, therefore, over the relevant justifying features and their relative weight.

Furthermore, the arithmetical form of Aristotle's principle suggests more precision than is possible. If Q is quantifiable (the number of hours worked, for example), then it is possible to say that A worked twice as many hours as B and hence ought to be paid twice as much. There is no reason, however, for insisting on strict ratios. A pay scale with higher wages for overtime might result in A's being paid *more* than twice as much, and a system in which workers are paid a bonus for exceeding a production quota might result in A's being paid more than B but *less* than twice as much. Some relevant differences are not quantifiable in ways that permit meaningful ratios. Education, for example, is a legitimate criterion for evaluating job applicants, and, other things being equal, an employer is justified in selecting an applicant with a 4.0 average over one with an average of 2.0. It does not follow, however, that the 4.0 applicant is twice as able and ought to be paid twice as much. Many relevant differences are simply incapable of being quantified and hence cannot be expressed in ratios at all. On what scale, for example, can we compute people's needs?

Despite these difficulties, Aristotle's principle of distributive justice contains much of value. The main shortcoming of the theory is that it is a purely *formal* principle, that is, the statement of the principle contains variables with no definite content. This defect could be remedied only by a principle that fully specifies, among other things, the relevant features that justify different treatment. The value of the principle lies chiefly in its insistence that different treatment be justified by citing relevant differences and that difference in treatment be in proportion to these differences.

UTILITY AND JUSTICE

Justice, along with rights, is commonly regarded as a stumbling block for utilitarianism. One problem is that the principle of utility seems to favor any redistribution that increases the total amount of utility, without regard for how it is distributed. Thus, in the following three distribution schemes, I and II are morally indistinguishable, as judged by the total amount of utility, and III is preferable to both I and II despite the apparent inequality.

I		**II**		**III**	
A	10	A	14	A	16
B	10	B	8	B	9
C	10	C	8	C	7
	30		30		32

We have seen, however, that justice does not require complete equality. In Aristotle's principle of justice, certain departures from I are justified if there are morally relevant differences between A, B, and C. So it might be that II and III are more just than I and that III is the most just distribution of all. This would be the case if A were more talented and hardworking, for example, and hence deserved more. The possibility of justified inequalities provides no defense for utilitarianism, however, because it might also be the case that there are no relevant differences between A, B, and C, so that the inequality is unjustified. The problem that justice poses for utilitarianism, then, is not merely that the theory seems to place no value on equality but also that it seems to make no allowance for justified claims for unequal treatment.

Utilitarians generally respond that any conflict between utility and justice is only apparent and that utilitarianism, properly understood, supports our judgments about justice.[11] A society with the greatest amount of utility, in other words, is also a just society, as justice is commonly understood.[12] The concepts of equality and desert, however, are not part of the meaning of utility; rather, the connection between them is purely a matter of fact, and if utility should ever conflict with equality or desert, the latter would have to give way. Utilitarians generally believe that conflicts of this kind rarely if ever occur, though.[13]

Diminishing Marginal Utility. The main argument for the convergence of utility and justice is that a system of maximizing utility has a tendency toward equality in distribution because of the phenomenon of *diminishing marginal utility*.[14] A basic tenet of economics is that the amount of utility we receive from a good decreases as the amount of the good increases. The first cookie in a box, for example, gives us much pleasure, especially if we are hungry; the second cookie, less so; and by the time we reach the last cookie, we are likely to be so full that it gives us no pleasure at all. The marginal utility of income is similar. We use the first few dollars we earn to satisfy our most basic needs and our strongest desires, with a great increase in utility. Succeeding dollars satisfy lesser felt needs and more ephemeral desires, and so they bring us correspondingly less utility. Each additional (marginal) dollar of a person's income, therefore, "buys" less utility than the one before it. A utilitarian decision maker with a dollar to give away would thus produce a greater amount of utility by giving it to a poor person than to a rich one. Similarly, practices and institutions that redistribute wealth from the rich to the poor, thereby producing greater equality, increase utility at the same time.

Although this argument is plausible, it has proven difficult in practice to verify the effect of income redistribution on the amount of utility—mainly because of the problem of measurement. Some economists argue that the marginal utility of income increases and decreases in stages, so that a person coming into great wealth might suddenly have a burning desire for a yacht and French Impressionist paintings which he did not have before.[15] Further, some inequality increases utility. A stock argument for putting money in the hands of the wealthy is that they will put it to more productive use than the poor and thereby increase the wealth of the whole society. The tendency of utilitarianism toward equality is offset, therefore, by a countertendency toward inequality, and the optimal level of equality (or inequality) is not easily determined. An additional complication is that diminishing marginal utility constitutes a powerful argument for satisfying the basic needs of all members of society, but it is less persuasive as an argument for equality once everyone attains a certain minimal level of welfare—that is, the wealthier a society becomes, the greater the inequality permitted by the principle of utility (as long as the basic needs of everyone are met).

Justice and Desert. A second argument for the convergence of utility and justice is that a system for maximizing utility also tends to reward people according to desert. The reason why it is considered just to distribute goods in proportion to ability, effort, accomplishment, or contribution is that doing so encourages people to develop their abilities, exert greater effort, accomplish more, and perform more valued services—all of which promote the welfare of society. We do not reward all abilities or efforts, for example, but only those that contribute to human well-being, and we do the same with accomplishments and contributions. The social benefit of these criteria can be obtained, moreover, only in a stable social order in which people are treated according to their deserts *consistently*. A company that discriminates against women, for example, is unlikely to motivate female employees to put forth their best efforts. A utilitarian objection to discrimination of all kinds, then, is that it reduces the overall level of utility.

Mill's Argument. Mill's position on the connection between justice and utility is that the standards of equality and desert are involved "in the very meaning of Utility." Mill's argument can be sketched as follows. One sense of equality, which is expressed in Aristotle's principle of justice, is a presumptive right of equal treatment that requires any inequalities in treatment to be justified. In Mill's words, "All persons are deemed to have a *right* to equality of treatment, except when some recognised social expediency requires the reverse." Mill characterized impartiality, which is a part of justice closely related to equality, as an obligation not to be swayed by irrelevant considerations in treating others. And this obligation, in turn, is part of "the more general obligation of giving to every one his right."

Mill's argument, in short, is that justice includes treating people according to their rights unless utility dictates otherwise, and that this obligation of justice is implicit in the meaning of the utilitarian principle. Mill thus held that equality is a part of the meaning of utility—all the while contending that it can be overridden by considerations of utility. "Each person maintains," Mill wrote, "that equality is a dictate of justice, except where he thinks that expediency requires inequality." This position is not contradictory, as some have argued.[16] But it does make equality superfluous. The meaning of "treat people equally except where utility requires unequal treatment" can be fully expressed by the simple "treat people as utility requires."

Mill did not succeed, then, in showing that equality and desert are involved in "the very meaning of Utility," and so we are left with the arguments for convergence, which are still relatively strong. Some philosophers remain unconvinced, however, and have sought other foundations for justice. One of these skeptics is John Rawls, who offers a highly influential egalitarian theory of justice as an alternative to utilitarianism.

THE EGALITARIAN THEORY OF JOHN RAWLS

In *A Theory of Justice* (1971), the contemporary American philosopher John Rawls offers two principles that he thinks express our considered views about justice. Although his theory is rather complex, the basic outlines, which were first presented in a 1958 article, "Justice as Fairness," are relatively easy to understand.[17]

Rawls's aim is to give an account of justice that embodies a Kantian conception of equality. His objection to utilitarianism, as we have already seen, is that it does not give adequate attention to the way in which utility is distributed among different individuals. "Utilitarianism," Rawls charges, "does not take seriously the difference between persons."[18] As an alternative to the utilitarian ideal of a society with the highest level of welfare, Rawls proposes a society that recognizes its members as free and equal moral persons, a concept he attributes to Kant.[19] For Rawls, questions of justice arise primarily when free and equal persons attempt to advance their own interests and come into conflict with others pursuing their self-interests.

The key to a well-ordered society is the creation of institutions that enable individuals with conflicting ends to interact in mutually beneficial ways. The principles of justice assist in this effort by assigning rights and duties in the basic institutions of society and distributing the benefits and burdens of mutual cooperation.[20] The focus of Rawls's theory, then, is on *social justice*, that is, on a conception of justice that is suited to a well-ordered society. Once we have determined what constitutes a just society, however, we can then apply the results to questions of justice in the political, legal, and economic spheres.

Rawls's Method

Rawls begins by asking us to imagine a situation in which free and equal persons, concerned to advance their own interests, attempt to arrive at unanimous agreement on principles that will serve as the basis for constructing the major institutions of society. He describes persons in this imaginary situation as self-interested agents, who evaluate principles according to whether they help or hinder them in achieving their ends. He assumes, moreover, that the people he is describing are rational in the sense that they conceive ends and act purposefully to achieve them and that they are willing to cooperate with others when this is possible and to abide by any agreements made.[21]

The approach taken by Rawls in *A Theory of Justice* is similar to traditional contract theories, which assume that if individuals in some hypothetical precontract situation would unanimously accept certain terms for governing their relations, then those terms are just and all people have an obligation to abide by them. Crucial to any contract theory is a description of the precontract situation, which is called the state of nature by Locke and others and the *original position* by Rawls. A distinctive feature of the original position, as described by Rawls, is the *veil of ignorance*.[22] The individuals who are asked to agree on the principles of justice must do so without knowing many facts about themselves and their situation. They do not know their social status or class, their natural assets or abilities, their intelligence or physical strength, their race or sex, or even their own conception of the good life. However, they have enough general knowledge about human psychology, social organization, political affairs, and the like to make informed choices about the principles of justice.

Rawls conceives of the process as a "bargaining game," in which people are free to offer proposals of their own and to reject those of others until unanimity is achieved.[23] If people in the original position know too much about themselves and how proposed principles would affect them personally, they might become dead-locked and prefer no agreement to one that puts them at a disadvantage. Behind the veil of ignorance, they are forced to be impartial and to view proposed principles from the perspective of all persons at once. Without any knowledge of their race or sex, for example, they are unlikely to advocate or support discriminatory principles, because they could be among the victims of discrimination. Certainly, no one could rationally choose a system of slavery without knowing who would be the masters and who the slaves.

The Principles of Justice

Now, what principles would rational, self-interested persons freely agree to in a position of equality behind a veil of ignorance? Rawls thinks that there are two, which he states as follows:

> 1. Each person is to have an equal right to the most extensive total system of basic liberties compatible with a similar system of liberty for all.
> 2. Social and economic inequalities are to be arranged so that they are both
>   a. to the greatest benefit of the least advantaged, and
>   b. attached to offices and positions open to all under conditions of fair equality of opportunity.[24]

The two principles are arranged by Rawls in order of priority. The task of ensuring that everyone has basic rights (the first principle) ought to be completed before any inequalities based on the second principle are permitted.[25] Further, liberty ought not to be traded for welfare; that is, no one following these principles would give up liberties for the sake of an increase in welfare.

The reasoning behind the first principle is that an equal share of whatever goods are available is the most that any person could reasonably expect, given the requirement of unanimous agreement.[26] No one would voluntarily accept less than an equal share if it were possible to share equally, because to do so would make that person comparatively worse off. Also, no one would be able to get more than an equal share by pressing for some special advantage, because someone else would have to agree to receive less than an equal share, and, as we have already seen, no one would voluntarily do that. Like the person who cuts the cake knowing that he will get the last piece, persons in the original position would opt for equal shares.

The second principle recognizes, however, that there are conditions under which rational, self-interested persons would make an exception to the first principle and accept less than an equal share of some primary goods. One such condition is that everyone would be better off with the inequality than without it.[27] If it is possible to increase the total amount of income, for example, but not possible to distribute it equally, then the resulting distribution is still just, according to Rawls, as long as the extra income is distributed in such a way that everyone benefits from the inequality.

To illustrate, consider the following distribution schemes:

I		II		III		IV	
A	10	A	14	A	16	A	14
B	10	B	8	B	9	B	11
C	10	C	8	C	7	C	11
	30		30		32		36

We have already seen that from the utilitarian point of view, I and II are morally indistinguishable; and III is superior to both I and II, despite the inequality, because of the increase in the total amount of utility. For Rawls, however, people in the original position would prefer I over II, because two people, B and C, are worse off in II. For the same reason, they would reject III, despite the greater amount of utility, because B and C are still worse off than they are in I. However, IV would be unanimously favored by people in the original position over I, because everyone is better off.

The Difference Principle and Equal Opportunity

In principle 2(a), commonly known as the *difference principle*, Rawls states another condition under which persons in the original position would be willing to accept inequalities. The condition is not that everyone be better off but that the *least-advantaged person* be better off.[28] What Rawls has in mind here is that in every distribution there is a worst-off person—although this is not necessarily the same person in each possible scheme. For example,

IV		V		VI		VII	
A	14	A	14	A	12	A	10
B	11	B	12	B	10	B	12
C	11	C	10	C	14	C	14
	36		36		36		36

The least-advantaged person in V, VI, and VII is C, B, and A, respectively. In V, VI, and VII, at least one person is better off than in IV. Still, IV is preferable, according to Rawls, because the worst-off persons, namely B and C, are better off than the worst-off person in each of V, VI, and VII. (A person in the original position would also choose IV over I, II, or III for the reason that the worst-off person in IV is still better off than the worst-off person in I, II and III.)

The difference principle accounts for our intuitive judgment that it would be wrong to enslave a portion of society even if the result were a vast increase in the total amount of utility. At the same time, it bars even the smallest sacrifice on the part of the least-advantaged person, no matter how great the increase in welfare to the rest of society. The difference principle is also compatible with a redistribution that considerably worsens the condition of people not too far above the least-advantaged person, as long as it improves the condition of the person at the bottom, even if only slightly. Thus, although Rawls's theory protects the least advantaged from a worsening of their position, it provides no corresponding protection to the more advantaged members of society.

Principle 2(b), the *principle of equal opportunity*, is similar to the view that careers should be open to all on the basis of talent. Whether a person gets a certain job, for example, ought to be determined by competence in that line of work and not by skin color, family connections, or any other irrelevant characteristic. In a just society, therefore, every effort should be made to eliminate differences that result from the accidents of birth and social conditions, in order to give people roughly equal prospects to fill positions in society. Natural differences that cannot be eliminated ought to be regarded, Rawls says, as "a common asset," to be used for the benefit of everyone.[29]

The Argument for the Difference Principle. Rawls's argument for the difference principle is that under conditions of uncertainty, a rational person would "play it safe" and choose the alternative in which the worst possible outcome is still better than the worst possible outcome of any other alternative. His argument thus employs a rule of rational choice drawn from game theory known as "maximin," which holds that it is rational to *maxi*mize the *min*imum outcome in choosing between different alternatives.[30] This is a rational rule to follow when, as Rawls says, your place will be determined by your worst enemy.

Whether Rawls's theory is superior to utilitarianism depends in large part on his success in defending maximin as a rule of rational choice.[31] Persons who have to choose principles behind a veil of ignorance might well "play it safe" and give exclusive priority to improving the condition of a representative least-advantaged person out of fear that they might be that person. But why is it not also rational for them to opt for a principle of maximizing average utility, especially if they have a propensity for risk?[32] By choosing the principle of maximum average utility, the chances of being better off are greater than under Rawls's principles. Further, they just might be one of the best-off people in the society that results.

Ultimately, the justification for Rawls's theory of justice is its congruence with our most deeply held beliefs about justice. Some of the consequences of the two principles fit better than those of utilitarianism; others do not fit as well. Although Rawls's list of basic liberties is relatively uncontroversial, he provides no priority among them when not all basic liberties can be realized or when they come into conflict with one another. There are other liberties, moreover, that might be added to the list, and Rawls provides no justification for excluding them. In addition, the priority of liberties over other primary goods is criticized for reflecting the bias of a well-to-do society whose material needs are already satisfied.[33] People in a struggling Third World country might prefer to sacrifice some liberties in order to improve their standard of living.

UTILITY AND THE MARKET SYSTEM

In a capitalist economy, major decisions about what goods and services to produce, how to manufacture and sell them, and so on, are primarily made through the impersonal forces of supply and demand in the marketplace. The principal aim of business firms in this system is to maximize the return on investment or, in other words, to make a profit. Although the market system operates largely without any explicit consideration of what is right or just, it does not follow that the system itself lacks an ethical justification. Indeed, we value our capitalist economy and hold it to be superior to other forms of economic organization, not only because of its ability to provide an abundance of goods and services at low prices but also because *we believe it to be morally justified.*

The market system is characterized by three main features: (1) *private ownership* of resources and the goods and services produced in an economy; (2) *voluntary exchange,* in which individuals are free to enter into mutually advantageous trades; and (3) the *profit motive,* whereby people engage in trading solely to advance their own well-being. In the market system described by Adam Smith and other classical economic theorists, individuals trade with each other, giving up things they own in exchange for other things they need or want. The motive of these economic actors is to improve their lives. This simple picture of the market is considerably complicated by neoclassical economics, in which the firm is the major unit of analysis. Business organizations of various kinds mobilize the productive resources of society, including labor, and transform them into goods and services to be sold in a mass market. Still, all three features are present in the description of the market system offered by neoclassical economists.

The arguments justifying the market system can be conveniently reduced to two: a utilitarian argument that a market system produces the highest possible level of welfare for society and a rights-based argument that a market system best protects our liberty, especially with respect to private property.[34] The first of these is derived from Adam Smith and his notion of an invisible hand that leads self-seeking individuals to promote a collective good. The second argument originated with John Locke's defense of property as a fundamental right, which today is generally advanced by conservative economists and libertarians. The utilitarian argument is examined in this section, and the rights-based argument is considered in the one following.

Adam Smith's "Invisible Hand"

Adam Smith (1723–1790) is generally regarded as the founder of modern economics, and his major work, *An Inquiry into the Nature and Causes of the Wealth of Nations,* published in 1776, is a landmark in the history of Western thought. One of Smith's key insights is the importance of trading in an economy. A central problem for any economic system is how to enlist the aid of others in satisfying our basic needs. Trading solves this problem by appealing to other people in terms of their own advantage. According to Smith,

> ... [M]an has almost constant occasion for the help of his brethren, and it is in vain for him to expect it from their benevolence only. He will be more likely to prevail if he can interest their self-love in his favour, and shew them that it is for their own advantage to do for him what he requires of them. Whoever offers to another a bargain of any kind, proposes to do this. Give me that which I want, and you shall have this which you want, is the meaning of every such offer; and it is in this manner that we obtain from one another the far greater part of those good offices which we stand in need of. It is not from the benevolence of the butcher, the brewer, or the baker, that we expect our dinner, but from their regard to their own interest. We address ourselves, not to their humanity but to their self-love, and never talk to them of our own necessities but of their advantages.

If two people want the same thing, conflict is inevitable, but if they want different things, and each person has what the other wants, then a trade is to the advantage of both. Therefore, trading is an instance of mutually advantageous cooperation. Whenever a trade takes place voluntarily, we can be sure that both parties believe themselves to be better off (or, at least, not worse off), because, by assumption, no one willingly consents to being worse off.

In an economy built on free trade, the same beneficial effects occur. Laborers, in search of the highest possible wages, put their efforts and skill to the most productive use. Buyers seeking to purchase needed goods and services at the lowest possible price force sellers to compete with one another by making the most efficient use of the available resources and keeping prices at the lowest possible level. The resulting benefit to society as a whole is due not to any concern with the well-being of others but solely to the pursuit of self-interest. By seeking only personal gain, each individual is, according to a famous passage in *The Wealth of Nations*, "led by an invisible hand to promote an end which was no part of his intention." He continued, "Nor is it always the worse for the society that it was no part of it. By pursuing his own interest he frequently promotes that of the society more effectually than when he really intends to promote it."

The invisible hand argument was developed further by the successors of Adam Smith. Using refined mathematical techniques, economic theorists are able to demonstrate that in a free market, individuals will continue to trade up to a point of equilibrium where no further mutually advantageous trades are possible. Thus, a system of free trade results in maximum efficiency. The details of this argument can be found in virtually any introductory economics textbook.[35]

Problems with the "Invisible Hand" Argument

There are many problems with the utilitarian argument in both the invisible hand version offered by Adam Smith and the more technical formulation of modern economic theory. First, the argument, strictly speaking, does not prove that free markets maximize *utility*; only that they are *efficient*, which is to say that they produce the greatest amount of output for the least amount of input. It stands to reason that a more productive economy has a higher level of utility, because more goods and services are available. But the connection is not assured unless the economy produces the goods and services that people want and manufactures and distributes them in ways that meet these needs.

Second, the argument that free markets are efficient presupposes *perfect competition.* This condition is satisfied when there are many buyers and sellers who are free to enter or leave the market at will and a large supply of relatively homogeneous products that buyers will readily substitute one for another. In addition, each buyer and seller must have complete knowledge of the goods and services available, including prices. In a market with these features, no firm is able to charge more than its competitors, because customers will purchase the competitors' products instead. Also, in the long run, the profit in one industry can be no higher than that in any other, because newcomers will enter the field and offer products at lower prices until the rate of profit is reduced to a common level.

Competition in actual markets is always imperfect to some degree. One reason is the existence of monopolies and oligopolies, in which one or a few firms dominate a market and exert an undue influence on prices. Competition is also reduced when there are barriers to market entry, when products are strongly differentiated, when some firms have information that others lack (about new manufacturing processes, for example), and when consumers lack important information. Competition is also reduced by *transaction costs*, that is, the expense required for buyers and sellers to find each other and come to an agreement.

Third, the argument that free markets are efficient makes certain assumptions about human behavior. It assumes, in particular, that the individuals who engage in economic activity are fully rational and act to maximize their own utility.[36] This construct, commonly called *economic man*, is faulty for at least two reasons. One is that people lack the ability to make all the calculations required to act effectively in their own interests. The other reason is that human motivation is much more complex than the simple view of economic theory. People often give money to the poor or return a lost wallet, for example, with no expectation of gain. Altruism and moral commitment play a prominent role in our economic life, along with self-interest, and yet economic theory gives them scant regard.[37]

Also, firms do not always act in the ways predicted by economic theory. Because of limitations on the ability of human beings to acquire and process the amount of information needed for corporate decision making, managers have what is described as *bounded rationality*. To compensate for these limitations, business organizations develop rules and procedures for decision making that substitute for the direct pursuit of profit. Firms also do not necessarily seek optimal outcomes, as economists assume, but, in the view of some organizational theorists, they settle for merely adequate solutions to pressing problems through a process known as *satisficing*. The immediate aim of firms, according to these theorists, is to achieve an *internal* efficiency, which is to say, the well-being of the firm, rather than *external* efficiency in the marketplace.[38]

Fourth, the invisible hand argument disregards the possibility of spillover effects or *externalities*. It assumes that all costs of production are reflected in the prices of goods and services and are not passed on to others. An externality is present when the manufacturer of a product is permitted to pollute a stream, for example, thereby imposing a cost on businesses downstream and depriving people of the opportunity to fish and swim in the stream. Other examples of externalities in present-day markets include inefficient use of natural resources (automobile drivers do not pay the full cost of the gasoline they use), occupational injuries, and accidents from defective products.

The task of dealing with externalities falls mainly to governments, which have many means at their disposal.[39] Polluters can be forced to internalize the costs of production, for example, by regulations that prohibit certain polluting activities (the use of soft coal, for example), set standards for pollution emissions, create tax incentives for installing pollution control devices, and so on. Some free market theorists have proposed solutions to the problem of externalities that make use of market mechanisms. Allowing firms that exceed pollution standards to "sell" the rights to pollute to those whose emissions violate the standards is one example.

A fifth and final objection to the invisible hand argument concerns the problem of *collective choice*.[40] In a market system, choices that must be made for a whole society—a transportation policy, for example—are made by aggregating a large number of individual choices. Instead of leaving to a central planner a decision about whether to build more roads or more airports, we allow individuals to decide for themselves whether to drive a car or to take an airplane to their destination, and through a multitude of such individual decisions, we arrive at a collective choice. The underlying assumption is that if each individual makes rational choices—that is, choices that maximize his or her own welfare—then the collective choice that results will also be rational and maximize the welfare of society as a whole.

Public Goods and the Prisoner's Dilemma

The validity of the assumption about collective choice is open to question, most notably, in the case of *public goods*. A market economy has a well-known bias in favor of private over public consumption, that is, the production of goods that can be owned and used by one person as opposed to goods that can be enjoyed by all.[41] Automobiles are an example of a private good. Roads, by contrast, are a public good in that their use by one person does not exclude their use by others. As a result of this bias in favor of private consumption, people spend large sums on their own cars but little to build and maintain a system of roads. Public parks, a free education system, public health programs, and police and fire protection are all examples of public goods that are relatively underfunded in an otherwise affluent society.[42]

The reason for this bias is simple: There is little profit in public goods. Because they cannot be packaged and sold like toothpaste, there is no easy way to charge for them. And although some people are willing to pay for the pleasure of a public park, for example, others, who cannot be excluded from enjoying the park as well, will be *free riders;* that is, they will take advantage of the opportunity to use a public good without paying for it. Indeed, if we assume a world of rational economic agents who always act in their own interest, then everyone would be a free rider, given the chance. To act otherwise would be irrational.[43] Consequently, public goods are ignored by the market and left for governments to provide, usually in a grudging manner. As the economist Joan Robinson has observed,

> When you come to think of it, what can easily be charged for and what cannot, is just a technical accident. Some things, such as drainage and street lighting, are so obviously necessary that a modicum is provided in spite of the fact that payment has to be collected through the rates, but it is only the most glaring necessities that are met in this way, together with some traditional amenities, like flower-beds in the parks, that are felt to be necessary to municipal self-respect.[44]

The Prisoner's Dilemma. The assumption that rational individual choices always result in rational collective choices is also brought into question by the *prisoner's dilemma.*[45] Suppose that two guilty suspects have been apprehended by the police and placed in separate cells where they cannot communicate. Unfortunately, the police have only enough evidence to convict them both on a minor charge. If neither one confesses, therefore, they will receive a light sentence of one year each. The police offer each prisoner the opportunity of going free if he confesses and the other does not. The evidence provided by the suspect who confesses will then enable the police to convict the other suspect of a charge carrying a sentence of 20 years. If they both confess, however, they will each receive a sentence of 5 years. A matrix of the four possible outcomes is represented in Figure 4–1.

Obviously, the best possible outcome—one year for each prisoner—is obtained when both do not confess. Neither one can afford to seek this outcome by not confessing, however, because he faces a 20-year sentence if the other does not act in the same way. Confessing, with the prospect of 5 years in prison or going scot-free, is clearly the preferable alternative. The rational choice for both prisoners, therefore, is to confess. But by doing so, they end up with the second-best outcome and are unable to reach the optimal solution to their problem.

The dilemma in this case would not be solved merely by allowing the prisoners to communicate, because the rational strategy for each prisoner in that case would be to agree not to confess and then turn around and break the agreement by confessing. The prisoner's dilemma is thus like the free-rider problem discussed earlier. If each prisoner has the opportunity to take advantage of the other's cooperation without paying a price, then it is rational to do so.[46] The true lesson of the

		Prisoner B	
		Confess	*Not Confess*
Prisoner A	Confess	A: 5 Years B: 5 Years	A: 0 Years B: 20 Years
	Not Confess	A: 20 Years B: 0 Years	A: 1 Year B: 1 Year

FIGURE 4–1   The Prisoner's Dilemma

prisoner's dilemma is that to reach the best possible outcome, each must be *assured* of the other's cooperation. The prisoner's dilemma is thus an assurance problem.[47] It shows that a rational collective choice can be made under certain circumstances only if each person in a system of cooperative behavior can be convinced that others will act in the same way.

Implications of the Dilemma. The prisoner's dilemma is not an idle intellectual puzzle. Many real-life situations involve elements of this problem.[48] Consider the following example. The factories located around a lake are polluting the water at such a rate that within a few years none will be able to use the water and they will all be forced to shut down or relocate. The optimal solution would be for each factory to install a water-purification system or take other steps to reduce the amount of pollution. It would not be rational for any one factory or even a few to make the investment required, however, because the improvement in the quality of the water would be minimal and their investment wasted. Without assurance that all will bear the expense of limiting the amount of pollution, each factory will continue to pollute the lake and everyone will lose in the end. The most rational decision for each factory individually will thus result in a disastrous collective decision.

The usual solution to prisoner's dilemma cases—along with those involving externalities and public goods—is government action. By ordering all the factories around the lake to reduce the amount of pollution and backing up that order with force, a government can assure each factory owner that the others will bear their share of the burden. As a result, they could achieve an end that they all desire but could not seek without this assurance. Regulation of this kind is not necessarily incompatible with the operation of a free market. Thomas C. Schelling points out that voluntarism versus coercion is a false dichotomy because coercion can enable firms to do what they want to do but could not do voluntarily.[49] Firms are not always averse to internalizing costs and providing public goods, Schelling observes, as long as they are not penalized more than their competitors.[50] This condition can also be secured by government regulation.

We have seen that the utilitarian argument for the market system, expressed primarily in Adam Smith's famous invisible hand metaphor, has a number of significant problems. These do not show that free markets are unjustified but only that they have limitations, or shortcomings. The proper response, therefore, is not to scrap the market system but to find ways of correcting the problems and helping markets achieve the goal of providing for our well-being. With an understanding of the problems of the utilitarian argument, we are in a better position to adopt measures that enable the market system to operate in a morally justified manner.

THE LIBERTARIAN JUSTIFICATION OF THE MARKET

Present-day libertarians are the intellectual successors to the laissez-faire tradition in economics. The distinctive feature of the libertarian philosophy is a commitment to individual liberty, conceived in the Lockean sense of a right to own property and to live as much as possible free from the interference of others. Two of the most prominent exponents of libertarianism are the Austrian-born economist Friedrich von Hayek and the American economist Milton Friedman, both of whom are strong defenders of the classical free market as the economic system that most fully supports their conception of individual liberty. The fullest statement of libertarianism by a contemporary philosopher is Robert Nozick's *Anarchy, State, and Utopia*, published in 1974.[51] The discussion that follows is mainly concerned, therefore, with the theory of justice that Nozick calls the entitlement theory.

Nozick's Entitlement Theory

The principles of justice in Nozick's theory differ from the principle of utility and Rawls's principles in two major respects. First, they are *historical* principles as opposed to nonhistorical or *end-state* principles.[52] Historical principles, Nozick explains, take into account the process by which a distribution came about, whereas end-state principles evaluate a distribution with regard to certain structural features at a given time. Second, the principles of justice in both utilitarianism and Rawls's theory are *patterned*.[53] A principle is patterned if it specifies some feature in a particular distribution and evaluates the distribution according to the presence or absence of that feature. Any principle of the form "Distribute according to _____," such as "Distribute according to IQ scores," is a patterned principle. Both Aristotle's principle and the principle of utility are patterned principles, as is the socialist formula "From each according to his abilities, to each according to his needs."

Nozick thinks that any acceptable principle of justice must be nonpatterned because any particular pattern of distribution can be achieved and maintained only by violating the right to liberty. Upholding the right to liberty, in turn, upsets any particular pattern of justice. He argues for this point by asking us to consider a case in which there is a perfectly just distribution, as judged by some desired pattern, and also perfect freedom. Now suppose that a famous athlete—Nozick suggests Wilt Chamberlain—will play only if he is paid an additional 25 cents for each ticket sold and that many people are so excited to see Wilt Chamberlain play that they will cheerfully pay the extra 25 cents for the privilege. Both Wilt Chamberlain and the fans are within their rights to act as they do, but at the end of the season, if one million people pay to see him play, then Wilt Chamberlain will have an additional income of $250,000, which is presumably more than he would be entitled to on a patterned principle of justice. By exercising their right to liberty, though, Wilt Chamberlain and the fans have upset the just distribution that formerly prevailed. In order to maintain the patterned distribution, it would be necessary to restrict the freedom of Wilt Chamberlain or the fans in some way, such as prohibiting the extra payment or taxing away the excess.

The entitlement theory can be stated very simply. A distribution is just, Nozick says, "if everyone is entitled to the holdings they possess."[54] Whether we are entitled to certain holdings is determined by tracing their history. Most of what we possess comes from others through transfers, such as purchases and gifts. Thus, we might own a piece of land because we bought it from someone, who in turn bought it from someone else, and so on. Proceeding backward in this fashion, we ultimately reach the original settler who did not acquire it through a transfer but by clearing the land and tilling it. As long as each transfer was just and the original acquisition was just, then our present holding is just. "Whatever arises from a just situation by just steps is itself just," Nozick writes.[55] In his theory, then, particular distributions are just not because they conform to some pattern (equality or social utility, for example) but solely because of antecedent events.

Two Principles. Nozick's theory requires at least two principles: a principle of just transfer and a principle of just original acquisition. Because holdings can be unjustly appropriated by force or fraud, a third principle, a principle of rectification, is also necessary, in order to correct injustices by restoring holdings to the rightful owners. If we rightfully possess some holding—a piece of land, for example, either by transfer or original acquisition—then we are free to use or dispose of it as we wish. We have a right, in other words, to sell it to whomever we please at whatever price that person is willing to pay, or we can choose to give it away. As long as the exchange is purely voluntary, with no force or fraud, the resulting redistribution is just. Any attempt to prevent people from engaging in voluntary exchanges in order to secure a particular distribution is a violation of liberty, according to the entitlement theory.

A world consisting only of just acquisitions and just transfers would be just, according to Nozick, no matter what the pattern of distribution. Some people, through hard work, shrewd trades, or plain good luck, would most likely amass great wealth, whereas others, through indolence, misjudgment, or bad luck, would probably end up in poverty. However, the rich in such a world would have no obligation to aid the poor,[56] nor would it be just to coerce them into doing so. Each person's share would be determined largely through his or her choices and those of others. Nozick suggests that the entitlement theory can be expressed simply as *From each as they choose, to each as they are chosen.*

The entitlement theory supports a market system with only the absolute minimum of government intervention, as long as the principles of just acquisition and just transfer are satisfied. The reason is that a system in which we have complete freedom to acquire property and engage in mutually advantageous trades (without violating the rights of another person, of course) is one in which our own rights are most fully protected. To critics who fear that unregulated markets would lead to great disparities between rich and poor and a lowering of the overall welfare of society, Nozick has a reply. The point of justice is not to promote human well-being or to achieve a state of equality; it is to protect our rights. Because a market system does this better than any other form of economic organization, it is just.

Objections to Nozick's Theory

The main difficulty with Nozick's theory is the lack of argument for the key assumption that liberty, conceived as the unhindered exercise of property rights, is a paramount value. First, there are rights, such as a right to a minimal level of welfare, that some consider to be at least as important as the rights of property owners. Yet, in Nozick's view, it would be unjust to provide for the well-being of the members of society if doing so involved forcing the well-to-do against their will to aid the poor. Taxation for this purpose, according to Nozick, is on a par with forced labor.[57] What is the justification, though, for allowing the poor to starve merely to protect the rich against being coerced in this way?

Second, Nozick seems to overlook the point that not all restrictions of liberty are due to interference by the state. By exercising their property rights, people sometimes effectively restrict the choices of others. Suppose that Wilt Chamberlain uses his wealth to buy the only factory in town and immediately cuts all wages. Libertarians hold that it would be an unjustified violation of property rights for the state to tax an individual $500, say, to finance a social security system, but according to Nozick, the liberty of the workers in the factory is not restricted by Wilt Chamberlain's action even if they are deprived by the same amount. Nozick appears to be inconsistent in treating the two cases differently.[58]

Third, justice in transfers and original acquisition depends on conditions that are scarcely ever satisfied in the actual world. Transfers are just, according to Nozick, if each person is entitled to the holdings he or she possesses. The present distribution of holdings is heavily determined, however, by forced takings—fraud, theft, wars, and the like. Correcting such injustices is a task for the principle of rectification, which is applied by attempting to predict what would have happened had the injustice not occurred and then enforcing the result. Although isolated injustices can be rectified in this manner,[59] what is the possibility of determining the course of events if Native Americans had not been deprived of their land, slavery had never occurred, and so on?[60] An adequate theory of justice should be capable of leading us from an unjust world to one a little more just, a test that Nozick's theory fails.

Conclusion

In this chapter, we have examined four theories of justice, namely, those of Aristotle, Mill, Rawls, and Nozick. The purpose of these theories is primarily to provide a means for evaluating existing and proposed institutional arrangements. On the question of the organization of our economic system, utilitarianism and libertarianism both support a system of free markets.[61] Insofar as the market system promotes utility *and* protects rights, it is doubly justified, so to speak. Some proponents of free markets employ both justifications. The two are not wholly compatible, however, and in practice we often face trade-offs between utility or welfare on the one hand and liberty or property rights on the other. When such situations occur, the holders of these two justifications are forced to choose between them.

Case 4.2 Executive Compensation

The news that Michael D. Eisner received $203 million in 1993 for his efforts on behalf of the Walt Disney Company was greeted with a predictable outcry about run-away executive pay. How could anyone be worth that much in any just economic system? Eisner's earnings reached that astronomical figure by a one-time exercise of accumulated stock options. In 1994, the highest paid executive received a paltry $25.9 million in salary, bonus, and long-term compensation, and Eisner ranked only sixteenth on that year's list of top-paid chief executives with a total package of $10.6 million.[62] The dollar value of executive compensation is difficult to gauge because of uncertainty in the pricing of stock options, and the ranking of executives varies from year to year as some executives exercise stock options and others defer them. Still, the average total compensation of CEOs in 1994 at 371 companies surveyed by *Business Week* was more than $2.8 million, and 25 executives each received more than $10 million.

Aside from the dollar amount of executive compensation, critics note that CEOs tend to receive their fat paychecks regardless of their company's performance. Louis V. Gerstner was rewarded with a 34 percent increase in his 1994 salary and bonus for raising the stock of IBM over 30 percent, but the CEO of AT&T, Robert E. Allen, received an even higher increase of 38 percent despite a drop of 4.3 percent in the price of Ma Bell's stock.[63] Some executives are compensated handsomely even as they order massive layoffs that create misery for many hard-working employees. One study found that compensation for the CEO rose in 1993 at 27 of the 30 companies with the largest staff reductions over a three-year period.[64] In 1993, Sears Roebuck increased by 198 percent the compensation of a CEO who presided over the elimination of 50,000 jobs.

By any measure, the CEOs of U.S. firms are well compensated. But are they overcompensated, as many contend? If Michael Eisner's $203 million is too much, how much is just right? How should executive pay be determined? One answer is that we should use the same standards for CEOs that are generally applied to lower-level managers and other employees—namely, the degree of responsibility involved; the knowledge, skill, and effort required; and success in meeting certain goals. A CEO's compensation should be higher for those who run larger companies and companies that face greater problems or risks. The reward for turning around a troubled giant like IBM should be greater than that for attending to everyday business at a heavily regulated utility, for example. Most of all, an executive should be rewarded for doing well, which may be reflected not only in the price of the stock but also by desired changes, such as improving customer relations or a company's environmental record. The demands on a CEO are heavy and the talents needed are in short supply, so high pay for the very best executives is to be expected.

That CEOs are paid more than most workers is no surprise. But how much more should they be paid? And what should their pay be compared with? Derek Bok, the former president of Harvard University, notes that he headed a very large and complex organization for 20 years with relatively modest compensation. Bok further observes that average CEO pay is two to three times the average in industrial nations, and some American CEOs are paid ten to twelve times their counterparts in Japan and Europe. "Are American CEOs ten or twelve, or even two or three times better?" he asks.[65] The average CEO in Japan and Europe earns ten to twelve times as much as the average manufacturing employee; in the United States the comparable figure is twenty-five times,[66] and among Fortune 500 companies, CEOs earn more than one hundred times the salary of the average employee.

Graef Crystal, a persistent critic of executive compensation, is less concerned about the amount of CEO pay than the lack of linkage to the usual standards.[67] Although sports and entertainment figures negotiate multimillion-dollar contracts, Crystal finds that at least 70 percent of those amounts can be explained by purely economic considerations. Thus, a star pitcher who negotiates a $10 million contract can usually be expected to produce at least $7 million in extra revenue for the team owner. Crystal's studies have never been able to account for more than 40 percent of CEO pay by such factors as company size, performance, business risk, government regulation, industry type, and location. Crystal concludes that the pay of top executives is highly arbitrary, with some CEOs receiving two to three times what his model indicates as "rational." The problem, according to Crystal, is that boards of directors, who hire the CEO, are not very good negotiators. Unlike team owners and movie producers, board members have little idea of what the talent they are buying is really worth. After a bad season, a highly paid sports star may receive much less the next year; this seldom happens with CEOs.

Another answer to the question of how executive pay should be determined looks to the benefit that attracting and motivating good talent can bring to the American economy. If high compensation succeeds in placing the most competent people in positions of business leadership, then everyone benefits. According to this view, companies bid against one another for the best among the available pool of managerial talent, and it may be worth $10 million to a company to hire the most capable person whose success may result in profits many times that amount. Compensation packages are also designed with a mixture of base pay, bonuses, and stock options in order to align the CEO's interests with those of the corporation. Some critics scoff at the idea that people of CEO caliber have to be coaxed with multimillion-dollar salaries to take the reins of a company and would not give the job their full attention without huge bonuses.

These objections are beside the point, however, if the need for high pay is not to attract a CEO from the pool of existing candidates but to attract the best to the field of management. If some other occupations, such as law and finance, attract the cream of the crop by their high pay, then American business and the whole of society will be deprived of their potential contribution. Michael C. Jensen and Kevin J. Murphy suggest that critics of executive pay should at least ask the right question: "Are current levels of CEO compensation high enough to attract the best and brightest individuals to careers in corporate management?" The answer, they say, is probably not.[68] If CEOs were paid for performance—which most people believe to be fair—then some would receive far more compensation than they currently do, and some would receive far less or be booted out the door. Why don't boards act in this way? One reason, according to Jensen and Murphy, is the public perception that executives are already overpaid. When boards are afraid to pay high performers what they are really worth because of popular pressure, they are also reluctant to punish poor performers. The result is a level of compensation that fails to attract the best people to management and keeps mediocre people in place. If Jensen and Murphy are right, then widespread concern about the justice of executive compensation is itself part of the problem.

Case 4.3   Merck and AIDS in South Africa

In April 2001, Merck & Company, a U.S.-based multinational pharmaceutical firm, had to decide whether to continue a legal battle against the South African government.[69] A suit by 39 drug manufacturers sought to overturn a 1997 law that, in Merck's view, threatened to undermine the international system of pharmaceutical patent protection that is essential for developing new life-saving products. The law was enacted to enable South Africa to deal with the AIDS crisis. With approximately 4.7 million people HIV-positive and more than 600,000 infected with AIDS, the country desperately needed cheap medicine. Merck had already reduced the prices on its two major AIDS drugs almost 90 percent. The question was whether to oppose a law that would permit the importation of even cheaper generic versions produced in violation of internationally recognized patents.

   Drug pricing pits market forces against human need. It costs on average $500 million to develop a new drug, and pharmaceutical companies must price their products so as to recoup their investment. Once a drug is on the market, it can be reproduced at relatively low cost. Consequently, patent protection, which confers exclusive control for a certain number of years, is vital. Without it, the incentives for engaging in costly research would be greatly reduced. The high prices of new drugs have little impact on poor countries, where medical treatment is generally provided by generic versions of basic pharmaceuticals. Patented drugs are also often sold at different prices in different countries, based on local market conditions. Such "tiered pricing" makes newer drugs more affordable in less developed parts of the world, although pharmaceutical companies are still accused of charging too much.

   AIDS in sub-Saharan Africa is different from other medical situations in three important respects. First, the few effective drugs are all recent discoveries; there are no alternative treatments for AIDS. As a result, there is no choice but to use the newly discovered costly AIDS drugs. The triple cocktail of drugs used in the United States costs more than $10,000 a year for each patient. Second, even much cheaper drugs would not benefit many AIDS victims in sub-Saharan Africa, where people can scarcely afford any medical treatment. In South Africa, the government spends approximately $10 per person on health care. Third, the human need is pressing. One quarter of South African adults are HIV-positive, and, in the year 2000, a quarter of a million people died from AIDS. A 15-year-old South African youth today faces a better than even chance of dying from AIDS. The United Nations estimates that in sub-Saharan Africa 25.3 million people are HIV-positive and 4.8 million have AIDS.

The legal suit, begun in 1998, challenged a 1997 law, the Medicines and Related Substances Control Act. This legislation permits the country's health minister to import generic versions of patented drugs from foreign suppliers and to allow domestic firms to produce the drugs (a step known as "compulsory licensing"). In 1993, South Africa signed the World Trade Organization Agreement on Trade-related Aspects of Intellectual Property Rights (TRIPS). This agreement pledged countries to respect recognized patents but allows an exception for pharmaceuticals in the event of a national emergency. South Africa was reluctant to declare such an emergency for fear of scaring off foreign investors. The suit charged, then, that the law gave the health minister power that is not permitted by the TRIPS agreement.

Although the law had yet to be invoked, the suit acquired renewed urgency when, in April 2001, an Indian company, Cipla, requested the health minister to authorize the production of generic versions of eight patented AIDS drugs. Many companies had already slashed prices on AIDS drugs. Merck offered to sell its two best-sellers at reduced prices—Crixivan for $600 per patient per year and Socrin for $500. In the United States, Crixivan sells for $5,000 per patient. Bristol-Myers Squibb reduced to $55 the price of Zerit, which sells in the United States for $3,400 a year. GlaxoSmithKline offered to cut the price on Combivir to $730 a year, which the company says is equal to its manufacturing costs and about one-tenth of the $7,000 per year cost in the United States.

However, Bombay-based Cipla said that it could provide all of these pharmaceuticals for about one-third of the reduced prices, so that a triple cocktail could be sold for about $600 per year. Even at that price, though, relatively few people in South Africa would receive treatment. Estimates are that the number of people on drug therapy, currently 10,000, could be increased to 100,000. The main impediments to AIDS treatment are the lack of funds and the poor health-care delivery system.

The stakes to Merck and other companies are very high. No money is lost by providing drugs at cost in Africa since they would not be sold otherwise. There is a risk, however, that reduced-price drugs sold in Africa could be diverted to more developed countries and interfere with sales there. Moreover, if South Africa permitted the production of generic versions of patented AIDS drugs, this practice could spread to other drugs and to other countries. South Africa is a relatively small market, accounting for less than 1 percent of global pharmaceutical sales, but great damage could be done if patent protection were weakened in such countries as Brazil, Thailand, or India, which are major producers of generic drugs. Finally, losing the court fight in South Africa might undermine patent protection in the United States. AIDS victims in the United States are guaranteed the necessary drugs, but reduced prices abroad could lead to a demand for lower prices in the United States for drugs that treat other illnesses. The chief executive of the Pharmaceutical Manufacturers Association said, "If we decided not to fight, the whole world could simply adopt laws undermining patents. We'd risk the whole market."

Postscript

On April 19, 2001, Merck and the 38 other companies that been suing South Africa dropped their opposition to the law permitting the importation of generic versions of patented drugs. The suit had been a public relations disaster for the pharmaceutical industry. The European Union and the World Health Organization were joined by AIDS activists and physician groups worldwide in calling for a halt to the suit. The United States government declared that it would not seek to enforce American patents as long as international agreements were observed. Although pressure forced the companies to capitulate, questions remain about the best policy. In the short run, more AIDS victims in South Africa will receive treatment as a result, but, in the long run, will the outcome of this case help or hinder the search for drugs to ease the plight of people in poor countries like South Africa?

NOTES

1. Material for this case is taken from Katherine Ellison, "The Two Valleys of the Green Giant," *West Magazine, San Jose Mercury News,* 16 June 1991, 6–11.

2. Robert B. Reich, "Who Is Them?" *Harvard Business Review,* 69 (March–April 1991), 77–78.

3. Aristotle makes the distinction between compensatory and retributive justice in terms of voluntary and involuntary relations. A contract is a voluntary arrangement between two people, whereas the victim of an assault enters into the relation involuntarily. Many commentators have found this a rather awkward way of making the distinction.

4. The distinction between comparative and noncomparative justice is discussed in Joel Feinberg, "Comparative Justice," *The Philosophical Review,* 83 (1974), 297–338. Reprinted in Joel Feinberg, *Rights, Justice, and the Bounds of Liberty* (Princeton, NJ: Princeton University Press, 1980), 265–306.

5. The distinction is discussed in Brian Barry, *Political Argument* (London: Routledge & Kegan Paul, 1965), 97–100.

6. For a discussion of the connection, see Gregory Vlastos, "Justice and Equality," in *Social Justice,* ed. Richard B. Brandt (Upper Saddle River, NJ: Prentice Hall, 1962), 31–72.

7. For a discussion of egalitarianism, see S. I. Benn and R. S. Peters, *Social Principles and the Democratic State* (London: Allen & Unwin, 1959), chap. 5. Also, H. J. McCloskey, "Egalitarianism, Equality, and Justice," *Australasian Journal of Philosophy,* 44 (1966), 50–69.

8. For a discussion of Aristotle's principle, see Hans Kelsen, "Aristotle's Doctrine of Justice," in Kelsen, *What Is Justice* (Berkeley and Los Angeles: University of California Press, 1957), 117–36; and Renford Bambrough, "Aristotle on Justice," in *New Essays on Plato and Aristotle,* ed. Renford Bambrough (London: Routledge & Kegan Paul, 1965), 159–74.

9. Taken from William K. Frankena, "Some Beliefs about Justice," in *Freedom and Morality,* ed. John Bricke (Lawrence: University of Kansas, 1976), 56.

10. For a discussion of this list, see Nicholas Rescher, *Distributive Justice* (Indianapolis, IN: Bobbs-Merrill, 1966), chap. 4.

11. For a discussion of utilitarian responses, see D. C. Emmons, "Justice Reassessed," *American Philosophical Quarterly,* 4 (1967), 144–51.

12. For an account of Bentham's theory of justice, see Hugo A. Bedau, "Justice and Classical Utilitarianism," in *Justice*, ed. Carl J. Friedrich and John W. Chapman (New York: Atherton Press, 1963), 284–305.

13. Contemporary philosophers who take this position are Jan Narveson, *Morality and Utility* (Baltimore, MD: Johns Hopkins University Press, 1967), 210–19; Rolf E. Sartorius, *Individual Conduct and Social Norms* (Encino, CA: Dickinson, 1975), 130–33; J. J. C. Smart, "Distributive Justice and Utilitarianism," in *Justice and Economic Distribution*, ed. John Arthur and William H. Shaw (Upper Saddle River, NJ: Prentice Hall, 1978), 103–15; and R. M. Hare, "Justice and Equality," in Arthur and Shaw, eds., *Justice and Economic Distribution*, 116–31. Richard B. Brandt argues for a convergence of justice with his version of utilitarianism as a pluralistic welfare-maximizing moral system in *A Theory of the Good and the Right* (Oxford: Oxford University Press, 1979), chap. 16.

14. A good statement of this argument is given in Richard B. Brandt, *Ethical Theory* (Upper Saddle River, NJ: Prentice Hall, 1959), 415–19. A more extended discussion is Brandt, *Theory of the Good and the Right*, 311–16.

15. On this point, see Brandt, *Ethical Theory*, 416–17; and C. Dyke, *The Philosophy of Economics* (Upper Saddle River, NJ: Prentice Hall, 1981), 38–39.

16. See Bedau, "Justice and Classical Utilitarianism," 303.

17. A good exposition is Robert Paul Wolff, *Understanding Rawls: A Reconstruction and Critique of A Theory of Justice* (Princeton, NJ: Princeton University Press, 1977). For a critical study, see Brian Barry, *The Liberal Theory of Justice* (Oxford: Oxford University Press, 1973). Among the many collections of articles are Norman Daniels, ed., *Reading Rawls: Critical Studies of A Theory of Justice* (New York: Basic Books, 1975); and H. Gene Blocker and Elizabeth H. Smith, eds., *John Rawls's Theory of Social Justice* (Athens: Ohio University Press, 1980).

18. John Rawls, *A Theory of Justice* (Cambridge, MA: Harvard University Press, 1971), 85–86.

19. See John Rawls, "A Kantian Conception of Equality," *Cambridge Review* (February 1975), 94.

20. Rawls, *Theory of Justice*, 4.

21. Ibid., 142–45.

22. Ibid., 136–42.

23. The term *bargaining game* is derived from game theory, which Rawls employs in *A Theory of Justice*. The classic work on game theory is R. D. Luce and Howard Raiffa, *Games and Decisions* (New York: John Wiley, 1957).

24. Rawls, *Theory of Justice*, 302.

25. Ibid., 243–51, 541–48.

26. The reasoning behind both principles is discussed in ibid., 150–61.

27. Rawls's formulation of the second principle in "Justice as Fairness" is "Inequalities are arbitrary unless it is reasonable to expect that they will work out for everyone's advantage"; John Rawls, "Justice as Fairness," *The Philosophical Review*, 67 (1958), 165. His initial statement in *A Theory of Justice* is that "social and economic inequalities are to be arranged so that they are both (a) reasonably expected to be to everyone's advantage, and (b) attached to positions and offices open to all;" Rawls, *Theory of Justice*, 60.

28. The reasons for the shift in the formulation of the difference principle involve rather technical problems concerning the comparison of alternatives. For an explanation, see Wolff, *Understanding Rawls*, 42–47, 63–65.

29. Rawls, *Theory of Justice*, 102.

30. Ibid., 152–56. For a full explanation, see Luce and Raiffa, *Games and Decisions*, chap. 13.

31. For further objections to the justification of maximin in the original position, see Thomas Nagel, "Rawls on Justice," *The Philosophical Review*, 82 (1973), 229–32.

32. Some critics of Rawls argue that a principle of maximum average utility is the only possible choice in the original position. See John C. Harsanyi, "Can the Maximin Principle Serve as a

Basis for Morality? A Critique of John Rawls' Theory," *American Political Science Review*, 69 (1975), 594–606. Also John C. Harsanyi, "Morality and the Theory of Rational Behavior," in *Utilitarianism and Beyond*, Amartya Sen and Bernard Williams, eds. (Cambridge: Cambridge University Press, 1982) 39–62.

33. On this problem, see Barry, *The Liberal Theory of Justice*, 59–82; and H. L. A. Hart, "Rawls on Liberty and Its Priority," *University of Chicago Law Review*, 40 (1973), 534–55.

34. A comprehensive survey of the economic and ethical arguments for and against the market system is Allen Buchanan, *Ethics, Efficiency, and the Market* (Totowa, NJ: Rowman and Little-field, 1988). See also the collection of essays in Gerald Dworkin, Gordon Bermant, and Peter G. Brown, eds., *Markets and Morals* (Washington, DC: Hemisphere, 1977); and Virginia Held, ed., *Property, Profits, and Economic Justice* (Belmont, CA: Wadsworth, 1980).

35. For an explanation, see A. Feldman, *Welfare Economics and Social Choice Theory* (Dordrecht: Martinus Nijhoff, 1980).

36. For a brief discussion of this assumption, see Amartya Sen, *On Ethics and Economics* (Oxford: Basil Blackwell, 1987), 10–28.

37. Some economists and behavioral scientists recognize the role of altruism and moral commitment and attempt to incorporate them into economic theory. See, for example, E. S. Phelps, ed., *Altruism, Morality, and Economic Theory* (New York: Russell Sage Foundation, 1975); Harvey Leibenstein, *Beyond Economic Man* (Cambridge, MA: Harvard University Press, 1976); David Collard, *Altruism and Economy: A Study in Non-Selfish Economics* (New York: Oxford University Press, 1978); Howard Margolis, *Selfishness, Altruism, and Rationality: A Theory of Social Choice* (Cambridge: Cambridge University Press, 1982); Amitai Etzioni, *The Moral Dimension: Towards a New Economics* (New York: Free Press, 1988); and Robert H. Frank, *Passions Within Reason: The Strategic Role of the Emotions* (New York: W. W. Norton, 1988).

38. On the concepts of bounded rationality and satisficing, see Herbert A. Simon, *Administrative Behavior: A Study of Decision Making Processes in Administrative Organization*, 3d ed. (New York: Free Press, 1976), originally published in 1947; James G. March and Herbert A. Simon, *Organizations* (New York: John Wiley, 1958); and Richard M. Cyert and James G. March, *A Behavioral Theory of the Firm* (Upper Saddle River, NJ: Prentice Hall, 1963).

39. See Buchanan, *Ethics, Efficiency, and the Market*, 24–25.

40. For an insightful study of this problem, see Amartya Sen, *Collective Choice and Social Welfare* (San Francisco: Holden Day, 1970). Also, Kenneth Arrow, *Social Choice and Individual Values*, 2d ed. (New York: John Wiley, 1963).

41. Joan Robinson, *Economic Philosophy* (London: C. A. Watts, 1962), 132.

42. The disparity between private and public consumption in the United States is the major theme in John Kenneth Galbraith, *The Affluent Society* (Boston: Houghton Mifflin, 1958).

43. If everyone attempts to be a free rider, however, then certain kinds of collective choices are impossible unless people are coerced in some way. See Mancur Olson, *The Logic of Collective Action* (Cambridge, MA: Harvard University Press, 1965), 44.

44. Robinson, *Economic Philosophy*, 132–33.

45. For discussion of the prisoner's dilemma, see any book on game theory, such as Luce and Raiffa, *Games and Decisions*.

46. This point is made by Russell Hardin, "Collective Action as an Agreeable *n*-Person Prisoners' Dilemma," *Behavioral Science*, 16 (1971), 472–79.

47. C. Ford Runge, "Institutions and the Free Rider: The Assurance Problem in Collective Action," *Journal of Politics*, 46 (1984), 154–81. Cited in Ian Maitland, "The Limits of Business Self-Regulation," *California Management Review*, 27 (Spring 1985), 134.

48. For the empirical significance of the prisoner's dilemma, see Anatol Rapoport and Albert M. Chammah, *The Prisoner's Dilemma* (Ann Arbor: University of Michigan Press, 1965); and Robert Axelrod, *The Evolution of Cooperation* (New York: Basic Books, 1984).

49. Thomas C. Schelling, "Command and Control," in *Social Responsibility and the Business Predicament,* ed. James W. McKie (Washington, DC: The Brookings Institution, 1974), 103–5.

50. Schelling, "Command and Control," 103.

51. Robert Nozick, *Anarchy, State, and Utopia* (New York: Basic Books, 1974).

52. Ibid., 153–55.

53. Ibid., 155–60.

54. Ibid., 151.

55. Ibid., 151.

56. Individuals could voluntarily agree with others to contribute to the relief of poverty, in which case an obligation would exist. See ibid, 265–68.

57. Ibid., 169.

58. James P. Sterba, *The Demands of Justice* (Notre Dame, IN: University of Notre Dame Press, 1980), 113.

59. For a discussion of some of the problems, see Lawrence Davis, "Comments on Nozick's Entitlement Theory," *The Journal of Philosophy,* 73 (1976), 839–42.

60. For one answer, see David Lyons, "The New Indian Claims and Original Rights to Land," *Social Theory and Practice,* 4 (1977), 249–72.

61. Rawls's theory also justifies the market system, although he distinguishes between free markets as a method of allocation and distribution and capitalism as an economic system that includes private ownership. Markets, according to Rawls, can be a feature of either capitalism or socialism, and whether an economic system with free markets is just depends a great deal on other background institutions. See Rawls, *Theory of Justice,* 270–74.

62. These figures were taken from "CEO Pay: Ready for Takeoff," *Business Week,* 24 April 1995, 88–119.

63. "Deliver or Else: Pay-for-Performance Is Making an Impact on CEO Paychecks," *Business Week,* 27 March 1995, 36–38.

64. Molly Baker, "I Feel Your Pain? When the CEO Orders Massive Layoffs, Should He Get a Pay Raise?" *Wall Street Journal,* 12 April 1995, A6.

65. "Former Harvard President Slams U.S. Exec Pay," *HR Focus* (December 1994), 11.

66. These figures are derived from data in Amanda Bennett, "Managers' Incomes Aren't Worlds Apart," *Wall Street Journal,* 12 October 1992, B1, B5.

67. Graef S. Crystal, *In Search of Excess* (New York: W. W. Norton, 1991).

68. Michael C. Jensen and Kevin J. Murphy, "CEO Incentives—It's Not How Much You Pay, but How, " *Harvard Business Review,* 68 (May–June 1990), 145.

69. This case is based on Kurt Shillinger, "Firms Fight South Africa to Keep Patents on AIDS Drugs," *Boston Globe,* 4 March 2001, A5; Jon Jeter, "Trial Opens in South Africa AIDS Drug Suit," *New York Times,* 6 March 2001, A1; Sheryl Gay Stolberg, "Africa's AIDS War," *New York Times,* 10 March, 2001, A1; Rachel L. Swarns, "AIDS Drug Battle Deepens in Africa," *New York Times,* 7 March 2001, A1; Denise Gellene, "AIDS Drug Pricing Controversy Opens Door to Wider Debate," *Los Angeles Times,* 25 March 2001, C1; Rachel L. Swarns, "Drug Makers Drop South Africa Suit of AIDS Medicine," *New York Times,* 19 April 2001, A1; Andrew Pollock, "Defensive Drug Industry: Fueling Clash over Patents," *New York Times,* 20 April, 2001, A6; Melody Petersen, "Lifting the Curtain on the Real Costs of Making AIDS Drugs," *New York Times,* 24 April 2001, C1.

5

Whistle-Blowing

Case 5.1   Two Whistle-Blowers

Chuck Atchinson

At the age of 40, Charles (Chuck) Atchinson had achieved a measure of success.[1] His job as a quality control inspector for Brown & Root, a construction company building a nuclear power plant for the Texas Utilities Electric Company, paid more than $1,000 a week. This was enough to provide a comfortable house for his wife and 13-year-old daughter, along with new cars, vacation trips, and a bounty of other luxuries. Four years later, the family was six months behind on the rent on a trailer home on a gravel street in Azle, Texas, near Fort Worth. Chuck Atchinson had been fired by Brown & Root and was unable to find work. The house was repossessed, and most of the family's furniture and other possessions had been sold to cover living expenses and mounting legal bills.

The source of Chuck Atchinson's misfortune was his inability to get his superiors at Brown & Root to observe safety regulations in the construction of the Comanche Peak nuclear power plant being built in Glen Rose, Texas, and to correct a number of potentially dangerous flaws. When his repeated complaints to the company got no response, he brought the situation to the attention of government regulators. Soon after that he was dismissed from his job. Brown & Root justified the dismissal on the grounds of poor performance as a safety inspector, and the company attempted to downplay the credibility of his charges. Around the time that he was testifying to government regulators, he received anonymous threatening telephone calls warning him to keep quiet, and he suspected that he was being followed and his telephone monitored.

Finding a new job was not easy. He worked for a while driving a wrecker, and out of desperation he even gathered cans along the highway for sale as scrap aluminum. After finally landing a job as an inspector at a plant in New Orleans, Chuck Atchinson was on the job for only a week before he was subpoenaed to give further testimony about the Comanche Peak plant. According to his own account, "When I got back, my boss called me in and fired me. He said I was a troublemaker." He was not even given the chance to assume a new job at a power plant in Clinton, Illinois. "Two days before I was to leave," he said, "they called and said they wouldn't take me, because I was a troublemaker. I tried other plants and I found that I was blacklisted."

Chuck Atchinson eventually found stable employment doing quality control work for the aerospace division of the LTV Corporation, and he is finally getting back on his feet financially. But the psychic scars still remain. Especially disturbing to the family was the loss of friends. "The whistle-blower has about the same image as the snitch does," he said. "Everyone thinks you're slime." Still he expresses few regrets for blowing the whistle and asserts that he would do it again if he had to. "I've got absolutely nothing in my hand to show as a physical effect of what I've done, except the losses I've had. But I know I was the cutting edge of the knife that prevented them from getting their license and sent them back to do repairs. I know I did right. And I know I'll always sleep right. I'll sleep like a baby."

Joseph Rose

In the course of his work as an in-house attorney for the Associated Milk Producers Incorporated (AMPI), Joseph Rose became aware of illegal political contributions to the Nixon reelection campaign.[2] He knew that the confidentiality of the attorney-client relation barred him from voluntarily releasing information about past contributions, even if they constituted a criminal conspiracy. At the same time, his legal training told him that his present activity—helping to cover up illegal activity—made him a co-conspirator in a crime. He first contacted the president of AMPI, a Wisconsin dairy farmer, and told him that he would not approve certain payments. Warned that he was now as guilty as anyone else, he concluded that there was no point in going to the other executives of AMPI, who were all deeply implicated in the illegal scheme, so he decided that he had to collect the incriminating evidence and present it to the board of directors. In his own words:

> I was never allowed to do that. My attempt [to talk to the board] happened on a weekend during their convention in Minneapolis. Labor Day followed, and then Tuesday I went to work. I found a guard posted at my door; locks had been changed. The general manager demanded to see me. My services had become very, very unsatisfactory. After I was fired, I felt virtually a sense of relief. I was glad to be out of it, and I planned to keep my mouth shut. Then I had a call from one of the lawyers involved in an antitrust case against AMPI. He said, "They are really slandering you—making some very vicious attacks on you." I had indicated to AMPI executives that if the board would not listen to me, I would go right to the dairy farmers and they obviously felt my career and credibility had to be completely destroyed to protect their own tails.

In the end, AMPI was fined $35,000 and forced to pay $2.9 million in back taxes. Two executives were convicted, and one received a prison sentence. (The other executive died while waiting to be sentenced.) Joseph Rose also paid a price in terms of the disruption of his career. Despite the fact that he had been compelled by law to testify before Congress and to the grand jury led by the Watergate special prosecutor, Joseph Rose was regarded by potential employers as someone who had been disloyal and was unreliable as an employee.

Joseph Rose was able to put his life back together. He has been able to establish a successful practice in San Antonio, Texas, and some of his clients come to him because of his reputation for integrity and toughness in the face of adversity. However, the experience has left him with a cynical view of American business.

> … [A]ll of the public utterances of corporations and indeed of our own government concerning "courage, integrity, loyalty, honesty, and duty" are nothing but the sheerest hogwash that disappear very rapidly when it comes to the practical application of these concepts by strict definition. The reason that there are very few … [whistle-blowers] is that the message is too clearly out in this society that white-collar crime, or nonviolent crime, should be tolerated by the public at large, so long as the conduct brings a profit or a profitable result to the institution committing it….

INTRODUCTION

There have always been informers, or snitches, who reveal information to enrich themselves or to get back at others. However, whistle-blowers like Chuck Atchinson and Joseph Rose are generally conscientious people who expose some wrongdoing, often at great personal risk. The term *whistle-blower* was first applied to government employees who "go public" with complaints of corruption or mismanagement in federal agencies. It is now used in connection with similar activities in the private sector.[3]

As the examples of Chuck Atchinson and Joseph Rose show, whistle-blowers often pay a high price for their acts of dissent. Retaliation is common and can take many forms—from poor evaluations and demotion to outright dismissal. Some employers seek to blacklist whistle-blowers so that they cannot obtain jobs in the same industry. Many whistle-blowers suffer career disruption and financial hardship resulting from the job dislocation and legal expenses, and there is severe emotional strain on whistle-blowers and their families as co-workers, friends, and neighbors turn against them. Given the high price that whistle-blowers sometimes pay, should people really be encouraged to blow the whistle? Is the exposure of corruption and mismanagement in government and industry the best way to correct these faults? Or are there more effective ways to deal with them without requiring individuals to make heroic personal sacrifices? Should whistle-blowers be protected, and if so, how can this best be done?

In addition to these practical questions, there are more philosophical issues about the ethical justification of whistle-blowing. Do employees have a right to blow the whistle? Although they usually act with the laudable aim of protecting the public by drawing attention to wrongdoing on the part of their organization, whistle-blowers also run the risk of violating genuine obligations that employees owe to employers. Employees have an obligation to do the work that they are assigned, to be loyal to their employer, and generally to work for the interest of the company, not against it. In addition, employees have an obligation to preserve the confidentiality of information acquired in the course of their work, and whistle-blowing sometimes involves the release of this kind of information. Cases of whistle-blowing are so wrenching precisely because they involve very strong conflicting obligations. It is vitally important, therefore, to understand when it is morally permissible to blow the whistle and when whistle-blowing is, perhaps, not justified. Our first task, though, is to develop a definition of whistle-blowing.

WHAT IS WHISTLE-BLOWING?

As a first approximation, whistle-blowing can be defined as the release of information by a member or former member of an organization that is evidence of illegal and/or immoral conduct in the organization or conduct in the organization that is not in the public interest. There are several points to observe in this definition.

First, blowing the whistle is something that can be done only by a member of an organization. It is not whistle-blowing when a witness of a crime notifies the police and testifies in court. It is also not whistle-blowing for a reporter who uncovers some illegal practice in a corporation to expose it in print. Both the witness and the reporter have incriminating information, but they are under no obligation that prevents them from making it public. The situation is different for employees who become aware of illegal or immoral conduct in their own organization. Whistle-blowing, therefore, is an action that takes place within an organization.

The difference is due to the fact that an employee is expected to work only as directed, to go through channels, and, especially, to act in all matters for the well-being of the organization. Also, the information involved is typically obtained by an employee in the course of his or her employment as a part of the job. Such information is usually regarded as confidential so that an employee has an obligation not to reveal it, especially to the detriment of the employer. To "go public" with information that is damaging to the organization is generally viewed as violating a number of obligations that an employee has as a member of the organization.

Second, there must be information. Merely to dissent publicly with an employer is not in itself to blow the whistle; whistle-blowing necessarily involves the release of nonpublic information. According to Sissela Bok, "The whistleblower assumes that his message will alert listeners to something they do not know, or whose significance they have not grasped because it has been kept secret."[4] A distinction can be made between *blowing the whistle* and *sounding the alarm*. Instead of revealing new facts, as whistle-blowers do, dissenters who take a public stand in opposition to an organization to which they belong can be viewed as trying to arouse public concern, to get people alarmed about facts that are already known.

Third, the information is generally evidence of some significant kind of misconduct on the part of an organization or some of its members. The term *whistle-*

blowing is usually reserved for matters of substantial importance. The illegal campaign contributions that AMPI funneled to the Nixon reelection campaign provide a good example because information about the contributions constituted grounds for legal action against the organization and several executives. An employee could also be said to blow the whistle about other practices that are legal but contrary to the public interest, such as a lobbying effort for maintaining artificially high prices for milk products. Information of this kind could alert the public and aid consumers and other interest groups in counteracting the lobbying effort, for example. However, merely exposing incompetent or self-serving management or leaking information to influence the course of events is not commonly counted as whistle-blowing. Lacking in these kinds of cases is a serious wrong that could be averted or rectified by whistle-blowing.

Fourth, the information must be released outside normal channels of communication. In most organizations, employees are instructed to report instances of illegal or improper conduct to their immediate superiors, and other means often exist for employees to register their concerns. Some corporations have an announced policy of encouraging employees to submit any suspicions of misconduct in writing to the CEO, with an assurance of confidentiality. Others have a designated official—often called an *ombudsman*—for handling employee complaints. Whistle-blowing does not necessarily involve "going public" and revealing information outside the organization. There can be *internal* as well as *external* whistle-blowing. However, an employee who follows established procedures for reporting wrongdoing is not a whistle-blower. Joseph Rose, by contrast, saw that it would be futile to confront the executives of AMPI with his evidence of their illegal campaign contributions and decided to go over their heads to the board of directors, thereby blowing the whistle.

A definition of whistle-blowing also needs to take into account *to whom* the whistle is blown. In both internal and external whistle-blowing, the information must be revealed in ways that can reasonably be expected to bring about a desired change. Merely passing on information about wrongdoing to a third party does not necessarily constitute whistle-blowing. Chuck Atchinson testified before a regulatory agency, for example, and Joseph Rose testified before a committee of Congress and the Watergate grand jury. Going to the press is often effective because the information ultimately reaches the appropriate authorities. Reporting to a credit-rating agency that a person faces bankruptcy, by contrast, would not usually be an instance of whistle-blowing but of ordinary snitching.

Fifth, the release of information must be something that is done voluntarily, as opposed to being legally required, although the distinction is not always clear. Joseph Rose did not volunteer his evidence of illegal campaign contributions outside the organization, for example; he had to be subpoenaed and legally forced to testify. And some of Chuck Atchinson's testimony was also the result of court orders. One thing that confuses the distinction is that whistle-blowers are often required by law to reveal information, but the call to testify comes only after they volunteer that they have incriminating evidence. However, in a state supreme court case, *Petermann* v. *International Brotherhood of Teamsters,* a treasurer for a union had no desire to be a whistle-blower, but he refused to perjure himself before a California state legislative body as he had been ordered to do by his employer.[5] Although Petermann acted with considerable courage, it is not clear whether he should be called a whistle-blower because he had little choice under the circumstances.

A sixth point is that whistle-blowing must be undertaken as a moral protest; that is, the motive must be to correct some wrong and not to seek revenge or personal advancement. This is not to deny that a person with incriminating evidence could conceivably be justified in coming forth, whatever the motive. People "go public" for all sorts of reasons—a common one being fear of their own legal liability—and by doing so, they often benefit society. Still, it is useful to draw a line between the genuine whistle-blower and corporate malcontents and intriguers. Because the motives of whistle-blowers are often misperceived in the organization, employees considering the act must carefully examine their own motivation.

Putting all these points together, a more adequate (but unfortunately long-winded) definition of whistle-blowing is as follows: Whistle-blowing is the voluntary release of nonpublic information, as a moral protest, by a member or former member of an organization outside the normal channels of communication to an appropriate audience about illegal and/or immoral conduct in the organization or conduct in the organization that is opposed in some significant way to the public interest.

THE JUSTIFICATION OF WHISTLE-BLOWING

The ethical justification of whistle-blowing might seem to be obvious in view of the laudable public service that whistle-blowers provide—often at great personal risk. However, whistle-blowing has the potential to do great harm to both individuals and organizations.

The negative case against whistle-blowing is given vigorous expression in a widely cited passage from a 1971 speech by James M. Roche, who was chairman of the board of General Motors Corporation at the time:

> Some critics are now busy eroding another support of free enterprise—the loyalty of a management team, with its unifying values of cooperative work. Some of the enemies of business now encourage an employee to be disloyal to the enterprise. They want to create suspicion and disharmony, and pry into the proprietary interests of the business. However this is labelled—industrial espionage, whistle blowing, or professional responsibility—it is another tactic for spreading disunity and creating conflict.[6]

A more temperate statement along the same lines is given by Sissela Bok:

> Furthermore, the whistleblower hopes to stop the game, but since he is neither referee or coach, and since he blows the whistle on his own team, his act is seen as a vio-

lation of loyalty. In holding his position, he has assumed certain obligations to his colleagues and clients. He may even have subscribed to a loyalty oath or a promise of confidentiality. Loyalty to colleagues and to clients comes to be pitted against loyalty to the public interest, to those who may be injured unless the revelation is made.[7]

As these remarks indicate, the main stumbling block in justifying whistle-blowing is the duty of loyalty that employees have to the organization of which they are a part. The public service that whistle-blowers provide has to be weighed against the disruptive effect that the disclosure of information has on bonds of loyalty. Does a person in a position to blow the whistle have a greater obligation to the public or to the organization? Where does the greater loyalty lie?

That we have an obligation to the public is relatively unproblematic; it is the obligation to prevent serious harm to others whenever this is within our power. An obligation of loyalty to an organization is more complex, involving, as it does, questions about the basis of such an obligation and the concept of loyalty itself. What does an employee owe an employer, and, more to the point, does the employment relation deprive an employee of a right to reveal information about wrongdoing in the organization? In order to answer these questions, let us begin with a commonly used argument against the right of an employee to blow the whistle.

The Loyal Agent Argument

According to one argument, an employee is an *agent* of an employer.[8] An agent is a person who is engaged to act in the interests of another person (called a *principal*) and is authorized to act on that person's behalf. This relation is typical of professionals, such as lawyers and accountants, who are called upon to use their skills in the service of a client. Employees are also considered to be agents of an employer in that they are hired to work for the benefit of the employer. Specifically, an employee, as an agent, has an obligation to work as directed, to protect confidential information, and, above all, to be loyal. All these are seemingly violated when an employee blows the whistle.

The loyal agent argument receives considerable support from the law, where the concept of agency and the obligations of agents are well developed. Although our concern is with the *moral* status of employees, the law of agency is a rich source of relevant insights about the employment relation.[9] According to one standard book on the subject, "an agent is a person who is authorized to act for a principal and has agreed so to act, and who has power to affect the legal relations of his principal with a third party."[10] Agents are employed to carry out tasks that principals are not willing or able to carry out for themselves. Thus, we hire a lawyer to represent us in legal matters where we lack the expertise to do the job properly.

The main obligation of an agent is to act in the interest of the principal. We expect a lawyer, for example, to act as we would ourselves, if only we had the same ability. This obligation is expressed in the *Second Restatement of Agency* as follows: "…

an agent is subject to a duty to his principal to act solely for the benefit of the principal in all matters connected with his agency."[11] The ethical basis of the duty of agents is a contractual obligation or an understood agreement to act in the interests of another person. Lawyers agree for a fee to represent clients, and employees are similarly hired with the understanding that they will work for the benefit of an employer.

Are Whistle-Blowers Disloyal Agents?

At first glance, a whistle-blower is a disloyal agent who backs out of an agreement that is an essential part of the employer-employee relation. A whistle-blowing employee, according to the loyal agent argument, is like a lawyer who sells out a client—clearly a violation of the legal profession's code of ethics. Closer examination reveals that the argument is not as strong as it appears. Although employees have an obligation of loyalty that is not shared by a person outside the organization, the obligation is not without its limits. Whistle-blowing is not something to be done without adequate justification, but at the same time, it is not something that can never be justified.

First, the law of agency does not impose an absolute obligation on employees to do whatever they are told. Rather, an agent has an obligation, in the words of the *Second Restatement*, to obey all *reasonable* directives of the principal. This is interpreted to exclude illegal or immoral acts; that is, employees are not obligated as agents to do anything illegal or immoral—even if specifically instructed by a superior to do so. Questions can arise, of course, about the legal and moral status of specific acts. Is an agent free to disobey an order to do something that is suspect but not clearly illegal or immoral, for example? Borderline cases are unavoidable, but in situations where a crime is being committed or people are exposed to the risk of serious injury and even death, the law of agency is clear: An employee has no obligation to obey.

The law of agency further excludes an obligation to keep confidential any information about the commission of a crime. Section 395 of the *Second Restatement of Agency* reads in part: "An agent is privileged to reveal information confidentially acquired ... in the protection of a superior interest of himself or a third person." The *Restatement* does not define what is meant by a "superior interest" except to note that there is no duty of confidentiality when the information is about the commission of a crime. "... [I]f the confidential information is to the effect that the principal is committing or is about to commit a crime, the agent is under no duty not to reveal it."[12] Protecting oneself from legal liability can reasonably be held to be a "superior interest," as can preventing some serious harm to others.

Second, the obligations of an agent are confined to the needs of the relation. In order for a lawyer to represent a client adequately, it is necessary to impose a strong obligation of loyalty, but the obligation of loyalty required for employees to do their job adequately is less stringent. The obligation of agents to follow orders exactly stems, in part, from the fact that they may be binding the principal to a contract or exposing the principal to tort liability. The duty of confidentiality is justified by the legitimate right of an employer to maintain the secrecy of certain vital information. Thus, a quality control inspector, such as Chuck Atchinson, has an obligation to perform his work as directed so that nuclear power plants are built safely, and he has an obligation of confidentiality because certain information could benefit a competitor. These obligations are essential to a quality control inspector's job.

Employees are hired for limited purposes, however. As Alex Michalos points out, a person who has agreed to sell life insurance policies on commission is committed to performing *that* activity as a loyal agent. "It would be ludicrous," he continues, "to assume that the agent has also committed himself to painting houses, washing dogs, or doing anything else that happened to give his principal pleasure."[13] Similarly, a quality control inspector is not hired to overlook defects, falsify records, or do anything else that would permit a dangerous plant to go into operation. Information about irregularities in safety matters is also not the kind that the employer has a right to keep confidential, because it is not necessary to the normal operation of the business of constructing nuclear power plants.

To conclude, the loyal agent argument does not serve to show that whistle-blowing can never be justified. The obligations that employees have as agents of an organization are of great moral importance, but they do have limits. Specifically, the agency relation does not require employees to engage in illegal or immoral activities or to give over their whole life to an employer.

The Meaning of Loyalty

The concept of loyalty itself raises some questions. One is whether whistle-blowing is always an act of disloyalty or whether it can sometimes be done out of loyalty to the organization. The answer depends, in part, on what we mean by the term *loyalty*. If loyalty means merely following orders and not "rocking the boat," then whistle-blowers are disloyal employees. But loyalty can also be defined as a commitment to the true interests or goals of the organization, in which case whistle-blowers are often very loyal employees. Thus, whistle-blowing is not necessarily incompatible with loyalty, and, indeed, in some circumstances, loyalty may require employees to blow the whistle on wrongdoing in their own organization.

All too often, the mistake of the whistle-blower lies not in being disloyal to the organization as such but in breaking a relation of trust with a few key members of an organization or with associates and immediate superiors. Insofar as an employee has a duty of loyalty, though, it cannot be merely to follow orders or to go along with others. Loyalty means serving the interests and goals of an organization, which can sometimes lead to divided loyalties and uncertainties about what is best for an organization.

The Sociological Evidence. Some evidence for the claim that whistle-blowers are often loyal—perhaps even too loyal—to the organizations they serve is provided by

Myron Glazer, a sociologist who interviewed 55 whistle-blowers in depth. One of his findings is that

> Virtually all of the ethical resisters ... had long histories of successful employment. They were not alienated or politically active members of movements advocating major changes in society. On the contrary, they began as firm believers in their organizations, convinced that if they took a grievance to superiors, there would be an appropriate response. This naiveté led them into a series of damaging traps. They found that their earlier service and dedication provided them with little protection against charges of undermining organizational morale and effectiveness.[14]

The irony of this finding is that whistle-blowers are often loyal employees who take the first steps toward whistle-blowing in the belief that they are doing their job and acting in the best interests of the company. This is true of Joseph Rose, who has written:

> I never set out to be a whistleblower; I merely tried to alert the appropriate officials at AMPI to the misconduct I became aware of—I felt that was my duty as AMPI's in-house counsel. Even though AMPI fired me abruptly for attempting to discharge my duty ... my personal set of ethics dictated that I attempt to shield the company because of the unsettled question of our attorney-client relationship. If I was a whistleblower, I became one reluctantly.[15]

Exit, Voice, and Loyalty. As further evidence that the relation between whistle-blowing and loyalty is far more complex than it first appears, the economist Albert O. Hirschman argues in a book entitled *Exit, Voice, and Loyalty* that members of organizations and people who deal with organizations, such as customers of a firm, can respond to dissatisfaction either by leaving the organization and having no further dealings with it (exit) or by speaking up and making the dissatisfaction known in the hope of bringing about change (voice). Loyalty is a factor that keeps people from exiting an organization; but, at the same time, it activates the voice option. According to Hirschman,

> ... the likelihood of voice increases with the degree of loyalty. In addition, the two factors are far from independent. A member with a considerable attachment to a product or organization will often search for ways to make himself influential, especially when the organization moves in what he believes is the wrong direction; conversely, a member who wields (or thinks he wields) considerable power in an organization and is therefore convinced that he can get it "back on the track" is likely to develop a strong affection for the organization in which he is powerful.[16]

On Hirschman's analysis, exit is a more extreme form of dissent than voice, but business firms do not usually regard an employee's departure as a form of disloyalty. In fact, whistle-blowers are often treated in ways designed to get them to leave voluntarily. It may benefit an organization in the short run to get rid of troublemakers, but Hirschman argues that in the long run, encouraging employees to use the exit option will harm the organization by depriving it of those people who can bring about healthy change.

As a result of loyalty, these potentially most influential customers and members will stay on longer than they would ordinarily, in the hope or reasoned expectation that improvement or reform can be achieved "from within." Thus, loyalty, far from being irrational, can serve the socially useful purpose of preventing deterioration from becoming cumulative, as it so often does when there is no barrier to exit.[17]

A further complication is the fact that employees typically have a number of loyalties, both inside and outside an organization, which can come into conflict. This point is well made by Daniel Ellsberg, the Defense Department employee who finally decided to disclose the so-called Pentagon Papers to the press.

> I think the principle of "company loyalty," as emphasized in the indoctrination within any bureaucratic structure, governmental or private, has come to sum up the notion of loyalty for many people. This is not a healthy situation, because the loyalty that a democracy requires to function is a somewhat varied set of loyalties which includes loyalty to one's fellow citizens, and certainly loyalty to the Constitution and to the broader institutions of the country. Obviously, these loyalties can come into conflict, and merely mentioning the word "loyalty" doesn't dissolve those dilemmas that one faces.[18]

Even if we limit loyalty to a specific employer, such as a government agency or a corporation, questions about what loyalty means still arise. *The Code of Ethics for Government Service*, for example, contains the following instruction for federal employees: "Put loyalty to the highest moral principles and to country above loyalty to persons, party, or government department." This lofty statement is a prescription for confusion when employees of an administration or an agency are called upon to be team players.

THE CONDITIONS FOR JUSTIFIED WHISTLE-BLOWING

The following are some questions that should be considered in deciding whether to blow the whistle in a specific case.[19]

Is the situation of sufficient moral importance to justify whistle-blowing? A cover-up of lethal side effects in a newly marketed drug, for example, is an appropriate situation for disclosure because people's lives are at stake. But situations are not always this clear. Is whistle-blowing warranted if the side effects are not lethal or debilitating but capable of causing temporary discomfort or pain? What if the drug is the most effective treatment for a serious medical problem, so that the harm of the side effect is outweighed by the benefit of using the drug? We need to ask, in such a case, how serious is the potential harm compared with the benefit of the drug and the trouble that would be caused by blowing the whistle. The less serious the harm, the less appropriate it is to blow the whistle.

In addition to the moral importance of the situation, consideration should also be given to the extent to which harm is a direct and predictable result of the activity that the whistle-blower is protesting. For example, a toy that might be hazardous under unusual circumstances warrants whistle-blowing less than one that poses a risk under all conditions. Sissela Bok contends that the harm should also be imminent. According to her, an alarm can be sounded about defects in a rapid-transit system that is already in operation or is about to go into operation, but an alarm should not be sounded about defects in a system that is still on the drawing boards and is far from being operational.[20]

Do you have all the facts and have you properly understood their significance?
Whistle-blowing usually involves very serious charges that can cause irreparable harm if they turn out to be unfounded or misinterpreted. A potential whistle-blower, therefore, has a strong obligation to the people who are charged with wrongdoing to make sure that the charges are wellfounded. The whistle-blower should also have as much documentation and other corroboration as possible. A whistle-blower's case is stronger when the evidence consists of verifiable facts and not merely hunches or rumors. Because whistle-blowing cases often end up in court, the proof should also be strong enough to stand up under scrutiny. The support for the charges need not be overwhelming, but it should meet the ordinary legal standard of a preponderance of evidence.

Employees often have access to only some of the facts of a case and are liable, as a result, to form false or misleading impressions. Would-be whistle-blowers must be careful, therefore, not to jump to conclusions about matters that higher-level managers, with a fuller knowledge of the situation, are in a better position to judge. Typically, employees have only one kind of expertise, so they are not able to make an accurate judgment when different kinds of knowledge are needed.

Have all internal channels and steps short of whistle-blowing been exhausted?
Whistle-blowing should be a last rather than a first resort. It is justified only when there are no morally preferable alternatives. The alternatives available to employees depend to a great extent on the provisions an organization makes for dissent, but virtually every organization requires employees to take up any matter of concern with an immediate superior before proceeding further—unless that person is part of the problem. Courts will generally not consider a complaint unless all possible appeals within an organization have been exhausted. Some progressive corporations have recognized the value of dissent in bringing problems to light and have set up procedures that allow employees to express their concern through internal chan-

nels. Steps of this kind reduce the need for whistle-blowing and the risks that external whistle-blowers take. It is possible to justify not using internal channels, however, when the whole organization is so mired in the wrongdoing that there is little chance that using them would succeed.

Another justification for "going public" before exhausting internal channels is if there is a need for a quick response and internal whistle-blowing would be too slow and uncertain. Two engineers at Morton Thiokol expressed concern to their superiors about the effects of low temperature on the O-rings on the booster rockets for the *Challenger* spacecraft, but their warning never reached the officials at NASA who were responsible for making the decision to go ahead with the launch. The engineers spoke out after the *Challenger* explosion—for which they were disciplined by Morton Thiokol—but their whistle-blowing was too late to avert the disaster. To be effective, they would have had to blow the whistle before the decision was made to launch the spacecraft. This would have required them to go outside the company and contact the officials at NASA directly.

What is the best way to blow the whistle? Once a decision is made to "go public," a host of other questions have to be answered. To whom should the information be revealed? How much information should be revealed? Should the information be revealed anonymously or accompanied by the identity of the whistle-blower? Often an anonymous complaint to a regulatory body, such as the Environmental Protection Agency or the Securities and Exchange Commission, is sufficient to spark an investigation. The situation might also be handled by contacting the FBI or a local prosecuting attorney or by leaking information to the local press. The less information that is revealed, the less likely an employee is to violate any duty of confidentiality. Employees can also reduce conflicts by waiting until they leave an organization to blow the whistle.

Whistle-blowing is also more likely to be effective when an employee presents the charge in an objective and responsible manner. It is especially important that a whistle-blower stick to the important issues and refrain from conducting crusades or making personal attacks on the persons involved. Organizations often seek to discredit whistle-blowers by picturing them as disgruntled misfits or crazy radicals; intemperate, wide-ranging attacks undermine the whistle-blower's own credibility. Many whistle-blowers recommend developing a clear plan of action. Do not blow the whistle impulsively, they advise, but think out each step and anticipate the possible consequences.[21]

What is my responsibility in view of my role in the organization? The justification for blowing the whistle depends not only on the wrongdoing of others, but also on the particular role that a whistle-blower occupies in an organization. Thus, an employee is more justified in blowing the whistle—and may even have an obligation to do so—when the wrongdoing concerns matters over which the employee has direct responsibility. Chuck Atchinson, for example, was not merely an employee of Brown & Root; he was a quality control inspector, and so the irregularities in the construction of the nuclear power plant involved matters that were part of his job. A company such as Brown & Root employs quality control inspectors in order to be sure that nuclear power plants are built to certain standards of safety. The welfare of the company and the public alike depend on people such as Chuck Atchinson doing their job conscientiously. In order to operate a nuclear power plant, moreover, a company must file an application under oath with the Nuclear Regulatory Commission (NRC), and it is a criminal offense for any employee to participate knowingly in the submission of an application that contains false information. Had Chuck Atchinson been called upon to verify that the work on the plant had been performed according to specifications, he would have faced the choice of refusing or becoming an accomplice in a crime.

When an employee is a professional, the question of whether to blow the whistle must be considered in the context of professional ethics. Professionals, such as lawyers, accountants, and engineers, have a greater obligation to blow the whistle under some circumstances and are restricted or prohibited from whistle-blowing under others. For example, Joseph Rose, the staff lawyer for Associated Milk Producers Incorporated, realized that he had an obligation as a lawyer for AMPI to seek to recover the missing funds, which is an obligation that an employee in a different position would not necessarily have. Yet, at the same time, he planned after leaving his position to keep quiet about what he knew, because he had also discharged his legal responsibility by informing his superiors. In addition, he had to be careful about violating the guarantee of confidentiality inherent in the lawyer-client relation. His situation was thus substantially different from that of other employees at AMPI.

What are the chances for success? Insofar as whistle-blowing is justified because of some good to the public, it is important to blow the whistle only when there is a reasonable chance of achieving that good. Whistle-blowing may be unsuccessful for many reasons. Sometimes the fault lies with the whistle-blower who fails to make a case that attracts widespread concern or to devise an effective plan of action; other times it is simply that the organization is too powerful or the public not sufficiently responsive.

IS THERE A RIGHT TO BLOW THE WHISTLE?

Even though whistle-blowing can be justified in some situations, the sad fact remains that courageous employees who perform a valuable public service are often subjected to harsh retaliation. Our reaction when this occurs is "There ought to be a law!" and, indeed, many have been proposed in Congress and various state legislatures.[22] Few have passed, however, and there are some strong arguments against providing legal protection for whistle-blowers. In this section we will examine the debate over the moral justification of laws to protect whistle-blowers against retaliation. It will be useful, first, to survey the existing legal protection.

Existing Legal Protection

Retaliation against federal employees who report instances of waste and corruption in government is prohibited by the Civil Service Reform Act of 1978, which also set up the Merit System Protection Board (MSPB) to receive and act on complaints of retaliation.[23] The provisions of this act were strengthened by the Whistleblower Protection Act of 1989, which designates the Office of Special Counsel as the recipient of whistle-blower reports, although federal workers are still protected if they disclose information about waste or corruption to "any person."

Some protection for whistle-blowers in both the public and private sectors exists in the antiretaliation provisions of various pieces of federal legislation. The National Labor Relations Act of 1935 (NLRA) forbids employers to retaliate against any employee who files a charge with the National Labor Relations Board (NLRB). Title VII of the 1964 Civil Rights Act protects employees who file a charge of discrimination, participate in an investigation or proceeding connected with a charge, or oppose an activity of a company that the employee believes is discriminatory. The Occupational Safety and Health Act of 1970 also prohibits retaliation against any employee who files a complaint with the Occupational Safety and Health Administration or testifies in a proceeding. Other federal acts with antiretaliatory provisions are the Surface Mining Act, the Railway Safety Act, the Surface Transportation Safety Act, the Safe Drinking Water Act, the Toxic Substance Control Act, the Clean Air Act, the Water Pollution Control Act, the Energy Reorganization Act, and the Solid Waste Disposal Act.

Perhaps the most effective federal statute for protecting whistle-blowers is the once-moribund Federal False Claims Act of 1863 (amended 1986). This act was originally passed by Congress to curb fraud during the post-Civil War reconstruction period by allowing private citizens who blow the whistle on government contractors to share in the financial recovery. The False Claims Act has been updated to encourage employees of defense industry contractors to report any fraud that they observe by entitling whistle-blowers to receive between 15 percent and 30 percent of the funds recovered in any suit.[24] The largest settlement to date was made to Douglas D. Keith, a former employee of United Technologies Corporation, who received $22.5 million of the $150 million that the company paid the government to settle a suit over billing practices in the building of helicopters for the Pentagon in the 1980s.[25]

More than two-thirds of the states have passed laws designed to protect whistle-blowers. Most of these apply only to government employees, but a few—Michigan's Whistle Blowers Protection Act, for example—extend more widely. Most of these state statutes specify the procedures that a whistle-blower must follow to receive protection and place requirements on the persons to whom the information is disclosed and on the kind of information that the whistle-blower discloses. Another source of protection for whistle-blowers is state court decisions limiting the traditional right of employers to fire at will. These decisions protect workers against retaliation for many reasons besides whistle-blowing, but they also leave some whistle-blowers unprotected. A further discussion of the issues in this kind of protection is in Chapter 10.

The Arguments against Whistle-Blower Protection

There are many problems with drafting legislation for protecting whistle-blowers. First, a law recognizing whistle-blowing as a right is open to abuse. Whistle-blowing might be used by disgruntled employees to protest company decisions or to get back at their employers. Employees might also find an excuse to blow the whistle in order to cover up their own incompetence or inadequate performance. Alan F. Westin notes, "Forbidding an employer to dismiss or discipline an employee who protests against illegal or improper conduct by management invites employees to take out 'antidismissal insurance' by lodging a whistle-blowing complaint."[26]

Second, legislation to protect whistle-blowers would encroach on the traditional right of employers to conduct business as they see fit and would add another layer of regulation to the existing legal restraints on business, thereby making it more difficult for managers to run a company efficiently. The courts would be called upon to review and possibly reverse a great many personnel decisions. The likely increase in employee litigation could also, according to Westin, "create an informer ethos at work that would threaten the spirit of cooperation and trust on which sound working relationships depend."[27]

Third, if whistle-blowing were protected by law, what should be the legal remedy for employees who are unjustly dismissed? Reinstatement in the workplace, which is the usual remedy in union contract grievance procedures, may not be feasible in the case of employees who are perceived as being disloyal. As an alternative to reinstatement, though, whistle-blowers could be offered a monetary settlement to compensate them for the losses suffered by being wrongly dismissed. An award could be arrived at by negotiation or arbitration, or it could result by allowing dismissed employees to sue for tort damages.

The Arguments for Whistle-Blower Protection

The main argument in defense of a law to protect whistle-blowers is a utilitarian one that rests on the contribution whistle-blowers make to society. There is a direct benefit in having instances of illegal corporate conduct, gross waste and mismanagement, and dangers to the public brought to light. This benefit can be achieved, the argument goes, only if whistle-blowers are encouraged to come forward and make their information known. Ralph Nader makes the further point that allowing employees greater freedom to speak out makes it easier to enforce existing laws and to bring about desirable changes in corporate behavior. He has observed:

> Corporate employees are among the first to know about industrial dumping of mercury or fluoride sludge into waterways, defectively designed automobiles, or undisclosed adverse effects of prescription drugs and pesticides. They are the first to grasp the technical capabilities to prevent existing product or pollution hazards. But they are very often the last to speak out, much less to refuse to be recruited for acts of cor-

porate or governmental negligence or predation. Staying silent in the face of a professional duty has direct impact on the level of consumer and environmental hazards.[28]

These benefits must be balanced against the undeniable harm that a greater incidence of whistle-blowing would have on business firms. Insofar as companies are less efficient—either because of the greater regulation or the loss of loyalty within organizations—a right to blow the whistle is not justified on utilitarian grounds.

A second argument for providing legal protection for whistle-blowers appeals to the First Amendment right of freedom of speech. A distinction needs to be made, though, between the appeal to freedom of speech as a legal argument and as a moral argument. Our rights under the Constitution protect us for the most part only against acts of government and not against those of private employers. Consequently, the freedom of speech that we have as a matter of legal right does not necessarily prevent corporations from retaliating against whistle-blowers, although it does confer some protection on government employees who speak out as citizens.

A teacher in Illinois, for example, wrote a letter to the editor of a local newspaper criticizing the school board for favoring athletics at the expense of the academic program. The teacher, named Marvin Pickering, was fired on the grounds that writing the letter was "detrimental to the efficient operation and administration of the schools of the district." Pickering charged in reply that writing the letter was an exercise of the First Amendment right of free speech that cannot be denied citizens just because they are government employees. The U.S. Supreme Court agreed with Pickering and thereby established whistle-blower protection for government employees.[29] The Court, in a 1975 decision, *Holodnak* v. *Avco Corporation*, extended the precedent set by *Pickering* to private employers who do extensive work for the federal government.[30] The Supreme Court further refined the *Pickering* precedent 15 years later in *Connick* v. *Myers*. In this case, the court ruled that the "speech" in question—which consisted of distributing a questionnaire to protest a personnel matter—did not address a matter of "public concern" sufficient to override the interest of the employer in maintaining control and efficiency in the workplace.[31]

Although the First Amendment right of free speech cannot be used as a *legal* argument for holding that whistle-blowing is a protected activity in the private sector, it can still be maintained that there is a *moral* right to freedom of speech and that (morally) there ought to be a law extending this right to whistle-blowers.[32] At least one writer has urged that we recognize a right that is broader than merely freedom of speech, namely, a right to follow one's own conscience. Whistle-blowers are often led to speak out not by a desire to serve the public good but to do what they feel is morally required of them. "Thus," this writer concludes, "the interests that weigh in favor of providing legal protection to the external whistleblower are not those embodied in an employee's obligation to society, but rather those embodied in his interest as an individual to act in accordance with the dictates of conscience."[33]

DEVELOPING A COMPANY WHISTLE-BLOWING POLICY

Companies have many incentives to develop a whistle-blowing policy.[34] No company is immune from wrongdoing, and an effective policy on whistle-blowing enables a company to deal with misconduct internally, thereby preventing embarrassing public disclosure. For a policy to be effective, however, employees must be assured that their reports will be taken seriously—which means that an investigation will be conducted and appropriate action taken. More importantly, employees must feel confident that they will not suffer any retaliation.

Benefits and Dangers of a Policy

Although companies might prefer to ignore some wrongdoing and to continue profitable but questionable practices, they can also benefit from learning about problems early and taking corrective action before the problems become public. The lack of a policy will not prevent whistle-blowing by a company's employees, and the increasing public acceptance of whistle-blowing combined with expanded legal protection makes whistle-blowing all the more likely. The aftermath of a whistle-blowing incident also creates problems that are best avoided. In particular, dismissing whistle-blowers with legitimate complaints sends the wrong signal to other employees, and yet allowing whistle-blowers to remain in the workplace may cause tension and strife. These equally undesirable alternatives can be avoided by eliminating the need for any employee to go outside of the normal channels of communication. An effective whistle-blowing policy can have the added benefit of affirming a company's commitment to good ethics and creating an ethical corporate climate.

Whistle-blowing policies also benefit employees by providing them with a channel of communication for responding to perceived wrongdoing in the organization. Employees are likely to welcome an opportunity to express their legitimate concerns without the risk of going public with damaging information. Whistle-blowing policies involve some dangers, however. Encouraging employees to report on each other can create an environment of mistrust and intimidation, especially if people feel vulnerable to the possibility of false accusations.

Components of a Whistle-Blowing Policy

A well-designed whistle-blowing policy should include the following components.

1. *An effectively communicated statement of responsibility.* Employees should understand that they have a responsibility to report all concerns about serious unethical or illegal conduct through the appropriate internal channels.

2. *A clearly defined procedure for reporting.* A procedure should be established that allows employees to report their concerns in a confidential manner. The procedure should specify the persons to whom reports are to be made and the proper

form, and employees should be made aware of the procedure. Some companies use an ethics "hot line" that allows employees to make a report by calling an 800 number; other companies insist that reports be made to a person's immediate superior unless that person is involved in the suspected wrongdoing. Multiple means of reporting concerns and the choice of anonymous reporting are available in some companies with whistle-blowing policies.

3. *Well-trained personnel to receive and investigate reports.* The success of a whistle-blowing policy depends heavily on the skill of the personnel who receive and investigate the reports from employees. Especially critical is the ability to maintain confidentiality and to conduct a fair and thorough investigation. For these reasons, the personnel should be well-trained and have sufficient authority within the organization, and the program should be evaluated periodically for effectiveness.

4. *A commitment to take appropriate action.* Employees must be assured that their reports of suspected wrongdoing will not be ignored or misused. Not only should the purposes of a whistle-blowing policy be effectively communicated to all employees, but the company must also assure employees by both word and deed that their reports will be used only for these purposes. The best policies also stipulate that reporting employees will be informed about the outcome of an investigation and the action taken.

5. *A guarantee against retaliation.* By far the most critical component in any whistle-blowing policy is the assurance that employees will not suffer retaliation for making reports in good faith. Retaliation can be prevented, however, only if the importance of the policy is effectively communicated to everyone in the organization and there is a credible commitment to the policy's success by top management. Companies must be on guard, of course, for employees who might abuse an ethics hot line or other reporting mechanisms for personal ends, but a fair and thorough investigation should reveal the facts of the case apart from the reporting employee's motives.

A whistle-blowing policy by itself will neither protect an organization from wrongdoing nor prevent whistle-blowing outside of prescribed channels. A poorly designed or implemented policy also runs the risk of doing more harm than good. Still, a policy with regard to whistle-blowing is worth considering by any company that is committed to ethical conduct.

Conclusion

Whether to blow the whistle on misconduct in an organization is the most difficult decision that some people ever have to make. The decision is wrenching personally because the stakes are so high. The lives of Chuck Atchinson and Joseph Rose were irreparably damaged, and yet many whistle-blowers say that they could not have lived with themselves if they had stayed silent. The decision is also difficult ethically, because whistle-blowing involves a conflict between two competing duties: to protect the public and to be loyal to an organization. Although loyalty is not always overriding, as the loyal agent argument holds, neither is it inconsequential. Deciding between these duties often requires that an employee exercise very careful judgment.

The one certain conclusion of this chapter is that whistle-blowing is ethically permissible under certain carefully specified conditions. (Whether it can ever be ethically required is a different question that seldom arises. Everyone has an obligation not to be a part of illegal and immoral activity, but exposing it at great risk to oneself is usually regarded as beyond what duty requires.) Blowing the whistle is only one response that an employee can make to corporate misconduct, however, and the act of whistle-blowing itself can take many different forms. So in addition to *whether* to become a whistle-blower, employees are faced with the further question of *how* to blow the whistle in a justified manner.

Finally, it is evident that employees who are justified in blowing the whistle ought not to suffer retaliation. What ought to be done to protect whistle-blowers from this fate is less clear. A plausible case can be made for legislation in this area, but the difficulty is drafting laws that achieve the desired result without interfering unduly in the legitimate conduct of business.

Case 5.2   A Whistle-Blower Accepts a "Deal"

As the head of corporate audit for a major pharmaceutical company, I was involved in the lengthy approval process that the Food and Drug Administration requires before a new drug can be brought to market.[35] The reviewer for the FDA was asking some tough questions about the data supporting our application to market a new drug. Although I managed to answer the reviewer's questions to his apparent satisfaction, doubts were beginning to form in my own mind about the reliability of the data I was defending, so I instructed my staff to get photocopies of the original research reports for me as soon as possible.

The photocopies provided evidence of "double books." The raw data in the original reports were entirely different from the data in our FDA application and showed the new drug failing every required test. I had heard rumors of other questionable conduct by the project director, and I suspected that he was implicated in the falsification of the data, although I had no proof for any accusations. I rejected the idea of blowing the whistle on the company by telling everything to the FDA and decided instead to follow the procedure outlined in the company's own whistle-blowing policy. Accordingly, I prepared a report stating only the facts that I could document, and I sent it to the next highest level above the person involved, which in this situation was the legal department of the corporation.

My internal whistle-blowing prompted a quick response. I was summoned to meet with the board of directors which had a team of lawyers from an outside firm present. The original research reports had apparently been destroyed, but there was no question about the authenticity of the photocopies that I still retained because the raw data were accompanied by the researchers' signatures and the dates of entry. After friendly but close questioning, the board of directors offered me a "deal." They would give me all of the resources that I needed to get the drug approved by the FDA, but they promised that the drug would never be marketed. The board intended to correct the problems within the company (and the project director soon resigned), but it wanted to avoid the embarrassment of public exposure. The board's plan was to request that approval of the drug be withdrawn afterward by telling the FDA that mistakes had been made in the marketing projections. I accepted the deal and succeeded in getting the drug approved. The board kept its word, and 10 years later the drug is still not on the market.

After my "deal" with the board, other changes were made. Corporate policy was revised so that I no longer had ready access to company records. The FDA has the authority to conduct "surprise" audits at any time, and the policy had been to allow my office to mimic FDA audits, so that the company would always be "FDA-ready." Under the new policy, audits must be prearranged with the department involved, and the department can stop an audit and reschedule it at any point. Finally, the department is allowed to review the audit report before it is submitted. To my knowledge, there has been no repetition of the events 10 years ago, but my ability to uncover such misconduct has been severely limited. Oftentimes I wonder whether I should have accepted that "deal."

Case 5.3 Better Late than Never?

Seven years ago, Rockland International adopted a whistle-blowing policy that requires employees to report any suspicions of unethical or illegal conduct. During that time, Ken Dryden had investigated many cases, but the thick folder on his desk presented him with his greatest challenge yet. As he thumbed through it once more, he reflected on the pledge that was the cornerstone of Rockland's policy: "No employee will suffer any adverse personnel action as a result of making a report in compliance with this policy, and the company will seek, as far as possible, to ensure that employees are protected against other forms of retaliation." He wondered now whether this guarantee could be kept.

The problem began when Ken followed up on a call to Rockland's 800 number "hot line" by seeking further information from the employee, Art Holmes. Art was a top-notch purchasing agent, who was usually cheerful and outgoing, but the prospect of impending layoffs had affected him along with many others in the company. He suspected—rightly as it turned out—that his job was slated for elimination.

Ken was surprised to be presented with photocopies of dozens of canceled checks and invoices that Art had meticulously gathered. A comparison of the invoices with shipments received revealed convincingly that payments had been made for goods not delivered, and the authorization for the checks was made in every case by the head of the purchasing department. The head of the department at first denied any knowledge of the discrepancy and then claimed that he had allowed the payment for undelivered goods because he wanted to keep a good supplier afloat during troubled times. Eventually he broke down and admitted that he had arranged a kick-back scheme with the supplier in which the two shared the amounts overpaid. The details of the kickback scheme—and Art's role in uncovering it—soon became known throughout the company. The head of purchasing made restitution to the company and hastily resigned.

So far, it would appear that the whistle-blowing policy was working as intended: Art suspected wrongdoing, documented his suspicions, and reported them to the appropriate office. What bothered Ken, however, was the fact that the checks and invoices in the photocopies were more than a year old. He surmised that Art had surreptitiously photocopied the evidence as a form of "insurance" and, now that he feared being laid off, was seeking the protective cover provided by the whistle-blow-ing policy. Art vehemently denied such motives and contended that he feared retalia-tion from the wrongdoer, who was his immediate boss. Now that he had less to lose, he felt safer in doing his duty according to the company's own policy. "Better late than never," he quipped.

If Art is let go, Ken thought to himself, then a lot of people in the company will attribute it to his whistle-blowing, and everything we have done to build trust at Rockland will go up in smoke. On the other hand, if employees get the idea that they can get job insurance by collecting evidence to present at an opportune time, the pol-icy will fail to encourage timely reporting of wrongdoing, and terminating anyone will become impossible. Closing the folder, he reminded himself that he not only had to make a decision in this case, but he also needed to learn how this mess had arisen and what could have been done to prevent it.

NOTES

1. The material for this case is taken from N. R. Kleinfield, "The Whistle Blowers' Morning After," *New York Times*, 9 November 1986, sec. 3, p. 1.

2. Material for this case is taken from Myron Glazer, "Ten Whistleblowers and How They Fared," *The Hastings Center Report*, 13 (December 1983), 33-41; and from Alan F. Westin, ed., *Whistle Blowing! Loyalty and Dissent in the Corporation* (New York: McGraw-Hill, 1981), 31-38.

3. For a discussion of the etymology of the word, see William Safire, *Safire's Political Dictionary*, 3d ed. (New York: Random House, 1978), 790.

4. Sissela Bok, "Whistleblowing and Professional Responsibility," in *Ethical Theory and Business*, 4th ed., ed. Tom L. Beauchamp and Norman E. Bowie (Upper Saddle River, NJ: Prentice Hall, 1993), 308.

5. *Petermann v. International Brotherhood of Teamsters*, 174 Cal. App. 2d 184, 344, P. 2d 25 (1959). This case is discussed further in Chapter 10 on unjust dismissal, because Petermann was fired for testifying.

6. James M. Roche, "The Competitive System to Work, to Preserve, and to Protect," *Vital Speeches of the Day* (May 1971), 445.

7. Bok, "Whistleblowing and Professional Responsibility," 307.

8. One form of this argument is examined in Alex C. Michalos, *A Pragmatic Approach to Business Ethics* (Thousand Oaks, CA: Sage Publications, 1995), 44-53.

9. The concept of agency is not confined to law but occurs in economics (especially the theory of the firm) and organizational theory. For a useful collection of articles exploring the ethical relevance of agency theory, see Norman E. Bowie and R. Edward Freeman, eds., *Ethics and Agency Theory* (New York: Oxford University Press, 1992).

10. R. Powell, *The Law of Agency* (London: Pitman and Sons, 1965), 7.

11. *Second Restatement of Agency*, Sec. 387. A Restatement is not a statute passed by a legislature but a summary of the law in a given area, written by legal scholars, which is often cited in court opinions. Other important Restatements are those on contracts and torts.

12. *Second Restatement of Agency*, Sec. 358, Comment f.

13. Michalos, *Pragmatic Approach to Business Ethics*, 51.

14. Myron Peretz Glazer and Penina Migdal Glazer, "Whistleblowing," *Psychology Today* (August 1986), 39. See also Myron Peretz Glazer and Penina Migdal Glazer, *The Whistle-Blowers: Exposing Corruption in Government and Industry* (New York: Basic Books, 1989).

15. Westin, *Whistle Blowing!* 36.

16. Albert O. Hirschman, *Exit, Voice, and Loyalty* (Cambridge, MA: Harvard University Press, 1970), 77.

17. Ibid., 79.

18. Charles Peters and Taylor Branch, *Blowing the Whistle: Dissent in the Public Interest* (New York: Praeger, 1972), 269.

19. For similar lists, see Richard T. DeGeorge, *Business Ethics*, 5th ed. (Upper Saddle River, NJ: Prentice Hall, 1995), 250–54; Gene G. James, "Whistle Blowing: Its Moral Justification," in *Business Ethics: Readings and Cases in Corporate Morality*, 3d ed., ed. W. Michael Hoffman and Robert E. Frederick (New York: McGraw-Hill, 1995), 290-301.

20. Bok, "Whistleblowing and Professional Responsibility," 307.

21. For a more thorough discussion of the practical aspects of whistle-blowing, see Peter Raven-Hansen, "Dos and Don'ts for Whistleblowers: Planning for Trouble," *Technology Review*, 83 (May 1980), 34-44.

22. See Martin H. Malin, "Protecting the Whistle-blower from Retaliatory Discharge," *Journal of Law Reform*, 16 (Winter 1983), 277-318; Douglas Massengill and Donald J. Petersen, "Whistle-blowing: Protected Activity or Not?" *Employee Relations Law Journal*, 15 (Summer 1989), 49-56; and Elleta Sangrey Callahan and Terry Morehead Dworkin, "Internal Whistleblowing: Protecting the Interests of the Employee, the Organization, and Society," *American Business Law Journal*, 37 (1991), 267-308.

23. Two studies of the MSPB in 1980 and 1983 showed that it had done little to encourage employees to report waste and corruption or to prevent retaliation against those who did. See Rosemary Chalk, "Making the World Safe for Whistleblowers," *Technology Review*, 91 (January 1988), 55.

24. See Elleta Sangrey Callahan and Terry Morehead Dworkin, "Do Good and Get Rich: Financial Incentives for Whistleblowing under the False Claims Act," *Villanova Law Review*, 37 (1992), 273-336.

25. Calvin Sims, "Trying to Mute the Whistle-Blowers," *New York Times,* 11 April 1984, C1, C3.

26. Westin, *Whistle Blowing!* 134.

27. Ibid., 136. The points in this paragraph are made by Westin.

28. Ralph Nader, Peter J. Petakas, and Kate Blackwell, eds., *Whistle Blowing: The Report of the Conference on Professional Responsibility* (New York: Grossman, 1972), 4.

29. *Pickering* v. *Board of Education*, 391 U.S. 563 (1968).

30. *Holodnak* v. *Avco Corporation*, 423 U.S. 892 (1975). In other Supreme Court decisions, the right to speak out on matters of public concern was upheld in cases involving a police officer, a firefighter, and a public health nurse. *Muller* v. *Conlisk*, 429 F. 2d 901 (7th Cir. 1970); *Dendor* v.

Board of Fire and Police Commissioners, 11 Ill. App. 3d 582, 297 N.E. 2d 316 (1973); *Rafferty* v. *Philadelphia Psychiatric Center,* 356 F. Supp. 500 (E.D.Pa. 1973).

31. *Connick* v. *Myers,* 461 U.S. 138 (1983).

32. This is advocated by Patricia H. Werhane, "Individual Rights in Business," in *Just Business: New Introductory Essays in Business Ethics,* ed. Tom Regan (New York: Random House, 1984), 114-18.

33. Malin, "Protecting the Whistleblower from Retaliatory Discharge," 309.

34. For discussions of whistle-blowing policies, see Tim Barnett, "Why *Your* Company Should Have a Whistleblowing Policy," *SAM Advanced Management Journal,* 57 (1992), 37-42; and Marcia P. Miceli and Janet P. Near, *Blowing the Whistle: The Organizational and Legal Implications for Companies and Employees* (New York: Lexington Books, 1992), chap. 7.

35. This case is based on an experience reported to Professor John T. Delaney, University of Iowa. Some details have been changed. Used with the permission of Professor Delaney.

6

Trade Secrets and Conflict of Interest

Case 6.1 The Aggressive Ad Agency

Rob Lebow was used to aggressive advertising agencies. As director of corporate communications for Microsoft Corporation, the giant computer software producer located in Redmond, Washington, Lebow helped to administer the company's $10 million advertising budget. So when it was announced in the fall of 1987 that Microsoft was conducting an agency review, putting its business up for grabs, he was prepared for a flood of calls and letters. One particular piece of mail that caught his eye was a specially prepared flier from a small agency in Boston named Rossin Greenberg Seronick & Hill (RGS&H).

Under the leadership of its president, Neal Hill, this five-year-old advertising agency had accounts totaling $26 million and a growth rate of 65 percent for the past year. Although its business was concentrated in New England, RGS&H was attempting to become a national force by going after high-tech industries. As part of this strategy, the agency recruited two talented people who had worked on an account for the Lotus Corporation at another firm. Jamie Mambro and Jay Williams, who were creative supervisors at Leonard Monahan Saabye in Providence, Rhode Island, joined RGS&H on November 2.

A few days later, Neal Hill read a news story in a trade publication about the agency review by the Lotus rival. Because Microsoft's new spreadsheet program, Excel, was competing directly against Lotus 1–2–3, the industry leader, this seemed to be an ideal opportunity for RGS&H.

The flier was sent by Neal Hill on November 20, after two previous letters and several telephone calls elicited no response from Microsoft. Included in the flier was a round-trip airline ticket from Seattle to Boston and an invitation that read in part:

> You probably haven't thought about talking to an agency in Boston.... But, since we know your competition's plans, isn't it worth taking a flier? ... You see, the reason we

125

know so much about Lotus is that some of our newest employees just spent the past year and a half working on the Lotus business at another agency. So they are intimately acquainted with Lotus' thoughts about Microsoft—and their plans to deal with the introduction of Excel.

In order to do an effective job for a client, advertising agencies must be provided with a certain amount of confidential information that would be of value to competitors. Many companies include a confidentiality clause in their contracts with advertising agencies, and Lotus had such an agreement with its agency, Leonard Monahan Saabye. Even in the absence of a confidentiality clause, however, advertising agencies generally recognize an obligation to preserve the confidentiality of sensitive information.

On the other hand, offering the experience of employees who have handled similar accounts is an accepted practice in the advertising industry. As the president of one firm observed, "There's a thin line between experience and firsthand recent knowledge." But, he continued, "I can't imagine a new-business presentation in which the agency didn't introduce people who worked on the prospect's kind of business."[1]

Rob Lebow was left to wonder: Was Neal Hill at RGS&H offering Microsoft the experience of two employees who had worked on the Lotus account, or was he offering to sell confidential information? In either event, what should Lebow do?

If the new employees at RGS&H had information about Lotus's advertising strategy for countering the introduction of Excel, this could be of considerable value to Microsoft. Anticipating the moves of rivals is often critical to the success of a campaign. However, moving even a part of Microsoft's business to another agency—especially to a small, untested agency like RGS&H—would surely attract the attention of Lotus. In the rumor-filled world of advertising, the presence of two employees who had formerly worked on a Lotus account would not go unnoticed. Therefore, any information that RGS&H had might be "too hot to touch."

Rob Lebow recognized that he could decline the offer in different ways. He could merely ignore the flier, or he could return it with the reply "Thanks but no thanks." Another possibility was to forward the flier to Lotus. Even the rumor that Microsoft had communicated with RGS&H could be damaging to the company, and so being open with Lotus would provide some protection. However, Lotus has a reputation within the industry of being quick to sue, and considerable harm could be done to RGS&H—and to the two new employees, Jamie Mambro and Jay Williams, who might be unaware of the offer made in the flier.

Thus, any decision that Rob Lebow made was bound to have significant ethical, legal, and practical implications.

INTRODUCTION

It is not surprising that corporations such as Microsoft and Lotus attempt to protect themselves against the loss of trade secrets and to utilize what they can learn about their competitors. Information is a valuable business asset that generally provides companies with a significant advantage over competitors who lack it. We need to ask, however, what rights do companies have in maintaining the secrecy of valuable information? And what corresponding obligations do employees have not to disclose company trade secrets to outsiders or use them for their own advantage? Because companies also seek to learn about each other through competitor intelligence gathering, the ethics of such activities is also a critical issue.

There is considerable justification for holding that companies have some rights with respect to trade secrets and other intellectual property, such as patents, copyrights, and trademarks. In general, employees have an obligation of confidentiality not to disclose or use information acquired during their employment. On the other hand, employees have the right to change jobs or to start up a business of their own using some of the skill and knowledge they have acquired while working for a former employer. Furthermore, companies have a right to use their own employees' skill and knowledge that have been legitimately acquired elsewhere and to gather legitimate competitor intelligence. The challenge for individuals and companies, as well as the law, is to balance all of these competing rights and obligations.

THE DEFINITION OF A TRADE SECRET

Trade secrets pose a complex set of problems about the rights and obligations of companies possessing valuable information as well as the rights and obligations of employees and competitors. The courts have long struggled with these problems without much success. Even what information constitutes a trade secret is a source of contention. A rough definition of a trade secret is that it is information used in the conduct of a business and is not commonly known by others. Section 757 of the *Restatement of Torts* defines a trade secret as follows:

> A trade secret may consist of any formula, pattern, device or compilation of information which is used in one's business, and which gives him an opportunity to obtain an advantage over competitors who do not know or use it.

Examples of trade secrets include the ingredients or chemical composition of a product, the design of a machine, the details of a manufacturing process, methods of quality control, results of marketing surveys, financial projections, and lists of customers and suppliers.

A distinction is made in the *Restatement* between trade secrets and confidential business information. The latter is information concerning specific matters, such as the salary of an employee, which is kept secret but not actually used to manufacture anything or provide a service. The amount of a specific bid is also not a trade secret, but the procedure of a company for calculating bids might be. A former employee who is knowledgeable about the bidding procedure of a company, for example, might be able to use that information to enter lower bids.

The *Restatement* admits that an exact definition is not possible, but it lists six factors that can be used to determine what information is protectable as a trade secret. These are

> (1) the extent to which the information is known outside his business; (2) the extent to which it is known by employees and others involved in his business; (3) the extent of measures taken by him to guard the secrecy of the information; (4) the value of the information to him and his competitors; (5) the amount of effort or money expended by him in developing the information; (6) the ease or difficulty with which the information could be properly acquired or duplicated by others.

Prior to 1996, trade secrets were protected only by state laws, except where government information was involved. Subsequently, Congress made the theft of trade secrets a federal offense by passing the Economic Espionage Act of 1996 (EEA). This act is intended to prevent the theft of trade secrets for the benefit of foreign governments, which has been estimated to cost U.S. firms tens of billions of dollars annually. The penalties for individuals convicted of foreign economic espionage include prison sentences up to 15 years and fines up to $500,000. The EEA also applies to domestic espionage, such as a theft of trade secrets involving two American firms. The fines for individuals convicted of domestic economic espionage range up to $250,000, and sentences, up to 10 years.

Although the EEA is aimed primarily at foreign espionage, some observers consider the domestic provisions of the act to have a greater impact.[2] The EEA defines theft very broadly, as the knowing misappropriation of a trade secret without the owner's consent. The act also defines a trade secret broadly as "All forms and types of financial, business, scientific, technical, economic, or engineering information … if (a) the owner thereof has taken reasonable measures to keep such information secret; and (b) the information derives independent economic value, actual or potential, from not being generally known to, and to being readily ascertainable through proper means by the public." Because of these definitions, many trade secret disputes between companies could be subject to criminal prosecution, and every company needs to be more careful in the acquisition of a competitor's information.

THE ARGUMENTS FOR TRADE SECRET PROTECTION

There are three major arguments for trade secret protection. One argument views trade secrets as a kind of *property* and attempts to apply common-law principles of property rights to them. In the second argument, cases involving trade secrets are considered in terms of the right to compete and the principles of *fair competition*. The third argument holds that employees who disclose trade secrets to others or who use them for their own gain violate an obligation of *confidentiality* that is part of the employer-employee relation.

Trade Secrets as Property

Imagine a lone inventor who, after years of hard work, develops an improved process for manufacturing a common product and builds a factory to turn out the product using the new process. Even if the innovations are not sufficiently original to be patentable, we can accept that he owns the results of his creative efforts, at least to the extent that it would be wrong for a worker in the factory to disclose the details of the manufacturing process to a competitor, especially if the employee had been sworn to secrecy.[3]

Trade secrets, along with patents, copyrights, and trademarks, are commonly regarded in the law as intellectual property that can be said to belong to an owner. Patents, copyrights, and trademarks, in particular, are like tangible property in that the owner has a right of exclusive use and the right to sell, license, or otherwise assign ownership to others. This right does not depend on keeping the information secret. Ownership of a trade secret, by contrast, does not confer a right of exclusive use but only a right not to have the secret misappropriated or wrongfully acquired by others. Once the information is widely known, it ceases to be a protectable trade secret. All forms of intellectual property are unlike tangible property, however, in that they are not inherently exclusive; that is, their use by one person does not preclude their use by another.

The question of who owns what becomes more complicated if the inventor is himself employed by a manufacturer of the product in question. As long as he gets his ideas while performing unrelated work for his employer, however, and conducts the experiments on his own time using his own materials and facilities, then it seems only right that he be recognized as the sole owner of the improved manufacturing process and be permitted—perhaps after leaving his present employer—to sell the secrets of the process to another manufacturer or to go into business for himself. If, on the other hand, he is hired as an inventor to develop improved methods of manufacture or if he does his creative work on his employer's time with the resources of his employer, then some or all the rights of ownership could reasonably be claimed to belong to the employer.

The Wexler Case. The case *Wexler* v. *Greenberg* is instructive in this regard. Alvin Greenberg was employed as chief chemist for the Buckingham Wax Company, which manufactured floor cleaners, polishes, and other maintenance materials. One of his tasks as chief chemist was to analyze the products of competitors and to use the results to develop new formulas. After eight years with the company, Greenberg left to join Brite Products, which had previously purchased exclusively from

Buckingham. With the formulas that Greenberg had developed while working for Buckingham, Brite was able to dispense with Buckingham as a supplier and become a manufacturer itself, whereupon Buckingham sued to prevent Greenberg and his new employer from using the formulas on the grounds that they were trade secrets which Greenberg had misappropriated.

According to the decision in this landmark case, an employer has the burden of showing two things: "(1) a legally protectable trade secret; and (2) a legal basis, either a covenant or a confidential relationship, upon which to predicate relief."[4] Information is protectable as a trade secret, in other words, only as long as it meets certain conditions, one of which is that it is genuinely a secret. Furthermore, the owner of a trade secret is protected against the use of this information by others only when it is disclosed by an employee in violation of an obligation of confidentiality, for example, or when a competitor obtains it by theft, bribery, industrial espionage, or some other impermissible means.

In overturning a lower court ruling that held that Greenberg had an obligation of confidentiality not to disclose the formulas, the Supreme Court of Pennsylvania ruled in favor of Greenberg, citing the fact that the supposed trade secrets had not been disclosed to Greenberg by his employer but had been developed by Greenberg himself. The court explained:

> The usual situation involving misappropriation of trade secrets in violation of a confidential relationship is one in which an employer *discloses to his employee* a pre–existing trade secret (one already developed or formulated) so that the employee may duly perform his work…. It is then that a pledge of secrecy is impliedly extracted from the employee, a pledge which he carries with him even beyond the ties of his employment relationship. Since it is conceptually impossible, however, to elicit an implied pledge of secrecy from the sole act of an employee turning over to his employer a trade secret which he, the employee, has developed, as occurred in the present case, the appellees must show a different manner in which the present circumstances support the permanent cloak of confidence cast upon Greenberg….

The formulas, moreover, were not significant discoveries on Greenberg's part but were merely the result of routine applications of Greenberg's skill as a chemist. As such, they were, in the court's view, the kinds of technical knowledge that any employee acquires by virtue of being employed. Even though the formulas are trade secrets, which the Buckingham Wax Company is permitted to use, they properly belong to Greenberg, who has a right to use them in his work for a new employer. Society also makes an investment in the development of information; it is not the exclusive property of an individual or a firm. Because patentable ideas and other innovations are generally built on foundations that have been laid by others, even companies that have spent a great deal for research cannot claim sole right of ownership.

The Basis for Property Rights. One source for the argument that patentable ideas, trade secrets, and the like are a form of property is the Lockean view that we own the results of our own labor.[5] Patent and copyright laws are based in part on the premise that inventors and writers who work with their minds and turn out such products as blueprints and novels should have the same right of ownership that is accorded to creators of more tangible objects. Insofar as intellectual property is created by individuals who have been hired by a company for that purpose and paid for their labor, then it follows, in the Lockean view, that the company is the rightful owner. Just as the products made on an assembly line belong to the company and not to the workers who make them, so too do inventions made by people who are hired to invent. The company has paid them for their efforts and provided them with the wherewithal to do their work.

In addition, there are good utilitarian reasons for holding that companies have property rights to certain kinds of information. First, society generally benefits from the willingness of companies to innovate, but without the legal protection provided by patent and trade secret laws, companies would have less incentive to make the costly investments in research and development that innovation requires. Second, patent and copyright laws encourage a free flow of information, which leads to additional benefits. Patent holders are granted a period of 17 years in which to capitalize on their discoveries, but even during the period of the patent, others can use the information in their research and perhaps make new discoveries.

The existence of legal protection for trade secrets, patents, and other forms of intellectual property also has its drawbacks. A patent confers a legal monopoly for a fixed number of years, which raises the price that the public pays for the products of patent holders during that time. Trade secrets permit a monopoly to exist as long as a company succeeds in keeping key information out of the hands of competitors. Because there is no requirement that patents be used, a company could conceivably patent a large number of processes and products that rival its own and thereby prevent competitors from using them.[6] The owner of copyrighted material can prevent the wide dissemination of important information either by denying permission to print it or by charging an exorbitant price.

These drawbacks can be minimized by the optimal trade-off between the advantages and disadvantages of providing legal protection for patents, trade secrets, and the like. This trade-off is achieved, in part, by the limits on what can be patented or copyrighted or protected as a trade secret. Other means for achieving the optimal trade-off include placing expiration dates on patents and copyrights and defining what constitutes infringements of patents and copyrights. Thus, the Copyright Act of 1976 includes a provision for "fair use" that permits short quotations in reviews, criticism, and news reports.

Clarifying the Ownership of Ideas. Many companies attempt to clarify the ownership of patentable ideas by requiring employees to sign an agreement turning over all patent rights to the employer. Such agreements are morally objectionable, however, when they give companies a claim on discoveries that are outside the scope of an employee's responsibilities and make no use of the employer's facilities and resources.[7] Courts in the United States have often invalidated agreements that force employees to give up the rights to inventions that properly belong to them. The laws in most of the other industrialized countries of the world provide for sharing the rights to employee inventions or giving additional compensation to employees, especially for highly profitable discoveries.[8]

The ownership of ideas is a difficult area, precisely because the contributions of employers and employees are so difficult to disentangle. Arguably, the law in the United States has tended to favor the more powerful party, namely, employers. Contracts or other agreements that spell out in detail the rights of employers and employees are clearly preferable to ambiguous divisions that often land in the courts. These arrangements must be fair to all concerned, however, and granting employees a greater share of the rewards might be a more just solution—and also one that benefits corporations in the long run, by motivating and retaining talented researchers.

Fair Competition

The second argument for trade secret protection holds that companies are put at an unfair competitive disadvantage when information they have expended resources in developing or gathering can be used without cost by their competitors. Even when the information is not easily classifiable as property and there is no contract barring disclosure or use of the information, it may still be protected on grounds of fairness in trade.

In *Wexler* v. *Greenberg*, the court considered not only who owns the formulas that Greenberg developed for the Buckingham Wax Company but also whether placing restrictions on Greenberg's use of the formulas in his work for another company unfairly deprived him of a right to compete with his former employer. According to the decision in *Wexler*,

> ... any form of post-employment restraint reduces the economic mobility of employees and limits their personal freedom to pursue a preferred course of livelihood. The employee's bargaining position is weakened because he is potentially shackled by the acquisition of alleged trade secrets; and thus, paradoxically, he is restrained because of his increased expertise, from advancing further in the industry in which he is most productive. Moreover ... society suffers because competition is diminished by slackening the dissemination of ideas, processes and methods.

The problem of trade secrets, in the view of the court, is one of accommodating the rights of both parties: "the right of a businessman to be protected against unfair competition stemming from the usurpation of his trade secrets and the right of an individual to the unhampered pursuit of the occupations and livelihoods for which he is best suited."

The Associated Press Case. A good illustration of the fair competition argument is provided by a 1918 case in which the Associated Press complained that a news service was rewriting its stories and selling them to newspapers in competition with the

Associated Press.[9] The defendant, International News Service, argued in reply that although the specific wording of a news story can be regarded as a form of property, like a literary work, which belongs to the writer, the content itself cannot belong to anyone. Further, there is no contract between the parties that International News Service had breached. In the words of Justice Louis D. Brandeis:

> An essential element of individual property is the legal right to exclude others from enjoying it.... But the fact that a product of the mind has cost its producer money and labor, and has a value for which others are willing to pay, is not sufficient to insure to it this legal attribute of property. The general rule of law is, that the noblest of human productions—knowledge, truths ascertained, conceptions, and ideas—become, after voluntary communication to others, free as the air to common use.

In this view, information that cannot be patented or copyrighted has the same legal status as trade secrets, so that a plaintiff must show that there is a breach of contract or some other wrongful means of acquisition. Accordingly, Brandeis continued:

> The means by which the International News Service obtains news gathered by the Associated Press is ... clearly unobjectionable. It is taken from papers bought in the open market or from bulletins publicly posted. No breach of contract, or of trust and neither fraud nor force, are involved. The manner of use is likewise unobjectionable. No reference is made by word or act to the Associated Press.... Neither the International News Service nor its subscribers is gaining or seeking to gain in its business a benefit from the reputation of the Associated Press. They are merely using its product without making compensation. That, they have a legal right to do; because the product is not property, and they do not stand in any relation to the Associated Press, either of contract or trust, which otherwise precludes such use.

A majority of the justices of the Supreme Court sided with the Associated Press, however, arguing that the case should be decided not on grounds of property rights or breach of contract but on considerations of fair competition. Although the public may make unrestricted use of the information contained in news stories, the two parties were direct competitors in a business in which the major stock in trade is news, a product that requires the resources and efforts of a news-gathering organization. In selling news stories based on dispatches from the Associated Press, the International News Service was, in the words of the majority opinion, "endeavouring to reap where it has not sown, and ... appropriating to itself the harvest of those who have sown." The opinion further held:

> We need spend no time, however, upon the general question of property in news matter at common law, or the application of the Copyright Act, since it seems to us the case must turn upon the question of unfair competition in business.... The underlying principle is much the same as that which lies at the base of the equitable theory of consideration in the law of trusts—that he who has fairly paid the price

should have the beneficial use of the property. It is no answer to say that complainant spends its money for that which is too fugitive or evanescent to be the subject of property. That might ... furnish an answer in a common-law controversy. But in a court of equity, where the question is one of unfair competition, if that which complainant has acquired fairly at substantial cost may be sold fairly at substantial profit, a competitor who is misappropriating it for the purpose of disposing of it to his own profit and to the disadvantage of complainant cannot be heard to say that it is too fugitive and evanescent to be regarded as property. It has all the attributes of property necessary for determining that a misappropriation of it by a competitor is unfair competition because contrary to good conscience.

Noncompetition Agreements. Because of the difficulty of imposing legal restraints on employees after they leave, many companies require employees to sign a noncompetition agreement when they are hired. These agreements typically restrict an employee from working for a competitor for a certain period of time or within a given geographical territory after leaving a company. Agreements not to compete are a common feature of the sale of a business, and the courts have generally not hesitated to enforce them.

But there is little justification for restricting employees in this way. Noncompetition agreements are almost entirely for the benefit of the employer and inflict a burden on employees that is out of proportion to any gain. At least 12 states consider them so unfair that they are prohibited entirely.[10] Where noncompetition agreements are permitted by law, the courts have generally imposed a number of tests to determine whether they are justified.[11] These tests are that the restrictions contained in an agreement (1) must serve to protect legitimate business interests; (2) must not be greater than that which is required for the protection of these legitimate interests; (3) must not impose an undue hardship on the ability of an employee to secure gainful employment; and (4) must not be injurious to the public. Legitimate business interests include the protection of proprietary information or customer relations, but the purpose of an agreement cannot be merely to protect an employer against competition.

In determining whether restrictions are greater than those required to protect the legitimate interests of an employer, three factors are important. These are the time period specified, the geographical area, and the kind of work that is excluded. The value of trade secrets is reduced over time, so that a noncompetition agreement designed to protect trade secrets can justifiably restrain an employee only during the time that they have value. Without a time limit on an agreement, an employee could be prevented from working for a competitor even after formerly proprietary information becomes common knowledge. Similarly, an employer with a legitimate interest in protecting the customers it serves in New York City, for example, might be justified in preventing a sales representative from working for a competitor in that area but not elsewhere.

Noncompetition agreements that specify the kind of work too broadly also run the risk of hampering an employee unduly. In one case, a woman in Georgia signed a contract with an employment agency in which she agreed not to work in any capacity for a period of one year for any competitor within a 25-mile radius. The Supreme Court of Georgia ruled that the time period and the area were reasonable but that the phrase "in any capacity" was unreasonably broad, because it would bar her from doing any work for a competitor and not merely the work that she had done for her former employer.[12] Generally, agreements prohibiting employees from working on a particular project or soliciting specific clients, for example, are less likely to be objectionable than vague restrictions such as writing computer programs or selling insurance.

The Confidentiality Argument

The third argument for trade secret protection is that employees who disclose trade secrets to others or use them themselves are guilty of violating an obligation of confidentiality. This argument is based on the view that employees agree as a condition of employment to become agents of an employer and be bound by the duty that agents have to preserve the confidentiality of certain information.[13] Section 395 of the *Restatement of Agency* states that an agent has an obligation

> … not to use or to communicate information confidentially given him by the principal or acquired by him during the course of or on account of his agency … to the injury of the principal, on his own account or on behalf of another … unless the information is a matter of general knowledge.

The obligation of confidentiality does not end with the employment relation but continues to exist after an employee leaves one job for another. Employees who sign an explicit confidentiality agreement may be bound by more stringent contractual obligations than those contained in the agency relation.

Companies can also have an obligation of confidentiality that prohibits them from misappropriating trade secrets. A company that inveigles trade secrets from another company under the guise of negotiating a license agreement or a merger, for example, might be charged with a violation of trade secret law, because the process of negotiation creates a relation of confidentiality. (It might also be charged with failing to negotiate in good faith.) It is also not uncommon for companies to reject ideas brought to them by outsiders, only to adopt them later as their own. The courts have ruled in many such instances that inventors and others have a right to expect that their ideas will be received in confidence and not misappropriated.

The argument for an obligation of confidentiality provides strong support for the right of employers to trade secret protection, but it too has a number of shortcomings. It assumes, for example, that the information was received from the employer and was not, as in the case of Alvin Greenberg, developed by the employee. The obligation of confidentiality is also limited by the right of employees to use their skill and knowledge in the pursuit of a trade or occupation.

Confidentiality Agreements. Many employees sign a confidentiality agreement which creates an explicit contractual obligation that is often more stringent than the obligation of confidentiality that employees ordinarily have as agents. Although confidentiality agreements have some advantages for both employers and employees, they are open to the same objections as agreements to assign patent rights to employers and to refrain from postemployment competition. Because they are usually required as a condition of employment, employees are effectively coerced into giving up rights to which they might otherwise be entitled.

By relying on an enforceable obligation of confidentiality, companies often place unnecessary restraints on employee mobility and career prospects. Michael S. Baram contends that litigation rarely preserves either the secrecy of company information or the liberty of employees and that both of these are better served by more sophisticated management.[14] Among the policies he suggests are improving security procedures in the workplace; securing the legal protection of patents, copyrights, and trademarks whenever possible; segmenting information so that fewer people know the full scope of a trade secret; limiting information to those with a need to know; and using increased pensions and postemployment consulting contracts to keep employees from taking competitive employment.

In addition, the incentive for employees to leave with valuable trade secrets can be reduced by greater recognition of employees for their contributions. Not infrequently, employees go to a competitor or set up a business of their own because of a feeling that they have not been fairly treated. Baram concludes that the key to protecting trade secrets lies in improved employee relations, in which both employers and employees respect the rights of the other and take their obligations seriously. And a key element in improving employee relations is an ethical climate of fair play. Employers might find that treating employees fairly provides more protection for trade secrets than relying on the law.

COMPETITOR INTELLIGENCE GATHERING

Not all use of a company's trade secrets and other confidential business information is unethical or illegal. The systematic collection and analysis of competitor intelligence has become an accepted practice in the corporate world, and companies that do not avail themselves of this valuable tool may find themselves at a disadvantage. This is especially true in a global environment where some of America's competitors have long-established and highly efficient intelligence units.

Computers have greatly facilitated competitor intelligence gathering, first, by making immense volumes of information available in open-access databases and, second, by enabling companies to store and sort through the information they have compiled. Much of the information used for intelligence purposes is publicly available from news sources, trade publications, court records, regulatory filings, and presentations at industry meetings, and some is also obtained from employees' own contacts with customers, suppliers, and even competitors themselves. The challenge is to piece the information together so that conclusions can be drawn.

Although competitor intelligence gathering has shed its unsavory cloak-and-dagger image, it still has ethical and legal limits that companies ignore at their peril. Unethical collection practices often lead to costly litigation and possibly to criminal prosecution under the Economic Espionage Act of 1996. The outcome of any legal action is uncertain because of confusion in the law. A lack of ethics in competitor intelligence gathering also creates a climate of mistrust that hampers normal business activity and forces companies to adopt costly defensive measures. Most importantly, companies that routinely cross ethical boundaries in gaining competitor intelligence can scarcely expect others to respect their own trade secrets and confidential business information.

The importance of ethics in competitor intelligence gathering has led some companies to adopt policies that give employees firm guidelines on acceptable practices and also set the tone for practices within their industries. Not only can a well-designed policy protect a company from the consequences of unethical or illegal intelligence gathering, but it can also enable a company to gain the maximum benefit from competitor intelligence by making the ethical and legal limits known to all employees. Companies can protect themselves from prosecution under the Economic Espionage Act by showing that any illegal conduct by an employee was in violation of an effective EEA-compliance program.

The Ethical and Legal Limits. The ethical and legal limits on competitor intelligence gathering are generally concerned with the *methods* used to acquire the information.[15] The importance of the method of acquisition is due to the point that trade secrets are protected, according to the *Wexler* decision, only if there is a legal basis "upon which to predicate relief," which means that some *duty* has been breached. The duties in question are most often breached by using improper methods to acquire information from a competitor. Thus, a company that carelessly allows a trade secret to become known has no right to prevent competitors from using it. Companies do have a right, however, to prevent the use of a trade secret that is sold by an employee, for example, or stolen by a competitor, because theft is present in both cases and there is a duty not to steal. The unethical methods for gathering competitor intelligence can be grouped under four headings, each of which involves a breach of a particular duty.

Theft and Receipt of Unsolicited Information. Theft of information, either by an employee or an outsider, is obviously an improper method for acquiring information because it involves a violation of *property rights*. Examples of employee theft include freely offering information to competitors to take revenge, selling it for monetary reasons, and taking it to a new job in order to advance one's career. Companies that receive the information, for whatever reason it is offered, are receiving, in effect, stolen property. More controversial, however, are cases in which an employee inadvertently leaves a document where it can be seen or taken by a competitor or carelessly discloses information in casual conversation. Suppose that a competitor's bid on an important project is accidentally enclosed along with the specifications that are sent by the customer. Would it be ethical to use that knowledge in preparing one's own bid? Or does the information still belong to the competitor?

Misrepresentation. To gain information under false pretenses is a form of deception that violates a duty to be *honest* in all dealings. Posing as a customer to obtain information from a competitor, for example, is an act of dishonesty. Other devious practices include asking consulting firms to solicit information from competitors under the guise of doing a study of the industry and getting friendly customers to make phony requests for bids from competitors, which might contain confidential technical information about the bidder's products. Because useful bits of information are sometimes picked up during job interviews with a competitor's employees, some companies have advertised and conducted interviews for positions that do not exist, in the hope that some applicants would inadvertently reveal trade secrets of their present employer.[16]

Improper Influence. The employment relation is built on *trust,* and to induce an employee to reveal information through bribery or some other means is to exert an improper influence that undermines that trust. An employee who accepts a bribe and turns over a company's secrets has broken a bond with the employer, but the company that offers the bribe has obtained those secrets by inducing that break. Improper influence can be exerted not only by bribery but also by promising or holding out the possibility of a job or some other opportunity. Offering to purchase from a supplier in return for a competitor's price list would be an example of improper influence. More direct would be plying a competitor's employee with drinks in order to make that person less discrete.

Covert Surveillance. Some methods for obtaining information intrude in ways that companies have not anticipated and taken steps to prevent. These can be said to violate a company's right to privacy. Employees who talk about confidential matters in a public place, for example, can have no expectation of privacy, but planting hidden microphones in a competitor's place of business is a form of espionage that intrudes into an area that is regarded as private. Virtually all of the high-tech gadgetry that government intelligence agencies use to spy on enemies abroad is available for competitor intelligence gathering at home. Whether corporations have a right to privacy is controversial, but if covert surveillance were to become an accepted practice, companies would be forced to take costly defensive measures. Respecting a company's reasonable expectations of privacy, then, is in everyone's best interests.

CONFLICT OF INTEREST

Among the many ethical problems in the collapse of Enron Corporation, conflict of interest looms large. A report of the Enron board of directors assigns much of the blame to a host of partnerships set up by Andrew S. Fastow, the company's former

Chief Financial Officer (CFO). These partnerships—which included LJM1, LJM2, Chewco, JEDI, Southhampton, and four Raptor partnerships—had the effect of removing from Enron's books unwanted assets and liabilities. They also greatly enriched Mr. Fastow along with other top Enron executives and favored investors. These supposedly independent entities contributed to Enron's demise because they created liabilities for the company should the price of Enron stock fall, as it did.

The conflict of interest for Mr. Fastow arose when he negotiated the terms of the deals on behalf of the partners with a company that he had a duty to serve. He was in effect bargaining with himself (or his subordinates) over matters in which he stood to gain. In some negotiations involving Enron payments to the partnerships, he reportedly did not reveal his stake or seek approval of transactions as the company's code of ethics required. In two instances the board waived the conflict of interest clause in the ethics code to permit the CFO's dual role, but the waivers themselves raise ethical concerns. Richard C. Breeden, a former chairman of the Securities and Exchange Commission, observed, "The very notion that the chief financial officer of a major corporation would have divided loyalties to this degree of magnitude is something I wouldn't have believed any board of directors would allow—or that any C.F.O. would accept." He added, "The C.F.O. is the financial conscience of the company, the guardian of the numbers. If he has a conflict, how can the system work?"[17]

Enron's public auditing firm, Arthur Andersen, has also been accused of conflict of interest. Although Andersen auditors were troubled by the partnerships—especially whether they were truly independent entities or merely accounting fictions—they apparently did not bring their concerns to the board, as would be expected. One possible reason for this failure is that Andersen also provided consulting services that were far more lucrative than auditing. When an accounting firm occupies such a double role, it has an incentive to ignore accounting irregularities in order to keep a consulting client. Moreover, Andersen auditors performed some of the company's internal bookkeeping, thus blurring the line between conducting an independent audit and managing a company's own financial operations. Auditor's who make money keeping a client's books are scarcely independent judges of the integrity of the information they contain.

In creating the partnerships, Enron engaged several major investment banks, which also encountered conflicts of interest. The conflicts here arise when a firm's analysts feel pressure to maintain "buy" recommendations in order to keep Enron's lucrative deal structuring business. As a result, many investors maintained confidence in Enron even as the company's troubles were becoming known to its investment advisers. Ironically, the Chinese walls that investment banks build between their analysis and advisory services in order to prevent conflict of interest may have prevented analysts from knowing about Enron's deteriorating condition.

Enron's law firm, Vinson & Elkins, was also accused of a conflict of interest when it was engaged to give a legal opinion after concerns were raised about certain deals in an anonymous letter to chairman Kenneth Lay. The writer of that letter, Sherron S. Watkins, wrote, "Can't use V&E due to conflict—they provided some true sale opinions on some of the deals." Vinson & Elkins was engaged, and the firm gave a clean bill of health to deals they helped develop. In addition, Enron's board of directors was accused of conflict of interest because the company contributed heavily to charities and institutions with which the members were involved and, in one instance, to the political campaign of a member's husband. The suspicion is that the board members' independence was undermined by the generosity of these gifts.

Companies and their employees have an obligation to avoid *conflict of interest* of the kinds illustrated by the Enron case. Virtually all corporate codes of ethics address conflict of interest because it interferes with the ability of employees to act in the best interests of a firm. Accepting gifts or lavish entertainment from suppliers, for example, is generally prohibited or strictly limited for the simple reason that the judgment of employees is apt to be compromised. Company codes usually contain guidelines on investing in customers, suppliers, and competitors of an employee's firm for the same reason.

Prohibitions on conflict of interest cannot be so extensive, however, as to prevent employees from pursuing unrelated business opportunities, taking part in community and political affairs, and generally acting as they see fit in matters outside the scope of their employment. One problem with conflict of interest is in drawing a line between legitimate and illegitimate activities of employees in the pursuit of their personal interests. A further problem is the large gray area that surrounds conflict-of-interest situations. Perhaps no other ethical concept in business is so elusive and subject to dispute. Many people charged with conflict of interest see nothing wrong with their behavior. It is important, therefore, to define the concept clearly and to understand the different kinds of conflict of interest.

What Is Conflict of Interest?

The Enron case features numerous individuals and firms with interests that conflict. It would be inaccurate, however, to define a conflict of interest as merely a clash between conflicting or competing interests because these are present in virtually every business relation. In the relation between buyer and seller, for example, each party strives to advance his or her own interest at the expense of the other, but neither party faces a conflict of interest as the term is commonly understood.

The conflict in a conflict of interest is not merely a conflict between conflicting interests, although conflicting interests are involved. The conflict occurs when a personal interest comes into conflict with an obligation to serve the interests of another. More precisely, we can say that a conflict of interest is a conflict that occurs when a personal interest interferes with a person's acting so as to promote the interests of another *when the person has an obligation to act in that other person's interest.* This obligation is stronger than the obligation merely to avoid harming a person and can arise only when the two persons are in a special relation, such as employer and employee.

Specifically, the kind of obligation described in this definition is that which characterizes an agency relation in which one person (an agent) agrees to act on behalf of another (the principal) and to be subject to that person's control. This fact explains why conflict of interest is most often encountered by professionals—lawyers, doctors, and accountants, for example—and among fiduciaries, such as executors and trustees. Employees of business firms are also in an agency relation in that they have a general obligation to serve the interests of an employer.

An important feature of an agency relation is its open-endedness. An agent is obligated to perform not merely this or that act but, in the words of the *Second Restatement of Agency*, "to act solely for the benefit of the principal *in all matters concerned with his agency*."[18] The duties of an agent are not determined solely by a list of moral rules but by the nature of the interests to be served. This open-ended character of the agency relation explains why it is a conflict of interest for an agent to acquire *any* interest that is contrary to that of a principal, because the kinds of situations in which an agent might be called upon to act in the interest of another are not easily anticipated.

To complete the definition of conflict of interest, some account should also be given of a personal interest. Roughly, a person has an interest in something when the person stands to gain a benefit or an advantage from that thing. "Having an interest" is not the same as "taking an interest." A person can take an interest in someone else's interest, especially when that person is a family member or a close associate. In that case, however, the benefit or advantage accrues to someone else. Furthermore, the benefit or advantage is usually restricted to a financial gain of some kind. Merely satisfying a desire, for example, would not seem to be enough, for otherwise a lawyer who secretly hopes that the client will be convicted would face a conflict of interest, as would a lawyer who prefers to play golf instead of spending the time adequately representing a client. The benefit or advantage would also have to be substantial enough to interfere significantly with a person's performance of an obligation.

Some Relevant Distinctions

All instances of conflict of interest are morally suspect, but some are more serious than others. In their rules on conflict of interest, company codes of ethics and codes for professionals, such as lawyers and accountants, contain a number of relevant distinctions that can aid us in understanding the concept of conflict of interest.

Actual and Potential Conflict of Interest. There is a distinction between *actual* and *potential* conflict of interest.[19] A conflict is actual when a personal interest leads a person to act against the interests of an employer or another person whose interests the person is obligated to serve. A situation constitutes a potential conflict of interest when there is the possibility that a person will fail to fulfill an obligation to act in the interests of another, even though the person has not yet done so.[20]

Andrew S. Fastow was apparently in an actual conflict of interest situation by virtue of having a duty to serve the interests of Enron at the same time that he stood to gain from negotiating favorable terms for the partnerships in which he had a stake. If another person at Enron were bargaining on behalf of the company, then he might have been free to serve the interests of himself and the other partners. However, it appears that he was attempting to serve both interests simultaneously.

Obviously, the categories of actual and potential conflict of interest involve subjective elements. A person of integrity might be able to have a strong personal interest in some matter and yet still serve the interests of another. Merely having a competing interest creates a potential conflict of interest, but determining whether an actual conflict of interest exists would require us to make a judgment about that person's objectivity. Similarly, whether an interest creates a potential conflict depends on the strength of the influence it exerts on a person. Owning a small amount of stock in a company, for example, is unlikely to influence anyone's conduct, and so most employers do not impose an absolute prohibition on investments. More often they place a dollar limit on outside financial interests, or else they require a disclosure of stock ownership so that the potential for conflict of interest can be evaluated in each case.

Personal and Impersonal Conflict of Interest. A second distinction can be made between *personal* and *impersonal* conflict of interest. The definition developed in the preceding section is phrased in terms of a personal interest that comes into conflict with the interests of another. A conflict can also arise when a person is obligated to act in the interests of two different persons or organizations whose interests conflict. Thus, a lawyer who represents two clients with conflicting interests may not stand to gain personally from favoring one or the other, and yet, according to Rule 1.7(a) of the American Bar Association's *Model Rules of Professional Conduct,* such an arrangement constitutes a conflict of interest.[21] A lawyer who has a personal interest that conflicts with the interests of a client has a personal conflict of interest, whereas a lawyer who represents two clients with conflicting interests faces an impersonal conflict of interest.

Insofar as Andrew Fastow stood to gain financially from the partnerships he headed, he faced a personal conflict of interest. However, even if he had no personal stake, there would still be an impersonal conflict of interest if he took an active role in the management of the partnerships. His role as the CFO of Enron commits him to acting in the best interests of the shareholders in all matters, and this duty cannot be fulfilled if he is also committed to serving the interests of the members of the partnerships. For example, deciding whether an unusually profitable investment opportunity should be allocated to Enron or to a partnership would require him to favor one set of interests over the other. This is an instance of the biblical injunction that a person cannot serve two masters.

Individual and Organizational Conflict. Third, conflict of interest can be either *individual* or *organizational*. In the agency relation, the agent is typically a person acting in the interests of a principal, which may be another person or an organization. However, organizations can be agents as well and hence parties to conflicts of interest. For example, many large accounting firms, like Arthur Andersen, provide management services to companies they also audit, and there is great concern in the profession that this dual function endangers the independence and objectivity of accountants.[22] Advertising agencies whose clients have competing products face a similar kind of conflict of interest. Investment banking houses have also been accused of conflict of interest for financing takeovers of companies with which they have had long-standing relations. Further, large law firms face the possibility of conflict of interest when they have clients with competing interests—even when the work is done by different lawyers in the firm.

For an accountant to provide management services to a company that he or she also audits—or for an individual ad person, banker, or lawyer to accept clients with conflicting interests—is a clear conflict of interest. But why should it be a conflict when these functions are performed by different persons in different departments of a firm? The answer is that an accounting firm, for example, also has an interest that is shared by every member of the organization, and the interests of the firm can affect decisions about individual clients. Thus, when management services are more lucrative than auditing, firms may have an incentive to concentrate on them to the detriment of other functions. They may also be tempted to conduct audits in ways that favor the clients to whom they provide management services.

Similarly, the creative work for competing advertising accounts is generally done by independent groups, but there is an incentive to commit greater resources and talent to more valuable accounts. In addition, when an organization such as an advertising agency takes on a client, there is an organizational commitment of loyalty that goes beyond merely delivering agreed-upon services. For an organization to work for and against a client at the same time is incompatible with this kind of organizational commitment. In addition, advertising campaigns involve sensitive information about product development and marketing strategies that is not easily kept confidential. Investment banks and large law firms encounter similar challenges to their ability to serve the interests of all clients to the fullest.

The Kinds of Conflict of Interest

The concept of conflict of interest is complex in that it covers several distinct moral failings that are often run together. It is important to separate them, though, in order to have a full understanding both of the definition of conflict of interest and of the reasons that it is morally wrong for a person to be in a conflict-of-interest situation. Briefly, there are four kinds of conflict of interest: (1) exercising biased judgment, (2) engaging in direct competition, (3) misusing a position, and (4) violating confidentiality. Each of these calls for some explanation.

Biased Judgment. The exercise of judgment is characteristic of professionals, such as lawyers, accountants, and engineers, whose stock in trade is a body of specialized knowledge that is used in the service of clients. Not only are professionals paid for using this knowledge to make judgments for the benefit of others but also part of the value of their services lies in the confidence that can be placed in a professional's judgment. Accountants do not merely examine a company's financial statement, for example; they also attest to the accuracy of that statement and to its compliance with generally accepted accounting principles, or GAAP. The National Society of Professional Engineers' *Code of Ethics for Engineers* stipulates that engineers shall not submit plans or specifications that are unsafe or not in conformity with accepted engineering standards.[23] So an engineer's signature on a blueprint is also a warrant of its quality.

Judgment is not exclusively a feature of professional work. Most employees are called upon to exercise some judgment in the performance of their jobs. Purchasing agents, for example, often have considerable latitude in choosing among various suppliers of a given product. The judgment of purchasing agents in all matters, however, should be used to make decisions that are in the best interests of the employing firm. For a purchasing agent to accept a bribe or kickback in return for placing an order constitutes a clear conflict of interest. The reason is simple. Bribes and kickbacks are usually intended to induce an employee to grant some favor for a supplier at the expense of the employer. Other factors that could influence the judgment of an employee include outside business interests, such as an investment in a competitor or a supplier, or dealings with businesses owned by family members.

Whether it is a potential conflict of interest for a purchasing agent to accept a gift from a supplier who expects favorable treatment in the future is less clear. An answer to this question depends largely on the value of the gift, the circumstances under which it is offered, the practice within the industry, and whether the gift violates any law. The code of ethics of a large bank, for example, states that employees should not accept gifts where the purpose is "to exert influence in connection with a transaction either before or after that transaction is discussed or consummated. Gifts, for any other purposes, should be limited to those of nominal value." "Gifts of nominal value," the code continues, "generally should be limited to standard advertising items displaying a supplier's logo." A maximum value of $25 is suggested as a guideline.

Direct Competition. For an employee to engage in direct competition with his or her employer is a conflict of interest. One reason, of course, is that an employee's judgment is apt to be impaired by having another interest. In addition, the quality of the employee's work might be reduced by the time and effort devoted to other activities. Unlike other kinds of outside business interests, however, direct competition is generally prohibited by companies even when it is disclosed and there is no danger of impaired judgment or diminished work performance. Consider this case, which is taken from a policy statement issued by the Xerox Corporation:

> The wife of a Xerox tech rep inherits money. They decide it would be profitable to open a copy shop with her money and in her name in a suburban city. The territory they choose is different from his. However, there are several other copy shops and an XRC [Xerox Resource Center] in the vicinity. She leases equipment and supplies from Xerox on standard terms. After working hours, he helps his wife reduce costs by maintaining her equipment himself without pay. He also helps out occasionally on weekends. His job performance at Xerox remains as satisfactory as before. One of the nearby competitive shops, also a lessee of Xerox equipment, writes to his manager complaining that the employee's wife is getting free Xerox service and assistance.

The conflict of interest in this case consists mainly in the fact that the employee's investment and work outside of his employment at Xerox place him in direct competition with the company. The territory is different from his own, and so he would never have to make decisions on the job that could be influenced by his wife's business. The outside interest also has no effect on the quality of his work for Xerox. Still, on the assumption that the husband benefits from his wife's business venture, he is competing directly with his employer, because Xerox operates an XRC in the area.

In addition, by maintaining the equipment in his wife's shop himself, the employee harms the company by depriving it of the potential for additional business. It would be a conflict of interest for the tech rep to do service for any copy shop using equipment of any make, as long as he is employed by Xerox. His skill as a technician, which is in part the result of company training, belongs in a sense to the company, and he would be free to exercise these skills only upon leaving the employment of Xerox. Finally, the employee is indirectly harming the interests of Xerox by upsetting the relations between the company and other lessees of Xerox equipment.

Misuse of Position. Misuse of position constitutes a third kind of conflict of interest. In April 1984, a reporter for the *Wall Street Journal* was fired for violating the newspaper's policy on conflict of interest. The firing occurred after R. Foster Winans, a contributor to the influential stock market column "Heard on the Street," admitted to his employer and investigators from the Securities and Exchange Commission (SEC) that he conspired over a four-month period, beginning in October 1983, with two stockbrokers at Kidder, Peabody & Company to trade on the basis of advance information about the content of the column. One of the charges against R. Foster Winans was that he misused his position as a *Wall Street Journal* reporter to enrich himself in violation of a provision in the newspaper's code of ethics that reads as follows:

> It is not enough to be incorruptible and act with honest motives. It is equally important to use good judgment and conduct one's outside activities so that no one— management, our editors, an SEC investigator with power of subpoena, or a political critic of the company—has any grounds for even raising the suspicion that an employee misused a position with the company.[24]

Consider the hypothetical case of a bank manager who, in the course of arranging home improvement loans, makes it a point to ask customers whether they have lined up a contractor. She casually drops the name of her brother who operates a general contracting business and mentions that a number of bank customers have been very satisfied with the work of his company. The bank manager's mention of her brother is clearly improper if she misuses her power to grant or deny a loan to induce customers to use him as a contractor. A conflict of interest is still present, though, even if she does not allow her personal interest to have any effect on the decisions she makes on behalf of her employer. There is no conflict between the interests of the manager and those of the bank, and the bank is not harmed in any significant way. Still, the manager has taken the opportunity to advance her personal interests while acting in her capacity as an official of the bank. Holding a position with a company or other organization gives a person powers and opportunities that would not be available otherwise, and an employee has an obligation not to use these powers and opportunities for personal gain.

Extortion also constitutes a misuse of position. Unlike bribery, with which it is often confused, extortion does not involve the use of a payment of some kind to influence the judgment of an employee. Rather, extortion in a business setting occurs when a person with decision-making power for a company demands a payment from another party as a condition for making a decision favorable to that party. For example, a purchasing agent who threatens a supplier with a loss of business unless the supplier agrees to give a kickback to the purchasing agent is engaging in extortion. Extorting money from a supplier in this way is a conflict of interest, even if the company is not directly harmed, because the purchasing agent is violating an obligation to act in the position solely for the interests of the employer.

Violation of Confidentiality. Finally, violating confidentiality constitutes, under certain circumstances, a conflict of interest. The duty of lawyers, accountants, and other professionals, for example, precludes the use of information acquired in confidence from a client to advance personal interests—even if the interests of the client are unaffected. Similarly, because a director of a company is privy to much information, it would be wrong to use it for personal gain or other business interests.

The case of R. Foster Winans also illustrates a conflict of interest involving a breach of confidentiality. A reporter with information prior to publication who attempts to capitalize on the expected results is using that information for his or her own personal gain. Specifically, the courts found Mr. Winans guilty of *misappropriating* confidential information that properly belonged to his employer. In the Supreme Court decision affirming the conviction of Mr. Winans, Justice Byron White observed:

> Confidential business information has long been recognized as property. "Confidential information acquired or compiled by a corporation in the course and conduct of its business is a species of property to which the corporation has the exclusive right and benefit. . . ."[25]

Justice White further noted:

> The District Court found, and the Court of Appeals agreed, that Winans had knowingly breached a duty of confidentiality by misappropriating prepublication information regarding the timing and the contents of the "Heard" column, information that had been gained in the course of his employment under the understanding that it would not be revealed in advance of publication and that if it were, he would report it to his employer.

MANAGING CONFLICT OF INTEREST

Conflict of interest is not merely a matter of personal ethics. A person in a conflict of interest, either potential or actual, may be in the wrong, but conflicts usually occur in the course of being a professional or a member of an organization. Often, these conflicts result from structural features of a profession or an organization and must be managed through carefully designed systems.

Professions, such as medicine, law, and accounting, are highly vulnerable to conflict of interest because of their strong duty to serve the interests of others as patients or clients. Business firms in particular industries also face conflicts because of their need to provide many different kinds of services to many different clients or customers. In some cases, both professional and organizational factors are involved. Accountants, for example, sometimes own stock in the companies that they audit, and the firm they work for may also provide consulting services to its auditing clients. Obviously, accounting firms need to employ a variety of means for managing these kinds of conflicts of interest.

Fortunately, there are many means for managing conflict of interest. Most corporations have a section in their code of ethics that specifically addresses the problem. In some industries, especially financial services, companies have comprehensive compliance programs for ensuring the utmost integrity. For example, an obvious conflict of interest exists when the portfolio manager of a mutual fund also engages in personal trading for his or her own account. Securities and Exchange Commission Rule 17j-1 requires mutual fund companies to develop policies and procedures to prevent inappropriate personal investing. In response, companies have adopted very extensive systems that prohibit certain kinds of trades, such as short-selling, require preclearance of other trades, and ban participation in Initial Public Offerings (IPOs). In addition, mutual funds closely monitor portfolio managers and prepare periodic reports of violations.

As these examples indicate, the management of conflict of interest requires a variety of approaches. The following is a list of the major means by which professional groups and business organizations can manage conflicts of interest.

1. *Objectivity.* A commitment to be objective serves to avoid being biased by an interest that might interfere with a person's ability to serve another. Virtually all professional codes require objectivity. Indeed, the code for certified public accountants, which requires objectivity and independence, identifies objectivity (the obligation to be impartial and intellectually honest) with avoiding actual conflicts of interest and independence (avoiding relations that would impair objectivity) with potential conflicts.

2. *Avoidance.* The most direct means of managing conflicts of interest is to avoid acquiring any interests that would bias one's judgment or otherwise interfere with serving others. Avoidance is easier said than done, however. First, it may be difficult to anticipate or identify a conflicting interest. For example, law firms typically conduct a review of new clients to avoid conflicts of interest, but when the number of

relations on both sides are numerous, such a review may miss some potentially conflicting interests.

Second, acquiring conflicting interests may be unavoidable due to the nature of the business. This is especially true of investment banking, where conflicts of interest are built into their structure. For example, a large investment bank routinely advises clients on deals that affect other companies which the bank also advises or whose securities the bank holds. Investment banks have been accused of modifying research reports on stocks in order to avoid antagonizing companies from which it solicits business. As one person notes, "The biblical observation that no man can serve two masters, if strictly followed, would make many of Wall Street's present activities impossible."[26]

Where adverse interests cannot be avoided, they can be countered by introducing new interests in a process known as *alignment*. For example, a problem in corporate governance is that CEOs, who are supposed to serve the interests of shareholders, have personal interests that often interfere. One solution is the use of pay-for-performance measures, such as stock options, that align the CEO's personal interest with that of shareholders. Stockbrokers are in a conflict of interest situation when their compensation is tied to the number of trades that a customer makes and not to the quality of these trades. The solution in this case is to base the broker's compensation on the customer's portfolio return, thus aligning the broker's interest more closely to the client's.

3. *Disclosure.* Disclosure serves to manage conflict of interest primarily because whoever is potentially harmed by the conflict has the opportunity to disengage or at least to be on guard. For example, a stock broker who is paid more to sell a firm's in-house mutual funds faces a conflict of interest in recommending a fund to a client. A client who knows of the potential bias can seek out another broker who is uninfluenced by the difference in compensation or can evaluate more carefully the broker's advice to detect any bias. In short, forewarned is forearmed. In legal ethics, there is no conflict of interest if the lawyer discloses the conflict and is confident that the client will be unaffected and the client accepts the lawyer's service under those conditions.[27]

In addition to adverse interests, disclosure may include all kinds of information. The greater the transparency—that is openness of information—the less opportunity there is for conflict of interest to occur. For example, conflict of interest in government is managed in part by requiring officials to disclose financial holdings, but disclosure in the press of officials' activities also reduces conflicts. Thus, we are better able to judge whether a legislator has a conflict of interest if we know not only how much stock he or she owns in a company affected by a bill but also how that person voted on the bill.

4. *Competition.* Strong competition provides a powerful incentive to avoid conflicts of interest, both actual and potential. For example, at one time commercial banks gave their brokerage business to firms that were already bank customers. This practice, know as reciprocation or "recip," has virtually disappeared because of the need for returns on trust accounts to compare favorably with alternative investments. Competition dictates that the allocation of brokerage commissions be based

on the "best execution" of trades and not on keeping bank customers happy. Of course, no firm would use increased competition as a means for managing conflict of interest, but industry regulators should recognize that the power of competition to reduce conflict of interest is another reason to encourage competition.

5. *Rules and Policies.* As already noted, most companies have policies concerning conflict of interest. These typically require employees to avoid acquiring adverse interests by not accepting gifts or investing in potential suppliers, for example. Rules and policies may also prohibit the kind of conduct that would constitute a conflict, as when a broker trades ahead of a large customer, a practice know as "frontrunning." Conflict of interest may be managed by other rules and policies that do not address conflict of interest directly and have other purposes. For example, controls on the flow of information that affect who has access to what information are necessary for many reasons, but the rules and policies in question also limit conflict of interest. Thus, a portfolio manager of a particular mutual fund who has no knowledge of pending purchases by other funds in the firm has fewer possibilities for conflict.

Priority rules are an especially useful means for managing conflict of interest. For example, an investment bank that advises outside investment funds faces a conflict of interest in deciding which investment opportunities to bring to each fund. This problem is especially acute if the bank also operates its own in-house funds. Generally, in such cases, the bank establishes priority rules so that each client knows the order of favor. A client who knows in advance that the better opportunities will be allocated to other funds cannot complain of unfair treatment.

6. *Independent Judgment.* Insofar as a conflict of interest results in biased judgment, the problem can be corrected by utilizing a third party who is more independent. In courts of law, a judge who, say, owns stock in a company affected by a case is generally obligated to *recuse* himself or herself and allow the decision to be rendered by other judges. Companies usually require an executive with a conflicting interest to pass the decision to the next level. Independent appraisers are often utilized in cases where an insider, such as an executive or a director, is engaging in a property transaction with a corporation. In firms with frequent conflict of interest problems, such as investment banks in which the conflict exists among various units, a standing independent advisory board is often formed to consider matters referred to it.

7. *Structural Changes.* Because conflicts of interest result from providing many different services to different customers or clients, they can be reduced by compartmentalizing these services. Advertising agencies, for example, form separate creative teams for each account; accounting firms separate auditing and advisory services; and commercial banks split trust management from the retail side of the business. Within multifunction institutions, conflicts can be reduced by strengthening the independence and integrity of each unit. For example, instead of treating the investment research division as an arm of their brokerage units, investment banks are being urged to upgrade their status and insulate them from pressure.

Some structural features of American business are dictated by law. Because of the potential conflicts of interest, Congress mandated that commercial banks could not also sell stocks or insurance, thereby making investment banking and insurance separate businesses. Pressure is building among federal regulators to force accounting firms to form separate auditing and consulting companies. Addressing the problem of conflict of interest by structural changes is probably unwise overall, however, because of the many advantages of combining different services in one firm. Separating the functions of an investment bank, for example, might reduce conflict of interest, but a firm that underwrites corporate securities needs the sales capacity of its brokerage unit and the skills of its research department. On the whole, we probably gain much more than we lose by having firms that provide multiple services.

Conclusion

Like whistle-blowing, trade secrets and conflict of interest involve a delicate balancing of the rights and interests of employers and employees, as well as the public at large. Especially in the case of trade secret protection, we see how different kinds of arguments—for property rights, fair competition, and a duty of confidentiality—underlie the law in this area and support our views about what is morally right. For the most part, the language of rights and the obligations of agents have dominated our discussion, although utilitarian considerations about the harm and benefit of protecting trade secrets have been introduced. In the next chapter we examine another right, namely, the right of employees and consumers to privacy. Utilitarian and Kantian arguments are used to explain the concept of privacy and provide a foundation for it. Unlike trade secrets and conflict of interest, privacy is an area where the law has yet to develop very far, and so we can focus mainly on ethical issues.

Case 6.2   The Conflict of an Insurance Broker

I work for an insurance brokerage firm, Ashton & Ashton (A&A), which is hired by clients to obtain the best insurance coverage for their needs.[28] To do this, we evaluate a client's situation, keep informed about insurance providers, negotiate on the client's behalf, and present a proposal to the client for approval. Our compensation comes primarily from a commission that is paid by the client as part of the premium. The commission is a percentage of the premium amount, and the industry average for commissions is between 10 and 15 percent. A secondary source of compensation is a contingency payment that is made annually by insurance providers; the amount of this payment is based on the volume of business during the past year.

One of our clients, a world-class museum in a major American city, has been served for years by Haverford Insurance Company. Haverford is a financially sound insurer that has provided the museum with reliable coverage at reasonable prices and has gone out of its way on many occasions to be accommodating. Haverford has also built good relations with A&A by allowing a 17 percent commission—a fact that is not generally known by the clients. When the museum's liability insurance policy was up for renewal, A&A was asked to obtain competitive proposals from likely insurers. We obtained quotations from four comparable insurance companies with annual premiums that ranged between $90,000 and $110,000. A fifth, unsolicited proposal was sent by a small, financially shaky insurance company named Reliable. The annual premium quoted by Reliable was $60,000.

There is no question that the museum is best served by continuing with Haverford, and our responsibility as an insurance broker is to place clients with financially sound insurers who will be able to honor all claims. The museum has a very tight operating budget, however, and funding from public and private sources is always unpredictable. As a result, the museum is forced to be extremely frugal in its spending and has always chosen the lowest bid for any service without regard for quality. The dilemma I faced, then, was: Do I present the Reliable bid to the museum? If I do, it will almost certainly accept it given its priority of saving money. Because the market indicates that the value of the needed policy is around $100,000, the Reliable proposal is definitely an attempt to "low-ball" the competition, and the company would probably raise the premium in future years. Is this honest competition? And if not, should A&A go along with it? Allowing a client to accept a low-ball bid might also jeopardize our relations with the reputable insurers who submitted honest proposals in good faith. If relations with Reliable are not successful, the museum is apt to blame us for not doing our job, which is not merely to pass along proposals but to evaluate them for suitability.

On the other hand, A&A will receive a higher commission and a larger contingency payment at the end of the year if the museum is presented with only the four solicited proposals and never learns of the Reliable bid. Because of our financial stake in the outcome, however, do we face a conflict of interest? Could we be accused of choosing a course of action that benefits us, even though in reality the client is also better served?

Case 6.3   Procter & Gamble Goes Dumpster Diving

According to *Competitive Intelligence Magazine,* John Pepper, the chairman of Procter & Gamble, told a group that competitive intelligence was "of singular importance" to a consumer products company and that P&G had shifted "from collecting, analyzing and dissemination information, to acquiring and using knowledge to create winning strategies."[29] Despite these strong words, Mr. Pepper was apparently alarmed to hear that competitive intelligence sleuths hired by P&G had obtained some documents from its European-based rival, Unilever, through questionable means. At least one person sorted through the trash bins at Unilever's Chicago office, a practice known as "dumpster diving." After learning how P&G's competitor intelligence had been obtained, Mr. Pepper informed Unilever of the misdeeds and personally called the Unilever chairman to settle the matter.

In the highly competitive business of shampoo and other hair care products, information about new lines, launch dates, pricing, advertising plans, and production figures is carefully guarded. Like many companies, P&G attempts to gather all publicly available information about their competitors' activities for what the company calls "competitive analysis." Competitive-analysis executives at P&G contracted with an outside firm, which in turn hired several subcontractors to do investigate competitors. A budget estimated at $3 million was allotted to the project, which began in the fall of 2000. The operation was run out of a safe house, called the "Ranch," located in P&G's home town of Cincinnati, Ohio. Among the secrets gained from dumpster diving in Chicago were detailed plans for a product launch in February. In addition to dumpster diving, which P&G admitted, Unilever believed that some rogue operators also misrepresented themselves to competitors in efforts to gain access, a charge that P&G denies.

Although P&G claims that nothing illegal was done, the dumpster diving violated the company's own code of ethics and its policies for competitive intelligence contractors. The ethics code of the Society of Competitive Intelligence Professionals also prohibits dumpster diving when the bins are on private property. In April 2001, when the company became aware that the spying operation had spun out of control, three executives overseeing the project were fired. Mr. Pepper then contacted Unilever with full disclosure and a promise not to use any of the information gained. P&G had, in effect, blown the whistle on itself. Mr. Pepper hoped perhaps that this gesture would put the matter to rest. However, Unilever had just begun to seek a settlement.

In the ensuing negotiations, Unilever proposed that P&G compensate Unilver between $10 million and $20 million for possible losses incurred from the unethical acquisition of information. In addition, Unilever wanted P&G to reassign key personnel in its hair-care division who had read the documents to other positions in which they could not utilize the information they had gained. Perhaps the most unusual remedy was that P&G allow an independent third party to investigate the company's hair-care business for several years and to report to Unilever any situations in which improperly gained information might have been used. Unilever suggested that if a satisfactory settlement could not be reached, then the company might sue in court, with uncertain results.

If John Pepper thought that notifying Unilever and firing the people involved were the right thing to do, then Unilever's proposals might seem to be an unwarranted punishment that would discourage others from being so forthright. On the other hand, aside from any monetary payment, P&G could continue to compete as vigorously as it would have had it not gained the information from dumpster diving. A settlement on Unilever's terms might effectively restore fair competition. On August 28, 2001, Mr. Pepper flew to London for final negotiations, knowing that he would soon have to make a decision.

NOTES

1. Cleveland Horton, "Ethics at Issue in Lotus Case," *Advertising Age,* 21 December 1987, 6.

2. Alan Farnham, "How Safe Are Your Secrets?" *Fortune,* 8 September 1997, 114–20; Chaim A. Levin, "Economic Espionage Act: A Whole New Ballgame," *New York Law Journal,* 2 January 1997, 5.

3. These are essentially the facts in the classic case *Peabody* v. *Norfolk,* 98 Mass. 452 (1868).

4. *Wexler* v. *Greenberg,* 160 A. 2d 430 (1960).

5. For a discussion of the Lockean view as well as the utilitarian argument discussed below, see Edwin C. Hettinger, "Justifying Intellectual Property," *Philosophy and Public Affairs,* 18 (1989), 36–51.

6. This point is made by Robert E. Frederick and Milton Snoeyenbos, "Trade Secrets, Patents, and Morality," in *Business Ethics* (Buffalo: Prometheus Books, 1983), 165–66.

7. For a discussion of the ethical issues, see Mark Michael, "Patent Rights and Better Mouse-traps," *Business and Professional Ethics Journal,* 3 (1983), 13–23.

8. For a summary of the law in other countries, see Stanley H. Lieberstein, *Who Owns What Is in Your Head?* (New York: Hawthorne Books, 1979), 225–32.

9. *International News Service* v. *Associated Press,* 248 U.S. 215 (1918).

10. See Kevin McManus, "Who Owns Your Brain?" *Forbes,* 6 June 1983, 178.

11. See Harlan M. Blake, "Employee Covenants Not to Compete," *Harvard Law Review,* 73 (1960), 625–91.

12. *Dunn* v. *Frank Miller Associates,* Inc., 237 Ga. 266 (1976). The case is cited by Lieberstein, "Who Owns What Is in Your Head?" 50–51.

13. All employees are regarded in law as agents, at least while acting within the scope of their assigned responsibilities, but their specific obligations, including those with respect to confidentiality, are determined by the amount of trust placed in them and any understandings, such as company policies or professional ethics.

14. Michael S. Baram, "Trade Secrets: What Price Loyalty?" *Harvard Business Review,* 46 (November–December 1968), 66–74.

15. Lynn Sharp Paine, "Corporate Policy and the Ethics of Competitor Intelligence Gathering," *Journal of Business Ethics,* 10 (1991), 423–36.

16. Much of the material in this paragraph and the one following is contained in Steven Flax, "How to Snoop on Your Competitors," *Fortune,* 14 May 1984, 28–33.

17. Diana B. Henriques with Kurt Eichenwald, "A Fog over Enron, and the Legal Landscape," *New York Times,* 27 January 2002.

18. *Second Restatement of Agency,* Sec. 385. Emphasis added.

19. This distinction is made in Thomas M. Garrett and Richard J. Klonosky, *Business Ethics,* 2d ed. (Upper Saddle River, NJ: Prentice Hall, 1986), 55; and in Manual G. Velasquez, *Business Ethics: Concepts and Cases,* 4th ed. (Upper Saddle River, NJ: Prentice Hall, 1998), 431.

20. Michael Davis, in "Conflict of Interest, " *Business and Professional Ethics Journal,* 1 (1982), 17–27, makes a threefold distinction between actual, latent, and potential conflicts of interest. Latent conflict of interest involves conflict situations that can reasonably be foreseen, whereas potential conflict of interest involves conflict situations that cannot reasonably be foreseen.

21. The rule reads: "A lawyer shall not represent a client if the representation of that client will be directly adverse to another client unless: (1) the lawyer reasonably believes the representation will not adversely affect the relationship with the other client; and (2) each client consents after consultation."

22. See Abraham J. Briloff, "Do Management Services Endanger Independence and Objectivity?" *CPA Journal,* 57 (August 1987), 22–29.

23. National Society of Professional Engineers, *Code of Ethics for Engineers*, 1987, III, 2(b).

24. "Media Policies Vary on Preventing Employees and Others from Profiting on Knowledge of Future Business Stories," *Wall Street Journal*, 2 March 1984, A12.

25. *Carpenter et al. v. U.S.*, 484 U.S. 19 (1987).

26. Warren A. Law, "Wall Street and the Pubic Interest," in *Wall Street and Regulation*, ed. Samuel L. Hayes (Boston, MA: Harvard Business School Press, 1987), 169.

27. American Bar Association, Model Rules of Professional Conduct, Rule 1.7.

28. This case is based on actual events, but the names of the companies have been disguised. The case was prepared by Michael Streett under the supervision of Professor John R. Boatright. Copyright 1995 by John R. Boatright.

29. This case is based on Andy Serwer, "P&G Comes Clean on Spying Operation," www.fortune.com, 30 August 2001; Julian E. Barnes, "Unilever Wants P&G Placed under Monitor in Spy Case," *New York Times*, 1 September 2001, C1; Andrew Edgecliffe-Johnson and Adam Jones, "Unilever Seeks Review after P&G 'Spying'," *Financial Times*, 1 September 2001, 14; Julian E. Barnes, "P&G Said to Agree to Pay Unilver $10 Million in Spying Case," *New York Times*, 7 September 2001, C7; Andy Serwer, "P&G's Covert Operation," www.fortune.com, 17 September 2001.

Case 7.1 Psychological Testing at Dayton Hudson

Answer each of the following questions True or False:

I feel sure there is only one true religion.

My soul sometimes leaves my body.

I believe in the second coming of Christ.

I wish I were not bothered by thoughts about sex.

I am very strongly attracted by members of my own sex.

I have never indulged in any unusual sex practices.

In April 1989, Sibi Soroka answered these questions satisfactorily and was hired as a Store Security Officer (SSO) at a Target store in California. Afterward, Soroka felt "humiliated" and "embarrassed" at having to reveal his "innermost beliefs and feelings." So he joined with two rejected job applicants in a class action suit, charging the Dayton Hudson Corporation, the owner of the Target store chain, with invasion of privacy.[1]

Psychological testing is one of many means for enabling employers to evaluate applicants and select the best employee for a job. In the 1920s, the owner of Frank Dry Goods Company in Fort Wayne, Indiana, noticed that some salesgirls sold two to four times as much merchandise as others.[2] Further investigation revealed that the top sellers came from large working-class families with savings accounts, whereas the low performers were from small families that did not need the money and were opposed to the employment. Accordingly, the company developed a set of test questions for job applicants that asked about family size, the occupations of family members, the amount of income needed for an average family, the attitude of the family about working in the store, and the existence of savings accounts.

Dayton Hudson defended the use of the psychological test, called Psychscreen, on the grounds that an SSO, whose main function is to apprehend suspected shoplifters, needs good judgment, emotional stability, and a willingness to take direction. Psychscreen is a combination of two standard tests that have been administered to applicants for such public safety positions as police officers, prison guards, air traffic controllers, and nuclear power plant operators. The completed Psychscreen test is interpreted by a firm of consulting psychologists which rates an applicant on five traits (emotional stability, interpersonal style, addiction potential, dependability, and rule-following behavior) and offers a recommendation on whether to hire the applicant. Dayton Hudson does not receive the answers to any specific questions.

Dayton Hudson admitted in court that it had not conducted any studies to show that Psychscreen was a reliable predictor of performance as a security officer, except to administer the test to 18 of its most successful SSOs. The company could not document any improvement in the performance of SSOs after adopting the test or any reduction in shoplifting. An expert witness for the plaintiffs contended that the test had not been proven to be reliable or valid for assessing job applicants in this particular setting. An expert witness for Dayton Hudson admitted that the use of Psychscreen resulted in a 61 percent rate of false positives. Thus, even if every unqualified applicant were identified by the test, more than six in ten qualified applicants would also be rejected as unfit.

Dayton Hudson conceded that the intimate questions in Psychscreen constitute an invasion of privacy but added that the intrusion was minor and was justified by the company's needs. Employment application forms ask for some job-related personal information. Even though questions about religion and sex are not themselves job-related, they enable the interpreters of the test to evaluate psychological traits that are related to the job. Dayton Hudson was no more interested than Frank Dry Goods in the personal life of its applicants for employment. The information gained by intimate questions was merely a means to an end. Left unanswered by this response are whether the company's need to administer the test offset the invasion of the applicants' privacy and if so, whether some less invasive means to achieve this end could have been found.

Some critics argue that psychological testing is an invasion of privacy not only because of the intimate nature of the questions but also because the tests seek personal information, namely psychological traits, in ways that the person does not understand and is unable to control. That is, not only the means but also the end are intrusive. Thus, even if a test could be constructed without questions about religion, sex, or any other intimate subject, these critics hold that the test would still be an invasion of privacy.

INTRODUCTION

Early in the century, the Ford Motor Company set up a "Sociological Department" in order to make sure that workers, in Henry Ford's words, were leading "clean, sober, and industrious" lives.[3] Company inspectors checked bank accounts to keep

Ford employees from squandering their munificent $5-a-day wages. They visited employees' living quarters to see that they were neat and healthful, and they interviewed wives and acquaintances about the handling of finances, church attendance, daily diet, drinking habits, and a host of other matters. Workers who failed to live up to Henry Ford's standards of personal conduct were dismissed.

Employers today would scarcely dare to intrude so openly into the private lives of their employees, but they possess less obvious means for acquiring the information sought by Ford's teams of snooping inspectors—and some information that Henry Ford could not have imagined! Among the tools available to present-day employers are quick and inexpensive drug tests, pencil-and-paper tests for assessing honesty and other personality traits of employees, extensive computer networks for storing and retrieving information about employees, and sophisticated telecommunication systems and concealed cameras and microphones for supervising employees' work activities. By administering medical insurance plans and providing on-site health care and counseling, employers are now in a position to know about employees' medical conditions. Some employers have also conducted genetic testing to screen employees for genes that make them more vulnerable to chemicals in the workplace.

Consumers have joined employees as targets for information gathering by American corporations. The same surveillance techniques that are used to monitor employees are now used to detect theft by store customers. Video cameras are commonplace in retail stores, and some retailers have installed hidden microphones as well. The main threat to consumer privacy comes from the explosive growth of database marketing. The countless bits of information that consumers generate in each transaction can now be combined in vast databases to generate lists for direct mail and telemarketing solicitations. Public records, such as automobile registrations and real estate transfers, are also readily available sources of information for the creation of specialized lists. The collection of information about users of the Internet, which is in its infancy, has immense potential for marketers.

Concern about privacy is a relatively recent occurrence. However, a 1979 public-opinion survey conducted by Louis Harris for the Sentry Insurance Company revealed that three out of four respondents believed that privacy should be regarded as a fundamental right akin to life, liberty, and the pursuit of happiness and that half of them fear that American corporations do not adequately safeguard the personal information they gather on individuals.[4] Over 90 percent of those who responded said that they favored safeguards to prevent the disclosure of personnel and medical files to outsiders. A law granting employees access to the information collected about them was favored by 70 percent, and 62 percent wanted Congress to pass a law regulating the kind of information that corporations may collect about individuals.

CHALLENGES TO PRIVACY

Privacy has become a major issue in government and business in recent years for many reasons. One is simply the vast amount of personal information that is collected by government agencies. The need to protect this information became especially acute after the passage of the Freedom of Information Act (FOIA) in 1966. Intended by Congress to make government more accountable for its actions, the act had the unforeseen consequence of compromising the confidentiality of information about private individuals. The Privacy Act of 1974 was designed in large part to resolve the conflict between government accountability and individual privacy. So great were the problems that Congress created the Privacy Protection Study Commission to investigate and make recommendations about further action. The National Labor Relations Board has long faced a similar problem with union demands for access to personnel files and other employee records. Unions claim that they need the information in order to engage in fair collective bargaining, but allowing unions to have unlimited access to this information without consent violates the employees' right of privacy.[5]

Employee Privacy

Government is not the only collector of information. Great amounts of data are required by corporations for the hiring and placement of workers, for the evaluation of their performance, and for the administration of fringe benefit packages, including health insurance and pensions. Private employers also need to compile personal information about race, sex, age, and handicap status in order to document compliance with the law on discrimination. In addition, workers' compensation law and occupational health and safety law require employers to maintain extensive medical records. Alan F. Westin, an expert on privacy issues, observes that greater concern with employee rights in matters of discrimination and occupational health and safety has had the ironic effect of creating greater dangers to employees' right of privacy.[6]

Workplace Monitoring. Monitoring the work of employees is an essential part of the supervisory role of management, and new technologies enable employers to watch more closely than ever before, especially when the work is done on telephones or computer terminals. Supervisors can eavesdrop on the telephone conversations of employees, for example, and call up on their own screens the input and output that appear on the terminals of the operators.[7] Hidden cameras and microphones can also be used to observe workers without their knowledge. A computer record can be made of the number of telephone calls, their duration, and their destination. The number of keystrokes made by a data processor, the number of errors and corrections made, and the amount of time spent away from the desk can also be recorded for use by management. Even the activities of truck drivers can be monitored by a small computerized device attached to a vehicle that registers speed, shifting, and the time spent idling or stopped.

Companies claim that they are forced to increase the monitoring of employees with these new technologies as a result of the changing nature of work. More complex and dangerous manufacturing processes require a greater degree of oversight by employers. The electronic systems for executing financial transactions and transferring funds used by banks and securities firms have a great potential for misuse and costly errors. In addition, employers are increasingly concerned about the use of drugs by workers and the high cost of employee theft, including the stealing of trade secrets. Employers also claim to be acting on a moral and a legal obligation to provide a safe workplace in which employees are free from the risk of being injured by drug-impaired co-workers.[8]

Even efforts to improve employees' well-being can undermine their privacy. Wellness programs that offer medical checkups along with exercise sessions result in the collection of medical data that can be used to terminate employees or defend against workplace injury claims. More than half of all U.S. employees have access to Employee Assistance Plans (EAPs) for help in handling personal problems and drug addictions. Although the information gained is generally held in confidence, it is available for company use when an employee files a workplace injury claim or sues for discrimination, wrongful discharge, or any other wrong. In some instances, employers have used the threat of revealing unrelated embarrassing information in court to dissuade employees from pressing a suit. Although the use of an EAP is usually voluntary, employees are often required to gain approval from an EAP counselor before seeking company-paid mental health care. Some employees thus face the choice of revealing their mental health condition to their company or paying for mental health care out-of-pocket.

Psychological Testing. One particular area of concern has been psychological testing of the kind conducted by Dayton Hudson (Case 7.1). Interest in psychological testing was spurred in the first half of the twentieth century by the "scientific management" ideas of Frederick Taylor and the development of the field of applied or industrial psychology. The massive testing programs of the armed forces in two world wars were carried over into civilian life by large American corporations. In the postwar period, American education became increasingly reliant on standardized testing for admission to colleges and universities, not only to identify qualified students but also to prevent discrimination. Tests that measure job-related abilities and aptitudes have raised little opposition. However, employers have increasingly come to recognize that an employee's psychological traits are important, not only for predicting successful job performance but also for identifying potentially dishonest and troublesome employees.

This latter goal is the appeal of integrity tests, which are sold by a handful of publishers and administered to an estimated five million job applicants annually. Use of the pencil-and-paper tests has been spurred by the banning of mechanical polygraph testing in 1988 and by the reluctance of former employers to reveal any but the most basic information. Studies by the congressional Office of Technology Assessment and the American Psychological Association have found that some tests have moderate predictive value but that others are virtually worthless. Large numbers of honest people are denied jobs and suffer a stigma because of faulty testing, and a few rogues slip through.

Some critics have charged that no one can pass an integrity test without a little dishonesty. Ironically, the highly honest may be among the most frequent victims of mistakes because they are more forthcoming in their answers. The use of integrity tests also assumes that people are honest or not and ignores the role of the work environment in promoting honesty—or dishonesty. Some employees steal when they believe that they are being cheated or abused, for example. One benefit of integrity tests, therefore, may be to enable employers to recruit a work force that will tolerate shabby treatment without retaliating.

Consumer Privacy

Concern about consumer privacy has focused primarily on the gathering and use of information in database marketing. Businesses have discovered that it pays to know their customers. For example, grocery stores that issue identification cards that are scanned along with the universal product code on each product are able to construct detailed profiles of each customer's purchasing preferences. This information may be used in many ways, including the making of offers that are tailored to appeal to specific customers. However, the main value of a database of consumer information lies in the capacity to generate customized mailing lists. If a company can identify the characteristics of potential customers by age, income, lifestyle, or other measures, then a mailing list of people with these characteristics would enable the seller to reach these customers at relatively low cost. Such targeted selling, known as direct mail, is also potentially beneficial to consumers, because a customized mailing list is more likely to produce offers of interest to consumers than is a random mailing.

The growth in database marketing has been facilitated by computer technology, which is able to combine data from many sources and assemble it in usable form. For example, by merging information about an individual with census data for that person's zip-code-plus-four area, it is possible to make reliable inferences about income, lifestyle, and other personal characteristics. Companies that specialize in data collection can provide direct marketers with customized mailing lists that target groups with the desired characteristics. Although American consumers are concerned about the threat to privacy posed by the use of personal information for this purpose, one survey showed that over two-thirds of respondents approved of the use of customized mailing lists to offer goods and services to people who are likely to be interested.[9] However, when Lotus Development Corporation announced plans in April 1990 for Lotus Marketplace:Households, a set of compact discs containing information on 120 million Americans (see Case 7.4), a storm of protest ensued. This episode showed that consumers believe that there are limits on the use of personal information for marketing purposes and that this product had crossed a line.

Issues in Consumer Privacy. One issue in the use of databases to generate mailing lists is the right of control over information. If we reveal some information about ourselves to a company, does that company "own" the information? For example, does a magazine have a right to sell a list of its subscribers to a direct marketer? We voluntarily provide our name and address to the magazine for the purpose of obtaining a subscription, just as we reveal our annual income to a bank in order to obtain a loan. These are examples of the *primary* use of information. The use of information for some other purpose is labeled *secondary*. Some privacy advocates hold that there should be no secondary use of information without a person's knowledge and consent. Thus, some magazines inform subscribers that they make their list available for direct mail and allow subscribers to "opt out" by removing their name and address from the list. In general, the secondary use of any information in a loan application is prohibited by law.

Other issues concern access to information and potential misuse. Although an individual's annual income is generally regarded as personal, people may not be upset to learn that this information is used to generate a mailing list—as long as no one has access to the information itself. A direct marketer has no interest in knowing a particular person's income but only whether that person is a likely prospect, and the fact that a person's name and address is on a list does not reveal to anyone that person's income. However, some information is considered too sensitive to be included in a marketing database. Health information has generally fallen into this category, but pharmaceutical companies now seek mailing lists of patients with particular conditions. For example, *Reader's Digest* succeeded in obtaining completed questionnaires on health problems from nine million subscribers, and the magazine is planning to make this information available for direct mail on specific pharmaceutical products.[10] Patients' records, prescription data from pharmacies, and even calls to the toll-free numbers of pharmaceutical companies are resources for information gatherers.[11] Companies that sell lists also have a responsibility to screen buyers to ensure that direct mailings are for legitimate purposes and do not involve consumer fraud.

Ethical questions about employee and consumer privacy are unavoidable because obtaining and using personal information is essential in both employment and marketing. But everyone has a legitimate interest in maintaining a private life that is free from unwarranted intrusion by business. Finding the right balance between the rights of everyone concerned is not a simple task. A set of guidelines or a company code on employee and consumer privacy must address an immense number of different questions. Before we make the attempt to find a balance between these competing rights, though, it is necessary for us to inquire into the meaning of privacy as an ethical concept.

THE CONCEPT OF PRIVACY

A definition of privacy has proven to be very elusive. After two years of study, the members of the Privacy Protection Study Commission were still not able to agree on one. Much of the difficulty is due to the diverse nature of the many different situations in which claims of a right of privacy are made. Even the narrower concept of privacy for employees and consumers is applied in such dissimilar circumstances that it is not easy to find a common thread running through them.

History of the Concept

As a legal concept, privacy dates only from the late nineteenth century. There is no mention of privacy in the original Constitution or the Bill of Rights. Although a number of rights related to privacy have long been recognized in American law, they have generally been expressed in terms of freedom of thought and expression, the right of private property, protection from "unreasonable searches and seizures," and other constitutional guarantees. The first sustained discussion of privacy occurred in an 1890 article in the *Harvard Law Review* written by two young attorneys, Samuel Warren and Louis Brandeis (who later became a famed Justice of the Supreme Court).[12]

The theory of privacy presented by Warren and Brandeis was slow to gain acceptance. It was rejected by the courts in a number of cases around the turn of the century in which the names and pictures of prominent persons were used to advertise products. The public uproar over one of these cases prompted the New York legislature to enact a law prohibiting the commercial use of a person's name or likeness without permission.[13] Gradually, most states followed the lead of New York in granting persons a right to be free of certain kinds of intrusion into their private life. But it was not until 1965 that the Supreme Court declared privacy to be a constitutionally protected right. The decision came in *Griswold* v. *Connecticut*, which concerned the right of married couples to be free of state interference in the use of contraceptives.[14]

Some philosophers and legal theorists have argued that the concept of privacy does not introduce any new rights into the law but merely expresses several traditional rights in a new way. Consequently, our legal system already contains the resources to protect individuals against these wrongs without creating a distinct right of privacy.[15] For example, disclosing embarrassing facts about a person or intruding into his or her solitude might be described as inflicting mental distress; and the publication of false accusations could be said to constitute libel or defamation of character. What, these critics ask, does the concept of privacy add to other, better-established rights?

Definitions of Privacy

The literature contains many attempts to elucidate privacy as an independent right that is not reducible to any other commonly recognized right. Three definitions in particular merit examination. One, which derives from Warren and Brandeis and finds expression in *Griswold* v. *Connecticut*, holds that privacy is the right to be left alone. Warren and Brandeis were concerned mainly with the publication of idle gossip in sensation-seeking newspapers. The aim of privacy laws, they thought, should be to protect "the privacy of private life" from unwanted publicity, and their proposals all deal with limits on the publication of information about the private lives of individuals. In his celebrated dissenting opinion in *Olmstead* v. *United States*, a 1928 case concerning the constitutionality of telephone wiretapping, Brandeis wrote that the right of privacy is "the right to be let alone—the most comprehensive of rights and the right most valued by civilized men."[16]

A similar view of privacy was expressed by the majority in *Griswold.* Laws governing the use of contraceptives intrude into an area of the lives of individuals where they have a right to be let alone. Justice William J. Brennan expressed the view in a subsequent birth control case that

> If the right to privacy means anything, it is the right of the individual, married or single, to be free from unwarranted government invasion into matters so fundamentally affecting a person as the decision whether to bear or beget a child.[17]

Many critics have pointed out that the phrase "to be let alone" is overly broad.[18] Individuals have a right "to be let alone" in matters of religion and politics, for example, but legal restrictions on religious practices, such as snake handling, or on political activities, such as the making of political contributions, do not involve violations of privacy. At the same time, the Warren and Brandeis definition is too narrow, because some violations of privacy occur in situations where there is no right to be let alone. Workers have no right to be free of supervision, for example, even though it can be claimed that their privacy is invaded by the use of hidden cameras to monitor their activity secretly.

These objections, in the view of critics, are merely symptoms of a deeper source of error in the Warren and Brandeis definition, which is the confusion of privacy with liberty. These examples show that a loss of liberty is neither a necessary nor a sufficient condition for a loss of privacy. Perhaps greater clarity is achieved by limiting the concept of privacy to matters involving information and not stretching the concept to include all manner of intrusions into our private lives. This suggests that smoking policies like those of the Ford Meter Box Company (Case 7.3) are better analyzed as limitations of liberty rather than invasions of privacy.

This suggestion is reflected in a second definition in which privacy is expressed in terms of control over information about ourselves.[19] The following definition by Alan F. Westin is typical: "Privacy is the claim of individuals ... to determine for themselves when, how, and to what extent information about them is communicated to others."[20] This definition is open to the same charge: It is at once too broad and too narrow. Richard B. Parker observes, "Not every loss or gain of control over information about ourselves is a gain or loss of privacy."[21] Furthermore, all definitions of privacy as exercising control flounder on the fact that individuals can relinquish their own privacy by voluntarily divulging all sorts of intimate details themselves.[22] There is a loss of privacy under such circumstances but not a loss of control. Therefore, privacy cannot be identified with control.

A third, more adequate definition of privacy holds that a person is in a state of privacy when certain facts about that person are not known by others. W. A. Parent, in an important 1983 article, "Privacy, Morality, and the Law," defines privacy as "the condition of not having undocumented personal knowledge about one possessed by others."[23] By the phrase "personal knowledge," Parent does not mean all information about ourselves but only those facts "which most individuals in a given society at any given time do not want widely known."[24] It is necessary that the definition be restricted to *undocumented* personal information, because some facts that individuals commonly seek to conceal are a matter of public record and can be known without prying into their private lives. A person does not suffer a loss of privacy, for example, when a conviction for a crime becomes known to others, because court records are public documents. Similarly, there is no loss of privacy when an easily observable fact, such as a person's baldness, is known to others, even though the person is sensitive about it and prefers that others not be aware of it.

In the remaining discussion, the concept of privacy is limited to matters involving information and, in particular, to the access of others to undocumented personal information, as described by Parent. The two other definitions—as a right to be let alone and to have control over information about ourselves—confuse privacy with other values. Having gained some understanding of the concept of privacy, we can now turn to the question of why privacy is a value.

THE VALUE OF PRIVACY

Why do we value privacy so highly and hold that it ought to be protected as a right? Certainly, we desire to have a sphere of our life in which others do not possess certain information about us. But the mere fact that we have this desire does not entail that we have a right of privacy. Nor does it tell us how far a right of privacy extends. Some arguments are needed, therefore, to establish the value of privacy and the claim that we have a right to it. Most of the arguments developed by philosophers and legal theorists fall into one of two categories. One category consists of utilitarian arguments that appeal to consequences, and the second is Kantian arguments that link privacy to being a person or having respect for persons. To a great extent, these two different kinds of arguments express a few key insights about privacy in slightly different ways.

Utilitarian Arguments

One of the consequences cited by utilitarians is that great harm is done to individuals when inaccurate or incomplete information collected by an employer is used as the basis for making important personnel decisions. The lives of many employees have been tragically disrupted by groundless accusations in their personnel record, for example, and the results of improperly administered polygraph and drug tests. Even factual information that ought not to be in an employee's file, such as the record of an arrest without a conviction, can cause needless harm. The harm from these kinds of practices is more likely to occur and to be repeated when employees are unable to examine their files and challenge the information (or misinformation) in them.

A drawback to this argument is that it rests on an unproved assumption that could turn out to be false. It assumes that on balance more harm than good will result when employers amass files of personal information, use polygraph machines, conduct drug tests, and so on. Whatever harm is done to employees by invading their privacy has to be balanced, in a utilitarian calculation, against the undeniable benefits that these practices produce for both employers and employees.

Furthermore, the argument considers only the possible harmful consequences of privacy invasions. However, some practices, such as observing workers with hidden cameras and eavesdropping on business conducted over the telephone, are generally considered to be morally objectionable in themselves, regardless of their consequences. Honest workers, for example, have nothing to fear from surveillance that is designed to protect against employee theft, and indeed the use of hidden cameras in a warehouse can even benefit those who are honest by reducing the possibility of false accusations. Still, workers have a right to complain that secret surveillance of their activities on the job violates the right of privacy. It is the fact that they are subjected to constant observation and not any possible consequence of being observed that is morally objectionable.

Expanding the Scope of Consequences. This objection is avoided by more sophisticated utilitarian arguments that do not locate the harmful consequences solely in the harm that occurs when information is misused. According to these arguments, a certain amount of privacy is necessary for the enjoyment of some activities, so that invasions of privacy change the character of our experiences and deprive us of the opportunity for gaining pleasure from them. Monitoring and surveillance in the workplace, for example, affect job satisfaction and the sense of dignity and self-worth of all workers. They send a message to employees that they are not trusted and respected as human beings, and the predictable result is a feeling of resentment and a decline in the satisfaction of performing a job.

An illustration of this point is provided by a truck driver with 40 years' experience with the Safeway Company who reports that he used to love his job because "you were on your own—no one was looking over your shoulder. You felt like a human being." After the company installed a computerized monitoring device on his truck, he decided to take early retirement. He complains, "They push you around, spy on you. There's no trust, no respect anymore." A directory-assistance operator reported, "I've worked all those years before monitoring. Why don't they trust me now? I will continue to be a good worker, but I won't do any more than necessary now."[25]

Privacy and Identity. Some writers argue that privacy is of value because of the role it plays in developing and maintaining a healthy sense of personal identity. According to Alan F. Westin, privacy enables us to relax in public settings, release pent-up emotions, and reflect on our experiences as they occur—all of which are essential for our mental well-being. A lack of privacy can result in mental stress and even a nervous breakdown.[26] Another common argument appeals to the importance of privacy in promoting a high degree of individuality and freedom of action among the members of a society. Critics of these arguments object, however, that there is little evidence that privacy has the benefits claimed for it or that the predicted harm would follow from limiting people's privacy.[27] Many societies function very well with less room for solitude than our own, and the experiences of human beings in prisons and detention camps are cited by critics to refute these arguments.

Kantian Arguments

Two Kantian themes that figure prominently in defense of a right to privacy are those of autonomy and respect for persons. Stanley I. Benn, for example, notes that utilitarian arguments for a right of privacy are not able to show what is morally wrong when a person is secretly observed without any actual harm being done. "But respect for persons," Benn claims, "will sustain an objection even to secret watching, which may do no actual harm at all." He continues:

> Covert observation—spying—is objectionable because it deliberately deceives a person about his world, thwarting ... his attempts to make a rational choice. One cannot be said to respect a man ... if one knowingly and deliberately alters his conditions of action, concealing the fact from him.[28]

Benn's argument thus appeals to both Kantian themes by arguing that invading a person's privacy violates the principle of respect for persons *and* prevents a person from making a rational choice as an autonomous being.

Hyman Gross argues in a similar vein that what is morally objectionable about being observed unawares through a hidden camera or having personal information in a data bank is that a person loses control over how he or she appears to others.[29] If people form incomplete or misleading impressions of us that we have no opportunity to correct, then we are denied the possibility of autonomous or self-directed activity, which is a characteristic of human beings. Hence, invasions of privacy diminish an essential condition for being human.

In a very influential discussion, Charles Fried argues that privacy is of value because it provides a "rational context" for some of our most significant ends, such as love, friendship, trust, and respect, so that invasions of privacy destroy our very integrity as a person.[30] The reason that privacy is essential for respect, love, trust, and so on is that these are intimate relations, and intimacy is created by the sharing of personal information about ourselves that is not known by other people. In a society without privacy, we could not share information with other people (because they would already know it), and hence we could not establish intimate relations with them. Thus, monitoring, in Fried's view, "destroys the possibility of bestowing the gift of intimacy, and makes impossible the essential dimension of love and friendship."[31] Similarly, trust cannot exist where there is monitoring or surveillance, because trust is the expectation that others will behave in a certain way without the need to check up on them.

The arguments of Benn, Gross, Fried, and others seize upon important insights about the value of privacy, but many critics have found flaws in the details of their arguments. Jeffrey H. Reiman, for one, objects that it is too strong to assert that all instances of watching a person unawares result in deceiving a person and depriving that person of a free choice. Otherwise, we would be violating a person's right of privacy by observing him or her strolling down a street or riding a bus.[32] Intimate relations such as love and friendship do not consist solely in the sharing of information but involve, as one writer says, "the sharing of one's total self—one's experiences, aspirations, weaknesses, and values."[33] Consequently, these relations can exist and even flourish in the absence of an exclusive sharing of information.

The Role of Privacy in Socialization

Several philosophers have suggested that the key to a more satisfactory theory of privacy can be constructed by understanding the way in which individuals are socialized in our culture.[34] Privacy, in the view of these philosophers, is neither a necessary means for realizing certain ends nor conceptually a part of these ends. Nevertheless, we are trained from early childhood to believe that certain things are shameful (for example, public nudity) and others strictly our own business (such as annual income). There is no intrinsic reason why our body or our financial affairs should be regarded as private matters. People at different times and places have been socialized differently with regard to what belongs to the sphere of the private, and we might even be better off if we had been socialized differently. Still, we have been socialized in a certain way. In our culture, certain beliefs about what ought to be private play an important role in the process by which a newborn child develops into a person and by which we continue to maintain a conception of ourselves as persons. In the words of Jeffrey H. Reiman:

> Privacy is an essential part of the complex social practice by means of which the social group recognizes—and communicates to the individual—that his existence is his own. And this is a precondition of personhood. To be a person, an individual must recognize not just his actual capacity to shape his destiny by his choices. He must also recognize that he has an exclusive moral right to shape his destiny. And this in turn presupposes that he believes that the concrete reality which he is, and through which his destiny is realized, belongs to him in a moral sense.[35]

This argument is broadly utilitarian. The consequences that it appeals to, however, are not the simple pleasures and pains of classical utilitarianism or even the notions of mental health and personal growth and fulfillment of more sophisticated utilitarian arguments. The argument goes deeper by appealing to the importance of privacy for personhood, a concept that is more commonly used by Kantian theorists. Unlike Kantian arguments, though, this one recognizes that privacy is not necessary for all people in all times and places but is merely a value specific to contemporary Western culture. There are societies that function very well with less privacy than we are accustomed to; however, given the role privacy plays in our socialization process, a certain amount is needed for us to develop as persons and have a sense of dignity and well-being.

Both utilitarian and Kantian arguments point to a key insight: Privacy is important in some way to dignity and well-being. They claim too much, however; privacy is not absolutely essential to either one, except insofar as we have come to depend on it. For better or worse, privacy has become an important value in our culture, and now that it has, it needs to be maintained. Privacy is like the luxury that soon becomes a necessity, but "necessary luxuries" are not less valuable just because we could formerly get by without them. The justification of privacy just offered is thus the most adequate one we have.

THE PRIVACY OF EMPLOYEE RECORDS

The arguments in the preceding section show that privacy is of such sufficient value that it ought to be protected. There are many instances, however, in which other persons and organizations are fully justified in having personal information about us and thereby in intruding into our private lives. The task of justifying a right of privacy, then, consists not only in demonstrating the value of privacy but also in determining which intrusions into our private lives are justified and which are not.[36]

As an example, consider the issues that must be addressed in developing the case for a right of privacy in employee records and in formulating a company privacy protection plan for these records. Among the issues are

1. The kind of information that is collected.
2. The use to which the information is put.
3. The persons within a company who have access to the information.
4. The disclosure of the information to persons outside the company.
5. The means used to gain the information.
6. The steps taken to ensure the accuracy and completeness of the information.
7. The access that employees have to information about themselves.

The first three issues are closely related, because the justification for an employer's possessing any particular kind of information depends, at least in part, on the purpose for which the information is gathered. Some information is simply of no conceivable use in company decision making and constitutes a gratuitous invasion of employee privacy. It is more often the case, however, that an employer has a need or an interest that provides some justification for intruding into the private lives of employees. An invasion of employee privacy is justified, however, only when the information is used for the intended purpose by the individuals who are responsible for making the relevant decisions.

Companies are generally justified in maintaining medical records on employees in order to administer benefit plans, for example, and to monitor occupational health and safety. If these are the purposes for which a company gathers this kind of information, then it follows that (1) only medical information that is essential for these purposes can be justifiably collected; (2) only those persons who are responsible for administering the benefit plans or monitoring the health and safety of employees are justified in having access to the information; and (3) these persons must use the information only for the intended purposes. There are three corresponding ways in which employees' right of privacy can be violated. These are when (1) personal information is gathered without a sufficient justifying purpose; (2) it is known by persons who are not in a position that is related to the justifying purpose; and (3) persons who are in such a position use the information for other, illegitimate purposes.

What Justifies a Purpose? Obviously, the notion of a justifying purpose plays a critical role in determining the exact scope of the right of privacy in employment. There is considerable room for disagreement on the questions of whether any given purpose is a legitimate one for a business firm to pursue, whether a certain kind of information is essential for the pursuit of a particular purpose, and whether the information is in fact being used for the intended purpose. Companies have an interest and, indeed, an obligation to ensure that employees are capable of performing physically demanding work and are not subjected to undue risk, for example. The purposes for which Henry Ford created the Sociological Department, however, went beyond this concern to include a paternalistic regard for the general welfare of his employees, which is not a legitimate purpose. Even to the extent that the work of the inspectors from the Ford Motor Company was justified by a legitimate purpose, there could still be an objection to the excessive amount of information they sought. Information about the handling of finances, church attendance, and eating and drinking habits is more than the company needs to know.

   Determining the purpose for which information is being used can raise difficult questions about intentions. A controversy was sparked in 1980, for example, when it became publicly known that the Du Pont Company was routinely screening black applicants at a plant in New Jersey for signs of sickle cell anemia. The company asserted that the purpose for conducting the screening was to protect black workers, because carriers of the disease, who are mostly black, were thought to be more vulnerable to certain chemicals used at the plant. Such a purpose is arguably legitimate, but some critics of Du Pont charged that the company was actually using genetic screening for another purpose, namely to prevent liability suits and to avoid having to protect workers from dangerous chemicals.[37]

Resolving Disagreements about Purpose. Is there any way in which the notion of a justifying purpose can be clarified so that such disagreements can be resolved? One possibility is to specify the conditions necessary for a business to conduct normal operations. In order to do this, a company must be able to assess the suitability of applicants for employment, supervise their work-related behavior, administer fringe benefit plans, and so on. In addition, employers must be left free to acquire the information necessary for complying with legal requirements about taxes, social security, discrimination, health and safety, and the like. As a result, employers are justified in asking potential employees about their educational background, past employment, and so on, but not, for example, about their marital status, because this information is not necessary in order to make a decision about hiring. Once employees are hired, a company may have a need to inquire about marital status in order to determine eligibility for medical benefits, but only if the employee in question chooses to participate in a medical insurance plan. Even then, this information should be used only for the purpose of determining eligibility for medical benefits.

Joseph R. DesJardins suggests that questions about the extent of the right of privacy in the workplace can be settled by appealing to a contract model of the employer-employee relation.[38] Viewing employment as a contractual relation between an employer and an employee provides a basis for granting a set of rights to both parties, because the validity of contracts requires that certain conditions be satisfied. Contracts are valid, first, only if they are free of force and fraud. As a result, an employer has a right to require applicants to provide enough information to make an informed decision about hiring and to submit to tests for measuring relevant aptitudes and skills. Once hired, employees have an obligation to permit employers to monitor work performance, for example, and to gather whatever information is necessary to maintain an ongoing contractual relation.

Second, valid contracts also require mutual voluntary consent, so a contract model of employment would not permit employers to collect information without the knowledge and permission of the employees affected. Covert searches, surveillance by hidden cameras, the use of private investigators, and so on would be incompatible with the view of employment as a contractual relation. Similarly, objections could be raised to employer demands that employees either submit to drug tests and interrogation with a polygraph machine or be dismissed, because an employee has little choice but to comply. Union contracts in which employees are able to exercise effective choice often contain provisions prohibiting such practices.

Disclosure to Outsiders

The fourth issue—concerning the disclosure of personal information to persons outside a company—arises because of the practice, once very common, of employers sharing the content of personnel files with landlords, lending agencies, subsequent employers, and other inquiring persons without the consent of the employees involved. Even when there is a legitimate purpose that would justify these various parties having the information, it can be argued that an employer has no right to provide it, because the employer is justified in collecting and using information only for purposes connected with the employer-employee relation. What is morally objectionable about an employer's disclosing personal information to an outside party, in other words, is not necessarily that the outside party is not justified in having it but that the employer has no justification for giving it to out.

Thus, medical records collected by a former employer ought not to be passed along to a subsequent employer without the employee's consent. The former employer presumably had a purpose that justified the gathering of that information, and the new employer might also have a similar purpose in gathering the same information. The justification in the case of the former employer, however, is related to *that* employment relation, and the information gathered can be justifiably used only for purposes connected with it. The subsequent employer, although perhaps justified in gathering the same information, must proceed in the same way as the former employer.

This argument points up an important difference between personal information and other kinds of corporate records. Databases of various kinds are generally regarded as resources that are *owned* by a company. Ownership, however, generally entails an exclusive and unrestricted right of access and control, which employers do not have with respect to personal information. A mailing list, for example, is a kind of property that a company can use in any way it pleases, with no restrictions. Medical records, by contrast, can be compiled by a company only for a specific purpose, and any use unrelated to this purpose is prohibited. The fact that employers bear a burden of proof for justifying the collection and use of personal information shows that the notion of ownership is inappropriate in this case.[39]

It is also not appropriate to describe the information in personnel files as belonging to employees either, because they relinquish some rights to it by virtue of entering into the employment relation. Neither an employer nor an employee, therefore, can be said to own the information in a company's personnel files. Such information is simply not property in the usual sense, unlike other kinds of data gathered by corporations. It is necessary, therefore, to develop a conceptual model for personal information other than that of ownership.

The Means Used to Gather Information

Justifying the means used to gather information, which is the fifth issue, involves a different set of considerations. Use of certain means may violate an employee's right of privacy, even when the information gathered is of a kind that an employer is fully justified in possessing. Examples of impermissible means are polygraph testing and pretext interviews. (Pretext interviews are inquiries made under false pretenses, as when an employer seeks information from an applicant's family while posing as a market researcher.) Even if employers are justified in asking certain questions on a job application, they are not, for that reason, justified in using a polygraph machine or a pretext interview to verify the accuracy of a person's responses.

A major consideration in evaluating the means used to gather information is whether less intrusive means are available. In general, less intrusive means are morally preferable to those that are more intrusive. Employers are justified in seeking information about drug use by employees in the workplace, for example, but such means as searches of lockers and desks, hidden cameras in rest rooms, random drug tests, and the like are not justified when sufficient information could be gathered by less intrusive means, such as closer observation of work performance and testing only for cause. (Some means are not justified, of course, even if less intrusive means are not available. Hidden cameras and random drug tests are possible examples.)

What makes some means more intrusive than others depends on several factors. Such practices as conducting strip searches and watching while a urine sample is produced involve an affront to human dignity. An objection to constant monitoring, personality tests, and the use of polygraph machines is that they collect more information than is necessary and that they collect it indiscriminately. Honesty tests, for example, often inquire into personal habits and interests, family relations, and sexual adjustment—matters that are extraneous to the ostensible purpose.[40] Improperly administered polygraph tests can easily become "fishing expeditions," which result in the revelation of information that an employer is not justified in having.

Another reason why some practices such as monitoring and surveillance by hidden cameras and polygraph testing are unusually intrusive is that they deprive persons of an opportunity to exercise control over how they appear to others, which is essential for being an autonomous individual. An employee who is unaware of being observed, for example, might be unwittingly led to reveal facts that he or she would otherwise keep from others. George G. Brenkert argues, very perceptively, that because a polygraph machine measures physical characteristics such as breathing rate, perspiration, and blood pressure over which we have little or no control, it "circumvents the person" and undercuts the "way by which we define ourselves as autonomous persons."[41] As a person, we can shape how we appear to others and create an identity for ourselves. A machine that registers our involuntary responses denies us the power to do that.

Accuracy, Completeness, and Access

The last two issues are concerned primarily with matters of fairness. If the information in personnel files and other corporate databases is going to be used to make critical decisions about wage increases, promotions, discipline, and even termination of employment, then it is only fair that the information be as accurate and complete as possible and that employees have access to their personnel files so that they can challenge the contents or at least seek to protect themselves from adverse treatment based on the information in them.

Employers who maintain inaccurate or incomplete files and deny employees access to them are not invading the privacy of their employees, as the concept of privacy is commonly defined. What is at issue is not the possession of personal information by an employer but its use in ways that are unfair to employees. The right that employers violate is a right of fair treatment, which is not the same as a right of privacy. Still, because these issues are involved in the handling of personal information, they must be considered in devising policies or laws dealing with employee privacy.

Another objection to drug tests and polygraph machines is their unreliability. A number of factors, including the use of prescription drugs and careless laboratory work, can result in false positives, especially in simpler, less-expensive drug tests. Polygraph machines are inherently unreliable, because they register only bodily responses and not the mental experience that triggers them. An investigator might conclude that a subject is lying when the responses recorded by the machine are actually due to a different kind of association. One study, in which 14 polygraphers were asked to evaluate the charts of 207 criminal suspects, found that 50 percent of the experts thought that innocent suspects gave deceptive answers and 36 percent of them considered the guilty suspects to be telling the truth.[42] After a review of the studies to date, the U.S. Office of Technology Assessment concluded in 1983 that polygraph testing was useless for screening in preemployment contexts.[43]

In summary, determining the exact limits of the right of employees to privacy in the workplace requires that we address a number of issues. Questions about four of these issues—those concerning the kind of information collected, the use to which it is put, and the persons both inside and outside the company who have access to it—can be answered largely by appealing to the notion of a legitimate purpose. The issue of the means used to gain information involves different questions about whether some means are inherently objectionable and whether others are objectionable because less intrusive means are available. Finally, the remaining issues involve the fair treatment of employees, which is not, strictly speaking, part of a right of privacy but is still related to the handling of personal information.

PRIVACY ON THE INTERNET

Imagine that most of the stores you entered created a record of your visit including not only your purchases but also what merchandise you looked at, how long you took, what route you followed through the store, what other stores you had visited, and what you bought there. Imagine further that, in many instances, the store could connect this information with your name, address, telephone number, and perhaps you age, income level, and lifestyle. You would probably have the feeling that your shopping activity was being closely scrutinized and that you lacked virtually any privacy while browsing.

This situation, which most people would find alarming in a shopping mall, is routine on the Internet. A 1999 study found that 92.8 percent of the Web sites surveyed collected at least one piece of personal information, such as name and e-mail address, and 56.8 percent collected at least one type of demographic data, such as age, gender, or zip code. Only 6.6 percent of these sites collected no information.[44] Some information is provided *overtly* by the user as a condition of making a purchase or gaining access to Web pages. Other information is obtained *covertly*, without the user's knowledge or consent.

The most common method for obtaining information covertly is the installation of a "cookie," which is a file placed on a user's hard drive that recognizes a repeat user and stores information from past visits. Cookies benefit users by eliminating the need to enter information each time, but they can also provide the site owner with "clickstream" data about what pages are visited and how much time is spent on each one. Because cookies identify only a user's computer (by tagging it with a unique number), this tool is considered to preserve anonymity. However, personal information can be obtained by combining cookie data with larger databases that identify and profile individuals. Once users are identified, site owners can share the information derived from cookies to form more complete profiles in a process know as "cookie synchronization."

A Boston technology company called Pharmtrak places cookies on the computers of visitors to the health information pages of pharmaceutical companies and records the kind of information they seek.[45] If this information could be combined with individual names, then drugs could be marketed to Web users with specific ailments. To date, Pharmatrak has not taken this extra, morally questionable step, but another company, DoubleClick, aroused a storm of protest when it proposed something similar. DoubleClick, which places banner ads on 1,500 Web sites using profiles based on information from cookies, purchased Abacus Direct Corporation, which has personal information on more than 80 million households. By merging the two databases, DoubleClick could tailor the banner ads on a Web site to match the user's purchasing preferences. Stung by vociferous protests, the company announced that it would wait until government and industry agreed on standards.[46] However, one observer said that although the industry was letting DoubleClick take the heat, the company is "not doing anything that anyone else isn't doing."[47]

The explosive growth of the Internet as a consumer marketplace is a benefit to consumers and businesses alike. The success of Web sites depends crucially on the collection of information. One reason is that anonymous sales with cash are not possible. Furthermore, sites that offer free content depend on advertising, and advertising space is much more valuable if it can be tailored to individual users. However, the collection of information on the Internet appears to pose threats to users' privacy. So we need to ask, first, what is the danger? What harm, if any, is done by information collecting by Web sites? Are any rights violated? Second, given the need for government regulation or industry self-regulation, what standards should be applied, and what should be the goal in setting standards? Finally, by what means should these standards be implemented? Already a variety of organizations have been formed to offer resources for developing privacy policies and to certify compliance by awarding seals of approval.

What's Wrong with Information Collection?

Although consumers may feel a lack of privacy when browsing the Internet, is this feeling well-grounded? Scott McNealy, the Chairman and CEO of Sun Microsystems, once remarked, "You have zero privacy anyway. Get over it!"[48] Much of the information compiled is publicly available; the Internet only makes its compilation easier and cheaper than in the past. Store owners, if they wish, could follow consumers around to see what merchandise they examined. Computers do not observe us unawares or intrude into our private lives the way that psychological tests or hidden cameras do. The Internet is arguably a public arena, so being online is like walking and talking in the town square. The use to which the information is put is primarily to sell us something. Although fraud is a serious concern, we seek mainly to avoid the annoyance of advertising on the Internet.

If privacy is defined as control over personal information or the dissemination of personal information without our consent, then the practices of Internet companies violate our rights. However, this position assumes that we "own" the information about ourselves and thus have a right of control. Moreover, we give up a great deal of information in order to enjoy the benefits of Internet commerce, and so perhaps some loss of privacy is a trade-off that we voluntarily make. So, following Scott McNealy's advice, should we get over it?

The noted expert Lawrence Lessig in his important book *Code and Other Laws of Cyberspace* raises two problems that are unique to computers and the Internet.[49] One risk for Lessig is that our initial contacts with information gatherers form a profile of who we are, and this profile will fit us into a particular mold, which may not be accurate to begin with and may inhibit our ability to change and grow. Lessig writes, "The system watches what you do; it fits you into a pattern; the pattern is then fed back to you in the form of options set by the pattern; the options reinforce the pattern; the cycle begins anew."[50] If we develop by selecting from the options available to us, then the choice of options is critical. In life apart from the Internet we can always seek out new options, but to the extent that we are bound on the Internet by the options presented to us, our possibilities for growth are limited.

Lessig's second risk is that information collection by computers, especially on the Internet, could undermine the traditional American value of equality. The American Revolution was in part a rejection of European society in which innumerable distinctions of rank divided people. According to Lessig, "An efficient and effective system for monitoring makes it possible once again to make these subtle distinctions of rank. Collecting data cheaply and efficiently will take us back to the past."[51] For example, by means of frequent flyer programs, airlines identify their better customers and offer them special treatment. Companies with 800-numbers recognize the telephone numbers of favored customers and put them at the head of the queue. As a result, some people suffer a form of discrimination in which they do not get a flight on standby or endure long waits on the telephone. Businesses have always provided better service to select customers, but any discrimination was limited by the cost of the information. Lessig observes, "Whereas before there was relative equality because the information that enabled discrimination was too costly to acquire, now it pays to discriminate."[52]

Neither of these concerns involves privacy per se; the first affects autonomy, the second, equality. Moreover, the effect may be slight and insignificant. Perhaps Lessig's greatest insight is that the way in which technology is configured, specifically the way in which computer code is written, has social consequences. Furthermore, to the extent that computer code has consequences, it can be rewritten so as to achieve more desirable results. One of the solutions discussed later is Lessig's suggestion that we develop special software that will act as an electronic butler, negotiating with Web sites the kind of privacy protection that we desire.

Principles for Protecting Privacy

The appropriate standards or principles that should be applied to information gathering on the Internet depend on what we want to achieve or avoid. Three camps can be identified.[53] Those who worry about a "dossier" society, in which every facet of our lives is available to those with power, want strict limits on the kinds and amounts of data collected and on the availability of this data. Those who view personal data as a kind of property that can be "traded" in a market for certain benefits want to ensure that consumers do not trade too cheaply and that this valuable commodity is fairly priced. In the view of this camp, personal information is currently too "cheap" and hence is being overutilized.

The dominant camp consists of people in industry, government, and public interest groups who want to balance people's concerns about privacy—well-founded or not—with the growth of the Internet as a consumer marketplace. They seek to provide consumers with a voice in the development of this important commercial medium. The danger is that the Internet will firmly fix some practices before the public is aware of what is happening. Their goal is primarily to prevent the most egregious abuses by developing standards or principles that safeguard consumers.

In 1972, the Department of Health, Education, and Welfare developed guidelines for its own handling of information called "fair information practices," which form the basis for much subsequent action. In 1980, the Organization for Economic Cooperation and Development (OECD) adopted a set of guidelines that underpin most international agreements and self-regulatory policies of multinational corporations. The European Parliament adopted the European Union Privacy Directive, which took effect on October 25, 1998. This law binds not only member countries but also nonmember states doing business in the European Union. Already some major American multinational corporations are being investigated for violations of the EU Privacy Directive. Although many American laws address various aspects of Internet privacy, the United States has preferred a piecemeal legal response instead of adopting an omnibus piece of legislation like the EU Privacy Directive. The Federal Trade Commission, in particular, has attempted to protect Internet privacy using various consumer protection laws. Principles of privacy on the Internet have also been developed by industry associations, such as the Online Privacy Alliance (OPA), and public interest groups, most notably the Electronic Privacy Information Center (EPIC).

Despite this great diversity of sources, a remarkably similar set of standards has emerged. The Federal Trade Commission list of five principles is representative of the many documents on Internet privacy.

1. *Notice/Awareness.* Disclose the identity of the collecting party, the information collected, the means for collecting it, and the uses to which the information will be put. This notice usually consists of a privacy policy that should be prominently displayed and easily understood. Ideally, the home page and every page that asks for information should include a link to the policy. Notice should also be given if the privacy policy is not the same for all linked sites or if data will be shared with other parties with different policies.

2. *Choice/Consent.* Provide a mechanism for choosing whether to allow information to be collected. The mechanism may either require explicit consent (opt in) or assume consent if a person takes no action (opt out). One could choose to permit the collection of some information (name and address, for example) but not other

(such as medical information), or one could consent to some uses of information (to select banner ads, for example) but not others (such as providing information to a third party).

3. *Access/Participation.* Allow consumers access to the information collected about them and the opportunity to contest the accuracy or completeness of the data. The right of access may exclude information that a company collects from sources other than the Web site and any results from processing Web site data.

4. *Integrity/Security.* Inform users of the steps taken to protect against the alteration, misappropriation, or destruction of data and of the action that will be taken in the event of a breach of security. Also, maintain information so that it is accurate and up-to-date.

5. *Enforcement/Redress.* Assure consumers that the company follows responsible information practices and that there are consequences for failing to do so. Consumers should also have some means for resolving disputes and for receiving an appropriate remedy. One way to ensure enforcement and redress is by contracting with an organization that monitors and certifies the information practices of Web sites.

Although substantial agreement exists on these five principles, much depends on their interpretation and implementation. In particular, how stringently should the principles be interpreted, and what are the most effective and efficient means for implementing them? Other questions include the responsibility of Internet Service Providers (ISPs). For example, Yahoo was criticized for revealing to the Navy the identity of a sailor who used the pseudonym "Boysrch" in gay chatrooms. (The Navy used this information in a attempt to oust the sailor from the service for homosexuality.) The principles do not specify whether they apply to information that Web sites acquire from sources other than the Internet which are then aggregated with data obtained from users. The most contentious issues are whether the weaker opt-out provision is satisfactory in most instances and in what cases, if any, opt-in ought to be required. Finally, few proposals have been developed for handling enforcement and redress.

Implementing Internet Privacy

Principles are of little value if they cannot successfully be implemented, and the Internet presents unique challenges for implementation. Its decentralized, democratic structure makes centralized, authoritarian approaches ineffective, as does its global reach. Because the Web is worldwide, so too must be any successful regulatory scheme. Although government regulation, as represented by the EU Privacy Directive, creates a powerful incentive to protect privacy, laws must still grapple with the difficult question of the appropriate means. In considering the problem of protecting Internet privacy, we must ask, first, who should be involved and, second, what means should be used.

Obviously, the principal parties are Internet firms (Web sites and Internet service providers), computer companies (both hardware and software suppliers), industry associations, governments and regulatory agencies, public interest groups, and, of course, individual users. The main approach to date has focused on government regulation and self-regulation by the industry, designed in large part to prevent further intrusion by government. Self-regulation has largely taken the form of developing privacy policies and, in some instances, creating the post of Chief Privacy Officer (CPO) to direct company efforts. In this task, Web sites have been aided by public interest groups which offer resources and certification. Organizations, such as TRUSTe and BBBOnline (a service of the Council of Better Business Bureaus), monitor a firm's compliance with its privacy policy and award a seal that can be displayed on its Web site.

The most effective solution to a problem created by runaway technology might very well be more technology. We can protect privacy through both *formal* and *material* means.[54] Regulation and certification as described above utilize rules or norms that are designed to influence behavior. Such formal means can be supplemented with changes in material conditions that prevent certain kinds of behavior. Although we need laws against theft (formal), we also protect property with locks (material). The suggestion, then, is that we develop technology that will enable Internet users to protect their privacy to the extent they desire. To be effective, this technology must be usable by even the most unsophisticated in order to overcome the problem of the "blinking twelve" (which refers to the number of people who cannot even set the clock on a VCR).

A material solution consists in the development of various Privacy Enhancing Technologies (PETs). Among such means are services that permit "proxy surfing" by hiding the identity of the user's computer and remailers that forward e-mail stripped of any identifying markers. Cookie-management software exists that can block or disable cookies. Intel caused controversy by encoding a unique Processor Serial Number(PSN)in its Pentium III processor, but the company later offered software that would enable a user to suppress this number.[55] These PETs are likely to be used, however, only by very sophisticated users and so encounter the "blinking twelve" problem.

The most promising technology follows Lessig's suggestion of creating an electronic butler or a Cyber-Jeeves. This is a software program that would allow a user to answer a few questions about the desired features of a Web site's privacy policy and then determine whether sites to be visited fit the user's preferences. Such software is the goal of the Platform for Privacy Project (P3P), which is being conducted by the World Wide Web Consortium. If installed on most personal computers, a Cyber-Jeeves would force Web sites to adopt the privacy policies that the majority of Internet users desire.

Conclusion

Although privacy is a relatively recent concept—dating in American law to the 1890s—public concern is clearly increasing, primarily in response to privacy-invading technologies. The problems facing employees, consumers, and Internet users are similar, as are the solutions. There is greater agreement, however, on the ends than on the means, but even the ends are in dispute. Americans say that they value privacy, and yet they give up a great deal for convenience and material gain. Without question, the technologies that threaten privacy have brought us many benefits. Finding the right means is a great challenge to business firms which must meet employee and consumer expectations as they utilize new technologies. More than many business ethics problems, protecting privacy requires a coordinated solution involving many parties. Until a solution is found, though, the focus of businesses will remain on developing and implementing privacy policies.

Case 7.2   Three Challenges to Employee Privacy

Dating at Wal-Mart

The company's position is crystal-clear: "Wal-Mart strongly believes and supports the 'family unit.'"[56] For this reason, the handbook that all newly hired employees must read and sign stipulates, "A dating relation between a married associate with another associate, other than his or her own spouse, is not consistent with this belief and is prohibited." This policy was designed in part to maintain businesslike relations among employees and to avoid the turmoil that extramarital affairs can cause in the workplace. An added, and perhaps not unintended, benefit of the policy is to reinforce the wholesome, down-home image that Wal-Mart carefully fosters. The policy was challenged, however, when the manager of a Wal-Mart store in Johnstown, New York, discovered that two sales associates—one a 20-year-old single man and the other a 23-year-old woman who was married but separated from her husband—were dating. The romantic tie between the two came to the manager's attention when the woman was served with custody papers at the store. The manager immediately fired them both. The woman could not remember reading the no-dating rule in the handbook but expressed her disagreement nonetheless. "I felt it was my personal life," she said.

Is E-Mail Private?

A reporter in the Moscow bureau of the *Los Angeles Times* guessed the passwords of his colleagues and read their e-mail, thereby gaining an advantage in office politics. He was disciplined.[57] Two women who worked for Nissan, training dealers in the use of the company's e-mail system, were fired after their supervisor overrode the passwords and read exchanges with the dealers they were training. The messages contained some sexually suggestive banter along with comments by the women disparaging the supervisor's own sexual prowess.[58] Did the reporter or the supervisor do anything unethical? A letter sent to a colleague by U.S. mail would be considered private, as would (perhaps) a sealed envelope delivered through a company's own mail system. E-mail messages are accessible only with a password known to the user, so they are also like letters stored in a locked desk drawer. The desk may belong to the company but not the contents. On the other hand, e-mail messages are also like notes posted on a company-owned bulletin board, because the technology is provided to the employee. Employers claim a right to monitor employee's messages to ensure that the system is not being misused. Company computers have been used to arrange drug deals, to operate office betting pools, and to engage in racial and sexual harassment. Some privacy experts contend that employers have a right to read employees' e-mail if they announce the policy in advance; others believe that reading electronic messages is always wrong, even if employees are put on notice.

Video Surveillance

In September 1997, an employee at the Consolidated Freightways truck terminal in Riverside County, California, noticed that the mirror above a sink in the men's rest room was askew.[59] When he went to adjust it, he came across a hidden video camera. A check by the local sheriff's office, which was called to investigate, uncovered hidden cameras in two of nine rest rooms and several recorded videotapes. A company representative explained that the cameras were aimed only at areas where drug dealing was suspected of taking place and that they were focused "nowhere near the urinals or the stall area." California state law forbids surveillance in areas where people have a "reasonable expectation" of privacy, and many of the 600 employees in the facility were outraged by the company's tactics. The company argued that it has a responsibility to protect the public, as well as its own employees and customers' freight, against the dangers of drug use by drivers. A suspected love tryst was the reason given for the installation of a video camera in a women's locker room at an Amoco chemical laboratory in Illinois.[60] In response to complaints that a male supervisor and a female worker were leaving their workstations to be together in the locker room, Amoco managers installed a hidden camera in the ceiling, trained on the door. After the existence of the camera was discovered, eight female employees, who would have been observed entering and leaving the locker room to change clothes, filed a lawsuit charging an invasion of privacy. The locker room was also used occasionally by visitors who needed to change clothes, and a female electrician, who was not an employee but who worked at the laboratory during the period of the surveillance, joined the women's suit.

Case 7.3   Ford Meter Box

In 1989, the Ford Meter Box Company adopted a policy of hiring only nonsmokers. Smoking had already been banned on the job by the Wabash, Indiana, manufacturer, but employees hired before the date of the policy's adoption were permitted to continue smoking away from the plant on their own time. However, newly hired employees were told that they could not smoke at all—anyplace, anytime. Janice Bone, a part-time clerk and a smoker, refrained from smoking at work and was unaffected by the new hiring policy . . . until she applied for a vacant full-time position. The job change would make her a "new hire," subject to the policy. To get the job she would have to give up smoking entirely. Janice Bone agreed to quit and got the job, but a drug test, routinely administered to all new employees after six weeks of service, revealed the presence of nicotine. Subsequently, Ford Meter Box terminated her employment.

Ford Meter Box is not alone in adopting a policy to hire only nonsmokers. An estimated 20 percent of employers give some preference to nonsmoking applicants,[61] and many provide some inducements for employees to stop smoking or impose penalties on those who continue to smoke. Turner Broadcasting is perhaps the best-known company that refuses to hire smokers, although previously hired smokers are permitted to smoke at work in designated areas. Employers' interest in their employees' smoking habits stem largely from the extra cost that smoking imposes. A study by Texas Instruments, Inc., found that the cost of a smoker's health care was 50 percent greater than that of a nonsmoker,[62] and a study by Du Pont in the late 1980s showed that smokers cost the company an additional $960 each in medical bills and days lost from work. Instead of forbidding employees to smoke, however, Texas Instruments charges employees an extra $10 a month in their health insurance premiums for each family member, including the employee, who smokes.

Critics of smoking bans and extra charges worry that employers will soon turn their attention to other unhealthy lifestyles and attempt to regulate employees' weight or alcohol consumption and dangerous hobbies, such as mountain climbing or bungee jumping. Employers' options are being limited, however, by state laws that protect employees in their pursuit of legal, off-duty activities. The new laws have been promoted by an unusual alliance between the tobacco industry lobby and the American Civil Liberties Union, each of which claims to be protecting employees' privacy and "right to choose."

Case 7.4   Lotus MarketPlace: Households

In April 1990, the Lotus Development Corporation announced plans for a revolutionary new product.[63] MarketPlace:Households would provide information about 80 million American households which would enable businesses to generate mailing lists for direct marketing. The product was developed jointly with Equifax, Inc., which collects credit information on consumers and provides credit reports and customized mailing lists to client businesses. The distinguishing feature of MarketPlace:Households was that businesses would have information that was formerly contained only in the Equifax mainframe database in a set of compact discs for use on a personal computer. For $695, any business could obtain CD-ROMs with consumers' names, addresses, and purchasing behavior and generate a 5,000-name mailing list with the desired demographic profile using special sorting software. Once a company owned the CD-ROMs and the software, additional lists could be generated at a cost of $400 for each list of 5,000 names.

The announcement of MarketPlace:Households was greeted with loud complaints by groups concerned with consumer privacy. The American Civil Liberties Union charged that the product blurred the line between information that is provided by consumers for obtaining credit and the use of that information for direct marketing. Many privacy advocates contend that people have a right to control the kind of information that they supply to obtain a mortgage or a credit card and that such information should not be used without their consent. When Lotus announced that consumers could have their names deleted from the CD-ROMs, the company was flooded with 30,000 requests. Critics of Lotus were also concerned that the product could be used by unscrupulous businesspersons to fleece consumers with bogus offers. Others raised questions about possible problems due to inaccurate data. Word of MarketPlace:Households spread on computer bulletin boards along with the e-mail address of Lotus CEO Jim Manzi. One irate message to Mr. Manzi read: "If you market this product, it is my sincere hope that you are sued by every person for whom your data is false, with the eventual result that your company goes bankrupt."

Lotus considered concerns about privacy in the development of MarketPlace:Households and engaged a respected privacy expert, Alan F. Westin of Columbia University, to advise the company. All personal information on the CD-ROMs was designed to be inaccessible to the user; only mailing labels in minimum lists of 5,000 names could be generated by the owner of the product. Lotus created a procedure for allowing individuals to request that their names be omitted from the compact disks, although critics pointed out that the names would still be included in CD-ROMs already sold. In addition, MarketPlace:Households would be available only to reputable businesses; the mailing lists generated by the product would be seeded with decoy names so that Lotus would receive copies of items mailed to consumers; and Lotus and Equifax were prepared to take legal action for misuse of the lists. Lotus maintained that MarketPlace:Households would merely provide small businesses with the same resources as large corporations that already buy mailing lists from companies like Equifax. Mr. Westin observed: "The aborigines used to believe that to take a picture of a man was to steal his soul. But the bottom line with Lotus Market-Place is that you'll get a few extra pieces of mail."

NOTES

1. *Soroka* v. *Dayton Hudson,* 1 Cal. Rptr. 2d 77 (Cal. App. 1 Dist. 1991).

2. Frank H. Williams, "Selecting Best Type of Salespeople Easy with This Set of Test Questions," *Dry Goods Economist,* 21 April 1923, 15.

3. For information on the Sociological Department, see Robert Lacey, *Ford: The Men and the Machine* (Boston: Little, Brown, 1986), 117–25.

4. See Al Noel, "Privacy: A Sign of Our Times," *Personnel Administrator,* 26 (March 1981), 59–62.

5. For a discussion of this problem, see Karl J. Duff and Eric T. Johnson, "A Renewed Employee Right to Privacy," *Labor Law Journal,* 34 (1983), 747–62.

6. Alan F. Westin, "The Problem of Privacy Still Troubles Management," *Fortune*, 4 June 1979, 120–26. See also Joyce Asher Gildea, "Safety and Privacy: Are They Compatible?" *Personnel Administrator*, 27 (February 1982), 80–83.

7. These new technologies are described in Gary T. Marx and Sanford Sherizen, "Monitoring on the Job: How to Protect Privacy as Well as Property," *Technology Review*, 89 (November–December 1986), 63–72. See also Gary T. Marx, "The New Surveillance," *Technology Review*, 88 (May–June 1985), 42–48.

8. See John C. North, "The Responsibility of Employers for the Actions of Their Employees: Negligent Hiring Theory of Liability," *Chicago-Kent Law Review*, 53 (1977), 717–30; and Marian M. Extejt and William N. Bockanic, "Theories of Negligent Hiring and Failure to Fire," *Business and Professional Ethics Journal*, 8 (Winter 1989), 21–34.

9. John A. Baker, "An Issue of Consumer Privacy ... and Corporate Retreat," *New York Times*, 31 March 1991, sec. 3, p. 9.

10. Sally Beatty, "Reader's Digest Targets Patients by Their Ailments," *Wall Street Journal*, 17 April 1998, B1, B3.

11. Michael W. Miller, "Patients' Records Are Treasure Trove for Budding Industry," *Wall Street Journal*, 27 February 1992, A1, A4.

12. Samuel Warren and Louis D. Brandeis, "The Right to Privacy," *Harvard Law Review*, 4 (1890), 193–220. Two earlier discussions are in James Fitzjames Stephen, *Liberty, Equality, and Fraternity* (New York: Henry Holt, 1873), and E. L. Godkin, "Rights of the Citizen, Part IV—To His Own Reputation," *Scribner's Magazine*, 8 (1890), 58–67.

13. The case is *Robertson* v. *Rochester Folding Box Co.*, 171 N.Y. 538 (1902).

14. *Griswold* v. *Connecticut*, 381 U.S. 479 (1965).

15. William L. Prosser, "Privacy," *California Law Review*, 48 (1960), 389. Other writers who argue that privacy is a complex of rights are Frederick Davis, "What Do We Mean by 'Right to Privacy'?" *South Dakota Law Review*, 4 (1959), 1–24; and Judith J. Thomson, "The Right to Privacy," *Philosophy and Public Affairs*, 4 (1975), 295–315. For criticism of Prosser, see Edward J. Bloustein, "Privacy as an Aspect of Human Dignity: An Answer to Dean Prosser," *New York University Law Review*, 39 (1964), 962–1007.

16. *Olmstead* v. *United States*, 277 U.S. 438 (1928).

17. *Eisenstadt* v. *Baird*, 405 U.S. 438 (1972).

18. See W. A. Parent, "Privacy, Morality, and the Law," *Philosophy and Public Affairs*, 12 (1983), 269–88; H. J. McCloskey, "Privacy and the Right to Privacy," *Philosophy*, 55 (1980), 17–38; and Joseph R. DesJardins, "Privacy in Employment," in *Moral Rights in the Workplace*, ed. Gertrude Ezorsky (Albany: State University of New York Press, 1987), 127–39.

19. See Charles F. Fried, *An Anatomy of Values* (Cambridge, MA: Harvard University Press, 1970), 141; Richard A. Wasserstrom, "Privacy," in *Today's Moral Problems*, 2d ed., ed. Richard A. Wasserstrom (New York: Macmillan, 1979), 393; Elizabeth L. Beardsley, "Privacy: Autonomy and Selective Disclosure," in *Privacy*, ed. J. Roland Pennock and John W. Chapman, (New York: Atherton Press, 1971), 65; and Arthur Miller, *The Assault on Privacy* (Ann Arbor: University of Michigan Press, 1971), 25.

20. Alan F. Westin, *Privacy and Freedom* (New York: Atheneum, 1967), 7.

21. Richard B. Parker, "A Definition of Privacy," *Rutgers Law Review*, 27 (1974), 279.

22. This point is made by Parent, "Privacy, Morality, and the Law," 273.

23. Ibid., 269.

24. Ibid., 269–70.

25. Both of these cases are contained in Marx and Sherizen, "Monitoring on the Job," 67.

26. Westin, *Privacy and Freedom*, 31–42.

27. For criticisms of these arguments, see McClosky, "Privacy and the Right to Privacy," 34–35; and Parent, "Privacy, Morality, and the Law," 275–76.

28. Stanley I. Benn, "Privacy, Freedom, and Respect for Persons," *Privacy*, 10–11.

29. Hyman Gross, "Privacy and Autonomy," in *Privacy*, 174.

30. Fried, *Anatomy of Values*, 9. A similar account is given in James Rachels, "Why Privacy Is Important," *Philosophy and Public Affairs*, 4 (Summer 1975), 295–333.

31. Fried, *Anatomy of Values*, 148.

32. See Jeffrey H. Reiman, "Privacy, Intimacy, and Personhood," *Philosophy and Public Affairs*, 6 (1976), 26–44.

33. Parent, "Privacy, Morality, and the Law," 275. A similar point is expressed in Ibid., Reiman, 33.

34. Arguments of this kind are presented in Wasserstrom, "Privacy," and Reiman, "Privacy, Intimacy, and Personhood."

35. Reiman, "Privacy, Intimacy, and Personhood," 39.

36. This point is made by Parent, "Privacy, Morality, and the Law," 280. Much of the argument in this section is derived from Parent's analysis of wrongful invasion of privacy.

37. See Thomas H. Murray, "Thinking the Unthinkable about Genetic Screening," *Across the Board*, 20 (June 1983), 34–39.

38. DesJardins, "Privacy in Employment."

39. This point is made by Donald Harris, "A Matter of Privacy," *Personnel*, 64 (February 1987), 38.

40. See Anne E. Libbin, Susan R. Mendelssohn, and Dennis P. Duffy, "The Right to Privacy, Part 5: Employee Medical and Honesty Testing," *Personnel*, 65 (November 1988), 47.

41. George G. Brenkert, "Privacy, Polygraphs and Work," *Business and Professional Ethics Journal*, 1 (1981), 30.

42. David T. Lykken, "Polygraphic Interrogation," *Nature*, 23 (February 1984), 681.

43. "Scientific Validity of Polygraph Testing: A Research Review and Evaluation," Technical Memorandum OTA-TM-H-15 (Washington, DC: Office of Technology Assessment, November 1983).

44. Mary J. Culnan, "Georgetown Internet Privacy Policy Survey: Report to the Federal Trade Commission," June 1999.

45. Marcia Stepanek, "Surf at Your Own Risk," *Business Week*, 30 October 2000, 143.

46. Hiawatha Bray, "DoubleClick Backs off on Net Data," *Boston Globe*, 3 March 2000, C1.

47. Matt Richtel, "Yahoo Says It Is Discussing Internet Privacy with the F.T.C.," *New York Times*, 31 March 2000, C5.

48. S. Sprenger, "Sun on Privacy: Get Over It'," *Wired News*, 26 January 1999.

49. Lawrence Lessig, *Code and Other Laws of Cyberspace* (New York: Basic Books, 1999).

50. Ibid., 154.

51. Ibid., 155.

52. Ibid., 156.

53. These camps are described and analyzed in Karl D. Belgum, "Who Leads at Half-time? Three Conflicting Views on Internet Privacy," *Richmond Journal of Law & Technology*, 6 (1999), http://www.richmond.edu/~jolt/.

54. Shawn C. Helms, "Translating Privacy Values with Technology," *Boston University Journal of Science and Technology Law*, 7 (2001), 156.

55. Declan McCullagh, "Intel Nixes Chip-Tracking ID," *Wired News*, 27 April 2000.

56. Material for this case is taken from Jacques Steinberg, "At Wal-Mart, Workers Who Dated Lose Jobs," *New York Times*, 14 July 1993, A16.

57. Calvin Sims, "Reporter Disciplined for Reading His Co-workers' Electronic Mail," *New York Times*, 6 December 1993, B9.

58. Glenn Rifkin, "Do Employees Have a Right to Electronic Privacy?" *New York Times*, 8 December 1991, sec. 3, p. 8.

59. Stuart Silverstein, "Employee Finds Hidden Camera in Bathroom," *Los Angeles Times*, 12 September 1997, D1.

60. *Brazinski* v. *Amoco Petroleum Additives Co.*, 6 F.3d 1176 (7th Cir. 1993).

61. "Company Policies Could Give Smokers Their Day in Court," *HRfocus*, 69 (August 1991), 1.

62. "If You Light Up on Sunday, Don't Come in on Monday," *Business Week*, 26 August 1991, 70.

63. Material for this case is taken from John R. Wilke, "Lotus Product Spurs Fears About Privacy," *Wall Street Journal*, 13 November 1990, B1; Alan Radding, "Lotus Stirs Research Marketplace," *Advertising Age*, 14 January 1991, 21; Michael W. Miller, "Lotus Likely to Abandon Consumer-Data Project," *Wall Street Journal*, 23 January 1991, B1; and Peter H. Lewis, "Why the Privacy Issue Will Never Go Away," *New York Times*, 7 April 1991, F4.

8

Discrimination and Affirmative Action

Case 8.1 Discrimination at Texaco

On November 4, 1996, the *New York Times* disclosed the contents of a secretly recorded conversation in which three senior Texaco executives discussed plans to destroy documents that were being sought in a class action lawsuit for racial discrimination.[1] Also on the tape were derogatory comments about black employees and ridicule of the company's efforts at diversity. One executive was heard to say, "This diversity thing. You know how all the jelly beans agree." Another said, "That's funny. All the black jelly beans seemed to be glued to the bottom of the bag." To this the first executive responded, "You can't just have black jelly beans and other jelly beans. It doesn't work."

The public reaction was fast and furious. Irate Texaco customers threatened to cut up their credit cards. The trustee of a large pension fund with 1.3 million Texaco shares wrote that the taped conversation suggests "a corporate climate of disrespect." Civil rights leaders, including the Reverend Jesse Jackson, threatened a nationwide boycott of Texaco service stations. Three days after the *New York Times* article appeared, Texaco CEO Peter I. Bijour issued an apology, saying, "The statements on the tapes arouse a deep sense of shock and anger among all the members of the Texaco family and decent people everywhere." Less than ten days later, Texaco abruptly settled the discrimination lawsuit, which had dragged on for two and one-half years. The company agreed to pay $141 million in compensation to 1,350 black employees and to spend another $35 million for improvements in the diversity program at Texaco.

The lawsuit was filed in March 1994 by six employees on behalf of 1,500 salaried African Americans employed by Texaco. Although these employees reported numerous examples of racist incidents, the suit focused on a pervasive pattern of discrimination in promotion and pay. The petroleum industry had never fully shed its "good old boy" culture that prevented the advancement of women and racial minorities, but an industrywide survey conducted annually showed that Texaco lagged behind every other major oil company. As a percentage of employees in each salary bracket in the survey, blacks at Texaco trailed the competition in every one, and the percentage of blacks declined more sharply than at other companies as the salary brackets increased. Of employees earning between $51,100 and $56,900 in 1993, 5.9 percent were black at Texaco versus 7.2 percent at the other major oil companies. In the highest bracket, above $128,800, only 0.4 percent of the income earners were black compared to 1.8 percent elsewhere. On average, blacks in each job category were paid 10 to 15 percent less than their white counterparts. Promotions were slower in each job category. A study by the U.S. Department of Labor in 1995 found that it took 6.1 years for minority employees to rise to the position of accountant and 4.6 years for employees at the other major oil companies to achieve the same position. Whites who were promoted to assistant accounting supervisor at Texaco took an average of 9.8 years, but blacks in that position had waited 15.0 years for promotion.

The plaintiffs decided to file the suit, *Roberts* v. *Texaco*, after they discovered striking similarities in the tactics that had been used to prevent their advancement. The lead plaintiff in the suit, Bari-Ellen Roberts, was a pension analyst who had been wooed from Chase Manhattan Bank where she supervised the Texaco pension account as a vice president. At Texaco, she quickly discovered that she had been hired mainly to improve the racial percentages. She once had a superior evaluation reduced to unsatisfactory because a higher executive had found her "uppity" for openly disagreeing in a meeting. When the position above her in the pension department became open, a white male with no experience in pensions was brought in with the explanation that "Bari will help train him." Another plaintiff complained that he was assigned less capable staff members, whose poor performance reduced his own evaluation. One member of his staff, a white male, was allowed to report directly to the plaintiff's superior to avoid reporting to a black.

Widespread discrimination flourished at Texaco despite an explicit company policy and an established diversity program. The booklet "Texaco's Vision and Values" states "Each person deserves to be treated with respect and dignity in appropriate work environments, without regard to race, religion, sex, age, national origin, disability or position in the company." The company had an affirmative action plan that set diversity goals and provided for diversity training. In fact, the idea of different-colored jelly beans was taken from a diversity training session attended by the executives in the taped conversation.

The problem, according to the observers of Texaco's culture, was the lack of high-level oversight. Implementation of the diversity program was left to middle- and low-level managers, with little guidance from above. Complaints of racist treatment were generally dismissed, and seldom was any action taken against the offenders. Texaco conducted no audits to measure the success of its own affirmative action plan, nor any studies to determine whether its personnel practices discriminated against women or racial minorities. The results of government investigations seldom reached top executives. Promotion was heavily determined by a secret list of "high-potential" employees, which was not formally scrutinized for its possible discriminatory effects. Indeed, no official criteria existed for the inclusion of people on the list—or their removal. (Bari-Ellen Roberts inadvertently discovered after her lowered evaluation that her name had also been removed from the high-potential list.)

The settlement ended Texaco's legal woes, but the task of changing the corporate climate remained. How should the company spend the $35 million that was committed to improving diversity? CEO Peter Bijour denied that the programs that Texaco had in place "were flawed in any way." The solution, in his view, was to expand and improve the initiatives already underway. These efforts included higher goals (but not quotas) for the percentages of black employees, more diversity training, greater emphasis on mentoring and career counseling, an increase in the use of minority suppliers, advertising in black publications, and support for black causes.

Critics contend that Texaco had incorrectly diagnosed the cause of the problems and consequently had failed to devise effective solutions. Some employees believe that the environment at the company is still racist and that such remedies as diversity training and mentoring simply increase the racial friction. Some other efforts are more show than substance. More attention should be given, they say, to punishing offenders and tying more of managers' bonuses to the achievement of goals. Determining who is right on these issues is of great importance for addressing the problems of discrimination in the workplace. In an era of setbacks for affirmative action, Texaco's actions will be closely watched by corporate America and by civil rights groups to gain some vision of the future.

INTRODUCTION

Many racial and ethnic groups have been subject to discrimination, and the treatment of women by American business constitutes another prominent form of discrimination. Discrimination is not simply a matter of the number of blacks, women, and other groups who are hired by an employer. In the early 1970s, for example, more than one-half of the employees of American Telephone & Telegraph Company (AT&T) were women, and racial minorities constituted over 10 percent of the AT&T workforce.[2] Women employees were largely concentrated in low-paying clerical and telephone operator jobs, however, and blacks, Hispanics, and members of other racial minorities were employed chiefly in unskilled job categories, such as maintenance workers and janitors. As a result, AT&T was charged with discriminating against women and racial minorities by using sex and race as factors in making job assignments. Eventually, the company agreed to increase the representation of these groups in job categories from which they had previously been excluded.

This chapter is concerned primarily with the steps that can be taken to prevent discrimination and rectify past wrongs. Some of these measures, such as nondiscriminatory hiring and promotion procedures, are widely implemented in the American workplace, but others, especially affirmative action, remain controversial. In order to address the ethical issues in discrimination and affirmative action, it is useful to begin with a definition of discrimination in its many forms and a discussion of the ethical arguments against it.

WHAT IS DISCRIMINATION?

The term *discrimination* describes a large number of wrongful acts in employment, housing, education, medical care, and other important areas of public life. Although discrimination in each of these areas takes different forms, what they have in common is that a person is deprived of some benefit or opportunity because of membership in some group toward which there is substantial prejudice. Discrimination in employment, which is our concern here, generally arises from the decisions employers make about hiring, promotion, pay, fringe benefits, and the other terms and conditions of employment that directly affect the economic interests of employees. There is nothing unjust about such decisions as long as they are made for reasons that are reasonably job-related, but singling out a person for adverse treatment merely because of that person's race or sex is generally an act of discrimination.

Although discrimination is a form of unequal treatment, not all unequal treatment is discrimination. An employer who shows favoritism in deciding on promotions, for example, is guilty of violating the principle of equality in dealing with employees but not necessarily of discriminating against them. Two further elements are necessary. First, discrimination involves decisions that directly affect the employment status of individuals or the terms and conditions of their employment; that is, discrimination occurs in what are generally regarded as *personnel* decisions, such as those involving hiring and firing, promotion, pay, advancement opportunities, and the like. Second, the unequal treatment results from prejudice or some other morally unjustified attitude against members of the group to which an individual belongs. In cases of discrimination, individuals are not treated on the basis of individual merit but on the basis of membership in a group.

The 1964 Civil Rights Act

These two elements can be observed in Title VII of the 1964 Civil Rights Act. Section 703(a) reads as follows:

> It shall be an unlawful employment practice for an employer—(1) to fail or refuse to hire or to discharge any individual, or otherwise to discriminate against any individual with respect to his compensation, terms, conditions, or privileges of employment, because of such individual's race, color, religion, sex, or national origin; or (2) to limit, segregate, or classify his employees or applicants for employment in any way which would deprive or tend to deprive any individual of employment opportunities or otherwise adversely affect his status as an employee, because of such individual's race, color, religion, sex, or national origin.[3]

Notice that Title VII first describes the kinds of employment decisions that are governed by the statute and then lists five factors—race, color, religion, sex, and national origin—that employers are not legally permitted to take into consideration. These factors define groups that are called in law *protected classes*. In subsequent legislation, Congress extended the list of protected classes in order to prevent discrimination against older people (Age Discrimination in Employment Act of 1967), the handicapped (Rehabilitation Act of 1973 and the Americans with Disabilities Act of 1990), and pregnant women (Pregnancy Discrimination Act of 1978).

The prejudice involved in discrimination is of several different kinds. One kind of prejudice behind much racial and ethnic discrimination consists of strong feelings of antipathy and intolerance. Some prejudice is based not on strong feelings but on misunderstandings, such as stereotypes about women, older workers, and the handicapped. Finally, some prejudice is due primarily to economic considerations. Employers are sometimes deterred from hiring pregnant women, older workers, and the handicapped merely because of the higher cost.

Is Intent Necessary? Among the questions to be addressed in defining discrimination is whether there must be an *intent* to discriminate. A charge of discrimination requires that the adverse treatment of a person be related in some way to membership in a certain group, but what is the nature of this relation? Section 703(a) of Title VII states that it is illegal for an employer to make decisions about hiring, discharge, compensation, and so on because of an individual's race, color, sex, religion, or national origin. The crucial phrase "because of" is not explained in the law, however.

One kind of discrimination occurs when there is an express intent to treat the members of certain groups differently. Thus, an employer who refuses to receive applications from blacks or has a policy of hiring blacks only for certain jobs would be guilty of discrimination of this kind. It would be discriminatory in a similar way for an employer to seek a "competent, experienced woman" for a secretarial position or a "recent college graduate" for an opening in sales. Such wording implies that sex and age will be considered in decisions about hiring.

In some cases, employers defend themselves against charges of discrimination by arguing that race, sex, or some other characteristic is relevant to the job. Although admitting an intent to treat their employees and applicants differently on the basis of some characteristic, these employers maintain that they are motivated not by racism or sexism or some other kind of prejudice but purely by business considerations. Some situations of this kind are relatively uncontroversial. Obviously, Catholic priests and Jewish rabbis must be of the religion that they serve. A woman's college could justifiably have a preference for a female president.

Exceptions to the Law. Section 703(e) of Title VII allows exceptions for sex, religion, and national origin when these are a "bona fide occupational qualification [BFOQ] reasonably necessary to the normal operation of that particular business or enterprise." Race and color are not included in Section 703(e) as a BFOQ and thus cannot be used legally to make distinctions for purposes of employment. The courts have interpreted the BFOQ exception very narrowly, so that employers must show that the qualification is absolutely essential for the conduct of business and not merely useful. Airlines are legally permitted to force pilots to stop flying at the age of 60 for safety reasons, but a federal appeals court rejected an airline's argument against male flight attendants on the grounds that the job of reassuring anxious passengers and giving courteous personalized service not only could be performed by men but were also peripheral to Pan Am's main business of transporting passengers safely.[4]

One of the few cases in which the Supreme Court has held sex to be a BFOQ concerned a rule adopted by the Alabama Board of Corrections excluding women from positions in a maximum-security male prison requiring close contact with the inmates.[5] The majority opinion argued, first, that women employees are likely to be the victims of sexual assaults in a prison characterized by "rampant violence" and a "jungle atmosphere." Second, the likelihood of sexual assaults would reduce the ability of a woman to maintain order in the prison, which is the main function of a prison employee. The presence of women in a male prison poses a threat, therefore, "not only to the victim of the assault but also to the basic control of the penitentiary and protection of its inmates and other security personnel."

Several members of the Court disagreed. It is not clear, they said, that women are exposed to significantly greater risk of attack than men who are employed in the same positions at the prison. Even if a job is more dangerous for some individuals than others, it is less discriminatory to allow individuals to decide voluntarily the degree of risk to assume rather than bar those at greater risk. Also, Justice Thurgood Marshall observed, "It is women who are made to pay the price in lost job opportunities for the threat of depraved conduct by male prison inmates." A better solution, perhaps, would be to make the workplace safer for women instead of limiting their employment opportunities because of the threatened conduct of others.

Disparate Treatment and Disparate Impact

Employment policies that do not explicitly involve classifying employees by race, sex, or other impermissible characteristics can still serve to exclude members of these groups in disproportionate numbers. In interpreting Title VII, the courts have generally held that employers are guilty of discrimination in the absence of any intent *when the effects are the same as if there had been an intent to discriminate*. Discrimination is thus not solely a matter of intention but also of consequences. A distinction is made in law between *disparate treatment,* which is discrimination of the first kind, involving an express intention, and *disparate impact,* which is discrimination of the second kind.

Griggs v. Duke Power. A landmark case in discrimination law that illustrates the distinction between disparate treatment and disparate impact is *Griggs* v. *Duke Power Company.*[6] Before the passage of Title VII of the Civil Rights Act of 1964, Duke Power Company openly practiced discrimination against blacks. At the Dan River Plant in Draper, North Carolina, blacks were employed only in the labor department, the lowest paying of the five operating divisions. In order to comply with Title VII, the company revised its hiring and promotion policies in 1965 so as to eliminate distinctions between blacks and whites. All applicants for jobs in any department except labor were now required to have a high school diploma and pass two standardized tests, the Wonderlic Personnel Test, which is designed to measure general intelligence, and the Bennett Mechanical Comprehension Test.

Thirteen black employees in the labor department brought suit against Duke Power Company, contending that the education and test requirements were discriminatory for two reasons. First, according to the 1960 census, 34 percent of white males in North Carolina had graduated from high school compared with only 12 percent for black males. The requirement of a high school diploma, therefore, served to exclude black applicants in proportionately greater numbers than white applicants. Second, the passing scores on the two standardized tests were set by the company at the national median of high school graduates, with the result that 58 percent of whites taking the test passed, whereas only 6 percent of the blacks succeeded in doing so. Again, a requirement imposed by the company had a disproportionate impact on blacks applying for employment.

The difference between blacks and whites in the percentages graduating from high school and the performance of the two groups on the standardized tests is largely attributable to the segregated school system in the state. Thus, the requirements, although ostensibly colorblind, served to perpetuate the effects of discrimination in schooling. Chief Justice Warren Burger asserted in the majority opinion that "practices, procedures, and tests neutral on their face, and even neutral in terms of intent, cannot be maintained if they operate to 'freeze' the status quo of prior discriminatory employment practices." Duke Power Company responded to the charge of discrimination by holding that Title VII does not require employers to treat workers without regard for qualifications. The requirement of a minimal educational attainment is reasonable, and intelligence and aptitude tests are specifically sanctioned by Section 703(h) of the Civil Rights Act. This section authorizes the use of "any professionally developed ability test" that is not "designed, intended, or used to discriminate because of race."

The Court Decision. The position of the Supreme Court was that neither requirement had been shown by the company to be related to successful job performance. According to the majority opinion:

> On the record before us, neither the high school completion requirement nor the general intelligence test is shown to bear a demonstrable relationship to successful performance of the jobs for which it was used.... The evidence, however, shows that employees who have not completed high school or taken the tests have continued to perform satisfactorily and make progress in departments for which the high school and test criteria are now used. The promotion record of present employees who would not be able to meet the new criteria thus suggests the possibility that the requirements may not be needed even for the limited purpose of preserving the avowed policy of advancement within the company....

The decision in *Griggs* v. *Duke Power Company* interprets Title VII as prohibiting employment practices that involve no intent to discriminate (disparate treatment) but still operate to exclude members of protected classes unnecessarily (disparate impact). Companies are free to hire and promote workers on the basis of defensible requirements. But in the words of the Court: "The touchstone is business necessity. If an employment practice which operates to exclude Negroes cannot be shown to be related to job performance, the practice is prohibited." Moreover, the burden of proof in showing business necessity rests on the employer.[7]

The Forms of Discrimination

A definition cannot answer all the difficult questions about discrimination in the workplace, and some further clarification is necessary for understanding each form of discrimination.

Discrimination on the Basis of Sex. In the interpretation of Title VII, sex discrimination is discrimination based on the fact that a person is male or female and not on sex-related matters, such as sexual orientation or marital status. Although some local governments have passed laws barring discrimination in employment and other matters against homosexuals, discrimination of this kind is not covered by the Civil Rights Act of 1964. Employers are permitted by Title VII to treat married and single employees differently as long as no distinction is made between men and women. An employer can give a preference in hiring to married applicants, for example, but it would be discriminatory to prefer married men and single women in filling jobs. Because of uncertainty over the legality of discrimination against pregnant women, Congress passed the Pregnancy Discrimination Act of 1978, which amends the phrase in Title VII "because of sex" to include decisions made on the basis of "pregnancy, childbirth, or related medical conditions." Sexual harassment has also been ruled by the courts to constitute a form of sex discrimination. These matters are treated in the following chapter.

Religious Discrimination. Religious discrimination is substantially different from discrimination based on race or sex. There are instances, to be sure, of religious discrimination in which employers refuse to hire or promote individuals simply because of prejudice against members of certain religious groups, such as Catholics and Jews. Most charges of religious discrimination in employment, however, involve conflicts between the religious beliefs and practices of employees and workplace rules and routines. Employees sometimes request revised work schedules for Sabbath observance or time off to observe religious holidays. Members of some religious groups have special dress or grooming requirements, such as a yarmulke for Jewish men and a turban and a beard for Sikh men.[8] Some employees have religious objections to performing certain kinds of work or to submitting to medical examinations; others request prayer breaks and special foods in the company cafeteria.

Religious discrimination also involves the violation of a right not to be adversely affected because of the religious beliefs and practices of employers or other employees. For example, a Texas woman who was an atheist resigned her job at a savings and loan association in order to avoid compulsory monthly staff meetings that began with a religious service.[9] Aggressive proselytizing on the job has occurred in many companies. There is a growing number of "Christian" companies that attempt to do business according to certain principles and to impose these principles on their employees. What makes these conflicts different from racial and sexual discrimination is that employees are not being treated differently because of their religion. The problem is quite the opposite: They are being treated like everyone else *when they have a right to be treated differently because of their religion or lack of one.*

In 1972, Congress amended Title VII by adding Section 701(j), which states that there is no religious discrimination if "an employer demonstrates that he is unable to reasonably accommodate an employee's or prospective employee's religious observance or practice without undue hardship on the conduct of the employer's business." As a result of this amendment, the bulk of the court cases involving charges of religious discrimination raise questions about what constitutes "reasonable accommodation" and "undue hardship." In addition, the courts have held that religious objections can be dismissed by an employer when they interfere with employee safety.

National Origin Discrimination. National origin discrimination overlaps discrimination based on race, color, and, to some extent, religion. It is conceptually distinct, however, because an employer could have employment policies that exclude Mexican immigrants but not other Hispanics, or Vietnamese but not other Asians. It is not discriminatory under Title VII for an employer to require U.S. citizenship as a condition for hiring or promotion as long as the requirement is reasonably job-related and is not a pretext for excluding members of some nationality group. Similarly, an employer is permitted by Title VII to impose a requirement that employees be fluent in English, even if it excludes recent immigrants, as long as the requirement is dictated by legitimate business reasons and is uniformly applied.

A curious unresolved question is whether there can be national origin discrimination against American nationals. As more businesses in the United States come under the control of Japanese and European parent companies, there is a growing tendency for them to draw plant supervisors and higher-level managers from their own countries. Is this discrimination on the basis of national origin against Americans by foreigners?

Age Discrimination. Age discrimination results largely from the benefits that employers perceive in shunting older employees aside to make room for younger employees who often have more up-to-date skills and innovative ideas. Younger employees are also less expensive to employ, because older employees generally have higher salaries and make more extensive use of fringe benefits. Youth is sometimes preferred by employers for marketing reasons. Three employees in their fifties were dismissed by I. Magnin Department Stores in 1978, for example, as part of what they claimed was a new strategy by the company to appeal to younger shoppers.[10]

The Age Discrimination in Employment Act (ADEA), passed by Congress in 1967, follows the form of Title VII in prohibiting employers from discriminating in the hiring, promotion, discharge, compensation, or other terms and conditions of employment because of age. Exceptions to the ADEA are permitted when age is a bona fide occupational qualification and in cases where a company has a bona fide seniority system. Highly paid corporate executives are also generally excluded from protection under the ADEA.

Discrimination against the Handicapped. In many respects, discrimination against the handicapped is like religious discrimination rather than discrimination on the basis of race or sex. Employing the handicapped often requires that they be treated differently in order to compensate for their disabilities. It may be argued that employers ought to be willing to make reasonable accommodations for the impairments or disabilities of the handicapped just as they are obligated to make reasonable accommodations for the religious beliefs of their employees.

Although some personnel decisions are obviously discriminatory, others are more subtle and even controversial. Moreover, employers are permitted to treat protected groups differently (except those based on race and color) if there is sufficient reason. Gender-based classifications can be justified by law, for example, for jobs in which sex is a BFOQ. Policies with a disparate impact are legal if they serve a legitimate business purpose (the "business necessity" test). The burden that businesses can be justifiably expected to bear also varies for different kinds of discrimination. Employers are required by law to go to great lengths to accommodate pregnant women and the handicapped, for example, but employees' religious beliefs are to be accommodated only if the cost is minimal. As our discussion shows, what constitutes discrimination is not always easy to determine.

ETHICAL ARGUMENTS AGAINST DISCRIMINATION

That discrimination is wrong can be shown by a variety of arguments. There are, first, straightforward utilitarian arguments that cite the ways discrimination harms individuals, business firms, and society as a whole. A second kind of argument appeals to the Kantian notions of human dignity and respect for persons. Arguments of a third kind are based on various principles of justice. Any one of these arguments is sufficient to establish the point, but it is still worthwhile to examine them all, because each brings out some important aspects of the problem of discrimination.

Utilitarian Arguments

One standard utilitarian argument favored by economists is that discrimination creates an economically inefficient matching of people to jobs. The productivity of individual businesses and the economy as a whole is best served by choosing the most qualified applicant to fill any particular position. When applicants are evaluated on the basis of characteristics, such as race and sex, that are not job-related, productivity suffers. Similarly, it is economically disadvantageous for employees to discriminate by refusing to work with blacks or women and for customers to discriminate by refusing to patronize minority-owned businesses.

There are a number of difficulties with this argument. First, not all forms of discrimination produce economic inefficiencies. This is especially true of religious discrimination and discrimination against the handicapped where complying with the law imposes some cost. It is often cheaper for employers to dismiss employees with troublesome religious beliefs and practices and to avoid hiring handicapped people who have special needs. Second, it is not clear that even racial and sexual discrimination are always inefficient. Under the assumptions of classical economic theory, employers who discriminate on the basis of race or sex are expressing a "taste for discrimination," which they pay for by imposing a higher cost on themselves.[11] For example, when a more productive black applicant is passed over by an employer who prefers to hire whites merely because of race, the output of that employer will be lower. The difference is a cost that the employer is presumably willing to assume in order to satisfy a preference for a white work force. In a free market, then, employers with a taste for discrimination are liable to be driven out of business, and discrimination should be reduced over time.[12] This theoretical result is not always borne out in practice, and economists have offered a variety of explanations for the discrepancy.[13]

Another utilitarian argument focuses on the harm that discrimination does to the welfare of society as a whole by perpetuating the effects of racism and sexism. When racial discrimination in employment is combined with discrimination in education, housing, medical care, and other areas of life, the result is poverty with all its attendant social ills. Sexism also serves to disadvantage women as a group and create social problems. Employers who discriminate on the basis of race and sex thus impose an external cost on society. An externality is also imposed when employers attempt to cut costs by refusing to hire the handicapped; the savings to employers may be more than offset by the cost to the handicapped themselves and to the society that is forced to care for them.

Kantian Arguments

From a nonconsequentialist point of view, discrimination can be shown to be wrong by appealing to the Kantian notions of human dignity and respect for persons. This is especially true of discrimination based on contempt or enmity for racial minorities or women. Discrimination of this kind typically involves a racist or sexist attitude that denies individuals in these groups the status of fully developed human beings who deserve to be treated as the equal of others. The victims of racial and sexual discrimination are not merely disadvantaged by being forced to settle for less desirable jobs and lower pay. They are also deprived of a fundamental moral right to be treated with dignity and respect.

This moral right is also denied when individuals are treated on the basis of group characteristics rather than individual merit. Much of the discrimination against women, older workers, and the handicapped does not result from the belief that they are less deserving of respect and equal treatment. It results instead from the stereotypes that lead employers to overlook significant differences among individuals. Stereotypes, which are a part of racism and sexism, clearly result in a denial of dignity and respect. But stereotyping by its very nature is morally objectionable because it leads employers to treat individuals only as members of groups.

Arguments Based on Justice

Perhaps the strongest arguments against discrimination are those that appeal to some principle of justice. Fundamental to many principles of justice is the requirement that we be able to justify our treatment of other people by giving good reasons, but to discriminate is to treat people differently when there is no good reason for doing so. According to Aristotle's principle of justice as proportional equality—that like cases should be treated alike, and unlike cases should be treated differently in proportion to the relevant differences—discrimination is unjust because characteristics such as race and sex are generally irrelevant to the performance of a job. Even when the differences between individuals constitute genuinely job-related characteristics, the difference in pay, for example, should still be in proportion to that difference.

The contract theory of John Rawls provides the basis for yet another argument against discrimination. One of the principles that would be adopted in the original position is described by Rawls as follows: "Social and economic inequalities are to be arranged so that they are ... attached to offices and positions open to all under conditions of fair equality of opportunity."[14] Even if it were to the advantage of everyone to exclude some groups from certain positions, such a denial of opportunity could not be justified because individuals would be deprived of an important human good, namely the opportunity for self-development.

AVOIDING DISCRIMINATION

Clara Watson felt fortunate in 1976 to be one of the few blacks ever employed as a teller by the Fort Worth Bank and Trust. Her ambition, though, was to become a supervisor in charge of other tellers at the bank despite the fact that only one other black person had ever held this position.[15] She applied to be a supervisor on four separate occasions, and each time she was denied the position. The bank claimed that all promotion decisions were based strictly on evaluations of fitness for the job and that race was not a factor in filling any of the vacancies for which Clara Watson applied. However, a study showed that during a four-year period, white supervisors at the bank hired 14.8 percent of the white applicants and only 3.5 percent of the black applicants. The same supervisors rated black employees 10 points lower than white employees on a scale used for annual salary evaluations. As a result, blacks were promoted more slowly from one salary grade to another and earned less. In 1981, Clara Watson left the bank, but not before she went to the Equal Employment Opportunity Commission and filed a charge of racial discrimination.

Although racial discrimination is illegal, the advancement of blacks and other racial minorities is still hampered by subtle forms of discrimination that are difficult to prove. How can it be determined, for example, whether Clara Watson was a victim of discrimination or simply less qualified than the whites who were promoted ahead of her? Is the statistical underrepresentation of blacks employed at the bank sufficient evidence to charge the bank with discrimination? What should the bank do to avoid a charge of discrimination?

Being a truly nondiscriminatory employer is not an easy task. In addition to good-faith compliance with the law, employers must be aware of some subtle and surprising sources of discrimination. This section discusses what is involved in pursuing a policy of nondiscrimination by examining three basic steps in the hiring and promotion process: analyzing the job to be performed, recruiting applicants, and assessing the applicants for suitability.

Job Analysis

In order to ensure that decisions on hiring and promotion consider only job-related characteristics and result in finding the best person for the job, it is necessary to conduct a *job analysis*. A job analysis consists of two parts: (1) an accurate *job description* that details the activities or responsibilities involved in a position, and (2) a *job specification* listing the qualifications required to perform the job as described. Virtually every job in any present-day corporation has been analyzed in this way, because job analysis is a standard management tool for organizing work and appraising performance.

Because a job description focuses on the specific activities or responsibilities of a position rather than on the people who have traditionally held it, certain kinds of work are less likely to be stereotyped as belonging to one group or another. Even when the qualifications for a job favor one sex over another, a job description that lists only the qualifications will not serve to exclude the members of the other sex who meet them. Because the qualifications must be related to the description, it is easier to determine whether they are really needed for the satisfactory performance of a job. A job analysis need not be confined to traditional job categories. If a job is unnecessarily identified with one sex or another, it can be redesigned, perhaps by combining the activities of one or more other jobs, so that the newly created job is attractive to both men and women. Jobs can also be narrowed so as to avoid excluding some groups unnecessarily. A desk job that involves some moving and lifting can be redesigned to exclude these tasks in order to accommodate the handicapped.

Recruitment and Selection

After a job analysis is done, a company is faced with the task of recruiting applicants in a nondiscriminatory manner. An obvious first step is to make sure that information about an opening is widely disseminated, especially to nontraditional groups. Employers who are serious about not discriminating will place listings of job opportunities with minority publications and educational institutions and employment agencies serving minorities. Also, applications from members of nontraditional groups are more likely to be received if significant numbers of minorities and women are involved in the company's recruitment effort and if a number of nontraditional applicants are hired at one time.[16]

After a sufficient number of applicants have been recruited, the next task is to select the person who is best suited to fill the job. Discrimination can enter into this stage of hiring in many different ways. The selection process itself, which often includes a battery of tests and rounds of interviews, can be discouraging for many nontraditional applicants. Employers can address this problem by simplifying the application procedure or providing instruction on how to proceed. Small differences in treatment can also make racial minorities and women feel uncomfortable. A company can further reduce the barriers to nontraditional applicants by reducing the number of promotion lines and increasing their flexibility so that minorities and women have more opportunities for advancement.

Objective Tests and Subjective Evaluations

Two other important sources of discrimination in both the hiring and the promotion processes are objective tests and subjective evaluations formed on the basis of personal interviews or the recommendations of supervisors. Three kinds of objective tests are commonly used to make decisions on hiring and promotion: (1) tests that measure specific knowledge and skills, such as those needed to be a bookkeeper or a typist; (2) tests that measure intelligence and general aptitude for performing certain kinds of work; and (3) tests that attempt to gauge an applicant's suitability for employment generally and the extent to which an applicant will fit into a specific work environment.

Objective tests of these kinds are permissible under Title VII of the 1964 Civil Rights Act as long as they are not used as a cover for discrimination. One condition laid down in *Griggs* v. *Duke Power Company*, however, is that a test not unnecessarily exclude a disproportionate number of members of protected classes, which is to say it should not have disparate impact. A second condition is that a test be validated; that is, an employer must be able to show that a test for any given job is a reliable predictor of successful performance in that job. The two tests administered by Duke Power Company, the Wonderlic Personnel Test and the Bennett Mechanical Comprehension Test, are professionally prepared instruments that presumably provide an accurate measure of general intelligence and mechanical ability, respectively. What the company failed to prove, however, is that passing scores on these tests are closely correlated with successful job performance. They failed, in other words, to validate the tests that they used.

The Supreme Court has held employers to very high standards of proof in validating tests. It is not sufficient merely to show that employees who successfully perform a certain job also attain high scores on any given test. An employer must be able to show, further, that applicants with lower scores would not be capable of performing just as well. A biased test that results in the exclusion of a substantial percentage of blacks or women, for example, might still be a reliable predictor of successful performance for those who pass but not a reliable predictor of the lack of success of those who fail the test. Further, comparing the scores of employees who are currently performing a job successfully with the scores of inexperienced applicants is not sufficient proof of the reliability of a test. In order to draw a significant conclusion, it would be necessary to know how the current job holders would have scored on the test before they were hired.[17]

Objective tests are ethically and legally permissible, then, as long as they do not have disparate impact and are validated. Do the same two conditions apply to subjective evaluations based on personal interviews or the recommendations of supervisors? On the one hand, evaluations of this kind are made by experienced employees who are well acquainted with the job to be filled and have an opportunity to assess qualities in an applicant that do not lend themselves to objective testing. On the other hand, the evaluations of interviewers and supervisors are apt to be influenced by irrelevant factors, such as a person's appearance or manner, and by conscious or unconscious prejudice. This is especially true when the evaluator is not well trained for the task.

The Clara Watson Case

The first opportunity for the Supreme Court to decide on the conditions for an acceptable subjective evaluation system came in 1988 with the suit of Clara Watson against the Fort Worth Bank and Trust Company which opened this section. The statistical evidence of discrimination was not in dispute, and so the principal issue in this case is not whether the procedures used by the bank had disparate impact (they did) but whether the fact that they involved subjective evaluations rather than objective tests made a difference in the interpretation of Title VII. A related issue is whether the requirement of validation, which applies to objective tests, should be extended to subjective evaluations.

The Fort Worth Bank and Trust Company used three common types of subjective evaluation procedures: interviews, rating scales, and experience requirements. Rating scales differ from interviews in that they record evaluations derived from observations made over a long period of time while an employee is actively at work. Typically, an evaluator is asked to rank an employee on a numerical scale with respect to certain qualities, such as drive and dependability. Experience requirements involve an inventory of specific jobs performed that provide a basis on which to make judgments about future performance.

The American Psychological Association (APA) submitted an *amicus curiae,* or "friend of the court," brief in the Watson case in order to support the claim that these three types of subjective evaluation procedures are capable of validation.[18] Each procedure is open to bias. The most common bias in interviews and rating scales is the "halo effect," in which a single trait exercises an inordinate influence on an evaluator. Closely related to the halo effect is stereotyping, in which assumptions about members of certain groups influence an evaluator. Interviewers are also subject to the "similar-to-me" phenomenon, in which they are inclined to be more favorable to people who have the same traits as themselves. Among the problems with rating scales are the tendencies of evaluators to place persons toward the center of a scale, thereby avoiding the extremes, and to be lenient, scoring most people favorably. The APA brief also cites considerable evidence to show that scores on rating scales are affected by racial factors.

All of these biases can be avoided by subjective evaluation procedures that are designed and carried out according to the APA's *Standards for Educational and Psychological Testing* and the *Principles for the Validation and Use of Personnel Selection Procedures.* The key in each type of procedure is to relate it to a thorough analysis of the job to be filled. The interview should be carefully structured with questions designed to elicit information that is relevant only to the qualifications and performance criteria of the job. The traits on the rating scale and the kinds of experience used as experience requirements should be similarly selected. As much as possible, the results of evaluation procedures should reflect the personal characteristics of the person being evaluated and not the person doing the evaluating, so that differences between evaluators are kept to a minimum. Interviewers, supervisors, and other persons involved in the process should be thoroughly trained in performing their roles in the hiring and promotion process.

The APA brief faults Fort Worth Bank and Trust for failing to meet the generally accepted standards for subjective evaluation procedures and for the lack of any validation of the procedures used. Interviews were conducted by only one person, a white male, and there is no evidence that the questions were carefully designed with job-related qualifications in mind. No job analysis was done in order to guide the selection of questions in the interview and the traits on the rating scale. Moreover, the traits on the rating scale were vaguely defined and not clearly related to job performance. The supervisors who performed the ratings were not specifically trained for that task, and no steps were taken to avoid the effect that race is known to have on the results of rating scales. Finally, it would be impossible without a job analysis of the position to determine what prior experience would enable her superiors to judge the success of Clara Watson in that position.

In an 8–0 decision, the Supreme Court found in favor of Clara Watson and established that the theory of disparate impact applied to subjective evaluation procedures as well as to objective tests of the kind at issue in *Griggs.*

AFFIRMATIVE ACTION

After the passage of the 1964 Civil Rights Act, employers scrutinized their hiring and promotion practices and attempted to eliminate sources of discrimination. However, even the best efforts of companies did not always succeed in increasing the advancement opportunities for women and racial minorities. As a result, many companies and other organizations established affirmative action plans in order to address the problem of discrimination more effectively. In some instances, these plans were adopted in order to be in compliance with Title VII; in others, the intent was to go beyond what the law requires.

Although the goal of eliminating discrimination in employment has generally been accepted in business, people are still divided over the appropriate means. Advocates of affirmative action argue that special programs are required as a matter of "simple justice." Victims of discrimination, they say, deserve some advantage. Preferential treatment is necessary to ensure equality of opportunity. Opponents counter that if it is unjust to discriminate against racial minorities and women on account of their race or sex, then it is similarly unjust to give them preference for the same reason. People who are passed over in favor of a black or a woman are victims of discrimination in reverse. Who is right in this debate?

Examples of Affirmative Action Plans

In 1974, at a plant operated by the Kaiser Aluminum Company in Grammercy, Louisiana, for example, only 5 skilled craft workers (out of 273) were black. Kaiser had long sought out qualified black workers, but few met the requirement of five years of prior craft experience, in part because of the traditional exclusion of blacks from craft unions. In an effort to meet this problem, Kaiser Aluminum and the local union, the United Steelworkers of America, developed an innovative program to train the company's own employees to become skilled craft workers. The plan set up a training program that admitted blacks and whites in equal numbers based on seniority until the proportion of blacks in the skilled craft category equaled the percentage of blacks in the area work force, which was approximately 39 percent.

In 1978, Santa Clara County in California undertook a similar effort. According to the 1970 census, women constituted 36.4 percent of local workers. Only 22.4 percent of county employees were women, and these were concentrated in two areas: paraprofessionals (90 percent female) and office and clerical workers (75.9 percent female). Out of 238 skilled craft workers, not one was a woman. In order to correct these imbalances, the county board adopted an Equal Employment Opportunity Policy that set broad goals and objectives for hiring and promotion in each agency. Specifically, the policy stated:

> It is the goal of the County and Transit District to attain a work force which includes in all occupational fields and at all employment levels, minorities, women and hand-

icapped persons in numbers consistent with the ratio of these groups in the area work force.

To carry out the board's policy, the Santa Clara Transportation Agency, the agency responsible for road maintenance in the county, developed a detailed plan for increasing the percentages of women and minorities in job categories where they were underrepresented.

The programs adopted by Kaiser Aluminum and Santa Clara County are examples of *affirmative action*. The idea behind affirmative action is that merely ceasing to discriminate is not enough. Employers, if they choose to do so, ought to be permitted to take more active steps to ensure a balanced work force, including plans that give preference to job applicants on the basis of race or sex. In response to the plan at Kaiser Aluminum, however, a white employee, Brian Weber, sued both the company and his own union.[19] Weber had insufficient seniority to be admitted as a white trainee, even though he had worked longer at the Kaiser plant than any of the blacks who were selected. He charged, therefore, that he himself was a victim of discrimination in violation of Title VII of the 1964 Civil Rights Act.

A suit also arose as a result of the plan adopted by the Transportation Agency of Santa Clara County when a white male, Paul Johnson, was passed over for promotion and the job went to a woman.[20] In December 1979, Paul Johnson and Diane Joyce applied along with ten other employees for promotion to the position of dispatcher for the road division of the transportation agency. Dispatchers, who assign road crews, equipment, and materials for road maintenance and keep records of the work done, are classified as skilled craft workers. Applicants for the position were required to have four years of dispatch or road maintenance experience with Santa Clara County. The eligible candidates were interviewed by two different boards. The first board, using a numerical scale, rated Johnson slightly above Joyce (75 to 73), and the second board, although rating both candidates "highly qualified," also recommended Johnson over Joyce. The director of the agency was authorized to select any eligible candidate, however, and in order to implement the affirmative action plan, he gave the nod to Joyce. Johnson, like Weber, believed himself to be a victim of discrimination and sued.

Another kind of affirmative action plan that has been challenged in the courts is a "set-aside" provision for minority contractors. The Public Works in Employment Act, passed by Congress in 1977, required that at least 10 percent of the funds granted to state and local governments for construction be set aside for "minority business enterprise." Some municipalities have enacted similar legislation. The city of Richmond, Virginia, for example, adopted a Minority Business Utilization Plan that required recipients of city construction contracts to subcontract at least 30 percent of the dollar amount of each contract to minority-owned businesses. Some white contractors have charged that set-aside provisions are an illegal form of discrimination.[21]

In *Weber* v. *Kaiser Aluminum* and *Johnson* v. *Transportation Agency*, the Supreme Court held that the affirmative action plans in question were not discriminatory under Title VII. Both Brian Weber and Paul Johnson thus lost their suits. After approving set-asides for minority-owned contractors in a 1980 decision, *Fullilove* v. *Klutznick*, the Court cast all such programs into doubt in *City of Richmond* v. *J. A. Croson* (1989) and subsequently *Adarand* v. *Peña* (1995).

The Compensation Argument

One argument for giving preferential treatment to members of certain groups is that it is owed to them as *compensation* for the injustice done by discrimination directed against them personally or against other members of a group to which they belong. The root idea in this argument derives from Aristotle's discussion in Book V of the *Nicomachean Ethics*, in which he distinguished between justice in the distribution of goods (distributive justice) and justice where one person has wrongfully inflicted some harm on another (corrective justice). In the latter case, justice requires that the wrong be corrected by providing compensation. For example, if A, while driving carelessly, crashes into B's new car, then B suffers a loss that is A's fault. An injustice has been done. It can be corrected, though, if A compensates B for the amount of the loss, say the cost of repairs plus any inconvenience. Similarly, if A has a right not to be discriminated against and B discriminates, thereby harming A unjustifiably, then it is only just that B compensate A to correct the harm done.

Aristotle's analysis perfectly fits the situation, for example, of an employer who has been found guilty of discriminating against women or racial minorities by assigning them to lower-paying jobs and bypassing them in promotions. The law in such cases is guided by the dictum "No right without a remedy," which is to say that a person cannot be said to have a right unless there is also some means of correcting a violation of that right. If we have a right not to be discriminated against, then the courts should be able to provide some remedy when that right is violated. The remedy is often to require the employer to pay the victims the difference between what they actually earned and what they would have earned had no discrimination taken place and to advance them to the positions that they would have attained.

Who Deserves Compensation? Critics of affirmative action charge that not all affirmative action plans are justified by the compensation argument because the individuals who are given preferential treatment are often not the same as those who are victims of discrimination. Affirmative action plans almost always single out persons as members of a group that has suffered discrimination without requiring any evidence that the persons themselves have been victimized in any way. A person may be a member of a disadvantaged group and yet lead a rather privileged life, relatively free of the effects of discrimination.

Defenders of affirmative action respond that racial and sexual discrimination have subtle psychological effects despite the profound changes that have taken place in our society, and racial and sexual discrimination affect all members of that group to some degree. Bernard R. Boxill observes that the critics' objection involves a non sequitur. From the premise that better-qualified blacks or women are less deserving of compensation than some who have been more severely handicapped by discrimination, we cannot conclude that no compensation is owed. "Because I have lost only one leg," he argues, "I may be less deserving of compensation than another who has lost two legs, but it does not follow that I deserve no compensation at all."[22] Nor does it follow, according to Boxill, that victims of discrimination who succeed in overcoming the harm done to them are any less deserving of compensation than those who are unable to succeed to the same degree.[23]

This response still does not answer the question, Why should preference not be given to those who most deserve compensation, the people who have most suffered the effects of racial discrimination and who are consequently among the least qualified? One reply is that giving preference in hiring is only one way of compensating individuals for past discrimination, and it is a way that is of greater help to those who are better qualified for the jobs available. Those who have been more disadvantaged by discrimination may be less able to benefit from affirmative action and may derive greater benefit from other forms of help, such as job-training programs.

A defender of affirmative action can also argue that claims of compensatory justice must be balanced against another principle of justice, the principle that hiring should be done on the basis of competence.[24] Giving preference to the best qualified of those who are deserving of compensation, according to this argument, is the best way of accommodating these two conflicting principles. Another argument cites the practical difficulty of evaluating each case to determine the extent to which an individual has been harmed by discrimination and hence deserves compensation. Giving preference to members of groups without regard for the particulars of individual cases, therefore, is a matter of administrative convenience.[25]

Does Compensation Punish the Innocent? A second objection to the compensation argument is that in affirmative action the burden of providing compensation often falls on individuals who are not themselves guilty of acts of discrimination. In the competition for admission into colleges and universities and for hiring and promotion in jobs, it is largely white males who are asked to pay the price of correcting injustices that are not of their making. Critics ask, why should a few white males bear such a disproportionate burden?

One answer to this question is that white males, even when they are not themselves guilty of discrimination, are still the beneficiaries of discrimination that has occurred and thus are merely being asked to give back some ill-gotten gain. This response does not fully meet the critics' point, however. Even if all white males have benefited to some degree, it still needs to be shown that what is given up by the few white males who are passed over when preference is given to others is equal to the benefit they have obtained by living in a discriminatory society. Furthermore, it would still seem to be unjust to place the full burden in such an arbitrary manner on a few when so many other members of society have also benefited from discrimination.[26]

A second answer to the critics' question is that white males are typically asked not to give up gains they have already made but to forgo a future benefit to which no one has an undisputed right.[27] Brian Weber and Paul Johnson, for example, were not deprived of any gains they had made but only of an opportunity for advancement. Although this is a real loss, neither one had a right to be selected but only a right not to be discriminated against in the selection process. If the compensation argument is correct, then those who were selected *deserved* the advantage given by their race or sex and hence no discrimination took place. (Ironically, the opportunity for Brian Weber to receive on-the-job training for a skilled craft position would not have existed had Kaiser Aluminum not adopted an affirmative action plan in order to hire more blacks.) Still, if the job prospects of white males are substantially reduced by affirmative action, then they have suffered a loss. There is surely some limit on the amount of compensation any individual or group of individuals can justifiably be required to pay.

Accordingly, the courts have laid down three conditions for permissible affirmative action plans to protect those adversely affected by them. They are (1) that a plan does not create an absolute bar to the advancement of any group, (2) that the plan does not unnecessarily trammel the rights of others, and (3) that it be temporary. The training program in the Weber case was open to black and to white workers in equal numbers. Although Brian Weber failed to gain admittance to the first class, his seniority would have assured him a place eventually. In addition, he had other opportunities for realizing his ambition of becoming a skilled craft worker. Finally, the training program was scheduled to terminate when the proportion of black craft workers reached 39 percent, the percentage of blacks in the local workforce.

Equality Arguments

We noted in Chapter 4 that, according to Aristotle, justice is a kind of equality. Whether affirmative action is just, therefore, can be decided, perhaps, by the principle that people ought to be treated equally or treated as equals. Two quite different concepts of equality are relevant to the debate over affirmative action, however. These are *equality of opportunity* and *equality of treatment.*

Justice, in the first interpretation of equality, requires that everyone have an equal opportunity to succeed in life and that no one be held back by arbitrarily imposed restraints or barriers. Better enforcement of the laws against discrimination can help to equalize the opportunities for everyone, but the effects of past discrimination also need to be neutralized in some way.

President Lyndon B. Johnson expressed the argument graphically in a 1965 speech on an executive order requiring every federal contractor to be an "equal opportunity employer":

> Imagine a hundred yard dash in which one of the two runners has his legs shackled together. He has progressed 10 yards, while the unshackled runner has gone 50 yards. How do they rectify the situation? Do they merely remove the shackles and allow the race to proceed? Then they could say that "equal opportunity" now prevailed. But one of the runners would still be forty yards ahead of the other. Would it not be the better part of justice to allow the previously shackled runner to make up

the forty yard gap; or to start the race all over again? That would be affirmative action towards equality.[28]

The equal opportunity argument addresses not only the harm done to individuals from past discrimination but also the barriers posed by discrimination in present-day society. Many fully qualified blacks and women have not been disadvantaged by past discrimination but are still at a competitive disadvantage because of lingering racism and sexism on the part of employers. Giving preferential treatment may be necessary under such circumstances simply to ensure that people are considered equally. Whether this is true depends, in part, on what we mean by equal opportunity.

The Meaning of Equal Opportunity. What does equal opportunity mean? To say that every child born today has an equal opportunity to become a surgeon, for example, has two distinct senses. One interpretation is that each child initially has an equal chance in the same way that every ticket holder is equally likely to win the lottery. The other is that the means for pursuing the career of a surgeon are open to all. Following Douglas Rae in his book *Inequalities,* let us call these two possibilities the *prospect-regarding* and the *means-regarding* interpretations of the concept of equal opportunity.[29] Prospect-regarding equality aims at eliminating *all* factors affecting the distribution of goods in a society except for mere chance. Means-regarding equality, by contrast, is compatible with considerable inequality of prospects. The only requirement for equality of opportunity in this latter interpretation is that the results reflect only differences in personal attributes and not differences in the means available to persons.

Equality of prospects is an ideal of egalitarians who want to minimize the role of "accidents" of birth on people's success in life. Just as our race or sex should have no bearing on what we are able to achieve, so too should it not matter whether we are born into wealth or poverty or whether we are born with certain mental or physical endowments. There are a number of difficulties with interpreting equal opportunity as equality of prospects. First, because people begin life with vastly different prospects, steps would have to be taken to equalize these prospects. Achieving an equality of prospects would entail considerable remedial education and a reallocation of resources to schools with disadvantaged students. Second, how are we to know when unequal prospects have been offset? One way might be to look at equality of outcomes, but equality of prospects need not result in equal outcomes if people make different choices. So there must be some way of determining when prospects are equal without looking at the outcomes.

The interpretation of equal opportunity as equality of means entails that rewards should be distributed on the basis of some relevant criteria. Artificial barriers to advancement, such as racial or sexual characteristics, are irrelevant and should be removed, but justice does not require the removal of inequalities in prospects resulting from differences in a person's various physical and mental characteristics. Equal opportunity, on this view, means a chance to compete under fair conditions.

One difficulty with this interpretation is that it requires only that all discrimination cease; it does nothing to address President Johnson's concern about the head start provided by discrimination in the past. A further difficulty is that the conditions for fair competition are highly suspect. What is commonly called talent is largely the acquisition of the expertise and skills provided by education and certified through formal procedures. If access to education or certification is affected by racial or sexual discrimination, then we can scarcely be said to have equal access to the means for achieving success in life.

Equal Opportunity or Equal Treatment? Some defenders of affirmative action have argued that the goal ought to be not equal opportunity but equal treatment. This concept, too, is ambiguous, with two distinct senses. Ronald Dworkin points out that when we say that certain white males have been denied a right to equal treatment, we might have in mind two different rights.[30]

> The first is the right to *equal treatment*, which is the right to an equal distribution of some opportunity or resource or burden.... The second is the right to *treatment as an equal*, which is the right, not to receive the same distribution of some burden or benefit, but to be treated with the same respect and concern as anyone else.[31]

The right to equal treatment in the sense of a right to receive an equal share applies only to a few things, such as the right that each person's vote shall count equally. In the distribution of most things, it is a right to the same respect and concern that is at stake. Affirmative action is objectionable, then, only if the alleged victims are not treated as equals.

Dworkin argues that any selection process advantages some people and disadvantages others. Admitting students to medical school on the basis of academic preparation, for example, serves to exclude some applicants. Such a selection process is justified, however, by a social good that outweighs any harm done to those who are turned away. Affirmative action plans are adopted to serve an important social good, namely overcoming the effects of racism. In so doing, it must be recognized that some white applicants who are denied admission to medical school would be admitted in the absence of a plan. As long as the interests of these unsuccessful applicants are taken into consideration with the same respect and concern as those of others in adopting rules for medical school admission that best meet the needs of society and, as long as these rules are applied impartially, showing the same respect and concern to every applicant, then no one has cause to complain. Everyone has been treated as an equal.

This argument is open to serious objections.[32] In Dworkin's interpretation, the right to equal treatment is the right to equal respect and concern as rational calculations are made about the social good. According to Robert L. Simon, "So understood, the right to treatment as an equal looks suspiciously like the utilitarian requirement that everyone count for one and only one in computing social benefits and burdens."[33] Simon suggests that placing more emphasis on *respect* for other persons rather than on concern that their *welfare* be given equal weight would make affirmative action programs less compatible with equal treatment interpreted as treatment as an equal.

The conclusion to be drawn from the discussion in this section is that the concept of equal opportunity is too vague and ambiguous to provide conclusive support for any particular position on the justification of affirmative action. Those who favor preferential treatment programs and those who oppose them can find a meaning of "equal opportunity" to fit their particular position. Equality of treatment, by contrast, provides a more solid basis for affirmative action. This principle demands, however, that we think carefully about the reasons for affirmative action and make sure that the goals to be achieved are worthwhile and cannot be attained by means that do not involve taking race or sex into account.

Utilitarian Arguments

Unlike the two previous arguments for affirmative action, arguments based on utility do not hold that programs of preferential treatment are morally *required* as a matter of justice but that we are morally *permitted* to use them as means for attacking pressing social problems. Utilitarian arguments stress that preferential treatment programs are necessary to eradicate lingering racial and sexual discrimination and to accelerate the pace of integrating certain groups into the mainstream of American society. The underlying assumption of these arguments is that racism and sexism are deeply embedded in the major social, political, and economic institutions of our society and in people's attitudes, expectations, and perceptions about social realities. If this assumption is correct, then antidiscrimination legislation addresses only the surface manifestations of racism and sexism and does not penetrate to the root causes. Action must be taken to change the institutions of society and the ways people think about themselves and their world. Otherwise, the goal of a discrimination-free society will come only slowly, if at all.

Preferential treatment programs serve to combat the effects of discrimination, first, by making more jobs available to racial minorities, women, and others through lowering the stated qualifications and formal accreditation required for hiring and promotion. Opportunities for groups subject to discrimination are further increased by breaking down stereotypes in the eyes of employers and the rest of society and by creating role models for people who would not otherwise consider certain lines of work. The long history of sexual and racial stereotyping of jobs in this country has hampered the acceptance of women and members of some racial minorities into desirable positions in our society, and this history has also affected the very people who were excluded by limiting their career aspirations. Finally, affirmative action increases opportunities by heightening awareness about discrimination and changing the hiring and promotion process. When business firms make a commitment to achieve a certain racial and sexual mix in their work force with established goals, the officials responsible for hiring and promoting employees cannot help but be sensitive to the issue of discrimination in every decision they make.

Affirmative action also provides a direct economic benefit to corporations themselves by increasing the pool of job applicants and generally improving community relations. Discrimination introduces inefficiency into the job market by excluding whole groups of people on the basis of race or sex, some of whom are highly qualified. The result is that people in these groups tend to be "underutilized" in jobs that do not make full use of their abilities and training, and employers are deprived of the best possible work force. The following statement by an executive of Monsanto Corporation testifies to the benefit that affirmative action can have for employers: "We have been utilizing affirmative action plans for over 20 years. We were brought into it kicking and screaming. But over the past 20 years we've learned that there's a reservoir of talent out there, of minorities and women that we hadn't been using before. We found that it works."[34]

Some Problems with Affirmative Action

Affirmative action has some significant undesirable consequences that must be balanced against the undeniable utilitarian benefits of preferential treatment programs. Three arguments in particular are commonly used by opponents of affirmative action. These are (1) that affirmative action involves hiring and promoting less qualified people and lowering the quality of the work force, (2) that it is damaging to the self-esteem of employees who are favored because of race or sex, and (3) that it produces race consciousness, which promotes rather than fights discrimination. Let us examine these in turn.

The Quality Argument. The first argument—the quality argument—can be expressed in the following way. The most qualified person for a position has no need for special consideration. Therefore, a person who is given preference on the basis of race or sex cannot be the most qualified person and cannot perform as well in a job as someone who is more qualified. The result is a decline in the quality of goods and services, which has an adverse affect on the whole of society.

A supporter of affirmative action can question, first, how much quality is given up. Preferential treatment does not involve the hiring or promotion of people who are *un*qualified but who are (at worst) *less* qualified to some degree. And the degree can be so slight as to be of no significance. Many jobs require only minimal qualifications and can readily be mastered by persons of normal abilities. Even occupations requiring considerable ability and expertise involve many tasks that can satisfactorily be performed by people who are not the best available. In many instances, "qualified" means "already trained," which brands as unqualified those people who are capable of being fully competent with some training.

Also, whether a person is qualified for a certain job depends on how qualifications are recognized or determined. Conventional measures, such as standardized tests and academic record, are often criticized for containing a bias against women and racial minorities. The credentials used to certify competence in various fields, such as licenses, certificates, union cards, diplomas, and the like, have been accused of containing a similar bias. More complications emerge when we ask, what are the relevant qualifications for the performance of any given job? It is sometimes argued, for example, that a black police officer can be more effective in a black community and that an applicant's race is, therefore, a legitimate consideration in the hiring of a police force.[35] In education, a largely male college faculty may not provide a learning environment that is as beneficial to women students as one with a substantial number of female professors.

The Injury Done by Affirmative Action. A second utilitarian argument against affirmative action is that it injures the very people it is designed to help. The effect of hiring and promoting blacks and women because of their race or sex is to draw attention to their lack of qualifications and create an impression that they could not succeed on their own. Another effect of affirmative action is to reduce the respect of society for the many hard-won achievements of blacks and women and to undermine their own confidence and self-esteem. The stigma attached by preferential treatment may even have the unintended consequence of impeding racial integration if qualified minority applicants avoid jobs where race is a factor in selection.

This is an argument to be taken seriously. It rests, however, on the questionable assumption that programs of preferential treatment have not significantly helped some people. Insofar as affirmative action has boosted some racial minorities and women into higher-level positions of prestige and responsibility, there is bound to be an increase in their pride and self-respect as well as their financial well-being. Success in life is often unearned, but there is little evidence that the beneficiaries of good fortune are psychologically damaged by it. Manuel Velasquez observes, "For centuries white males have been the beneficiaries of racial and sexual discrimination without apparent loss of their self-esteem."[36]

The Importance of Race. The third and final utilitarian argument against preferential treatment programs is that they increase rather than decrease the importance of race and other factors in American society. If the ideal of an equal society is one in which no one is treated differently because of color, ethnic origin, religion, or any other irrelevant factor, then preferential treatment defeats this ideal by heightening our consciousness of these differences. To some critics of affirmative action, all uses of racial, ethnic, and religious classifications are abhorrent and ultimately destructive of the fabric of a society.

One response by proponents of affirmative action is that the use of racial classifications is a temporary expedient, necessary only to eradicate racism before we can realize the ideal of an equal society. Justice Harry Blackmun wrote in an opinion:

> I suspect that it would be impossible to arrange an affirmative action program in a racially neutral way and have it successful. To ask that this be so is to demand the impossible. In order to get beyond racism, we must first take account of race. There is no other way. And in order to treat some persons equally, we must treat them differently.[37]

The Supreme Court has long held that distinctions based on race and ethnic origin are "by their very nature odious to a free people whose institutions are founded upon the doctrine of equality."[38] Nevertheless, they are permissible when the conditions warrant their use.

Others argue that there is nothing inherently wrong with race consciousness and the awareness of sexual, religious, ethnic, and other differences. What makes any of these wrong is their use to degrade and oppress people with certain characteristics. There is a great difference between the racial distinctions that were an essential element of the institution of slavery, for example, and the race consciousness that is a part of the present-day attack on racism. Any utilitarian analysis of affirmative action must take into account the history of racial minorities and women in this country and current social realities. All things considered, the race consciousness engendered by affirmative action may be socially beneficial.

Conclusion

The ethical issues surrounding discrimination and affirmative action are very problematical. Rights figure prominently in these issues—both the rights of people who have been victimized by discrimination and the rights of people who now bear the burden of correcting past wrongs. Considerations of justice also play a role. Justice requires that people who have been wronged be compensated in some way and that all people be treated equally, but the concepts of just compensation and of equal opportunity or equal treatment are subject to differing interpretations. Finally, arguments based on utility provide strong support for antidiscrimination and affirmative action policies, although the benefits of any given policy must be weighed against the harms with uncertain results. The ideal of a nondiscriminatory society is clear, but the pathway to it is strewn with formidable obstacles.

Case 8.2   The Alaskan Salmon Cannery

You have been sent from the head office of a large food-packing company to a salmon cannery in Alaska with instructions to investigate charges of discrimination that have been made by a group of nonwhite workers.[39] Your assignment is especially critical in view of the fact that the protesting cannery workers have threatened to bring a lawsuit against the company for its alleged discriminatory hiring and promotion practices.

This company-owned salmon cannery on a remote section of the Alaskan coast operates for only a few months each year during the salmon run. The entire work force must be transported to the unpopulated location and housed in specially constructed barracks. There are two kinds of jobs at the cannery: unskilled jobs on the cannery line and both skilled and unskilled noncannery jobs. The cannery workers, who are mostly Filipino and Native American, prepare the fish and pack it into four-ounce cans. Afterward, the cans are sealed and the fish is cooked under carefully controlled conditions. The Filipino workers are recruited through an agreement with a major labor union, whereas the Native Americans live in villages near the cannery. The noncannery employees are mostly skilled and include machinists, engineers, carpenters, cooks, bookkeepers, and office workers. A few noncannery jobs are classified as unskilled; examples of these are kitchen help, table servers, janitors, laundry workers, night watchmen, and deckhands.

The nonwhite cannery workers have complained that the unskilled noncannery jobs generally pay more than similar unskilled cannery jobs and are filled almost entirely by whites. The existence of two classes of unskilled workers, one consisting mainly of lower-paid nonwhites and another consisting largely of higher-paid whites, is prima facie evidence, they claim, of disparate impact. As a result, the company can legally be required in court to show that the selection criteria for both kinds of jobs are reasonably necessary for the operation of the business. (This is called the "business necessity" test.) The cannery workers also charge that certain employment practices prevent them from filling unskilled noncannery jobs. These include nepotism, a preference for rehiring previous employees, not posting notices of openings at the cannery, and an English proficiency requirement. Finally, the two classes of unskilled workers are housed separately and rarely mingle.

The managers you talk to admit the truth of the workers' complaints but differ in their interpretation of the facts. There is no discrimination, they contend, because the unskilled noncannery workers are recruited from the company's base of operations in the Pacific Northwest, and the racial composition of the unskilled noncannery category is roughly equal to the racial composition in that region of the country and also to the composition of the people who apply for the jobs. "It is a mistake," one manager says, "to compare the two unskilled job categories at the cannery and conclude that there is a statistical imbalance. The proper comparison is with the available work force for temporary labor in the Pacific Northwest. And there we compare very favorably."

Upon returning to company headquarters, you consult with a staff lawyer about the situation. She tells you that the courts have indeed held disparate impact to be determined by comparing the racial or sexual composition of an employer's work force to the composition of the available, qualified pool of applicants in the local area. The fact that workers must be recruited elsewhere and transported to the work site makes this case unusual, she admits. "If the workers succeed in bringing suit," she continues, "we might still win using some untested legal arguments. We could argue, for example, that the workers need to show more than a racial imbalance in the work force; that they need to show that our employment practices are responsible for that imbalance. That would be very difficult for them to do, but the requirement is a reasonable one. Otherwise every employer could be brought into court to justify their practices as a matter of business necessity merely because of a statistical imbalance, and this could force employers to adopt strict quotas in order to avoid baseless suits, which is not what Congress intended in passing the Civil Rights Act."

Having heard from all sides in this dispute, you now sit down at your desk and begin drawing up a list of recommendations.

Case 8.3 The Walkout at Wilton's

News of the walkout greeted Sam Hilton as he settled into his office chair on Monday morning. As the director of human resources for Wilton's Department Stores, he prided himself on the company's progressive personnel relations, but a policy that he thought a success was backfiring on him. Eight of the twenty drivers in the Wilton's warehouse were sitting in their trucks in the parking lot, refusing to enter the warehouse building where the merchandise is loaded.

The walkout occurred when the drivers learned that Roy Stone had returned to work from a six-month medical leave of absence. A year ago, Roy revealed to his supervisor, who had demanded an explanation for repeated requests to take time off for doctors' appointments, that he had ARC, or AIDS-Related Complex. ARC is a less severe condition than AIDS that is caused by the Human Immunodeficiency Virus (HIV), which is also responsible for AIDS, and about 50 percent of ARC patients eventually develop AIDS. Although Roy had insisted that his revelation remain absolutely confidential, the supervisor felt compelled to explain the situation to his own superior, and within weeks the information had spread to Roy's co-workers. A few were sympathetic, but others avoided him; derogatory comments began appearing on bulletin boards and bathroom walls; and some employees avoided using the sinks and toilets in the warehouse building.

Sam had been preparing a company policy statement on AIDS and decided that now was the time to act. The Vocational Rehabilitation Act of 1973 and the 1990 Americans with Disabilities Act, which became effective in 1992, protect employees with handicaps and disabilities from discrimination. Whether AIDS would be considered a disability by the courts was uncertain, but many companies were already preparing themselves for the difficult challenges that the inevitable rise of AIDS would create. Relying on the most up-to-date information from the U.S. Centers for Disease Control, which indicates that AIDS cannot be transmitted through ordinary casual contact, Sam and his staff completed the draft of an AIDS policy statement that sought to protect the rights of everyone in the organization.

One provision of the policy statement says: "Wilton's recognizes that employees who have or may be perceived as having AIDS may wish to continue in their normal work activities as long as their physical condition allows them to do so. No difference in treatment should be accorded these individuals so long as they are able to meet work standards and so long as medical evidence indicates that their condition is not a threat to themselves or others."[40] In addition, the policy states that the company will make reasonable accommodation for AIDS-infected employees to the extent that they would for any handicapped or disabled employee, and that the confidentiality of all medical information provided to the responsible company officials will be preserved. After the adoption of the policy, Roy Stone requested and received a medical leave of absence, a benefit to which all employees are entitled for a maximum of one year.

After six months, Roy felt strong enough to return to work full time, although by this time he had developed AIDS, a fact that he shared with the company doctor. He requested a position on the loading dock instead of the work as a driver that he had done previously. Sam and his staff met with all employees in the warehouse to prepare them for Roy's return. Through presentations and pamphlets, the employees were informed that HIV could be transmitted only through the sharing of bodily fluids such as occurs in sexual intercourse and the sharing of needles. Employees in the ordinary course of work are at virtually no risk.

The Monday of Roy's return did not go smoothly. In addition to the walkout of the eight drivers, several employees were threatening a lawsuit under a provision in their labor contract that obligates the company to take reasonable steps to assure the health and safety of all employees. The supervisor of the warehouse, who reported the walkout to Sam, told him, "These guys aren't troublemakers. They're really scared about breathing the same air and touching what he's touched. They don't believe that everything is known about AIDS, and they don't want to be guinea pigs for the company's sake, not when they have wives and children to protect. One guy who just got married and really needs this job was crying hysterically this morning."

As the news sank in, Sam thought that maybe the AIDS policy was a mistake, or maybe it could have been handled better. At any rate, Sam had to go out now and speak with the warehouse personnel again. But what should he say to them? And what should he say to Roy?

NOTES

1. Material for this case is drawn from Bari-Ellen Roberts, *Roberts* vs. *Texaco* (New York: Avon Books, 1998); Amy Myers Jaffe, "At Texaco, the Diversity Skeleton Still Stalks the Halls," *New York Times*, 11 December 1994, Sec. 3, p. 5; Peter Fritsch, "Trustee of Big Fund with Texaco Stock Says Tape Shows 'Culture of Disrespect,'" *Wall Street Journal*, 6 November 1996, A15; Kurt Eichenwald, "Texaco Punishes Executives for Racial Comments and Plans to Destroy Papers," *New York Times*, 7 November 1996; Kurt Eichenwald, "The Two Faces of Texaco," *New York Times*, 10 November 1996, sec. 3, pp. 1, 10-11; Adam Bryant, "How Much Has Texaco Changed?" *New York Times*, 2 November 1997, sec. 3, pp. 1, 16-17.

2. The information on AT&T is taken from Earl A. Molander, "Affirmative Action at AT&T," in *Responsive Capitalism: Case Studies in Corporate Conduct* (New York: McGraw-Hill, 1980), 56–70.

3. 42 U.S.C. 2000e-2.

4. *Diaz* v. *Pan American World Airways, Inc.*, 442 F. 2d 385 (1971).

5. *Dothard* v. *Rawlinson*, 433 U.S. 321 (1977).

6. *Griggs* v. *Duke Power Company*, 401 U.S. 424 (1970). Some material is also taken from a case prepared by Nancy Blanpied and Tom L. Beauchamp, *Ethical Theory and Business*, 3d ed., ed. Tom L. Beauchamp and Norman E. Bowie (Upper Saddle River, NJ: Prentice Hall, 1988), 383–85.

7. The precedent of *Griggs* was altered by several subsequent court decisions, most notably *Wards Cove Packing Co.* v. *Antonio*, 490 U.S. 642 (1988), which made it more difficult for employees to sue for discrimination. A 1991 civil rights bill largely restored the interpretation of *Griggs* that had prevailed before.

8. For a discussion of these and other problems, see James G. Frierson, "Religion in the Workplace," *Personnel Journal*, 67 (July 1988), 60-67; and Douglas Massengill and Donald J. Petersen, "Job Requirements and Religious Practices: Conflict and Accommodation," *Labor Law Journal*, 39 (July 1988), 402-10.

9. *Young* v. *Southwestern S & L Association*, 509 F. 2d 140 (5th Cir. 1975).

10. Tony Mauro, "Age Bias Charges: Increasing Problem," *Nation's Business* (April 1983), 46.

11. This analysis and the phrase "taste for discrimination" are due to Gary S. Becker, *The Economics of Discrimination*, 2d ed. (Chicago: University of Chicago Press, 1971).

12. This point is also made in Milton Friedman, *Capitalism and Freedom* (Chicago: University of Chicago Press, 1962), 109–10. Ironically, Friedman uses the analysis to argue *against* legislation curbing discrimination on the grounds that competition alone is sufficient to bring discrimination to an end.

13. See Kenneth J. Arrow, "The Theory of Discrimination," in *Discrimination in Labor Markets*, ed. Orely Ashenfelter and Albert Rees (Princeton, NJ: Princeton University Press, 1973), 3–33; Lester C. Thurow, *Generating Inequality* (New York: Basic Books, 1975); Dennis J. Aigner and Glen G. Cain, "Statistical Theories of Discrimination in Labor Markets," *Industrial and Labor Relations Review*, 30 (1977), 175–87; and Barbara R. Bergmann and William Darity, Jr., "Social Relations, Productivity, and Employer Discrimination," *Monthly Labor Review*, 104 (April 1982), 47–49.

14. John Rawls, *A Theory of Justice* (Cambridge, MA: Harvard University Press, 1971), 83.

15. *Clara Watson* v. *Fort Worth Bank and Trust*, 487 U.S. 977 (1988).

16. Rosabeth Moss Kanter, in *Men and Women of the Corporation* (New York: Basic Books, 1977), chap. 8, proposes "batch" promotions of two or more individuals from excluded groups so that they can support each other and break down barriers.

17. These points are made in *Albemarle Paper Company* v. *Moody*, 422 U.S. 405 (1975).

18. "In the Supreme Court of the United States: *Clara Watson* v. *Fort Worth Bank and Trust*," *American Psychologist*, 43 (1988), 1019–28. See also an accompanying explanation by Donald N. Bersoff, "Should Subjective Employment Devices Be Scrutinized? It's Elementary, My Dear Ms. Watson," *American Psychologist*, 43 (1988), 1016–18.

19. *United Steelworkers and Kaiser Aluminum* v. *Weber*, 443 U.S. 193 (1979).

20. *Johnson* v. *Transportation Agency, Santa Clara County*, 480 U.S. 616 (1987).

21. The landmark cases are *Fullilove* v. *Klutznick*, 448 U.S. 448 (1980); *City of Richmond* v. *J. A. Croson Company*, 488 U.S. 469 (1989); and *Adarand* v. *Peña*, 115 S. Ct. 896 (1995).

22. Bernard R. Boxill, "The Morality of Preferential Hiring," *Philosophy and Public Affairs*, 7 (1978), 247.

23. Ibid., 248.

24. For a defense of this principle, see Alan H. Goldman, "Justice and Hiring by Competence," *American Philosophical Quarterly*, 14 (1977), 17–26.

25. For an argument of this kind, see James Nickel, "Classification by Race in Compensatory Programs," *Ethics*, 84 (1974), 147–48.

26. These points are made by George Sher, "Preferential Hiring," in *Just Business: New Introductory Essays in Business Ethics*, ed. Tom Regan (New York: Random House, 1984), 48.

27. Boxill, "Morality of Preferential Hiring," 266.

28. Quoted in Lewis D. Solomon and Judith S. Heeter, "Affirmative Action in Higher Education: Towards a Rationale for Preference," *Notre Dame Lawyer*, 52 (October 1976), 67.

29. Douglas Rae, *Inequalities* (Cambridge, MA: Harvard University Press, 1981), 66–68.

30. Ronald Dworkin, *Taking Rights Seriously* (Cambridge, MA: Harvard University Press, 1978), 223–39; and Ronald Dworkin, *A Matter of Principle* (Cambridge, MA: Harvard University Press, 1985), 293–331.

31. Dworkin, *Taking Rights Seriously*, 227.

32. For criticism of Dworkin's argument, see Robert L. Simon, "Individual Rights and 'Benign' Discrimination," *Ethics*, 90 (1979), 88–97.

33. Ibid., 91.

34. Quoted in Peter Perl, "Rulings Provide Hiring Direction: Employers Welcome Move," *Washington Post*, 3 July 1986, A1.

35. The example and much that follows is taken from Alan Wertheimer, "Jobs, Qualifications, and Preferences," *Ethics*, 94 (1983), 99–112. A different kind of argument for the relevance of race as a qualification is presented by Michael Davis, "Race as Merit," *Mind*, 42 (1983), 347–67. For a criticism of the kind of argument presented by Wertheimer, see Robert K. Fullinwider, *The Reverse Discrimination Controversy: A Moral and Legal Analysis* (Totowa, NJ: Rowman & Littlefield, 1980), 78–86; and Alan H. Goldman, *Justice and Reverse Discrimination* (Princeton, NJ: Princeton University Press, 1979), 167–68.

36. Manuel G. Velasquez, *Business Ethics: Concepts and Cases*, 4th ed. (Upper Saddle River, NJ: Prentice Hall, 1998), 405.

37. *Regents of the University of California* v. *Bakke*, 438 U.S. 265 (1978).

38. *Hirabayashi* v. *United States*, 320 U.S. 81 (1943).

39. This case is adapted from *Wards Cove Packing Co., Inc. et al.* v. *Antonio*, 490 U.S. 642 (1988).

40. Adapted from a proposed AIDS policy statement in Rose Knotts and J. Lynn Johnson, "AIDS in the Workplace: The Pandemic Firms Want to Ignore," *Business Horizons* (July–August 1993), 5–9.

9

Women and Family Issues

Case 9.1   Jacksonville Shipyards

Lois Robinson was a first-class welder at Florida-based Jacksonville Shipyards, Inc.(JSI).[1] Women in any skilled craft job are a rarity in the largely men's world of shipbuilding and repair. JSI records show that less than 5 percent of shipyard workers between 1980 and 1987 were female, and no woman had ever held a supervisory or executive position at the company. Starting out as a third-class welder in 1977, Lois Robinson had steadily increased her skill so that she was the equal of any male welder. Still, she never quite fit in at JSI, which has been characterized as "a boys club," where a woman could be admitted only as a sex object. She could not be accepted merely as a good welder.

None of Lois Robinson's co-workers or supervisors had ever solicited her for sex, nor had any of them offered some benefit for her sexual favors or threatened to retaliate if she refused. Lois Robinson was occasionally ridiculed, as when one co-worker handed her a pornographic magazine while those around laughed at her response, or when another co-worker passed around a picture of a nude woman with long blond hair and a whip. (Because she has long blond hair and uses a welding tool known as a whip, she thought that the picture was being displayed to humiliate her.) It was not these incidents that infuriated her, however; it was rather the pervasive presence of calendars, magazines, pictures, graffiti, and other visual displays of nude women that she found intolerable.

The workplace was plastered with pinup calendars from suppliers that featured nude or partially clad women in sexually submissive poses, often with breasts and genital areas exposed. The suppliers' calendars were distributed by JSI to its employees with permission to display them wherever they pleased. Employees were required to get permission to post any other material in the workplace—and permission was denied in some instances for requests to post material of a commercial or political nature—but pictures of nude women from magazines or other sources were displayed with the full knowledge of management, from the president of JSI down.

The pictures observed by Lois Robinson included one with a woman's pubic area exposed and a meat spatula pressed against it and another of a nude woman in full-frontal view and the words "USDA Choice." A drawing on a dartboard pictured a woman's breast with the nipple as the bull's eye. Lois Robinson also became aware that the sexually suggestive comments increased when her male co-workers noticed that she had seen one of the pornographic pictures. Although crude sexual jokes were sometimes told in her presence, she was often warned to "take cover" or leave so that the men could exchange jokes out of her hearing.

In January 1985, Lois Robinson complained to JSI management about the visual displays. Afterward, the pictures became more numerous and more graphic and the number of sexually suggestive comments to her and the other women increased. The complaints to her supervisors were apparently passed to higher levels of management, and a few pictures were removed only to be replaced by others. Some of the pictures to which she objected were in the shipfitter's trailer, where she and other workers reported to receive instructions, and she sometimes entered the trailer to check on paperwork. One day the words "Men Only" appeared on the door of the trailer, and though the sign was soon painted over, the words could still be observed. One supervisor pointed out that the company had no policy against the posting of pictures and claimed that the men had a constitutional right to do so. The supervisor's superior declined to order the pictures removed. Another supervisor suggested that Ms. Robinson "was spending too much time attending to the pictures and not enough time attending to her job."

As a federal contractor (JSI performed repairs on ships for the U.S. Navy), the company is obligated by presidential order to be nondiscriminatory and to have an affirmative action plan. In 1980, JSI adopted a policy entitled "Equal Employment Opportunity." The policy stated in part:

> ... we should all be sensitive to the kind of conduct which is personally offensive to others. Abusing the dignity of anyone through ethnic, sexist, or racist slurs, suggestive remarks, physical advances or intimidation, sexual or otherwise, is not the kind of conduct that can be tolerated.

The policy asked that any violations be reported to the Equal Employment Opportunity (EEO) coordinator at the facility. The policy was not generally known to the supervisors at the shipyards, nor was it incorporated in the standard JSI rule book. The supervisors received no training on how to deal with reports of sexual harassment or other problems, and the name of the EEO coordinator was not given in the policy and was not widely known to employees in the company. In any event, the experience of Lois Robinson was not likely to encourage any victim of harassment to make a report to anyone at JSI.

On September 2, 1986, Lois Robinson filed a suit against Jacksonville Shipyards, Inc., for sexual harassment. In the suit she cited the pervasive presence of sexually explicit pictures, the sexually suggestive and humiliating comments of her male co-workers, and the "Men Only" sign on the shipfitter's trailer.

INTRODUCTION

Women populate the American workplace. According to 1990 data, women hold 46 percent of the jobs in this country, and the figure is expected to rise to 50 percent by 2001. A closer look, however, reveals that women are more highly represented in traditionally female job categories, which carry less pay, prestige, and potential for promotion. The result is a "pay gap" between men and women, which has narrowed in recent years, but women still receive only 71 cents for every dollar earned by men. Although women have entered the ranks of management in greater numbers, a 1994 *Wall Street Journal* study of the 200 largest U.S. firms found that women held only a quarter of all management positions and that these were mainly at lower levels.[2] Women constituted a mere 5 percent of the vice presidents surveyed, and the number of women presidents, CEOs, and board directors was exceedingly small. A "glass ceiling" appears to exist in corporate America, preventing the promotion of women beyond a certain level.

In addition to the pay gap and the glass ceiling, women experience other adversities in the workplace from which men are largely spared. Women who enter male-dominated lines of work are often made to feel unwelcome, either because their presence breaks accepted routines or because of attitudes about women's roles. Sexual stereotypes about women's commitment to work or their capacity for leadership also interfere with their advancement up the corporate ladder. In addition, women are compelled to deal with the complications of sexual attraction in the workplace, which range from office romances to sexual harassment. Although sexual harassment is thought by some to be merely unwanted sexual interest, it is viewed by others to be an abuse of power and an assertion of existing power relations by men over women.

Finally, women more than men face the challenges of combining full-time work with family obligations. Two-thirds of men in the work force have wives who also work, thus creating so-called two-earner families, and more than half of all employed women have children under the age of six. Although responsibility for children in two-earner families could be shared 50-50, this is seldom the case. The sociologist Arlie Hochschild has called work at home "the second shift," and she notes that it constitutes roughly an extra month of 24-hour days each year. Working this second shift takes its toll on women who attempt to balance the conflicting demands, and some women drop out of the work force for varying periods of time in order to fulfill family obligations. Many companies have responded to the problems of working mothers and two-earner families by a set of family-friendly programs that includes maternity leaves, flexible working hours, and child-care centers.

This chapter examines three issues that mainly affect women in the workplace. These are sexual harassment as a form of sex discrimination, comparable worth as a solution for the problem of the pay gap, and the family-friendly programs that help employees balance the demands of family and work life.

SEXUAL HARASSMENT

Improper sexual conduct in the workplace—which includes lewd and suggestive comments, touching and fondling, persistent attention, and requests for sexual favors—has long been a problem for women, and occasionally for men. All too often, such sexual harassment has been regarded by employers as a personal matter beyond their control or as an unavoidable part of male-female relations. However, increased attention to the problem and developments in the law have made employers aware of their responsibilities—and women, of their rights!

That sexual harassment is morally wrong is not in dispute. The main questions are: What is sexual harassment? How serious is the problem? Who is responsible for preventing it? A further, legal question is whether sexual harassment is prohibited by the 1964 Civil Rights Act. Although Title VII does not mention sexual harassment, the courts have ruled that it is a form of discrimination and, hence, an illegal employment practice.

What Is Sexual Harassment?

Surveys of employee attitudes reveal substantial agreement on some of the activities that constitute sexual harassment and differences on others. In particular, most of the respondents in a 1980 poll conducted by *Harvard Business Review* and *Redbook* magazine consistently rated a supervisor's behavior as more serious than the same action by a co-worker, thereby recognizing that sexual harassment is mainly an issue of power.[3] Barbara A. Gutek has found that over 90 percent of both men and women consider socializing or sexual activity as a job requirement to be sexual harassment. However, 84 percent of the women surveyed, but only 59 percent of the men, identified "sexual touching" as sexual harassment.[4] In general, women are more likely than men to label the same activity as sexual harassment.

In 1980, the Equal Employment Opportunity Commission (EEOC) issued guidelines on sexual harassment that included the following definition:

> Unwelcome sexual advances, requests for sexual favors, and other verbal or physical conduct of a sexual nature constitute sexual harassment when (1) submission to such conduct is made either explicitly or implicitly a term or condition of an individual's employment, (2) submission to or rejection of such conduct by an individual is used as the basis for employment decisions affecting such individual, or (3) such conduct has the purpose or effect of unreasonably interfering with an individual's work performance or creating an intimidating, hostile, or offensive working environment.

This definition makes a distinction between two kinds of harassment. One is *quid pro quo* harassment, in which a superior, who is usually a man, uses his power to grant or deny employment benefits to exact sexual favors from a subordinate, who is usually a woman. The other kind is *hostile working environment* harassment, in which the sexual nature of the conduct of co-workers and others causes a woman (or a man) to be very uncomfortable. What constitutes discomfort is not easy to specify, but the judge in the *Jacksonville Shipyards* case ruled that the display of pinup calendars and pornographic pictures constitutes an unrelenting "visual assault on the sensibilities of female workers" and that such a situation constitutes sexual harassment under the "hostile working environment" provision.

Whether a work environment is hostile or offensive is not easily determined. Much depends on the attitudes of the employees involved and the response of management to employee concerns. The two situations described in Case 9.2, for example, are debatable. The manager of the Dairy Mart store was not subjected to harassing treatment by particular persons, nor was she exposed to the contents of the magazines in question; she merely objected to the fact that the magazines were being sold in her store. The readers of the magazine were customers, not Dairy Mart employees, and she would have no further contact with the readers after they left the store. Unlike the women at Stroh Brewery, she did not blame any actual harassing treatment on fellow employees who were influenced by the objectionable material. Both Dairy Mart and Stroh Brewery were responding to demonstrable consumer interest in the kinds of material that some employees found offensive.

Sexual Harassment as a Form of Sex Discrimination

Title VII of the 1964 Civil Rights Act and other legislation protect women against many forms of discrimination. The Equal Pay Act of 1963 forbids an employer to offer different wages to men and women who perform the same or substantially similar work unless the difference is based on some valid factor other than sex, such as seniority or productivity. Unlike race and color, sex can be a Bona Fide Occupational Qualification (BFOQ). Many of the problems about what constitutes sexual discrimination arise in cases where a person's sex can arguably be taken into consideration, such as in hiring guards for a male prison. In some other cases, sex is not a BFOQ, but the stated qualifications serve to exclude virtually all women. Examples are tests for police officers and firefighters that require considerable strength and endurance. Whether the qualifications are discriminatory depends largely on whether they are reasonably necessary for the performance of the job.

The Pregnancy Discrimination Act. Does the Title VII prohibition against discrimination because of sex protect women who are pregnant? Congress passed the Pregnancy Discrimination Act in 1978 to resolve this question. The main impetus came from a case in which a woman challenged a policy at General Electric Company forcing her to take unpaid maternity leave with the loss of all fringe benefits.[5] After a miscarriage, the woman suffered a further medical problem unrelated to the pregnancy. Because she lost all fringe benefits due to her pregnancy, the company refused to cover the cost of the unrelated medical problem. The Supreme Court ruled that the policy at General Electric did not constitute sexual discrimination under Title VII but challenged Congress, if it disagreed, to clarify whether the law on sexual discrimination covers pregnancy. Congress accepted the challenge and declared that it did.

The Pregnancy Discrimination Act amends the phrase in Title VII "because of sex" to include decisions made on the basis of "pregnancy, childbirth, or related medical conditions." The act mainly affects fringe benefits of two kinds: hospital and major medical plans and policies on temporary disability and sick leave. All such plans and policies are required by the act to treat pregnancy like any other condition. Employers are not required to grant maternity leave or pay for the medical care associated with pregnancy, but if they allow an employee with a broken leg, for example, to have a paid leave of absence or they pay for the cost of a broken leg under the company's medical plan, then a woman's pregnancy must be treated in exactly the same way. Men are also protected by the Pregnancy Discrimination Act in that the medical coverage for the pregnant wife of a male employee must be the same as that provided for the husband of a female employee.[6]

Quid Pro Quo Harassment. Quid pro quo harassment clearly violates the Title VII provision that men and women should not be treated differently in their "compensation, terms, conditions, or privileges of employment." A woman who is promised a promotion or a raise—or threatened with demotion, termination, or loss of pay—based on whether she submits to the sexual demands of her boss is being held to a different standard, merely because of her sex.

Some observers contend that quid pro quo harassment, while unfortunate, is not sexual discrimination but merely a wrongful act committed by one employee against another. It is not uncommon for workers of both sexes to encounter personal problems on the job, and harassment, in this view, is one of these personal problems. However, Catharine A. MacKinnon has argued that sexual harassment in the workplace is more than "personal"; it has a connection to "the female condition as a whole."

> As a practice, sexual harassment singles out a gender-defined group, women, for special treatment in a way which adversely affects and burdens their status as employees. Sexual harassment limits women in a way men are not limited. It deprives them of opportunities that are available to male employees without sexual conditions. In so doing, it creates two employment standards: one for women that includes sexual requirements, one for men that does not.[7]

In *Meritor Savings Bank* v. *Vinson* (1986), the U.S. Supreme Court declared that "without question" both quid pro quo harassment and hostile working environment harassment constitute sexual discrimination under Title VII.[8] The decision in the *Jacksonville Shipyards* case further upheld the EEOC view that a hostile working environment constitutes sexual harassment. Even when there is no demand for sexual favors, conditions in a workplace can produce a form of stress that interferes with a person's ability to work and erodes that person's sense of well-being. Not only does a visual display of pornographic pictures produce stress, but the need to be diligent to avoid the next incident may induce more stress; and the feeling that their complaints will not produce any change further compounds the stress that harassed women have.

Hostile Working Environment Harassment. In reaching its decision in *Jacksonville Shipyards*, the court relied on testimony about sexual (and racial) stereotyping. Stereotyping is likely to occur when members of a group are few in number and when members of another group are in power. The stereotypes in sexual harassment cases are those that prevail outside the workplace where some men view women as sex objects. The conditions for stereotyping thus permit "sex role spillover," in which women's roles outside of employment "spill over" or become central in an environment where other roles, such as the job to be performed, ought to be the only ones relevant.[9] A good welder who is also a husband and father can be treated on the job only as a good welder, whereas a woman like Lois Robinson cannot escape the stereotypes that the men bring with them to the workplace. She cannot be, in their eyes, only a good welder. Stereotyping becomes more prevalent when there are "priming" elements, such as pictures that create a stimulus for harassing treatment. One effect of stereotyping is selective interpretation, whereby complaints may be perceived in accord with a stereotype, such as that women are "overly emotional." The failure of Lois Robinson's supervisors to take her complaints seriously may have been due to that effect.

Hostile working environment harassment is both more pervasive and more difficult to prove. Studies have shown that quid pro quo harassment is relatively rare, but in surveys about one-third of working women (33 percent) report incidents of sexual remarks and jokes and around a fourth cite staring and suggestive leers (27 percent) and unwanted sexual touching (24 percent).[10] Not all of this conduct is considered to be sexual harassment, however, even by the women who report it. Still, a line must be drawn somewhere. One possibility is a *reasonable person* standard, whereby conduct that is offensive to a person of average sensibilities would be impermissible. However, one court has rejected this approach on the grounds that it "tends to be male-biased and tends to systematically ignore the experiences of women." This court has proposed, instead, a *reasonable woman* standard, which requires that the alleged harassment be judged from the recipient's point of view.[11]

Further Issues. Initially, the courts were reluctant to recognize sexual harassment as discrimination unless a woman suffered some economic loss, such as a reduction in pay or the loss of her job. If this position is accepted, however, then any amount of harassment is legal as long as the woman's employment status is not affected. In 1981, though, a court held that sexual harassment is illegal even when there is no economic loss, as long as there is psychological harm. No woman, the court declared, should be forced to endure the psychological trauma of a sexually intimidating workplace as a condition of employment.[12] This position was expanded by the decision in *Harris* v. *Forklift Systems, Inc.* (1993), in which the victim could not establish even psychological harm.

Theresa Harris's employer made disparaging comments about women, suggested that she negotiate a raise at a local motel, publicly announced (falsely) that she had slept with a client to get an account, and required her to retrieve change from his pants pocket. The employer claimed that he was only joking. A lower court found that the employer, the president of a Nashville-based truck-leasing company, was "a vulgar man" but contended that his behavior was not so egregious as to seriously affect her "psychological well-being."[13] In *Harris* v. *Forklift Systems, Inc.,* the U.S. Supreme Court ruled that no psychological harm needs to be shown as long as a reasonable person would find the conduct offensive.[14] In the words of one observer, "You don't have to have a nervous breakdown, but one joke does not make a case."[15] The highest court had an opportunity in the *Harris* case to affirm the reasonable *woman* standard, but the justices relied instead on the reasonable *person* standard.

The most intractable issue for the courts has been the responsibility of an employer for the conduct of an employee, especially when the employer is unaware of the harassment by an employee. In *Meritor,* Mechelle Vinson charged that her supervisor, Sidney Taylor, made repeated sexual advances and raped her on several occasions, but she did not report this to anyone at the bank or use the bank's formal complaint procedure. The bank held that it was not responsible, therefore, because of the lack of knowledge. The Supreme Court disagreed, however, and held that an employer has a responsibility to ensure that the workplace is free of sexual harassment. But how far does this responsibility extend?

In two 1998 cases, *Burlington Industries* v. *Ellerth* and *Faragher* v. *City of Boca Raton*, the U.S. Supreme Court established a two-step test.[16] First, if the harassment is by a superior and results in a "tangible employment action, such as discharge, demotion, or undesirable assignment," then the employer is liable, regardless of whether the employer knew about the harassing activity. Second, if there is no "tangible employment action," the employer is still liable unless the employer can show (1) that reasonable care was exercised to prevent and correct sexual harassment and (2) that the employee unreasonably failed to take advantage of the opportunities provided by the employer to correct or avoid the harassing conduct. In the decision, the Court declined to consider who is at fault in cases of harassment and focused instead on how to prevent them. Employers are now on notice that they must anticipate the possibility of harassment and take demonstrable steps to address the problem. Employees have also been told that they have a responsibility to use whatever means an employer has made available for dealing with harassment.

Preventing Sexual Harassment

Although sexual harassment is usually committed by one employee against another, employers bear both a legal and an ethical obligation to prevent harassment and to act decisively when it occurs. Harassment is more likely to occur when management has not prescribed clear policies and procedures with regard to conduct of a sexual nature. Employers who display an insufficient concern (a "head-in-the-sand" attitude) or have inadequate procedures for detecting harassment in the workplace bear some responsibility for individuals' harassing conduct. In addition, companies cannot fully evade responsibility by blaming the victim for not reporting sexual harassment in accord with established procedures. The way in which employers respond to claims of sexual harassment sends a powerful message about the seriousness with which management takes it own policies and procedures. The legal duty of an employer also extends to harassment by nonemployees, such as customers and clients.

Aside from the law—including the cost of litigating and paying settlements—employers have strong financial incentives to avoid sexual harassment. Tolerating sexual harassment can result in hidden costs to companies. A 1988 study of 160 Fortune 500 companies calculated that sexual harassment costs an average company with about 24,000 employees a stunning $6.7 million.[17] This figure does not include litigation and settlement costs but merely the losses that result from absenteeism, low morale, and employee turnover. According to the study, productivity suffers when women are forced to waste time avoiding uncomfortable situations or to endure the stress of coping with them. The stress induced by sexual harassment leads to health problems, loss of self-confidence, and a lack of commitment, all of which may reduce career prospects and deprive employers of valuable talent. Women who have been harassed are more likely to seek transfers or to quit, thereby increasing the cost of employee training.

Sexual Harassment Programs. Most corporations have recognized the cost of sexual harassment and accepted their responsibility to prevent it by establishing programs to deal with sexual harassment on the job. The major features of these programs are (1) developing a firm policy against harassment; (2) communicating this policy to all employees and providing training, where necessary, to secure compliance; (3) setting up a procedure for reporting violations and investigating all complaints thoroughly and fairly; and (4) taking appropriate action against the offenders.

1. *A Sexual Harassment Policy.* The first step in a corporate program to eliminate sexual harassment is a firm statement from a high level in the organization that certain conduct will not be tolerated. The policy statement should not only convey the serious intent of management but also describe the kinds of actions that constitute sexual harassment. A good policy should educate as well as warn.

2. *Communicating the policy.* No policy can be effective unless it is effectively communicated to the members of the organization, and effective communication is not merely a matter of making the policy known but of gaining understanding and acceptance of the policy. Many corporations include sexual harassment awareness in their initial training and ongoing education programs, often utilizing videos of situations and simulation games to heighten employee sensitivity to the issues.

3. *Setting up procedures.* A complete policy should include a well-publicized procedure for handling incidents of sexual harassment with assurances of nonretaliation against an accuser. Employees should be informed of the procedure to follow in making a complaint, including the specific person or office to which complaints should be made and preferably offering several alternatives for making complaints. In addition, those who handle complaints should be aware of the procedure they

should follow. The policy should assure all parties—the accuser as well as the accused—of confidentiality. The investigation itself should seek to ascertain all the relevant facts and to observe the rules of due process, especially in view of the harm that could result from false accusations. Although companies should have a formal complaint procedure, some also make use of an informal process through which a situation may be resolved to the victim's satisfaction. An informal procedure is well suited for less serious, infrequent incidents among peers where there is some misunderstanding or insensitivity; it is inappropriate for repeat offenses with multiple victims and for harassment by a victim's superior.

   4. *Taking appropriate action.* Any disciplinary action—which may include a reprimand, job transfer, demotion, pay reduction, loss of a bonus, or termination—should aim, at a minimum, to deter the offender and perhaps to deter others in the organization (although the deterrent effect on others will depend on publicizing the penalty). Because the victims of harassment may have suffered some job loss or been deprived of some opportunities, a proper resolution may also include compensating the victims for any harm done.

Case 9.2   Sexual Harassment or Business as Usual?

Selling *Playboy* at Dairy Mart

Dolores Stanley, a 33-year-old churchgoing mother of three, was promoted in 1990 to the position of manager of a Dairy Mart store in rural Toronto, Ohio, after ten years with the convenience store chain. One of her first acts as a manager was to remove all "adult" magazines, including *Playboy* and *Penthouse*, from behind the counter where they were displayed in opaque plastic wrappers. Her Dairy Mart superiors insisted that the magazines be put back on display in the store, saying that it was not the company's role to censor their customers' reading. Dolores Stanley replied that she could not participate in the selling of material that was offensive to her personally and degrading for all women. Forcing her to sell the magazines constituted sexual harassment, she claimed.

The Swedish Bikini Team

In a TV ad for Old Milwaukee beer, the scantily clad Swedish Bikini Team descends by parachute into the campsite of several awed young males. The voiceover to the women's bump-and-grind routine is "It just doesn't get any better than this." Things got a lot worse for five women employees of Stroh Brewery, the maker of Old Milwaukee beer, who claim that their harassment by fellow male employees was encouraged by the company's advertising. The Stroh executives conceded that the women had been subjected to lewd comments and displays of pornographic pictures but pointed to the company's own sexual harassment policy as evidence that such conduct is not condoned. They labeled as "preposterous" the claim that the Swedish Bikini Team ads contributed in any way to the harassment.[18] The women argued that "these ads tell Stroh's male employees that women are stupid, panting playthings."[19] One woman said of the men in her office, "When they are getting feedback from the top of the company that women are bimbos and that's OK, that's why I'm getting treated the way I'm getting treated."[20] Shortly after the women filed suit, the Swedish Bikini Team was featured in *Playboy* magazine.

COMPARABLE WORTH

In 1981, a strike was called by a union representing about half of the 4,000 municipal workers in San Jose, California. One issue in the strike was the union's insistence that the city spend $3.2 million over four years to upgrade the salaries of nonmanagement employees in female-dominated jobs. A study recently conducted by the city with the aid of a national consulting firm revealed considerable differences in salary between men and women in comparable positions. Using a point-factor evaluation system, with points being assigned for knowledge, problem solving, accountability, and working conditions, the study showed that Senior Chemists, (501 points) for example, received $29,094 annually, whereas the salary for the comparable female-dominated job of Senior Librarian (496 points) was $23,348. The male-dominated job category of Painter (173 points) was held by the study to be comparable to that of Secretary (177 points), but the pay for painters and secretaries was $24,518 and $17,784, respectively. The nine-day strike ended after the city council pledged to spend $1.45 million over two years to upgrade salaries in the lowest-paid female-dominated positions and to bargain in good faith in subsequent years to close the wage gap further. Mayor Janet Gray Hayes hailed the settlement as "the first giant step toward fairness in the workplace for women."

The situation in San Jose is remarkable only for the response of the city to the problem of the undervaluing of jobs held predominantly by women. It is well documented that women earn less than men. The figure at the end of 1989 was that the median income for women was 68 percent of the earnings for men. This earnings gap has been relatively constant over the past 50 years with the difference in pay between men and women dipping in recent years before rising again. In 1939, women who were employed full time, year-round earned 63.6 percent of the wages of men who were similarly employed, and this figure remained above 63 percent during the middle of the 1950s. In 1972, the earnings of women as a percentage of men's fell to a record low of 56.6, and by 1981 the figure had risen only to 59.2 percent.

Women are not the only group with lower earnings. The income of racial minorities is also substantially less than that of whites. With increasing access to education and the elimination of discrimination in hiring and promotion, though, black men have made significant economic gains. In 1959, the mean salary of black men was 71 percent of the earnings of the average white man. Ten years later the figure had risen to 75 percent, and by 1975 the percentage had increased to 85. Conventional remedies for discrimination have thus proven effective in reducing wage differences between blacks and whites but not those between men and women. As a result, advocates of women's rights have sought new means of attacking this form of discrimination.

The Principle of Comparable Worth

The means chosen is the principle of *comparable worth*. This principle holds that dissimilar jobs can still be compared with respect to certain features and that jobs that are similar with respect to these features ought to be paid the same. The job of a secretary, for example, is quite different from that of a painter, but if it can be shown that the two jobs require a similar degree of skill and effort, then secretaries and painters should be paid at the same rate. Expressed in this way, the principle of comparable worth provides a method for setting wages that is an alternative to market forces operating according to the laws of supply and demand.

On the standard economic view, wages are determined largely by the contribution workers make to *productivity*. More technically, the pay of any given worker is a function of the net addition that worker makes to the revenues of the employer. The value of a job, then, on the economic view, is what the work of an employee is worth to an employer, and this value is measured largely by the price a worker can command in the marketplace. In a competitive labor market in which the workers are free to move from one job to another and employers are forced to compete for their services, the laws of supply and demand will result in a state of equilibrium in which the wages of workers match their productivity.

Advocates of the principle of comparable worth hold the view that compensation ought to be based on *job content*. This is typically done by a process known as job evaluation, which measures and compares the features of a job for which a worker ought to be compensated. The general features considered in most job evaluation systems are skill, effort, responsibility, and working conditions. Each job is assigned a certain number of points from a range for each feature; these points are added together to arrive at a total; different jobs are then ranked according to their total number of points. Compensation is based on these rankings so that jobs with the same number of points are paid the same and jobs with more points carry higher rates of pay.

The use of job evaluations to set wages is a familiar practice in American business. Since the 1920s, companies have employed methods for measuring the content of jobs in order to provide a rational means for setting wages where market forces alone are inadequate for the task. Although the laws of supply and demand determine the overall level of wages for different kinds of work, they are not suited for making fine distinctions between the multitude of jobs in large organizations, especially where the productivity of individual workers is difficult to measure. This method also gives managers a great deal of control over decisions about wages and the deployment of labor, which is especially valuable in dealing with a unionized work force.

Job evaluation studies are used extensively in the public sector, because governments are largely insulated from market forces. In order to attract and retain competent personnel and to maintain equity among different jobs, civil service systems on the local, state, and federal levels evaluate the content of jobs and set wages according to the value of comparable jobs in the private sector. Prior to the study of nonmanagement workers in San Jose, for example, the city conducted a job evaluation of salaried managers in the municipal government. The purpose was merely to ensure that all salaries were competitive in order to prevent other cities in the area from luring managers away with offers of higher pay. Washington, which was the first state to conduct a comparable worth study to determine the extent of discrimination against women, had been required since 1960 to pay workers the prevailing "market rate," which was determined by an elaborate salary survey of state jobs and jobs in business and industry. No attempt was made to compare every state job with jobs in the private sector. Rather, the compensation for a few "benchmark" jobs was set by determining the wages that work of that kind would command in a competitive market, and the pay for other state jobs was set by comparing them with the "benchmark" jobs.

Two Versions of the Principle

There are two different versions of the principle of comparable worth. In one version, the value of a job is determined by features of the work performed—by the content of the job, in other words, as opposed to supply and demand. Unlike the value of other goods, then, the value of labor is not the same as the market price, and the possibility exists that the wages resulting from market forces undervalue (or in some cases overvalue) the work performed. The principle of comparable worth, in this version, is offered as a morally preferable alternative to the market, especially when the market price does not accurately reflect the value of the work performed.

A second version of the principle of comparable worth does not maintain that there is a concept of value other than the price that a worker can command in a competitive market. Because of discrimination, however, the wages that are actually paid to some workers are not the same as what they would be paid if there were no discrimination. Because the skill and effort required to perform a job are largely the features that enable workers to command a certain price for their labor in a free market, a comparison of jobs according to their content provides a means for determining whether discrimination exists and what wages would be without the presence of discrimination. In this version, comparable worth is a method for detecting discrimination in the labor market and correcting it. It is not an alternative to the market but an adjunct that frees the market from the distorting force of discrimination.

The difference between the two versions of the principle of comparable worth can be expressed in the language of rights. The first version claims that each worker has a right to be paid according to the content of the job performed and that this right is violated when one worker is paid less than another for performing comparable work, whatever the reason for the difference in pay. The right involved in the second version is simply the right not to be discriminated against in the setting of wages, that is, the right not to be paid lower wages simply because the work is performed predominantly by persons of a certain sex or race. The principle "equal pay for comparable work," in this version, is thus a means for protecting an already existing right rather than the creation of a new right.

In order to justify the principle of comparable worth in the first version, it is necessary to demonstrate that the only morally relevant features for setting wages are those that concern the character of the work performed and that all other differences, such as those resulting from the forces of supply and demand, are morally irrelevant. This is by no means an easy task. The justification of the second version of the principle is much easier. A defender need only show that the market has operated in a discriminatory manner and that setting wages on the basis of job content is better suited than any other means for a nondiscriminatory labor market. Because the second version is what most advocates intend, it is the one considered in the remainder of this discussion.

The Equal Pay Act and Title VII

The principle of comparable worth ("equal pay for comparable work") is not the same as the principle embodied in the Equal Pay Act of 1963 (EPA). Passed by Congress as an amendment to the Fair Labor Standards Act, the EPA prohibits an employer from paying men and women in the same establishment different wages for jobs that require "equal skill, effort, and responsibility" and are performed under the same conditions. Exceptions are allowed by the act for differences due to any factor other than sex, such as seniority, a merit system, and compensation based on productivity. As interpreted by the courts, the EPA applies to employees who perform substantially the same but not necessarily identical work.[21] The principle embodied in the EPA can be expressed, therefore, as "equal pay for the *same* work" rather than "equal pay for *comparable* work."

Although Title VII of the 1964 Civil Rights Act provides more extensive protection against wage discrimination than the EPA, the courts have not ruled that the failure of employers to pay equal wages for work of comparable worth is a violation of Title VII. The issue was raised in a suit filed by a woman, Alberta Gunther, and three co-workers who were employed as jail matrons in Washington County, Oregon. Because the matrons were responsible for fewer inmates than male guards and performed some clerical duties that were not shared by their male counterparts, the jobs were held by the courts to be dissimilar. Consequently, there was no violation of the EPA. Still, the women argued that the difference in pay was the result of intentional discrimination on the basis of sex. In support of their argument, they cited a county-commissioned study that evaluated the job of jail matron at 95 percent of the market value of the guards' job. While the male guards received the full recommended wages, the women were paid only 70 percent of theirs. As a result, the women earned one-third less than the men. In a 5–4 decision, the Supreme Court ruled that Alberta Gunther and her co-workers could sue under Title VII for equal pay for work that is comparable to but not substantially the same as that performed by men. The decision emphasized, however, that the matrons' suit was "not based on the controversial concept of 'comparable worth,'" but that they "seek to prove by direct evidence, that their wages were depressed because of *intentional sex discrimination.*"[22]

Are Wage Differences Due to Discrimination?

One important issue in the debate over comparable worth is whether the difference in earnings between men and women is due to discrimination or whether it results from the impersonal workings of the labor market. If discrimination is not the cause of the earnings gap, then there is no substance to the claim that women are being unfairly compensated for the work they do and hence there is no need to adopt remedial measures, such as a system of compensation based on the principle of comparable worth. Of course, a comparable worth system could still be advanced as a morally preferable method of setting wages (the first version of the principle) or as a pragmatic step toward reducing the disparity in earnings between the sexes as a socially desirable goal.

The Evidence of Discrimination. The main evidence for discrimination in compensation is statistical. Unfortunately, the statistical evidence—which indisputably documents the existence of an earnings gap between men and women—is explainable in a variety of ways, not all of which point to discrimination as the culprit. One reason why men earn more is that they work more hours, and so a distinction has to be made between income (the amount actually received in a paycheck) and wages (the amount earned for each hour worked). Earnings also increase with experience, and women, who voluntarily leave the work force more often to raise a family and pursue other interests, have a shorter work history on the average than men. When comparisons are made between male and female workers with the same number of years of experience or between never married women and men, the wage gap narrows, though not by much. By 1980, women who had never married, for example, still earned only 65 percent as much as men their own age.

Success also depends on a willingness to acquire an education and specialized training, to accept positions of greater responsibility, to relocate when necessary, and to make other sacrifices. These are examples of what economists call *human capital*, which is the amount of resources invested in order to increase the productivity of individual workers. If men "invest" more in themselves and thereby increase their value to employers, then they can rightly expect to be paid more. It can be objected, however, that insofar as women leave the work force more often and are less willing to do what is necessary to build a career, their choices simply reflect the lower return that their investment in human capital brings in a job market that discriminates against them. The result is a "chicken-and-egg" question: Do women earn less because they have less human capital, or do women rationally choose to invest less in themselves because of low pay?

Job Segregation. By far the most significant factor in explaining the lower wages of women is the segregation of jobs according to sex. Women are crowded into traditionally female occupations that are not highly valued by the market in the first place, and the large number of women competing for jobs further depresses wages. Of the 553 occupations listed in the 1970 census, 310 were 80 percent or more male-dominated, while women held 80 percent or more of the jobs in 50 of the occupations. Further, 70 percent of men and 54 percent of women were in occupations where workers of their own sex held more than 70 percent of the jobs. Twenty-five percent of women workers were employed in occupations that were 95 percent female.

Experts are sharply divided on whether women are forced into traditionally female occupations because they are the only jobs available or whether they freely choose them because of a preference for certain kinds of work. Advocates of the latter view cite gender differences in the factors that contribute to job satisfaction. According to some studies, women are willing to sacrifice some income for clean, safe, comfortable work, and they place more value than men on interpersonal relations, the opportunity to serve others, and the intrinsic interest of a job.[23] As a result, some of the gap between the earnings of men and women may be due to the higher wages employers must offer to induce workers (who are predominantly men) to perform less desirable work, while women compete among themselves for more attractive jobs.

Realignment as a Solution. If the explanation of the pay gap is that women are confined by discrimination to female-dominated jobs, then one solution is to remove the barriers to the advancement of women into traditionally male lines of work by more rigorous enforcement of Title VII. This is an alternative to comparable worth known as *realignment,* which involves reducing the extent of job segregation according to sex. In the view of many experts, however, realignment does not address the source of the problem, which can be succinctly stated as follows:

> Women are paid less because they are in women's jobs, and women's jobs are paid less because they are done by women. The reason is that women's work—in fact, virtually anything done by women—is characterized as less valuable.[24]

As an illustration, the formerly male-dominated job of bank teller has been largely taken over by women. The result has not been an increase in the opportunities of women for higher pay but a decrease in the relative earning power (as well as the prestige) of this kind of work. If the undervaluing of work done by women is a significant factor in explaining why men earn more, then comparable worth is likely to be a more effective strategy than realignment.

To summarize this portion of the discussion, some of the differences in male and female earnings are due to the choices women make. These are (perhaps) the result of inherent differences between men and women, and certainly they result from social forces that operate outside the job market. Many structural features of the job market also play a role. Among these are differences between industries and firms within an industry, regional differences, age distribution in the work force, unionization, and a host of other variables that have not been adequately studied. Whatever residual amount is left unaccounted for by these factors is due to discrimination against women. The only conclusion to emerge consistently from the many studies on the question, though, is that we do not know how much that residual amount is.

Measuring Job Content

A second issue in the comparable worth debate is whether it is possible to measure job content in such a way that the resulting comparisons are meaningful and reliable. Some critics use the analogy of comparing apples and oranges to support their contention that no significant comparison can be made between extremely dissimilar jobs. Other opponents of comparable worth argue that job evaluations are not wholly objective and are potentially discriminatory. The evaluator must use judgment, first, in determining the features to be taken into account and their relative weight and, second, in deciding the number of points to assign for each feature in evaluating a job. Studies have shown that there is considerable variation in the results obtained by different evaluators. There is a tendency, for example, for the judgment of an evaluator to reflect the prevailing status and pay of the jobs being evaluated, with the result that the evaluation simply ratifies the existing practices of an employer. This is a problem known as "policy capturing."

A study of comparable worth conducted by the National Academy of Sciences for the Equal Employment Opportunity Commission also cites the problem that the judgment of evaluators may also introduce a systematic bias against women.[25] The potential for introducing bias is vividly illustrated by an evaluation of the job of bindery workers in the U.S. Government Printing Office.[26] Not only did the evaluator assign women fewer points for performing work with the same features (lifting heavy objects and handling confidential material, for example), but the skill and effort required for the work done by women were consistently underestimated. The evaluator gave no points to the women for stitching bindings, for example, because sewing is a skill possessed by most women! The response of supporters of comparable worth is that there are carefully designed, generally reliable techniques for conducting job evaluations that are not open to the critics' charge. The reasons that some workers deserve to be paid more than others refer to factors that are inherent in all jobs. The high pay of truck drivers is deserved because of the long grueling hours and the heavy responsibility of hauling valuable cargo. But nursing is also exhausting work on which the lives of patients often depend. So these two jobs have much in common despite the obvious differences.

The reliability of a comparable worth study depends on how it is conducted. Experts in the field recommend that a committee of employees be formed, so that the choice of factors and the assignment of weights are done by people who are thoroughly familiar with the jobs being evaluated. In the San Jose study, the city's personnel department distributed questionnaires to employees in order to gather firsthand information on each job. After interviews with approximately 20 percent of the respondents, detailed job descriptions were written, and a committee was formed to study and evaluate every job. This committee consisted of an equal number of men and women, and before a decision was made, seven of the ten members had to agree on the number of points to be assigned for each factor. Furthermore, the factors used and their weight can be validated by applying a job evaluation system only to white males, where discrimination is not present, and comparing actual earnings with those predicted by the system.

The Effect on the Labor Market

A third issue in the debate is the charge of many opponents of comparable worth that ignoring market forces will undermine the ability of the market to price and allocate labor in an efficient manner, with a resultant lowering of productivity. One beneficial effect of market forces, for example, is to provide incentives for workers to leave crowded areas with low-productivity jobs for which there is a declining demand and to prepare themselves for more productive work in newer areas where the demand is greater. Paying workers on the basis of job content will remove this valuable incentive. The introduction of comparable worth, moreover, is likely to be accompanied by a complex administrative structure that will increase the hand of government in business decision making and further impair productivity.

A major difficulty with arguments based on the virtues of a free market is that they assume that wages are now set in an efficient manner by the workings of an impersonal economy, when in fact the actual wage setting practices of employers are highly arbitrary and discriminatory. The issues in this dispute are highly complex and form the basis for competing theories in labor economics. In opposition to mainline economists who hold that wages are determined largely by human capital in a competitive market, a growing number of economists, commonly known as "institutionalists," stress the importance of structural features of labor markets and features that are peculiar to specific industries.[27]

One of these structural features is the existence of *internal* labor markets. Most jobs are filled by the transfer or promotion of workers already employed, so the practices and relationships within a firm, including collective bargaining agreements, play a much more important role than the supply of labor on the outside. In addition, some jobs involve unique firm-specific tasks for which employees receive extensive on-the-job training. In the absence of an external market for these jobs, wages cannot be set by the laws of supply and demand, and other means must be used. Institutionalists further cite the existence of *dual* labor markets. Alongside a primary market consisting of "good" jobs, with high status and pay, fringe benefits, and advancement opportunities, there exists a secondary market of low-paying, dead-end, "bad" jobs. The primary market is largely the preserve of white males, while women and minorities are heavily represented in the secondary market.

Institutionalist theories of labor markets suggest strategies for reducing discrimination that are quite different from those of the standard economic view. Instead of encouraging women to prepare themselves for entering male-dominated lines of work (the market solution), it may be more effective to scrutinize the personnel practices of firms and especially the ways in which employees are selected in the internal market for on-the-job training and other opportunities. If the dual market theory is correct, then discrimination is likely to persist unless more "good" jobs are created and upward mobility from "bad" jobs is increased. Compensation systems based on the principle of comparable worth can serve to enhance both of these strategies.

The Issue of Cost

A fourth and final issue in the debate is the cost of implementing a system of compensation based on comparable worth. In addition to the indirect cost of reduced efficiency, there is also the direct cost of raising the wages of women (and some men) in undervalued jobs. Faced with the need to meet the increased payroll cost, employers may be led to increase prices, thus producing inflation, or to lower the wages of the men (and a few women) in predominantly male lines of work. The $1.45 million committed by the city of San Jose to implement comparable worth over two years was still considerably short of the union's estimate of $5.4 million over four years. The Minnesota legislature set aside $21.8 million to be spent in 1983-1984 to eliminate about half of the wage gap that was estimated to be due to discrimination against women among the 9,000 state employees. In 1984, Iowa began the first phase of a comparable worth plan at a cost of $10 million. Estimates of the cost, both direct and indirect, of eliminating wage discrimination nationwide range between $2 billion and $150 billion.

The National Academy of Sciences report concludes that "because of the complexity of market processes, actions intended to have one result may well turn out to have other, even perverse, consequences."[28] One of these unintended consequences might be to harm the women who are supposed to be the beneficiaries. This is because another way for employers to reduce their payroll costs is by increasing capital investment so as to cut down on the number of workers in less productive jobs. And the workers thrown out of jobs will tend to be women with less training and experience. According to one writer, "Nine secretaries working at word processors might become more cost-effective than twelve secretaries working at typewriters."[29] The nine secretaries who are still employed will benefit by higher wages, but the other three will lose.

Advocates of comparable worth respond that the estimates of cost and predictions of unintended consequences are overstated. Helen Remick and Ronnie J. Steinberg point out that

> ... the assumption underlying these estimates is that all wage discrimination in *all* work organizations is going to be rectified all at once and tomorrow.... Most legal reforms that impact upon the labor market have been implemented in stages: either the scope of coverage is initially restricted and gradually expanded to cover a larger proportion of employees over time, or the legal standard is introduced in steps.[30]

Furthermore, if increased pay is owed to women in undervalued jobs as a matter of right, then justice requires that society be willing to bear the burden, no matter what the cost. With regard to the impact on the employment of women, evidence one way or the other is scarce, because comparable worth has not been extensively implemented in the United States. The only available evidence comes from the experience of Australia. Between 1969 and 1975, a system of comparable worth was adopted in stages. In the decade of the 1970s, the ratio of women's earnings to those of men rose from 65 percent to 86 percent. Experts who have studied the Australian experience agree that the economic impact has been slight. One study reports that while the number of women employed continued to rise, due to an expanding economy, the rate was one-third less than it would have been otherwise, and the unemployment among women increased by 0.5 percent.[31] These consequences, while not insignificant, fall short of the dire predictions made before Australia adopted a comparable worth system.

Summary

The comparable worth debate teaches many valuable lessons. It shows us, first, the importance of an accurate understanding of the problem at hand. The existence of a wage gap is a fact, but a great deal of careful analysis is necessary to determine whether the differences in pay are due to discrimination or other causes. Second, the debate forces us to address a fundamental issue in any economic system, namely, how should wages be determined? What is a just wage? One version of the comparable worth principle challenges the basic tenet of our economic system, that our labor is worth what it can command in the marketplace. The content of a job and not supply and demand, its advocates say, should determine what we receive. Even if we accept the version of the principle that seeks to use comparable worth as a remedy for discrimination in the market, we still have to ask, is this the best solution to the problem? Insofar as the wage gap is due to discrimination, it ought to be eliminated. About this there can be no dispute. But many strategies have been proposed for dealing with the problem, and more, perhaps, will be developed. Each solution involves a host of factors that need to be considered before we settle on any one.

Case 9.3 The Mommy Track

Memo

From: Robert C. Begley, President and CEO
To: Maria L. Mendoza, Vice President for Human Resources

Your innovative programs to recruit and promote women in every department of our company have surpassed all expectations. Our recent listing among the top ten of the best companies for working women is a tribute to your dedication and management skill. After studying some recent figures on turnover among the managers in the company and the demographics of those in higher-level positions, however, I see some problems, and I would like your response to some ideas I have for addressing these problems.

First, despite our nationally recognized family-friendly environment, the turnover rate among women managers continues to be two and one-half times the rate for men, which is little different from the experience of other companies. I need not remind you that every employee who leaves is money out the window because of the need to train his or her replacement. As a result, the cost of employing women in our company is greater than the cost of employing men. Second, we have a higher percentage of women executives than most other companies, but a majority of them, over 60 percent, have no children, while 95 percent of male executives do. This suggests to me that although the overall numbers look good, we have not been successful in enabling women with children to reach the executive ranks.

Both of these problems might be due to the family-friendly programs of which we are so proud. We encourage women to take advantage of extended leaves, flexible scheduling, and flextime, and we have permitted some women to share jobs and to work at home through telecommuting. The result has been that women managers have been pleased with the work environment, but too many may have been content to stay at the lower and middle layers of management and not strive for the highest positions. Some women managers are the equal of their male colleagues and are future COO, CFO, and even CEO material. They will be held back in their progress toward these ranks, however, if they take time out or fail to get critical experience.

These thoughts were inspired by reading an article in the *Harvard Business Review* by Felice Schwartz, the founder and president of Catalyst, a research organization that works with business to improve the situation of women.[32] In this article, "Management Women and the New Fact of Life," she identifies two types of women managers in corporations: "career-and-family women," who are "willing to trade some career growth and compensation for freedom from the constant pressure to work long hours," and high-potential "career-primary women," who put their careers first and are willing to make whatever sacrifices are required to reach the top. Ms. Schwartz recommends that women in this latter group be identified early and given the same opportunities as men to gain experience and develop skills. They should also be asked to travel and relocate and to work the same long hours as their male colleagues.

In this vein, I have been considering whether we should recognize two tracks, a "fast track" and a so-called "mommy track." Should we explain to women in management positions the realities of what it takes to get ahead and the consequences of using "family-friendly" programs? We are already doing what Ms. Schwartz recommends to enable career-and-family women to balance conflicting demands. But are we shortchanging career-primary women or women who ought to be encouraged in this direction? I recognize some of the objections to a two-track approach.[33] Ms. Schwartz has been widely criticized for giving male executives a ready excuse to put most women in a second-class, nonpromotable category and for giving women only two inflexible alternatives that preclude shifting gears at different points in a career.

Her proposals have also been charged with placing full responsibility for child rearing on women and trying to keep traditional patterns of promotion.

Still, I think we ought to consider the problems raised above seriously. Please give me the benefit of your thoughts as soon as possible.

FAMILY AND WORK

The increasing prevalence of single parents and two-earner couples with children has led many American companies to assess the impact of this development on their own business and to develop programs that ease the demands on their employees. Traditional attitudes are still present in the American workplace, however, as evidenced by the following two comments. First, from a woman office manager with a working husband and four children:

> When you punch a time clock, your children don't exist. You don't go to the supervisor and tell him that you must go home to care for a sick or injured child. You punch out in the face of many complaints; you will not be paid for the time; and if you do this too often, you will be looking for another job. Things are changing slowly. But when you get down into most companies' production lines, the script still says, "If you want to work, be here and you will get paid."[34]

The second statement comes from a lawyer on the staff of a high-tech manufacturing company who has a working wife and a two-year-old daughter.

> I left private practice and entered the corporate world, in part, to get more control over my life so that I could spend more time with my family.... [When the babysitter has a crisis in her own life] we have to decide who can take the morning off ... or who can take the baby to the office and for how long.... Many friends tell me that their companies are not receptive to the idea of fathers who want to take time off to go to the school-appointed pediatrician. The "that's women's work" attitude persists.[35]

This traditional attitude is changing in response to new realities. Companies that were once reluctant to hire women now find that they have little choice. In addition to antidiscrimination legislation, which provides legal protection for women, the demographics of the work force make the employment of women a practical necessity. An estimated two-thirds of the entrants into the work force for the rest of the century will be women, and about three-fourths of them will have children during some portion of their working career.[36] The cost to employers of ignoring the family burdens of their employees includes lost productivity due to stress, the loss of the investment in training when experienced workers leave, and the failure to attract the best talent in the first place. Many companies are discovering that a family-friendly environment is good business and have developed a range of programs that enable employees to balance family and work life.

Further impetus for change has come from the Family and Medical Leave Act (FMLA) which became effective in August 1993. The FMLA requires companies with 50 or more employees to allow employees up to 12 weeks of unpaid leave for the birth or adoption of a child or for the serious illness or injury of a family member, with the guarantee of the same or an equivalent job upon the employee's return and continued medical benefits. Passage of the FMLA was vigorously opposed by many American businesses, despite the fact that the United States was the last major industrial country to provide some form of maternity leave, and many employers have failed to abide fully by the bill's provisions.[37] The main concern of opponents is the cost of unpaid leave, especially for small companies, but studies suggest that permitting leaves for childbirth is statistically less costly than the alternative of recruiting and training a replacement.[38] Whether employees will still be set back in their careers if they take advantage of the FMLA and other family-friendly programs is also a source of concern for employees and employers alike.

The Problems Facing Employees

The advancement of women in the workplace has been impeded by at least three different factors. First is the unalterable biological function of motherhood; and as Felice Schwartz observes, "Maternity is not simply childbirth but a continuum that begins with an awareness of the ticking of the biological clock, proceeds to the anticipation of motherhood, includes pregnancy, childbirth, physical recuperation, psychological adjustment, and continues on to nursing, bonding, and child rearing." Second, the responsibility of running a household—the "second shift" in Arlie Hochschild's phrase— falls primarily on women in our society, so that women, more than men, find it difficult to be both a parent and a full-time worker. Third are the stereotypes about why women work and what they want in a job. Even women who are fully committed to a career are often assumed to be working temporarily, until the time comes for motherhood; or, if they have a family, they are frequently thought to be less committed than a man because of their dual role. As a result, they are often passed over for promotion or denied the experiences that would lead to advancement.

Motherhood can be accommodated—and historically has been accommodated—by women leaving and reentering the work force. The role of homemaker can be similarly accommodated when women accept less demanding jobs that allow flexibility and do not disrupt family life. Such accommodation rules out "fast track" jobs that involve long hours, unpredictable schedules, and a willingness to relocate. The drawback to such solutions is that women pay a price in career advancement and earning potential. Much of the pay gap is due to interruptions in women's work histories which are incompatible with the expectations for promotion in most companies. Traditionally, careers have been built with a single employer in a slow, steady rise to the top, and to reenter the workforce after an absence is often to start the ascent over again.

Both men and women find it difficult to balance family and work life when emergencies arise, as when a family member suffers a serious illness or is injured in an accident. As the American population ages, caring for the elderly will impose a greater burden on whole families. One specialist on aging predicts, "In the coming years, elder care will have a greater impact on the workplace than child care."[39] Whereas child care generally affects working people in the early years of a career, the responsibility of caring for elderly parents usually falls during people's peak working years, with a significant impact on both employers and employees.

The Cost of Employing Women. Maternity and home responsibilities thus impose a cost, and until recently, this cost has been borne mostly by women themselves. Viewed from an employer's perspective, retaining women in management positions and allowing them to advance is possible only if employers assume some portion of this cost. This is the reality that prompted Felice Schwartz to make a controversial statement in her *Harvard Business Review* article "Management Women and the New Facts of Life." "The cost of employing women in management is greater," she announced, "than the cost of employing men."[40] She immediately added, however, that we must draw the right conclusions from this fact.

First, some of the cost of employing women arises from corporate practices that serve little useful purpose and can easily be altered. "Business as usual" in American corporations reflects to some extent a past era in which men worked and women stayed home. That managers should work from eight o'clock to five and always be present in an office, for example, is a work pattern that was made possible by nonworking wives, but modern communications make flexible working hours and telecommuting possible. Changes like these can reduce costs overall and benefit both women and employers. Other corporate practices result from male perceptions and expectations that are now counterproductive. Schwartz cites as an example the masculine conception of a career as "either an unbroken series of promotions and advancements toward CEOdom or stagnation and disappointment."[41] Women bring similar counterproductive perceptions and expectations to the workplace as a result of social conditioning. As the socialization of both men and women changes, corporations have an opportunity to change those practices that have arisen from the differing perceptions and expectations of both men and women.

Second, much of the cost involved in employing women in management positions is offset by the value that they bring to an employer. Some of the cost of employing women cannot be cut merely by changing corporate practices, but women constitute a valuable resource that employers cannot afford to neglect. Family-friendly programs that enable women to remain employed and advance involve an unavoidable cost to employers, but the money may be well spent when one considers the alternatives. In addition to the cost of recruiting and retraining when women leave, employers incur a cost when they do not compete for the best talent available. Employers who want to hire from the top 10 percent of college graduates or MBAs, now find that about half of the people in this select group are women, and that proportion is likely to increase.

Third, limited career opportunities constitute a form of discrimination that reduces women's choice and their freedom of action. Both men and women are concerned about the impact of work on family life and would prefer employment that enables them to balance family and work more successfully. In one study, 60 percent of employees described the effect on family life as very important in deciding to take a job, and 46 percent cited family-friendly policies as a very important factor in their decision.[42] The strain of balancing family and work life takes a toll on all employees and affects other family members. Among the results of family-friendly programs could be improvements in family well-being and a corresponding reduction in poverty, crime, and other social ills. Aside from considerations of cost, there are substantial moral reasons, rooted in freedom of choice and individual and social welfare, for changing corporate practices that pose problems for employees with families.

Evaluating Family-Friendly Programs

Family-friendly programs take a variety of forms, and companies vary in the mix of programs that they adopt. In general, the programs fall into three groups: (1) resource and assistance programs, such as providing child care, giving guidance about care of the elderly, and counseling employees with family problems; (2) programs for emergencies and occasional needs, such as sick-child care, leaves for childbirth and adoption that go beyond legal requirements, and financial aid for family emergencies; and (3) programs involving flexible working arrangements, including flexible hours, temporary part-time status, working at home by means of telecommuting, and job sharing (in which two people fill a single full-time job together).

Although a few companies have highly commended programs and consistently rank in lists of America's most family-friendly companies, the evidence suggests that most companies have yet to take effective steps.[43] The Families and Work Institute, which publishes *The Corporate Reference Guide*, a benchmarking guide for companies with family-friendly programs, has developed a three-stage classification. Companies at Stage I have several policies but no comprehensive program; at Stage II, companies have developed their policies into a coordinated program; and Stage III companies have begun to change the corporate culture. In the 1991 edition of *The Corporate Reference Guide*, only four companies were identified as being at Stage III.

Employers face a number of problems in becoming family-friendly. Resource and assistance programs are relatively easy to establish, but programs that require changes in work patterns often impose hardships. Allowing a key member of a project to take an extended leave of absence or keeping a job open for an employee on leave may be difficult. Flexible working arrangements may not be suitable for jobs that demand an employee's presence during certain hours or that require the coordination of several employees' schedules. Job sharing may be resisted by clients, for example, who want a single person to handle their business. Employers are also concerned about employee abuse of family-friendly policies and fear that voluntary policies might come to be perceived as entitlements for which they could be sued.[44] Family-friendly policies are apt to be resented by single employees and others without family obligations (see Case 9.4). In addition, the Family and Medical Leave Act of 1993 leaves open questions about what illnesses qualify, who counts as a family member, and what constitutes a calendar year.

Family-friendly programs also create some uncertainties and dangers for employees. Employees who exercise their rights under the FMLA or who take advantage of flextime or other company benefits run the risk of alienating some superiors or of being considered uncommitted. In organizations that have not made changes in the culture, family-friendly policies are often utilized only by women, and the women who utilize them may be identified as being on the "mommy track." Much of the critical reaction to Felice Schwartz's article, "Management Women and the New Facts of Life" (see Case 9.3), was due to the fear that without changes in the corporate culture, her proposals for enabling women to balance family and work life would result in two classes of female employees. Evidence that such a development is already taking place is provided by a study that finds that companies with the best family-friendly programs have some of the worst records for promoting women.[45] This study confirms the suspicion of many women managers that taking advantage of family benefits is dangerous to their careers, and it suggests that programs that concentrate on training and succession planning are more effective in boosting women into top management positions.

Conclusion

Major structural changes in the economy, such as the entry of women into the work force and the rise of the two-earner family, involve costs that must be distributed in some way and require accommodation by both employers and employees. However, the need to balance family and work life raises issues beyond the distribution of costs and the means of accommodation. The central ethical issues are the freedom of employees, men and women alike, to integrate work into a satisfying whole life and the interest of society in minimizing the ills that result when work affects the quality of family life. The political response to these issues has consisted mainly in the enactment of the Family and Medical Leave Act. American corporations have also responded to these issues by developing family-friendly programs, which benefit both employers and their employees. The evaluation of these programs suggests that they cannot be fully effective until significant changes are made in corporate cultures. Until family-friendly programs are perceived as benefits for both men and women and as a normal part of corporate routine, they will fail to address the issues of family and work.

Case 9.4   Is Family-Friendly Always Fair?

Martha Franklin had never seen Bill so angry. It was past 7:30 on a Friday evening, and both were tired from a long day. The November sales report had been due at four o'clock, but Martha's assistant, Janet, had taken the last two days off to be with her ten-year-old daughter, who had undergone emergency surgery to remove an abdominal obstruction. Martha had attempted to pull the figures together herself but was slowed by her unfamiliarity with the new computerized sales-reporting system. As the regional sales director, the report was her responsibility, but she generally relied on others to generate the numbers. Just before lunch, she asked Bill Stevens, one of three district sales managers, to help out until the job was finished. As he dropped the completed report on her desk, he slumped down in a chair and began to complain, calmly at first and then with increasing agitation.

"Don't get me wrong," he said. "I'm willing to do my part, and it's great that Janet was able to spend this time with her daughter. Many employers are not as caring as we are here. But every time someone in this office gets time off to care for a family member, one of us single people takes up the slack. I feel that I'm doing my own job and a bit of everyone else's. If you recall, I spent half the day on Thanksgiving straightening out a billing problem for Frank, so that he wouldn't have to disrupt his family's plans. Many people in the office jealously guard their time, leaving at five sharp in order to attend a son's Little League game or get their children to a birthday party.

"This is a very family-friendly place. But what about those of us without families? It's as if we're expected to be married to the company. No one considers that we have a life to lead too. Also, no one wanted to be transferred to the office in Omaha, but Susan was selected to go because she had no family to relocate. She's been transferred three times while most people with families have managed to stay put. And most of the fringe benefits are for families, so we lose out yet again. We've got a great child-care center but no workout room. It's unfair. This company is discriminating against single people and childless couples, and a lot of us are beginning to resent the unequal treatment."

As Martha heard Bill out, she sympathized with his complaints and wondered what could be done. She supported the family-friendly programs for which the company had received national recognition. Was Bill describing the inevitable trade-off, or could the company treat everyone fairly and yet differently?

NOTES

1. Material for this case is taken from *Lois Robinson* v. *Jacksonville Shipyards, Inc.*, No. 86–927–Civ–J–12 (1991).
2. Rochelle Sharpe, "Women Make Strides, but Men Stay Firmly in Top Company Jobs," *Wall Street Journal*, 29 March 1994, A1.

3. Eliza G. C. Collins and Timothy B. Blodgett, "Sexual Harassment: Some See It … Some Won't," *Harvard Business Review*, 59 (March–April 1981), 76–95.

4. Barbara A. Gutek, *Sex and the Workplace* (San Francisco: Jossey-Bass, 1985), 43–44.

5. *General Electric Co. v. Gilbert*, 419 U.S. 125 (1976).

6. *Newport News Shipbuilding and Dry Dock Co. v. EEOC*, 462 U.S. 669 (1983). For a discussion, see Michael A. Mass, "The Pregnancy Discrimination Act: Protecting Men from Pregnancy-Based Discrimination," *Employee Relations Law Journal*, 9 (1983), 240–50.

7. Catharine A. MacKinnon, *Sexual Harassment of Working Women* (New Haven, CT: Yale University Press, 1979), 193.

8. *Meritor Savings Bank v. Vinson*, 477 U.S. 57 (1986).

9. The expert witness in *Robinson v. Jacksonville Shipyards, Inc.* who made this point was Susan Fiske. The term *sex role spillover* was developed by Veronica F. Nieva and Barbara A. Gutek, *Women and Work: A Psychological Perspective* (New York: Praeger, 1981).

10. Robert C. Ford and Frank McLaughlin, "Sexual Harassment at Work: What Is the Problem?" *Akron Business and Economic Review*, 20 (Winter 1988), 79–92.

11. *Ellison v. Brady*, 924 F. 2d 872 (1991). See also Howard A. Simon, "*Ellison v. Brady*: A 'Reasonable Woman' Standard for Sexual Harassment," *Employee Relations Law Journal*, 17 (Summer 1991), 71–80.

12. *Bundy v. Jackson*, 641 F. 2d 934 (D.C. Cir. 1981).

13. Linda Greenhouse, "High Court to Decide Burden of Accusers in Harassment Cases," *New York Times*, 2 March 1993, A1.

14. *Harris v. Forklift Systems, Inc.*, 510 U.S. 17 (1993).

15. Barbara Presley Noble, "Little Discord on Harassment Ruling," *New York Times*, 13 November 1993, sec. 5, p. 25.

16. *Burlington Industries v. Ellerth*, No. 97–569, and *Faragher v. City of Boca Raton*, No. 97–282.

17. The study by Freada Kelin is reported in Susan Crawford, "A Wink Here, a Leer There: It's Costly," *New York Times*, 28 March 1993, sec. 5, p. 17.

18. "Battling the Bimbo Factor," *Time*, 28 November 1991, 70.

19. Ibid.

20. Martha T. Moore, "Taste Test: Debate Brews over Selling Beer with Sex," *USA Today*, 15 November 1991, 1B.

21. The landmark case is *Schultz v. Wheaton Glass Company*, 21 F. 2d 259 (3d Cir. 1970), *cert. denied* 398 U.S. 905 (1970).

22. *Gunther v. County of Washington*, 452 U.S 967 (1981).

23. See, for example, Guiseppi A. Forgionne and Vivian E. Peters, "Differences in Job Motivation and Satisfaction among Male and Female Managers," *Human Relations*, 35 (1982), 101–18.

24. Sharon Toffey Shepela and Ann T. Viviano, "Some Psychological Factors Affecting Job Segregation and Wages," in *Comparable Worth and Wage Discrimination*, ed. Helen Remick (Philadelphia: Temple University Press, 1984), 47.

25. Donald J. Treimam and Heidi I. Hartmann, eds., *Women, Work, and Wages: Equal Pay for Jobs of Equal Value* (Washington, DC: National Academy Press, 1981).

26. *Thompson v. Boyle*, 499 F. Supp. 1147 (D.D.C. 1980).

27. See P. B. Doeringer and M. J. Piore, *Internal Labor Markets and Manpower Analysis* (Lexington, MA: D. C. Heath, 1971); and Francine Blau, *Equal Pay in the Office* (Lexington, MA: Lexington Press, 1978).

28. Treiman and Hartmann, *Women, Work, and Wages*, 65–66.

29. Michael Evan Gold, *A Dialogue on Comparable Worth* (Ithaca, NY: ILR Press, 1983), 55.

30. Helen Remick and Ronnie J. Steinberg, "Technical Possibilities and Political Realities: Concluding Remarks," in *Comparable Worth and Wage Discrimination*, 290.

31. Robert G. Gregory and Robert C. Duncan, "Segmented Labor Market Theories and the Australian Experience of Equal Pay for Women," *Journal of Post Keynesian Economics*, 3 (1981), 403–28.

32. Felicia Schwartz, "Management Women and the New Facts of Life," *Harvard Business Review*, 67 (January–February 1989), 65–76.

33. "The Mommy Track," *Business Week*, 20 March 1989, 126–34; and reader response in "Is the Mommy Track a Blessing—Or a Betrayal?" *Business Week*, 15 May 1989, 98–99.

34. *Work and Family Policies: The New Strategic Plan* (New York: The Conference Board, 1990), 13.

35. Ibid.

36. Barbara Presley Noble, "The Family Leave Bargain," *New York Times*, 7 February 1993, sec. 3, p. 25.

37. Sue Shellenbarger, "Many Employers Flout Family and Medical Leave Law," *Wall Street Journal*, 26 July 1994, B1, B7.

38. Noble, "Family Leave Bargain."

39. Sue Shellenbarger, "The Aging of America Is Making 'Elder Care' a Big Workplace Issue," *Wall Street Journal*, 16 February 1994, A1.

40. Schwartz, "Management Women and the New Facts of Life," 65.

41. Ibid., 67.

42. Charlene Marmer Solomon, "Work/Family's Failing Grade: Why Today's Initiatives Aren't Enough," *Personnel Journal*, May 1994, 82.

43. Sue Shellenbarger, "If You Want a Firm That's Family Friendly the List Is Very Short," *Wall Street Journal*, 6 September 1995, B1.

44. Sue Shellenbarger, "How Accommodating Workers' Lives Can Be a Business Liability," *Wall Street Journal*, 4 January 1995, B1.

45. Rochelle Sharpe, "Family Friendly Firms Don't Always Promote Females," *Wall Street Journal*, 29 March 1994, B1, B5.

On May 22, 1987, Robert Greeley was abruptly dismissed from his job as a laborer at Miami Valley Maintenance Contractors, Inc., in Hamilton, Ohio.[1] This was a blow not only to the 30-year-old, recently divorced father of two young children, but also to his ex-wife, who was relying on this job for child support payments. Three weeks earlier, a county court judge had ordered that Mr. Greeley's employer withhold the payments from his paycheck as permitted under Ohio law, but his bosses at Miami Valley Maintenance Contractors decided that the bookkeeping involved was too much trouble. Firing him was much easier.

Divorced fathers often fail to make court-ordered child support payments, and judges and legislators have few means for making deadbeat fathers pay. To address this problem, the U.S. Congress enacted the Child Support Enforcement Amendments of 1984 which require states to provide income withholding as a means of collecting payments. The federal law also mandates that states make provisions for fining employers who refuse to withhold such payments. The Ohio General Assembly complied with this federal law by passing legislation the following year. An employer who violates the Ohio law is subject to a $500 fine.

Miami Valley Maintenance Contractors readily admitted that it fired Mr. Greeley to avoid complying with the Ohio law, and it did not contest the $500 fine. The company contended, however, that Robert Greeley, who was not a union member under a contract, was an at-will employee. Accordingly, he could leave his employment at any time, for any reason, and his employer could terminate him with the same ease.

The law in some states prohibits employers from firing for certain kinds of reasons—such as for refusing to break the law or for serving on a jury—because permitting them to do so conflicts with important matters of public policy. However,

Ohio was, at the time, a strict employment-at-will state. Employers could hire and fire at will, with virtually no legal restrictions. In 1986, for example, the state supreme court upheld the firing of a Toledo-area chemist who reported illegal dumping of toxic wastes, even though the employer was eventually found guilty and fined $10 million by the Ohio Environmental Protection Agency. By comparison, Mr. Greeley's employer got off cheaply: The company had to pay a paltry $500 fine for the privilege of firing him.

INTRODUCTION

At first glance, there is nothing remarkable about this case. In the United States, employers are generally regarded as having the right to make decisions about hiring, promotion, and discharge as well as wages, job assignments, and other conditions of work. Employees have a corresponding right to accept or refuse work on the terms offered and to negotiate for more favorable terms. But in the absence of a contract that spells out the conditions under which employment can be terminated, employees can be legally dismissed for any reason—or for no reason at all.

The moral and legal basis for this particular assignment of rights for employers and employees is a doctrine known as *employment at will.* Employment, according to this doctrine, is an "at-will" relation that comes into existence when two parties willingly enter into an agreement, and the relation continues to exist only as long as both parties will that it do so. Employers and employees both have the right to enter into any mutually agreeable arrangement without outside interference. Each party is also free to end an arrangement at any time without violating the rights of the other, as long as doing so is in accord with the terms that they have agreed on. Employment at will is a common-law doctrine long embodied in American labor practice.

The first explicit statement that employment is an at-will relation occurred in an 1877 work by H. G. Wood entitled *A Treatise on the Law of Master and Servant.*[2] The doctrine was first given legal force by an 1884 Tennessee Supreme Court decision in the case *Paine* v. *Western & A.R.R.* In an often-quoted sentence, the court declared, "All may dismiss their employee(s) at will, be they many or few, for good cause, for no cause, or even for a cause morally wrong."[3] Other state courts followed the example of Tennessee, as did the U.S. Supreme Court, so that shortly after the turn of the century, the doctrine of employment at will was firmly established in American law.

A typical decision from the turn of the century is *Lochner* v. *New York* (1905).[4] At issue in this case was an 1897 New York statute limiting the work of bakers to 10 hours a day and 60 hours a week. The law was intended to protect the health of bakers, which was being undermined by the long, exhausting hours they were required to work. This piece of protective legislation was struck down, however, on the grounds that it violated the right of bakers and bakery owners alike to contract on mutually agreeable terms. According to the majority opinion, "the freedom of master and employee to contract with each other in relation to their employment … cannot be prohibited or interfered with, without violating the Federal Constitution." More specifically, the Court ruled that the law violated the due process clause of the Fourteenth Amendment, which stipulates that no state shall "deprive any person of life, liberty, or property, without due process of law."

The reasoning of *Lochner* was embraced in a long series of decisions extending into the 1930s, striking down worker protection legislation and minimum wage laws as similar violations of the due process clause. But in the changed climate of the Depression and the New Deal, the Court abruptly reversed itself, for reasons that are described later in this chapter. The doctrine of employment at will, however, remained firmly entrenched, and it has been only in the past 30 years that some state courts have begun to make exceptions to it.

The task of this chapter is to examine the justification of this doctrine in order to determine the rights that employees have against unjust dismissal and other adverse treatment at the hands of employers. Three arguments are commonly used to justify employment at will. One argument holds that the doctrine is entailed by the rights of property owners; the second argument appeals to the notion of freedom of contract; and the third argument is based on considerations of efficiency. Each of these arguments upholds certain rights of employers, but they can also be used to make a strong case for the rights of employees and to provide greater protection from unjust dismissal than the prevailing legal interpretation of employment at will.

PROPERTY RIGHTS AND EMPLOYMENT AT WILL

One argument for employment at will—the property rights argument—begins with the assumption that both employers and employees have property of some kind. The owner of a factory, for example, owns the machinery and raw materials for the manufacture of a product, along with a certain amount of money for wages. The only resource lacking is labor for operating the machinery and turning the raw materials into a finished product. Labor, or more precisely the productivity of labor, thus has an economic value and can be said to be a kind of "property" that is "owned" by the worker. Employment can be described, therefore, as an exchange of a worker's productive power for the wages that are given out in return by the factory owner.

In this exchange, both parties are free to exercise the rights of property ownership. The owner of the factory is free to utilize the productive resources of the factory and to pay out money as wages in any way that workers are willing to accept. The workers are free to accept work under the conditions and at the wages offered or to seek work elsewhere on more favorable terms. It follows that any restriction on the kinds of agreements that employers and employees can make is a violation of the property rights of one or both parties. Just as consumers are under no obligation to continue buying a product, employers are free to stop "buying" the labor of an employee. Although the loss of a job may create some hardship for the person dismissed, no rights are violated.

The historical roots of the property rights argument are contained in John Locke's idea that there is a *natural* right to property, by which he meant a morally fundamental right that exists apart from any particular legal system. Accepting the biblical belief that God gave the bounty of the earth to all persons in common for the purposes of life, Locke went on to observe that we can make use of this bounty only by appropriating it and making it our own. The fruit of a tree cannot nourish us, for example, until we pluck and eat it, but when one person eats a piece of fruit, that person deprives another of its use. Locke's argument for property as a natural right is based, therefore, on the role that property, including labor, plays in satisfying human needs.

How Important Are Property Rights?

The property rights argument seems to confer a right on the owners of businesses to hire and fire at will. All rights are limited, however, for the simple reason that they inevitably come into conflict with each other and with important societal interests. Government is permitted under the Constitution to place legal restrictions on the use of property as long as "due process" is observed. The doctrine of employment at will is supported by the property rights argument, then, only when it is further assumed that the right to property is of such importance that it takes precedence over all other competing values.

This was the view of the Supreme Court in the decisions of the *Lochner* era when considerations of adequate pay, worker health and safety, the right to organize, and the like were held to be of little account. The *Lochner* era came to an end in 1937 with the decision in *West Coast Hotel* v. *Parrish*, which upheld the constitutionality of a Washington State law setting a minimum wage for women and children.[5] In the majority opinion, Chief Justice Charles Evans Hughes observed that

> … the Constitution does not recognize an absolute and uncontrollable liberty. Liberty in each of its phases has its history and connotation. But the liberty safeguarded is liberty in a social organization which requires the protection of law against the evils which menace the health, safety, morals and welfare of the people.

He continued:

> In dealing with the relation of employer and employed, the legislature has necessarily a wide field of discretion in order that there may be suitable protection of health and safety, and that peace and good order may be promoted through regulations designed to insure wholesome conditions of work and freedom from oppression.[6]

Problems with Locke's Theory

A further objection to an unlimited right to property can be constructed using Locke's own premises. Property rights are fundamental in Locke's political theory because of the role they play in satisfying our basic needs and securing liberty. It can be argued, however, that instead of serving these Lockean ends, the doctrine of employment at will has the opposite effect, namely, the impoverishment of a substantial portion of society and their subjugation to the will of others. Philip J. Levine has observed, for example,

> The notion of an absolute right to property, although originally conceived of as the basis of liberty, has resulted in the subjugation of the working class. It has allowed employers to exercise dominion over their employees, without any corresponding protection of the employee's interests. This freedom from governmental restraint has enabled employers to place unconscionable conditions on employment.
>
> The seriousness of the situation is compounded by the importance of employment to the individual employee. The essential elements of his life are all dependent on his ability to derive income. His job is the basis of his position in society, and, therefore, may be the most meaningful form of wealth he possesses.[7]

If this argument is correct, then something has gone seriously wrong. The explanation, according to C. B. Macpherson, is to be found in two factors, neither of which was anticipated by Locke. One is a change in the conception of property.

> As late as the 17th century, it was quite usual for writers to use the word in what seems to us an extraordinarily wide sense. John Locke repeatedly and explicitly defined men's properties as their lives, liberties, and estates. … One's own person, one's capacities, one's rights and liberties were regarded as individual property. They were even more important than individual property in material things and revenues, partly because they were seen as the source and justification of individual material property.
>
> That broad meaning of property was lost in the measure that modern societies became fully market societies. Property soon came to have only the narrower meaning it generally has today: property in material things or revenues.[8]

The second factor is the unequal distribution of property that prevails in modern society. A certain amount of inequality is justified by Locke insofar as it reflects the differing efforts of individuals. What Locke failed to anticipate is that those who have only their labor to "sell" accelerate the process of transferring the wealth of society to those who are in a position to employ them. As a result of these two factors, Macpherson has claimed:

> Those who have to pay for access to the means of using their capacities and exerting their energies, and pay by making over to others both the control of their capacities

and some of the product of their energies—those people are denied equality in the use and development and enjoyment of their own capacities. And in a modern market society, that amounts to most people....[9]

PROPERTY RIGHTS AND DEMOCRACY

In holding that property rights enable individuals to be free and equal members of a state, Locke failed to recognize another aspect of property rights, pointed out in a classic essay, "Property and Sovereignty" by Morris R. Cohen. Property makes it possible for individuals to be citizens in a democracy, in Locke's view, by providing them with a base of individual power that enables them to stand up against the sovereign power of the state. Without a right to the means for making a living by developing and exercising our own capacities, human beings would be in the position of supplicants who receive the necessities of life as gifts from the state. What Cohen observed is that property gives individuals not only power against the state but also power against each other. The owner of property has not only a right to the exclusive use of some material thing but also a means for gaining control over fellow human beings. Property itself thus constitutes a kind of sovereignty in addition to that of the state.

According to Cohen,

> The extent of the power over the life of others which the legal order confers on those called owners is not fully appreciated by those who think of the law as merely protecting men in their possession. Property law does more. It determines what men shall acquire. Thus, protecting the property rights of a landlord means giving him the right to collect rent, protecting the property of a railroad or a public-service corporation means giving it the right to make certain charges....
>
> From this point of view it can readily be seen that when a court rules that a gas company is entitled to a return of 6 percent on its investment, it is not merely protecting property already possessed, it is also determining that a portion of the future social produce shall under certain conditions go to that company. Thus not only medieval landlords but the owners of all revenue-producing property are in fact granted by the law certain powers to tax the future social product. When to this power of taxation there is added the power to command the services of large numbers who are not economically independent, we have the essence of what historically has constituted political sovereignty.[10]

Because of the many problems with Locke's theory, there is ample reason to limit property rights—and with them the doctrine of employment at will—by recognizing counterbalancing rights of employees in the employment relation. If the justification for property rights is the securing of liberty, and if these rights, when applied to the employment relation, fail to do that, then we are no longer justified in holding property rights as absolute.

Property Rights in a Job

For some philsosophers and legal theorists, the problem lies not with the notion of property rights but with the limited way property is conceived. Employees, according to Locke, have property in their own labor, but it is argued that they also have property rights in a job and in the education and special skills that they put at the service of an employer. If employees have property rights of this kind, then the right of an employer to dismiss "at will," instead of being a right entailed by the employer's property rights, should be viewed as a violation of the property rights of an employee. The property rights argument can thus be turned on its head to undermine rather than to support the doctrine of employment at will. Before jumping to this conclusion, however, it is necessary to defend the claim that employees can be said to have property rights in a job.[11]

Consider the case of Robert Sindermann, who had been a teacher for ten years with the Texas state college system, rising to the position of professor of government and social science at Odessa Junior College. After four years in this position, he was not offered a contract for the next academic year. The reason cited by the college's board of regents was Sindermann's "insubordination." But no specific reasons were given, and his request for a hearing to challenge the dismissal was refused by the regents. During the past academic year, however, Sindermann had served as president of the Texas Junior College Teachers Association, and in this capacity he left the classroom on several occasions to testify before committees of the Texas State Legislature. His name also appeared in a newspaper advertisement that was highly critical of the board of regents.

In the resulting lawsuit, *Perry* v. *Sindermann*, Sindermann claimed that in addition to violating his right to freedom of speech, the board of regents had violated his rights under the Fourteenth Amendment by depriving him of property without due process of law.[12] Because the college did not have a tenure system, Sindermann had no contractual guarantee of continued employment. The college had nevertheless given assurances that any faculty member with seven or more years of service "may expect to continue in his academic position unless adequate cause for dismissal is demonstrated in a fair hearing, following established procedures of due process." Sindermann contended that this promise of continued employment constitutes property, so that he was at least entitled to a hearing of the charges against him.

What Is Property?

A crucial issue in evaluating Sindermann's claim is, what is property? As Morris R. Cohen noted, "Any one who frees himself from the crudest materialism readily recognizes that as a legal term 'property' denotes not material things but certain rights."[13] In a famous passage, R. H. Tawney wrote:

> Property is the most ambiguous of categories. It covers a multitude of rights which have nothing in common except that they are exercised by persons and enforced by

the State. Apart from these formal characteristics, they vary indefinitely in econom-ic character, in social effect, and in moral justification. They may be conditional like the grant of patent rights, or absolute like the ownership of ground rents, ter-minable like copyright, or permanent like a freehold, as comprehensive as sover-eignty or as restricted as an easement, as intimate and personal as the ownership of clothes and books, or as remote and intangible as shares in a gold mine or rubber plantation.[14]

The economic assets of most employees do not consist of land, machines, and the like, which are commonly thought of as productive property. Rather, they consist in the possession of an education and certain specialized skills, which are often certified by a diploma or a license of some kind. The most valuable economic asset of a medical doctor is likely to be the M.D. degree and the license from a state, which permit the doctor to practice. Charles A. Reich has observed that government bestows many entitlements and that these are steadily taking the place of more tra-ditional forms of property.[15] Unless we recognize that the wealth of many employees consists of their education, skills, and government entitlements, these valuable assets will not be protected as well as traditional forms of property.

In *Perry* v. *Sindermann*, the Supreme Court ruled for the first time that a job is a form of property for purposes of the Fourteenth Amendment, so that Sinder-mann and employees in a similar situation are entitled by the due process clause to a fair hearing. In the majority opinion, Justice Potter Stewart, wrote:

> We have made it clear … that "property" interests subject to procedural due process protection are not limited by a few rigid, technical forms. Rather, "property" denotes a broad range of interests that are secured by "existing rules or understandings." A person's interest in a benefit is a "property" interest for due process purposes if there are such rules or mutually explicit understandings that support his claim of entitlement to the benefit and that he may invoke at a hearing.

There are some legal obstacles, however, to the claim that the due process clause applies to cases of private employment where an employee is dismissed with-out a just cause or a fair hearing. *Perry* v. *Sindermann* and similar cases have involved only government and not private employment. The due process clause protects an individual only when deprivation of property occurs by the action of a state, not by a private employer. Furthermore, there must be some basis other than the Constitu-tion for asserting that a property right exists. The Supreme Court has clearly stated that the Constitution does not create property rights but merely extends "various procedural safeguards" to protect them.[16]

Still, a powerful moral argument can be constructed to support the claim that employees such as Sindermann have interests in a job that (morally) ought to be protected. If we accept such an argument, then these moral rights of employees must be weighed against the property rights that have traditionally been accorded to the owners of a business. The right of employees to be protected from unjust dismissal cannot be denied merely by appealing to the property rights of corporations. Employees have property rights, too, and these are just as deserving of protection as the property rights of employers.

THE FREEDOM OF CONTRACT ARGUMENT

Employment can be viewed as a contractual arrangement between employers and employees. This arrangement arises in some instances from an *explicit* contract, a legal document signed by both parties, in which a business firm states the terms under which it is willing to hire a person and that person signifies by his or her acceptance a willingness to work under those terms. Union employees are typically covered by a companywide contract that is agreed to by both the management of a company and the union rank and file. In the absence of an explicit contract, we can still understand the employment relation as involving an *implicit* contract insofar as the conditions of employment are understood and tacitly accepted by both parties.

To place a limit, then, on the kinds of agreements that can be made between an employee and an employer is to limit the right of contract of both parties. Just as it would be a violation of rights to force an employee to remain in a job, so it would be a violation of rights to prevent an employer from terminating an employee who voluntarily entered into an at-will employment relation. This was the reasoning of the Supreme Court in *Lochner*. A more explicit statement of employment at will as a consequence of freedom of contract occurs in a decision upholding the right of an employer to fire an employee for belonging to a labor organization. The majority opinion in this case, *Adair v. United States* (1907), held:

> ... [I]t is not within the functions of government—at least in the absence of contract between the parties—to compel any person, in the course of his business and against his will, to accept or retain the personal services of another, or to compel any person, against his will to perform personal services for another. The right of a person to sell his labor upon such terms as he deems proper is, in its essence, the same as the right of the purchaser of labor to prescribe the conditions upon which he will accept such labor from the person offering to sell it. So the right of the employee to quit the service of the employer for whatever reason is the same as the right of the employer, for whatever reason, to dispense with the services of such employee.... In all such particulars the employer and the employee have equality of right, and any legislation that disturbs that equality is an arbitrary interference with the liberty of contract which no government can legally justify in a free land.[17]

In another case the Court held, "This right is as essential to the laborer as to the capitalist, to the poor as to the rich; for the vast majority of persons have no other honest way to begin to acquire property, save by working for money."[18]

The Philosophical Basis for Freedom of Contract

In the British and American legal traditions, the philosophical basis for the freedom of contract argument, as for the property rights argument, derives from John Locke, who considered the exercise of property rights to be only one part of a more general freedom of action. On the Continent, however, the philosophical basis for freedom of contract derives not from Locke but from Immanuel Kant and his concept of autonomy.

The Kantian argument can be sketched briefly as follows. Autonomy involves the capacity and opportunity to make meaningful choices about matters that bear most significantly on our lives. That is, we are autonomous insofar as it is *we* who make the important decisions affecting our lives and not *others*. An essential part of acting autonomously in this sense is the possibility of making mutually binding voluntary agreements. Therefore, autonomy entails freedom of contract.

Use of the freedom of contract argument to support the doctrine of employment at will is problematic because of the immense difference in bargaining power that usually prevails between employers and employees. Bargaining almost always takes place between parties of different strengths, and the stronger side usually gains at the expense of the weaker. The outcome need not be unjust for this reason. But is there some point at which employers ought not to be permitted to take advantage of their superior bargaining position? The decision in *Lochner* v. *New York* denied that there was any morally significant difference in bargaining strength between bakery workers and their employers. The majority opinion held, "There is no contention that bakers as a class are not equal in intelligence and capacity to men in other trades … or that they are not able to assert their rights and care for themselves without the protecting arm of the state."

As mentioned earlier, the *Lochner* era came to an end in 1937 with the decision in *West Coast Hotel* v. *Parrish*. Chief Justice Hughes, who delivered the majority opinion, cited "an additional and compelling consideration which recent economic experience has brought into a strong light." This consideration is the "exploitation of a class of workers who are in an unequal position with respect to bargaining power and are thus relatively defenseless against the denial of a living wage." The doctrine of employment at will cannot be justified by a right to freedom of contract, according to Chief Justice Hughes, when the result is to deprive employees of the ability to protect their most vital interests. This decision was followed by a series of federal and state worker protection laws that addressed the most serious abuses of employment at will.

Restrictions on Freedom of Contract

Freedom of contract, like the right to property, is not absolute. Outside the employer-employee relation, there are a number of moral and legal limits on the right of individuals to enter into contractual relations. Children and people who are mentally incompetent cannot be parties to a contract, for example. Contracts can be invalidated by showing that they were made under duress or under threats of harm or that they involve misrepresentation or fraud. The courts sometimes refuse to recognize contracts that exploit people's inexperience or ignorance.

The reason for these restrictions is that contracts are valid only when they are made with the genuine consent of both parties. Children and the retarded and insane, for example, do not have the mental or emotional capacity to understand the terms of a contract. Hence, they are not able to give their consent in a meaningful way. A person who is coerced into accepting the terms of a contract through the use of intimidation or force "consents" only in a very artificial sense of the term, because the person lacks any significant degree of choice. When a contract is secured by misrepresentation or fraud—as is the case when a buyer of a house is not told about hidden structural defects or termite damage—the victim's "consent" is also not consent in the full meaning of the term, because the person lacks full knowledge of what is being agreed to.[19] These restrictions are compatible with, and indeed even required by, Locke's conception of freedom.

The same restrictions on freedom of contract are justified when freedom of contract is based on Kant's conception of autonomy. Children and the mentally incompetent lack the capacity for autonomous action to a sufficient degree. And when a person is coerced into agreeing to the terms of a contract—whether it be by physical compulsion or deception of some kind—that person cannot be said to be acting autonomously. Autonomy, along with Lockean freedom, requires that our consent be voluntarily given.

An Autonomy Argument

Although autonomy can be used to support freedom of contract—and with it the doctrine of employment at will—the same concept can also serve as the basis for an argument *against* employment at will.[20] A sketch of the argument is as follows.

If people are to have autonomy, they must then have not only the capacity for autonomous action, but also an acceptable range of alternatives from which to choose. What makes us autonomous is not the mere fact that we are able to make a choice, no matter what the alternatives are. A meaningful account of autonomy must also consider the number and desirability of alternatives available to us. When we choose to do one thing and not another, the alternatives are largely fixed by a social, political, and economic order over which we have little control. A student who wants to become a doctor, for example, cannot merely *choose* to do so. For some, a career in medicine is not a viable alternative. The student may be limited by his or her own abilities and education in comparison with those who are also competing for entry into a medical school. Once trained, a doctor must practice within a system of medical care that is constrained by the resources available, by the demand for services, and by the priorities of the society.

Some people have more alternatives from which to choose than others. A person with few skills and limited resources might "choose" to accept menial, exhausting, and dangerous work at low pay when the alternative is no work at all. But the choices these people make are scarcely the ones they would *like* to make. They would like to be able to choose from a more desirable list of alternatives. To some extent the degree of freedom of choice each of us has is a matter of our own endowments, but it is also due to forces outside ourselves. Complete freedom of choice, in the sense that we are free to make any choice we please from an unlimited list of alternatives, is an unrealistic ideal. And this cannot be what autonomy requires. At the other extreme, it would be absurd to say that slaves are free simply because they can choose to obey their master or accept the consequences. There must be some point between these two extremes where we are justified in saying that a person is acting autonomously.

Once we recognize that the autonomy of a person cannot be separated from the alternatives available to that person, it becomes obvious that employment cannot be analyzed merely as a contractual arrangement between two parties. Any agreement between an employer and an employee takes place in a larger setting that profoundly influences the relative bargaining strength of the two parties and consequently the outcome of their negotiations. Among people with relatively equal bargaining strength with respect to employment, there might be no need for any restrictions on the agreements that are made. But this is seldom the case. To require employees to negotiate a contract with regard to the terms of their employment under conditions of decidedly unequal bargaining strength, then, is not to enhance their autonomy but to diminish it.

How should the employment relation be constructed in a society that values and wishes to advance the autonomy of its members? Adina Schwartz suggests, "A crucial implication is that respecting people as autonomous is not equivalent, as commonly assumed, to leaving them as free as possible from social influences. Instead, a society respects all its members as autonomous to the extent that it assists them in leading a certain kind of life."[21] A kind of life that embodies a realistic view of autonomy, according to Schwartz, is one that allows for "shaping one's circumstances by planning rationally to achieve some over-all conception of one's goals" and "shaping one's goals by rationally criticizing and changing one's over-all conception."[22] In short, a society fosters autonomy to the extent that it makes it possible for people to decide for themselves how they want to live and enables them to implement their decisions effectively.

The two arguments for freedom of contract discussed in this section lead to the same conclusion. Whether the right of employers and employees to contract freely is based on a Lockean conception of freedom or a Kantian conception of autonomy, certain conditions must be satisfied in order for the resulting agreements to be just. There is ample room for controversy over what these conditions are. But to the extent that the necessary conditions are not satisfied, the doctrine of employment at will cannot be supported by the freedom of contract argument.

Efficiency and Employment at Will

The third argument for employment at will is a utilitarian one that relies not on property rights or the freedom of contract but on the importance of this doctrine for the efficient operation of business. The success of any business enterprise depends on the efficient use of all resources, including the resource of labor. For this reason, employers are generally accorded considerable leeway to determine the number of workers needed, to select the best workers available, to assign them to the jobs for which they are best suited, and to discipline and dismiss workers who perform inadequately. The intrusion of factors other than the most efficient allocation of resources into business decision making can only impair efficiency, according to this argument, and thereby harm everyone concerned. Furthermore, legal limitations on the commonly accepted prerogatives of employers puts legislatures and courts in the position of making vital business decisions.

Public Policy Exceptions

Although employment at will has undeniable benefits, most of these go to employers. A utilitarian justification does not allow us to look merely at the consequences of this doctrine for employers but requires that we also consider the consequences for workers and for the whole of society. One of the objections made by Chief Justice Hughes in *West Coast Hotel* v. *Parrish* is that the exploitation of workers that the doctrine of employment at will makes possible is "not only detrimental to their health and well-being but casts a direct burden for their support upon the community. What these workers lose in wages the taxpayers are called upon to pay." When the benefits to employers are balanced against the cost to workers and the other members of society, it may be that employment at will is not an efficient but a very inefficient way of using human resources.

The case of Dr. Grace Pierce is instructive here. As an employee of the Ortho Pharmaceutical Corporation, she was forced to resign after she refused to proceed with tests of a new drug containing what she considered to be a dangerously high level of saccharin. In remanding the case for trial, an appeals court judge ruled that there was a clear and well-defined public policy at issue. Members of the public have a right to be protected against the testing of potentially unsafe drugs, and the public interest is not served by allowing the company to pressure Dr. Pierce to act against her best professional judgment. The trial judge stated that it may be that

> ... public policy will develop to a degree that professionals, even though employees at will, will be permitted to resist what they consider to be a professionally unsound and unethical decision without fear of demotion or discharge.

Dr. Pierce eventually lost her case in the New Jersey Supreme Court, however.[23]

Robert Greeley won his case, and with the decision in *Greeley* v. *Miami Valley Maintenance Contractors, Inc.*, Ohio joined the growing list of states that recognize a public policy exception to the doctrine of employment at will. Ensuring that the children of divorced parents are properly supported is an important matter of public policy, and the means devised by the Ohio state legislature—namely, ordering employers to withhold child support payments from a parent's paycheck—is a reasonable means of achieving a policy objective. Allowing employers to ignore a court order merely by paying a small fine would undermine the child support enforcement mechanism created by the state legislature. The state legislature established a policy, and the state supreme court's justices said, "It is our job to enforce, not frustrate, that policy."

The efficiency argument, then, not only provides some grounds for justifying employment at will but also justifies some limitations for reasons of public policy. Merely admitting that there are some limits, however, does not enable us to determine exactly what these limits are. There are still some good reasons for proceeding cautiously in this area. These are well expressed by the appeals judge in *Pierce* v. *Ortho Pharmaceutical Corporation*:

> We note that a public policy exception would represent a departure from the well-settled common law employment-at-will rule. If such a departure is to be made, care is required in order to insure that the reasons underlying the rule will not be undermined. Most notably in this regard, the employer's legitimate interests in conducting his business and employing and in retaining the best personnel available cannot be unjustifiably impaired. Thus, it cannot change the present rule which holds that just or good cause for the discharge of an employee at will or the giving of reasons therefore are not required. In addition the exception must guard against a potential flood of unwarranted disputes and litigation that might result from such a doctrine, based on vague notions of public policy. Hence, if there is to be such an exception to the at-will employment rule, it must be tightly circumscribed so as to apply only in cases involving truly significant matters of clear and well-defined public policy and substantial violations thereof. If it is to be established at all, its development must be on a case-to-case basis.

Examples of Public Policies. What are some of the "truly significant matters of clear and well-defined public policy" that justify making exceptions to the doctrine of employment at will? First, employees ought not to be subject to demotion, discipline, or discharge for refusing to violate the law. If anything is contrary to public policy, it is a doctrine that permits employers to use the threat of dismissal to force an employee to commit illegal acts. In a 1959 California case, *Petermann* v. *International Brotherhood of Teamsters*, Petermann, the business manager for a local of the Teamsters union, was fired by the union after he refused an order to commit perjury before an investigative committee of the state legislature.[24] The California Court of Appeals held that in order to make the law against perjury effective, some restriction had to be placed on an employer's right of discharge. Another employee was dismissed for being away from work to serve on a jury.[25] Although jury duty is a legal obligation for the persons selected, the court in Oregon did not find that the employer was prevented by any statute from dismissing the employee. Nevertheless, the court ruled in favor of the employee on the grounds that the discharge was for "a socially undesirable motive" that tended to "thwart" the jury system.

Second, employers ought not to prevent employees from receiving the full benefit of their legal rights and entitlements relating to employment. The legislation creating many employee rights include antiretaliation provisions, so there is no need for the courts to create a separate public policy exception. Thus, the National Labor Relations Act and the Occupational Safety and Health Act, among others, not only forbid retaliation but also provide for legal remedies. In many instances, employees have been dismissed for filing workers' compensation claims for benefits that are guaranteed as a matter of right for injuries suffered on the job. In one such case, the court ruled: "When an employee is discharged solely for exercising a statutorily conferred right, an exception to the general rule [of employment at will] must be recognized."[26]

Aside from public policy, the courts have made exceptions under two other conditions. One is the existence of an *implied contract* and the other is the presence of *bad faith and malice.*

An Implied Contract to Continued Employment

In some instances, prospective employees are given assurances in job interviews that dismissal is only for cause, that attempts are made to work through any problems before the company resorts to dismissal, and that due process is followed in all cases. These assurances are conveyed in other instances by employee manuals, policy statements, personnel guidelines and procedures, and other company documents. The claim that an implied contract exists as a result of different kinds of assurances makes an appeal not to a utilitarian justification based on public policy but to the law of contracts. Employee suits have not always been successful, however, as witness the following case.

Walton L. Weiner was hired away from his job at Prentice Hall with promises of secure employment.[27] He was told in a preemployment interview that it was the policy at McGraw-Hill not to dismiss an employee without just cause, and the company's personnel handbook contained a procedure for help in overcoming any problem with an employee's performance. When he was summarily fired after eight years at McGraw-Hill, without just cause and without following the prescribed procedure, he sued his employer for breach of contract. However, the court ruled against Weiner's claim that the company's verbal and written assurances constituted an implied contract. The personnel handbook, for example, was not itself a contract, because it did not spell out the terms of employment, such as salary and duration, and it could be modified at any time by the company. In a dissenting opinion, however, one judge asserted:

> ... I cannot agree that an employee handbook on personnel policies and procedures is a corporate illusion, "full of sound ... signifying nothing."
>
> The application form presented to the employee which required his signature prior to the employment, stated that employment would be subject to handbook

rules. An employee should be able to rely thereon, perhaps to his detriment. The employer should be estopped from acting other than with respect thereto.

In two Michigan cases that were decided together, *Toussaint* v. *Blue Cross and Blue Shield of Michigan* and *Ebling* v. *Masco Corporation,* the plaintiffs were given assurances of job security by their employers as long as they performed satisfactorily.[28] Charles Toussaint testified that he was told by his employer that he would be with the company until the mandatory retirement age of sixty five "as long as I did my job." The supervisory manual at Blue Cross and Blue Shield stipulated that employees could be dismissed only for just cause and that specific disciplinary proceedings were to be used. The court found that his supervisor, in asking him to resign, had not observed these provisions in the manual. Furthermore, the court held:

> While an employer need not establish personnel policies or practices, where an employer chooses to establish such policies and practices and makes them known to its employees, the employment relation is presumably enhanced. The employer secures an orderly, cooperative and loyal work force, and the employee the peace of mind associated with job security and the conviction that he will be treated fairly.

Walter Ebling was given similar assurances of job security at the Masco Corporation as long as he performed satisfactorily, and he testified that he was told that he could be discharged only for just cause and after a review of the case by the company's executive vice president. Ebling claimed that he was discharged merely to prevent him from exercising a stock option that had greatly appreciated in value. The Michigan Supreme Court held that the oral assurances and the company manual create a commitment on the part of the employer with regard to the terms of employment and thus constitute an implied contract.

Bad Faith and Malice

Even without an implied contract, a commonly accepted principle in business is acting in good faith. This concept is applied widely as both a moral and a legal requirement in collective bargaining, contract negotiations, consumer relations, and indeed virtually all commercial dealings. The Uniform Commercial Code, for example, requires that all sales be in good faith, which is defined as "honesty in fact and the observance of reasonable commercial standards of fair dealing in trade." The UCC also holds that an "unconscionable" sales contract is unenforceable. Although the same requirements have not commonly been applied by the courts to the employment relation, there are good moral reasons for doing so.

An example of conspicuous *bad* faith is the case of a 25-year veteran employee of the National Cash Register Company, named Fortune, who was dismissed the next business day after he had secured an order for $5 million worth of equipment to be delivered over the next four years. The court found that the dismissal was motivated by a desire to deprive Fortune of the very substantial commission he would receive as the equipment was sold. The Massachusetts court held that Fortune had an implied contract with his employers and stated that:

> … in every contract there is an implied covenant that neither party shall do anything which will have the effect of destroying or injuring the right of the other party to receive the fruits of the contract, which means that in every contract there exists an implied covenant of good faith and fair dealing.[29]

A California court issued a similar ruling in the case of *Cleary* v. *American Airlines*.[30] Cleary claimed that his dismissal, after eighteen years of satisfactory service, was based on a false accusation of work rule violations and was not in accord with the company's personnel policies. In agreeing with Cleary, the court held that the dismissal was contrary to the "implied-in-law covenant of good faith and fair dealing contained in all covenants." At about the same time, a court in New York ruled in favor of an employee who was fired after thirteen years of service for no reason except to deprive him of his pension benefits, which would be vested in two more years.[31] Although New York State law had no provision for wrongful discharge, the court held that the employer's action "smacks of the unconscionable."

PROTECTING AGAINST UNJUST DISMISSAL

Although the courts have recognized limited exceptions to employment at will, American workers lack the broad protection of a law against unjust dismissal. In this absence of legislation prohibiting termination without sufficient cause, the United States stands alone among virtually all industrialized countries. Union contracts generally contain provisions governing dismissal, and some companies have voluntarily granted employees certain rights. However, an estimated 70 percent of American workers are at-will employees who can be dismissed for any reason, or even no reason. Accordingly, there have been calls for a law to provide at-will employees with the same kind of protection now enjoyed by unionized workers and workers in other industrialized countries.

Is a Law against Unjust Dismissal Needed?

Because some legal protection against unjust dismissal is currently available, the case for a new law depends on the adequacy of the present system, which is based primarily on the right of contract and considerations of public policy. Terminating employees without cause violates a contractual right when they are promised secure employment as long as they perform satisfactorily. In addition, every contract is assumed to include an implied covenant of good faith and fair dealing. It might be argued, therefore, that a right not to be dismissed without cause is already contained in the implied covenant of good faith and fair dealing, but this view has been adopted by the courts only in cases of egregious misconduct. Dismissals that conflict with public policy, such as the firing of Robert Greeley (see Case 10.1) do not violate any employee right; they are legally prohibited because of their impact on society at large.

The effect of a law against unjust dismissal, then, would be to introduce the right not to be dismissed without cause into every employment relation. This right would be guaranteed for all employees and would not depend on an explicit contract or the implied covenant of good faith and fair dealing. Moreover, such a right is meaningful only if there are mechanisms in place for hearing employee complaints and providing a remedy. The remedy should not only provide full compensation for an employee's loss but also constitute an effective deterrent for employers to refrain from dismissing employees unjustly. In short, a right against unjust dismissal is effective only if any wrong committed in discharging an employee is rectified and the incidence of such wrongdoing is minimized.

What is the justification for such a right? One possible argument is that a terminated employee suffers some substantial harm that ought not to be inflicted without an adequate reason. If comparable jobs were readily available, then no employee would be harmed by being forced to change, but when jobs are scarce and the alternatives are less desirable, then the loss of a job is a significant financial blow. The cost of an unjust dismissal is borne not only by the affected individual but also by the whole of society. In addition, our social standing and self-esteem are closely linked to our work, so that a job loss may also result in great psychological harm.

The importance of a job to our welfare is eloquently expressed in the following often-cited passage:

> We have become a nation of employees. We are dependent upon others for our means of livelihood, and most of our people have become completely dependent upon wages. If they lose their jobs they lose every resource, except for the relief supplied by the various forms of social security. Such dependence of the mass of the people upon others for all of their income is something new in the world. For our generation, the substance of life is in another man's hands.[32]

Because a job is so essential to our well-being, it should not be subject to the arbitrary power of an employer. Employers have great power over us, and this power should be exercised responsibly. A law that requires employers to have a good reason for any dismissal is reasonable, then, in view of power that they hold.

There is no need for a new law if the present system works. However, critics charge that current law does not provide predictable and uniform protection for all employees.[33] Employees win a high percentage of wrongful discharge suits that reach a jury; in California the figure exceeds 70 percent. The damage awards in these cases average between $300,000 and $500,000 and often exceed $1 million.[34] But the majority of the successful plaintiffs in these suits are middle- and upper-level managers. Rank-and-file workers win less often, and their awards are much smaller. Protection through the courts is rather like a lottery in which a few employees are big winners and the rest receive little. Employers suffer as well when the outcome of trials is so unpredictable and the cost of losing is so great. In addition, many unjustly dismissed employees do not seek protection in the court because of the high standard in current law. It has been estimated that of the two million employees who are terminated annually, approximately 150,000 to 200,000 would have a legitimate complaint under a "good cause" standard.[35]

The Model Employment Termination Act

The Model Employment Termination Act has been proposed as a guide for the development of state laws. Since its drafting in 1991, no state has chosen to follow the lead of this document. Still, it can serve to illustrate the difficulties involved in legislating a right against unjust dismissal.

The key proposal in the Model Employment Termination Act is that an employer may not terminate the employee without "good cause," which is defined as:

> (i) a reasonable basis for the termination of an individual's employment in view of the relevant factors and circumstances, which may include the individual's conduct, job performance and employment record; and the appropriateness of termination for the conduct involved; or (ii) the good faith exercise of business judgment, which may include setting economic goals and determining methods to achieve those goals, organizing or reorganizing operations, discontinuing or divesting operations or parts of operations, determining the size and composition of the work force, and determining and changing performance standards for positions.[36]

This definition conforms with the concept of "good cause" that has evolved in decades of labor union arbitration, and it does not interfere with decision making on the basis of legitimate business considerations, as long as these are in good faith. Among the potential difficulties of the definition is determining whether an exercise of business judgment is in good faith. For example, an employer might raise the standards of performance merely in order to dismiss a particular employee. In a unionized setting, any raising of standards would be subject to negotiation, but a nonunion worker would have no similar protection.

The model act also contains a waiver provision under which an employee can waive the right not to be dismissed without cause in exchange for the employer's agreement to make a severance payment equal to one month's salary for every year of service.[37] The danger of this provision is that an employer might require all employees to sign a waiver as a condition of employment. Doing so would effectively deprive employees of the option to pursue legitimate termination complaints in court, with the possibility of obtaining a high award, and, at the same time, it would protect employers from court litigation and the risk of paying high awards.

Two final issues concern the method of resolving disputes and the remedy. The Model Employment Termination Act recommends arbitration as the preferred dispute resolution method and reinstatement with lost pay as the preferred remedy. Arbitration works well in a unionized setting, where it often serves as an extension of union-management negotiation and involves the terms of a master contract. Without this context, arbitrators would have little guidance for resolving disputes. Reinstatement, too, is an appropriate remedy only in a unionized setting where the union is able to protect reinstated employees from subsequent retaliation. In addition, reinstatement with lost pay would not constitute a significant deterrent to employers. An alternative to state laws that follow the model act is a federal statute modeled on Title VII and other antidiscrimination laws, which provides for court action and monetary compensation.[38]

Conclusion

This discussion of unjust dismissal is of vital importance for all employees and not just those unfortunate enough to be fired without cause. The reason is that the pivotal doctrine of employment at will pervades the whole employer-employee relation. To be an at-will employee is to be at all times subject to the power of an employer to grant or deny something that is essential for our well-being, namely, a job. The alternative is not necessarily a lifetime guarantee of employment but an assurance that we will be treated fairly.

Employment at will is, fortunately, an idea whose time is past, although it still retains a strong hold in the laws of many states. The main supports, historically, have been property rights and the right of contract, which we have examined in depth. These arguments have been rebutted not by denying these rights but by showing their limits. In particular, the imbalance that exists between employer and employee makes a mockery of the idea that the rights of both are being preserved. When employers can dismiss at will, legitimate property rights of employees and the right of employees to contract freely are severely eroded.

The present-day debate revolves mainly around utilitarian issues.[39] To what extent is the welfare of society advanced by preserving or limiting the traditional prerogatives of employers? Employers typically favor employment at will not because they want to fire without cause but because they would rather avoid the need to account for their personnel decisions in court and face the possibility of stiff punitive awards. Even advocates of greater employee protection recognize the dangers of the courts becoming too deeply involved in business decision making.

Case 10.2   Waiving the Right to Sue

Seven years ago, David Parker thought that he made the right decision. He now realizes that it was the biggest mistake of his life, and he hopes that it can be undone. But the decision, in any event, was a tough one.

At the time, the company was reducing its work force because of declining sales, and a group of dismissed employees had filed a class action suit, alleging that they were terminated in violation of the procedures in the company's own personnel handbook. Previously, the company had been sued by disgruntled former employees for wrongful discharge on a variety of grounds. Although no suit had been successful, the cost of defending the company in court, not to mention the cost in management's time, was substantial. In order to protect the company against employee suits, an offer was made to all senior-level employees: If they would waive their right to sue the company in the event of termination, then the company would guarantee them an enhanced severance package. Dismissed employees regularly receive six months' pay with no continued benefits; employees who signed a waiver releasing the company from any other claims would receive full pay and all benefits for one year. The company emphasized the significance of signing and encouraged employees to take their time and to consult with an attorney. No pressure was placed on anyone to sign the waiver, and employees were assured that the fact that they did or did not sign would not be used in any decision to terminate employment.

David Parker finally decided to sign the waiver. Without consulting an attorney, he concluded that if he were laid off, the enhanced severance package would enable him to maintain his present standard of living until he found a new job, and it would soften the blow of a job loss on his wife and children. He never doubted his ability to land a new job fairly quickly. Fortunately, he had kept his job through previous turbulent times, but now, seven years later, he found himself dismissed at the age of 58 in yet another round of cutbacks. This time, however, finding a new job had proven difficult, and he was still unemployed.

Although his position was officially eliminated, the major responsibilities were assumed by a 38-year-old woman, whom he had trained as his assistant. Five other employees, all over the age of 40, were terminated at the same time and also found their jobs turned over to younger colleagues. After consulting with a lawyer, a decision was made to sue the company for violation of the Age Discrimination in Employment Act of 1967, which protects employees over the age of 40 from dismissal solely on the grounds of age. The six employees, including David Parker, held that the company had no reason to believe that they were any less capable than the employees who remained. The lawyer thought that they had a good chance of prevailing in court, but the waivers posed a problem. Two of the other five employees had also signed waivers not to sue the company; two had joined the company within the past five years and had not been asked to sign; and one refused to sign a waiver seven years ago. The only hope for those who signed a waiver, the lawyer explained, would be for the court to hold they were improperly obtained by the company. They could argue that because of the unequal bargaining power between employers and employees, their consent was not wholly voluntary. Even if they knew what they were signing at the time, they could not anticipate their situation in the future, so asking them to sign away their rights like that is unfair.

Case 10.3 A "State-of-the-Art" Termination

Monday had been the most humiliating day of Bill Collins's life. Rumors of downsizing had been swirling for months, and every computer analyst in Bill's department knew that the ax would fall on some of them. Bets had even been taken on who would stay and who would go. When the news was finally delivered, Bill was not surprised. He also understood the necessity of reducing the computer support staff in view of the merger that had made many jobs redundant, and he felt confident that he would find a new job fairly quickly. What upset him was the manner in which he had been terminated.

Bill arrived in the office at eight o'clock sharp to find a memo on his desk about a 9:30 meeting at a hotel one block away. Because this site was often used for training sessions, he gave the notice little thought. Bill decided to arrive a few minutes early in order to chat with colleagues, but he found himself being ushered quickly into a small conference room where three other people from his department were already seated. His greeting to them was cut short by a fourth person whom Bill had never seen before. The stranger explained that he was a consultant from an outplacement firm that had been engaged to deliver the bad news and to outline the benefits the company was providing for them. Once he started talking, Bill felt relieved: The package of benefits was greater than he had dared hope. All employees would receive full salary for six months plus pay for accrued vacation time; medical insurance and pension contribution would be continued during this period; and the outplacement firm would provide career counseling and a placement service that included secretarial assistance, photocopying and fax service, and office space. The consultant assured the four longtime employees that the company appreciated their years of service and wanted to proceed in a caring manner. It was for this reason that they hired the best consulting firm in the business, one that had a reputation for a "state-of-the-art" termination process.

Bill's relief was jolted by what came next. The consultant informed the four that they were not to return to their office or to set foot inside the corporate office building again; nor were they to attempt to contact anyone still working for the company. (At this point, Bill suddenly realized that he had no idea how many employees might be in other four-person groups being dismissed at the same time.) The contents of their desks would be boxed and delivered to their homes; directories of their computer files would be provided, and requests for any personal material would be honored after a careful review of their contents to make sure that no proprietary information was included. The consultant assured them that all passwords had already been changed, including the password for remote access. Finally, they were instructed not to remain at the hotel but to proceed to a service exit where prepaid taxis were stationed to take them home.

Bill regretted not being able to say goodbye to friends in the office. He would have liked some advance warning in order to finish up several projects that he had initiated and to clear out his own belongings. The manner in which he had been terminated was compassionate up to a point, Bill admitted, but it showed that the company did not trust him. A few days later, Bill understood the company's position better when he read an article in a business magazine that detailed the sabotage that had been committed by terminated employees who had continued access to their employer's computer system. Some disgruntled workers had destroyed files and done other mischief when they were allowed to return to their offices after being informed of their termination. One clever computer expert had previously planted a virtually undetectable virus that remained dormant until he gained access long enough through a co-worker's terminal to activate it. The advice that companies were receiving from consulting firms that specialize in termination was be compassionate, but also protect yourself. Good advice, Bill thought, but the humiliation was still fresh in his mind.

NOTES

1. *Greeley* v. *Miami Valley Maintenance Contractors, Inc.*, 49 Ohio St. 3d 228 (1990); 551 N.E. 2d 981 (1990).

2. H.G. Wood, *A Treatise on the Law of Master and Servant* (Albany, NY: John D. Parsons, Jr., 1877), 134. Cited in Patricia H. Werhane, *Persons, Rights, and Corporations* (Upper Saddle River, NJ: Prentice Hall, 1985), 82.

3. *Paine* v. *Western & A.R.R.*, 81 Tenn. 507, 519–20 (1884).

4. *Lochner* v. *New York*, 198 U.S. 45, 25 S.Ct. 539 (1905).

5. *West Coast Hotel* v. *Parrish*, 300 U.S. 379 (1937).

6. In a landmark decision, *Nebbia* v. *New York*, 291 U.S. 502 (1934), the Supreme Court had previously ruled that states have the power to regulate business—and thereby limit the property rights of owners—for the sake of the public welfare. The decision in *West Coast Hotel* thus extends the precedent of *Nebbia*, which concerned the setting of prices, to matters of employment.

7. Philip J. Levine, "Towards a Property Right in Employment," *Buffalo Law Review*, 22 (1973), 1084.

8. C. B. Macpherson, "Human Rights as Property Rights," *Dissent*, 24 (Winter 1977), 72.

9. Ibid., 74.

10. Morris R. Cohen, "Property and Sovereignty," in *Law and the Social Order* (New York: Harcourt, Brace & Co., 1933), 47.

11. For discussions of a job as property, see Barbara A. Lee, "Something Akin to a Property Right: Protections for Employee Job Security," *Business and Professional Ethics Journal*, 8 (Fall 1989), 63–81; and William B. Gould, IV, "The Idea of the Job as Property in Contemporary America: The Legal and Collective Bargaining Framework," *Brigham Young Law Review* (1986), 885–918.

12. *Perry et al.* v. *Sindermann*, 408 U.S. 593 (1971). See also *Arnett* v. *Kennedy*, 416 U.S. 134 (1974); *Cleveland Board of Education* v. *Loudermill*, 470 U.S. 532 (1985); and *Foley* v. *Interactive Data Corporation*, 47 Cal. 3rd 654, 254 Cal. Rptr. 211 (1988).

13. Cohen, "Property and Sovereignty," 45.

14. R. H. Tawney, *The Acquisitive Society* (New York: Harcourt, Brace & World, 1920), 53–54.

15. Charles A. Reich, "The New Property," *Yale Law Review*, 73 (1964), 733.

16. *Leis* v. *Flynt*, 439 U.S. 438 (1979). There is at least one writer, though, who believes that the due process clause, when combined with all the protective legislation that applies to private employment, can be interpreted to make interests in a job a form of property so that employees can be protected against unjust discharge. See Cornelius J. Peck, "Unjust Discharges from Employment: A Necessary Change in the Law," *Ohio State Law Journal*, 40 (1979), 1–49. This view remains untested in the courts, however.

17. *Adair* v. *United States*, 208 U.S. 161, 28 S.Ct. 277 (1907).

18. *Coppage* v. *Kansas*, 26 U.S. 1, 35 S.Ct. 240 (1914).

19. For a discussion of these restrictions, see Anthony T. Kronman, "Contract Law and Distributive Justice," *Yale Law Journal*, 89 (1980), 472–97.

20. The argument developed here is largely derived from Adina Schwartz, "Autonomy in the Workplace," in *Just Business: New Introductory Essay in Business Ethics*, ed. Tom Regan, (New York: Random House, 1984), 129–66.

21. Ibid., 151.

22. Ibid., 150.

23. *Pierce* v. *Ortho Pharmaceutical*, 84 N.J. 58, 417 A. 2d 505 (1980).

24. *Petermann* v. *International Brotherhood of Teamsters*, 174 Cal. App. 2d 184, 344 P. 2d 25 (1959).

25. *Nees* v. *Hocks*, 272 Or. 210, 536 P. 2d 512 (1975).

26. *Frampton* v. *Central Indiana Gas Company*, 260 Ind. 249, 297 N.E. 2d 425 (1973). In addition, see *Kelsay* v. *Motorola*, 74 Ill. 2d 172, 384 N.E. 2d 353 (1978); *Sventko* v. *Kroger Co.*, 69 Mich. App. 644, 245 N.W. 2d 151 (1976).

27. *Weiner* v. *McGraw-Hill*, 83 A.D. 2d 810 (1981).

28. *Toussaint* v. *Blue Cross and Blue Shield of Michigan* and *Ebling* v. *Masco Corporation*, 408 Mich. 579, 272 N.W. 2d 880 (1980).

29. *Fortune* v. *National Cash Register Company*, 364 N.E. 2d 1251 (1977).

30. *Cleary* v. *American Airlines*, 168 Cal. Rptr. 722 (1980).

31. *Savodnick* v. *Korvettes, Inc.*, 488 F. Supp. 822 (1980).

32. Frank Tannenbaum, *A Philosophy of Labor* (New York: Alfred A. Knopf, 1951), 9.

33. For criticism, see Kenneth A. Strang, "Beware the Toothless Tiger: A Critique of the Model Employment Termination Act," *The American University Law Review*, 43 (1994), 849-924.

34. *Uniform Commissioners' Model Employment Termination Act, Uniform Laws Annotated*, 7A Cumulative Annual Pocket Part (St. Paul, MN: West Publishing, 1997), Prefatory Note.

35. Ibid.

36. Ibid.

37. Ibid.

38. Strang, "Beware the Toothless Tiger," 921-23.

39. One exception is provided by Richard A. Epstein, who argues for retaining contract theory as a basis for the employment relation. See "In Defense of Contract at Will," *University of Chicago Law Review*, 51 (1984), 947–82.

11

Marketing, Advertising, and Product Safety

Case 11.1   Dow Corning's Breast Implants

On May 15, 1995, Dow Corning Corporation filed for federal bankruptcy protection.[1] Founded in 1943 as a joint venture of Dow Chemical Company and Corning Glass Works (later changed to Corning, Inc.), Dow Corning had prospered by making lubricants, sealants, electrical insulators, and other products from silicone. One product, however, silicone breast implants, now threatened to destroy the company.

During the 1980s, as an estimated two million women received silicone breast implants, thousands of women began to experience severe headaches, unexplained rashes, pain in the joints and muscles, weight loss, and extreme fatigue. The women and their doctors were baffled, until a common thread began to emerge: the presence of silicone. The implants themselves often caused a painful hardening of the surrounding tissues, known as capsular contracture. The semiliquid filling in the implants was suspected of bleeding through the pouchlike outer covering, and some women's implants ruptured, releasing silicone that eventually lodged in the liver, spleen, and other internal organs. A variety of autoimmune diseases—in which the body attacks its own tissue as if it were a foreign object—were believed by doctors to result from the bleeding and ruptured implants.

In December 1991, Dow Corning was found guilty in a San Francisco court of manufacturing a defective product and of fraudulently concealing evidence of safety problems. The jury awarded the woman in the case, Mariann Hopkins, $7.3 million (a figure that was upheld on appeal). Dow Corning repeated its contention that scientific evidence showed the implants to be safe, and the company denounced the verdict as a politicization of the issue. At the time, the Food and Drug Administration (FDA) was deciding whether to continue the unrestricted use of breast implants. Before 1976, medical devices of all kinds were unregulated, but a law was passed that year by Congress requiring manufacturers to prove the safety of their products. In

273

April 1992, the FDA ruled that sufficient proof had not been provided, and breast implants were ordered off the market (except for special cases of reconstructive surgery, which were to be closely monitored).

More suits followed, and in April 1994, Dow Corning agreed to contribute $2 billion to a $4.23 billion global settlement fund for women around the world who claimed to suffer because of breast implants. At the time of Dow Corning's bankruptcy filing, 410,000 women had submitted claims for compensation from the fund, and the company faced between 5,000 and 15,000 suits from women who elected not to join the settlement. The bankruptcy filing put the global settlement fund and all unsettled suits in jeopardy. The two parent corporations, Dow Chemical and Corning, had to write off their $374 million investment in the joint venture.

Dow Corning introduced the first version of a breast implant in 1963 for use in reconstructive surgery for breast cancer victims following a mastectomy. The prototype had been developed by Dr. Thomas D. Cronin, a surgeon at Baylor University in Texas. Dow Corning performed no tests on the safety of the implant prior to the introduction, relying instead on a limited study done by Dr. Cronin on dogs. However, the company believed silicone to be biologically inert on the basis of other animal studies and experience with a few other silicone medical devices. In 1964, Dow Corning commissioned an independent laboratory to do a study on the safety of the medical-grade silicone used in all implants. Researchers discovered that silicone injected into dogs caused "persistent, chronic inflammation," but the results were dismissed by Dow Corning officials as a typical foreign body reaction that is not specific to silicone. A 1969 study revealed that injected silicone eventually lodged in the vital organs of test animals, but this result too was not regarded as significant because the company did not advocate the injection of silicone directly into the body.

In 1975, Dow Corning introduced a new type of breast implant that was softer and more pliable. This second-generation version was the result of a crash program in response to the entry of competitors into the market, several of which were aided by ex-employees who built on Dow Corning's existing technology. These upstart companies reduced Dow Corning's market share to 35 percent by providing plastic surgeons with an implant that produced a more natural look and feel. The fact that the newer implants were also suited for breast enlargement and not merely reconstructive surgery fueled an increase in purely cosmetic uses. Within four months the redesigned breast implant was ready to demonstrate to plastic surgeons, and full production was achieved by September. No additional testing was done for the new product even though the changed design used a more liquid form of silicone. Some Dow Corning scientists voiced concerns that greater silicone bleed through the outer covering was possible and constituted an unknown risk. These fears were confirmed when the samples displayed to plastic surgeons became oily to the touch and stained the velvet showcases that were used. Sales personnel were instructed to change the sample often and to wipe them before they were presented to the surgeons.

In news articles based on internal company documents, Dow Corning was charged with conducting inadequate research and with suppressing memos and research reports that questioned the safety of breast implants. One indignant salesperson wrote that selling possibly defective implants "has to rank right up there with the Pinto gas tank." Another person in marketing reported that he had assured a group of doctors "with crossed fingers" that Dow Corning had safety studies underway. A *New York Times* columnist described Dow Corning as "a company adrift without a moral compass." One of many cartoons in the press showed a Dow Corning representative explaining to a group of women, "We're testing breast implants on you to see if they're safe for guinea pigs."

Several recent studies involving large populations of women over extended periods of time have found no increase in autoimmune diseases or any other illnesses among women with implants. Some doctors question whether the health problems of women with implants are due to silicone, because a certain number of women in any population will develop autoimmune diseases. The prestigious *New England Journal of Medicine*, which published one of the studies, questioned the need for the FDA's decision to pull breast implants from the market. If breast implants eventually prove to be safe, this fact will not satisfy critics who fault Dow Corning for going ahead with the product without having sufficient evidence. Nor will it save Dow Corning from its current financial crisis. Ironically, in no year did breast implants account for more than 1 percent of Dow Corning's profits.

INTRODUCTION

Virtually all aspects of marketing—from the development of new products to pricing, promotion, and sales—raise ethical questions that do not always have an easy answer. Advertising, in particular, raises numerous ethical concerns, as does safety in the development of new products, which is at issue in the Dow Corning case. This chapter begins with an examination of marketing in general, with an emphasis on the rights of consumers and anticompetitive marketing practices. Many of these practices have been addressed for more than a century by antitrust law, although the consumer movement that began in the 1960s has produced additional legislation dealing with a broader range of marketing activities. The chapter continues with a section on advertising that examines the ethics of persuasion and the definition of deception. The final section on product safety presents three rival theories concerning the responsibility of manufacturers to compensate victims of defective products.

ETHICAL ISSUES IN MARKETING

Marketing, according to one often-cited definition, "consists of the performance of business activities that direct the flow of goods and services from producer to consumer or user."[2] Within this broad characterization, there are a number of distinct functions, including product development, distribution, pricing, promotion, and sales. Although many of the activities of marketers are relatively free of ethical problems, others have drawn extensive criticism and given rise to a demand for an expanded list of consumer rights.

Consumer Rights

Traditionally, producers are regarded as having the following rights in a free market system:

1. The right to make decisions regarding the products offered for sale, such as their design and style.
2. The right to set the price for products and all other terms of sale, including warranties.
3. The right to determine how products will be made available to consumers (that is, the right to make decisions about distribution).
4. The right to promote products in any way that they choose, including the use of any truthful advertising message.

These rights are limited by the usual rules of fair market exchanges. Thus, products must be as represented; producers must live up to the terms of the sales agreement; and advertising and other information about products must not be deceptive. Except for these restrictions, however, producers are free, according to free market theory, to operate pretty much as they please.

Consumers, on the other hand, are traditionally recognized as having only the right not to buy a product that is offered for sale. The opportunity for consumers to satisfy their needs and desires is thus restricted to a "veto" option. If the goods that consumers want are provided by producers, then all is well, but the rights of consumers are not violated if producers fail for any reason to provide consumers with goods they desire. Moreover, the burden of protecting the interests of consumers falls primarily on consumers themselves. In particular, they have the responsibility for acquiring the information needed to make rational choices. The number-one rule in market exchanges is thus caveat emptor, or buyer beware.

In 1962, President John F. Kennedy proclaimed a four-point bill of rights for consumers:

1. The right to be protected from harmful products.
2. The right to be provided with adequate information about products.
3. The right to be offered a choice that includes the products that consumers truly want.
4. The right to have a voice in the making of major marketing decisions.

These rights are needed, according to consumer advocates, because the right not to buy provides inadequate opportunities for consumers to satisfy their needs and desires. Also, the burden of protecting their own interests is too heavy for consumers to bear, especially in view of the unequal relation between buyers and sellers in present-day markets.

The first two of these rights are now embodied to some extent in consumer protection legislation. The Consumer Product Safety Act (1972), for example, created an independent regulatory body, the Consumer Product Safety Commission, which has the power to issue standards, require warnings, and even ban dangerous products entirely. The Fair Packaging and Labeling Act (1966) requires that containers disclose the ingredients of the product, the amount, and other pertinent information, including nutritional content in the case of food. The Magnuson-Moss Warranty Act (1975) specifies the information and the minimum conditions that must be included in a full warranty and requires that all warranties be written in comprehensible language. The latter two rights, especially the right to a voice, have not been addressed by the law and remain unrealized goals of the consumer movement.

Packaging and Labeling

Consumers need a certain amount of information to make rational choices, and often this information is not easily obtained. Consider the plight of a consumer examining a frozen apple pie in a sealed, opaque cardboard box. Without information on the label, consumers have no practical means for determining the size of the frozen pie, the ingredients used, the nutritional content, or the length of time the product has been sitting in the freezer case. Health-conscious consumers are especially disadvantaged by the welter of claims about low fat and salt content and the unregulated use of words such as *light* and *healthy*. Certainly, the more information consumers have, the better they can protect themselves in the marketplace. The ethical question, though, is, how much information is a manufacturer obligated to provide? To what extent are consumers responsible for informing themselves about the products for sale?

The Fair Packaging and Labeling Act was passed by Congress in 1966 to enable consumers to make meaningful value comparisons. Specifically, the act requires that each package list the identity of the product; the name and location of the manufacturer, packer, or distributor; the net quantity; and, as appropriate, the number of servings, applications, and so on. There are detailed requirements for many specific kinds of products in the Fair Packaging and Labeling Act and other statutes.

The Nutrition Labeling and Education Act (NLEA) of 1990 further requires that the labels on packaged food products contain information about certain ingredients expressed by weight and as a percentage of the recommended daily diet in a standardized serving size. The total number of calories and the number of calories from fat must also be listed along with the percentage of the recommended daily intake of certain vitamins and minerals. In addition, the NLEA lists the health claims that are permissible and defines such terms as *low fat, light,* and *healthy.* Previously, food manufacturers were able to manipulate the information on labels. The amount of fat or salt could be reduced, for example, merely by decreasing the listed serving size. A product labeled "light" could contain a substantial amount of fat.

Manufacturers offer a number of reasons for not providing more information. A detailed listing of amounts of ingredients might jeopardize recipes that are trade secrets; listing the kind of fat would prevent them from switching ingredients to take advantage of changes in the relative prices of different oils; product dating is often misunderstood by consumers, who reject older products that are still good; and packaging has to be designed with many considerations in mind, such as ease in filling, the protection of goods in transit, the prevention of spoilage, and so on. Therefore, the objectives of the Fair Packaging and Labeling Act and the Nutrition Labeling and Education Act, manufacturers argue, need to be balanced against a number of practical constraints.

Pricing

The question of how much information sellers are obligated to provide arises not only in packaging and labeling but also in pricing. The proliferation of products at different prices makes it difficult for consumers to compare even those from the same manufacturer. Price codes that can be understood only by sales personnel put consumers at a disadvantage. The use of a Universal Product Code (UPC) that can be machine read has raised concern about the accuracy of posted prices, and some retailers attempt to reduce costs by not marking prices on individual packages. As a result, some local and state governments now require retailers to mark the price on each product, and the FTC has investigated the accuracy of the price posted on products.

Also, some products have hidden costs. The price of tires, for example, often excludes mounting, balancing, extended warranties, and other extras, which are often mentioned to consumers after a decision has been made to buy (a sales technique known as "low-balling"). Consumers cannot compare two air conditioners without knowing the cost of operating them, because a cheaper but less efficient air conditioner can cost more in the long run. Manufacturers of electrical household appliances are now required by law, therefore, to disclose the amount of energy used in a year and the range of energy consumption for products of the same kind. Comparisons are facilitated by standard units for measuring the relevant factors. Thus, tires are required by federal law to be graded with a tread-wear index and a rating for traction, and insulating materials have an R-value that enables consumers to compare different grades of insulation.

The Disclosure of Information

What is the justification for the Fair Packaging and Labeling Act, the Nutrition Labeling and Education Act, and various other statutes governing price displays, hidden costs, and standardized units of measurement? Consumer protection legislation of this kind seems, at first glance, to run counter to classical economic theory. In a free market system, buyers bear the primary burden of informing themselves about the products offered for sale. Generally, sellers are not obligated to provide complete information but only to avoid misrepresentation, although buyers are entitled to rely on any representations that are made and to make minimal assumptions about the quality of goods and their suitability. These are referred to in law as implied warranties of merchantability and fitness for use. However, beyond the obligations to be truthful and fulfill warranties, both expressed and implied, caveat emptor is the rule of the marketplace.

An alternative rule that underlies much consumer protection legislation is that manufacturers have an obligation to provide relevant information that consumers cannot reasonably obtain for themselves. The rationale for abandoning caveat emptor is that in a modern industrial economy, it is impossible for consumers to acquire sufficient information to protect themselves in the marketplace. A division of the burden on buyers and sellers that might have been appropriate to an earlier age is impractical, therefore, under present-day economic conditions. Because manufacturers already have the information and because they make the key decisions about products, it is more cost effective to place the burden of providing information on them.

Deceptive and Manipulative Marketing Practices

Roughly, marketing practices are deceptive when consumers are led to hold false beliefs about a product. Examples of some common deceptions are markdowns from a "suggested retail price" that is never charged, "introductory offers" that incorrectly purport to offer a savings, and bogus clearance sales in which inferior goods are brought in. Packaging and labeling are deceptive when the size or shape of a container, a picture or description, or the use of terms such as *economy size* and *new and improved* mislead consumers in some significant way. Warranties that cannot easily be understood by the average consumer are also deceptive.

Manipulation is distinguished from deception in that it typically involves no false or misleading claims. Instead, it consists of taking advantage of consumer psychology to make a sale. More precisely, manipulation is noncoercively shaping the alternatives open to people or their perception of those alternatives so that they are effectively deprived of a choice.[3] Examples of relatively harmless forms of manipulation include multiple pricing, such as "3 for $1" and "buy two, get one free," and odd-even pricing, $2.99 instead of $3.00. When customers are accustomed to paying a certain price for a product, such as a candy bar, manufacturers often reduce the amount in order to maintain the same price, a practice known as customary pricing. In 1991, the size of one popular brand of tuna, for example, was reduced without any fanfare from the industry standard of 6 $\frac{1}{2}$ ounces to 6 $\frac{1}{8}$ ounces, a nearly invisible 5.8 percent price increase.[4]

A more objectionable form of manipulation is "bait and switch," a generally illegal practice in which a customer is lured into a store by an advertisement for a low-cost item and then sold a higher priced version. Often the low-cost item is not available, but even if it is, the advertised product may be of such low quality that customers are easily "switched" to a higher priced product. Bait and switch is manipulative not only because consumers are tricked into entering the store but because they enter in a frame of mind to buy. Manipulation can also take place when salespeople use high-pressure tactics. The sales force of one encyclopedia company used deception to gain entry to the homes of prospects by claiming to be conducting advertising research (the questionnaires were thrown away afterward). Another company offered to place a set of encyclopedias "free," provided the family bought a yearly supplement for a certain number of years at a price that exceeded the cost of the encyclopedia set alone. Some groups of people are more vulnerable to manipulation than others, most notably children, the elderly, and the poor.[5] Special care needs to be taken, therefore, in marketing aimed at those groups.

The moral case against deceptive and manipulative marketing needs little explanation, because it rests on the requirement that markets be free of force and fraud. The difficult ethical questions in this area concern the definition of deception and manipulation and the dividing line between acceptable and unacceptable marketing practices. Virtually all companies that market to children, for example, concede the special vulnerability of this group but vigorously deny that their own marketing practices are deceptive or manipulative. Personal selling requires an understanding of consumer psychology and skill in using persuasive techniques. When does effective selling, however, cross the line and become ethically objectionable manipulation?

Marketing Research

Some of the objections to deceptive and manipulative marketing practices apply to marketing research, which also raises questions about privacy of the kind discussed in Chapter 7. Corporations engage in a great amount of systematic information gathering about consumers to aid them in developing new products and planning marketing strategies. Data are collected using all the techniques of social science research, including in-depth surveys, field studies, and controlled laboratory experiments.

One set of problems for marketing research conducted by outside agencies concerns the relation between researchers and clients, including integrity in undertaking research assignments and honesty in interpreting data and presenting results.[6] Another set concerns the treatment of research subjects or respondents. These include manipulating persons into participating in research projects, deceiving them about the purpose of a study, and invading their privacy by use, for example, of one-way mirrors during interviews.[7] Marketing research has been misused to make sale pitches or to generate lists of sales prospects in a practice known as *sugging* (from the acronym for selling under the guise of marketing research). A further threat to privacy comes from the use of research data in the growing field of database marketing, in which retailers, through credit-card records and other information, are able to construct detailed profiles of individual customers.

Anticompetitive Marketing Practices

Most anticompetitive marketing practices are also illegal under the Sherman Act (1890), the Clayton Act (1914), the Federal Trade Commission Act (1914), and the Robinson-Patman Act (1936). Many states also have antitrust statutes that prohibit the same practices. Our task, therefore, is to examine the ethical arguments for prohibiting certain anticompetitive marketing practices. The major anticompetitive marketing practices follow.

Price Fixing. Price fixing is an agreement among two or more companies operating in the same market to sell goods at a set price. Such an agreement is contrary to the usual practice, whereby prices are set in a free market by arm's-length transactions. Most commonly, price fixing is *horizontal*, among different sellers at the same level of distribution, but price fixing can also be *vertical*, when it occurs between buyers and sellers at different levels, such as an agreement between a manufacturer and a wholesaler. Price fixing occurs not only when there is an explicit agreement among competitors to charge similar prices but also when the same result is achieved by other means. Among these are a tacit agreement to follow an industry standard (parallel pricing), the lead of a dominant seller (price leadership), a situation in which one company effectively controls the prices of competitors (administered price), and market allocation, in which competitors agree not to compete in certain geographical areas or with certain buyers.

Resale Price Maintenance. This is a practice whereby products are sold on the condition that they be resold at a price fixed by the manufacturer or distributor. Resale price maintenance is thus a form of vertical price fixing, as described above. There are various reasons for imposing resale price maintenance on retailers, including fostering a prestige image, enabling a larger number of retailers to carry a product, and providing an adequate margin for promotion or service. As a form of price fixing, resale price maintenance prevents prices from being set by the forces of a competitive market.

Price Discrimination. Sellers engage in price discrimination when they charge different prices or offer different terms of sale for goods of the same kind to different buyers. Often this occurs when buyers are located in different geographical regions or vary in size or their access to other sellers. Thus, a seller who gives a discount to large buyers solely by virtue of their size is guilty of discriminating against small buyers. However, bulk discounts and other price differences are legal as long as the same terms are available to all buyers. Price discrimination can be practiced not only by sellers but also by large buyers. The Robinson-Patman Act prevents large buyers, such as chain stores, from demanding and receiving preferential treatment from manufacturers and wholesalers to the detriment of smaller buyers.

Reciprocal Dealing, Tying Arrangements, and Exclusive Dealing. Reciprocal dealing involves a sale in which the seller is required to buy something in return, as when an office supply firm agrees to buy a computer system only on the condition that the computer firm agrees to purchase supplies from the office supply firm. A tying arrangement exists when one product is sold on the condition that the buyer purchase another product as well. An example of a tying arrangement is an automotive supply firm that requires as a condition for selling tires to a service station that the buyer also purchase batteries from the seller. In an exclusive dealing agreement, a seller provides a product—a brand of sportswear, for example—on the condition that the buyer not handle competing brands.

The overall ethical objection to these practices is that they are not fair forms of competition. In particular, anticompetitive practices distort prices by preventing the market from serving as a mechanism for setting prices fairly. The concept of a fair price is subject to varying interpretations, but a price arrived at by mutual agreement under competitive conditions at arm's-length is generally considered to be fair. Classical economic theory assumes that competitive markets are characterized by a number of buyers and sellers who are free to enter or leave the market at any time and by largely undifferentiated products that buyers are willing to substitute one for another. Under such conditions, the forces of supply and demand ensure that goods are fairly priced.

Even when prices are not distorted, anticompetitive marketing practices are objectionable insofar as they lead to monopoly. Where there is a single seller of a product, the forces of supply and demand no longer work and the monopolist is free to manipulate the supply to obtain a higher price than could be achieved under conditions of perfect competition. The primary objective of antitrust legislation, therefore, is to prevent the formation of monopolies, which are clearly incompatible with the operation of a free market. Thus, horizontal price fixing and market allocation permit the creation of monopoly conditions, despite the existence of many sellers, because they act in concert and thus become, in effect, one. Even when monopoly is not the result, price discrimination enables stronger sellers to inflict harm on weaker competitors and thereby gain an additional advantage.

The remaining practices—resale price maintenance, reciprocal dealing, tying arrangements, and exclusive dealing—are objectionable for the reason that they result in market transactions that are less efficient than they would be otherwise. They do not necessarily lead to monopoly, but they still distort the market mechanism for setting prices. The assumption of the courts is that there would be no need for these practices if any of the exchanges involved would be made under perfectly competitive conditions. For example, the need for the automotive supply firm to insist that buyers of tires also purchase batteries would not exist if the batteries represented as great a value as the tires. Similarly, an exclusive dealing agreement would not be necessary if the retailer would handle only one line of sportswear given the opportunity to choose.

ADVERTISING

Advertising pervades our lives. It is impossible to read a newspaper or magazine, watch a television show, or travel the streets of our cities without being bombarded by commercial messages. Although some ads may be irritating or offensive, the better efforts of Madison Avenue provide a certain amount of entertainment. We also derive benefit from information about products and from the boost that advertising gives to the economy as a whole. On the other side of the fence, companies with products or services to sell regard advertising as a valuable, indeed indispensable marketing tool. Approximately 2 percent of the gross national product is currently devoted to advertising. So whether we like it or not, advertising is a large and essential part of the American way of doing business.

A typical definition of advertising, from a marketing textbook, is that it is "a paid nonpersonal communication about an organization and its products that is transmitted to a target audience through a mass medium."[8] So defined, advertising is only one kind of promotional activity. The others are *publicity* (press releases and other public relations efforts that do not involve the purchase of air time or space in the mass media), *sales promotion* (contests, coupons, free samples, and so on, which are not, strictly speaking, forms of communication), and *personal selling* by shop clerks and telephone solicitors (which, of course, is not impersonal and also does not take place through a mass medium). Although most advertising is for a product or a service, some of it is devoted to enhancing the image of a corporation or advancing some issue or cause. Thus, a distinction is commonly made between product advertising on the one hand and corporate or advocacy advertising on the other.[9]

Advertising is widely criticized. Exaggerated claims and outright falsehoods are the most obvious targets for complaints, followed closely by the lack of taste, irritating repetition, and offensive character of many ads. More recently, questions have been raised about the morality of specific kinds of advertising, such as advertising for alcohol and tobacco products and ads aimed at children. Particular ads are also faulted for their use of excessive sex or violence or for presenting negative stereotypes of certain groups. Other critics complain about the role advertising plays in creating a culture of consumerism. Advertising encourages people not only to buy more but also to believe that their most basic needs and desires can be satisfied by products. Finally, there is great concern about the potential of advertising for behavior control.

These objections to advertising have led to calls for government regulation and industry self-regulation. Deceptive advertising is subject to control by the Federal Trade Commission (FTC), but many questions still arise about the definition of deception in advertising. The American Association of Advertising Agencies has adopted a code of ethics that addresses more subtle issues, such as the disparagement of competing products and the use of testimonials and authoritative sources. Although admitting that taste is subjective, the code of ethics also commits agencies to avoid ads that are in poor taste. Perhaps the greatest deterrent to offensive ads is public opinion, which has led to the quick removal of more than one questionable effort by Madison Avenue.

This section examines two ethical issues in advertising, namely, the charge that advertising uses unacceptable means of persuasion and the problem of defining and preventing deceptive advertising.

Persuasion and Behavior Control

In 1957, Vance Packard frightened Americans with his best-selling book *The Hidden Persuaders*, which revealed how advertisers were turning to motivational research to discover the subconscious factors that influence human action. A pioneer in this area, Dr. Ernest Dichter, declared in 1941 that advertising agencies were "one of the most advanced laboratories in psychology" and that a successful advertiser "manipulates human motivations and desires and develops a need for goods with which the public has at one time been unfamiliar—perhaps even undesirous of purchasing."[10] The key to success in advertising, according to Dr. Dichter, is to appeal to feelings "deep in the psychological recesses of the mind" and to discover the right psychological "hook."[11]

These claims are disturbing because of the possibility that advertisers have means of influence that we are powerless to resist. We now know that advertising and propaganda—advertising's political cousin—have limited power to change people's basic attitudes. Still, there is evidence that the techniques of modern advertising are reasonably successful in playing on natural human desires for security, acceptance, self-esteem, and the like so as to influence consumer choices. Advertisers have also discovered that visual images are more powerful than the written word, in part because they bypass our rational thought processes. Finally, constant exposure to advertising in general is bound to have some cumulative psychological effect in creating a consumer society.

Although there is no consensus on the nature and extent of advertising's power to persuade, there is no disputing the fact that this power is substantial. The major difference between critics of advertising and defenders is whether there is anything morally objectionable about the use of this power. What, if anything, is wrong with using the methods of Madison Avenue to bring about changes in consumer behavior? There are two closely related answers to this question. One is given primarily by economists and the other by philosophers. Both of these answers contend that certain kinds of advertising cross the boundary between legitimate persuasion and unacceptable forms of behavior control.

The Dependence Effect. The economist John Kenneth Galbraith coined the term *dependence effect* to describe the fact that present-day industrial production is concerned not merely with turning out goods to satisfy the wants of consumers but also with creating the wants themselves. He has written:

> As a society becomes increasingly affluent, wants are increasingly created by the process by which they are satisfied. This may operate passively. Increases in consumption … act by suggestion or emulation to create wants. Or producers may proceed actively to create wants through advertising and salesmanship. Wants thus come to depend on output.[12]

The dependence effect, in turn, is a consequence of the increasingly planned nature of the American economy. New products with a proven consumer demand, such as a pain reliever, require advertising only for the purpose of competing with products already available. But the introduction of an unfamiliar product, such as (at one time) mouthwash, must be preceded by a campaign designed to ensure a receptive market. People who never before worried about "bad breath" now need to be made to feel that they have a problem that only gargling with mouthwash will solve.

The main significance of the dependence effect lies in the challenge it poses to a fundamental tenet of economic theory which is that production is justified because it satisfies consumer demand. The defense offered by corporate leaders when they are criticized for applying their immense resources to seemingly nonessential consumer goods while ignoring pressing needs is, "We only give the public what it wants." These words are a hollow, self-serving excuse if, as Galbraith and others claim, these same corporations have a good deal to do with determining what the public wants.

More technically, economic theory presupposes that (1) the value of goods arises from their role in satisfying the needs and desires of consumers, and (2) consumers themselves are the best judges of what will best satisfy their needs and desires. (This latter assumption is known as the principle of consumer sovereignty.) It follows that the optimal satisfaction of everyone's needs and desires will be achieved by allowing individuals to express their preferences in a free market. Producers of goods, therefore, do not need to ask whether people would be better off with more books and fewer television sets, because this question is best answered by the free choices of individual consumers.

Galbraith contends that there is a flaw in this argument:

> If the individual's wants are to be urgent they must be original with himself. They cannot be urgent if they must be contrived for him. And above all they must not be contrived by the process of production by which they are satisfied. For this means that the whole case for the urgency of production, based on the urgency of wants, falls to the ground. One cannot defend production as satisfying wants if that production creates the wants.... Production only fills a void that it has itself created.[13]

Furthermore, if Galbraith is right about the dependence effect, one cannot argue for the production of more goods on the grounds that the members of society will be better off. "The higher level of production," Galbraith says, "has, merely, a higher level of want creation, necessitating a higher level of want satisfaction."[14]

The significance of the dependence effect is not confined wholly to economic theory. The creation of wants, aided by advertising, is responsible for a conspicuous feature of American society, in Galbraith's view, namely, the imbalance between the private production of consumer goods and the level of public services. When advertising instills a strong desire for products that can be packaged and sold for a profit by a business firm but not for services—such as road maintenance, recreational areas, public health, police protection, and education—that are typically provided by government, the inevitable result is an abundance of the former and a dearth of the latter.

What Wants Are Worth Satisfying? Galbraith's criticism of advertising has not gone unchallenged. The dependence effect, in Galbraith's formulation, involves a distinction between wants that originate in a person and those that are created by outside forces. F. A. von Hayek has pointed out that almost all wants beyond the most primitive needs for food, shelter, and sex are the result of cultural influences.[15] Thus, desires for art, music, and literature are no less *created* than desires for any consumer product. The creation of the former desires, moreover, is due in part to efforts by painters, composers, and novelists to earn a living. It is a complete non sequitur, therefore, to hold that wants that are created by the forces that also satisfy them are less urgent or important for that reason.

The point that Galbraith is trying to make is that some wants are more worth satisfying than others. What Hayek convincingly demonstrates is that the worth of a want cannot depend on its *source*—that is, on whether it originates within a person or is created by outside forces. A distinction between wants of greater and lesser worth, therefore, cannot be made in the way Galbraith proposes. It may still be possible, however, to rescue Galbraith's point—and with it, his criticism of advertising—by drawing the distinction in another way.

If some wants are more worth satisfying than others, then we need a criterion for making such a distinction. One such criterion is whether a desire is *rational*. Rationality is often described as a matter of the suitability of a desire for achieving some end or purpose. Consider, for example, the desires that advertising creates for expensive brands of liquor or designer clothing by appealing to people's yearning for status. Any clearheaded person should see the absurdity of thinking that status could be achieved merely by what one drinks or wears. A defender of advertising can reply that people do not really believe (irrationally) that they are "buying" status in making certain consumer purchases. Rather, advertising has succeeded in surrounding some products with an aura of status, so that people derive a certain satisfaction from purchasing and using those products.

Consumer behavior suggests, however, that people really do make certain purchases because they want status. Vance Packard reported that in the 1950s people expressed reluctance to buy small cars because they were less safe. Research showed, however, that a process of rationalization was taking place.[16] People wanted large cars for reasons of status but disguised their true motivation as a concern for safety. Today, many ads for luxury cars stress safety so that buyers can assure themselves of the rationality of their decisions, even though status is uppermost in their minds. (The headline of one ad asked, "How important is the elegance of Chrysler Fifth Avenue if it can't protect you in an emergency?")

Rational Persuasion. The main concern of philosophers with advertising is whether the influence it exerts on consumers is consistent with a respect for personal freedom or autonomy. Persuasion is a broad category that ranges from the laudable (such as guidance by parents and teachers) to the sinister (psychoactive drugs, psychosurgery, and torture, for example). Advertising does not involve such extreme methods, of course. Still, advertising that cynically exploits deep-seated emotions or short-circuits logical thought processes can be criticized on the ground that it wrongfully deprives people of a certain amount of freedom in the making of consumer choices.

An advertising technique that might be faulted for this reason is subliminal communication. There is a story, possibly apocryphal, of an experiment in which a movie theater in New Jersey boosted sales of ice cream by flashing split-second messages on the screen during the regular showing of a film.[17] Several other studies have reported a decrease in shoplifting in department stores when exhortations against stealing were mixed with the background music being piped over speakers.[18] A related form of unconscious, if not subliminal, communication is the conspicuous placement of brand-name products in movies, a practice known as *product placement.*[19] Although many people believe that subliminal communication is a commonly used technique in advertising, there is little evidence to establish either its frequency or its effectiveness.[20]

The ethical argument against the use of subliminal communication in advertising—assuming that it is effective in influencing consumer behavior—is quite simple. Richard T. DeGeorge expressed it in the following way:

> Subliminal advertising is manipulative because it acts on us without our knowledge, and hence without our consent. If an ad appears on TV, we can tune it out or change stations if we do not want to be subject to it. If an ad appears in a magazine, we are not forced to look at it. In either case, if we do choose to look and listen, we can consciously evaluate what we see and hear. We can, if we wish, take a critical stance toward the advertisement. All of this is impossible with subliminal advertising, because we are unaware that we are being subjected to the message. The advertiser is imposing his message on us without our knowledge and consent.[21]

A similar argument can be made against product placement. Because moviegoers are unaware that advertising is being directed at them, they may not be prepared to evaluate it critically. Ads in newspapers and magazines and on television are clearly identified as such, so that we can separate them from news, entertainment, and other elements and treat them accordingly. Plugs in movies, under the guise of entertainment, catch us unawares, without our critical faculties at work, so to speak. We are not able to subject them to the same scrutiny as other ads because we do not recognize them for what they are.[22]

In both of these arguments, the main complaint is that certain advertising techniques—namely, subliminal communication and product placement—do not allow people to use their capacity for critical evaluation. The significance of this capacity lies in the role that it plays in freedom of choice. In the view of many philosophers, a choice is free to the extent that a person makes it on the basis of reasons that are considered by that person to be good reasons for acting. Freedom, in this view, is compatible with persuasion, but only as long as the techniques used do not undermine the ability of people to evaluate reasons for or against a course of action.

Defenders of advertising point out that there is some argumentation in advertising, especially in the print media. Ads that cite good reasons for buying a product (by showing its uses, for example) or preferring one product over another (such as low cost or better quality) make a rational appeal to consumers and permit them to evaluate the reasons and decide for themselves. Most attempts at persuasion, including those that involve rational argument, make some emotional or other non-rational appeal. A suitor is unlikely to win the heart of his beloved with logical arguments alone; a romantic setting with candlelight and soft music improves the chances of success. Courtroom lawyers do not rely solely on strong legal arguments to win cases but also on their ability to play on the feelings of jurors.[23] Similarly, good advertising appeals on many levels; it is aesthetically pleasing, intellectually stimulating, and often humorous or heartwarming. In many ads, both rational and nonrational elements are combined for greater effect without reducing people's freedom of choice.

What kind of persuasion, then, would deprive a person of freedom of choice? One possibility put forth by several philosophers is that it is nonrational persuasion that a person could not reasonably be expected to resist.[24] This is a very promising approach, but more needs to be said about what constitutes nonrational persuasion and what a person can reasonably be expected to resist. Even more difficult, though, is drawing the line between resistible and irresistible influences. Stanley I. Benn holds that one test is whether people can be aware of what is happening to them.[25] By this test, he suggests, subliminal communication is probably objectionable because it operates subconsciously, as is advertising directed at children, who have not formed the capacity for critical judgment. Beyond these obvious cases, however, the test is not specific enough to be useful. How much do people have to be able to know, for example, about what is happening to them? For the test suggested by Benn to be accepted, more psychological research would have to be done into the effects of advertising in order to say with any assurance what people can and cannot resist.

Finally, any criterion of rational persuasion must consider the ends for which techniques of persuasion are used and not merely the techniques themselves, considered as *means*. Campaigns against smoking and drug use that employ slick Madison Avenue ads, for example, are not usually thought to be objectionable. But if ads to induce people to smoke and antismoking ads both employ the same nonrational techniques of persuasion—including some that people cannot reasonably be expected to resist—then the moral difference between them cannot be solely a matter of the morality of the means. Although some techniques, such as subliminal communication, may be morally unacceptable means no matter how worthy the end, most of the techniques of advertisers can be put to morally acceptable uses. So the morality of advertising depends, at least in part, on the ends for which certain techniques are used.

Deceptive Advertising

Consider the following examples:

- An ad claims that a children's cereal contains less sugar than an apple. The cereal is 40 percent sugar, and its calories have little nutritional value, whereas an apple contains many nutrients that the cereal lacks.

- A reformulated gasoline is advertised as a "breakthrough in technology" that provides the "highest performance." A rival has used the same process for more than ten years and produces gasoline that is just as good. The company claims that "highest performance" means "the highest performance of which the engine is capable" and does not imply that the gasoline is better than the competitor's.
- Airfares are often advertised for one-way travel, but this price is available only with a round-trip purchase. The advertised airfare also omits applicable airport taxes. The round-trip condition and the possibility of taxes are revealed in small print elsewhere in the ad.
- An ad for a brand of disposable diapers shows a handful of dirt and proclaims, "Ninety days ago this was a disposable diaper." The ad does not reveal that the diaper is recyclable only if the plastic liner is removed, and that no facilities exist in most metropolitan areas to recycle the product.
- A popular fashion magazine produces a separate advertising supplement (called an "outsert") that is packaged with copies of the magazines. The supplement, commissioned by a cosmetic company, uses the magazine's typeface and layout with photographs by the magazine's contributing photographers, but the copy is written by an ad agency instead of the magazine's editorial staff.

Whether these ads are deceptive is not easy to determine because the term *deception* does not have a clear, settled meaning. As a first approximation, something is deceptive if it has a tendency to deceive. In this definition, the deceptiveness of an ad does not depend solely on the truth or falsity of the claims made in it but also on the impact the ad has on the people who see or hear it. It is possible for advertising to contain false claims without being deceptive and for advertising to be deceptive without containing any false claims.

A patently false claim—for a hair restorer, for example—might not actually deceive anyone or even have the potential to do so. Furthermore, there are other advertising claims that are false if taken literally but are commonly regarded as harmless exaggerations or bits of puffery.[26] Every razor blade, for example, gives the closest, most comfortable shave; every tire, the smoothest, safest ride; and every pain reliever, the quickest, gentlest relief. Similarly, we are not misled by suggestive rhetorical expressions, such as AT&T's claim "It's all within your reach" or Apple's ungrammatical "Think different." Some ad copy has no determinate meaning at all and cannot be characterized as either true or false.

Some writers even defend the literal falsehoods and meaningless babble of advertising as legitimate and even socially desirable. Perhaps the best known of these defenders is Theodore Levitt, who compares advertising to poetry:

> Like advertising, poetry's purpose is to influence an audience; to affect its perceptions and sensibilities; perhaps even to change its mind…. [P]oetry's intent is to convince and seduce. In the service of that intent, it employs without guilt or fear of criticism all the arcane tools of distortion that the literary mind can devise. Keats does not offer a truthful engineering description of his Grecian urn. He offers, instead,… a lyrical, exaggerated, distorted, and palpably false description. And he is

thoroughly applauded for it, as are all other artists, in whatever medium, who do precisely this same thing successfully.

Commerce, it can be said without apology, takes essentially the same liberties with reality and literality as the artist, except that commerce calls its creations advertising.... As with art, the purpose is to influence the audience by creating illusions, symbols, and implications that promise more than pure functionality....[27]

In order to see that true claims can still be deceptive, consider an ad for Anacin that prompted a complaint by the FTC in 1973. The ad asserted that Anacin has a unique painkilling formula that is superior to all other nonprescription analgesics. Anacin is composed of two active ingredients, aspirin (400 milligrams) and caffeine (32.5 milligrams), but the sole pain-relieving component is aspirin. Aspirin itself is unique in the way that all chemical compounds are different from each other, and aspirin was superior to any other pain reliever available at that time without a prescription. Therefore, it is literally true that Anacin contains a unique and superior painkilling formula: aspirin. The impression that the ad conveyed, however, was that only Anacin has this superior pain-relieving ingredient (false) and that consequently Anacin itself is superior to competing brands of analgesics containing aspirin (also false).

The basis of the FTC complaint, therefore, was not that the claims made for Anacin are literally false but that they gave rise to, or were likely to give rise to, false beliefs in the minds of consumers. Ads for Anacin also claimed that it causes less frequent side effects. The position of the FTC is that the deceptiveness of this claim does not depend solely on whether it is true or false but also on whether the manufacturer, American Home Products, had sufficient evidence to back it up. That is, unsupported claims that turn out to be true are still deceptive, because, in the words of the court, "a consumer is entitled to assume that the appropriate verification has been performed." Even if Anacin does cause less frequent side effects, the consumer is deceived by being led to believe that there is evidence for the claim when there is not. Whether a claim is deceptive, therefore, depends, in some cases, on the strength of the evidence for it.

The Definition of Deception. The rough definition of deception that has been developed so far—namely, an ad is deceptive if it has a tendency to deceive— is not adequate, either for increasing our understanding of the ethical issues in deceptive advertising or for enforcing a legal prohibition against it. Unfortunately, the FTC has yet to offer a precise legal definition, and none of the attempts by marketing theorists and others to define deception in advertising has been entirely successful.[28] An adequate definition of deception must overcome several obstacles.

First, we need to consider whether the deception is due to the ad or the person. Is an ad deceptive if it creates a false belief in relatively few, rather ignorant consumers or only if it would deceive more numerous, reasonable consumers? Ivan Preston recounts the story of a customer who failed to catch the joke in an ad for a novelty beer, Olde Frothingslosh, that proclaimed it to be the only beer with the foam on the bottom and was outraged to discover that the foam was on the top, like all other beers.[29] Or consider whether it was deceptive for Clairol to advertise in the 1940s that a dye will "color hair permanently."[30] Only a few ignorant people would fail to realize the need to dye new growth. Yet the FTC, employing an ignorant consumer standard, found this claim deceptive. The FTC was called upon more recently to make a ruling in a case against the Ford Motor Company in which data on a mileage test were accompanied by the following qualification:

> You yourself might actually average less, or for that matter more. Because mileage varies according to maintenance, equipment, total weight, driving habits and road conditions. And no two drivers, even cars, are exactly the same.[31]

Would a reasonably intelligent, well-informed person reading this carefully qualified claim conclude that the data given in the advertisement described the mileage for an *average* driver? The FTC, employing the more restrictive reasonable consumer standard, decided that it did and asked Ford for substantiation.

Second, an ad may not actually create a false belief but merely take advantage of people's ignorance. Consider health claims in food advertising. The word *natural*, which usually means the absence of artificial ingredients, evokes images of wholesomeness in the minds of consumers. Yet many food products advertised as natural contain unhealthy concentrations of fat and sugar and are deficient in vitamins and minerals.[32] The makers of some brands of peanut butter advertise their products as cholesterol-free even though cholesterol is present only in animal fats, so no brand of peanut butter contains any cholesterol. Is it deceptive for food advertising to make use of terms such as *natural* and *no cholesterol?* Even though the advertisers' claims may not create false beliefs, they still depend for their effect on people's lack of full understanding.

Central to any definition of deception in advertising is the concept of rational choice. Deception is morally objectionable because it interferes with the ability of consumers to make rational choices, which generally depend on adequate information. But advertising is not intended to produce knowledgeable consumers, and so it should not be faulted for every failure to do so. Also, not every false belief is of such importance that consumers should be protected from it. But there is still a certain standard of rational consumer behavior, and advertising is deceptive when it achieves its effect by false beliefs that prevent consumers from attaining this standard. A proposed definition of deception is the following:

> Deception occurs when a false belief, which an advertisement either creates or takes advantage of, substantially interferes with the ability of people to make rational consumer choices.

Whether an ad *substantially interferes* with the ability of people to make rational consumer choices assumes some view of what choices they would make if they were not influenced by an ad. At least two factors are relevant to the notion of *substantial interference*. One is the ability of consumers to protect themselves and make rational choices despite advertising that creates or takes advantage of false beliefs.

Thus, claims that are easily verified or not taken seriously by consumers are not necessarily deceptive. The second factor is the seriousness of the choice that consumers make. False beliefs that affect the choices we make about our health or financial affairs are of greater concern than false beliefs that bear on inconsequential purchases. Claims in life insurance advertising, for example, ought to be held to a higher standard than those for chewing gum.

Application of the Definition. Both of these factors can be observed in two cases involving the Campbell Soup Company. Campbell ran afoul of the FTC in 1970 when it ran television ads showing a bowl of vegetable soup chock-full of solids.[33] This effect was achieved by placing clear-glass marbles on the bottom of the bowl to hold the solids near the surface. How does this case differ from one in which clear-plastic cubes are used instead of real ice in ads for cold drinks? In each case, false beliefs are created in consumers' minds. The false beliefs that viewers have about the contents of a glass of iced tea as a result of using plastic cubes have no bearing on a decision to purchase the product (an iced-tea mix, for example), whereas a decision to purchase a can of Campbell's vegetable soup can definitely be influenced by the false belief created by the glass marbles. The consumers who buy the soup in the belief that the bowl at home will look like the one in the ad will be disappointed, but not the consumers who buy the iced-tea mix. The soup ads have the potential, therefore, to interfere with the ability of consumers to make rational choices.

In 1991, the Campbell Soup Company was charged again by the FTC for ads stressing the low-fat, low-cholesterol content of some of its soups and linking these qualities to a reduced risk of heart disease.[34] The soups in question have reduced amounts of fat and cholesterol, but the ads failed to mention that they are high in sodium, which increases the risk of some forms of heart disease. In the FTC's judgment, Campbell was implying that its soups could be part of a diet that reduces heart disease, while at the same time refusing to tell consumers how much sodium the soups contain or that salt should be avoided by people concerned about heart disease. Consumers who are unaware of the salt content of canned soups might purchase Campbell products as part of a diet aimed at reducing the risk of heart disease. In so doing, they would be better off buying these products than high-salt soups that are also high in fat and cholesterol. But health-conscious consumers who are aware of the salt content might well make different, more rational consumer purchases instead.

Although these ads may not directly cause consumers to have false beliefs about certain Campbell products, the campaign depends for its success on consumer ignorance about the salt content of its soups and the link between salt and heart disease and thus takes advantage of this ignorance. Whether Campbell would have an obligation to reveal the sodium content of its soups if it did not make health claims is debatable, but having made claims designed to lead people concerned about heart disease to buy its products, Campbell definitely has such an obligation. (Campbell eventually agreed to reveal the sodium content in ads for soups with more than 500 milligrams of sodium in an eight-ounce serving.)[35] The health claims made on behalf of some Campbell soups also involve the two factors that are a part of substantial interference. The salt content of a soup, unlike the amount of vegetable solids, cannot easily be verified by consumers, and the decisions consumers make to protect their health are of great importance. Accordingly, the FTC rigorously scrutinizes health claims in ads and holds them to a higher standard.

PRODUCT LIABILITY

The right of consumers to be protected from harmful products raises innumerable problems for manufacturers. Many products can injure and even kill people, especially if the products are used improperly. Every dangerous product can be made safer at some cost, but is there a limit to the safety improvements that a manufacturer ought to provide? Do manufacturers also have a responsibility to ensure that a product is safe before it is placed on the market? Even if silicone breast implants (see Case 11.1) are eventually shown to be safe, did Dow Corning fail in a responsibility to conduct the appropriate tests? Three theories are commonly used to determine when a product is defective and what is owed to the victims of accidents caused by defective products.[36] Each of these theories appeals to a different ground for its ethical justification, and as legal doctrines, they each have a different source in the law.

The Due Care Theory

This theory holds that manufacturers ought to exercise *due care*. Their obligation is to take all reasonable precautions to ensure that products they put on the market are free of defects likely to cause harm. According to this theory, manufacturers are liable for damages only when they fail to carry out this obligation and so are at fault in some way. One ethical justification for this view is the Aristotelian principle of corrective justice: Something is owed by a person who inflicts a wrongful harm upon another. By failing to exercise due care, a manufacturer is acting wrongly and hence ought to pay compensation to anyone who is injured as a result.

The legal expression of this theory is the view in the law of torts that persons are liable for acts of negligence. Negligence is defined in the *Second Restatement of Torts* (Section 282) as "conduct which falls below the standard established by law for the protection of others against unreasonable risk of harm." The usual standard established by law is the care that a "reasonable person" would exercise in a given situation. Accordingly, a reckless driver is negligent (because a reasonable person would not drive recklessly), and the driver can be legally required to pay compensation to the victims of any accident caused by the negligent behavior.[37] In the case of persons with superior skill or knowledge, the standard is higher. A manufacturer can be assumed to know more than the average person about the product and, hence, can be legally required to exercise a greater degree of care.

The Due Care Standard. The standard of due care for manufacturers or other persons involved in the sale of a product to a consumer, including wholesalers and retailers, covers a wide variety of activities. Among them are the following.

1. *Design.* The product ought to be designed in accord with government and industry standards to be safe under all foreseeable conditions, including possible misuse by the consumer. A toy with small parts, for example, that a child could choke on, or a toy that could easily be broken by a child to reveal sharp edges is badly designed. Similarly, due care is not taken in the design of a crib or a playpen with slats or other openings in which a child's head could become wedged.

2. *Materials.* The materials specified in the design should also meet government and industry standards and be of sufficient strength and durability to stand up under all reasonable use. Testing should be done to ensure that the materials withstand ordinary wear and tear and do not weaken with age, stress, extremes of temperature, or other forces. The wiring in an appliance is substandard, for example, if the insulation cracks or peels, posing a risk of electrical shock.

3. *Production.* Due care should be taken in fabricating parts to specifications and assembling them correctly, so that parts are not put in the wrong way or left out. Screws, rivets, welds, and other ways of fastening parts should be properly used, and so on. Defects due to faulty construction can be avoided, in part, by giving adequate training to employees and creating conditions that allow them to do their job properly. Fast assembly lines, for example, are an invitation to defects in workmanship.

4. *Quality control.* Manufacturers should have a systematic program to inspect products between operations or at the end to ensure that they are of sufficient quality in both materials and construction. Inspections may be done either by trained personnel or by machines. In some programs every product is inspected, whereas in others inspection is done of samples taken at intervals. Records should be maintained of all quality control inspections, and the inspectors themselves should be evaluated for effectiveness.

5. *Packaging, labeling, and warnings.* The product should be packaged so as to avoid any damage in transit, and the packaging and handling of perishable foodstuffs, for example, should not create any new hazard. Also, the labels and any inserts should include instructions for correct use and adequate warnings in language easily understood by users.

6. *Notification.* Finally, the manufacturers of some products should have a system of notifying consumers of hazards that only become apparent later. Automobile manufacturers, for example, maintain lists of buyers, who can be notified of recalls by mail. Recalls, warnings, and other safety messages are often conveyed by paid notices in the media.

One question that arises in the due care theory is whether manufacturers have an obligation to ensure that a product is safe to use as intended or to anticipate all the conditions under which injury could occur. The driver of a 1963 Chevrolet Corvair, for example, was severely injured in a head-on collision when the steering column struck him in the head. In the model of the car he was driving, the steering column was a rigid shaft that extended to the front end of the car. Although this design did not cause the accident, the victim claimed that his injuries were greater as a result of it. General Motors contended that its cars were intended to be used for driving on streets and highways and not for colliding with other objects. Consequently, it had no obligation to make them safe for this latter purpose. A U.S. court of appeals held, however, that due care includes "a duty to design the product so that it will fairly meet any emergency of use which can reasonably be anticipated."[38]

The Problem of Misuse. This duty also extends to foreseeable misuse by the consumer. The owner's manual for the 1976 Mercury Cougar, for example, explicitly stated that the original equipment Goodyear tires should not be used "for continuous driving over 90 miles per hour."[39] A U.S. court of appeals determined that the tread separation on the right-rear tire of a Cougar being driven in excess of 100 miles per hour was not the result of any flaw in the tire. However, the Ford Motor Company should have known that a car designed for high performance and marketed with an appeal to youthful drivers would occasionally be driven at speeds above the safe operating level of the tires. Accordingly, Ford should have warned owners of the Cougar more effectively or else equipped the car with better tires.

Some courts have held companies responsible not only for foreseeable misuse but also for misuse that is actively encouraged in the marketing of a product.[40] General Motors, for example, marketed the Pontiac Firebird Trans Am by entering specially reinforced models in racing competitions and by featuring the car in crash scenes in "antihero scofflaw" motion pictures. A promotion film that had spliced together stunt scenes from these movies was used for promotions in dealers' showrooms. In a suit brought on behalf of the driver of a 1978 Trans Am who was injured when the car went out of control while traveling more than 100 miles per hour, the court found for the plaintiff by invoking a doctrine of "invited misuse."[41]

The Concept of Negligence. The major difficulty with the due care theory is establishing what constitutes due care. Manufacturers have an obligation to take precautions that are more stringent than the "reasonable person" standard, but no means exist for determining exactly how far the obligation of manufacturers extends. The courts have developed a flexible standard derived from Justice Learned Hand's famous formulation of the negligence rule.[42] In this rule, negligence involves the interplay of three factors: (1) the probability of harm, (2) the severity of the harm, and (3) the burden of protecting against the harm. Thus, manufacturers have a greater obligation to protect consumers when injury in an accident is more likely to occur, when the injury is apt to be greater, and when the cost of avoiding injury is relatively minor. These are relevant factors in formulating a standard of due care, but they are not sufficient by themselves to decide every case.

Some standards for design, materials, inspection, packaging, and the like have evolved through long experience and are now incorporated into engineering practice and government regulations. However, these standards reflect the scientific knowledge and technology at a given time and fail to impose an obligation to guard against hazards that are discovered later. Asbestos companies claimed that the danger of asbestos exposure was not known until the 1960s, at which time they instituted changes to make handling asbestos safer. This so-called "state-of-the-art" defense—in which a company contends that it exercised due care as defined by the scientific knowledge and technology at the time—was flatly rejected by the New Jersey Supreme Court decision in *Bashada* v. *Johns-Manville Products Corp.* in 1982.[43] The court reasoned that Johns-Manville ought to have made a greater attempt to discover the hazards of asbestos. In the words of the court: "Fairness suggests that manufacturers not be excused from liability because their prior inadequate investment in safety rendered the hazards of their products unknowable."

As a legal doctrine, the due care theory is difficult to apply. The focus of the theory is on the *conduct* of the manufacturer rather than on the *condition* of the product. So the mere fact that a product is defective is not sufficient for holding that a manufacturer has failed in an obligation of due care; some knowledge is needed about specific acts that a manufacturer performed or failed to perform. Lawsuits based on the theory thus require proof of negligence, which the victims of accidents caused by defective products are often not able to provide. In addition, common law allows for two defenses under the due care theory: contributory negligence and assumption of risk. Just as a manufacturer has an obligation to act responsibly, so too does a consumer. Similarly, if consumers know the dangers posed by a product and use it anyway, then to some extent they assume responsibility for any injury that results.

The Contractual Theory

A second theory is that the responsibility of manufacturers for harm resulting from defective products is that specified in a sales *contract*. The relation between buyer and seller is viewed in this theory as a contractual relation, which is subject to the terms of a contract. Even in the absence of an explicit, written contract, there may still be an implicit, understood contract between the two parties that is established by their behavior. This fact is recognized by the Uniform Commercial Code (UCC), Section 2-204(1), which states that "A contract for sale of goods may be made in any manner sufficient to show agreement, including conduct by both parties which recognizes the existence of such a contract."

One of the usual understandings is that a product be of an acceptable level of quality and fit for the purpose for which it is ordinarily used. These implicit contractual provisions are part of what is described in Section 2-314 of the UCC as an *implied warranty of merchantability*. Manufacturers have both a moral and a legal obligation, therefore, by virtue of their contractual relation, to offer only products free from dangerous defects. A person who buys a new automobile, for example, is entitled to assume that it will perform as expected and that nothing in the design makes it especially hazardous in the event of an accident.

There is also an *implied warranty of fitness* for a particular purpose when the buyer is relying on the seller's expertise in the selection of the product. In addition, an *express warranty* is created, according to Section 2-313 of the UCC, as follows:

> Any affirmation of fact or promise made by the seller to the buyer which relates to the goods and becomes part of the basis of the bargain creates an express warranty that the goods shall conform to the affirmation or promise.

The notion of an affirmation is very broad and includes any description or illustration on a package or any model or demonstration of the product being used in a certain way.

The ethical basis for the contractual theory is fairness in commercial dealings. Agreements to buy or sell a product are fair only when they are entered into freely by the contracting parties. Freedom in such agreements entails, among other things, that both buyers and sellers have adequate information about the product in question. Consumers know that the use of many products involves some danger, and they voluntarily assume the risk when the nature and extent of the hazards are revealed to them. Manufacturers may not take unfair advantage of consumers by exposing them to the risk of harm from hazards that are not disclosed. Selling a product that the manufacturer knows to be dangerous, without informing consumers, is a form of deception, because crucial information is either suppressed or misrepresented. Even when the manufacturer is unaware of a defect, the cost of any accident caused by a defective product still ought to be borne by the manufacturer, because the product was sold with the understanding that it posed no hazards except those already revealed to consumers.

Objections to the Contractual Theory. One objection to the contractual theory is that the understandings in a sales agreement, which are the basis for implied and express warranties, are not very precise. Whether a product is of an acceptable level of quality or is fit for the purpose for which it is ordinarily used is an extremely vague standard. In practice, the theory leaves consumers with little protection, except for grossly defective products and products for which the manufacturer makes explicit claims that constitute express warranties.

Second, a sales agreement may consist of a written contract with language that sharply limits the right of an injured consumer to be compensated. If buyers and sellers are both free to contract on mutually agreeable terms, then the sales agreement can explicitly disclaim all warranties, express or implied. Section 2-316 of the Uniform Commercial Code provides for the exclusion or modification of an implied warranty of merchantability as long as (1) the buyer's attention is drawn to the fact that no warranty is being given, with expressions such as "with all faults" or "as is"; (2) the buyer has the opportunity to examine the goods; and (3) the defect is one that can be detected on examination. If a consumer signs a contract with limiting language or explicit disclaimers, then, according to the contractual theory, the terms of that contract are binding.

Henningsen v. Bloomfield Motors. Both of these objections are illustrated in the classic court case in warranty law *Henningsen* v. *Bloomfield Motors, Inc.*[44] Claus Henningsen purchased a new 1955 Plymouth Plaza "6" Club Sedan for use by his wife,

Helen. Ten days after taking delivery of the car from a Chrysler dealer in Bloomfield, New Jersey, Mrs. Henningsen was traveling around 20 miles per hour on a smooth road when she heard a loud noise under the hood and felt something crack. The steering wheel spun in her hands as the car veered sharply to the right and crashed into a brick wall. Mrs. Henningsen was injured and the vehicle was declared a total wreck by the insurer. At the time of the accident, the odometer registered only 468 miles.

In the sales contract signed by Mr. Henningsen, the Chrysler Corporation offered to replace defective parts for 90 days after the sale or until the car had been driven 4,000 miles, whichever occurred first, "if the part is sent to the factory, transportation charges prepaid, and if examination discloses to its satisfaction that the part is defective." The contract further stipulated that the obligation of the manufacturer under this warranty is limited to the replacement of defective parts, which is "in lieu of all other warranties, expressed or implied, and all other obligations or liabilities on its part." By this language, liability for personal injuries was also excluded in the contract.

The question, as framed by the court, is simple:

> In return for the delusive remedy of replacement of defective parts at the factory, the buyer is said to have accepted the exclusion of the maker's liability for personal injuries arising from the breach of warranty, and to have agreed to the elimination of any other express or implied warranty. An instinctively felt sense of justice cries out against such a sharp bargain. But does the doctrine that a person is bound by his signed agreement, in the absence of fraud, stand in the way of any relief?

In giving an answer, the court decided that considerations of justice have greater force than an otherwise valid contract. Furthermore, the main conditions for a valid contract—namely, that the parties have roughly equal bargaining power and are able to determine the relevant facts for themselves—are absent in this case.

First, there is a gross inequality of bargaining power between consumers and manufacturers. Virtually all American cars at the time were sold using a standardized form written by the Automobile Manufacturers Association which the dealer was prohibited from altering. Due to the lack of competition among manufacturers with respect to warranties, consumers had no choice but to buy a car on the manufacturer's terms—or else do without, which is not a genuine alternative in a society where an automobile is a necessity. Hence, consumers did not have freedom of choice in any significant sense, and manufacturers were not offering consumers what they truly wanted. Consumers would most likely prefer to buy cars with better warranties.

Second, consumers are also at a profound disadvantage in their ability to examine an automobile and determine its fitness for use. They are forced to rely, for the most part, on the expertise of the manufacturer and the dealer to ensure that a car is free of defects. Further, the relevant paragraphs in the contract itself were among the hardest to read, and there was nothing in them to draw the reader's attention. "In fact," the court observed, "a studied and concentrated effort would have to be made to read them."

The Strict Liability Theory

A third theory, now gaining wider acceptance in the courts, holds that manufacturers are responsible for all harm resulting from a dangerously defective product even when due care has been exercised and all contracts observed. In this view, which is known in law as *strict liability*, a manufacturer need not be negligent nor be bound by any implied or express warranty to have responsibility. The mere fact that a product is put into the hands of consumers in a defective condition that poses an unreasonable risk is sufficient for holding the manufacturer liable.

A more precise account of the theory of strict liability is given in Section 402A of the *Second Restatement of Torts* as follows:

1. One who sells any product in a defective condition unreasonably dangerous to the user or consumer or to his property is subject to liability for physical harm thereby caused to the ultimate user or consumer, or to his property, if (a) the seller is engaged in the business of selling such a product, and (b) it is expected to and does reach the user or consumer without substantial change in the condition in which it is sold.
2. The rule stated in Subsection (1) applies although (a) the seller has exercised all possible care in the preparation and sale of his products, and (b) the user or consumer has not bought the product from or entered into any contractual relation with the seller.

Is Privity Necessary? The provision of 2(b) addresses an important legal issue in both the due care and the contract theories. Generally, lawsuits under either theory have required that the victim of an accident be in a direct contractual relation with the manufacturer. This relation is known in law as *privity*. Suppose an accident is caused by a defective part that is sold to a manufacturer by a supplier, and the finished product is sold to a wholesaler, who sells it to a retailer. The consumer, under a requirement of privity, can sue only the retailer, who can sue the wholesaler, who in turn can sue the manufacturer, and so on.

The requirement of privity developed as a way of placing reasonable limits on liability, because the consequences of actions extend indefinitely. In a simpler age when goods were often bought directly from the maker, this rule made sense. With the advent of mass production, however, most goods pass through many hands on the way to the ultimate consumer, and the requirement of a direct contractual relation greatly restricts the ability of consumers to collect compensation from manufacturers. In the landmark case *MacPherson* v. *Buick Motor Company* (1916), the New York State Court of Appeals ruled that privity was not necessary when there is negligence.[45] Negligence was present, according to the decision, because the defect in the wooden wheel supplied by another manufacturer should have been detected during the assembly of the car.

The main blow to privity in the contractual theory came in *Baxter* v. *Ford Motor Company* (1934).[46] The Supreme Court of Washington held that a driver who was injured by flying glass when a pebble struck the windshield had a right to compensation because all Ford cars were advertised as having Triplex shatterproof glass—"so made that it will not fly or shatter under the hardest impact." Because the truth of this claim could not easily be determined by an ordinary person, buyers have a right to rely on representations made by the Ford Motor Company. Hence, the wording of Ford's advertisements creates a warranty, in the view of the court, even without a direct contractual relation.

Legal Issues in Strict Liability. Strict liability as a legal doctrine did not make much headway in the courts until 1963, when the California State Supreme Court ruled in *Greenman* v. *Yuba Power Products.*[47] The relevant facts are that for Christmas 1955, Mr. Greenman's wife gave him a multipurpose power tool, called a Shopsmith, which could be used as a saw, a drill, and a lathe. Two years later, while using the machine as a lathe, the piece of wood he was turning flew out of the machine and struck him on the forehead. Expert witnesses testified that some of the screws used to hold parts of the machine together were too weak to withstand the vibration.

The court declined to consider whether Yuba Power Products was negligent in the design and construction of the Shopsmith or whether it breached any warranties, either express or implied. The only relevant consideration, according to the decision, was the fact that the tool was unsafe to use in the intended way. Specifically, the court held:

> To establish the manufacturer's liability it was sufficient that the plaintiff proved that he was injured using the Shopsmith in a way it was intended to be used as a result of a defect in design and manufacture of which plaintiff was not aware that made the Shopsmith unsafe for its intended use.

Section 402A was formulated a year later in 1964. Since that time, all 50 states and the District of Columbia have adopted the doctrine of strict liability as expressed in the *Second Restatement of Torts.*

The wording of Section 402A raises two questions of definition: What is a "defective condition" and what does it mean to say that a product is "unreasonably dangerous"? Generally, a product is in a defective condition either when it is unsuitable for use as it is intended to be used or when there is some misuse that can reasonably be foreseen and steps are not taken to prevent it. A ladder that cannot withstand the weight of an ordinary user is an example of the first kind of defect; a ladder without a label warning the user against stepping too high is an example of the second. A defect in a product can include a wide range of problems—from poor design and manufacture to inadequate instructions or warnings.

The definition of "unreasonably dangerous," offered in a comment on Section 402A, is "The article sold must be dangerous to an extent beyond that which would be contemplated by the ordinary consumer who purchases it, with the ordinary knowledge common to the community as to its characteristics." This definition is inadequate, however, because it implies that a product is not unreasonably dangerous if most consumers are fully aware of the risks it poses. All power lawn mowers are now required by federal law to be equipped with a "kill switch," which stops the engine when the handle is released. Although the dangers of power mowers are obvious to any user, a machine without a "kill switch" is (arguably) unreasonably dangerous.

The Ethical Arguments for Strict Liability. The ethical arguments for strict liability rest on the two distinct grounds of efficiency and equity. One argument is purely utilitarian and justifies strict liability for securing the greatest amount of protection for consumers at the lowest cost. The second argument is that strict liability is the fairest way of distributing the costs involved in the manufacture and use of products.

Both of these arguments recognize that there is a certain cost in attempting to prevent accidents and in dealing with the consequences of accidents that do occur. Preventing accidents requires that manufacturers expend greater resources on product safety. Consumers must also expend resources to avoid accidents by learning how to select safe products and how to use them correctly. Insofar as manufacturers avoid the cost of reducing accidents and turn out defective products, this cost is passed along to consumers who pay for the injuries that result. A manufacturer may save money, for example, by using a cheaper grade of steel in a hammer, but a user who suffers the loss of an eye when the head chips ends up paying instead. When product safety is viewed as a matter of cost, two questions arise: (1) How can the total cost to both manufacturers and consumers be reduced to the lowest possible level? (2) How should the cost be distributed between manufacturers and consumers?

The *efficiency argument* holds, in the words of one advocate, that "responsibility be fixed wherever it will most effectively reduce the hazards to life and health inherent in defective products that reach the market."[48] By this principle, manufacturers ought to bear this responsibility, because they possess greater expertise than consumers about all aspects of product safety. They also make most of the key decisions about how products are designed, constructed, inspected, and so on. By giving manufacturers a powerful incentive to use the advantages of their position to ensure that the products they turn out are free of dangerous defects, strict liability protects consumers at a relatively low cost. The alternatives, which include placing primary responsibility on government and consumers, generally involve comparatively higher costs.

The principle involved in the *equity argument* is expressed by Richard A. Epstein as follows: "[T]he defendant who captures the entire benefit of his own activities should … also bear its entire costs."[49] Insofar as manufacturers are the beneficiaries of their profit-making activity, it is only fair, according to this principle, that they be forced to bear the cost—which includes the cost of the injuries to consumers as a result of defective products. Much of the benefit of a manufacturer's activity is shared by consumers, however. But they also share the cost of compensating the victims of accidents through higher prices, and it is also just that they do so insofar as they reap some benefit. The distribution of the cost of compensating the victims of product-related injuries is fair, then, if this cost is distributed among all who benefit in the proportion that they benefit, so that it is not borne disproportionately by accident victims.

The Problem of Fault. The major stumbling block to the acceptance of strict liability is that the theory ignores the element of fault, which is a fundamental condition for owing compensation on the Aristotelian conception of compensatory justice.[50] We all benefit from automotive travel, for example, but we can justly be required to pay only for accidents that are our fault. Any system of liability that makes us pay for the accidents of others is unjust—or so it seems. Similarly, it is unjust to hold manufacturers liable to pay large sums to people who are injured by defective products in the absence of negligence or a contractual obligation to compensate. It is equally unjust to force consumers to pay indirectly through higher prices the settlements in product liability suits.

      The response of some advocates of strict liability is that it is not unjust to require those who are faultless to pay the cost of an activity *if everyone benefits by the use of an alternative method of paying compensation.* After all, the victims of accidents caused by defective products are not necessarily at fault either, and everyone is potentially a victim who deserves to be compensated for injuries received from defective products. Thus, those who "pay" under a system of strict liability are also protected. Furthermore, if manufacturers were not held strictly liable for the injuries caused by defective products, then they would take fewer precautions. As a result, everyone would have to spend more to protect themselves—by taking more care in the selection of products, by using them more carefully, and perhaps by taking out insurance policies—and to make up the losses they suffer in product-related accidents where no one is at fault. In either case, the lower prices that consumers pay for products under a negligence system based on the due care theory would not be sufficient to offset the higher cost of insurance, medical care, and so on.

      Under a system of strict liability, consumers give up a right they have in the due care theory—namely, the right not to be forced to contribute to the compensation of accident victims when they (the consumers) are not at fault. Prices are also higher under a strict liability system in order to cover the cost of paying compensation. But consumers gain more than they lose by not being required to spend money protecting themselves and making up their own losses. They also acquire a new right: the right to be compensated for injuries from defective products without regard to fault. Thus, everyone is better off under a strict liability system than under a negligence system.

Objections to Strict Liability. Critics reject many key assumptions in the two arguments for strict liability. First, product liability covers many different kinds of accidents, and the most efficient or equitable system for one kind may not be efficient or equitable for another. Careful studies need to be made of the consequences of competing theories for each kind of accident. Some proposals for reform have recommended strict liability for defects in construction and a negligence system for design defects, for example.[51]

Second, the view that corporations are able to distribute the burden of strict liability to consumers effortlessly is not always true. Multimillion-dollar awards in product liability suits and the high cost of insurance premiums place a heavy burden on manufacturers, driving some out of business and hindering the ability of others to compete. Other complaints of critics are that the threat of liability suits stifles innovation, because new and untested products are more likely to be defective, and that a patchwork of state laws with differing theories and standards creates uncertainty for manufacturers. For these reasons, many business leaders have pressed for uniform product liability laws, upper limits on awards, and other steps to ease the impact of product liability on manufacturers.

Conclusion

Which theory of liability is applied by the courts is of immense importance to manufacturers and consumers. Although this is a matter to be decided, in part, by legal and political considerations, there are also important ethical issues in the debate. The theories rest on different ethical foundations. The due care theory is based on the Aristotelian principle of compensatory justice; the contractual theory, on freedom of contract; and strict liability, largely on utilitarian considerations. Each one embodies something we consider morally fundamental, and yet the three theories are ultimately incompatible. The contractual theory is the least satisfactory because of the power of manufacturers to write warranties and other agreements to their own advantage and to offer them to consumers on a "take it or leave it" basis. The main shortcoming of the due care theory is the difficulty of deciding what constitutes due care and whether it was exercised. Strict liability, despite the absence of fault, is arguably the best theory. It provides a powerful incentive for manufacturers to take extreme precautions and creates a workable legal framework for compensating consumers who are injured by defective products. For strict liability to be just, however, the costs have to be properly distributed, so that they are fair to all parties.

Case 11.2   Volvo's "Bear Foot" Misstep

The television ad showed a monster truck riding atop the roofs of cars lined up in its path.[52] The truck, named "Bear Foot" because of its oversized, 6-foot tires, crushed every car but one—a Volvo station wagon. The scene of devastation around the still-standing Volvo vividly illustrated the company's advertising message of strength and safety. The TV and print ads both appeared in October 1990 and received immediate critical acclaim for their effectiveness. The monster truck campaign was quickly dropped, however, amid charges of deceptive advertising.

The idea for the ad came from a monster truck rally in Vermont in which a Volvo was the only survivor of a similar stunt. In re-creating the scene at a Texas arena, the production crew employed by the advertising agency Scali, McCabe, Sloves reinforced the roof of the Volvo with lumber and steel and partially sawed through the roof supports of the other cars. When word leaked out, the attorney general of Texas began an investigation that confirmed the rigging and led to a lawsuit for consumer fraud. Volvo quickly settled the suit by running corrective ads and by reimbursing the state of Texas for the cost of the investigation and the legal fees incurred. Scali, McCabe, Sloves also resigned its Volvo account, which generated $40 million a year in revenues.

In apologizing for the ads, Volvo insisted that the company was unaware of the rigged demonstration but defended the rigging all the same. The reasons for the alterations to the cars, the company explained, were to enable the production crew to conduct the demonstration and to film it safely and to allow the Volvo to withstand the repeated runs of the monster truck that were required for filming. The claim being made was not false: Volvo engineers had determined that the roof could withstand the weight of a 5-ton monster truck. The mistake was in not revealing to consumers that the ad was not an actual demonstration but a dramatization of the event in Vermont.

This was not the first time that Volvo and Scali, McCabe, Sloves had been criticized for questionable ads. The year before, an ad was produced that showed a large truck perched atop a Volvo with the tag line "How well does your car stand up to heavy traffic?" This ad was similar to one from the 1970s showing a Volvo withstanding the weight of six other Volvos stacked one on top of another. In both ads, the Volvo on the bottom was supported by jacks. The reason for the jacks, according to the company, was that the ads were intended to show only the strength of the main body; no claim was being made about the tires and suspension system, which, in any event, could not withstand such a load. The tires would blow out and the suspension system would collapse.

Case 11.3   The Target Marketing of Cigarettes

Target marketing is taught in business schools as an effective means for increasing sales by tailoring products to the needs and wants of specific consumers. A widely used marketing textbook compares target marketing to using a rifle rather than taking a shotgun approach.[53] This analogy is frighteningly apt, however, when the product is one that kills. As a critic of target marketing by the tobacco industry observed: "When you target for marketing [of cigarettes], you target for death."[54]

The R. J. Reynolds Tobacco Company, a division of RJR Nabisco, was criticized in 1990 for the development of two new cigarette brands that were aimed at narrow market segments. One brand, to be called Uptown, was designed to appeal to blacks, the other, named Dakota, was intended to attract young, blue-collar white women. The introductions of both brands were scuttled by R. J. Reynolds after protests from outraged civil rights groups and women's organizations. The Secretary of Health and Human Services, Dr. Louis W. Sullivan, who is a black physician, charged that Uptown was "deliberately and cynically targeted toward black Americans," and he urged the company to cancel plans to test-market the new brand. "At a time when our people desperately need the message of health promotion," he said, "Uptown's message is more disease, more suffering, and more death for a group already bearing more than its share of smoking-related illness and mortality."

The development of Uptown was based on extensive marketing research. A light menthol flavor was selected because 69 percent of black smokers prefer menthol-flavored cigarettes compared with 27 percent for all smokers. Many blacks open a package from the bottom, and so the cigarettes were to be packed with the filter end pointing down. Researchers discovered that the name Uptown, which evokes images of sophisticated nightlife, drew the most favorable response from blacks in test groups, and the theme of elegance was reinforced by lettering in black and gold, which were chosen instead of the green that is more commonly used for menthol brands. The market testing for Uptown, which was scheduled to begin on February 5, 1990, in Philadelphia, involved print ads in black magazines and newspapers and billboards and point-of-sale displays in black neighborhoods. The target group for Dakota was young white women, 18 to 20 years of age, with a high school education or less, who held an "entry-level service or factory job." The profile of the potential Dakota smoker also mentioned that she spent her free time accompanying her boyfriend to tractor pulls and car races. The test marketing for this product was planned for April 1990.

The groups targeted by Uptown and Dakota reflect market realities. As more smokers quit, the tobacco companies can best expand market share by luring customers away from rival brands, and groups with the lowest rate of quitting are more likely prospects. In 1990, 34 percent of adult blacks smoked whereas the rate among the whole adult population was 27 percent. In the period from 1965 to 1987, the number of white smokers declined 28 percent compared with a drop of only 21 percent among blacks. Although more men than women smoke (32 percent versus 27 percent in 1965), the percentage of men who quit smoking (37 percent) during the 1965 to 1987 period was much higher than the figure for women (16 percent). The decline in smoking among college-educated people was 52 percent, which contrasts sharply with a quit rate of 2 percent for high school dropouts. The data show that the best markets for tobacco products are blacks, women, and the uneducated.

The three largest-selling menthol brands on the market held 53 percent of the black market, attesting to the appeal of menthol-flavored cigarettes among blacks, but R. J. Reynold's product in this market, Salem, ranked behind the other two brands with a market share of only 15 percent. Another menthol brand could increase Reynold's share of the black market. The most popular brand for teenage women was Marlboro, manufactured by rival Philip Morris. R. J. Reynolds could not easily challenge the dominance of this brand among all smokers, but the company apparently thought that many women in that age group did not identify fully with the "Marlboro man" and could be enticed by a cigarette for women only. Thus, the "Dakota woman" was conceived as a feminine counterpart of the "Marlboro man."

The marketing of cigarettes (and alcohol) to blacks is nothing new. Ads for cigarettes contribute between 7 and 10 percent of the advertising revenue in black-oriented publications. Inner-city areas contain more billboards than do suburbs, and more of the billboards in the inner city advertise cigarettes and alcohol. A 1987 survey in St. Louis, for example, revealed that 62 percent of the billboards in predominantly black neighborhoods advertised cigarettes and alcohol compared with 36 percent in white neighborhoods.[55] R. J. Reynolds denied that it was attempting to attract new smokers among blacks and maintained that the company was merely trying to take away business from its competitors. After canceling the test marketing for Uptown, an executive from Reynolds remarked, "Maybe in retrospect we would have been better off not saying we were marketing to blacks. But those were the smokers we were going after, so why shouldn't we be honest about it?"

NOTES

1. This case is based on material in John A. Byrne, *Informed Consent* (New York: McGraw-Hill, 1996); Anne T. Lawrence, "Dow Corning and the Silicone Breast Implant Controversy," *Case Research Journal*, 13, no. 4 (1993), 87–112; Philip J. Hilts, "Maker Is Depicted as Fighting Tests on Implant Safety," *New York Times*, 13 January 1993, A1; Steven Fink, "Dow Corning's Moral Evasions," *New York Times*, 16 February 1992, F13; Gina Kolata, "Study Finds Nothing to Link Implants with any Diseases," *New York Times*, 16 June 1994, A8.

2. *Marketing Definitions: A Glossary of Marketing Terms* (Chicago: American Marketing Association, 1960), 15.

3. See Tom L. Beauchamp, "Manipulative Advertising," in *Ethical Theory and Business*, 4th ed., Tom L. Beauchamp and Norman E. Bowie, eds. (Upper Saddle River, NJ: Prentice Hall, 1993), 475–83.

4. John B. Hinge, "Critics Call Cuts in Package Size Deceptive Move," *Wall Street Journal*, 5 February 1991, B1.

5. Manipulative sales practices that exploit the poor are well documented in David Caplovitz, *The Poor Pay More* (New York: Free Press, 1963) and the report by the National Advisory Commission on Civil Disorders, better known as the Kerner Commission, in 1968. Another study is Alan R. Andreasen, *The Disadvantaged Consumer* (New York: Free Press, 1975).

6. See, for example, Leo Bogart, "The Researcher's Dilemma," *Journal of Marketing*, 26 (January 1962), 6–11; A. B. Blankenship, "Some Aspects of Ethics in Marketing Research," *Journal of Marketing Research*, 1 (May 1964), 26–31; Shelby D. Hunt, Lawrence B. Chonko, and James B. Wilcox, "Ethical Problems of Market Researchers," *Journal of Marketing Research*, 21 (1984), 309–24.

7. See Alice M. Tybout and Gerald Zaltman, "Ethics in Marketing Research: Their Practical Relevance," *Journal of Marketing Research*, 11 (November 1974), 357–68; Kenneth C. Schneider, "Subject and Respondent Abuse in Marketing Research," *MSU Business Topics* (Spring 1977), 13–20; and Del I. Hawkins, "The Impact of Sponsor Identification and Direct Disclosure on Respondent Rights on the Quantity and Quality of Mail Survey Data," *Journal of Business*, 52 (1979), 577–90.

8. William M. Pride and O. C. Ferrell, *Marketing: Basic Concepts and Decisions*, 10th ed. (Boston: Houghton Mifflin, 1997), 40.

9. See H. L. Darling, "How Companies Are Using Corporate Advertising," *Public Relations Journal*, 31 (November 1975), 26–29; W. S. Sachs, "Corporate Advertising: Ends, Means, Prob-

lems," *Public Relations Journal*, 37 (November 1981), 14–17; and S. Prakash Sethi, "Institutional/Image Advertising and Idea/Issue Advertising as Marketing Tools: Some Public Policy Issues," *Journal of Marketing*, 47 (January 1983), 68–78.

10. Vance Packard, *The Hidden Persuaders* (New York: David McKay, 1957), 20–21.

11. Ibid., 25.

12. John Kenneth Galbraith, *The Affluent Society* (New York: Houghton Mifflin, 1958), 128.

13. Ibid., 124–25.

14. Ibid., 128.

15. F. A. Von Hayek, "The Non Sequitur of the 'Dependence Effect'," *Southern Economic Journal*, 27 (April 1961), 346–48. Similar points are made in Theodore Levitt, "The Morality (?) of Advertising," *Harvard Business Review*, 48 (July–August 1970), 84–92.

16. Packard, *The Hidden Persuaders*, 110.

17. Cited in Ibid., 35.

18. For one example, see "Secret Voices: Messages That Manipulate," *Time*, 10 September 1979, 71.

19. See Michael Schudson, *Advertising, The Uneasy Persuasion: Its Dubious Impact on Society* (New York: Basic Books, 1984), 102–3.

20. A researcher who claims to find evidence for both the frequency and the effectiveness of subliminal advertising is Wilson Brian Key, who has written three books: *Subliminal Seduction* (Upper Saddle River, NJ: Prentice Hall, 1974); *Media Sexploitation* (Upper Saddle River, NJ: Prentice Hall, 1976); and *The Clam-Plate Orgy and Other Subliminal Techniques for Manipulating Your Behavior* (Upper Saddle River, NJ: Prentice Hall, 1980). A generally negative appraisal of the effectiveness of the technique is offered in N. F. Dixon, *Subliminal Perception* (London: McGraw-Hill, 1971). For evidence specifically on advertising, see Stephen G. George and Luther B. Jennings, "Effect of Subliminal Stimuli on Consumer Behavior: Negative Evidence," *Perceptual and Motor Skills*, 41 (1975), 847–54; Del I. Hawkins, "The Effects of Subliminal Stimulation on Drive Level and Brand Preference," *Journal of Marketing Research*, 8 (1970), 322–26; and Timothy E. Moore, "Subliminal Advertising: What You See Is What You Get," *Journal of Marketing*, 46 (Spring 1982), 38–47.

21. Richard T. DeGeorge, *Business Ethics*, 4th ed. (Upper Saddle River, NJ: Prentice Hall, 1995), 262.

22. This is not the only ground for objecting to product placement. Critics also cite the element of deception and the corrupting effect of product placement on the artistic integrity of movies.

23. Although this practice is generally accepted as legitimate persuasion, an interesting issue is posed by the increasing use of market research in jury selection and the formation of legal strategy. If lawyers start adopting the techniques of advertisers, then they could open themselves up to many of the same criticisms. See Scott M. Smith, "Marketing Research and Corporate Litigation ... Where Is the Balance of Ethical Justice?" *Journal of Business Ethics*, 3 (1984), 185–94.

24. This criterion is advanced by Stanley I. Benn, "Freedom and Persuasion," *Australasian Journal of Philosophy*, 45 (1967), 267.

25. Ibid., 269.

26. For a thorough study of puffery, see Ivan L. Preston, *The Great American Blow-Up: Puffery in Advertising and Selling* (Madison: University of Wisconsin Press, 1975).

27. Levitt, "Morality (?) of Advertising," 85.

28. For examples, see David M. Gardner, "Deception in Advertising: A Conceptual Approach," *Journal of Marketing*, 39 (January 1975), 40–46; and Thomas L. Carson, Richard E. Wokutch, and James E. Cox, Jr., "An Ethical Analysis of Deception in Advertising," *Journal of Business Ethics*, 4 (1985), 93-104.

29. Ivan L. Preston, "Reasonable or Ignorant Consumer? How the FTC Decides," *Journal of Consumer Affairs*, 8 (1974), 132.

30. *Gelb* v. *FTC*, 144 F. 2d 580 (2d Cir. 1944).

31. Cited in James C. Miller, "Why FTC Curbs Are Needed," *Advertising Age*, 22 March 1982, 83.

32. See Bonnie Liebman, "Nouveau Junk Food: Consumers Swallow the Back-to-Nature Bunk," *Business and Society Review*, 51 (Fall 1984), 47–51.

33. *Campbell Soup*, 77 FTC 664 (1970).

34. See Jeanne Saddler, "Campbell Soup Will Change Ads to Settle Charges," *Wall Street Journal*, 9 April 1991, B6.

35. "F.T.C. in Accord with Campbell," *New York Times*, 10 April 1991, sec. 4, p. 4.

36. For the sake of simplicity, the discussion in this section focuses only on manufacturers, but a responsibility for product safety extends to wholesalers, distributors, franchisers, and retailers, among others, although their responsibility is generally less than that of manufacturers.

37. The conditions under which a negligent act is a cause of injury to another person (known in law as *proximate* cause) are complicated. See any standard textbook on business law for an explanation.

38. *Larsen* v. *General Motors Corporation*, 391 F. 2d 495 (8th Cir. 1968).

39. *LeBouef* v. *Goodyear Tire and Rubber Co.*, 623 F. 2d 985 (5th Cir. 1980).

40. See Ed Timmerman and Brad Reid, "The Doctrine of Invited Misuse: A Societal Response to Marketing Promotion," *Journal of Macromarketing*, 4 (Fall 1984), 40–48.

41. *Commercial National Bank of Little Rock, Guardian of the Estate of Jo Ann Fitzsimmons* v. *General Motors Corporation*, U.S. Dist. Ct. E.D.Ark., No. LR-C-79-168 (1979).

42. *United States* v. *Carroll Towing Co.*, 159 F. 2d 169 (2d Cir. 1947).

43. *Bashada* v. *Johns-Manville Products Corp.*, 90 N.J. 191, 447 A. 2d 539 (1982). The "state-of-the-art" defense has been accepted in other cases. See, for example, *Boatland of Houston, Inc.* v. *Bailey*, 609 S.W. 2d 743 (Tex. 1980). For an overview, see Jordan H. Leibman, "The Manufacturer's Responsibility to Warn Product Users of Unknowable Dangers," *American Business Law Journal*, 21 (1984), 403–38.

44. *Henningsen* v. *Bloomfield Motors, Inc. and Chrysler Corporation*, 161 A. 2d 69 (1960).

45. *MacPherson* v. *Buick Motor Company*, 217 N.Y. 382 (1916).

46. *Baxter* v. *Ford Motor Company*, 168 Wash. 456, 12 P. 2d 409 (1932); 179 Wash. 123, 35 P. 2d 1090 (1934).

47. *Greenman* v. *Yuba Power Products*, 59 Cal. 2d 57, 377 P. 2d 897, 27 Cal. Rptr. 697 (1963). The theory was enunciated 19 years earlier in *Escola* v. *Coca-Cola Bottling Co.*, 24 Cal. 2d 453, 150 P. 2d 436 (1944).

48. *Escola* v. *Coca-Cola Bottling Co.*

49. Richard A. Epstein, *Modern Products Liability Law* (Westport, CT: Quorum Books, 1980), 27. Epstein holds, however, that this principle has limited application in product liability cases.

50. George P. Fletcher, "Fairness and Utility in Tort Theory," *Harvard Law Review*, 85 (1972), 537–73; Richard A. Epstein, "A Theory of Strict Liability," *Journal of Legal Studies*, 2 (1973), 151–204; and Jules L. Coleman, "The Morality of Strict Tort Liability," *William and Mary Law Review*, 18 (1976), 259–586. Richard Posner argues that the Aristotelian principle of compensatory justice in a negligence system, correctly understood, is the same as the principle of efficiency that underlies strict liability. See "A Theory of Negligence," *Journal of Legal Studies*, 1 (1972), 29–96; and "The Concept of Corrective Justice in Recent Theories of Tort Law," *Journal of Legal Studies*, 10 (1981), 187–206.

51. This proposal was contained in S. 44, an unsuccessful bill introduced in the 98th Congress.

52. Material for this case is from Barry Meier, "For This Pounding, Volvo Had Help," *New York Times*, 6 November 1990, D1, D17; Kim Folz, "Scali Quits Volvo Account, Citing Faked Commercial," *New York Times*, 14 November 1990, D1; Krystal Miller and Jacqueline Mitchell, "Car Marketers Test Gray Area of Truth in Advertising," *Wall Street Journal*, 19 November 1990, B1, B6.

53. Philip Kotler, *Marketing Management: Analysis, Planning, Implementation, and Control*, 9th ed. (Upper Saddle River, NJ: Prentice Hall, 1997), 249.

54. Anthony Ramirez, "New Cigarette Raising Issue of Target Marketing," *New York Times*, 18 February 1990, A28. Other sources used in the preparation of this case are Anthony Ramirez, "A Cigarette Campaign Under Fire," *New York Times*, 12 January 1990, D1, D32; Philip J. Hilts, "Health Chief Assails Reynolds Co. for Ads That Target Blacks," *New York Times*, 19 January 1990, A1, A13; Anthony Ramirez, "Reynolds, After Protests, Cancels Cigarette Aimed at Black Smokers," *New York Times*, 20 January 1990, A1, A11.

55. "Media Uproar Over Billboards in Poor Areas," *New York Times*, 1 May 1989, D32.

12

Occupational Health and Safety

Case 12.1 The Regulation of Benzene

Benzene is a colorless, vaporous liquid widely used as a solvent in the printing, rubber, paint, and dry-cleaning industries. It is a raw material or an intermediate product in the synthesis of many chemicals and a constituent of many petroleum products. Benzene is also highly toxic. Contact with the skin causes irritation and blistering, and the consequences of breathing benzene vapors in high concentrations include headache, fatigue, dizziness, convulsions, and even death from paralysis of the nervous system. The most serious effects of benzene exposure result from damage to the blood-cell-forming system of the bone marrow. Victims of benzene poisoning experience a variety of blood disorders that often lead to debilitating aplastic anemia and potentially fatal infections. Chronic exposure to high levels of benzene has also been linked to leukemia.

Before the dangers of benzene were known, the exposure of workers was routine. According to a study done in 1939, three printing plants in New York City, employing 350 men, used about 50,000 gallons of benzene a month, mostly as an ink solvent.[1] The workers were exposed to fumes from open troughs of ink on the printing presses, from accidental spills on the floor, and from ink drying on the paper. Benzene was also used to clean the machines, and workers cleaned themselves with it at the end of the day. An examination of 332 workers found that 130 of them had benzene poisoning in varying degrees, and further tests on a group of 102 workers revealed 22 severe cases, including 6 requiring hospitalization. The use of benzene in printing has long since been discontinued, but it was replaced at first with methanol, also known as wood alcohol, and carbon tetrachloride, both of which damage the central nervous system and internal organs, including the liver and kidneys. Methanol can also produce blindness.

Because of the toxicity of benzene, a Permissible Exposure Limit (PEL) of 10 parts per million (ppm) was set by the Occupational Safety and Health Administration (OSHA), the federal regulatory body created in 1970 to protect Americans from workplace hazards. The assumption that exposure below the level of 10 ppm is safe was challenged in 1977 when a disproportionate number of leukemia deaths occurred at two rubber pliofilm plants in Ohio.[2] On the evidence contained in a report by the National Institute for Occupational Safety and Health (NIOSH), OSHA declared benzene to be a leukemia-causing agent and issued an emergency temporary standard ordering that the PEL for benzene in most work sites be reduced to 1 ppm until a hearing could be conducted on setting a new limit. OSHA was acting under a section of the law that requires the PEL for a known carcinogen (cancer-causing agent) to be set at the lowest technologically feasible level that will not impair the viability of the industries being regulated.

In the resulting uproar, the American Petroleum Institute, a trade association of domestic oil companies, contended that the evidence linking benzene to leukemia was not conclusive and that the exposure standard should take into account the cost of compliance. Previous studies had documented the incidence of leukemia only at exposures above 25 ppm. One study of exposure below 10 ppm, conducted by Dow Chemical Company, found 3 leukemia deaths in a group of 594 workers, where 0.2 deaths would be expected, but it was impossible to rule out other causes, because the workers who developed leukemia had been exposed to other carcinogens during their careers. OSHA was unable to demonstrate, therefore, that exposure to benzene below the level of 10 ppm had ever caused leukemia.

According to OSHA figures, complying with the 1-ppm standard would require companies to spend approximately $266 million in capital improvements, $187 million to $205 million in first-year operating costs, and $34 million in recurring annual costs. The burden would be least in the rubber industry, where two-thirds of the workers exposed to benzene are employed. The petroleum-refining industry, by contrast, would be required to incur $24 million in capital costs and $600,000 in first-year operating expenses. The cost of protecting 300 petroleum refinery workers would be $82,000 each, compared with a cost of only $1,390 per worker in the rubber industry.

INTRODUCTION

No one questions the hazards posed by benzene and the necessity to limit the exposure of workers. Workers have a right to be protected from excessive exposure to toxic chemicals in the workplace. It may be asked, however, whether a reduction to the lowest technologically feasible level is justified, especially in view of the lack of conclusive evidence of the hazards at low levels of exposure. Also, should the cost of complying with the reduced standard be a factor in determining the acceptable level of safety in the workplace?

These are some of the questions that are posed by the benzene case. Questions can also be raised about the right of employees to be given information about the workplace hazards to which they are exposed and their right to refuse to perform dangerous work without fear of dismissal or other reprisals. An especially difficult kind of case is posed by the fact that certain jobs pose a health threat to the fetus of a pregnant woman and to the reproductive capacities of both men and women. Some pregnant women and women of childbearing age are demanding the right to transfer out of jobs thought to pose reproductive hazards. On the other hand, employers who exclude pregnant women or women of childbearing age from certain jobs because of reproductive hazards are open to charges of illegal sexual discrimination, especially when they do not show an equal concern for the reproductive risk to men.

This chapter is concerned with determining how questions about the rights of workers in matters of occupational health and safety ought to be answered. At issue in these questions is not only the obligation of employers with respect to the rights of workers, but also the justification for government regulation of the workplace, especially by the Occupational Safety and Health Administration. As a result, many of the questions discussed in this chapter deal with specific regulatory programs and policies which are the subject of intense controversy.

THE SCOPE OF THE PROBLEM

Many Americans live with the possibility of serious injury and death every working day. For some workers, the threat comes from a major industrial accident, such as the collapse of a mine or a refinery explosion, or from widespread exposure to a hazardous substance, such as asbestos, which is estimated to have caused more than 350,000 cancer deaths since 1940.[3] The greatest toll on the work force is exacted, however, by little-publicized injuries to individual workers, some of which are gradual, such as hearing loss from constant noise or nerve damage from repetitive motions. Some of the leading causes of death, such as heart disease, cancer, and respiratory conditions, are thought to be job-related, although causal connections are often difficult to make. Even stress on the job is now being recognized as a workplace hazard that is responsible for headaches, back and chest pains, stomach ailments, and a variety of emotional disorders.

The Distinction between Safety and Health

Although the term *safety* is often used to encompass all workplace hazards, it is useful to make a distinction between *safety* and *health*.[4] Safety hazards generally involve loss of limbs, burns, broken bones, electrical shocks, cuts, sprains, bruises, and impairment of sight or hearing. These injuries are usually the result of sudden and often violent events involving industrial equipment or the physical environment of the workplace. Examples include coming into contact with moving parts of machinery or electrical lines, getting hit by falling objects or flying debris, chemical spills and explosions, fires, and falls from great heights.

Health hazards are factors in the workplace that cause illnesses and other conditions that develop over a lifetime of exposure. Many diseases associated with specific occupations have long been known. In 1567, Paracelsus identified pneumoconiosis, or black lung disease, in a book entitled *Miners' Sickness and Other Miners' Diseases*. Silicosis, or the "white plague," has traditionally been associated with stone cutters. Other well-known occupational diseases are caisson disease among divers, cataracts in glassblowers, skin cancer among chimney sweeps, and phosphorus poisoning in match makers. Mercury poisoning, once common among felt workers, produces tremors, known as "the hatters' shakes," and delusions and hallucinations, which gave rise to the phrase "mad as a hatter."

In the modern workplace, most occupational health problems result from routine exposure to hazardous substances. Among these substances are fine particles, such as asbestos, which causes asbestosis, and cotton dust, which causes byssinosis; heavy metals, such as lead, cadmium, and beryllium; gases, including chlorine, ozone, sulphur dioxide, carbon monoxide, hydrogen sulfide, and hydrogen cyanide, which damage the lungs and often cause neurological problems; solvents, such as benzene, carbon tetrachloride, and carbon disulfide; and certain classes of chemicals, especially phenols, ketones, and epoxies. Pesticides pose a serious threat to agricultural workers, and radiation is an occupational hazard to X-ray technicians and workers in the nuclear industry.

Because occupationally related diseases result from long-term exposure and not from identifiable events on the job, employers have generally not been held liable for them, and they have not, until recently, been recognized in workers' compensation programs. The fact that the onset of many diseases occurs years after the initial exposure—30 or 40 years in the case of asbestos—hides the causal connection. The links are further obscured by a multiplicity of causes. The textile industry, for example, claims that byssinosis among its workers results from their own decision to smoke and not from inhaling cotton dust on the job.[5] Lack of knowledge, especially about cancer, adds to the difficulty of establishing causal connections.

Regulation of Occupational Health and Safety

Prior to the passage of the Occupational Safety and Health Act (OSH Act) in 1970, government regulation of occupational health and safety was almost entirely the province of the states. Understaffed and underfunded, the agencies charged with protecting workers in most states were not very effective.[6] Only a small percentage of workers in many states were even under the jurisdiction of regulatory agencies; often, powerful economic interests were able to influence their activities. Because the agencies lacked the resources to set standards for exposure to hazardous substances, they relied heavily on private standard-setting organizations and the industries themselves. The emphasis in most states was on education and training, and prosecutions for violations were rare. State regulatory agencies were also concerned almost exclusively with safety rather than with health.

States still play a major role in occupational health and safety through workers' compensation systems, but in 1970, primary responsibility for the regulation of working conditions passed to the federal government. The "general duty clause" of the OSH Act requires employers "to furnish to each of his employees employment and a place of employment which are free from recognized hazards that are causing or are likely to cause death or serious injury."[7] In addition, employers have a specific duty to comply with all the occupational safety and health standards that OSHA is empowered to make. Employees also have a duty, under Section 5(b), to "comply with occupational safety and health standards and all rules, regulations, and orders issued pursuant to this Act which are applicable to his own actions and conduct." OSHA regulates occupational health and safety primarily by issuing standards, which are commonly enforced by workplace inspections. Examples of standards are Permissible Exposure Limits (PELs) for toxic substances and specifications for equipment and facilities, such as guards on saws and the height and strength of railings.

THE RIGHT TO A SAFE AND HEALTHY WORKPLACE

At first glance, the right of employees to a safe and healthy workplace might seem to be too obvious to need any justification. This right—and the corresponding obligation of employers to provide working conditions free of recognized hazards— appears to follow from a more fundamental right, namely, the right of survival. Patricia H. Werhane writes, for example, "Dangerous working conditions threaten the very existence of employees and cannot be countenanced when they are avoidable." Without this right, she argues, all other rights lose their significance.[8] Some other writers base a right to a safe and healthy workplace on the Kantian ground that persons ought to be treated as ends rather than as means. Mark MacCarthy has described this view as follows:

> People have rights that protect them from others who would enslave them or otherwise use them for their own purposes. In bringing this idea to bear on the problem of occupational safety, many people have thought that workers have an inalienable right to earn their living free from the ravages of job-caused death, disease, and injury.[9]

Congress, in passing the OSH Act granting the right to all employees of a safe and healthy workplace, was apparently relying on a cost-benefit analysis, balancing the cost to industry with the savings to the economy as a whole. Congress, in other words, appears to have been employing essentially utilitarian reasoning. Regardless of the ethical reasoning used, though, workers have an undeniable right not to be injured or killed on the job.

It is not clear, though, what specific protection workers are entitled to or what specific obligations employers have with respect to occupational health and safety. One position, recognized in common law, is that workers have a right to be protected against harm resulting directly from the actions of employers where the employer is at fault in some way. Consider the case of the owner of a drilling company in Los Angeles who had a 23-year-old worker lowered into a 33-foot-deep, 18-inch-wide hole that was being dug for an elevator shaft.[10] No test was made of the air at the bottom of the hole, and while he was being lowered, the worker began to have difficulty breathing. Rescue workers were hampered by the lack of shoring, and the worker died before he could be pulled to the surface. The owner of the drilling company was convicted of manslaughter, sentenced to 45 days in jail, and ordered to pay $12,000 in compensation to the family of the victim. A prosecutor in the Los Angeles County district attorney's office explained the decision to bring criminal charges with the words, "Our opinion is you can't risk somebody's life to save a few bucks. That's the bottom line."

Few people would hesitate to say that the owner of the company in this case violated an employee's rights by recklessly endangering his life. In most workplace accidents, however, employers can defend themselves against the charge of violating the rights of workers with two arguments. One is that their actions were not the *direct cause* of the death or injury, and the other is that the worker *voluntarily assumed the risk*. These defenses are considered in turn.

The Concept of a Direct Cause

Two factors enable employers to deny that their actions are a direct cause of an accident in the workplace.[11] One factor is that industrial accidents are typically caused by a combination of factors, frequently including the actions of workers themselves. When there is such a multiplicity of causes, it is difficult to assign responsibility to any one person.[12] The legal treatment of industrial accidents in the United States incorporates this factor by recognizing two common-law defenses for employers; A workplace accident was caused in part by (1) lack of care on the part of the employee (the doctrine of "contributory negligence") or by (2) the negligence of co-workers (the "fellow-servant rule"). As long as employers are not negligent in meeting minimal obligations, they are not generally held liable for deaths or injuries resulting from industrial accidents.

The second factor is that it is often not practical to reduce the probability of harm any further. It is reasonable to hold an employer responsible for the incidence of cancer in workers who are exposed to high levels of a known carcinogen, especially when the exposure is avoidable. But a small number of cancer deaths can statistically be predicted to result from very low exposure levels to some widely used chemicals. Is it reasonable to hold employers responsible when workers contract cancer from exposure to carcinogens at levels that are considered to pose only a slight risk? The so-called Delaney amendment, for example, forbids the use of any food additive found to cause cancer.[13] Such an absolute prohibition is practicable for food additives, because substitutes are usually readily available. But when union and public-interest groups petitioned OSHA in 1972 to set zero tolerance levels for ten powerful carcinogens, the agency refused on the ground that workers should be protected from carcinogens "to the maximum extent practicable *consistent with continued use.*"[14] The position of OSHA, apparently, was that it is unreasonable to forgo the benefit of useful chemicals when there are no ready substitutes and the probability of cancer can be kept low by strict controls. This is also the position of philosopher Alan Gewirth, who argues that the right of persons not to have cancer inflicted on them is not absolute. He concluded, "Whether the use of or exposure to some substance should be prohibited should depend on the degree to which it poses the risk of cancer.... If the risks are very slight ... and if no substitutes are available, then use of it may be permitted, subject to stringent safeguards."[15]

The Benzene Case. This issue arose again in 1977 in the benzene case that begins this chapter. The main legal issue faced by the Supreme Court is the authority of the secretary of labor under the OSH Act to set standards. An "occupational safety and health standard" is defined in Section 3(8) of the OSH Act as "a standard which requires conditions, or the adoption or use of one or more practices, means, methods, operations, or processes, reasonably necessary or appropriate to provide safe or healthful employment and places of employment." Section 6(b)(5) states:

> The Secretary, in promulgating standards dealing with toxic materials or harmful physical agents under this subsection, shall set the standard which most adequately assures, to the extent feasible, on the basis of the best available evidence, that no employee will suffer material impairment of health or functional capacity even if such employee has regular exposure to the hazard dealt with by such standard for the period of his working life. Development of standards under this subsection shall be based upon research, demonstrations, experiments, and such other information as may be appropriate. In addition to the attainment of the highest degree of health and safety protection for the employee, other considerations shall be the latest available scientific data in the field, the feasibility of the standards, and experience gained under this and other health and safety laws.

Two specific issues raised by the language of the OSH Act are (1) What is meant by "to the extent feasible"? Should all technologically possible steps be taken? Or should some consideration be given to the economic burden on employers? Should the agency, in determining what is "reasonably necessary and appropriate," also weigh costs and benefits? (2) What amount of evidence is required? To what extent should the agency be required to provide valid scientific studies to prove that a standard addresses a genuine problem and will, in fact, achieve the intended result? On the one hand, the power to set emergency standards on the basis of flimsy evidence that turns out later not to warrant action exposes businesses to the possibility of heavy and unpredictable expenses. On the other hand, adequate scientific research is very costly and time-consuming. Lack of funding for research and the time lag between suspicion and proof that a substance is carcinogenic, for example, could result in the needless deaths of many workers. A high level of proof exposes workers to preventable harm, therefore, but at the price of a possibly unnecessary cost to employers. Where does the balance lie?

A plurality of justices agreed with a lower court opinion that OSHA does not have "unbridled discretion to adopt standards designed to create absolutely risk-free workplaces regardless of the cost." But no consensus was achieved on how the level of risk is to be determined and, in particular, whether OSHA is required to use cost-benefit analysis in setting standards. By a bare majority, though, the Supreme Court struck down the 1-ppm standard because the agency was unable to prove that exposure to benzene below concentrations of 10 ppm is harmful. Ample evidence existed at the time to justify a 10-ppm standard, but little evidence indicated that a lower standard would achieve any additional benefits. The Court's answer to the second question, then, is that a rather high level of proof is needed to impose a highly restrictive standard.

In a subsequent case, however, the Court ruled that OSHA is not required to take cost into account but only technological feasibility. The case, *American Textile Manufacturers Institute Inc.* v. *Raymond J. Donovan, Secretary of Labor*, concerned an established standard for cotton dust, which is the primary cause of byssinosis, or "brown lung" disease. The evidence linking cotton dust to byssinosis was not in dispute, but OSHA relied on cost-benefit studies that the agency itself admitted underestimated the cost of compliance. The reason given for not requiring OSHA to consider costs and benefits is that Congress in passing the OSH Act had already made a cost-benefit analysis.[16] According to the majority opinion,

> Not only does the legislative history confirm that Congress meant "feasible" rather than "cost-benefit" when it used the former term, but it also shows that Congress understood that the Act would create substantial costs for employers, yet intended to impose such costs when necessary to create a safe and healthful working environment.... Indeed Congress thought that the *financial costs* of health and safety problems in the workplace were as large as or larger than the *financial costs* of eliminating these problems.

Cost-benefit analysis is the proper instrument for striking a balance on issues of worker health and safety, in other words, but this instrument was used by Congress as a justification for empowering OSHA to set standards without regard for costs. OSHA, therefore, should base its decisions solely on technological feasibility.

As a postscript to the benzene case, OSHA announced in March 1986 a new standard based on subsequent research. The new standard, which took effect in February 1988, set a permissible exposure limit of 1 ppm for an eight-hour workday, with a short-term limit of 5 ppm. By this time, however, improvements in equipment and the monitoring of airborne pollutants required by the Environmental Protection Agency (EPA) had already reduced the level of exposure to benzene close to the 1-ppm standard. Consequently, the initial capital expenditures for meeting the new standard were considerably less than in 1977, ten years earlier, when the controversy first erupted.

The Voluntary Assumption of Risk

A further common-law defense is that employees voluntarily assume the risk inherent in work. Some jobs, such as coal mining, construction, longshoring, and meatpacking, are well known for their high accident rates, and yet some individuals freely choose these lines of work even when safer employment is available. The risk itself is sometimes part of the allure, but more often the fact that hazardous jobs offer a wage premium in order to compensate for the greater risk leads workers to prefer them to less hazardous, less well-paying jobs. Like people who choose to engage in risky recreational activities, such as mountain climbing, workers in hazardous occupations, according to the argument, knowingly accept the risk in return for benefits that cannot be obtained without it. Injury and even death are part of the price they may have to pay. And except when an employer or a fellow employee is negligent in some way, workers who have chosen to work under dangerous conditions have no one to blame but themselves.

A related argument is that occupational health and safety ought not to be regulated because it interferes with the freedom of individuals to choose the kind of work that they want to perform. Workers who prefer the higher wages of hazardous work ought to be free to accept such employment, and those with a greater aversion to risk ought to be free to choose other kinds of employment or to bargain for more safety, presumably with lower pay. To deny workers this freedom of choice is to treat them as persons incapable of looking after their own welfare.[17] W. Kip Viscusi, who served as a consultant to OSHA during the Reagan administration, adds an extra twist by arguing that programs designed to keep workers from being maimed and killed on the job are a form of class oppression. He has written:

> Efforts to promote present risk regulations on the basis that they enhance worker rights are certainly misguided. Uniform standards do not enlarge worker choices; they deprive workers of the opportunity to select the job most appropriate to their own risk preferences. The actual "rights" issue involved is whether those in upper income groups have a right to impose their job risk preferences on the poor.[18]

The argument that employees assume the risk of work can be challenged on several grounds. First, workers need to possess a sufficient amount of information about the hazards involved. They cannot be said to assume the risk of performing dangerous work when they do not know what the risks are. Also, they cannot exercise the right to bargain for safer working conditions without access to the relevant information. Yet, employers have generally been reluctant to notify workers or their bargaining agents of dangerous conditions or to release documents in their possession. Oftentimes, hazards in the workplace are not known by the employer or the employee until after the harm has been done. In order for employers to be relieved of responsibility for injury or death in the workplace, though, it is necessary that employees have adequate information *at the time they make a choice.*

Second, the choice of employees must be truly free. When workers are forced to perform dangerous work for lack of acceptable alternatives, they cannot be said to assume the risk. For many people with few skills and limited mobility in economically depressed areas, the only work available is often in a local slaughter-house or textile mill, where they run great risks. Whether they are coerced into accepting work of this kind is a controversial question. Individuals are free in one sense to accept or decline whatever employment is available, but the alternatives of unemployment or work at poverty-level wages may be so unacceptable that people lack freedom of choice in any significant sense.

Risk and Coercion

In order to determine whether workers assume the risk of employment by their free choice, we need some account of the concept of coercion. A paradigm example is the mugger who says with a gun in hand, "Your money or your life." The "choice" offered by the mugger contains an undesirable set of alternatives that are imposed on the victim by a threat of dire consequences. A standard analysis of coercion that is suggested by this example involves two elements: (1) getting a person to choose an alternative that he or she does not want, and (2) issuing a threat to make the person worse off if he or she does not choose that alternative.

Consider the case of an employer who offers a worker who already holds a satisfactory job higher wages in return for taking on new duties involving a greater amount of risk.[19] The employer's offer is not coercive because there is no threat involved. The worker may welcome the offer, but declining it leaves the worker still in possession of an acceptable position. Is an employer acting like a mugger, however, when the offer of higher pay for more dangerous work is accompanied by the threat of dismissal? Is "Do this hazardous work or be fired!" like or unlike the "choice" offered by the mugger? The question is even more difficult when the only "threat" is not to hire a person. Is it coercive to say, "Accept this dangerous job or stay unemployed!" because the alternative of remaining out of work leaves the person in exactly the same position as before? Remaining unemployed, moreover, is unlike getting fired, in that it is not something that an employer inflicts on a person.

In order to answer these questions, the standard analysis of coercion needs to be supplemented by an account of what it means to issue a threat. A threat involves a stated intention of making a person worse off in some way. To fire a person from a job is usually to make that person worse off, but we would not say that an employer is coercing a worker by threatening dismissal for failure to perform the normal duties of a job. Similarly, we would not say that an employer is making a threat in not hiring a person who refuses to carry out the same normal duties. A person who turns down a job because the office is not provided with air conditioning, for example, is not being made worse off by the employer. So why would we say that a person who chooses to remain unemployed rather than work in a coal mine that lacks adequate ventilation is being coerced?

The answer of some philosophers is that providing employees with air conditioning is not morally required; however, maintaining a safe mine is. Whether a threat is coercive because it would make a person worse off can be determined only if there is some baseline that answers the question, worse off compared with what? Robert Nozick gives an example of an abusive slave owner who offers not to give a slave his daily beating if the slave will perform some disagreeable task the slave owner wants done.[20] Even though the slave might welcome the offer, it is still coercive, because the daily beating involves treating the slave in an immoral manner. For Nozick and others, what is *morally required* is the relevant baseline for determining whether a person would be made worse off by a threatened course of action.[21]

It follows from this analysis that coercion is an inherently ethical concept that can be applied only after determining what is morally required in a given situation.[22] As a result, the argument that the assumption of risk by employees relieves employers of responsibility involves circular reasoning. Employers are freed from responsibility for workplace injuries on the ground that workers assume the risk of employment only if they are not coerced into accepting hazardous work. But whether workers are coerced depends on the right of employees to a safe and healthy workplace—and the obligation of employers to provide it.

In conclusion, the right of employees to a safe and healthy workplace cannot be justified merely by appealing to a right not to be injured or killed. The weakness of this argument lies in the difficulty of determining the extent to which employers are *responsible* for the harm that workers suffer as a result of occupational injuries and diseases. The argument applies only to dangers that are directly caused by the actions of employers; however, industrial accidents result from many causes, including the actions of co-workers and the affected workers themselves. The responsibility of employers is also problematical when the probability of harm from their actions is low. Moreover, the responsibility of employers is reduced insofar as employees voluntarily assume the risk inherent in employment. Whether the choice to accept hazardous work is voluntary, though, depends in part on difficult questions about the concept of coercion, which, on one standard analysis, can be applied only after the rights of employees in matters of occupational health and safety have been determined.

Case 12.2   Whirlpool Corporation

The Whirlpool Corporation operates a plant in Marion, Ohio, for the assembly of household appliances.[23] Components for the appliances are carried throughout the plant by an elaborate system of overhead conveyors. To protect workers from the objects that occasionally fall from the conveyors, a huge wire mesh screen was installed approximately 20 feet above the floor. The screen is attached to an angle-iron frame suspended from the ceiling of the building. Maintenance employees at the plant spend several hours every week retrieving fallen objects from the screen. Their job also includes replacing paper that is spread on the screen to catch dripping grease from the conveyors, and occasionally they do maintenance work on the conveyors themselves. Workers are usually able to stand on the frame to perform these tasks, but occasionally it is necessary to step on to the screen.

In 1973, several workers fell partway through the screen, and one worker fell completely through to the floor of the plant below but survived. Afterward, Whirlpool began replacing the screen with heavier wire mesh, but on June 28, 1974, a maintenance employee fell to his death through a portion of the screen that had not been replaced. The company responded by making additional repairs and forbidding employees to stand on the angle-iron frame or step on to the screen. An alternative method for retrieving objects was devised using hooks.

Two maintenance employees at the Marion plant, Virgil Deemer and Thomas Cornwell, were still not satisfied. On July 7, 1974, they met with the maintenance supervisor at the plant to express their concern about the safety of the screen. At a meeting two days later with the plant safety director, they requested the name, address, and telephone number of a representative in the local office of the Occupational Safety and Health Administration. The safety director warned the men that they "had better stop and think about what they were doing," but he gave them the requested information. Deemer called the OSHA representative later that day to discuss the problem.

When Deemer and Cornwell reported for the night shift at 10:45 P.M. the next day, July 10, they were ordered by the foreman to perform routine maintenance duties above an old section of the screen. They refused, claiming that the work was unsafe whereupon the foreman ordered the two employees to punch out. In addition to losing wages for the six hours they did not work that night, Deemer and Cornwell received written reprimands, which were placed in their personnel files.

THE RIGHT TO KNOW ABOUT AND REFUSE HAZARDOUS WORK

The Whirlpool case illustrates a cruel dilemma faced by many American workers. If they stay on the job and perform hazardous work, they risk serious injury and even death. On the other hand, if they refuse to work as directed, they risk disciplinary action, which can include loss of wages, unfavorable evaluation, demotion, and dismissal. Many people believe that it is unjust for workers to be put into the position of having to choose between safety and their job. Rather, employees ought to be able to refuse orders to perform hazardous work without fear of suffering adverse consequences. Even worse are situations in which workers face hazards of which they are unaware. Kept in the dark about dangers lurking in the workplace, employees have no reason to refuse hazardous work and are unable to take other steps to protect themselves.

Features of the Right to Know and Refuse

The right to refuse hazardous work is different from a right to a safe and healthy workplace. If it is unsafe to work above the old screen, as Deemer and Cornwell contended, then their right to a safe and healthy workplace was violated. A right to refuse hazardous work, however, is only one of several alternatives that workers have for securing the right to a safe and healthy workplace. Victims of racial or sexual discrimination, for example, also suffer a violation of their rights, but it does not follow that they have a right to disobey orders or to walk off the job in an effort to avoid discrimination. Other means are available for ending discrimination and for receiving compensation for the harm done. The same is true for the right to a safe and healthy workplace.

The right to know is actually an aggregation of several rights. Thomas O. McGarity classifies these rights by the correlative duties that they impose on employers. These are (1) the duty to *reveal* information already possessed; (2) the duty to *communicate* information about hazards through labeling, written communications, and training programs; (3) the duty to *seek out* existing information from the scientific literature and other sources; and (4) the duty to *produce* new information (for example, through animal testing) relevant to employee health.[24] Advocates of the right of workers to know need to specify which of these particular rights are included in their claim.

Disagreement also arises over questions about what information workers have a right to know and which workers have a right to know it. In particular, does the information that employers have a duty to reveal include information about the past exposure of workers to hazardous substances? Do employers have a duty to notify past as well as present employees? The issue at stake in these questions is a part of the "right to know" controversy commonly called *worker notification.*

The main argument for denying workers a right to refuse hazardous work is that such a right conflicts with the obligation of employees to obey all reasonable directives from an employer. An order for a worker to perform some especially dangerous task may not be reasonable, however. The foreman in the *Whirlpool* case, for example, was acting contrary to a company rule forbidding workers to step on the screen. Still, a common-law principle is that employees should obey even an improper order and file a grievance afterward, if a grievance procedure is in place, or seek whatever other recourse is available. The rationale for this principle is that employees may be mistaken about whether an order is proper, and chaos would result if employees could stop work until the question is decided. It is better for workers to obey now and correct any violation of their rights later.

The fatal flaw in this argument is that later may be too late. The right to a safe and healthy workplace, unlike the right not to be discriminated against, can effectively provide protection for workers only if violations of the right are prevented in the first place. Debilitating injury and death cannot be corrected later; neither can workers and their families ever be adequately compensated for a loss of this kind. The right to refuse hazardous work, therefore, is necessary for the existence of the right to a safe and healthy workplace.

The Justification for Refusing Hazardous Work

A right to a safe and healthy workplace is empty unless workers have a right in some circumstances to refuse hazardous work, but there is a tremendous amount of controversy over what these circumstances are. In the *Whirlpool* case, the Supreme Court cited two factors as relevant for justifying a refusal to work. These are (1) that the employee reasonably believes that the working conditions pose an imminent risk of death or serious injury, and (2) that the employee has reason to believe that the risk cannot be avoided by any less disruptive course of action. Employees have a right to refuse hazardous work, in other words, only as a last resort—when it is not possible to bring unsafe working conditions to the attention of the employer or to request an OSHA inspection. Also, the hazards that employees believe to exist must involve a high degree of risk of serious harm. Refusing to work because of a slight chance of minor injury is less likely to be justified. The fact that a number of workers had already fallen through the screen at the Whirlpool plant, for example, and that one had been killed strengthens the claim that the two employees had a right to refuse their foreman's order to step on to it.

The pivotal question, of course, is the proper standard for a reasonable belief. How much evidence should employees be required to have in order to be justified in refusing to work? Or should the relevant standard be the actual existence of a workplace hazard rather than the belief of employees, no matter how reasonable? A minimal requirement, which has been insisted on by the courts, is that employees act in *good faith*. Generally, acting in good faith means that employees have an honest belief that a hazard exists and that their only intention is to protect themselves from the hazard. The "good faith" requirement serves primarily to exclude refusals based on deliberately false charges of unsafe working conditions or on sabotage by employees. Whether a refusal is in good faith does not depend on the reasonableness or correctness of the employees' beliefs about the hazards in the workplace. Thus, employees who refuse an order to fill a tank with a dangerous chemical in the mistaken but sincere belief that a valve is faulty are acting in good faith, but employees who use the same excuse to conduct a work stoppage for other reasons are not acting in good faith, even if it should turn out that the valve is faulty.

Three Standards for Reasonable Belief. Three standards are commonly used for determining whether a good faith refusal is based on a reasonable belief. One is the *subjective* standard, which requires only that employees have evidence that they sincerely regard as sufficient for their belief that a hazard exists or that most workers in their situation would regard as sufficient. At the opposite extreme is the *objective* standard. This standard requires evidence that experts regard as sufficient to establish the existence of a hazard. In between these two is the *reasonable person* standard, which requires that the evidence be strong enough to persuade a reasonable person that a hazard exists.

The subjective standard provides the greatest protection for worker health and safety. Employees cannot be expected to have full knowledge of the hazards facing them in the workplace. They may not be told what chemicals they are using, for example, or what exposure levels are safe for these chemicals. Safe exposure levels for the chemicals may not even have been scientifically determined. Employees may also not have the means at the moment to measure the levels to which they are exposed. Yet, the objective standard forces employees to bear the consequences if their beliefs about the hazards present in the workplace cannot be substantiated. The reasonable person standard is less exacting, because it requires only that employees exercise reasonable judgment. Still, this standard places a strong burden of proof on workers who have to make a quick assessment under difficult circumstances. A wrong decision can result in the loss of a job or possibly the loss of a worker's life.

On the other hand, the subjective standard allows employees to make decisions that are ordinarily the province of management. Usually management is better informed about hazards in the workplace, along with other aspects of the work to be performed, and so its judgment should generally prevail. To allow workers to shut down production on the basis of unsubstantiated beliefs, and thereby to substitute their uninformed judgment for that of management, is likely to result in many costly mistakes. The subjective standard creates no incentive for workers to be cautious in refusing hazardous work because the cost is borne solely by the company. The reasonable person standard, therefore, which places a moderate burden of proof on employees, is perhaps the best balance of the competing considerations.

The Justification of a Right to Know

Unlike the right to refuse hazardous work, the right to know about workplace hazards is not necessary for the right to a safe and healthy workplace. This latter right is fully protected as long as employers succeed in ridding the workplace of significant hazards. Some argue that the right to know is still an effective, if not an absolutely essential, means for securing the right to a safe and healthy workplace. Others maintain, however, that the right to know is not dependent for its justification on the right to a safe and healthy workplace; that is, even employees who are adequately protected by their employers against occupational injury and disease still have a right to be told what substances they are handling, what dangers they pose, what precautions to take, and so on.

The Argument from Autonomy. The most common argument for the right to know is one based on autonomy. This argument begins with the premise that autonomous individuals are those who are able to exercise free choice in matters that affect their welfare most deeply.[25] Sometimes this premise is expressed by saying that autonomous individuals are those who are able to *participate* in decision making about these matters. One matter that profoundly affects the welfare of workers is the amount of risk that they assume in the course of earning a living. Autonomy requires, therefore, that workers be free to avoid hazardous work, if they so choose, or have the opportunity to accept greater risks in return for higher pay, if that is their choice. In order to choose freely, however, or to participate in decision making, it is necessary to possess relevant information. In the matter of risk assumption, the relevant information includes knowledge of the hazards present in the workplace. Workers can be autonomous, therefore, only if they have a right to know.

In response, employers maintain that they can protect workers from hazards more effectively than workers can themselves without informing workers of the nature of those hazards. Such a paternalistic concern, even when it is sincere and well-founded, is incompatible, however, with a respect for the autonomy of workers. A similar argument is sometimes used to justify paternalism in the doctor–patient relation. For a doctor to conceal information from a patient even in cases where exclusive reliance on the doctor's greater training and experience would result in better medical care is now generally regarded as unjustified. If paternalism is morally unacceptable in the doctor–patient relation where doctors have an obligation to act in the patient's interest, then it is all the more suspect in the employer–employee relation where employers have no such obligation.[26]

Although autonomy is a value, it does not follow that employers have an obligation to further it in their dealings with employees. The autonomy of buyers in market transactions is also increased by having more information, but the sellers of a product are not generally required to provide this information except when concealment constitutes fraud.[27] The gain of autonomy for employees must be balanced, moreover, against the not inconsiderable cost to employers of implementing a "right to know" policy in the workplace. In addition to the direct cost of assembling information, attaching warning labels, training workers, and so on, there are also indirect costs. Employees who are aware of the risk they are taking are more likely to demand higher wages or else safer working conditions. They are more likely to avail themselves of workers' compensation benefits and to sue employers over occupational injury and disease. Finally, companies are concerned about the loss of valuable trade secrets that could occur from informing workers about the hazards of certain substances.

Bargaining over Information. An alternative to a right to know policy that respects the autonomy of both parties is to allow bargaining over information. Thomas O. McGarity has described this alternative in the following way:

> Because acquiring information costs money, employees desiring information about workplace risks should be willing to pay the employer (in reduced wages) or someone else to produce or gather the relevant information. A straightforward economic analysis would suggest that employees would be willing to pay for health and safety information up to the point at which the value in wage negotiations of the last piece of information purchased equaled the cost of that additional information.[28]

Although promising in theory, this alternative is not practical. It creates a disincentive for employers, who possess most of the information, to yield any of it without some concession by employees, even when it could be provided at little or no cost. Bargaining is feasible for large unions with expertise in safety matters, but reliance on it would leave members of other unions and nonunionized workers without adequate means of protection. In the absence of a market for information, neither employers nor employees would have a basis for determining the value of information in advance of negotiations. Finally, there are costs associated with using the bargaining process to decide any matter—what economists call "transaction costs"—and these are apt to be quite high in negotiations over safety issues. It is unlikely, therefore, that either autonomy or worker health and safety would be well served by the alternative of bargaining over matters of occupational health and safety.

Utilitarian Arguments for a Right to Know

There are two arguments for the right to know as a means to greater worker health and safety. Both are broadly utilitarian in character. One argument is based on the plausible assumption that workers who are aware of hazards in the workplace will be better equipped to protect themselves. Warning labels or rules requiring protective clothing and respirators are more likely to be effective when workers fully appreciate the nature and extent of the risks they are taking. Also, merely revealing information about hazardous substances in the workplace is not apt to be effective without extensive training in the procedures for handling them safely and responding to accidents. Finally, workers who are aware of the consequences of exposure to hazardous substances will also be more likely to spot symptoms of occupational diseases and seek early treatment.

The second utilitarian argument is offered by economists who hold that overall welfare is best achieved by allowing market forces to determine the level of acceptable risk. In a free market, wages are determined in part by the willingness of workers to accept risks in return for wages. Employers can attract a sufficient supply of workers to perform hazardous work either by spending money to make the workplace safer, thereby reducing the risks, or by increasing wages to compensate workers for the greater risks. The choice is determined by the marginal utility of each kind of investment. Thus, an employer will make the workplace safer up to the point that the last dollar spent equals the increase in wages that would otherwise be required to induce workers to accept the risks. At that point, workers indicate their preference for accepting the remaining risks rather than suffering a loss of wages in return for a safer workplace.

Unlike the autonomy argument, in which workers bargain over risk information, this argument proposes that workers bargain over the trade-off between risks and wages. In order for a free market to determine this trade-off in a way that achieves overall welfare, it is necessary for workers to have a sufficient amount of information about the hazards in the workplace. Thomas O. McGarity has expressed this point as follows:

> A crucial component of the free market model of wage and risk determination is its assumption that workers are fully informed about the risks that they face as they bargain over wages. To the extent that risks are unknown to employees, they will undervalue overall workplace risks in wage negotiations. The result will be lower wages and an inadequate incentive to employers to install health and safety devices. In addition, to the extent that employees can avoid risks by taking action, uninformed employees will fail to do so. Society will then under invest in wages and risk prevention, and overall societal wealth will decline. Moreover, a humane society is not likely to require diseased or injured workers to suffer without proper medical attention. In many cases, society will pick up the tab....[29]

Although these two utilitarian arguments provide strong support for the right to know, they are both open to the objection that there might be more efficient means, such as more extensive OSHA regulation, for securing the goal of worker health and safety. Could the resources devoted to complying with a right-to-know law, for example, be better spent on formulating and enforcing more stringent standards on permissible exposure limits and on developing technologies to achieve these standards? Could the cost of producing, gathering, and disseminating information be better borne by a government agency than by individual employers? These are difficult empirical questions for which conclusive evidence is largely lacking.

The Problem of Trade Secrets.   One issue that remains to be addressed is the protection of trade secrets. Occasionally, information about the chemical composition of the substances used in industrial processes cannot be disclosed to workers without compromising the ability of a company to maintain a legally protected trade secret. When this is the case, is it justifiable to place limits on the right to know? These two rights—the right of workers to know about the hazards of substances that they use and the right of employers to maintain trade secrets—are not wholly incompatible, however. So it may be possible to implement the right to know in a way that maximizes the amount of information available to workers while at the same time minimizing the risk of exposing trade secrets.

Before limiting the right of employees to know, we need to be sure, first, that a genuine trade secret is involved and that it cannot be protected by any less restrictive means, such as obtaining a patent. Also, employers do not have a right to conceal *all* information connected with a trade secret but only the portion that cannot be revealed without jeopardy. Even when an employer is justified in not disclosing the chemical identity of a critical ingredient in a formula or a process, for example, there may be no justification for not providing workers with information about some of the physical characteristics and special hazards of that ingredient. Finally, information can sometimes be revealed in ways that preserve secrecy. It might be possible, for example, to reveal information about hazardous substances to union representatives or to the employees' own private physicians under a pledge of confidentiality without endangering valuable trade secrets.

Case 12.3   Johnson Controls, Inc.

Based in Milwaukee, Wisconsin, Johnson Controls is the largest manufacturer of automobile batteries for the United States replacement market.[30] Known chiefly for the production of control instruments for heating, lighting, and other electrical functions, Johnson Controls is also the largest independent supplier of automobile seating, and the company provides many small components for cars and light trucks. In 1978, Johnson Controls purchased Globe Union, Inc., a battery manufacturer. By 1990, the Globe Battery Division of Johnson Controls operated 14 plants nationwide and employed approximately 5,400 people. Batteries account for roughly 18 percent of Johnson Controls' sales and 17 percent of operating income.

Lead plates, which are essential for an automobile battery, are formed by compressing a paste of lead oxide. In this process, lead dust and lead vapor are released into the work area. Lead has been known for centuries to cause extensive neurological damage, and recent studies have shown that it affects the body's cardiovascular system, leading to heart attacks and strokes. Children who are exposed to lead, through eating peeling lead-based paint, for example, exhibit hyperactivity, short attention span, and learning difficulties. Lead in a pregnant woman's bloodstream can affect the neurological development of an unborn child, resulting in mental retardation, impaired motor control, and behavioral abnormalities. Pregnant women exposed to lead also run an increased risk of spontaneous abortion, miscarriage, and stillbirth. Although the effects of lead on men is less well understood, studies have shown some genetic damage to sperm that might cause birth defects.

Prior to Johnson Controls' purchase of Globe Union, the battery manufacturer had instituted a comprehensive program to minimize lead exposure in the workplace and to keep employees from carrying lead home on their body and clothing. Although no legal standards for lead exposure existed at the time, Globe Union routinely tested employees' blood lead levels and transferred employees with readings of 50 micrograms per deciliter of blood to other jobs without loss of pay until their levels had dropped to 30 µ/dl. In 1978, the Occupational Safety and Health Administration (OSHA) set a permissible exposure limit for lead of 50 µg/dl. OSHA did not establish a separate standard for pregnant women but recommended that both men and women who planned to conceive maintain blood levels below 30 µg/dl. OSHA concluded that there is no reason to exclude women of childbearing age from jobs involving lead exposure.

In 1977, as more women began working in battery production, Globe Union informed women employees of the hazards of lead and asked them to sign a statement that they had been told of the risks of having a child while exposed to lead in the workplace. Between 1979 and 1983, after Johnson Controls acquired Globe Union, eight women with blood lead levels in excess of 30 micrograms per deciliter became pregnant. In response, Johnson Controls changed its policy in 1982 to exclude fertile women from all jobs where lead is present. Specifically, the policy stated: "It is [Johnson Controls'] policy that women who are pregnant or who are capable of bearing children will not be placed into jobs involving lead exposure or which could expose them to lead through the exercise of job bidding, bumping, transfer or promotion rights." The policy defined women "capable of bearing children" as "all women except those whose inability to bear children is medically documented." In defending this policy, Johnson Controls maintained that a voluntary approach had failed and that to permit lead poisoning of unborn children was "morally reprehensible." Undoubtedly, the company was also concerned with its legal liability.

In April 1984, a class action suit was filed by several workers and their union, the United Auto Workers, charging that Johnson Controls' fetal-protection policy violated the Title VII prohibition against sex discrimination. The policy is discriminatory, the employees complained, because it singled out fertile women for exclusion, when evidence indicates that lead also poses a hazard to the reproductive capacities of men. Although Title VII permits exceptions for Bona Fide Occupational Qualifications (BFOQs), the inability to bear children has no relevance to the job of making a battery and therefore cannot be a legitimate BFOQ. Johnson Controls' fetal-protection policy applied to women who had no intention of becoming pregnant and those who might choose to accept the risk for the sake of keeping their jobs. The policy also offered women a choice of becoming sterile or losing their job, which some regarded as coercive. Among the employees suing were a woman who had chosen to be sterilized in order to keep her job, a 50-year-old divorcée who had been forced to accept a demotion because of the policy, and a man who had requested a leave of absence in order to reduce his blood lead level before becoming a father.

THE PROBLEM OF REPRODUCTIVE HAZARDS

The problem of reproductive hazards puts companies such as Johnson Controls in a very difficult position. On the one hand, fetal-protection policies are open to the charge of being discriminatory because they limit the job opportunities of fertile women while ignoring the substantial evidence of risk to men. Employees also claim that the policies are adopted by employers as a quick and cheap alternative to cleaning up the workplace. On the other hand, employers have a responsibility not to inflict harm, and this obligation extends, presumably, to the fetus of a pregnant employee. Asking employees to sign waivers releasing a company from responsibility is no solution, however, because employees cannot waive the right of their future children to sue for prenatal injuries. Employers seem to face a choice, therefore, between discriminating against their female employees and allowing fetuses to be harmed, with potentially ruinous consequences for the company.

The Issue of Who Decides

A number of employers have resolved this dilemma by adopting fetal-protection policies similar to that of Johnson Controls. However, many women are claiming the right to decide for themselves whether to work at jobs that involve reproductive risks. Although they have a strong maternal interest in the health of an unborn child, they are also concerned about their own economic well-being and want to be free to choose the level of risk to themselves and their offspring that they feel is appropriate. Ronald Bayer has observed that these responses to the problem of reproductive hazards involve a reversal of the usual positions of employers and employees:

> Typically, workers and their representatives have pressed management for the most extensive reductions in exposure levels to toxic substances. Further, they have argued that uncertainty requires the most cautious assumptions about the possibility of harmful consequences. Corporations have responded by arguing that a risk-free environment is chimerical and that uncertainty requires a willingness to tolerate levels of exposure that have not been proven harmful. Yet in relation to reproductive hazards, and especially fetal danger, it is labor that has tended to view with some skepticism the data on potential risk. Corporations, on the other hand, have adopted an almost alarmist posture.[31]

This reversal is not hard to understand. Protecting unborn children from harmful chemicals in the workplace and coping with the consequences of occupationally related birth defects involve substantial costs. The struggle between employers and employees is over who will exercise the choice about assuming the risk of reproductive hazards and who will bear the burden of responsibility. Fetal-protection policies are adopted by corporate managers who assume the right to make crucial decisions about how the fetus of an employee will be protected. The cost of these decisions is borne largely by women, however, who find that their economic opportunities are sharply limited. According to estimates by the federal government in 1980, at least 100,000 jobs were closed to women because of fetal-protection policies already in place,[32] and as many as 20 million jobs would be closed if women were excluded from all work involving reproductive hazards.[33] The women most affected by the reduction in the number of jobs available are those with the fewest skills and the least education, who are already near the bottom of the economic ladder. Because they bear the heavy cost, women argue that they should be the ones to make the decisions involved in protecting their own offspring.

Are Women Forced to Undergo Sterilization?

The debate over fetal-protection policies takes on a tragic dimension when women undergo sterilization for fear of losing a job. When five women were laid off by the Allied Chemical Company in 1977 because of concern for fetal damage by a substance known as Fluorocarbon 22, two of them underwent surgical sterilization in order to return to work. Shortly afterward, the company determined that Fluorocarbon 22 posed no threat to a developing fetus, so the women's loss of fertility was needless.[34] When the American Cyanamid Company announced the adoption of a fetal-protection policy at a plant in Willow Island, West Virginia, in 1978, five women between the ages of 26 and 43 submitted to sterilization in order to retain jobs that involved exposure to lead chromate, an ingredient of paint pigments. Two years later, American Cyanamid stopped producing paint pigments because of a decreased demand for the lead-based product.[35]

Certainly, no person, man or woman, should be put in the position of having to choose between holding a job and being able to bear children unless there is no acceptable alternative. Employers insist that they do not encourage women to take the drastic step of undergoing sterilization, but they also maintain that they have no control over such an intimate decision by employees or women seeking employment and that there is no reason to exclude women who undergo sterilization from jobs involving exposure to reproductive hazards. According to critics of fetal-protection policies, however, companies that exclude fertile women from such jobs effectively force sterilization on those who have no other satisfactory employment opportunities. The problem is one of intentions versus results. Although there is no intention to force women to become sterile, employers create situations that have precisely this result.

Issues in the Charge of Sex Discrimination

Whether employers have a right to adopt fetal-protection policies depends in part on whether excluding fertile women from certain jobs is a form of sex discrimination. The point at issue is not whether women are vulnerable to reproductive hazards (they certainly are), but whether men are vulnerable as well. If they are, then it is discriminatory for employers to adopt a policy that applies only to women and not to men. In order to evaluate the scientific evidence, it is necessary to understand more about the nature of reproductive hazards.

The Scientific Background. Substances harmful to a fetus are of three kinds. First, there are *fetotoxins*. These are toxic substances that affect a fetus in the same way that they affect an adult, although a fetus, because of the smaller size, may be harmed by exposure to substances below the permissible limits for adults. *Teratogens*, the second kind of substances, interfere with the normal development of the fetus in utero. These may pose no danger to a fully developed person outside the womb. Finally, some substances are *mutagens*, which damage genetic material before conception. The effects of fetotoxins, teratogens, and mutagens are similar. They include spontaneous abortion, miscarriage, stillbirth, and congenital defects. Some defects, such as deformities, are visible at birth, whereas others may be latent conditions that manifest themselves later, as in the case of childhood cancers.

Fetotoxins and teratogens (many substances are both) must be transmitted to the fetus through the mother. This can occur, however, when the father is exposed to a hazardous substance in the workplace. Studies show that the nonworking wives of men exposed to lead, beryllium, benzene, vinyl chloride, and anesthetic gases, for example, have higher-than-normal rates of miscarriage. Children of fathers exposed to asbestos and benzene are shown in studies to have a greater incidence of cancer. The most likely explanation for these correlations is that the hazardous substances are brought home on the father's body or on his clothing and other belongings.

The main reproductive hazard to men is posed by mutagenic substances, because these are capable of altering the chromosomal structure of both the ovum and the sperm. Although the mutagenic effect of many suspected substances is not firmly established, some researchers theorize that most teratogens are also mutagens and that there is a strong connection between the three phenomena: teratogenesis, mutagenesis, and carcinogenesis (that is, the development of cancers). The reason is that all three operate on the cellular level by altering the DNA molecule.[36] If these relationships exist, then virtually any substance that poses a reproductive hazard to a woman is also hazardous to the reproductive capacity of a man and ultimately the health of a fetus.

Are Fetal-Protection Policies Discriminatory? The research on reproductive hazards, although inconclusive, suggests that fetal harm can result when either parent is exposed to hazardous substances. If that is in fact the case, then fetal-protection policies should apply to both sexes. Wendy W. Williams argues:

> There is simply no basis for resolving doubts about the evidence by applying a policy to women but not to men on the unsubstantiated generalization that the fetus is placed at greatest risk by workplace exposure of the pregnant woman. Indeed, the limited scientific evidence available requires that doubts about the evidence be resolved in favor of an assumption of equal jeopardy.[37]

Fetal-protection policies are also discriminatory if they are applied only to women who occupy traditionally male jobs and not to women in female-dominated lines of work where the hazard is just as great. Some critics charge that fetal-protection policies have been used to discriminate by reinforcing job segregation through their selective application to women with jobs in areas formerly dominated by men, whereas the reproductive risks to other women have been ignored.

American Cyanamid, for example, identified five substances in use at the Willow Island plant as suspected fetotoxins and notified 23 women in production jobs that they were exposed to reproductive hazards and would be transferred if they became pregnant. However, the fetal-protection policy was applied only to the nine women who held comparatively high-paying jobs in the paint pigments department, which had formerly been reserved for men. Barbara Cantwell, one of the four women who submitted to sterilization in order to retain her job, remarked, "I smelled harassment when the problem was suddenly narrowed to the area where women worked."[38]

Among the women in traditionally female lines of work who are exposed to reproductive hazards are nurses in operating rooms, who have twice as many miscarriages as other women because of anesthetic gases; female X-ray technicians, who are twice as likely to bear defective children; and women who work in dry-cleaning operations, where petroleum-based solvents are used.[39] Yet there has been no movement in the industries employing these women to implement fetal-protection policies similar to those in the male-dominated chemical, petroleum, and heavy-manufacturing industries.

Defenses to the Charge of Sex Discrimination

Distinctions based on sex are not always discriminatory. They are morally permissible if they have an adequate justification, and Title VII of the 1964 Civil Rights Act recognizes this by allowing employers two defenses. These are the claims that a sex-based policy serves a proper business purpose (the business necessity defense) and that a person's sex is a Bona Fide Occupational Qualification (the BFOQ defense). A bona fide occupational qualification is defined as a qualification "reasonably necessary to the normal operation" of a business. As we saw in Chapter 8 on discrimination, this exception has been narrowly limited by the courts. The business necessity defense is less stringent. Generally, an employer must establish that a policy is needed for achieving a proper business objective in a safe and efficient manner and that the objective cannot reasonably be achieved by less discriminatory means.

The lower courts ruled that the fetal-protection policy adopted by Johnson Controls was permissible under Title VII. The Court of Appeals for the Seventh Circuit held that the policy passed a three-step business necessity defense because the company had established (1) there is a substantial risk to a fetus; (2) this risk occurs only through women; and (3) there is no less discriminatory alternative. Although not required to do so, the court addressed the question of whether sex is a BFOQ in this case and concluded that it is. The crucial phrase in the BFOQ defense "reasonably necessary for the normal operation" of a business has been interpreted by the courts to include "ethical, legal, and business concerns about the effects of an employer's activities on third parties." Simply put, a qualification is a BFOQ if it is necessary for conducting business without greatly endangering other people. Under this interpretation, the courts had ruled that age was a BFOQ for being an airline flight engineer because the safety of passengers depended on that person's performance.[40] By the same line of reasoning, sex is a BFOQ because a woman working in a battery factory is liable to expose a third party—namely, a fetus—to harm.

The U.S. Supreme Court in a landmark 9–0 decision overruled the lower courts and held Johnson Controls' fetal-protection policy to be discriminatory in violation of Title VII. The policy is discriminatory for the reason that it excludes women based only on their childbearing capacity and ignores the risk to men. Thus, the policy does not protect "effectively and equally" the offspring of all employees. Further, the Court ruled that sex is not a BFOQ in this case. All evidence indicates that fertile and even pregnant women are as capable of manufacturing batteries as anyone else. According to the majority opinion, the claim that the safety of third parties ought to be taken into consideration is inapplicable because this exception concerns only the ability of an employee to perform a job in a safe manner, and a fetus, in this case, is not endangered by the manner in which the job is performed. The workers at Johnson Controls, in other words, are not manufacturing batteries in ways that are unsafe for children. "No one can disregard the possibility of injury to future children," the opinion states, but the BFOQ defense "is not so broad that it transforms this deep social concern into an essential aspect of batterymaking."[41]

Remaining Issues

The decision in *Johnson Controls* clearly establishes that fetal-protection policies constitute illegal sex discrimination, and the ruling is a victory for working women. It gives women the right not to have their job opportunities limited because of their ability to conceive and bear children. Still, two important issues remain. One is how to balance this right with the desirable social goal of fetal protection. The other is whether the right established in *Johnson Controls* conflicts with another desirable goal, namely, protecting corporations against liability in the event that a person is harmed before birth from exposure to reproductive hazards in the workplace.

The ruling in *Johnson Controls* does not leave the unborn without protection. In the words of the Court, "Decisions about the welfare of future children must be left to the parents who conceive, bear, support, and raise them rather than to the employers who hire those parents." There is little reason to believe that employees are any less concerned than employers about the well-being of their offspring. Indeed, they have a far more compelling interest. This is not to say that some parents-to-be (both men and women) will not continue to choose work that exposes them to reproductive hazards. However, they, rather than their employer, will be making the crucial decision about the reasonableness of that choice. Employees may decide—all things considered—that the risk is worth the price. Parents also choose where to live with their children, thereby deciding whether they can afford to live farther away from sources of pollution and other hazards. Because parents in our society make choices in other matters that bear on the welfare of their children, why should an exception be made in the case of reproductive hazards? What is unfortunate is that any parents are required to choose between making a living and protecting their children.

Whether the *Johnson Controls* ruling exposes corporations to heavy tort liability is an open question. The view expressed in the majority opinion is that the prospect is "remote at best," provided that employers (1) fully inform employees of the risks they face and (2) do not act negligently. If employees are to make rational choices in cases of exposure to reproductive hazards, then they must have sufficient information, which employers have an obligation to provide. Employers can protect themselves from suits by cleaning up the workplace and reducing the risk as much as possible. The issue of tort liability, then, is hypothetical but nonetheless important for that reason.

NOTES

1. May Meyers et al., "Benzene (benzol) Poisoning in the Rotogravure Printing in New York City," *Journal of Industrial Hygiene and Toxicology* (October 1939), 395–420. The study is cited in Joseph A. Page and Mary-Win O'Brien, *Bitter Wages* (New York: Grossman, 1973), 37–38.

2. The details in this case are contained in *Industrial Union Dept., AFL-CIO v. American Petroleum Institute*, 448 U.S. 607 (1980).

3. The estimate is made in W. J. Nicholson, "Failure to Regulate—Asbestos: A Lethal Legacy," U.S. Congress, Committee of Government Operations, 1980.

4. Much of the following description of health and safety problems is adapted from Nicholas Askounes Ashford, *Crisis in the Workplace: Occupational Disease and Injury* (Cambridge, MA: MIT Press, 1976), chap. 3.

5. For a review of the controversy that discredits the industry claim, see Robert H. Hall, "The Truth about Brown Lung," *Business and Society Review*, 40 (Winter 1981), 15–20.

6. For a critical assessment of state efforts, see Page and O'Brien, *Bitter Wages*, 69–85.

7. Sec. 5(a)(1). This clause is the subject of much legal analysis. See Richard S. Morley, "The General Duty Clause of the Occupational Safety and Health Act of 1970," *Harvard Law Review*, 86 (1973), 988–1005.

8. Patricia H. Werhane, *Persons, Rights, and Corporations* (Upper Saddle River, NJ: Prentice Hall, 1985), 132.

9. Mark MacCarthy, "A Review of Some Normative and Conceptual Issues in Occupational Safety and Health," *Environmental Affairs*, 9 (1981), 782–83.

10. The case is related in Stephen G. Minter, "Are Prosecutors Stepping in Where OSHA Fears to Tread?" *Occupational Hazards*, 49 (September 1987), 101–2.

11. These two factors are discussed by Alan Gewirth, who cites them as exceptions to his claim that persons have a right not to have cancer inflicted on them by the actions of others. See Alan Gewirth, "Human Rights and the Prevention of Cancer," in *Human Rights: Essays on Justification and Applications* (Chicago: University of Chicago Press, 1982), 181–96. For an evaluation, see Eric Von Magnus, "Rights and Risks," *Journal of Business Ethics*, 2 (February 1983), 23–26.

12. See H. L. A. Hart and A. M. Honore, *Causation in the Law* (Oxford: Oxford University Press, 1959), 64–76.

13. U.S. Code 21, 348(c)(3).

14. *Federal Register*, 39, no. 20 (29 January 1974), 3758. Emphasis added.

15. Gewirth, "Human Rights and Prevention of Cancer," 189.

16. *American Textile Manufacturers Institute Inc.* v. *Raymond J. Donovan, Secretary of Labor*, 452 U.S. 490 (1981).

17. For an argument of this kind, see Tibor R. Machan, "Human Rights, Workers' Rights, and the Right to Occupational Safety," in *Moral Rights in the Workplace*, ed., Gertrude Ezorsky (Albany: State University of New York Press, 1987), 45–50.

18. W. Kip Viscusi, *Risk by Choice* (Cambridge, MA: Harvard University Press, 1983), 80.

19. The following argument is adapted from Norman Daniels, "Does OSHA Protect Too Much?" in *Moral Rights in the Workplace*, 51–60.

20. Robert Nozick, "Coercion," in *Philosophy, Science and Method*, ed., Sidney Morgenbesser, Patrick Suppes, and Morton White (New York: St. Martin's Press, 1969), 440–72.

21. In considering whether a person voluntarily chooses undesirable work when all of the alternatives are even worse as a result of the actions of other people, Nozick says that the answer "depends upon whether these others had the right to act as they did." Robert Nozick, *Anarchy, State, and Utopia* (New York: Basic Books, 1974), 262.

22. Some philosophers have attempted to give a morally neutral analysis of coercion that involves no assumptions about what is morally required. See David Zimmerman, "Coercive Wage Offers," *Philosophy and Public Affairs*, 10 (1981), 121–45.

23. Information on this case is contained in *Whirlpool Corporation* v. *Marshall*, 445 U.S. 1 (1980).

24. Thomas O. McGarity, "The New OSHA Rules and the Worker's Right to Know," *The Hastings Center Report*, 14 (August 1984), 38–39.

25. For a version of this argument see ibid. Much of the following discussion of the autonomy argument is adapted from this article.

26. This point is made in Ruth R. Faden and Tom L. Beauchamp, "The Right to Risk Information and the Right to Refuse Health Hazards in the Workplace," in *Ethical Theory and Business*, 4th ed., ed., Tom L. Beauchamp and Norman E. Bowie (Upper Saddle River, NJ: Prentice Hall, 1993), 205.

27. This exception suggests a further argument for the right to know based on fairness. Employers who knowingly place workers at risk are taking unfair advantage of the workers' ignorance. See McGarity, "New OSHA Rules and Worker's Right to Know," 40.

28. Ibid.

29. Ibid., 41.

30. This case is based on *United Automobile Workers, et al.* v. *Johnson Controls, Inc.*, 449 U.S. 187, 111 S.Ct. 1196 (1991). Other sources include "Court Backs Right of Women to Jobs with Health Risks," *New York Times*, 21 March 1991, sec. 1, p. 1; Stephen Wermiel, "Justices Bar 'Fetal Protection' Policies," *Wall Street Journal*, 21 March 1991, B1; and Anne T. Lawrence, "Johnson Controls and Protective Exclusion from the Workplace," *Case Research Journal*, 13 (1993), 1-14.

31. Ronald Bayer, "Women, Work, and Reproductive Hazards," *The Hastings Center Report*, 12 (October 1982), 18.

32. Wendy W. Williams, "Firing the Woman to Protect the Fetus: The Reconciliation of Fetal Protection with Employment Opportunity Goals under Title VII," *The Georgetown Law Journal*, 69 (1981), 647.

33. Equal Employment Opportunity Commission and Office of Federal Contract Compliance Programs, "Interpretive Guidelines on Employment Discrimination and Reproductive Hazards," *Federal Register* (1 February 1980), 7514.

34. Gail Bronson, "Issue of Fetal Damage Stirs Women Workers at Chemical Plants," *Wall Street Journal*, 9 February 1979, A1.

35. "Company and Union in Dispute as Women Undergo Sterilization," *New York Times*, 4 January 1979, sec. 1, p. 1; "Four Women Assert Jobs Were Linked to Sterilization," *New York Times*, 5 January 1979, sec. 1, p. 1; Philip Shabecoff, "Job Threats to Workers' Fertility Emerging as Civil Liberties Issue," *New York Times*, 15 January 1979, sec. 1, p. 1; and Bronson, "Issue of Fetal Damage Stirs Women Workers at Chemical Plants."

36. See Williams, "Firing the Woman to Protect the Fetus," 659–60.

37. Ibid., 663.

38. Cited in Bronson, "Issue of Fetal Damage Stirs Women Workers at Chemical Plants."

39. These examples are cited in Lois Vanderwaerdt, "Resolving the Conflict between Hazardous Substances in the Workplace and Equal Employment Opportunity," *American Business Law Journal*, 21 (1983), 172–73.

40. *Western Airlines* v. *Criswell*, 472 U.S. 400 (1985).

41. The Supreme Court did not consider the business necessity defense, because only the BFOQ defense was held to be applicable in this case. The reason is that the fetal-protection policy adopted by Johnson Controls was considered by the court to be "facial" discrimination, which is to say that a distinction was made explicitly on the basis of sex, and "facial" discrimination requires the more demanding BFOQ defense rather than the weaker business necessity defense.

13

Ethics in Finance

Case 13.1 Pacific Lumber Company

Since its founding in the nineteenth century, Pacific Lumber Company had been a model employer and a good corporate citizen. As a logger of giant redwoods in northern California, this family-managed company had long followed a policy of perpetual sustainable yield. Cutting was limited to selected mature trees, which were removed without disturbing the forests, so that younger trees could grow to the same size. Employees—many from families who had worked at Pacific Lumber for several generations—received generous benefits, including an overfunded company-sponsored pension plan. With strong earnings and virtually no debt, Pacific Lumber seemed well-positioned to survive any challenge.

However, the company fell prey to a hostile takeover. In 1986, financier Charles Hurwitz and his Houston-based firm Maxxam, Inc., mounted a successful $900 million leveraged buyout of Pacific Lumber. By offering $40 per share for stock that had been trading at $29, Hurwitz gained majority control. The takeover was financed with junk bonds issued by Drexel Burnham Lambert under the direction of Michael Milken, the junk-bond king. Hurwitz expected to pare down the debt by aggressive clear-cutting of the ancient stands of redwoods that Pacific Lumber had protected and by raiding the company's overfunded pension plan.

Using $37.3 million of $97 million that Pacific Lumber had set aside for its pension obligations, Maxxam purchased annuities for all employees and retirees and applied more than $55 million of the remainder toward reducing the company's new debt. The annuities were purchased from First Executive Corporation, a company that Hurwitz controlled. First Executive was also Drexel's biggest junk bond customer, and the company purchased one-third of the debt incurred in the takeover of Pacific Lumber. After the collapse of the junk bond market, First Executive failed in 1991 and was taken over by the State of California in a move that halted pension payments to Pacific Lumber retirees. Today, Charles Hurwitz and Maxxam are mired in lawsuits by former stockholders, retirees, and environmentalists.

INTRODUCTION

Some cynics jokingly deny that there is any ethics in finance, especially on Wall Street. This view is expressed in a thin volume, *The Complete Book of Wall Street Ethics*, which claims to fill "an empty space on financial bookshelves where a consideration of ethics should be."[1] Of course, the pages are all blank! However, a moment's reflection reveals that finance would be impossible without ethics. The very act of placing our assets in the hands of other people requires immense trust. An untrustworthy stockbroker or insurance agent, like an untrustworthy physician or attorney, finds few takers for his or her services. Financial scandals shock us precisely because they involve people and institutions that we should be able to trust.

Finance covers a broad range of activities, but the two most visible aspects are financial markets, such as stock exchanges, and the financial services industry, which includes not only commercial banks, but also investment banks, mutual fund companies, pension funds, both public and private, and insurance. Less visible to the public are the financial operations of a corporation, which are the responsibility of the Chief Financial Officer (CFO). This chapter focuses first on ethics in financial markets and the financial services industry by examining market regulation, which is the subject of securities law and regulation by the Securities and Exchange Commission (SEC), and the treatment of customers by financial services firms. Because Wall Street was shaken in the 1980s by instances of insider trading by prominent financiers and by hotly contested battles for corporate control by some of the same financiers, the chapter also covers the topics of insider trading and hostile takeovers.

FINANCIAL SERVICES

The financial services industry still operates largely through personal selling by stockbrokers, insurance agents, financial planners, tax advisers, and other finance professionals. Personal selling creates innumerable opportunities for abuse, and although finance professionals take pride in the level of integrity in the industry, misconduct still occurs. However, customers who are unhappy over failed investments or rejected insurance claims are quick to blame the seller of the product, sometimes with good reason.

For example, two real estate limited partnerships launched by Merrill Lynch & Co. in 1987 and 1989 lost close to $440 million for 42,000 investor-clients.[2] Known as Arvida I and Arvida II, these highly speculative investment vehicles projected double-digit returns on residential developments in Florida and California, but both stopped payments to investors in 1990. At the end of 1993, each $1,000 unit of Arvida I was worth $125, and each $1,000 unit of Arvida II, a mere $6.

The Arvida partnerships were offered by the Merrill Lynch sales force to many retirees of modest means as safe investments with good income potential. The brokers themselves were told by the firm that Arvida I entailed only "moderate risk," and company-produced sales material said little about risk while emphasizing the projected performance. Left out of the material was the fact that the projections included a return of some of the investors' own capital, that the track record of the real estate company was based on commercial, not residential projects, and that eight of the top nine managers of the company had left just before Arvida I was offered to the public.

This case raises questions about whether investors were *deceived* by the brokers' sales pitches and whether material information was *concealed*. In other cases, brokers have been accused of *churning* client accounts in order to generate higher fees and of selecting *unsuitable* investments for clients. Other abusive sales practices in the financial services industry include *twisting*, in which an insurance agent persuades a policy holder to replace an older policy with a newer one that provides little if any additional benefit but generates a commission for the agent, and *flipping*, in which a loan officer persuades a borrower to repay an old loan with a new one, thereby incurring more fees. In one case, an illiterate retiree, who was flipped ten times in a four-year period, paid $19,000 in loan fees for the privilege of borrowing $23,000.

This section discusses three objectionable practices in selling financial products to clients, namely deception, churning, and suitability.

Deception

The ethical treatment of clients requires salespeople to explain all of the relevant information truthfully in an understandable, nonmisleading manner. One observer complains that brokers, insurance agents, and other salespeople have developed a new vocabulary that obfuscates rather than reveals.

> Walk into a broker's office these days. You won't be sold a product. You won't even find a broker. Instead, a "financial adviser" will "help you select" an "appropriate planning vehicle," or "offer" a menu of "investment choices" or "options" among which to "allocate your money." ... [Insurance agents] peddle such euphemisms as "private retirement accounts," "college savings plans," and "charitable remainder trusts." ... Among other linguistic sleights of hand in common usage these days: saying tax-free when, in fact, it's only tax-deferred; high yield when it's downright risky; and projected returns when it's more likely in your dreams.[3]

Salespeople avoid speaking of commissions, even though they are the source of their compensation. Commissions on mutual funds are "front-end" or "back-end loads"; and insurance agents, whose commissions can approach 100 percent of the first year's premium, are not legally required to disclose this fact—and they rarely do. The agents of one insurance company represented life insurance policies as "retirement plans" and referred to the premiums as "deposits."[4]

Deception is often a matter of interpretation. Promotional material for a mutual fund, for example, may be accurate but misleading if it emphasizes the strengths of a fund and minimizes the weaknesses. Figures of past performance can carefully be selected and displayed in ways that give a misleading impression. Deception can also occur when essential information is not revealed. Thus, an investor may be deceived when the sales charge is rolled into the fund's annual expenses, which may be substantially higher than the competition's, or when the projected hypothetical returns do not reflect all charges. As these example suggest, true claims may lead a typical investors to hold a mistaken belief.

Deception aside, what information *ought* to be disclosed to a client? The Securities Act of 1933 requires the issuer of a security to disclose all material information, which is defined as information about which an average prudent investor ought reasonably to be informed or to which a reasonable person would attach importance in determining a course of action in a transaction. The rationale for this provision of the Securities Act is both fairness to investors, who have a right to make decisions with adequate information, and the efficiency of securities markets, which requires that investors be adequately informed. Most financial products, including mutual funds and insurance policies, are accompanied by a written prospectus that contains all of the information that the issuer is legally required to provide.

The analysis of deception by a financial service provider is similar to that provided for deceptive advertising in Chapter 11. In general, a person is deceived when that person is unable to make a rational choice as a result of holding a false belief that is created by some claim made by another. That claim may be either a false or misleading statement or a statement that is incomplete in some crucial way. Consider two cases of possible broker (mis)conduct:

1. A brokerage firm buys a block of stock prior to issuing a research report that contains a "buy" recommendation in order to ensure that enough shares are available to fill customer orders. However, customers are not told that they are buying stock from the firm's own holdings, and they are charged the current market price plus the standard commission for a trade.
2. A broker assures a client that an Initial Public Offering (IPO) of a closed-end fund is sold without a commission and encourages quick action by saying that after the IPO is sold, subsequent buyers will have to pay a 7 percent commission. In fact, a 7 percent commission is built into the price of the IPO, and this charge is revealed in the prospectus but will not appear on the settlement statement for the purchase.

In the first case, one might argue that if an investor decides to purchase shares of stock in response to a "buy" recommendation, it matters little whether the shares are bought on the open market or from a brokerage firm's holdings. The price is the same. An investor might appreciate the opportunity to share any profit that is realized by the firm (because of lower trading costs and perhaps a lower stock price before the recommendation is released), but the firm is under no obligation to share any profit with its clients. On the other hand, the client is buying the stock at the current market price and paying a fee as though the stock were purchased at the order of the client. The circumstances of the purchase are not explained to the client, but does the broker have any obligation to do so? And would this knowledge have any effect on the client's decision?

In the second case, however, a client might be induced to buy an initial offering of a closed-end mutual fund in the mistaken belief that the purchase would avoid a commission charge. The fact that the commission charge is disclosed in the prospectus might ordinarily exonerate the broker from a charge of deception except that the false belief is created by the broker's claim, which, at best, skirts the edge of honesty. Arguably, the broker made the claim with an intent to deceive, and a typical, prudent investor is apt to feel that there was an attempt to deceive.

Churning

Churning is defined as excessive or inappropriate trading for a client's account by a broker who has control over the account with the intent to generate commissions rather than to benefit the client. Although churning occurs, there is disagreement on the frequency or the rate of detection. The brokerage industry contends that churning is a rare occurrence and is easily detected by firms as well as clients. No statistics are kept on churning, but complaints to the SEC and various exchanges about unauthorized trading and other abuses have risen sharply in recent years.

The ethical objection to churning is straightforward: It is a breach of a fiduciary duty to trade in ways that are not in a client's best interests. Churning, as distinct from unauthorized trading, occurs only when a client turns over control of an account to a broker, and by taking control, a broker assumes a responsibility to serve the client's interests. A broker who merely recommends a trade, is not acting on behalf of a client or customer and is more akin to a traditional seller, but a broker in charge of a client's portfolio thereby pledges to manage it to the best of his or her ability.

Although churning is clearly wrong, the concept is difficult to define. Some legal definitions offered in court decisions are: "excessive trading by a broker disproportionate to the size of the account involved, in order to generate commissions,"[5] and a situation in which "a broker, exercising control over the frequency and volume of trading in the customer's account, initiates transactions that are excessive in view of the character of the account."[6] The courts have held that for churning to occur a broker must trade with the *intention* of generating commissions rather than benefiting the client. The legal definition of churning contains three elements, then: (1) The broker controls the account; (2) the trading is excessive for the character of the account; and (3) the broker acted with intent.

The most difficult issue in the definition of churning is the meaning of "excessive trading." First, whether trading is excessive depends on the character of the account. A client who is a more speculative investor, willing to assume higher risk for a greater return, should expect a higher trading volume. Second, high volume is not the only factor; pointless trades might be considered churning even if the volume is relatively low. Third, churning might be indicated by a pattern of trading that consistently favors trades that yield higher commissions. Common to these three points is the question of whether the trades make sense from an investment point of view. High-volume trading that loses money might still be defended as an intelligent but unsuccessful investment strategy, whereas investments that represent no strategy beyond generating commissions are objectionable, no matter the amount gained or lost.

A 1995 SEC report concluded that the compensation system in brokerage firms was the root cause of the churning problem.[7] The report identified some "best practices" in the industry that might prevent churning, including ending the practice of paying a higher commission for a company's own products, prohibiting sales contests for specific products, and tying a portion of compensation to the size of a client's account, regardless of the number of transactions. However, an SEC panel concluded that the commission system is too deeply rooted to be significantly changed and recommended better training and oversight by brokerage firms.

Suitability

In general, brokers, insurance agents, and other salespeople have an obligation to recommend only suitable securities and financial products. However, suitability, like churning, is difficult to define precisely. The rules of the National Association of Securities Dealers include the following:

> In recommending to a customer the purchase, sale, or exchange of any security, a member shall have reasonable grounds for believing that the recommendation is suitable for such customer upon the basis of the facts, if any, disclosed by such customer as to his other security holding and as to his financial situation and needs.[8]

The most common causes of unsuitability are (1) *unsuitable types of securities*, that is, recommending stocks, for example, when bonds would better fit the investor's objectives; (2) *unsuitable grades of securities*, such as selecting lower-rated bonds when higher-rated ones are more appropriate; (3) *unsuitable diversification*, which leaves the portfolio vulnerable to changes in the markets; (4) *unsuitable trading techniques*, including the use of margin or options, which can leverage an account and create greater volatility and risk; and (5) *unsuitable liquidity*. Limited partnerships, for example, are not very marketable and are thus unsuitable for customers who may need to liquidate the investment.

The critical question, of course, is, when is a security unsuitable? Rarely is a single security unsuitable except in the context of an investor's total portfolio. Investments are most often deemed to be unsuitable because they involve excessive risk, but a few risky investments may be appropriate in a well-balanced, generally conservative portfolio. Furthermore, even an aggressive, risk-taking portfolio may include unsuitable securities if the risk is not compensated by the expected return.

Ensuring that a recommended security is suitable for a given investor thus involves many factors, but people in the financial services industry offer to put their specialized knowledge and skills to work for us. We expect suitable recommendations from physicians, lawyers, and accountants. Why should we expect anything less from finance professionals?

FINANCIAL MARKETS

Financial transactions typically take place in organized markets, such as stock markets, commodities markets, futures or options markets, currency markets, and the like. These markets presuppose certain moral rules and expectations of moral behavior. The most basic of these is a prohibition against fraud and manipulation, but, more generally, the rules and expectations for markets are concerned with fairness, which is often expressed as a level playing field. The playing field in financial markets can become "tilted" by many factors, including unequal information, bargaining power, and resources.

In addition to making one-time economic exchanges, participants in markets also engage in financial contracting whereby they enter into long-term relations. These contractual relations typically involve the assumption of fiduciary duties or obligations to act as agents, and financial markets are subject to unethical conduct when fiduciaries and agents fail in a duty. In the standard model of contracting, the terms of a contract specify the conduct required of each party and the remedies for noncompliance. In short, there is little "wiggle room" in a well-written contract. However, many contractual relations in finance and other areas fall short of this ideal, because actual contracts are often vague, ambiguous, incomplete, or otherwise problematic. The result is uncertainty and disagreement about what constitutes ethical (as well as legal) conduct.

Much of the necessary regulatory framework for financial markets is provided by law. The Securities Act of 1933 and the Securities Exchange Act of 1934 with their many amendments and the rules adopted by the Securities and Exchange Commission (SEC) constitute the main regulatory framework for markets in securities, and particular financial investment institutions, such as banks, mutual funds, and pension and insurance companies, are governed by industry-specific legislation.

Equity and Efficiency

The main aim of financial market regulation is to ensure *efficiency*, but markets can be efficient only when people have confidence in their *fairness* or *equity*. Efficiency is itself an ethical value because achieving the maximum output with the minimum input—which is a simple definition of efficiency—provides an abundance of goods and services and thereby promotes the general welfare. A society is generally better off when capital markets, for example, allocate the available capital to its most productive uses. People will participate in capital markets, however, only if the markets are perceived to be fair, that is fairness has value as a means to the end of efficiency.

We also value fairness as an end in itself, and because fairness can conflict with efficiency, some choice or trade-off between the two must often be made. This unfortunate fact of life is commonly described as the *equity/efficiency trade-off*. Painful choices between efficiency and fairness (or equity), or between economic and social well-being, are at the heart of many difficult public policy decisions, but we should not lose sight of the fact that fairness contributes to efficiency even as the two conflict.

Fairness in Markets

What constitutes fairness in financial markets?[9] Fairness is not a matter of preventing losses. Markets produce winners and losers, and in many cases the gain of some persons comes from an equal loss to others (although market exchanges are typically advantageous to both parties). In this respect, playing the stock market is like playing a sport: The aim is not to prevent losses but only to ensure that the game is fair. Still, there may be good reasons for seeking to protect individual investors from harm, even when the harm does not involve unfairness. Just as bean balls are forbidden in baseball (but playing hardball is okay!), so too are certain harmful practices prohibited in the financial marketplace.

The regulation of financial markets protects not only individual investors, but also the general public. The stock market crash of 1929, which prompted the first securities legislation, profoundly affected the entire country. Everyone is harmed when financial markets do not fill their main purpose but become distorted by speculative activity or disruptive trading practices. The deleterious effect of stock market speculation is wryly expressed by John Maynard Keyne's famous quip: "When the capital development of a country becomes a by-product of the activities of a casino, the job is likely to be ill-done."[10] More recently, the question of whether junk bonds or program trading pose risks to the stability of the financial markets has been a subject of dispute.

The possible ways in which individual investors and members of society can be treated unfairly by the operation of financial markets are many, but the main kinds of unfairness are the following.

Fraud and Manipulation. One of the main purposes of securities regulation is to prevent fraudulent and manipulative practices in the sale of securities. The common-law definition of fraud is the willful misrepresentation of a material fact that causes harm to a person who reasonably relies on the misrepresentation. Section 17(a) of the 1933 Securities Act and Section 10(b) of the 1934 Securities Exchange Act both prohibit anyone involved in the buying or selling of securities from making false statements of a material fact, omitting a fact that makes a statement of material facts misleading, or engaging in any practice or scheme that would serve to defraud.

Investors—both as buyers and as sellers—are particularly vulnerable to fraud because the value of financial instruments depends almost entirely on information that is difficult to verify. Much of the important information is in the hands of the issuing firm, and so antifraud provisions in securities law place an obligation not only on buyers and sellers of a firm's stock, for example, but also on the issuing firm. Thus, a company that fails to report bad news may be committing fraud, even though the buyer of that company's stock buys it from a previous owner who may or may not be aware of the news. Insider trading is prosecuted as a fraud under Section 10(b) of the Securities Exchange Act on the grounds that any material nonpublic information ought to be revealed before trading.

Manipulation generally involves the buying or selling of securities for the purpose of creating a false or misleading impression about the direction of their price so as to induce other investors to buy or sell the securities. Like fraud, manipulation is designed to deceive others, but the effect is achieved by the creation of false or misleading appearances rather than by false or misleading representations.

Fraud and manipulation are addressed by mandatory disclosure regulations as well as by penalties for false and misleading statements in any information released by a firm. Mandatory disclosure regulations are justified, in part, because they promote market efficiency: Better informed investors will make more rational investment decisions, and they will do so at lower overall cost. A further justification, however, is the prevention of fraud and manipulation under the assumption that good information drives out bad. Simply put, fraud and manipulation are more difficult to commit when investors have easy access to reliable information.

Equal Information. A "level playing field" requires not only that everyone play by the same rules, but also that they be equally equipped to compete. Competition between parties with very unequal information is widely regarded as unfair because the playing field is tilted in favor of the player with superior information. When people talk about equal information, however, they may mean that the parties to a trade actually *possess* the same information or have equal *access* to information.

That everyone should posses the same information is an unrealizable ideal, and actual markets are characterized by great information asymmetries. The average investor cannot hope to compete on equal terms with a market pro, and even pros often possess different information that leads them to make different investment decisions. Moreover, there are good reasons for encouraging people to acquire superior information for use in trade. Consider stock analysts and other savvy investors who spend considerable time, effort, and money to acquire information. Not only are they ordinarily entitled to use this information for their own benefit (because it represents a return on an investment), but they perform a service to everyone by ensuring that stocks are accurately priced.

The possession of unequal information strikes us as unfair, then, only when the information has been illegitimately acquired or when its use violates some obligation to others. One argument against insider trading, for example, holds that an insider has not acquired the information legitimately but has stolen (or "misappropriated") information that rightly belongs to the firm. In this argument, the wrongfulness of insider trading consists not in the possession of unequal information but in violating a moral obligation not to steal or a fiduciary duty to serve others. Insider trading can also be criticized on the grounds that others do not have the same access to the information, which leads us to the second sense of equal information, namely equal access.

The trouble with defining equal information as having equal access to information is that the notion of equal access is not absolute but relative. Any information that one person possesses could be acquired by another with enough time, effort, and money. An ordinary investor has access to virtually all of the information that a stock analyst uses to evaluate a company's prospects. The main difference is that the analyst has faster and easier access to information because of an investment in resources and skills. Anyone else could make the same investment and thereby gain the same access—or a person could simply "buy" the analyst's skilled services. Therefore, accessibility is not a feature of information itself but a function of the investment that is required in order to obtain the information.

We also hold that some information asymmetries are objectionable to the extent that they reduce efficiency. In particular, markets are more efficient when information is readily available, so we should seek to make information available at the lowest cost. To force people to make costly investments in information—or to suffer loss from inadequate information—is a deadweight loss to the economy if the same information could be provided at little cost. Thus, the requirement that the issuance of new securities be accompanied by a detailed prospectus, for example, is intended not only to prevent fraud through the concealment of material facts but also to make it easier for buyers to gain certain kinds of information, which benefits society as a whole.

Although efficiency and fairness both support attempts to reduce information asymmetries in financial markets, exactly what fairness or justice requires is not easy to determine. Consider, for example, whether a geologist, who concludes after careful study that a widow's land contains oil, would be justified in buying the land without revealing what he knows.[11] A utilitarian could argue that without such opportunities, geologists would not search for oil, and so society as a whole is better off if such advantage taking is permitted. In addition, the widow herself, who would be deprived of a potential gain, is better off in a society that allows some exploitation of superior knowledge. A difficult task for securities regulation, then, is drawing a line between fair and unfair advantage taking when people have unequal access to information.

Equal Bargaining Power. Generally, agreements reached by arm's-length bargaining are considered to be fair, regardless of the actual outcome. A trader who negotiates a futures contract that results in a great loss, for example, has only himself or herself to blame. However, the fairness of bargained agreements assumes that the parties have relatively equal bargaining power. Unequal bargaining power can result from many sources—including unequal information, which is discussed above—but other sources include the following factors.

1. *Resources.* In most transactions, wealth is an advantage. The rich are better able than the poor to negotiate over almost all matters. Prices of groceries in low-income

neighborhoods are generally higher than those in affluent areas, for example, because wealthier customers have more options. Similarly, large investors have greater opportunities. They can be better diversified; they can bear greater risk and thereby obtain higher leverage; they can gain more from arbitrage through volume trading; and they have access to investments that are closed to small investors.

2. *Processing Ability.* Even with equal access to information, people vary enormously in their ability to process information and to make informed judgments. Unsophisticated investors are ill-advised to play the stock market and even more so to invest in markets that only professionals understand. Fraud aside, financial markets can be dangerous places for people who lack an understanding of the risks involved. Securities firms and institutional investors overcome the problem of people's limited processing ability by employing specialists in different kinds of markets, and the use of computers in program trading enables these organizations to substitute machine power for gray matter.

3. *Vulnerabilities.* Investors are only human, and human beings have many weaknesses that can be exploited. Some regulation is designed to protect people from the exploitation of their vulnerabilities. Thus, consumer protection legislation often provides for a "cooling off" period during which shoppers can cancel an impulsive purchase. The requirements that a prospectus accompany offers of securities and that investors be urged to read the prospectus carefully serve to curb impulsiveness. Margin requirements and other measures that discourage speculative investment serve to protect incautious investors from overextending themselves, as well as to protect the market from excess volatility. The legal duty of brokers and investment advisers to recommend only suitable investments and to warn adequately of the risks of any investment instrument provides a further check on people's greedy impulses.

Unequal bargaining power that arises from these factors—resources, processing ability, and vulnerabilities—is an unavoidable feature of financial markets, and exploiting such power imbalances is not always unfair. In general, the law intervenes when exploitation is unconscionable or when the harm is not easily avoided, even by sophisticated investors. The success of financial markets depends on reasonably wide participation, and so if unequal bargaining power were permitted to drive all but the most powerful from the marketplace, then the efficiency of financial markets would be greatly impaired.

Efficient Pricing. Fairness in financial markets includes efficient prices that reasonably reflect all available information. A fundamental market principle is that the price of securities should reflect their underlying value. The mandate to ensure "fair and orderly" markets—set forth in the Securities Exchange Act of 1934—has been interpreted to authorize invention to correct volatility or excess price swings in stock markets. Volatility that results from a mismatch of buyers and sellers is eventually self-correcting, but in the meantime, great harm may be done by inefficient pricing. Individual investors may be harmed by buying at too high a price or selling at two low a price during periods of mispricing. Volatility also affects the market by reducing investor confidence, thus driving investors away, and some argue that the loss of confidence artificially depresses stock prices. At its worst, volatility can threaten the whole financial system, as it did in October 1987.

INSIDER TRADING

Insider trading is commonly defined as trading in the stock of publicly held corporations on the basis of material, nonpublic information. In a landmark 1968 decision, executives of Texas Gulf Sulphur Company were found guilty of insider trading for investing heavily in their own company's stock after learning of the discovery of rich copper ore deposits in Canada.[12] The principle established in the *Texas Gulf Sulphur* case is that corporate insiders must refrain from trading on information that significantly affects stock price until it becomes public knowledge. The rule for corporate insiders is, reveal or refrain!

Much of the uncertainty in the law on insider trading revolves around the relation of the trader to the source of the information. Corporate executives are definitely "insiders," but some "outsiders" have also been charged with insider trading. Among such outsiders have been a printer who was able to identify the targets of several takeovers from legal documents that were being prepared; a financial analyst who uncovered a huge fraud at a high-flying firm and advised his clients to sell; a stockbroker who was tipped off by a client who was a relative of the president of a company and who learned about the sale of the business through a chain of family gossip; a psychiatrist who was treating the wife of a financier who was attempting to take over a major bank; and a lawyer whose firm was advising a client planning a hostile takeover.[13] The first two traders were eventually found innocent of insider trading; the latter three were found guilty (although the stockbroker case was later reversed in part). From these cases a legal definition of insider trading is slowly emerging.

The key points are that a person who trades on material, nonpublic information is engaging in insider trading when (1) the trader has violated some legal duty to a corporation and its shareholders; or (2) the source of the information has such a legal duty and the trader knows that the source is violating that duty. Thus, the printer and the stock analyst had no relation to the corporations in question and so had no duty to refrain from using the information that they had acquired. The stockbroker and the psychiatrist, however, knew or should have known that they were obtaining inside information indirectly from high-level executives who had a duty to keep the information confidential. The corresponding rule for outsiders is, Don't trade on information that is revealed in violation of a trust. Both rules are imprecise, however, and leave many cases unresolved.

Arguments against Insider Trading

The difficulty in defining insider trading is due to disagreement over the moral wrong involved. Two main rationales are used in support of a law against insider trading. One is based on *property rights* and holds that those who trade on material, nonpublic information are essentially stealing property that belongs to the corporation. The second rationale is based on *fairness* and holds that traders who use inside information have an unfair advantage over other investors and that, as a result, the stock market is not a level playing field. These two rationales lead to different definitions, one narrow, the other broad. On the property rights or "misappropriation" theory, only corporate insiders or outsiders who bribe, steal, or otherwise wrongfully acquire corporate secrets can be guilty of insider trading. The fairness argument is broader and applies to anyone who trades on material, nonpublic information no matter how it is acquired.

Inside Information as Property. One difficulty in using the property rights or misappropriation argument is determining who owns the information in question. The main basis for recognizing a property right in trade secrets and confidential business information is the investment that companies make in acquiring information and the competitive value that some information has. Not all insider information fits this description, however. Advance knowledge of better-than-expected earnings would be an example. Such information still has value in stock trading, even if the corporation does not use it for that purpose. For this reason, many employers prohibit the personal use of any information that an employee gains in the course of his or her work. This position is too broad, however, since an employee is unlikely to be accused of stealing company property by using knowledge of the next day's earning report for any purpose other than stock trading.

A second difficulty with the property rights argument is that if companies own certain information, they could then give their own employees permission to use it, or they could sell the information to favored investors or even trade on it themselves to buy back stock. Giving employees permission to trade on insider information could be an inexpensive form of extra compensation that further encourages employees to develop valuable information for the firm. Such an arrangement would also have some drawbacks; for example, investors might be less willing to buy the stock of a company that allowed insider trading because of the disadvantage to outsiders. What is morally objectionable about insider trading, according to its critics, though, is not the misappropriation of a company's information but the harm done to the investing public. So the violation of property rights in insider trading cannot be the sole reason for prohibiting it. Let us turn, then, to the second argument against insider trading, namely the argument from fairness.

The Fairness Argument. Fairness in the stock market does not require that all traders have the same information. Indeed, trades will take place only if the buyers and sellers of a stock have different information that leads them to different conclusions about the stock's worth. It is only fair, moreover, that a shrewd investor who has spent great time and money studying the prospects of a company should be able to exploit that advantage. Otherwise there would be no incentive to seek out new information. What is objectionable about using inside information is that other traders are barred from obtaining it no matter how diligent they may be. The information is unavailable not for lack of *effort* but for lack of *access*. Poker also pits card players with unequal skill and knowledge without being unfair, but a game played with a marked deck gives some players an unfair advantage over others. By analogy, then, insider trading is like playing poker with a marked deck.

The analogy may be flawed, however. Perhaps a more appropriate analogy is the seller of a home who fails to reveal hidden structural damage. One principle of stock market regulation is that both buyers and sellers of stock should have sufficient information to make rational choices. Thus, companies must publish annual reports and disclose important developments in a timely manner. A CEO who hides bad news from the investing public, for example, can be sued for fraud. Good news, such as an oil find, need not be announced until a company has time to buy the drilling rights, and so on; but to trade on that information before it is public knowledge might also be described as a kind of fraud.

Insider trading is generally prosecuted under SEC Rule 10b-5, which merely prohibits fraud in securities transactions. In fraudulent transactions, one party, such as the buyer of the house with structural damage, is wrongfully harmed for lack of knowledge that the other party concealed. So too—according to the fairness argument—are the ignorant parties to insider trading transactions wrongfully harmed when material facts, such as the discovery of copper ore deposits in the *Texas Gulf Sulphur* case, are not revealed.

The main weakness of the fairness argument is determining what information ought to be revealed in a transaction. The reason for requiring a homeowner to disclose hidden structural damage is that doing so makes for a more efficient housing market. In the absence of such a requirement, potential home buyers would pay less because they are not sure what they are getting, or they would invest in costly home inspections. Similarly—the argument goes—requiring insiders to reveal before trading makes the stock market more efficient.

The trouble with such a claim is that some economists argue that the stock market would be more efficient *without* a law against insider trading.[14] If insider trading were permitted, they claim, information would be registered in the market more quickly and at less cost than the alternative of leaving the task to research by stock analysts. The main beneficiaries of a law against insider trading, critics continue, are not individual investors but market professionals who can pick up news "on the street" and act on it quickly. Some economists argue further that a law against insider trading preserves the illusion that there is a level playing field and that individual investors have a chance against market professionals.

Economic arguments about market efficiency look only at the cost of registering information in the market and not at possible adverse consequences of legalized insider trading, which are many. Investors who perceive the stock market as an unlevel playing field may be less inclined to participate or will be forced to adopt costly defensive measures. Legalized insider trading would have an effect on the treatment of information in a firm. Employees whose interest is in information that they can use in the stock market may be less concerned with information that is useful to the employer, and the company itself might attempt to tailor its release of information for maximum benefit to insiders. More importantly, the opportunity to engage in insider trading might undermine the relation of trust that is essential for business organizations.[15] A prohibition on insider trading frees employees of a corporation to do what they are supposed to be doing—namely, working for the interests of the shareholders—not seeking ways to advance their own interests.

The harm that legalized insider trading could do to organizations suggests that the strongest argument against legalization might be the breach of fiduciary duty that would result. Virtually everyone who could be called an "insider" has a fiduciary duty to serve the interests of the corporation and its shareholders, and the use of information that is acquired while serving as a fiduciary for personal gain is a violation of this duty. It would be a breach of professional ethics for a lawyer or an accountant to benefit personally from the use of information acquired in confidence from client, and it is similarly unethical for a corporate executive to make personal use of confidential business information.

The argument that insider trading constitutes a breach of fiduciary duty accords with recent court decisions that have limited the prosecution of insider trading to true insiders who have a fiduciary duty. One drawback of the argument is that "outsiders," whom federal prosecutors have sought to convict of insider trading, would be free of any restrictions. A second drawback is that insider trading, on this argument, is no longer an offense against the market but the violation of a duty to another party. And the duty not to use information that is acquired while serving as a fiduciary prohibits more than insider trading. The same duty would be violated by a fiduciary who buys or sells property or undertakes some other business dealing on the basis of confidential information. That such breaches of fiduciary duty are wrong is evident, but the authority of the SEC to prosecute them under a mandate to prevent fraud in the market is less clear.

The* O'Hagan *Decision. In 1997, the U.S. Supreme Court ended a decade of uncertainty over the legal definition of insider trading. The SEC has long prosecuted insider trading using the misappropriation theory, according to which an inside trader breaches a fiduciary duty by misappropriating confidential information for personal trading. In 1987, the high court split 4–4 on an insider trading case involving a reporter for the *Wall Street Journal* and thus left standing a lower court decision that found the reporter guilty of misappropriating information.[16] However, the decision did not create a precedent for lack of a majority. Subsequently, lower courts rejected the misappropriation theory in a series of cases in which the alleged inside trader did not have a fiduciary duty to the corporation whose stock was traded. The principle applied was that the trading must itself constitute a breach of fiduciary duty. This principle was rejected in *U.S.* v. *O'Hagan.*

James H. O'Hagan was a partner in a Minneapolis law firm that was advising the British firm Grand Metropolitan in a hostile takeover of Minneapolis-based Pillsbury Company. O'Hagan did not work on Grand Met business but allegedly tricked a fellow partner into revealing the takeover bid. O'Hagan then reaped $4.3 million by trading in Pillsbury stock and stock options. An appellate court ruled that O'Hagan did not engage in illegal insider trading because he had no fiduciary duty to Pillsbury, the company in whose stock he traded. Although O'Hagan misappropriated confidential information from his own law firm—to which he owed a fiduciary duty—trading on this information did not constitute a fraud against the law firm or against Grand Met. Presumably, O'Hagan would have been guilty of insider trading only if he were an insider of Pillsbury.

In a 6–3 decision, the Supreme Court reinstated the conviction of Mr. O'Hagan and affirmed the misappropriation theory. According to the decision, a person commits securities fraud when he or she "misappropriates confidential information for securities trading purposes, in breach of a fiduciary duty owed to the source of the information." Thus, an inside trader need not be an insider (or a temporary insider, like a lawyer) of the corporation whose stock is traded. Being an insider in Grand Met is sufficient in this case to hold that insider trading occurred. The majority opinion observed that "it makes scant sense" to hold a lawyer like O'Hagan to have violated the law "if he works for a law firm representing the target of a tender offer, but not if he works for a law firm representing the bidder." The crucial point is that O'Hagan was a fiduciary who misused information that had been entrusted to him. This decision would also apply to a person who receives information from an insider and who knows that the insider source is violating a duty of confidentiality. However, a person with no fiduciary ties who receives information innocently (by overhearing a conversation, for example) would still be free to trade.

HOSTILE TAKEOVERS

Hostile takeovers—which are acquisitions opposed by the management of the target corporation—appear to violate the accepted rules for corporate change. Peter Drucker observed that the hostile takeover "deeply offends the sense of justice of a great many Americans."[17] An oil industry CEO charged that such activity "is in total disregard of those inherent foundations which are the heart and soul of the American free enterprise system."[18] Many economists—most notably Michael C. Jensen—defend hostile takeovers on the grounds that they bring about needed changes that cannot be achieved by the usual means.[19]

The ethical issues in hostile takeovers are threefold. First, should hostile takeovers be permitted at all? Insofar as hostile takeovers are conducted in a market through the buying and selling of stocks, there exists a "market for corporate control." So the question can be expressed in the form, should there be a market for corporate control? Or should change of control decisions be made in some other fashion? Second, ethical issues arise in the various tactics that have been used by raiders in launching attacks as well as by target corporations in defending themselves. Some of these tactics are criticized on the grounds that they unfairly favor the raiders or incumbent management, often at the expense of shareholders. Third, hostile takeovers raise important issues about the fiduciary duties of officers and directors in their responses to takeover bids. In particular, what should directors do when an offer that shareholders want to accept is not in the best interests of the corporation itself? Do they have a right, indeed a responsibility, to prevent a change of control?

The Market for Corporate Control

Defenders of hostile takeovers contend that corporations become takeover targets when incumbent management is unable or unwilling to take steps that increase shareholder value. The raiders' willingness to pay a premium for the stock reflects a belief that the company is not achieving its full potential under the current management. "Let us take over," the raiders say, "and the company will be worth what we are offering." Because shareholders often find it difficult to replace the current managers through traditional proxy contests, hostile takeovers are an important means for shareholders to realize the value of their investment. Although restructurings of all kinds cause some hardships to employees, communities, and other groups, society as a whole benefits from the increased wealth and productivity.

Just the threat of a takeover serves as an important check on management, and without this constant spur, defenders argue, managers would have less incentive to secure full value for the shareholders. With regard to the market for corporate control, defenders hold that shareholders are, and ought to be, the ultimate arbiters of who manages the corporation. If the shareholders have a right to replace the CEO, why should it matter when or how shareholders bought the stock? A raider who bought the stock yesterday in a tender offer has the same rights as a shareholder of long standing. Any steps to restrict hostile takeovers, the defenders argue, would entail an unjustified reduction of shareholders' rights.

Critics of hostile takeovers challenge the benefits and emphasize the harms. Targets of successful raids are sometimes broken up and sold off piecemeal or downsized and folded into the acquiring company. In the process, people are thrown out of work and communities lose their economic base. Takeovers generally saddle companies with debt loads that limit their options and expose them to greater risk in the event of a downturn. Critics also charge that companies are forced to defend themselves by managing for immediate results and adopting costly defensive measures.

The debate over hostile takeovers revolves largely around the question of whether they are good or bad for the American economy. This is a question for economic analysis, and the evidence, on the whole, is that takeovers generally increase the value of both the acquired and the acquiring corporation.[20] However, there is little evidence that newly merged or acquired firms outperform industry averages in the long run.[21] The effect on the economy aside, the benefit of hostile takeovers must be viewed with some caution.

First, not all takeover targets are underperforming businesses with poor management. Other factors can make a company a takeover target. The "bust up" takeover operates on the premise that a company is worth more sold off in parts than retained as a whole. Large cash reserves, expensive research programs, and other sources of savings enable raiders to finance a takeover with the company's own assets. The availability of junk bond financing during the 1980s permitted highly leveraged buyouts with levels of debt that many considered to be unhealthy for the economy. Finally, costly commitments to stakeholder groups can be tapped to finance a takeover. Thus, Pacific Lumber's pension plan and cutting policy constituted commitments to employees and environmentalists respectively. Both commitments were implicit contracts that had arguably benefited shareholders in the past but that could be broken now with impunity.

Second, some of the apparent wealth that takeovers create may result from accounting and tax rules that benefit shareholders but create no new wealth. For example, the tax code favors debt over equity by allowing a deduction for interest payments on debt while taxing corporate profits. Rules on depreciation and capital gains may result in tax savings from asset sales following a takeover. Thus, taxpayers provide an indirect subsidy in the financing of takeovers. Some takeovers result in direct losses to other parties. Among the losers in hostile takeovers are bondholders, whose formerly secure, investment grade bonds are sometimes downgraded to speculative, junk bond status.

Takeover Tactics

In a typical hostile takeover, an insurgent group—often called a "raider"—makes a *tender offer* to buy a controlling block of stock in a target corporation from its present shareholders. The offered price generally involves a premium, which is an amount in excess of the current trading price. If enough shareholders tender their shares in response to the offer, the insurgents gain control. In the usual course of events, the raiders replace the incumbent management team and proceed to make substantial changes in the company. In some instances, a tender offer is made directly to the shareholders, but in others, the cooperation of management is required.

The officers and directors of firms have a fiduciary duty to consider a tender offer in good faith. If they believe that a takeover is not in the best interests of the shareholders, then they have a right, even a duty, to fight the offer with all available means. Corporations have many resources for defending against hostile takeovers. These tactics—collectively called "shark repellents"—include poison pills, white knights, lockups, crown-jewel options, the Pac-Man defense, golden parachutes, and greenmail (see Exhibit 13.1). Some of the defensive measures (such as poison pills and golden parachutes) are usually adopted in advance of any takeover bid, while others (white knights and greenmail) are customarily employed in the course of fighting an unwelcome offer. Many states have adopted so-called antitakeover statutes that further protect incumbent management against raiders. Because of shark repellents and antitakeover statutes, a merger or acquisition is virtually impossible to conduct today without the cooperation of the board of directors of the target corporation.

All takeover tactics raise important ethical issues, but three, in particular, have elicited great concern. These are unregulated tender offers, golden parachutes, and greenmail.

Tender Offers. Ethical concern about the tactics of takeovers has focused primarily on the defenses of target companies, but unregulated tender offers are also potentially abusive. Before 1968, takeovers were sometimes attempted by a so-called "Sat-

---

Exhibit 13.1
Takeover Defenses

Crown Jewel Option   A form of lockup in which an option on a target's most valuable assets (crown jewels) is offered to a friendly firm in the event of a hostile takeover.

Golden Parachute   A part of the employment contract with a top executive that provides for additional compensation in the event that the executive departs voluntarily or involuntarily after a takeover.

Greenmail   The repurchase by a target of an unwelcome suitor's stock at a premium in order to end an attempted hostile takeover.

Lockup Option   An option given to a friendly firm to acquire certain assets in the event of a hostile takeover. Usually, the assets are crucial for the financing of a takeover.

Pac-Man Defense   A defense (named after the popular video game) in which the target makes a counteroffer to acquire the unwelcome suitor.

Poison Pill   A general term for any devise that lowers the price of a target's stock in the event of a takeover. A common form of poison pill is the issuance of a new class of preferred stock that shareholders have a right to redeem at a premium after a takeover.

Shark Repellant   A general term for all takeover defenses.

White Knight   A friendly suitor which makes an offer for a target in order to avoid a takeover by an unwelcome suitor.

---

urday night special," in which a tender offer was made after the close of the market on Friday and set to expire on Monday morning. The "Saturday night special" was considered to be coercive because shareholders had to decide quickly whether to tender their shares with little information.[22] Shareholders would generally welcome an opportunity to sell stock that trades at $10 a share on a Friday afternoon for, say, $15. If, on Monday morning, however, the stock sells for $20 a share, then the shareholders who tendered over the weekend gained $5 but lost the opportunity to gain $10. With more information, shareholders might conclude that $15 or even $20 was an inadequate price and that they would be better off holding on to their shares—perhaps in anticipation of an even better offer.

Partial offers for only a certain number or percentage of shares and two-tier offers can also be coercive. In a two-tier offer, one price is offered for, say, 51 percent of the shares and a lower price is offered for the remainder. Both offers force shareholders to make a decision without knowing which price they will receive for their shares or indeed whether their shares will even be bought. Thus, tender offers can be structured in such a way that shareholders are stampeded into tendering quickly lest they lose the opportunity. The payment that is offered may include securities—such as shares of the acquiring corporation or a new merged entity—and the value of these securities may be difficult to determine. Without adequate information, shareholders may not be able to judge whether a $15 per share noncash offer, for example, is fairly priced.

Congress addressed these problems with tender offers in 1968 with the passage of the Williams Act. The guiding principle of the Williams Act is that shareholders have a right to make important investment decisions in an orderly manner and with adequate information. They should not be stampeded into tendering for fear of losing the opportunity or forced to decide in ignorance. Under Section 14(d) of the Williams Act, a tender offer must be accompanied by a statement detailing the bidder's identity, the nature of the funding, and plans for restructuring the takeover target. A tender offer must be open for 20 business days, in order to allow shareholders sufficient time to make a decision, and tendering shareholders have 15 days in which to change their minds—thereby permitting them to accept a better offer should one be made. The Williams Act deals with partial and two-tier offers by requiring proration. Thus, if more shares are tendered than the bidder has offered to buy, then the same percentage of each shareholder's tendered stock must be purchased. Proration ensures the equal treatment of shareholders and removes the pressure on shareholders to tender early.

Golden Parachutes. At the height of takeover activities in the 1980s, between one-quarter and one-half of major American corporations provided their top executives with an unusual form of protection—golden parachutes.[23] A golden parachute is a provision in a manager's employment contract for compensation—usually, a cash settlement equal to several years' salary—for the loss of a job following a takeover. In general, golden parachutes are distinct from severance packages because they become effective only in the event of a change of control and are usually limited to the CEO and a small number of other officers.

The most common argument for golden parachutes is that they reduce a potential conflict of interest. Managers who might lose their jobs in the event of a takeover cannot be expected to evaluate a takeover bid objectively. Michael C. Jensen observes, "It makes no sense to hire a realtor to sell your house and then penalize your agent for doing so."[24] A golden parachute protects managers' futures, no matter the outcome, and thus frees them to consider only the best interests of the shareholders. In addition, golden parachutes enable corporations to attract and retain desirable executives because they provide protection against events that are largely beyond managers' control. Without this protection, a recruit may be reluctant to accept a position at a potential takeover target, or a manager might leave a threatened company in anticipation of a takeover bid.

Critics argue, first, that golden parachutes merely entrench incumbent managers by raising the price that raiders would have to pay. In this respect, golden parachutes are like poison pills; they create costly new obligations in the event of a change of control. All such defensive measures are legitimate if they are approved by the shareholders, but golden parachutes, critics complain, are often secured by executives from compliant boards of directors that they control. If golden parachutes are in the shareholders' interests, then executives should be willing to obtain shareholder approval. Otherwise, they appear to be self-serving defensive measures that violate a duty to serve the shareholders.

Second, some critics object to the idea of providing additional incentives to do what they are being paid to do anyway.[25] Philip L. Cochran and Steven L. Wartick observe that managers are already paid to maximize shareholder wealth. "To provide additional compensation in order to get managers to objectively evaluate takeover offers is tantamount to management extortion of the shareholders."[26] One experienced director finds it "outrageous" that executives should be paid *after* they leave a company. Peter G. Scotese writes: "Why reward an executive so generously at the moment his or her contribution to the company ceases? The approach flies in the face of the American work ethic, which is based on raises or increments related to the buildup of seniority and merit."[27]

The principle for justifying golden parachutes is clear, even if its application is not. The justification for all forms of executive compensation is to provide incentives for acting in the shareholders' interests. If golden parachutes are too generous, they entrench management by making the price of a takeover prohibitive—or else they motivate managers to support a takeover against the interests of shareholders. In either case, the managers enrich themselves at the shareholders' expense. The key is to develop a compensation package with just the right incentives, which, as Michael Jensen notes, will depend on the particular case.[28]

Greenmail. Unsuccessful raiders do not always go away empty-handed. Because of the price rise that follows an announced takeover bid, raiders are often able to sell their holdings at a tidy profit. In some instances, target corporations have repelled unwelcome assaults by buying back the raiders' shares at a premium. After the financier Saul Steinberg accumulated more than 11 percent of Walt Disney Productions in 1984, the Disney board agreed to pay $77.50 per share, a total of $325.3 million, for stock that Steinberg had purchased at an average price of $63.25. As a reward for ending his run at Disney, Steinberg pocketed nearly $60 million. This episode and many like it have been widely criticized as *greenmail.*

The play on the word "blackmail" suggests that there is something corrupt about offering or accepting greenmail. A more precise term that avoids this bias is *control repurchase.* A control repurchase may be defined as a "privately negotiated stock repurchase from an outside shareholder, at a premium over the market price, made for the purpose of avoiding a battle for control of the company making the repurchase."[29] Although control repurchases are legal, many people think that there ought to be a law. So we need to ask first why control repurchases are considered to be unethical.

There are three main ethical objections to control repurchases.[30] First, control repurchases are negotiated with one set of shareholders, who receive an offer that is not extended to everyone else. This is a violation, some say, of the principle that all shareholders should be treated equally. The same offer should be made to all shareholder—or none. To buy back the stock of raiders for a premium is unfair to other shareholders.

This argument is easily dismissed. Managers have an obligation to treat all shareholders according to their rights under the charter and bylaws of the corporation and the relevant corporate law. This means one share, one vote at meetings and the same dividend for each share. There is no obligation for managers to treat shareholders equally otherwise. Moreover, paying a premium for the repurchase of stock is a use of corporate assets that presumably brings some return to the shareholders, and the job of managers is to put all corporate assets to their most productive use. If the $60 million that Disney paid to Saul Steinberg, for example, brings higher returns to the shareholders than any other investment, then the managers have an obligation *to all shareholders* to treat this one shareholder differently.

Second, control repurchases are criticized as a breach of the fiduciary duty of management to serve the shareholders' interests. One critic of greenmail makes the case as follows:

> Say you owned a small apartment building in a distant city, and you hired a professional manager to run it for you. This person likes the job, and when someone—an apartment "raider"—sought to offer you a good price for the building, the manager does everything to prevent you from being able to consider the offer.... When all else fails, the manager takes some of your own money and pays the potential buyer greenmail to look elsewhere.[31]

If managers use shareholders' money to pay raiders to go away merely to save their own jobs, they have clearly violated their fiduciary duty. However, this may not be the intent of managers in all cases of greenmail.

Managers of target corporations may judge that an offer is not in the best interests of shareholders and that the best defensive tactic is a repurchase of the raiders' shares. With $60 million, Disney might have made another movie that would bring a certain return. However, Disney executives might also have calculated that the costs to the company of continuing to fight Saul Steinberg—or allowing him to gain control—would outweigh this return. If so, then the $60 million that Disney paid in greenmail is shareholder money well spent. So there is no reason to believe that greenmail or control repurchases necessarily involve a breach of fiduciary duty.

Third, some critics object to greenmail or control repurchases on the grounds that the payments invite *pseudobidders* who have no intention of taking control and mount a raid merely for the profit.[32] The ethical wrong, according to this objection, lies with the raiders' conduct, although management may be complicitous in facilitating it. At a minimum, pseudobidders are engaging in unproductive economic activity, which benefits no one but the raiders themselves; at their worst, pseudobidders are extorting corporations by threatening some harm unless the payments are made.

Is pseudobidding for the purpose of getting greenmail a serious problem? The effectiveness of pseudobidding depends on the credibility of the threatened takeover. No raider can pose a credible threat unless an opportunity exists to increase the return to shareholders. Therefore, the situations in which pseudobidders are likely to emerge are quite limited. Even if a pseudobidder or a genuine raider is paid to go away, that person has pointed out some problem with the incumbent management and paved the way for change. Unsuccessful raiders who accept greenmail may still provide a service for everyone.[33]

The Role of the Board of Directors

In 1989, Paramount Communications made a tender offer for all outstanding stock in Time Incorporated. Many Time shareholders were keen to accept the all-cash, $175-per-share bid (later raised to $200 per share), which represented about a 40 percent premium over the previous trading price of Time stock. However, the board of directors refused to submit the Paramount offer to the shareholders. Time and Warner Communications, Inc., had been preparing to merge, and the Time directors believed that a Time-Warner merger would produce greater value for the shareholders that an acquisition by Paramount. Disgruntled Time shareholders joined Paramount in a suit that charged the directors with a failure to act in the shareholders' interests.

This case raises two critical issues. First, who has the right to determine the value of a corporation in a merger or acquisition? Is this a job for the board of directors and their investment banking firm advisors? Both boards and their advisors have superior information about a company's current financial status and future prospects, but they also have a vested interest in preserving the status quo. Should the task of evaluation be left to the shareholders, whose interests are the ultimate arbiter but whose knowledge is often lacking? Some of the shareholders are professional arbitragers, who are looking merely for a quick buck. Second, does the interest of the shareholders lie with quick, short-term gain or with the viability of the company in the long run? Acceptance of the Paramount offer would maximize the immediate stock price for Time shareholders but upset the long-term strategic plan that the board had developed.

The Delaware State Supreme Court decision in *Paramount Communications, Inc.* v. *Time Inc.* addressed both issues by ruling that the Time board of directors had a right to take a long-term perspective in evaluating a takeover bid and had no obligation to submit the Paramount proposal to the shareholders.[34] The court recognized that increasing shareholder value in the long run involves a consideration of interests besides those of current shareholders, including other corporate constituencies, such as employees, customers, and local communities. One concern of the Time directors was to preserve the "culture" of *Time* magazine because of the importance of editorial integrity to the magazine's readers and journalistic staff.

The *Paramount* decision is an example of a so-called other constituency statute. A majority of states have now adopted (either by judicial or legislative action) laws that permit (and, in a few states, require) the board of directors to consider the impact of a takeover on a broad range of nonshareholder constituencies.[35] Other constituency statutes reflect a judgment by judges and legislators that legitimate nonshareholder interests are harmed by takeovers and that directors faced with a takeover do not owe allegiance solely to the current shareholders.[36] As a result of other constituency statutes, decisions about the future of corporations depend more on calm deliberations in boardrooms and less on the buying and selling of shares in a noisy marketplace.

Conclusion

Ethical issues in finance are important because they bear on our financial well-being. Ethical misconduct, whether it be by individuals acting alone or by financial institutions, has the potential to rob people of their life savings. Because so much money is involved in financial dealings, there must be well-developed and effective safeguards in place to ensure personal and organizational ethics. Although the law governs much financial activity, strong emphasis must be placed on the integrity of finance professionals and on ethical leadership in our financial institutions. Some of the principles in finance ethics are common to other aspects of business, especially the duties of fiduciaries and fairness in sales practices and securities markets. However, such activities as insider trading and hostile takeover raise unique issues that require special consideration.

Case 13.2   E. F. Hutton

In 1980, E. F. Hutton, a respected brokerage firm with more than 400 offices worldwide, was searching for ways to increase its income.[37] Brokerage commissions, which had accounted for 68 percent of revenues ten years earlier, had fallen to 38 percent. In a period of high inflation, with interest rates hovering around 18 percent, fewer investors were interested in the stock market. However, one source of income was producing remarkable results: the interest from the firm's own checking accounts with banks. Little did anyone at Hutton suspect at the time that actively pursuing these returns would seriously tarnish the company's reputation.

A firm that handles a lot of cash needs a cash management system in order to put money to its best use. This system may be as simple as a store depositing each day's receipts in an interest-bearing account. Any delay may not only lose the opportunity to earn interest but also prevent the store from having access to funds needed for other purposes. In developing its own cash management system, E. F. Hutton sought to move money quickly from branch office bank accounts to regional office bank accounts, and from there to national headquarters. The cash on hand in branch offices came from the firm's own funds (called general ledger accounts) and from client accounts. Clients were required to deposit money when trading on margin or short selling, and some clients left funds from stock sales on deposit pending another purchase. Once the money on hand reached national headquarters it could be loaned or invested by the firm.

In order to move money quickly to national headquarters, Hutton would write a check on a local office bank account as soon as possible and deposit it in the bank of a regional office. Regional offices would repeat this process in sending funds to national headquarters. By drawing down accounts so quickly, Hutton was depriving banks of the opportunity to use these funds, which is a major source of compensation to banks for providing checking service. However, Hutton was doing nothing illegal, and banks can protect themselves by charging fees or requiring minimum deposits or both.

At some point, cashiers at E. F. Hutton discovered that money could be moved even more quickly by writing checks on anticipated deposits. According to the legal principle of "means of payment," one should not write a check for which one does not have sufficient funds on deposit. Thus, one should not write a check on Monday, planning to go to the bank on Tuesday to deposit enough to cover the check, even though the check might not clear until Wednesday. However, if the check is not cleared by the bank until Wednesday, then the bank has the benefit of the money for two days. The time that it takes for checks to clear is known as the "float," and the float is an additional source of income for banks. By writing checks on one day in the amount of the next day's anticipated deposit, Hutton in effect captured the float and gained an extra day's interest on the money.

The float on interest-bearing checking accounts is a quirk of the check clearing system that would be eliminated if all transactions were simultaneous electronic transfers. As long as funds are credited when they are deposited and debited when a check clears, the float will remain. With interest rates unusually high, the management of E. F. Hutton saw a great opportunity. Cashiers were instructed how to use a formula for determining how large a check could be written in anticipation of the next day's receipts. The formula included a multiplier to take account of weekends and days on which checks took longer to clear. One cashier inadvertently added an extra zero to an overdraft, making it $9 million instead of $900,000. When he noticed that the bank did nothing, he began to overdraft the account more aggressively.

Hutton cashiers were encouraged in these efforts by the president, George Ball, who wrote in a memo dated August 5, 1981, "We have certainly had the luxury of high interest profits, profits which may be importantly lower in the year ahead. Our corporate goal is to earn in excess of $100,000,000 [in interest]. Together we can do it, but it will take a mighty push." How to increase interest income became a subject of many meetings. Branches were compared on their level of interest income, and tips on more aggressive use of the formula were passed from one branch to another.

One large overdraft seemed unusual and caught the attention of the Justice Department. A bank in Batavia, New York, was presented with a check for $8 million drawn on that bank that had been deposited in a New York City bank. The deposits for Hutton in the Batavia bank for covering this check consisted of a check drawn on a bank in Wilkes-Barre, Pennsylvania, that had not yet cleared. Thus, the Batavia bank was being asked to transfer funds to the New York City bank that it had not yet collected from the Wilkes-Barre bank. The bank manager in Batavia called the bank in Wilkes-Barre and was informed that the deposits for backing the check consisted of checks payable to Hutton that had not yet cleared. It appeared that Hutton was not merely using the unavoidable float that is built into the banking system but was also creating additional float by writing checks in such quick succession. Creating additional float in this manner is known as "chaining." The practice of chaining developed in Hutton's Atlantic region and had been adopted elsewhere.

The Justice Department estimated that between July 1980 through February 1982, Hutton managed to earn daily interest on $1 billion dollars more than they actually had on deposit in banks. Whether the Justice Department could successfully prosecute Hutton was open to question. Hutton's hope of prevailing in court were dashed, however, by the discovery of an internal memo, in which a branch manager defended these practices. This manager wrote:

> I believe those activities are encouraged by the firm and are in fact identical to what the firm practices on a national basis. Specifically we will from time to time draw down not only depositions plus anticipated depositions, but also bogus deposits.

The memo, described as "the nail in the coffin," forced Hutton in May 1985 to accept a settlement in which the company pleaded guilty to 2,000 counts of fraud and agreed to pay a $2 million fine plus $750,000 for the costs of the investigation. That the deal would enable the firm to put its troubles behind it proved to be wishful thinking. E. F. Hutton never recovered from the scandal. In 1987, the firm was bought by Shearson Lehman Brothers, which later put the distinguished name to rest.

Case 13.3 Salomon Brothers

> At Salomon Brothers, trading has always been a form of war in which the opponent is entitled to no pity and rules are viewed as impediments to be sidestepped, if possible. Now that attitude threatens to destroy the firm.[38]

In early August 1991, John Gutfreund, the CEO of Salomon Brothers struggled with a serious problem. Gutfreund (pronounced like "good friend") had been informed on April 29 that Paul Mozer, who directed Salomon's trading in U.S. Treasury securities, had violated the Treasury Department's rules in several auctions of Treasury bonds. Although Gutfreund took the matter seriously, no decisive action had been taken—until now, when events were coming to a head.

Among investment banks, Salomon Brothers was unusual in the percentage of its business devoted to trading for its own account, as opposed to providing services for clients. Because of this reliance on trading profits, Salomon had attracted skilled, aggressive traders and built a culture that formed and supported them. Salomon traders were known on Wall Street for their no-holds-barred, bare-knuckled style. Gutfreund had deliberately encouraged this culture by giving each department wide latitude in trading and allowing its members to share in the trading profits. The compensation in successful departments could run into the tens of millions of dollars. Upper management did not closely monitor the trading activity, and this highly decentralized structure generally paid off in huge profits.

Salomon Brothers had managed to create opportunities in staid securities. The firm had pioneered the practice of bundling together home mortgage loans into securities that could be traded. The bond arbitrage unit used sophisticated mathematics to identify small differences in pricing that could be exploited. Salomon was able to make money on Treasury bonds by buying a large portion of each issue at a relatively good price and then reselling the bonds at a slightly higher price to its own customers and to other firms that needed them to fill customer orders.

The Treasury Department sells securities at weekly auctions, in which a few major firms—called primary dealers—submit sealed bids for a certain percentage of the offering at a particular price. (The price, which is less than the face value of the bonds, determines the interest rate.) The key to success is bidding high enough to get the desired percentage of the bonds being offered and yet low enough to get a favorable price. Bid too high and there's no profit; bidding too low results in getting no bonds. The margin for error is very small, and mistakes can result in huge losses. Successful bidding requires a firm to have access to the best information. By handling orders from many clients, Salomon traders were able to gauge the demand for each issue and bid accordingly.

The Treasury Department is concerned not only with selling each issue at the lowest interest rate but also with ensuring an orderly market in the long run. An orderly market requires that enough primary dealers are successful so that healthy competition is maintained. Everyone involved must also perceive the auctions as fair. In particular, foreign investors, on whom the United States depends to buy much of its debt, would especially be reluctant to participate in a rigged market. In some instances, a firm was able to obtain a high percentage of a issue in demand and thereby execute a "squeeze." As a result, firms with commitments to clients were forced to pay a high price to obtain the needed securities.

In order to prevent squeezes, the Treasury Department issued a rule that no firm could submit a bid for more than 35 percent of the securities in any auction. The rules for Treasury auctions were kept rather informal in the belief that precise rules increase costs and interfere with the clubby atmosphere of the auctions. New rules were usually made in response to problems and were issued in press releases. The rule limiting bids to 35 percent resulted from concern about some of Salomon's aggressive trading tactics. The announcement angered Paul Mozer, who complained that the rule was made without consultation and was unfair to the large traders. The controversy produced strained relations between Mozer and Treasury officials.

What Gutfreund learned in April was that on February 21, Mozer had submitted a bid for 35 percent of a $9 billion five-year note auction for Salomon and an unauthorized bid for the same amount in the name of a Salomon customer. Afterward, Mozer ordered that the notes received from the bogus bid be credited to Salomon's account and that the usual confirmation to the customer not be sent.

On May 22, Mozer executed a more successful squeeze. In an auction for $12 billion two-year notes, Salomon entered a bid for 34 percent of the notes along with bids for two customers, Tiger Investments and Quantum Fund, for an additional 52 percent. The bids offered the best price to the Treasury, so that Salomon and its customers received the full amount sought and ended up controlling more than 86 percent of the issue. Because of the high price, the government saved an estimated $5 million in interest. However, dealers who had made commitments to their customers were forced to borrow or buy the notes from Salomon at a substantial cost.

In May, Gutfreund asked a law firm to investigate Mozer's trading activities. The lawyers reported that unauthorized bids in the name of a customer had been submitted in December 1990 and again in April 1991. They also discovered that in the February 21 auction, Mozer had submitted unauthorized bids for not one but two customers and, as a result, had received 57 percent of the notes.

By August pressure was mounting. The Justice Department had opened an investigation, and some executives were urging a meeting of the Board of Directors. Gutfreund needed to decide immediately how to respond to these disclosures. In the long run, he—or his successor, if he were forced to leave—would have to find ways of preventing a recurrence of these trading violations. Any changes, however, might threaten the formula for success that had made Salomon Brothers and its traders very wealthy.

NOTES

1. Jay L. Walker [pseudonym], *The Complete Book of Wall Street Ethics* (New York: William Morrow and Company, 1987).

2. "Burned by Merrill," *Business Week*, April 25, 1994, 122-25.

3. Ellen E. Schultz, "You Need a Translator for Latest Sales Pitch," *Wall Street Journal*, February 14, 1994, C1.

4. Michael Quint, "Met Life Shakes Up Its Ranks," *New York Times*, October 29, 1994, C1.

5. *Marshak* v. *Blyth Eastman Dillon & Co., Inc.*, 413 F. Supp. 377, 379 (1975).

6. *Kaufman* v. *Merrill Lynch, Pierce, Fenner & Smith*, 464 F. Supp. 528, 534 (1978).

7. *Report of the Committee on Compensation Practices*, issued by the Securities and Exchange Commission, 10 April 1995.

8. *NASD Rules of Fair Practice*, art. III, sec. 2.

9. This question is addressed in Hersh Shefrin and Meir Statman, "Ethics, Fairness and Efficiency in Financial Markets," *Financial Analysts Journal* (November-December 1993), 21-29, from which portions of this section are derived.

10. John Maynard Keynes, *The General Theory of Employment, Interest, and Money* (New York: Harcourt Brace & Co., 1936), 159.

11. The example is taken from Anthony Kronman, "Contract Law and Distributive Justice," *Yale Law Journal*, 89 (1980), 472-79.

12. *SEC* v. *Texas Gulf Sulphur*, 401 F.2d 19 (1987).

13. *Chiarella* v. *U.S.*, 445 U.S. 222 (1980); *Dirks* v. *SEC*, 463 U.S. 646 (1983); *U.S.* v. *Chestman*, 903 F.2d 75 (1990); *U.S.* v. *Willis*, 737 F.Supp. 269 (1990); *U.S.* v. *O'Hagen*, No. 96-842.

14. Henry Manne, *Insider Trading and the Stock Market* (New York: Free Press, 1966).

15. This point is argued in Jennifer Moore, "What Is Really Unethical about Insider Trading?" *Journal of Business Ethics*, 9 (1990), 171-82.

16. *Carpenter et al.* vs. *U.S.*, 484 U.S. 19 (1987).

17. Peter Drucker, "To End the Raiding Roulette Game," *Across the Board* (April 1986), 39.

18. Michel T. Halbouty, "The Hostile Takeover of Free Enterprise," *Vital Speeches of the Day* (August 1986), 613.

19. See Michael C. Jensen, "The Takeover Controversy," *Vital Speeches of the Day* (May 1987), 426-29; "Takeovers: Folklore and Science," *Harvard Business Review* (November-December 1984), 109-21.

20. Jensen, "Takeovers"; Michael C. Jensen and Richard S. Ruback, "The Market for Corporate Control: The Scientific Evidence," *Journal of Financial Economics*, 11 (1983), 5-50; and Douglas H. Ginsburg and John F. Robinson, "The Case Against Federal Intervention in the Market for Corporate Control," *The Brookings Review* (Winter-Spring 1986), 9-14.

21. F. M. Scherer, "Takeovers: Present and Future Dangers," *The Brookings Review* (Winter-Spring 1986), 15-20.

22. For a discussion of coercion in tender offers see John R. Boatright, "Tender Offers: An Ethical Perspective," in *The Ethics of Organizational Transformation: Mergers, Takeovers, and Corporate Restructuring*, W. M. Hoffman, R. Frederick, and E. S. Petry, Jr., eds. (New York: Quorum Books, 1989), 167-81.

23. Philip L. Cochran and Steven L. Wartick, "'Golden Parachutes': A Closer Look," *California Management Review*, 26 (1984), 113.

24. Michael C. Jensen, "The Takeover Controversy: Analysis and Evidence," in *Knights, Raiders, and Targets: The Impact of the Hostile Takeover*, John C. Coffee, Jr., Louis Lowenstein, and Susan Rose-Ackerman, eds. (New York: Oxford University Press, 1988), 340.

25. Peter G. Scotese, "Fold Up Those Golden Parachutes," *Harvard Business Review* (March-April 1985), 170.

26. Cochran and Wartick, "Golden Parachutes," 121.

27. Scotese, "Fold Up Those Golden Parachutes," 168.

28. Jensen, "The Takeover Controversy," 341.

29. J. Gregory Dees, "The Ethics of 'Greenmail'," in *Business Ethics: Research Issues and Empirical Studies,* William C. Frederick and Lee E. Preston, eds. (Greenwich, CT: JAI Press, 1990), 254.

30. These arguments are developed and evaluated in Dees, "Ethics of Greenmail."

31. Quoted in Robert W. McGee, "Ethical Issues in Acquisitions and Mergers," *Mid-Atlantic Journal of Business,* 25 (March 1989), 25.

32. John C. Coffee, Jr., "Regulating the Market for Corporate Control: A Critical Assessment of the Tender Offer's Role in Corporate Governance," *Columbia Law Review,* 84 (1984), 1145-1296.

33. This is argued in Roger J. Dennis, "Two-Tiered Tender Offers and Greenmail: Is New Legislation Needed?" *Georgia Law Review,* 19 (1985), 281-341.

34. *Paramount Communications, Inc.* v. *Time Inc.,* 571 A.2d 1140 (1990).

35. See Eric W. Orts, "Beyond Shareholders: Interpreting Corporate Constituency Statutes," *George Washington Law Review,* 61 (1992), 14-135.

36. Roberta S. Karmel, "The Duty of Directors to Non-shareholder Constituencies in Control Transactions—A Comparison of U.S. and U.K. Law," *Wake Forest Law Review,* 25 (1990), 68.

37. This case is adapted from Joanne Ciulla, "When E. F. Hutton Speaks...," in *Cases in Ethics and the Conduct of Business* John R. Boatright, ed. (Upper Saddle River, NJ: Prentice Hall, 1995), 274-288.

38. Floyd Norris, "Looking Out for No. 1," *New York Times,* 19 August 1991, C1.

14

Ethics and Corporations

Case 14.1 The Nun and the CEO

In 1996, a nun from Philadelphia sparked a debate with a Silicon Valley CEO over corporate social responsibility.[1] Sister Doris Gormley of the Sisters of St. Francis, whose retirement fund held 7,000 shares of Cypress Semiconductor stock, wrote a letter to the company protesting the lack of women and minorities on the board of directors. Speaking for a religious congregation of approximately 1,000 women, Sister Gormely wrote, "We belive that a company is best represented by a Board of qualified Directors reflecting the equality of the sexes, races and ethnic groups.... Therefore our policy is to withhold authority to vote for nominees of a Board of Directors that does not include women and minorities."

T. J. Rodgers, the CEO of Cypress, an international distributor of semiconductors with about $600 million in annual sales, composed a six-page response in which he labeled the Sister's view "immoral" and urged her to get down from her "moral high horse."[2] His letter reads in part:

> Thank you for your letter criticizing the lack of racial and gender diversity of Cypress's Board of Directors.... The semiconductor business is a tough one with significant competition....For [this] reason, our Board of Directors is not a ceremonial watchdog, but a critical management function. The essential criteria for Cypress board membership are as follows:
>
> - Experience as a CEO of an important technology company.
> - Direct expertise in the semiconductor business based on education and management experience.
> - Direct experience in the management of a company that buys from the semiconductor industry.

A search based on these criteria usually yields a male who is 50-plus years old, has a Master's degree in an engineering science, and has moved up the managerial ladder to the top spot in one or more corporations. Unfortunately, there are currently few minorities and almost no women who chose to be engineering graduate students thirty years ago.... Bluntly stated, a "woman's view" on how to run our semiconductor company does not help us, unless that woman has an advance technical degree and experience as CEO.... Therefore, not only does Cypress not meet your requirements for boardroom diversification, but we are unlikely to, because it is very difficult to find qualified directors, let alone directors that also meet investors' racial and gender preferences....

I presume you believe your organization does good work and that the people who spend their careers in its service deserve to retire with the necessities of life assured. If your investment in Cypress is intended for that purpose, I can tell you that each of the retired Sisters of St. Francis would suffer if I were forced to run Cypress on anything but a profit-making basis. The retirement plans of thousands of other people also depend on Cypress stock.... Any choice I would make to jeopardize retirees and other investors from achieving their lifetime goals would be fundamentally wrong.... If all companies in the U.S. were forced to operate according to some arbitrary social agenda, rather than for profit, all American companies would operate at a disadvantage to their foreign competitors, all Americans would become less well off (some laid off), and charitable giving would decline precipitously. Making Americans poorer and reducing charitable giving in order to force companies to follow an arbitrary social agenda is fundamentally wrong....

You have voted against me and the other directors of the company, which is your right as a shareholder. But here is a synopsis of what you voted against:

- Employee ownership. Every employee of Cypress is a shareholder and every employee of Cypress—including the lowest-paid—receives new Cypress stock options every year....
- Excellent pay. Our employees in San Jose averaged $78,741 in salary and benefits in 1995....
- A significant boost to our economy. In 1995, our company paid out $150 million to its employees. That money did a lot of good: it bought a lot of houses, cars, movie tickets, eyeglasses, and college education.
- A flexible health-care program. A Cypress-paid health-care budget is granted to all employees to secure the health-care options they want....
- Profit sharing. Cypress shares it profits with its employees. In 1995, profit sharing added up to $5,000 per employee, given in equal shares, regardless of rank or salary....
- Charitable work. Cypress supports Silicon Valley. We support the Second Harvest Food Bank ... I was chairman of the 1993 food drive, and Cypress has won the good-giving title three years running.... We also give to the Valley Medical Center, our Santa Clara-based public hospital....

I believe you should support management teams that hold our values and have the courage to put them into practice. So, that's my reply. Choosing a Board of Directors based on race and gender is a lousy way to run a company. Cypress will never do it. Furthermore, we will never be pressured into it, because bowing to well-meaning, special-interest groups is an immoral way to run a company, given all the people it would hurt. We simply cannot allow arbitrary rules to be forced on us by organizations that lack business expertise. I would rather be labeled as a person who is unkind to religious groups than as a coward who harms his employees and investors by mindlessly following high-sounding, but false, standards of right and wrong.... With regard to shareholders who exercise their right to vote according to a social agenda, we suggest that they reconsider whether or not their strategy will do net good—after all of the real costs are considered.

T. J. Rodgers asserted that "Cypress is run under a set of carefully considered moral principles," although one of these principles is that making a profit is the primary objective of the company. He disagreed with Sister Doris on whether morality requires a company to have a diverse board of directors. Such diversity is, in his view, more a matter of "political correctness." Sister Doris claimed, however, that her letter had nothing to do with "political correctness" but simply expressed "concern for the social integrity of business." After the exchange of letters, Sister Doris discovered that the order's retirement fund had sold its 7,000 shares of Cypress stock for unrelated reasons, thus bringing to an end the debate between the nun and the CEO.

INTRODUCTION

Although corporations are primarily business organizations run for the benefit of shareholders, they have a wide-ranging set of responsibilities—to their own employees, to customers and suppliers, to the communities in which they are located, and to society at large. Most corporations recognize these responsibilities and make a serious effort to fulfill them. Often, these responsibilities are set out in formal statements of a company's principles or beliefs. Corporations do not always succeed in fulfilling the responsibilities they acknowledge, however, and disagreements inevitably arise over the responsibilities of corporations in particular situations.

For example, T. J. Rodgers believes that Cypress Semiconductor is socially responsible, as witness the good deeds he describes. He contends, however, that profit must come first, as the fuel for doing good, and he objects to pressure from outside sources, especially when they do not understand business. He admits, though, that Sister Doris has a right as a shareholder to vote according to her views, but do not other members of society not have a right to make demands on corporations through their economic and political choices? At issue, then, are two questions: What is the social responsibility of business, and who has the right to make this determination?

There are no easy answers to such questions, but some help can be obtained from the theories of corporate social responsibility examined in this chapter. These range from a very restricted position—that corporations have no social responsibility beyond making a profit—to the position that corporations ought to assume a more active role in addressing major social problems. The social responsibility of corporations cannot be understood without an examination of the nature of corporations and their objectives. The fundamental question of corporate governance is, whose interests ought corporations to serve? This question is examined in a section on corporate governance. Finally, this chapter considers the ethics programs that many corporations have adopted to ensure ethical conduct by their employees.

CORPORATE SOCIAL RESPONSIBILITY

The concept of corporate social responsibility originated in the 1950s when American corporations rapidly increased in size and power. The concept continued to figure prominently in public debate during the 1960s and 1970s as the nation confronted pressing social problems such as poverty, unemployment, race relations, urban blight, and pollution.[3] Corporate social responsibility became a rallying cry for diverse groups demanding change in American business. In the last two decades of the twentieth century, corporations generally recognized a responsibility to society, but that responsibility was weighed against the demands of being competitive in a rapidly changing global economy. Pressure for improved performance was exerted by institutional investors, especially mutual and pension fund managers, who have a fiduciary duty to their investors to push for maximum return. During this period, the wealth of many American households was closely tied to the stock market, and so increasing stock price became a strong imperative for corporate managers.

The Debate over Social Responsibility

Some contend that corporate social responsibility is altogether a pernicious idea. The well-known conservative economist Milton Friedman writes in *Capitalism and Freedom*, "Few trends could so thoroughly undermine the very foundations of our free society as the acceptance by corporate officials of a social responsibility other than making as much money for their stockholders as possible."[4] He continues:

> The view has been gaining widespread acceptance that corporate officials ... have a "social responsibility" that goes beyond serving the interest of their stockholders.... This view shows a fundamental misconception of the character and nature of a free economy. In such an economy, there is one and only one social responsibility of business—to use its resources and engage in activities designed to increase its profits so long as it stays within the rules of the game, which is to say, engages in open and free competition, without deception or fraud.... It is the responsibility of the rest of us to establish a framework of law such that an individual pursuing his own interest is, to quote Adam Smith ..., "led by an invisible hand to promote an end which was no part of his intention."[5]

At the other extreme are critics who would like corporations to be more socially responsible but are mistrustful. They consider talk about corporate social responsibility to be a public relations ploy designed to legitimize the role of corporations in present-day American society and to divert attention away from the destructive social consequences of corporate activity. Even those who are more favorably disposed to the idea have reservations about the ability of corporations, especially as they are currently structured, to respond effectively to social issues. Businesses are single-purpose institutions, conceived, organized, and managed solely in order to engage in economic activity. As such, they lack the resources and the expertise for solving major social problems, and some add that they lack the legitimacy as well. Corporate executives are not elected officials with a mandate from the American people to apply the resources under their control to just any ends that they deem worthwhile.

Furthermore, the idea that corporations should be more socially responsible fails to give adequate ethical guidance to the executives who must decide which causes to pursue and how much to commit to them. This problem is especially acute in view of the fact that all choices involve trade-offs. A program to increase minority employment, for example, might reduce efficiency, thereby reducing wages for employees or raising prices for consumers. Or such a program might be adopted at the expense of achieving a greater reduction in the amount of pollution. Corporations committed to exercising greater social responsibility need more specific moral rules or principles to give them reasons for acting in one way rather than another.

The Definition of the Concept

All accounts of corporate social responsibility recognize that business firms have not one but many different kinds of responsibility, including economic and legal responsibilities. Corporations have an *economic* responsibility to produce goods and services and to provide jobs and good wages to the work force while earning a profit. Economic responsibility also includes the obligation to seek out supplies of raw materials, to discover new resources and technological improvements, and to develop new products. In addition, business firms have certain *legal* responsibilities. One of these is to act as a fiduciary, managing the assets of a corporation in the interests of shareholders, but corporations also have numerous legal responsibilities to employees, customers, suppliers, and other parties. The vast body of business law is constantly increasing as legislatures, regulatory agencies, and the courts respond to greater societal expectations and impose new legal obligations on business.

The concept of corporate social responsibility is often expressed as the voluntary assumption of responsibilities that go beyond the purely economic and legal responsibilities of business firms.[6] More specifically, social responsibility, according to some accounts, is the selection of corporate goals and the evaluation of outcomes not solely by the criteria of profitability and organizational well-being but by ethical standards or judgments of social desirability. The exercise of social responsibility, in this view, must be consistent with the corporate objective of earning a satisfactory level of profit, but it implies a willingness to forgo a certain measure of profit in order to achieve noneconomic ends.

Archie B. Carroll views social responsibility as a four-stage continuum.[7] Beyond economic and legal responsibilities lie ethical responsibilities, which are "additional behaviors and activities that are not necessarily codified into law but nevertheless are expected of business by society's members."[8] At the far end of the continuum are discretionary responsibilities. These responsibilities are not legally required or even demanded by ethics; but corporations accept them in order to meet society's expectations. S. Prakash Sethi notes that social responsibility is a relative concept: What is only a vague ideal at one point in time or in one culture may be a definite legal requirement at another point in time or in another culture. In most of the advanced nations of the world, fulfilling traditional economic and legal responsibilities is no longer regarded as sufficient for legitimizing the activity of large corporations. Corporate social responsibility can thus be defined as "bringing corporate behavior up to a level where it is congruent with the prevailing social norms, values, and expectations of performance."[9]

In 1971, the Committee for Economic Development issued an influential report that characterized corporate social responsibility in a similar fashion but without an explicit mention of legal responsibilities. The responsibilities of corporations are described in this report as consisting of three concentric circles.

> The *inner circle* includes the clear-cut basic responsibilities for the efficient execution of the economic function—products, jobs, and economic growth.
>
> The *intermediate circle* encompasses responsibility to exercise this economic function with a sensitive awareness of changing social values and priorities: for example, with respect to environmental conservation; hiring and relations with employees; and more rigorous expectations of customers for information, fair treatment, and protection from injury.
>
> The *outer circle* outlines newly emerging and still amorphous responsibilities that business should assume to become more broadly involved in actively improving the social environment. Society is beginning to turn to corporations for help with major social problems such as poverty and urban blight. This is not so much because the public considers business singularly responsible for creating these problems, but because it feels large corporations possess considerable resources and skills that could make a critical difference in solving these problems.[10]

Examples of Social Responsibility. Although there are some disagreements about the meaning of corporate social responsibility, there is general agreement on the types of corporate activities that show social responsibility. Among these are

1. Choosing to operate on an ethical level that is higher than what the law requires.
2. Making contributions to civic and charitable organizations and nonprofit institutions.
3. Providing benefits for employees and improving the quality of life in the workplace beyond economic and legal requirements.
4. Taking advantage of an economic opportunity that is judged to be less profitable but more socially desirable than some alternatives.

5. Using corporate resources to operate a program that addresses some major social problem.

Although these activities are all beyond the economic and legal responsibilities of corporations and may involve some sacrifice of profit, they are not necessarily antithetical to corporate interests. For example, corporate philanthropy that makes the community in which a company is located a better place to live and work results in direct benefits. The "goodwill" that socially responsible activities create makes it easier for corporations to conduct their business. It should come as no surprise, then, that some of the most successful corporations are also among the most socially responsible. They are led by executives who see that even the narrow economic and legal responsibilities of corporations cannot be fulfilled without the articulation of noneconomic values to guide corporate decision making and the adoption of nontraditional business activities that satisfy the demands of diverse constituencies.

Going beyond Social Responsibility. An important aspect of corporate social responsibility is the responsiveness of corporations—that is, the ability of corporations to respond in a socially responsible manner to new challenges.[11] William C. Frederick explains that the concept of *corporate social responsiveness* "refers to the capacity of a corporation to respond to social pressures."[12] The emphasis of corporate social responsiveness, in other words, is on the *process* of responding or the readiness to respond, rather than on the *content* of an actual response. Thus, a socially responsive corporation uses its resources to anticipate social issues and develop policies, programs, and other means of dealing with them. The management of social issues in a socially responsive corporation is integrated into the strategic planning process, instead of being handled as an ad hoc reaction to specific crises.

The *content* of a response is also important because it represents the outcome of being socially responsible. Donna Wood has combined all three elements—the *principle* of being socially responsible, the *process* of social responsiveness, and the socially responsible *outcome*—in the concept of *corporate social performance.*[13] Using the example of environmental concerns about packaging, the *principles* of corporate social responsibility would be those that lead the company to recognize an obligation to change its packaging in order to protect the environment. The *processes* might consist of establishing an office of environmental affairs or working with environmentalists to develop new packaging. The *outcomes* could include the switch to environmentally responsible packaging and perhaps building facilities to recycle the packaging.

The Classical View

The dominant conception of the corporation, at least in the United States, is called the *classical view*. This view, which prevailed in the nineteenth century, is still very influential today, especially among economists. It is expressed by James W. McKie in three basic propositions.

1. Economic behavior is separate and distinct from other types of behavior, and business organizations are distinct from other organizations, even though the same individuals may be involved in business and nonbusiness affairs. Business organizations do not serve the same goals as other organizations in a pluralistic society.
2. The primary criteria of business performance are economic efficiency and growth in production of goods and services, including improvements in technology and innovations in goods and services.
3. The primary goal and motivating force for business organizations is profit. The firm attempts to make as large a profit as it can, thereby maintaining its efficiency and taking advantage of available opportunities to innovate and contribute to growth.[14]

In the classical view, corporations should engage in purely economic activity and be judged in purely economic terms. Social concerns are not unimportant, but they should be left to other institutions in society.

The classical view is part of a larger debate about the legitimate role of the corporation in a democracy. In his introduction to the influential volume *The Corporation in Modern Society*, Edward Mason described the problem of the modern corporation as follows: America is a "society of large corporations ... [whose] management is in the hands of a few thousand men. Who selected these men, if not to rule over us, at least to exercise vast authority, and to whom are they responsible?"[15] The classical view is a response to this problem that recognizes that corporate power must be harnessed to a larger social good if it is not to become tyrannical. Confining corporations to economic ends is intended, in part, to limit their role in society so as to preserve other kinds of institutions, both public and private.

Arguments for Social Responsibility

Business activity, in the classical view, is justified partly on the ground that it secures the well-being of society as a whole. The crux of this argument is the efficacy of Adam Smith's invisible hand in harmonizing self-interested behavior to secure an end that is not a part of anyone's intention. This justification also depends on the ability of the rest of society to create the conditions necessary for the invisible hand to operate and to address social problems without the aid of business. The debate over the workings of the invisible hand cannot be settled here (it is examined at some length in Chapter 4), but the invisible hand argument, upon which the classical view depends, is not incompatible with certain arguments for corporate social responsibility.

The Moral Minimum of the Market. First, a certain level of ethical conduct is necessary for the invisible hand to operate, or indeed for business activity to take place at all. Milton Friedman speaks of the "rules of the game," by which he means "open and free competition, without deception or fraud." Theodore Levitt, in his article "The Dangers of Social Responsibility," says that, aside from seeking material gain, business has only one responsibility, and that is "to obey the elementary canons of everyday face-to-face civility (honesty, good faith, and so on)."[16] The "rules of the game" and "face-to-face civility" impose not inconsequential constraints on business. Presumably, the prohibition against deception and fraud obligates corporations to deal fairly with employees, customers, and the public and to avoid sharp sales practices, misleading advertising, and the like.

The moral minimum of the market also includes an obligation to engage in business without inflicting injury on others. Critics of Levitt's position observe:

> ... Levitt presents the reader with a choice between, on the one hand, getting involved in the management of society ... and, on the other hand, fulfilling the profit-making function. But such a choice excludes another meaning of corporate responsibility: the making of profits in such a way as to minimize social injury. Levitt at no point considers the possibility that business activity may at times injure others and that it may be necessary to regulate the social consequences of one's business activities accordingly.[17]

Thus, corporations in a free market have an obligation not to pollute the environment and to clean up any pollution they cause.

It may also be in the best interests of a corporation to operate above the moral minimum of the market. Corporations that adhere only to the moral minimum leave themselves open to pressure from society and regulation by government. One of the major reasons advanced for corporations to exercise greater social responsibility is to avoid such external interference. By "internalizing" the expectations of society, corporations retain control over decision making and avoid the costs associated with government regulation.

Power and Responsibility. Second, corporations have become so large and powerful that they are not effectively restrained by market forces and government regulation, as the invisible hand argument assumes. Some self-imposed restraint in the form of a voluntary assumption of greater social responsibility is necessary, therefore, for corporate activity to secure the public welfare. Keith Davis expressed this point succinctly in the proposition "*social responsibility arises from social power.*"[18] He also cited what he calls the Iron Law of Responsibility: "In the long run, those who do not use power in a manner which society considers responsible will tend to lose it."[19] The need for greater social responsibility by corporations, then, is an inevitable result of their increasing size and influence in American society.

Holders of the classical theory argue in reply that precisely because of the immense power of corporations, it would be dangerous to unleash it from the discipline of the market in order to achieve vaguely defined social goals.[20] Kenneth E. Goodpaster and John B. Matthews, Jr., concede that this is a matter for serious concern but argue in response:

> What seems not to be appreciated is the fact that power affects when it is used as well as when it is not used. A decision by [a corporation] ... not to exercise its economic influence according to "non-economic" criteria is inevitably a moral decision and

just as inevitably affects the community. The issue in the end is not whether corporations (and other organizations) should be "unleashed" to exert moral force in our society but rather how critically and self-consciously they should choose to do so.[21]

Giving a Helping Hand to Government. Third, the classical view assumes that business is best-suited to provide for the economic well-being of the members of a society, whereas noneconomic goals are best left to government and the other noneconomic institutions of society. This sharp division of responsibility is true at best only as a generalization, and it does not follow that corporations have *no* responsibility to provide a helping hand. Corporations cannot attempt to solve every social problem, of course, and so some criteria are needed for distinguishing those situations in which corporations have an obligation to assist other institutions. John G. Simon, Charles W. Powers, and Jon P. Gunnemann propose the following four criteria:[22]

1. The urgency of the need;
2. The proximity of a corporation to the need;
3. The capability of a corporation to respond effectively;
4. The likelihood that the need will not be met unless a corporation acts.

Accordingly, a corporation has an obligation to address social problems that involve more substantial threats to the well-being of large numbers of people, that are close at hand and related in some way to the corporation's activity, that the corporation has the resources and expertise to solve, and that would likely persist without some action by the corporation.

Friedman's Argument against Social Responsibility

Perhaps the best-known critic of corporate social responsibility is Milton Friedman. Friedman's main argument against corporate social responsibility is that corporate executives, when they are acting in their official capacity and not as private persons, are agents of the stockholders of the corporation. As such, executives of a corporation have an obligation to make decisions in the interests of the stockholders, who are ultimately their employers. He has asked:

> What does it mean to say that the corporate executive has a "social responsibility" in his capacity as businessman? If this statement is not pure rhetoric, it must mean that he is to act in some way that is not in the interest of his employers. For example, that he is to refrain from increasing the price of the product in order to contribute to the social objective of preventing inflation, even though a price increase would be in the best interests of the corporation. Or that he is to make expenditures on reducing pollution beyond the amount that is in the best interests of the corporation or that is required by law in order to contribute to the social objective of improving the environment. Or that, at the expense of corporate profits, he is to hire "hardcore" unem-

ployed instead of better-qualified available workmen to contribute to the social objective of reducing poverty.

In each of these cases, the corporate executive would be spending someone else's money for a general social interest. Insofar as his actions in accord with his "social responsibility" reduce returns to stockholders, he is spending their money. Insofar as his actions raise the price to customers, he is spending the customers' money. Insofar as his actions lower the wages of some employees, he is spending their money.[23]

When corporate executives act in the way Friedman describes, they take on a role of imposing taxes and spending the proceeds that properly belongs only to elected officials. They become, in effect, civil servants with the power to tax, and as civil servants, they ought to be elected through the political process instead of being selected by the stockholders of private business firms.[24]

Criticism of Friedman's Argument. The classical view does not sanction an unrestrained pursuit of profit. Friedman himself acknowledges that business must observe certain essential limitations on permissible conduct, which he describes as the "rules of the game." Presumably, he would also grant the necessity of government with limited powers for setting and enforcing rules. Business activity requires, in other words, a minimal state in order to prevent anticompetitive practices and to enforce the basics of commercial law. Friedman recognizes, further, that many supposed socially responsible actions are really disguised forms of self-interest. Contributions to schools, hospitals, community organizations, cultural groups, and the like are compatible with the classical view insofar as corporations receive indirect benefits from the contributions. All Friedman asks is that corporations recognize these as effective means for making a profit and not as philanthropic activities.

In addition, holders of the classical view generally admit the legitimacy of three other functions of government that place limits on business activity.[25] First, business activity generates many *externalities*, that is, social harms, such as worker injury, which result indirectly from the operation of business firms. In order to prevent these harms or to correct them after they occur, it is proper for government to act—by requiring safer working conditions, for example, or by taxing employers to fund workers' compensation programs.[26] Second, the operation of a free market economy results in considerable *inequalities* in the distribution of income and wealth. Insofar as it is desirable as a matter of public policy to reduce these inequalities, it is appropriate for government to undertake the task by such means as progressive taxation and redistribution schemes. It is the job of government, in other words, and not business, to manage the equity/efficiency trade-off.[27] Third, free markets are prone to *instability* that manifests itself in inflation, recessions, unemployment, and other economic ills. Individual firms are too small to have much effect on the economy as a whole, and so government must step in and use its powers of taxation, public expenditure, control of the money supply, and the like to make the economy more stable.

The classical view is compatible, then, with some intervention in business activity by government in order to secure the public welfare. The important point to recognize is that the restraints are almost entirely *external.* The primary burden for ensuring that corporations act in a way that is generally beneficial rests on society as a whole, which is charged by Friedman with the task of creating a framework of law that allows business firms to operate solely in their self-interest. The classical theory, therefore, does not permit corporations to *act* in a socially irresponsible manner; it only relieves them of the need to *think* about matters of social responsibility. In a well-ordered society, corporations attend to business while government and other institutions fulfill their proper roles.

The "Taxation" Argument. The objections to Friedman's argument against corporate social responsibility discussed so far do not address his point that in exercising corporate social responsibility, managers are spending someone else's money. Investors, according to the "taxation" argument, entrust their money to the managers of corporations in order to make profits for the shareholders. Spending money to pursue social ends is thus a form of taxation.

Many things are wrong with the "taxation" argument. To say, as Friedman does, that corporate assets belong to the shareholders, that it's *their* money, is not wholly accurate. The role of shareholders in corporate governance is a rather complex issue, however, that is examined in the next section. Even if Friedman's assumption is accepted, it does not follow that corporations have no social responsibility.

First, managers of a corporation do not have an obligation to earn the greatest amount of profit for shareholders without regard for the means used. A taxi driver hired to take a passenger to the airport as fast as possible, for example, is not obligated to break traffic laws and endanger everyone else on the road. Similarly, money spent on product safety or pollution control may reduce the potential return to shareholders, but the alternative is to conduct business in a way that threatens the well-being of others in society. Friedman would insist, of course, that managers carry out their responsibility to shareholders within the rules of the game, but the moral obligation of managers to be sensitive to the social impact of their actions is more extensive than the minimal restraints listed by Friedman.

Second, the obligation of managers is not merely to secure the maximum return but also to preserve the equity invested in a corporation. Securing the maximum return for shareholders consistent with the preservation of invested capital requires managers to take a long-term view that considers the stability and growth of the corporation. For corporations to survive, they must satisfy the legitimate expectations of society and serve the purposes for which they have been created. Friedman admits the legitimacy of acts of social responsibility as long as they are ultimately in the self-interest of the corporation. The main area of disagreement between proponents and critics of social responsibility is, how much socially responsible behavior is in a corporation's long-term self-interest?

Third, the interests of shareholders are not narrowly economic; corporations are generally expected by their owners to pursue some socially desirable ends. Shareholders are also consumers, environmentalists, and citizens in communities. Consequently, they are affected when corporations fail to act responsibly. In fact, shareholders may be morally opposed to some activities of a corporation and in favor of some changes. One writer contends that "there are conventionally motivated investors who have an interest in the social characteristics of their portfolios *as well* as dividends and capital gains."[28] If so, managers who exercise social responsibility are not "taxing" shareholders and spending the money contrary to their interests but quite the opposite; managers who do not act in a socially responsible manner are using shareholders' money in ways that are against the interests of their shareholders. Friedman's response is, if shareholders want certain social goals, let them use their dividends for that purpose. However, it may be more efficient for corporations to expend funds on environmental protection, for example, than for shareholders to spend the same amount in dividends for the same purpose.

For these reasons, then, the "taxation" argument against corporate social responsibility is not very compelling. Although the rights of shareholders place some limits on what businesses can justifiably do to address major social concerns, they do not yield the very narrowly circumscribed view of Friedman and others. However, the issues raised by the debate over corporate social responsibility on the classical view is only part of a larger controversy over the governance of the modern corporation. Who should control a corporation? Whose interests should the corporation serve? To these questions of corporate governance we now turn.

CORPORATE GOVERNANCE

A corporation brings together many different groups—most notably managers, employees, suppliers, customers, and, of course, investors—for the purpose of conducting business. Because these various corporate constituencies have different and sometimes conflicting interests, the question arises: In whose interest should the corporation be run?

Debate has long raged over the nature of the corporation. Is the corporation the private property of the stockholders who choose to do business in the corporate form, or is the corporation a public institution sanctioned by the state for some social good?[29] In the former view, which may be called the *property rights theory*, the right to incorporate is an extension of the property rights and the right of contract that belong to all persons.[30] The latter view—let us call it the *social institution theory*—holds that the right to incorporate is a privilege granted by the state and that corporate property has an inherent public aspect.[31] A third view is the *contractual theory* of the firm. In the contractual theory, shareholders, along with other investors, employees, and the like, each own assets that they make available to the firm. Thus, the firm results from the property rights and the right of contract of every corporate constituency and not from those of shareholders alone.

Whether corporations ought to serve the interests of shareholders alone or the interests of a wider range of constituencies depends on the theory of the firm that we accept. Even though holders of all three theories generally conclude that the interests of shareholders are primary, the arguments that they provide are different, and it is important to understand the logic of each argument.

The Property Rights and Social Institution Theories

The original form of the modern corporation was the joint stock company in which a small group of wealthy individuals pooled their money for some undertaking that they could not finance alone. In the property rights theory, this corporate form of business organization is justified on the grounds that it represents an extension of the property rights and the right of contract enjoyed by everyone. Just as individuals are entitled to conduct business with their own assets, so, too, have they a right to contract with others for the same purpose. Although individual shareholders in a joint stock company or a corporation have exchanged their personal assets for shares of stock, they jointly own the common enterprise, and as owners they are entitled to receive the full proceeds.

The social institution theory emphasizes that a corporation is not merely a *private* association created for the purpose of personal enrichment but also a *public* enterprise that is intended to serve some larger social good. The earliest joint stock companies were special grants that kings bestowed on favored subjects for specific purposes. Today, corporations are chartered by states, so that the opportunity for individuals to do business in the corporate form is a state-granted privilege. The courts have also held in decisions such as *Munn* v. *Illinois* (1876) that corporate property is "affected with a public interest" so that states have a right to regulate its use.[32] Corporations are thus not wholly private; they have an inherent public aspect.

In a pure expression of the property rights theory, the Michigan State Supreme Court ruled in 1919 that the Ford Motor Company could be forced to pay more dividends to the shareholders in spite of Henry Ford's view that the company had made too much profit and ought to share some of it with the public by reducing prices. In *Dodge* v. *Ford Motor Co.*, the court declared, "A business corporation is organized and carried on primarily for the profit of the stockholders."[33] The profit-making end of a corporation is set forth in its charter of incorporation, which represents a contract among the shareholders who have invested their money, and Henry Ford had no right to substitute another end by using corporate resources for an essentially philanthropic purpose.

The decision in *Dodge* v. *Ford Motor Co.* assumed that shareholders are the owners of a corporation. This assumption was true as long as corporations had relatively few shareholders who actively controlled the business. However, in 1932, a book by Adolf A. Berle, Jr., and Gardiner C. Means, *The Modern Corporation and Private Property*, documented a dramatic shift that had occurred in American business.[34] Stock ownership in large corporations had become dispersed among numerous investors who had little involvement in corporate affairs, and the actual control of corporations had passed to a class of professional managers. The result was a separation of ownership and control, and with this separation came a change in the nature of corporate property.

The Nature of Corporate Property. Strictly speaking, property is not a tangible thing like land, but a bundle of rights that defines what an owner is entitled to do with a thing. A property right, in the full sense of the term, involves control over the thing owned and an assumption of responsibility. In the separation of ownership and control, shareholders had relinquished both control and responsibility. As a result, shareholders of large publicly held corporations had ceased to be owners in the full sense and had become merely one kind of provider of the resources needed by a corporation.

According to Berle and Means, "The property owner who invests in a modern corporation so far surrenders his wealth to those in control of the corporation that he has exchanged the position of independent owner for one in which he may become merely recipient of the wages of capital."[35] They continued:

> ... [T]he owners of passive property, by surrendering control and responsibility over the active property have surrendered the right that the corporation should be operated in their sole interest —they have released the community from the obligation to protect them to the full extent implied in the doctrine of strict property rights. At the same time, the controlling groups ... have in their own interest broken the bars of tradition which require that the corporation be operated solely for the benefit of the owners of passive property.[36]

Because of the separation of ownership and control, managers have assumed the position of trustee for the immense resources of a modern corporation, and in this new position, they face the question: For whom are corporate managers trustees?

The Berle-Dodd Debate. In a famous 1932 exchange with Berle in the *Harvard Law Review*, E. Merrick Dodd, Jr., argued that the corporation is "an economic institution which has a social service as well as a profit-making function."[37] According to Dodd, the modern corporation had become a public institution, as opposed to a private activity of the shareholders, and as such, it had a social responsibility that could include the making of charitable contributions.[38] Because a corporation is property only in a "qualified sense," it may be regulated by society so that the interests of employees, customers, and others are protected. Corporate managers have a right, even a duty, to consider the interests of all those who deal with the corporation.[39]

Berle cautioned against Dodd's position because of the dangers posed by unrestrained managerial power. In a response to Dodd in the *Harvard Law Review*, Berle wrote, "When the fiduciary obligation of the corporate management and 'control' to stockholders is weakened or eliminated, the management and 'control' become for all practical purposes absolute."[40] It would be unwise, in Berle's estimation, for the law to release managers from a strict accountability to shareholders, not out of respect for their property rights as owners of a corporation but as a matter of sound public policy. He wrote:

> Unchecked by present legal balances, a social-economic absolutism of corporate administrators, even if benevolent, might be unsafe; and in any case it hardly affords the soundest base on which to construct the economic commonwealth which industrialism seems to require.[41]

Corporate law has evolved effective means for restraining managerial power by directing managers to act in the interests of the shareholders, but we lack effective means for ensuring that managers serve the interests of society as a whole. There is no guarantee that managers will exercise their newly acquired control in any interests but their own. Berle described the rise of corporate managers as a "seizure of power without recognition of responsibility—ambition without courage."[42] But in the absence of effective restraints on managerial power, Berle concludes, "... [W]e had best be protecting the interests we know, being no less swift to provide for the new interests as they successively appear."[43]

Eventually, the law loosened the restraints that were imposed in *Dodge* v. *Ford Motor Co.* and allowed corporations to expend some corporate funds for the good of society. In *A. P. Smith Manufacturing Co.* v. *Barlow* (1953), the New Jersey State Supreme Court ruled that the managers of the company were permitted by law to give $1,500 to Princeton University despite shareholder objections on the grounds that to bar corporations from making such contributions would threaten our democracy and the free enterprise system. The court agreed with the testimony of a former chairman of the board of United States Steel Company that if American business does not aid important institutions such as private universities, then it is not "properly protecting the long-range interests of its stockholders, its employees and its customers." After the decision in *A. P. Smith Manufacturing Co.* v. *Barlow*, Berle conceded defeat in his debate with Dodd. Public opinion and the law had accepted Dodd's contention that corporate powers ought to be held in trust for the whole of society.[44]

Although the separation of ownership and control documented by Berle and Means undermined the property rights theory, a fully developed social institution theory did not replace it. Instead, a conception of the corporation as a quasi-public institution emerged, in which managers have limited discretion to use the resources at their command for the good of employees, customers, and the larger society. In a world of giant corporations, managers are called upon to balance the interests of competing corporate constituencies, and in order to fill this role they have developed a sense of management as a profession with public responsibilities. Managers ceased being the exclusive servants of the stockholders and assumed the mantle of public-spirited leaders, albeit of profit-making business organizations.

In the last several decades, another theory of the firm has emerged that now dominates thinking in financial economics and corporate law. This is the contractual theory, in which the firm is viewed as a nexus of contracts among all corporate constituencies.

The Contractual Theory

The origin of the contractual, or nexus-of-contracts, theory is the work of the economist Ronald Coase, who claimed that firms exist as less costly alternatives to market transactions.[45] In a world where market exchanges could occur without any costs (what economists call *transaction costs*), economic activity would be achieved entirely by means of contracting among individuals in a free market. In the actual world, the transaction costs involved in market activity can be quite high, and some coordination can be achieved more cheaply by organizing economic activity in firms. Thus, there are two forms of economic coordination—firms and markets—and the choice between them is determined by transaction costs.

In the Coasean view, the firm is a market writ small in which parties with economic assets contract with the firm to deploy these assets in productive activity. Generally, an individual's assets are more productive when they are combined with the assets of others in joint or team production. Individuals will choose to deploy their assets in a firm instead of the market when the lower transaction costs of a firm combined with the benefits of team production yield them a higher return.

Deploying assets in a firm involves some risks, however, when those assets are *firm-specific*. Consider the situation of an employee who acquires skills that are needed by a particular employer. A worker with such skills will generally earn more than one with only generic skills, but only a few employers will value those special skills and be willing to pay the higher wages. An employee with special skills is also tied more closely to the firm because the skills in question are not easily transferable, and the employee would likely suffer a loss if forced to move to another employer. The assets of an employee that can be more profitably employed with one or a few firms are thus firm-specific. Firm-specific assets enable a worker to create more wealth, which makes possible the higher pay, but this wealth can also be appropriated by the firm itself, leaving the employee without adequate compensation for acquiring special skills.

Not only employees but also investors, suppliers, customers, and other groups have firm-specific assets, and these groups will make their assets available to a firm only with adequate safeguards against misappropriation. That is, each group will seek guarantees to ensure that they are adequately compensated for any assets that cannot easily be removed from joint productive activity. Most groups protect themselves by means of contracts. These may be either explicit, legally enforceable contracts or implicit contracts that have no legal standing. Thus, employees are often protected by employment contracts, suppliers by purchase contracts, consumers by warranties, and so on. An implicit contract is created, for example, when an employer creates an expectation about the conditions of employment. In the nexus-of-contracts firm, managers coordinate these contracts with the various corporate constituencies. The contracts with most corporate constituencies are relatively unproblematical, but one group raises special problems, namely shareholders.

The Role of Shareholders. In the usual interpretation of the contractual theory, shareholders along with bondholders and other investors, provide capital, but, more significantly, they also assume the residual risk of conducting business, which is the risk that remains after a firm has fulfilled all of its legal obligations. Residual risk could be borne by every group that contracts with a firm, but risk bearing can also be a specialized role in which one group bears the preponderance of risk. In the large, publicly held corporation, shareholders fill the role of residual risk bearer, and it is this role, rather than the role of capital provider, that sets them apart from other groups in the nexus of contracts.

The position of residual risk bearer is difficult to protect by ordinary contractual provisions, such as those available to bondholders, employees, customers, and other constituencies. Some protection is provided by the opportunity to diversify and by the limited liability that shareholders enjoy. Shareholders are also rewarded for their investment with prospects of a higher return than that assured to secured creditors such as bondholders. However, the most important protection for shareholders is corporate control. Corporate control is a package of rights that includes the right to select the board of directors and approve important changes. In addition, shareholders have a claim on management's allegiance. This is commonly expressed by saying that management has a fiduciary duty to operate the corporation in the shareholders' interest and that the objective of the firm is to maximize shareholder wealth.

The contractual theory of the firm itself does not assign this right to shareholders—or to any other group for that matter. Rather, corporate control is a benefit to be bargained for in the nexus-of-contracts firm, and through bargaining any constituency group could conceivably assume the right of control. Indeed, other groups take control when a corporation becomes employee-owned or customer-owned (a co-op, for example). Employees have also successfully bargained for representation on boards of directors, and bond indentures sometimes give bondholders the right to vote on certain risky ventures. When corporations are in bankruptcy, creditors take control from shareholders and the creditors' interests become primary until the firm recovers.

Still, the current system of corporate governance assigns control, for the most part, to shareholders. Some further argument is needed, however, for this assignment of rights. Why should shareholders, as residual risk bearers, control a corporation?

The Argument for Shareholder Control. The starting point of the argument for shareholder control is that this right is of greater value to residual claimants than to other constituencies. First, control is better suited than other kinds of protection for the special situation of shareholders. That is, control is a fitting solution for the problem of protecting the firm-specific assets of residual risk bearers. Second and more importantly, the return to shareholders is wholly dependent on the profitability of a firm, and so they value control as a means to spur a firm to the highest possible level of performance. By contrast, nonshareholder constituencies have little to gain from control because their fixed claims will be satisfied as long as the firm remains solvent. The right of control, therefore, is more valuable to whichever group settles for residual rather than fixed claims.

The contractual theory envisions the corporation as a nexus of contracts that is formed by bargaining among all corporate constituencies. In an actual or hypothetical bargaining situation, shareholders as residual claimants would insist on corporate control and be more willing to pay for this right. Moreover, other constituencies would agree to this arrangement. The reason is that bondholders, employees, and other groups would prefer different contractual terms. Employees, for example, would opt for contracts that assure them a specific wage and other benefits. Bargaining for control rights, such as the right to a seat on the board of directors, would require employees to give up something else, and the reluctance of employees to bargain for such rights suggests that the gain is not worth the price.

The contractual theory also holds that society benefits when shareholders control a corporation. Assuming that maximum wealth creation is the goal of business activity, control should go to the group with the appropriate incentives for making wealth-maximizing decisions, and this group—according to the argument—is the shareholders. It has already been observed that bondholders, employees, and other corporate constituencies with fixed claims tend to favor decisions that secure their claims and no more. Therefore, some profitable investment opportunities might not be pursued if these groups had control. Managers, too, lack the incentives to pursue all profitable ventures, especially those that would reduce their power or place them at risk. Wealth-maximizing decisions are more likely to be made by the residual risk bearers, because they incur the marginal costs and gain the marginal benefits of all new ventures. Shareholder control, therefore, is the ideal arrangement for the whole of society.

The contractual theory argument for shareholder control, then, is that corporations will create more wealth if decisions are made with only the shareholders' interests in mind. What is important to note about this argument is that the *ultimate* objective of the firm is the maximization of wealth for the whole of society. The objective of shareholder wealth maximization is merely a means to this larger end. This argument does not neglect other constituencies, such as employees, customers, suppliers, and the larger community. It assumes, rather, that these groups are well protected by other kinds of contracts and that, on the whole, they are well served by shareholder control. In other words, this arrangement is optimal not merely for shareholders but for society as a whole.

Frank H. Easterbrook and Daniel R. Fischel make the point in the following way:

> A successful firm provides jobs for workers and goods and services for consumers. The more appealing the goods to consumers, the more profit (and jobs). Prosperity for stockholders, workers, and communities goes hand in glove with better products for consumers. Other objectives, too, come with profit. Wealthy firms provide better working conditions and clean up their outfalls; high profits produce social wealth that strengthens the demand for cleanliness.... [W]ealthy societies purchase much cleaner and healthier environments than do poorer nations—in part because well-to-do citizens want cleaner air and water, and in part because they can afford to pay for it.[46]

As a result, they claim, "maximizing profit for equity investors assists the other 'constituencies' automatically."[47]

Criticism of the Contractual Theory

The contractual theory argument for shareholder control is similar to Adam Smith's famous invisible hand argument according to which each self-interested individual is, to quote from the *Wealth of Nations,* "led by an invisible hand to promote an end which was no part of his intention." Just as the invisible hand argument is subject to some well-known objections, so, too, is the contractual theory argument.

Externalities. One problem is posed by externalities, such as pollution. The welfare of society is not promoted, for example, when corporations make a profit for shareholders by polluting a stream. The idea that managers ought to consider only the interests of shareholders appears to invite, indeed require, actions that impose harms on other constituencies. Easterbrook and Fischel recognize this possibility but deny that it has any bearing on the argument for shareholder control. They write:

> We do not make the Panglossian claim that profit and social welfare are perfectly aligned. When costs fall on third parties—pollution is a common example—firms do injury because harm does not come back to them as private cost.... Users of the stream impose costs on the firm (and its consumers) as fully as the firm imposes costs on the users of the stream. No rearrangement of corporate governance structures can change this. The task is to establish property rights so that the firm treats the social costs as private ones, and so that its reactions, as managers try to maximize profits given these new costs, duplicate what all of the parties (downstream users and customers alike) would have agreed to were bargaining among all possible without cost.[48]

Easterbrook and Fischel contend that clear assignments of property rights would force firms to internalize what would otherwise be external costs.[49] Pollution can also be handled by other means, such as government regulation. At issue, then, is whether externalities such as pollution are due to shareholder control and hence whether any change in control would provide a solution.

Implicit Contracts. The contractual theory argument assumes that nonshareholder constituencies are well served by their various contractual arrangements. Explicit contracts, such as an employee contract or a sales agreement, are legally enforceable. However, many contracts are implicit, so that they depend on the goodwill of management. Although an employer may create an expectation of secure employment, for example, this implicit contract can be violated without any legal consequences. When longtime employees are terminated in corporate restructurings or downsizings, the charge is often made that the shareholders gain by violating the implicit contracts with those who are forced to leave. Similarly, companies that close a plant are often accused of violating an implicit contract with a community that has provided support. Critics of hostile takeovers have argued that one source of financing for the raid on a target company is the value of the implicit contracts with various constituencies that shareholders are able to capture. Thus, the takeover of a lumber company that has made an implicit contract with environmentalists to limit logging can be financed, in part, by increasing the number of trees felled after the takeover.

The defense against these charges is that the affected groups are still better off with implicit claims that can be violated. Without the freedom to restructure or downsize a work force, for example, corporations would be less able to compete and hence to offer high wages. Employees could gain greater job security only by settling for lower wages, and communities could keep plants only by settling for lesser benefits. Faced with these alternatives, each constituency would negotiate contracts that rely to some extent on vague promises and nonbinding commitments instead of precisely written, legally enforceable contracts.

Still, implicit contracts are a kind of promise, and it seems unfair for shareholders to benefit by, in effect, going back on their word. For this reason, Congress has addressed the plight of communities by requiring prior notification of plant closings,[50] and some states have sought to limit the impact of hostile takeovers by so-called "other constituency" statutes that permit directors to consider the interests of nonshareholder groups.[51] Congress has also imposed a fiduciary duty on corporations to serve the interests of employees in the management of pension funds. The effect of these measures is to limit the freedom of corporate managers to violate various implicit contracts with nonshareholder groups.

Residual Risk. The contractual theory assumes that shareholders bear all residual risk, but other constituencies, most notably employees and suppliers, bear some. In truth, well-diversified individual shareholders bear relatively little risk, but the relevant point is that their claim is on the residual returns of a firm. By the logic of the argument for shareholder control, however, any group that bears some residual risk should have some say in major corporate decisions. It has been argued, for example, that highly skilled employees who develop valuable firm-specific human capital may, in fact, assume considerable residual risk.[52] This is especially true in start-up companies that can attract employees only with promises of higher wages in the future. These promised wages are contingent on the performance of the firm, so that employees, in the words of one writer, "are also 'residual claimants,' who share in the business risk associated with the enterprise."[53] For this reason, many start-up companies offer their employees shares of stock or stock options. When other groups besides shareholders bear residual risk, they are sometimes accorded a seat on the board of directors.

Distribution and Power. Finally, the contractual theory takes no account of the fact that the contracting groups have unequal bargaining power, and this imbalance results in correspondingly unequal distribution of the wealth created by a firm. As in any market, those with the greatest resources will reap the lion's share of goods. If workers, for example, are weak and disempowered, they will not be able to bargain effectively in a nexus-of-contracts firm. Unequal bargaining power is a serious problem in most societies. Some of the inequality in assets comes from accidents of birth, as when one person is more intelligent or more talented than another. The main sources of inequality, however, are social and political in nature. Holders of the contractual theory maintain that all kinds of inequality can be corrected only by social and political change. No change in corporate governance can improve the condition of any corporate constituency without some alteration in its underlying bargaining power.

Stakeholder Theory

The contractual theory generally supports a stockholder-centered conception of the corporation. A much-discussed alternative to this view is *stakeholder theory*. The central claim of the stakeholder approach is that corporations are operated or ought to be operated for the benefit of all those who have a *stake* in the enterprise, including employees, customers, suppliers, and the local community. A stakeholder is variously defined as "those groups who are vital to the survival and success of the corporation"[54] and as "any group or individual who can affect or is affected by the achievement of the organization's objectives."[55] Although the relation of each stakeholder group to the corporation is different, each of these constituencies is integral to the operation of a corporation, and its role must be taken into account by managers.

Serious attempts have been made to develop the stakeholder concept into a full-blown view of the corporation that might replace the stockholder-centered conception. In the standard stockholder view, business is primarily an economic activity in which economic resources are marshaled for the purpose of making a profit. Employees are critical to this enterprise as a source of labor, but they are merely one *input* that can be "bought" in the market. Customers are also critical, and they receive the *output* of a corporation's activity, namely some good or service. But what customers give and receive is also the result of market exchanges. The resulting view of the corporation is the input–output model displayed in Figure 14.1.

The concept of a stakeholder highlights the fact that corporate activity is not solely a series of market transactions but also a cooperative (and competitive)

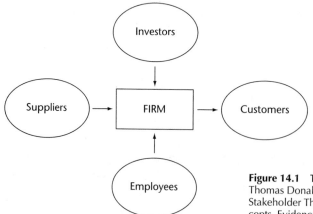

Figure 14.1   The Input–Output Model   *Source:* Thomas Donaldson and Lee E. Preston, "The Stakeholder Theory of the Corporation: Concepts, Evidence, and Implications," *Academy of Management Review,* 20 (1995), 68.

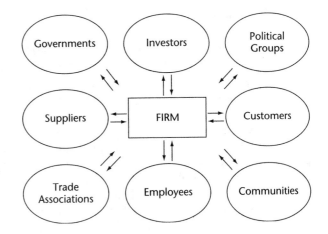

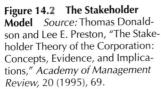

Figure 14.2   The Stakeholder Model   *Source:* Thomas Donaldson and Lee E. Preston, "The Stakeholder Theory of the Corporation: Concepts, Evidence, and Implications," *Academy of Management Review,* 20 (1995), 69.

endeavor involving large numbers of people organized in various ways. The corporation or firm is an organizational entity through which many different individuals and groups attempt to achieve their ends. A firm interacts continually with its stakeholder groups, and much of the success of a firm depends on how well all of these stakeholder relations are managed. Managing stakeholder relations, rather than managing inputs and outputs, may provide a more adequate model for understanding what people in corporations actually do as well as what they ought to do. Such a stakeholder model of the corporation is displayed in Figure 14.2.

Thomas Donaldson and Lee E. Preston have distinguished three uses of the stakeholder model: descriptive, instrumental, and normative.[56] First, the model can be used as a *description* of the corporation that can enable us to understand the corporation better. That is, a researcher who believes that the stakeholder model accurately describes corporations can use it to answer questions about how corporations are organized and managed or about how people in corporations think about their roles. The belief that the stakeholder model is an accurate description can be confirmed to the extent that these answers are put to a test and empirically verified.

Second, the stakeholder model can be used *instrumentally* as a tool for managers. Even if making a profit is the ultimate goal of corporate activity, this goal does not provide much help in the daily conduct of business. By contrast, telling managers to handle stakeholder relations well is a more practical action guide that may actually lead to greater profit. Certainly, many companies that care deeply about their employees, customers, suppliers, and other affected groups are also highly profitable.

Third, the stakeholder model can be used as a *normative* account of how corporations *ought* to treat their various stakeholder groups. The descriptive and instrumental uses of the stakeholder model suggest that corporations must deal with their stakeholders as a matter of practical necessity. Used normatively, the stakeholder model would have managers recognize the interests of employees, customers, and others as worth furthering for their own sakes. As Donaldson and Preston explain, "The interests of all stakeholders are of '*intrinsic value.*'"[57]

Criticism of Stakeholder Theory

Some writers reject the stakeholder model in its normative use on the ground that the interests of all groups other than shareholders constitute *constraints* on corporate activities rather than *goals*. Holders of the stockholder view are well aware that employees, customers, suppliers, and the general public are important to the operation of a corporation. However, satisfying these groups is necessary only as a means for achieving the end of making a profit. Igor Ansoff, for example, in his classic 1965 book *Corporate Strategy*, contends that "responsibilities" and "objectives" are not the same; the former are obligations that limit the achievement of the main objectives of a firm.[58] The stakeholder view, then, confuses the responsibilities of a corporation (which include the obligations to stakeholder groups) with its objectives (one of which is to make a profit for shareholders).

A more crucial objection to the stakeholder model is its shortcomings as an action guide for business. Even managers committed to honoring obligations to all stakeholders will find that the stakeholder model leaves many questions unanswered. Many difficult corporate decisions involve trade-offs in which a benefit to one group must be balanced against a loss to another. Thus, increasing the pension benefits of employees might result in higher prices to consumers, or a switch to environmentally safer packaging might lead to the termination of a supplier. A company that protects one community by keeping an unprofitable factory open might become uncompetitive and be forced eventually to close more factories in other communities.

Finally, the implications of stakeholder theory for corporate governance are unclear. In keeping with Berle's concern about the dangers of unrestrained managerial power, a stakeholder corporation would need to be structured so as to ensure the well-being of corporate constituencies. This is no easy task. To date, no stakeholder theorist has offered a detailed proposal for changes in corporate governance that would result in a stakeholder corporation. The systems of corporate governance in some foreign countries provide examples of some possibilities for changes. For example, workers in Germany and Japan have a larger role in strategic decision making than their counterparts in the United States. These systems reflect cultural and historical differences, however, and it is questionable whether they could be adopted in other settings.

Despite these objections, the stakeholder model remains a promising alternative to the stockholder view. The concept of a stakeholder is a valuable device for identifying and organizing the multitude of obligations that corporations have to different groups. At the present time, however, the theory is only a framework to help us get started on this very difficult task.

Case 14.2   Bath Iron Works

On May 17, 1991, a quick decision by CEO William E. Haggett almost destroyed Bath Iron Works, the largest private employer in Maine.[59] Founded in 1884, Bath Iron Works (BIW) is a major shipbuilder for the U.S. Navy with 10,400 employees. As one of two companies with the capability to build Aegis naval destroyers worth $250 million each, BIW was competing fiercely for contracts with its rival, Ingalls Shipbuilding in Mississippi. At 5:30 that morning, a janitor found a 67-page document stamped "Business Sensitive" in a conference room that had been used the previous day for a meeting with Navy officials. Two vice presidents who examined the document realized that it contained a detailed comparison of BIW's and Ingalls's costs for building the Aegis destroyer. They delivered the document to Mr. Haggett at 9:00 A.M. The CEO, who was leaving the office to deliver a luncheon speech, examined it for 15 minutes before making a decision. He ordered the two vice presidents to copy the document, return the original to the conference room, and meet with him late in the afternoon to discuss how they should handle the situation.

During the next few hours, the two executives analyzed the information and did some computer modeling based on it. At 2:15 they decided to notify the president of BIW, Duane D. "Buzz" Fitzgerald, who had a reputation for impeccable integrity. Mr. Fitzgerald immediately recognized that the federal Procurement Integrity Act requires defense contractors to certify that they have not been in unauthorized possession of any propriety information. In addition, Bath Iron Works is a signatory to the Defense Industry Initiative on Business Ethics and Conduct (DII), which was formed in 1986 in response to revelations by the Packard Commission of irregularities in defense industry contracting. The six principles of the DII require not only that signatories adopt a written code of ethics, engage in ethics training, and provide mechanisms for internal reporting of possible misconduct, but also that they take responsibility for any violation of law. Principle 4 states, "Each company has the obligation to self-govern by monitoring compliance with federal procurement laws and adopting procedures for voluntary disclosure of violations of federal procurement laws and corrective actions taken." Mr. Fitzgerald ordered that all copies be shredded and all data erased from the computer. Upon his return, Mr. Haggett agreed with the action taken and admitted that he had made an "inappropriate business-ethics decision." The CEO personally delivered the original document to Navy officials on site. However, Mr. Haggett decided not to reveal that copies had been made but to admit only that "no copies existed."

The Navy launched its own investigation and concluded that the bidding process had not been compromised. An adverse decision could have resulted in suspension or debarment as a government contractor, which would have jeopardized the survival of the firm with devastating consequences for its employees and the surrounding community. As part of the settlement with the Navy, BIW agreed to establish an ethics program headed by an ethics officer, expand ethics training, create a board committee for ensuring compliance, and report to the Navy for three years on the implementation of this agreement. BIW was still competing for contracts to build at least two new Aegis destroyers, and many at the company feared that lingering suspicion about the use of a competitor's information would be an impediment. To allay this concern, the two vice presidents who first handled the discovery of the document were asked to leave the company and William Haggett resigned as CEO. He later severed all connections with BIW, thus ending a 28-year career with a company where his father had worked as a pipe fitter. He lamented that 15 minutes of ethical uncertainty had cost him his job. Buzz Fitzgerald became the new CEO and immediately declared that BIW "must meet the highest ethical standards and avoid even the appearance of impropriety."

CORPORATE ETHICS PROGRAMS

Corporations are increasingly paying attention to ethics in the conduct of employees at all level of the organization. Unlike the emphasis on corporate social responsibility, which focuses on the impact of business activity on society at large, the corporate ethics movement addresses the need to guide individual decision making and to develop an ethical workplace environment. Much of the impetus in the United States has come from a recognition of the dangers posed by individual misconduct. However, unethical business practices are seldom due to a lone rogue employee but usually result from factors in the organization.[60] Ethics programs are designed, therefore, to create an organization that fosters ethical conduct. The case of Bath Iron Works shows that no program can prevent momentary lapses of judgment, much less intentional wrongdoing. The incident occurred in spite of an existing ethics program. The consequences of this incident might have been far worse, though, had the company not implemented the principles of the Defense Industry Initiative. Significantly, the chosen remedy was a strengthening of the program in place in order to prevent a recurrence.

This section examines corporate ethics programs. Specifically, what are the components of an ethics program? What leads corporations to adopt a program, and what do they expect to achieve? Some companies have adopted ethics programs in response to serious scandals, whereas others seek to prevent scandals before they occur. In particular, the corporate ethics movement has been spurred by the Federal Sentencing Guidelines, which offer lenient treatment for convicted organizations with an effective ethics program. These are primarily defensive strategies aimed at legal compliance. However, many corporations strive for a higher level of conduct in the belief that a reputation for integrity provides a strategic advantage. Like all other corporate initiatives, though, ethics programs represent an investment that must be justified, and so we need to take a critical look at their benefits and also at possible objections to them.

The Components of an Ethics Program

Every organization has an ethics program of some kind, although it may not be recognized as such.[61] In the broadest sense, an ethics program consists of the rules and policies of an organization and the procedures and systems for motivating and monitoring ethical performance. Rules and policies include the culture and values of an organization and formal documents, such as mission statements, codes of ethics, policy and personnel manuals, training materials, and management directives. Compliance with rules and policies is secured by various procedures and systems for orientation, training, compensation, promotion, auditing, and investigation. These procedures and systems are essential functions in any business organization. Companies with an identifiable ethics program are distinguished by the emphasis that they place on these functions and the manner in which they address them.

The components of a corporate ethics program generally include a code of ethics, ethics training for employees, means for communicating with employees about matters of ethics, a reporting mechanism for enabling employees to report alleged wrongdoing, an audit system for detecting wrongdoing, and a system for conducting investigations and taking corrective action. In addition, more than 500 U.S. corporations have established the position of ethics officer to oversee all aspects of an ethics program. Many companies without an ethics officer still assign the main responsibilities to one or more high-level executives.

This list of components does not reveal the range of corporate ethics programs. At one end of the spectrum are programs designed merely to secure compliance with the law and with the company's own rules and policies. The goals are to prevent criminal conduct and violation of government regulations on the one hand and to protect the company from self-interested action by employees on the other. Compliance of this kind is essential in any organization, but some corporations take a broader view of ethics. At the other end of the spectrum are ethics programs that communicate the values and vision of the organization, seek to build relations of trust with all stakeholder groups, and emphasize the responsibility of each employee for ethical conduct.

Lynn Sharp Paine describes this latter kind of program as an *integrity strategy* in contrast to the *compliance strategy* that is represented by the former kind.[62] Whereas a compliance approach imposes standards of conduct on employees and attempts to compel acceptable behavior, a program guided by integrity aligns the standards of employees with those of the organization and enables them to act ethically. An integrity strategy seeks to create conditions that foster right action instead of relying on deterrence and detection. These conditions are created by the whole management team rather than being relegated to lawyers or others in compliance, and by employing the whole resources of the organization. In particular, the full range of procedures and policies, the accounting and control systems, and the decision-making structures of the corporation are utilized for the end of fostering right conduct. An integrity strategy also attempts to motivate employees by appealing to their values and ideals, rather than relying solely on material incentives.

The Benefits of an Ethics Program

The main benefit of an ethics program is to prevent ethical misconduct by employees, which is costly to companies not only in direct losses but also in those sustained from a tarnished reputation. The total cost to Sears, Roebuck and Company for settling suits nationwide over allegations that its Sear Auto Center made unnecessary repairs (see Case 14.3) has been estimated to be $60 million. In addition, the trust of consumers that enabled the company to enter the competitive auto repair market was seriously damaged. The falsification of records by defense contractors, for which the companies were fined and forced to make restitution, led to development of the Defense Industry Initiative.

The financial services industry has produced some examples of very costly misconduct. A bond trading scandal at Salomon Brothers in 1991 cost the firm almost $1 billion, and in 1994 Prudential Securities agreed to pay fines and penalties in excess of $700 million for crimes committed in the sale of limited partnerships in the 1980s. A Japanese copper trader hid losses estimated at $2.6 billion from his employer, Sumitomo Corporation; and Nicholas Leeson, a 29-year-old, Singapore-based trader for Barings Bank, destroyed this venerable British firm by losing more than $1 billion in unauthorized trading. In some instances, the main loss from employee misconduct has been the company's reputation. For example, NYNEX adopted an ethics program after learning that between 1984 and 1988, its purchasing unit had hosted an annual convention in Florida for suppliers and company employees featuring strippers and prostitutes. The public exposure of these events—dubbed "pervert conventions" by the press—came at the same time that the struggling company was seeking an unpopular $1.4 billion rate increase from the New York State Public Service Commission.

Second, ethics programs provide a managerial tool for adapting the organization to rapid change. Among the factors that have led corporations to adopt ethics programs are increased competition, the development of new technologies, increased regulation, recent mergers and acquisitions, and the globalization of business. The problems at NYNEX, for example, were not confined to risqué parties. The breakup of AT&T in 1984 had forced NYNEX and all Baby Bells to compete in an unfamiliar environment that required new ways of doing business. NYNEX needed to provide individual guidance to employees during a period in which all the rules were being rewritten. Mergers and acquisitions also disrupt familiar routines and create the need to develop new ones rapidly. Finally, a formerly domestic company that becomes a global enterprise must not only set the rules for its own employees' behavior abroad but also mesh its conduct with that of foreign customers, suppliers, and joint venture partners.

A third benefit of ethics programs is managing relations with external constituencies. An ethics program serves to reassure customers, suppliers, investors, and the general public of the serious intent of a corporation to adhere to high ethical standards. It is no accident that the first ethics programs were developed by defense contractors, which have only one customer, namely the Department of Defense. The Defense Industry Initiative, which commits each signatory to develop an ethics program, was an attempt to assure this all-important customer of its trustworthiness. Problems often develop when a company and its suppliers or vendors operate by different standards, and so a company's ethics program helps make its standards known. For example, some companies notify suppliers of their policy on gift giving and ask them to respect it.

The existence of an ethics program is an assurance not only to socially responsible investors, who look for such indicators, but also to shareholders generally, who want to avoid the cost of major scandals. The shareholders of Caremark International, Inc., a health-care provider, sued the individual members of the board of directors for failing to prevent criminal violations that cost the company $260 million. In deciding this case, the Delaware State Supreme Court held in 1996 that directors have a fiduciary duty to the shareholders to ensure that the corporation's reporting systems are reasonably well designed to provide management with sufficient information to detect violations of law.[63] The *Caremark* decision has been described as a "wake-up call" to directors that they may be personally liable for their failure to ensure that a corporation has an adequate compliance system in place.[64] Of course, an ethics program is only one means for securing compliance. However, the *Caremark* decision, combined with the Federal Sentencing Guidelines, provides a strong incentive for developing one.

The Federal Sentencing Guidelines

In 1984, Congress created the United States Sentencing Commission in order to bring greater uniformity and effectiveness to the sentences that judges impose for federal crimes. In developing new guidelines, the Sentencing Commission departed from prevailing practices by holding organizations responsible for the conduct of individual decision makers and creating incentives for organizations to prevent misconduct by their members. Under the Federal Sentencing Guidelines for Organizations, which took effect in 1991, the sentence for an organization that has been convicted of a federal crime depends, in part, on the effort that has been made to prevent and detect criminal wrongdoing, including the adoption of an effective ethics program.

The Federal Sentencing Guidelines have the dual aim of imposing a just sentence on any convicted organization and influencing the conduct of all organizations. The former aim is achieved by guidelines that base the penalty on the seriousness of the offense and the culpability or blameworthiness of the organization. The guidelines provide not only for fines that punish an organization but also for restitution to the victims, and the fines can be set so as to wipe out any gain for an organization from its criminal activity. In addition, the severity of the fines, which can reach $290 million or the higher of the gain to the organization or the loss to the victims, provides a powerful incentive for organizations to take preventive steps and to cooperate in an investigation.

How the Guidelines Work. The first step in applying the guidelines is determining a base fine. This amount is generally taken from a table that ranks crimes according to their seriousness and assigns a monetary amount to each level. The base fine ranges from $5,000 for a level 6 offense or lower (embezzlement, theft, bribery of a public official) to $72.5 million for a level 38 offense or higher. Commercial bribery, for example, is a level 8 offense with a $10,000 base fine, and money laundering is a level 20 offense with a base fine of $650,000.

The base fine may then be either increased or decreased by a multiplier based on a "culpability score." This score, which ranges from 1 to 10, is determined by starting with five points and by subtracting or adding points for certain factors. Five points are added if high-level personnel were involved in the wrongdoing; three points are added if the organization obstructed the investigation or prosecution of the crime; and two points are added if similar misconduct had occurred before. Up to five points are subtracted for reporting an offense, cooperating with the investigation, and accepting full responsibility. The significant factor for the development of an ethics program is that a sentencing judge subtracts three points if the offense occurred "despite an effective program to prevent and detect violation of the law." (This provision does not apply if high-level personnel were involved or if they delayed reporting the offense after becoming aware of it.)

For each culpability score, there is a range from which a judge can choose a multiplier that either reduces or increases the base fine. The minimum multiplier is 0.05, which reduces the fine imposed to 5 percent of the base fine. The maximum multiplier is 4.00, which quadruples the base fine. Hence, the highest fine is $72.5 million (the highest base fine) multiplied by 4.00 (the highest multiplier), or $290 million. The highest fine imposed so far by using the Federal Sentencing Guidelines is $340 million levied against Daiwa Bank in New York, even though the bank was a victim of unauthorized trading by an employee that cost the bank $1.1 billion.[65] The charge against the bank was that officials had failed to inform U.S. officials within 30 days of the discovery of the loss as required by law. The bank officials engaged in the cover-up in part to avoid a decline in the bank's stock price but also because they felt that they needed time to understand what had happened and because the Japanese Ministry of Finance feared that disclosure would have an adverse impact on markets in Japan. However, if the bank had adopted an adequate compliance program, perhaps the loss would have been detected earlier with less severe consequences.

An Effective Ethics Program. The Federal Sentencing Guidelines define an effective ethics program as one "that has been reasonably designed, implemented, and enforced so that it generally will be effective in preventing and detecting criminal conduct."[66] The program need not prevent or detect every instance of wrongdoing, but the organization must have practiced "due diligence," which involves the following steps.[67]

1. The organization must have established compliance standards and procedures that are reasonably capable of reducing misconduct.
2. Specific high-level personnel must have been assigned responsibility for overseeing compliance with the standards and procedures.
3. The organization must take due care not to assign substantial discretionary authority to individuals with a propensity to engage in illegal behavior.
4. Standards and procedures must have been communicated to all employees and agents through such means as publications and training programs.

5. The organization must have taken reasonable steps to ensure compliance by using monitoring and auditing systems and a reporting system that employees may use without fear of retaliation.
6. The standards must have been consistently enforced through appropriate disciplinary measures, including, as appropriate, the punishment of employees who fail to detect an offense by others.
7. After an offense has been detected, the organization must have taken all reasonable steps to respond appropriately and to prevent further similar offenses.

The specific actions involving these steps will depend on many factors, including the size of the organization, the nature of the industry, and the organization's prior history.

Although the Federal Sentencing Guidelines provide a strong incentive for corporations to establish ethics programs and contain a good definition of an effective ethics programs, questions have been raised about the overall approach of the guidelines and specific features of the definition. First, there is no solid evidence that ethics programs are more effective than other kinds of compliance systems in preventing illegal behavior. Some evidence indicates that misconduct occurs not because of ignorance about the standards for acceptable conduct but because of organizational pressures and the actions of peers.[68] To be effective, therefore, an ethics program must go beyond setting and enforcing rules and include the goal-setting and reward systems of an organization. Second, to the extent that ethics programs are not effective, the guidelines may encourage corporations to create highly visible "window dressing" programs at the expense of more substantive initiatives. Moreover, most large corporations already have compliance programs that would satisfy the guideline's requirements, so that little is to be gained by offering a reduction in any fine. But small firms may be penalized for investing their more limited resources in a formal ethics program when other systems of control might be more effective in preventing misconduct.

Codes of Ethics

The first step in developing an ethics program, and the only step that some companies take, is a code of ethics. Codes of ethics vary widely, falling into three main types. The most common is a statement of specific rules or standards for a variety of situations. These are most often called codes of conduct or statements of business standards or practices. A second type is a statement of core values or the vision of an organization, sometimes called a credo or mission statement. These statements frequently include affirmations of the commitments of a company to key stakeholders, such as customers, employees, and the community. Third are corporate philosophies that describe the beliefs guiding a particular company. Perhaps the best-known of these is Hewlett Packard's "The HP Way". Corporate philosophy statements are generally written by the founders of businesses in emerging industries, such as computers, where new ways of doing business are needed.

Most codes of ethics combine elements of the first two types, but at least one firm, Levi Strauss & Company, has adopted all three kinds of statements. An Aspiration Statement describes what kind of company its members want it to be. A Code of Ethics explains the values and ethical principles that guide action. And finally, Levi Strauss has adopted a document entitled "Global Sourcing and Operating Guidelines," which contains very specific rules on working with business partners and choosing countries for operations. A few weeks before the guidelines were officially adopted, Levi Strauss canceled a contract with a supplier in Saipan (a U.S. territory) after reports of human rights violations. Subsequently, the U.S. Department of Labor charged that the contractor worked the employees, mostly Chinese women, up to 11 hours a day in guarded compounds and paid them well below the Saipan minimum wage. The contractor settled the charges for $9 million. One Levi Strauss manager observed, "If anyone doubted the need for guidelines, this convinced them."

In addition to company codes of ethics, there are many industry codes, generally adopted by a trade organization. These include organizations for the advertising, banking, direct marketing, franchising, insurance, and real estate industries. Because a commitment to high ethical standards and self-regulation is integral to a profession, most professional groups have also developed ethics codes to which their members are generally required to subscribe. Among professions with codes of ethics are physicians, lawyers, accountants and auditors, architects, engineers, financial planners, public administrators, consultants, and journalists. Unlike company and industry codes of ethics, which are of recent origin, some professional codes are as old as the profession, as witness the Hippocratic oath for physicians, which dates from the fourth century B.C.

The development of ethics codes for corporations is a relatively recent phenomenon, with most written since 1970.[69] In many instances, these codes replaced other kinds of documents, such as policy manuals, executive directives, and customary practices. An early prominent code of ethics was "The Penney Idea", a set of seven principles set forth by the merchandising pioneer J.C. Penney in 1913. A major impetus for the development of corporate ethics codes was provided by the influential National Commission on Fraudulent Financial Reporting (the "Treadway Report"), issued in October 1987. This report recommended:

> Public companies should develop and enforce written codes of corporate conduct. Codes of conduct should foster a strong ethical climate and open channels of communication to help protect against fraudulent financial reporting. As a part of its ongoing oversight of the effectiveness of internal controls, a company's audit committee should review annually the program that management establishes to monitor compliance with the code.[70]

Until recently, codes of ethics have primarily been adopted by American companies. A study in 1987 revealed that more than three-quarters of U.S. respondents had a code of ethics, but less than half of the responding European corporations had one.[71] The number of companies abroad with a code of ethics is increasing, however, in part because of the rise of mergers and joint ventures between American and foreign firms.

The Reasons for Adopting a Code. The reasons for adopting a code of ethics include those that lead companies to develop ethics programs. Even without a program, a code of ethics serves a number of valuable functions. A written document enables an organization to clarify standards that may otherwise be vague expectations left to individual interpretation. Where there is disagreement on the appropriate standards, codes can achieve a measure of consensus, and where standards are lacking or in need of revision, codes enable an organization to create new ones. This is especially true for American corporations with foreign operations and relationships, although a code of ethics may need to be modified when applied abroad. Codes of ethics are an effective means for disseminating standards to all employees in an easily understood form. Finally, an effective code of ethics that is enforced in an organization provides employees with a tool for resisting pressure to perform unethical or illegal actions. A code of ethics may enable employees to do what they believe to be right.

Even well-written codes of ethics have limitations, and badly written ones may have some unintended consequences. An emphasis on rules may create a rigid literalness that discourages judicious discretion. An especially dangerous situation is created when employees conclude that whatever is not prohibited is permitted. Some codes focus primarily on employee misconduct that can harm the company, which may lead to cynicism about the purpose of ethics. Some companies do not adopt a code of ethics because they believe that their way of doing business is best achieved by maintaining a strong culture and leading by example. Other companies believe that a code is inappropriate to their situation because extensive government regulation and internal auditing are sufficient to deter both unethical and illegal behavior.

Studies of which companies adopt codes of ethics reflect these advantages and disadvantages. Codes are more prevalent in large companies, in companies with more complex structures, especially those that have grown rapidly or recently merged, and in companies that have high visibility and depend on their reputation.[72] Codes of ethics are less common among financial firms—investment banks, for example—in part because of the extensive government regulation, but also because of the strong incentive to monitor employee behavior closely.

Writing a Code. There is no blueprint for writing a code of ethics. Both the procedure and content must arise from specific features of the company in question. However, some values, such as respect of the individual, fair treatment, honesty, integrity, responsibility, trust, teamwork, and quality, are included in typical codes, as are such topics as conflict of interest, use of company resources, gifts and entertainment, confidentiality of information, and workplace behavior. A few valuable guidelines for developing a code of ethics are offered by W. Michael Hoffman:

- Be clear about the objectives the code is intended to accomplish.
- Try to get support and ideas for the code from all levels of the organization.
- Be aware of the latest developments in the laws and regulations affecting your industry.

- Write as simply and clearly as possible. Be sure the code is legally defensible, but avoid legal jargon.
- Try to give reasons for the various provisions of the code.
- Devise a concrete program for communicating the code and for educating employees about the code and all programs designed to support it.
- Devise a concrete and responsible program for enforcing the code.
- Select competent persons to administer the code and give them the time and resources to get the job done.
- Make sure to provide for changing the code to meet new situations and challenges.[73]

There is one common trait of all *successful* codes of ethics, namely, that they have the clear support of top-level management. A code is unlikely to be successful, though, if it is imposed from the top down. Ideally, everyone in a company should have "ownership" of the code.

Case 14.3   Sears Auto Centers

On June 11, 1992, the CEO of Sears, Roebuck and Company, Edward A. Brennan, learned that the California Department of Consumer Affairs (DCA) was seeking to shut down the 72 Sears Auto Centers in that state.[74] A year-long undercover investigation by the DCA had found numerous instances in which Sears employees had performed unnecessary repairs and services. Officials in New Jersey quickly announced similar charges against six local Sears Auto Centers, and several other states, including Florida, Illinois, and New York, opened their own probes into possible consumer fraud. In the wake of this adverse publicity, revenues from the auto centers fell 15 percent, and the public's trust in Sears was badly shaken.

Sears Auto Centers, which were generally connected with a Sears department store, concentrated on basic "undercar" services involving tires, brakes, mufflers, shock absorbers, and steering mechanisms. Investigators from the DCA's Bureau of Automotive Repair purchased old vehicles in need of minor repairs and disassembled the brakes and suspension systems. After examining and photographing each part, the investigators towed the automobiles to a shop where they requested a brake inspection. In 34 of 38 instances, Sears employees recommended unnecessary repairs and services, and some auto centers charged for parts that were not installed or work that was not performed. The average overcharge was $235, but in two cases the amount overcharged exceeded $500.

Brennan had been notified in December 1991 of early results from the investigation, and Sears executives negotiated for six months with California officials. The company objected to the state's position that no part should be replaced unless it had failed and claimed that many repairs were legitimate preventive maintenance. For example, there is disagreement in the industry on whether brake calipers should be reconditioned whenever the pads are replaced. In addition, some of the automobiles used in the investigation showed sign of damage from worn parts that had already been replaced, thus leading mechanics to believe that repairs were needed. The DCA moved to revoke the licenses of all Sears Auto Centers in the state after the negotiations broke down over details of the financial settlement.

California officials charged that the problems at the Sears Auto Centers were not confined to a few isolated events but constituted systemic consumer fraud. According to a deputy attorney general, "There was a deliberate decision by Sears management to set up a structure that made it totally inevitable that the consumer would be oversold." Until 1991, service advisers, who make recommendations to customers, were paid a flat salary, but subsequently their compensation included a commission incentive. The service advisers were also required to meet quotas for a certain number of parts and services in a fixed period of time. The new incentive system also affected the mechanics, who perform the work on the customers' automobiles. Instead of an hourly wage, they were now compensated by a lower base hourly figure plus a fixed dollar amount based on the time required to install a part or perform a service. Under this system, a mechanic would receive the former hourly wage only by doing an hour's worth of work, but a mechanic could also earn more by working faster.

Commissions and quotas are commonly used in competitive sales environments to motivate and monitor employees. However, critics of Sears charge that there were not enough safeguards to protect the public. One former auto center manager in Sacramento complained that quotas were not based on realistic activity and were constantly escalating. He said that "sales goals had turned into conditions of employment" and that managers were "so busy with charts and graphs" that they could not properly supervise employees. A mechanic in San Bruno, California, alleged that he was fired for not doing 16 oil changes a day and that his manager urged him to save his job by filling the oil in each car only halfway. This illustrated, he said, the "pressure, pressure, pressure to get the dollar."

The changes in the compensation system at Sears Auto Centers were part of a company wide effort to boost lagging performance. In 1990, net income for all divisions, including Allstate (insurance), Coldwell Banker (real estate), and Dean Witter (brokerage), dropped 40 percent. Net income for the merchandising group, which included the department stores and the auto centers, fell 60 percent. Brennan, CEO since 1985, was under strong pressure to cuts costs and increase revenues. Some dissident shareholders were urging the board of directors to spin off the more profitable insurance, real estate, and brokerage divisions and focus on the ailing merchandising group. Brennan's response was to cut jobs, renovate stores, and motivate people. The overall thrust, according to a story in *Business Week*, was to "make every employee, from the sales floor to the chairman's suite focus on profits." Some critics of Sears attribute the problems at the auto centers to an unrealistic strategic plan that sought to wring more revenue out of the auto repair business than was possible. Robert Monk, who unsuccessfully sought a seat on the company's board, said, "Absent a coherent growth strategy, these sorts of things can happen."

At a press conference on June 22, 1992, Edward Brennan announced that, effective immediately, Sears would eliminate its incentive compensation system for automotive service advisers and all product-specific sales goals. Although he admitted that the company's compensation program "created an environment where mistakes did occur," Brennan continued: "We deny allegations of fraud and systemic problems in our auto centers. Isolated errors? Yes. But a pattern of misconduct? Absolutely not." He reaffirmed his belief that the California investigation was flawed and that Sears was practicing responsible preventive maintenance. He further announced that the company would retain an independent organization to conduct random "shopping audits" to ensure that no overcharging would occur. Sears also paid $8 million to settle claims in California and gave auto center customers $50 coupons that were expected to cost the company another $3 million. The total cost, including legal bills and lost sales, is estimated to be $60 million.

On September 30, 1992, Sears revealed plans to spin off its three nonretail divisions, Allstate, Coldwell Banker, and Dean Witter, and to reorganize the merchandising group. A new CEO, Arthur C. Martinez, succeeded Brennan and began a turnaround of the company. In describing his vision, Martinez said, "I want to revisit and intensify the theme of our customer being the center of our universe." A cornerstone of Martinez's strategy, according to the *New York Times*, was "clean business ethics."

Case 14.4   Campbell Soup Company

In 1985, the Campbell Soup Company faced a challenge from an Ohio-based group that was fighting for increased wages and improved living and working conditions for migrant farmworkers.[75] The group, known as the Farm Labor Organizing Committee (FLOC), was threatening to disrupt the annual shareholders' meeting with a demonstration in support of its cause. Since 1979, the company had been the target of a nationwide boycott of its products instigated by FLOC. Under the dynamic leadership of Baldemar Velasquez, the group gained support for the boycott from the major Protestant and Catholic church organizations in Ohio and from many influential national groups. In July 1983, about one hundred members and supporters of FLOC marched 560 miles from FLOC headquarters in Toledo, Ohio, to the headquarters of the Campbell Soup Company in Camden, New Jersey, to dramatize the boycott and to press their demand for a labor contract.

At the urging of the National Council of Churches, which had been persuaded by FLOC to intervene in the dispute, representatives from the two sides sat down and negotiated an agreement that set up procedures for conducting elections to determine whether the farmworkers wanted to be represented by FLOC. The agreement was signed in May 1985, and the voting took place in September, during the first month of the tomato harvest. The election process broke down, however, amid allegations by FLOC of unfair labor practices, including charges that the growers brought in local laborers on the day of the election and prevented some migrant farmworkers from voting. The possibility now existed that the National Council of Churches would support the boycott, a move that would generate considerable adverse publicity for Campbell Soup. FLOC's plan to stage a demonstration at the upcoming shareholders' meeting would create additional bad press.

The plight of the migrant farmworkers is truly deplorable. Many live in crowded camps, without electricity, fresh drinking water, or adequate toilets. Poor nutrition, communicable diseases, and constant exposure to pesticides result in a high infant mortality rate and a life expectancy 25 years below the average. FLOC claimed that laborers received an average hourly wage of less than $2.00, although Campbell Soup disputed this figure and insisted they were paid at least the minimum wage of $3.35 an hour. In addition, migrant farmworkers generally lack health insurance and the other benefits that most American workers take for granted.

When the National Labor Relations Act was passed in 1935, giving most workers the right to organize and engage in collective bargaining, farm laborers were deliberately excluded. So the company was under no legal obligation to recognize the existence of FLOC. The migrant farmworkers, moreover, were not Campbell employees but were hired by independent growers, who held contracts with the company to devote a specified number of acres to tomatoes that were sold to the company at a fixed rate. Executives of Campbell Soup did not believe that they had a right to negotiate over the heads of the growers or to dictate to the growers how they should treat their employees.

FLOC decided to launch a campaign against Campbell Soup in the belief that the price the company set for tomatoes did not allow enough profit margin for the growers to increase wages and improve working conditions even if they wanted to. FLOC disregarded the claim that Campbell Soup did not employ the migrant farmworkers, because the company provided seedlings to the growers and dictated how they were to be raised and picked. In response to a strike organized by FLOC in 1978, the company ordered the growers to switch to mechanical harvesters. Thus, the control that Campbell Soup exercised over the growers was greater than that normally exercised by a company over suppliers. The high visibility of the company and its carefully cultivated reputation also made it vulnerable to attack and contributed to FLOC's decision to make Campbell Soup the target of its campaign.

The charge of allowing child labor was one that deeply stung company officials. The basis of the accusation was a 1983 ruling by a federal district court that a pickle grower, who employed workers on a piece-rate basis, had no legal obligation to prevent children from working alongside their parents in the fields, because the workers, in the view of the court, are independent, self-employed contractors.[76] Although Campbell Soup also markets pickles through its Vlasic subsidiary, the company contended that the decision did not apply to tomato pickers, who are paid hourly rates because of the different method of harvesting tomatoes. The company denied, moreover, that it approved of children helping their parents in the tomato fields. In fact, Campbell Soup created the post of Ombudsman for Migrant Affairs in part to promote the availability of day-care facilities and to ensure that children of migrant farmworkers attend classes during the school year.

Campbell Soup took other steps to improve the living conditions of the workers who picked tomatoes. After the passage of an Ohio law upgrading the minimum standards for migrant housing, the duties of the ombudsman were expanded to include inspection of facilities on farms with company contracts. The company also offered to finance half the cost of new housing and to provide low-interest loans for the balance. In addition, a pilot health-care project was started.

R. Gordon McGovern, president and CEO of Campbell Soup Company, believed that the company had done all it could reasonably be expected to do. At his urging, Campbell executives refused to budge on FLOC's main demand for union recognition and a labor contract. As the day of the annual meeting approached, however, Mr. McGovern realized that he would soon have to face not only the company's shareholders inside the meeting hall but also the protesters from FLOC on the outside.

NOTES

1. Ellen Joan Pollock, "CEO Takes on a Nun in a Crusade against 'Political Correctness'," *Wall Street Journal*, 15 July 1991, A1.
2. Letter dated May 23, 1996, distributed to all Cypress Semiconductor shareholders.
3. For a comprehensive account of the historical development of the concept of corporate social responsibility, see Morrell Heald, *The Social Responsibilities of Business: Company and Community, 1900–1960* (Cleveland: Case Western Reserve University Press, 1970). Brief accounts are given in Clarence C. Walton, *Corporate Social Responsibilities* (Belmont, CA: Wadsworth, 1967), 21–53; and James W. McKie, "Changing Views," in *Social Responsibility and the Business Predicament*, ed. James McKie (Washington, DC: The Brookings Institution, 1974), 17–40.
4. Milton Friedman, *Capitalism and Freedom* (Chicago: University of Chicago Press, 1962), 133.
5. Ibid., 133. The quotation by Adam Smith is from *The Wealth of Nations*, Book IV, chap. II. This famous paragraph concludes: "It is an affectation, indeed, not very common among merchants, and very few words need be employed in dissuading them from it."
6. Joseph W. McGuire, *Business and Society* (New York: McGraw-Hill, 1963), 144. For the point that the assumption of responsibility must be voluntary, see Henry Manne and Henry C. Wallich, *The Modern Corporation and Social Responsibility* (Washington, DC: American Enterprise Institute for Public Policy Research, 1972), 5.
7. Archie B. Carroll, "A Three-Dimensional Conceptual Model of Corporate Performance," *Academy of Management Review*, 4 (1979), 497–505.
8. Ibid., 500.
9. S. Prakash Sethi, "Dimensions of Corporate Social Performance: An Analytical Framework for Measurement and Analysis," *California Management Review*, 17 (Spring 1975), 62. Emphasis in original omitted.
10. *Social Responsibilities of Business Corporations* (New York: Committee for Economic Development, 1971), 15.
11. The concept of corporate social responsiveness is developed in Robert W. Ackerman and Raymond A. Bauer, *Corporate Social Responsiveness: The Modern Dilemma* (Reston, VA: Reston, 1976).
12. William C. Frederick, "From CSR1 to CSR2: The Maturing of Business-and-Society Thought," *Business & Society Review*, 33 (1994), 150-64.
13. Donna J. Wood, "Corporate Social Performance Revisited," *Academy of Management Review*, 16 (1991), 693.
14. McKie, "Changing Views," 18–19.

15. Edward S. Mason, ed., *The Corporation in Modern Society* (New York: Atheneum, 1974), 5.

16. Theodore Levitt, "The Dangers of Social Responsibility," *Harvard Business Review*, 36 (September–October 1958), 49.

17. John G. Simon, Charles W. Powers, and Jon P. Gunnemann, "The Responsibilities of Corporations and Their Owners," in *The Ethical Investor: Universities and Corporate Responsibility* (New Haven, CT: Yale University Press, 1972), 16–17.

18. Keith Davis, "Five Propositions for Social Responsibility," *Business Horizons*, 18 (June 1975), 20. Italics in the original.

19. Keith Davis and Robert L. Blomstrom, *Business and Society: Environment and Responsibility*, 3d ed. (New York: McGraw-Hill, 1975), 50.

20. This objection is formulated in Kenneth E. Goodpaster and John B. Matthews, Jr., "Can a Corporation Have a Conscience?" *Harvard Business Review*, 60 (January–February 1982), 139–40.

21. Ibid., 140.

22. Simon, Powers, and Gunnemann, "Responsibilities of Corporations and Their Owners."

23. Milton Friedman, "The Social Responsibility of Business Is to Increase Its Profits," *New York Times Magazine*, 13 September 1970, 33.

24. Ibid., 122.

25. See, for example, Richard Musgrave, *The Theory of Public Finance* (New York: McGraw-Hill, 1959), in which the three functions of securing efficiency (which includes considerations of externalities), equity, and stability properly belong to government.

26. Holders of the classical view generally favor market solutions over government action in the belief that many externalities result from a lack, rather than an excess, of free market forces and that regulation is often ineffective. Still, they usually admit the principle that government regulation is appropriate in some instances to deal with externalities.

27. For an explanation and discussion of this concept, see Arthur M. Okun, *Equality and Efficiency: The Big Tradeoff* (Washington, DC: The Brookings Institution, 1975).

28. Marvin A. Chirelstein, "Corporate Law Reform," in *Social Responsibility and the Business Predicament*, 55.

29. The distinction between the property rights and the social institution conceptions of the corporation is due to William T. Allen, "Our Schizophrenic Conception of the Business Corporation," *Cardozo Law Review*, 14 (1992), 261–81. See also, William T. Allen, "Contracts and Communities in Corporate Law," *Washington and Lee Law Review*, 50 (1993), 1395–1407.

30. Because the right to incorporate is alleged to "inhere" in the right to own property and to contract with others, this view is also known as the *inherence theory*.

31. The view that incorporation is a privilege "conceded" by the state in order to achieve some social good is also known as the *concession theory*.

32. *Munn v. Illinois*, 94 U.S. 113, 24 L. Ed. 77 (1876).

33. *Dodge v. Ford Motor Co.*, 170 N.W. 668, 685 (1919).

34. Adolf A. Berle, Jr., and Gardiner C. Means, *The Modern Corporation and Private Property* (New York: Macmillan, 1932).

35. Ibid., 3.

36. Ibid., 355.

37. E. Merrick Dodd, "For Whom Are Corporate Managers Trustees?" *Harvard Law Review*, 45 (1932), 1148.

38. Ibid., 1161.

39. Ibid., 1162.

40. Adolf A. Berle, Jr., "For Whom Corporate Managers Are Trustees: A Note," *Harvard Law Review*, 45 (1932), 1367.

41. Ibid., 1372.

42. Ibid., 1370.

43. Ibid., 1372.

44. Adolf A. Berle, Jr., *The 20th Century Capitalist Revolution* (New York: Harcourt, Brace & World, 1954), 169.

45. Ronald M. Coase, "The Nature of the Firm," *Economica*, N.S., 4 (1937), 386–405. The contractual theory has been developed by economists using an agency or transaction cost perspective. See Armen A. Alchian and Harold Demsetz, "Production, Information Costs, and Economic Organization," *American Economic Review*, 62 (1972), 777–95; Benjamin Klein, Robert A. Crawford, and Armen A. Alchian, "Vertical Integration, Appropriable Rents, and the Competitive Contracting Process," *Journal of Law and Economics*, 21 (1978), 297–326; Michael C. Jensen and William H. Meckling, "Theory of the Firm: Managerial Behavior, Agency Costs, and Ownership Structure," *Journal of Financial Economics*, 3 (1983), 305–60; Eugene F. Fama and Michael C. Jensen, "Separation of Ownership and Control," *Journal of Law and Economics*, 26 (1983), 301–25; Steven N. S. Cheung, "The Contractual Theory of the Firm," *Journal of Law and Economics*, 26 (1983), 1–22; and Oliver E. Williamson, *The Economic Institutions of Capitalism* (New York: Free Press, 1985). An authoritative development of the theory of the firm in corporate law is Frank H. Easterbrook and Daniel R. Fischel, *The Economic Structure of Corporate Law* (Cambridge, MA: Harvard University Press, 1991). See also William A. Klein, "The Modern Business Organization: Bargaining under Constraints," *Yale Law Journal*, 91 (1982), 1521–64; Oliver Hart, "An Economist's Perspective on the Theory of the Firm," *Columbia Law Review*, 89 (1989), 1757–73; and Henry N. Butler, "The Contractual Theory of the Firm," *George Mason Law Review*, 11 (1989), 99–123.

46. Easterbrook and Fischel, *The Economic Structure of Corporate Law*, 38.

47. Ibid., 38.

48. Ibid., 39.

49. This reasoning follows from the Coase Theorem, which is presented in a highly influential paper by Ronald Coase, "The Problem of Social Cost," *Journal of Law and Economics*, 3 (1960), 1–44.

50. Worker Adjustment and Retraining Notification Act of 1988 (WARN).

51. See Eric W. Orts, "Beyond Shareholders: Interpreting Corporate Constituency Statutes," *George Washington Law Review*, 61 (1992), 14–135.

52. Margaret M. Blair, *Ownership and Control: Rethinking Corporate Governance for the Twenty-first Century* (Washington, DC: The Brookings Institution, 1995).

53. Ibid., 257.

54. William M. Evan and R. Edward Freeman, "A Stakeholder Theory of the Modern Corporation: Kantian Capitalism," in *Ethical Theory and Business*, 4th ed., ed. Tom L. Beauchamp and Norman E. Bowie (Upper Saddle River, NJ: Prentice Hall, 1993), 79. See also R. Edward Freeman and D. Reed, "Stockholders and Stakeholders: A New Perspective on Corporate Governance," in *Corporate Governance: A Definitive Exploration of the Issues*, C. Huizinga, ed., (Los Angeles: UCLA Extension Press, 1983).

55. R. Edward Freeman, *Strategic Management: A Stakeholder Approach* (Boston: Pitman, 1984), 46. This book provides a useful discussion of the history of the stakeholder concept and the literature on it.

56. Thomas Donaldson and Lee E. Preston, "The Stakeholder Theory of the Corporation: Concepts, Evidence, and Implications," *Academy of Management Review*, 20 (1995), 65–91.

57. Ibid., 67.

58. Igor Ansoff, *Corporate Strategy* (New York: McGraw-Hill, 1965), 38.

59. Material for this case is taken from Rushworth M. Kidder, *How Good People Make Tough Choices* (New York: Fireside, 1995), 35–38; Glenn Adams, "Bath Iron Works CEO Admits Ethics

Breach, Steps Down," The Associated Press, 16 September 1991; Jerry Harkavy, "Haggett Severs Ties with Shipyard in Aftermath of Photocopy Scandal," The Associated Press, 25 September 1991; Suzanne Alexander, "Bath Iron Works Says Haggett Quit as Chief," *Wall Street Journal*, 17 September 1991, B9; and Joseph Pereira and Andy Pasztor, "Bath Chairman, 2 Vice Presidents Quit under Navy Pressure over Secret Data," *Wall Street Journal*, 26 September 1991, A4.

60. Lynn Sharp Paine, "Managing for Organizational Integrity," *Harvard Business Review*, 72 (March–April 1994), 106.

61. Steven N. Brenner, "Ethics Programs and Their Dimensions," *Journal of Business Ethics*, 11 (1992), 391.

62. Paine, "Managing for Organizational Integrity."

63. *In re Caremark International Inc.* Derivative Litigation. Civil Action No. 13670 (Del. Ch. 1996).

64. Dominic Bencivenga, "Words of Warning: Ruling Makes Directors Accountable for Compliance," *New York Law Journal*, 13 February 1997, 5.

65. For an explanation, see Jeffrey M. Kaplan, "Why Daiwa Bank Will Pay $340 Million under the Sentencing Guidelines," *Ethikos*, 9 (May–June 1996).

66. United States Sentencing Commission, *Federal Sentencing Guidelines Manual*, §8A1.2, Commentary, Application Note 3(k).

67. Ibid., Application Note 3(k)(1-7).

68. O. C. Farrell, Debbie Thorne LeClair, and Linda Ferrell, "The Federal Sentencing Guidelines for Organizations: A Framework for Ethical Compliance," *Journal of Business Ethics*, 17 (1998), 353-63.

69. *Corporate Ethics*, The Conference Board, Research Report No. 900 (1987), 14.

70. National Commission on Fraudulent Financial Reporting, *Report of National Commission on Fraudulent Financial Reporting*, 1987, p. 35.

71. *Corporate Ethics*, 13.

72. Messod D. Benish and Robert Chatov, "Corporate Codes of Conduct: Economic Determinants and Legal Implications for Independent Auditors," *Journal of Accounting and Public Policy*, 12 (1993), 3-35.

73. W. Michael Hoffman, "A Blueprint for Corporate Ethical Development," in *Business Ethics: Readings and Cases in Corporate Morality*, 3d ed., ed., W. Michael Hoffman and Robert E. Frederick (New York: McGraw-Hill, 1995), 580–81.

74. Material for this case is taken from "Sears Auto Centers (A), Harvard Business School, 1993; Kevin Kelly, "How Did Sears Blow This Gasket?" *Business Week*, 29 June 1992, 38; Julia Flynn, "Did Sears Take Other Customers for a Ride?" *Business Week*, 3 August 1992, 24-25; Judy Quinn, "Repair Job," *Incentive*, October 1992, 40-46; Jennifer Steinhauer, "Time to Call a Sears Repairman," *New York Times*, 15 January 1998, sec. 3, pp. 1, 3; News from Sears, Roebuck and Co. [press release], 22 June 1992.

75. Information on this case is taken from "Campbell Soup Company," in *Cases in Ethics and the Conduct of Business*, ed., John R. Boatright (Upper Saddle River, NJ: Prentice Hall, 1995), 314–30.

76. *Brandel v. United States*, U.S. District Court, Western District of Michigan, Southern Division, 83–1228 (1984).

15

International Business Ethics

Nike is a leader in the sports shoe industry, with sales of $9.5 billion in 1998 and a 40 percent share of the American sneaker market. This Oregon-based company has also become a lightening rod for worldwide protests over alleged "sweatshop" conditions in factories across Southeast Asia. In a May 1998 speech, Phil Knight, the founder and CEO, admitted that "the Nike product has become synonymous with slave wages, forced overtime, and arbitrary abuse."[1] How did a prominent company, whose "swoosh" logo is a symbol for the "Just Do It" spirit, come to be associated with deplorable labor practices?

Nike's phenomenal success is due to a visionary strategy which was developed by Phil Knight during his student days at the Stanford Business School. The strategy involves outsourcing all manufacturing to contractors in low-wage countries and pouring the company's resources into high-profile marketing. One Nike vice president observed, "We don't know the first thing about manufacturing. We are marketers and designers."[2] Central to Nike's marketing effort is placing the Nike "swoosh" on the uniforms of collegiate and professional athletes and enlisting such superstars as Michael Jordan and Tiger Woods.

When Nike was founded in 1964, the company contracted with manufacturers in Japan, but as wages in that country rose, Nike transferred production to contractors in Korea and Taiwan. By 1982, more than 80 percent of Nike shoes were made in these two countries, but rising wages there led Nike to urge its contractors to move their plants to Southeast Asia. By 1990, most Nike production was based in Indonesia, Vietnam, and China.

In the early 1990s, young Indonesian women working in Korean-owned plants under contract with Nike started at 15 cents an hour. With mandatory over-

411

time, which was often imposed, more experienced workers might make $2 for a grueling 11-hour day. The Indonesian minimum wage was raised in 1991 from $1.06 for a 7-hour day to $1.24, only slightly above the $1.22 that the government calculated as necessary for "minimum physical needs." The women lived in fear of their brutal Korean managers, who berated them for failing to meet quotas and withheld pay to enforce discipline. Indonesian labor laws, lax to begin with, were flouted with impunity by contractors, since the government was eager to attract foreign investment. Workers often toiled in crowded, poorly ventilated factories, surrounded by machinery and toxic chemicals. There was little effective union activity in Indonesia, and labor strikes were firmly suppressed by the army.

Nike's initial response to growing criticism was to deny any responsibility for the practices of its contractors. These are independent companies from which Nike merely buys shoes. The workers are not Nike employees, and their wages are above the legal minimum and the prevailing market rate. When asked about labor strife in some factories supplying Nike, John Woodman, the company's general manager for Indonesia, said that he did not know the causes and added, "I don't know that I need to know."[3] Mr. Woodman defended Nike by arguing, "Yes, they are low wages. But we've come in here and given jobs to thousands of people who wouldn't be working otherwise."[4] He might have added that the company had also given additional employment to Michael Jordan, whose reported $2 million dollar fee in 1992 was larger than the payroll that year for all Nike production in Indonesia.[5]

INTRODUCTION

Increasingly, business is being conducted across national boundaries. Large Multinational Corporations (MNCs) that have long operated in other countries are being joined by smaller domestic firms going abroad for the first time. Intense competition is forcing companies worldwide to enter the global marketplace, whether they are ready or not.

This development presents a host of ethical problems that managers are often unprepared to address. Some of these problems arise from the diversity of business standards around the world and especially from the lower standards that generally prevail in Less Developed Countries (LDCs). Companies such as Nike (Case 15.1) are able to pay wages and impose working conditions that are shockingly low by U.S. standards, and yet they usually operate well above the standards of local firms. Environmental standards in LDCs are also invariably lower than those of more developed countries. And in countries with pervasive corruption it may be difficult to conduct business without paying bribes.

Additional problems result from the power of multinational corporations to affect the development of LDCs. MNCs often exploit the cheap labor and the natural resources of LDCs without making investments that would advance economic development. These problems are exacerbated when companies sucessfully avoid paying their fair share of taxes. Even though less developed countries invariably benefit to some extent from the activities of MNCs, the distribution of the gains is usually unequal. Critics ask whether it is fair for corporations from developed countries to return so little to the less developed parts of the world from which they derive so much.

Operations in foreign countries also raise questions about the proper role of corporations in political affairs. Most multinationals consider themselves to be guests in host countries and refrain from influencing local governments. However, Shell Oil Company (Case 15.3) was widely criticized for failing to intervene in a case of human rights abuse by the Nigerian government. MNCs have an opportunity to play a constructive role in countries making the transition from a socialist, planned economy to a free market. The high level of corruption in some of these countries, though, presents special challenges.

This chapter begins with the problem of determining the appropriate standards for operations in less developed countries. Applying home country standards in all parts of the world is usually not morally required, but adopting host country standards for wages, working conditions, and other matters may be morally impermissible. What principles can guide us in finding a happy medium? These principles are then applied to the issue of wages and working conditions that are raised in the Nike case. Although bribery is universally recognized as wrong, it, too, is a practice that is viewed differently around the world. Accordingly, a section is devoted to understanding this critical problem in international business. Finally, this chapter examines the difference in the ethical outlook of business people in different parts of the world, especially in Asia, and concludes with developing of codes of ethics that attempt to guide responsible business conduct worldwide.

WHAT TO DO IN ROME

The main charge against multinational corporations is that they adopt a double standard, doing in less developed, Third World countries what would be regarded as wrong if done in the developed First World. However, many criticized practices are legal in the countries in question and are not considered to be unethical by local standards. Should MNCs be bound by the prevailing morality of the home country and, in the case of American corporations, act everywhere as they do in the United States? Should they follow the practices of the host country and adopt the adage "When in Rome, do as the Romans do"? Or are there special ethical standards that apply when business is conducted across national boundaries? If so, what are the appropriate standards for international business?

Unfortunately, there are no easy answers to these questions. In some cases, the standards contained in American law and morality ought to be observed beyond our borders; in other cases, there is no moral obligation to do so. Similarly, it is morally permissible for managers of MNCs to follow local practice and "do as the Romans do" in some situations but not others. Even if there are special ethical standards for international business, these cannot be applied without taking into account differences in cultures and value systems, the levels of economic development, and the social, political, and legal structures of the foreign countries in which MNCs operate.

Absolutism versus Relativism

In answer to the question "When in Rome, do what?" there are two extremes. The absolutist position is that business ought to be conducted in the same way the world over with no double standards. In particular, U.S. corporations ought to observe a single code of conduct in their dealings everywhere. This view might be expressed as "When in Rome or anywhere else, do as you would at home."[6] The opposite extreme is relativism, which may be expressed in the familiar adage, "When in Rome do as the Romans do." That is, the only guide for business conduct abroad is what is legally and morally accepted in any given country where a company operates.

Neither of these positions can be adopted without exception. The generally high level of conduct that follows from "When in Rome, do as you would at home" is not morally required of MNCs in all instances, and they should not be faulted for every departure from home country standards in doing business abroad. However, "When in Rome, do as the Romans do" is not wholly justified either. The mere fact that a country permits bribery, unsafe working conditions, exploitive wages, and violations of human rights does not mean that these practices are morally acceptable, even in that country. The debate over absolutism and relativism revolve around four important points.

Morally Relevant Differences. First, some conditions in other countries, especially those in less developed parts of the world, are different in morally relevant ways. As a result, different standards may be morally permitted, indeed required. If Rome is a significantly different place, then standards that are appropriate at home do not necessarily apply there.

For example, pharmaceutical companies have been criticized for adopting a double standard in promoting drugs in less developed countries with more indications for their use and fewer warnings about side effects. Although such practices may be designed solely to promote sales, some drugs may be medically appropriate in an LDC for a wider range of conditions. With regard to one powerful but dangerous antibiotic, which is prescribed in the United States only for very serious infections, doctors in Bolivia claim that this limited use is a luxury that Americans can afford because of generally better health. "Here," they say, "the people's general health is so poor that one must make an all out attack on illness."[7] Thus, an antibiotic that should be marketed in the United States with one set of indications might be justifiably sold abroad with a more extensive list.

More generally, the relative level of economic development must be taken into account in determining the appropriate standards for different countries. For example, health and safety standards in the developed world are very stringent, reflecting greater affluence and a greater willingness to pay for more safety. The standards of these countries are not always appropriate in poorer, less developed countries with fewer resources and more pressing needs. The United States made different trade-offs between safety and other values at earlier stages of the country's economic development. At the present time, less developed countries might prefer lower safety standards in return for more jobs. On the other hand, exposing people in LDCs to unreasonable dangers may be a violation of basic human rights.

The Variety of Ethical Outlooks. Second, the absolutist position assumes that one country's standards are correct and that they should be imposed on people elsewhere, perhaps in conflict with their own moral values and principles. Acting on these assumptions ignores the wide variety of ethical outlooks in the world. Although some bedrock conceptions of right and wrong exist among people everywhere, many variations occur due to cultural, historical, political, and economic factors.

Cultural differences are important because they may affect the meaning of acts performed. For example, lavish gifts that would be considered bribes or kickbacks in the United States are an accepted and expected part of business in Japan and some other Asian countries. This difference in perception is due, in part, to the role that gift giving plays in building relationships, which are more critical in Asian business. Thus, giving gifts in Japan and China is usually viewed not as an attempt to improperly influence a person's judgment but rather as a means to cement a legitimate relationship.[8] Similarly, whistle-blowing, which is generally viewed favorably in the United States as a moral protest, is regarded unfavorably in Japan as an act of disloyalty. In both cases, culture plays a powerful role in the interpretation of what a person has done in giving a gift or blowing the whistle.

The impact of historical, political, and economic differences can be seen in Russian views of business ethics. The collapse of communism and the chaotic development of free markets in the former Soviet Union have created great uncertainty about ethical business behavior.[9] Although Russians and Americans agree on many matters, such as the importance of keeping one's word, paying debts, competing fairly, and avoiding extortion, they still differ in their ethical assessment of certain other matters. For example, less stigma is attached in Russia to making payments for favors (*blat*), falsifying information, and coordinating prices because of the prevalence of these practices in the previous planned economy. The lack of a workable legal system forces Russian managers to ignore senseless and contradictory regulations. Unfortunately, a certain amount of lawlessness is necessary for operating in the current business environment. On the other hand, Russia's socialist tradition leads them to criticize America's tolerance for exorbitant pay differentials and massive layoffs. As in Japan, whistle-blowing is viewed with suspicion, but for a different reason: It reminds Russians of the informer ethos that existed during the communist era.

The Right of People to Decide. Third, the absolutist position denies the right of the people who are affected to decide on important matters of business conduct. The primary responsibility for setting standards should rest on the government and the people of the country in which business is being conducted. The argument that the people affected have a right to decide is not a form of ethical relativism. Just because people approve of a certain practice does not make it right. The argument is rather an expression of respect for the right of people to govern their own affairs, rightly or wrongly. Imposing the standards of a developed, First World country in the Third World is criticized by some as a form of "ethical imperialism."

Avoiding "ethical imperialism" and allowing the people affected to decide must be approached cautiously. A respect for the right of people to set their own standards does not automatically justify corporations in inflicting grave harm on innocent people, for example, or violating basic human rights. Furthermore, it may be difficult to determine what people have decided. Some countries lack the capacity to regulate effectively the activities of multinational corporations within their own borders. The governments of developing countries are, in many instances, no less committed than those in the United States and Western Europe to protecting their people against harm, but they do not always have the resources—the money, skilled personnel, and institutions—to accomplish the task.

Some countries with the capacity to regulate multinationals lack the necessary will. MNCs, through the exercise of economic power, including bribery, are able to influence regulatory measures. The governments of less developed countries are also careful not to offend the developed countries on which they depend for aid. Furthermore, the absence of laws against unethical business practices is sometimes part of a pattern of oppression that exists within the country itself, so that MNCs are taking advantage of the immorality of others when they follow the law of countries with corrupt governments.

Consequently, we need to ask whether a standard in a host country, if it is lower than that at home, truly represents the considered judgment of the people in question. Does it reflect the decision that they would make if they had the capacity to protect their own interests effectively? A genuine respect for the right of people to determine which standards to apply in their own country requires a careful and sympathetic consideration of what people would do under certain hypothetical conditions rather than what is actually expressed in the law, conventional morality, and commonly accepted practices.

Required Conditions for Doing Business. Fourth, some practices may be justified where local conditions require that corporations engage in them as a condition of doing business. This point may be expressed by saying, "We don't agree with the Romans, but find it necessary to do things their way." American firms with contracts for projects in the Middle East, for example, have complied in many instances with requests not to station women and Jewish employees in those countries. Although discrimination of this kind is morally repugnant, it is (arguably) morally permissible when the alternative is to risk losing business in the Arab world. A more complicated case is posed by the Arab boycott of Israel, which was begun by the countries of the Arab League in 1945. In order to avoid blacklisting that would bar them from doing business with participating Arab countries, many prominent U.S. companies cooperated by avoiding investment in Israel. Other firms, however, refused to cooperate with the boycott for ethical reasons. (Congress addressed this issue in 1977 by amending the Export Administration Act to prohibit American corporations from cooperating with the Arab boycott against Israel.)

As with the other arguments, "There is no other way of doing business" cannot be accepted without some qualifications. The alternative is seldom to cease doing business; rather, the claim that a practice is "necessary" often means merely that it is the most profitable way of doing business. For example, the Arab embargo against Israel greatly complicated the problem of doing business in the Middle East, but some companies were able to avoid cooperating with the boycott and still have business relations in Arab countries. Similarly, during the period of apartheid in South Africa, some American companies defied the government and integrated their work forces. There are some situations, however, in which a company is morally obligated to withdraw if there is no other way to do business. Some companies have refused to do business in certain countries because they believe that involvement in an immoral system cannot be justified. For example, Robert Haas, the CEO of Levi Strauss, made the decision in 1993 to discontinue relations with suppliers in China and to defer any direct investment because of "pervasive violations of basic human rights." (Five years later, in 1998, Levi Strauss decided to resume sourcing in China and to sell clothing there.)

Guidelines for Multinationals

If neither home country nor host country standards provide complete guidance, what principles or rules should multinational corporations follow? Three kinds of guidelines have been offered, based on considerations of human rights, welfare, and fairness or justice. All of these are relevant moral concepts; the challenge is determining exactly how they apply to international business.

Human Rights. Thomas Donaldson has proposed that corporations have an obligation to respect certain rights, namely those that ought to be recognized as fundamental international rights.[10] MNCs are not obligated to extend all the rights of U.S. citizens to people everywhere in the world, but there are certain basic rights that no person or institution, including a corporation, is morally permitted to violate. Fundamental international rights are roughly the same as natural or human rights, discussed in Chapter 3, and some of these are given explicit recognition in such documents as the United Nations Universal Declaration of Human Rights, the International Covenant on Social, Economic and Cultural Rights, and the International Covenant on Civil and Political Rights.

The main problem with a principle to respect fundamental international rights (or fundamental rights, for short) is specifying the rights in question. Even undeniable human rights that create an obligation for some person or institution, such as the government of a country, are not always relevant to a multinational corporation. Moreover, observing a right ranges from merely not depriving people from some protection to ensuring the fulfillment of a right. For example, everyone has a right to subsistence, but a corporation may be under no obligation to feed the hungry in a country where it operates, especially if doing so has no relation to its business activity. It has an obligation, however, not to contribute directly to starvation by, say, destroying farm land. In general, Donaldson claims, a corporation is morally bound only by those minimal duties such that "the persistent failure to observe [them] would deprive the corporation of its moral right to exist" and not by maximal duties whose fulfillment would be "praiseworthy but not absolutely mandatory."[11]

Donaldson suggests the following fundamental rights as a moral minimum:

1. The right to freedom of physical movement
2. The right to ownership of property
3. The right to freedom from torture
4. The right to a fair trial
5. The right to nondiscriminatory treatment
6. The right to physical security
7. The right to freedom of speech and association
8. The right to minimal education
9. The right to political participation
10. The right to subsistence[12]

Sample applications of these rights, according to Donaldson, include failing to provide safety equipment to protect employees from serious hazards (the right to physical security); using coercive tactics to prevent workers from organizing (the right to freedom of speech and association); employing child labor (the right to minimal education); and bribing government officials to violate their duty or seeking to overthrow democratically elected governments (the right to political participation).

Donaldson recognizes that it may be impossible to observe all these rights, especially in less developed countries where human rights violations are routine. However, insofar as the acceptance of a practice in a host country is due to its low level of economic development, we can ask ourselves: Would we, in our home country, regard the practice as morally permissible under conditions of similar economic development? He calls this the *rational empathy test*, which he describes as putting ourselves in the shoes of a foreigner:

> To be more specific, it makes sense to consider ourselves and our own culture at a level of economic development relevantly similar to that of the other country. And, if, having done this, we find that under such *hypothetically altered social circumstances* we ourselves would accept the lower risk standards, then it is permissible to adopt the standards that appear inferior.[13]

Although Donaldson's list of fundamental rights sets some minimal conditions for ethical behavior, it is not a complete guide for managers. First, the bearing of these rights on controversial questions is not wholly clear. For example, no one disputes that causing starvation by destroying farm land is a human rights violation.

But what does the right to subsistence tell us about cases in which multinationals convert land from the production of domestic crops to foods for export? Even though the MNC is acting within its rights within a free market as a property owner, and even though the country may benefit from a more productive use of the land, the ability of local people to feed themselves may be severely curtailed. Has the multinational violated the right of these people to subsistence? To this kind of question, Donaldson's rights-based approach offers little guidance.

Second, many of the most difficult moral questions in international business do not involve rights at all. Although rights violations by corporations receive great public attention, they are relatively infrequent. The critical issues at the forefront of global business today focus more on abuses of power by multinationals and on the failure of MNCs to aid developing countries. For example, the OECD Guidelines for Multinational Enterprises, adopted by the Organization for Economic Cooperation and Development (OECD), covers such matters as competing fairly, disclosing information, paying taxes, considering countries' balance of payment and credit policies, utilizing appropriate technologies, and aiding economic development. In general, the goal of the OEDC guidelines is to achieve a smoothly functioning global economic system that spreads the benefits widely, rather than to protect people's rights.

In sum, guidelines based on human rights provide a bedrock moral minimum. However, the application of rights-based guidelines are uncertain in more controversial situations where we are most in need of guidance, and they are inapplicable to many other pressing matters.

Welfare. Richard DeGeorge offers seven basic guidelines for multinational corporations that cover a variety of moral considerations, including rights. However, several of these rules concern avoiding harm and providing benefits. His guidelines are

1. Multinationals should do no intentional direct harm.
2. Multinationals should produce more good than harm for the host country.
3. Multinationals should contribute by their activity to the host country's development.
4. Multinationals should respect the human rights of their employees.
5. To the extent that local culture does not violate ethical norms, multinationals should respect the local culture and work with and not against it.
6. Multinationals should pay their fair share of taxes.
7. Multinationals should cooperate with the local government in developing and enforcing just background institutions.

The first three of these guidelines express in different ways a duty to consider the welfare of people in a host country. The first, do no intentional direct harm, is vacuous if it excludes all actions with a legitimate business purposes. No company intends to do harm; any harm results rather from its regular business activity. However, a company can engage in such activity knowing that harm will result. In some instances, this harm may be wrong, as when a firm produces oil in such a way that the land is polluted. On the other hand, a firm might open a plant that, because of its efficiency, will drive local competitors out of business. This latter result, unlike the former, is not wrong but is merely the working of market forces. In a well-functioning economy, more efficient producers should replace the less efficient. For the same reason, polluting oil production (Is there any other kind?) may be acceptable if the benefits outweigh the costs. What DeGeorge's first guideline presumably excludes is pollution that results from unjustifiably low standards. What standards ought to be adopted, however, is a question not merely about the harm done but about the benefit gained—and also about how these ought to be distributed.

The same problems afflict DeGeorge's second guideline, do more good than harm. Economic theory tells us that all voluntary exchange results in more good than harm for the reason that no one makes a trade that is not beneficial. Thus, as long as a multinational corporation offers employment in a factory and workers in a developing country are willing to accept, then everyone is better off. Controversial cases like that of Nike in Indonesia (Case 15.1) raise two questions. First, how much more good than harm should a multinational bring to a country like Indonesia? And, second, does it matter that multinationals are often able to use their market power to reap most of the benefit, leaving the host country with a narrow balance of good over harm? The answers to these questions depend not whether more good than harm is produced but on whether the outcome is obtained justly.

Multinationals are criticized primarily in cases where they take more than a fair share by exploiting their superior position in an imperfect market. A developed country, such as the United States, attempts to maintain perfect markets by preventing monopolies and other conditions that reduce fair competition. However, the marketplace in which multinationals encounter less developed countries is highly imperfect. Under such circumstances, some outcomes may be criticized for resulting from unfair competition. As Manuel Velasquez observes, DeGeorge's approach "fails to take seriously the importance of justice in evaluating the activities of multinationals."[14]

Justice. Much of the criticism of multinational corporations rests on considerations of justice. Even when MNCs respect human rights and produce more good than harm, their activities may still be criticized for being unfair or unjust. This is true even for the outcomes of voluntary market exchanges when they occur in imperfect markets, as when a multinational exploits a monopoly position.

One kind of unfairness cited by critics is the often one-sided division of the benefits from foreign investment. Certainly, the gap between rich and poor countries is an urgent moral concern, and multinational corporations have much to offer. Thus, the third of DeGeorge's guidelines is that "[m]ultinationals should contribute by their activity to the host country's development." The main questions, however, are who should act and what should be done? National governments and world organizations are the primary actors, and it is questionable what role multinationals should play. What should they do to aid development other than engage in business activity?

The answer to this question depends, in part, on how we answer the second question about what should be done. What is the most effective strategy for aiding developing countries? The main approach being taken today is to increase foreign investment and export production in an increasingly integrated world economy. If this development—generally called *globalization*—is the most effective strategy, then multinational corporations can contribute best by being efficient—but responsible!—businesses. Indeed, if MNCs are expected to expend resources on development, they may then choose not to invest in poorer countries, thus depriving them of any aid. However, opponents of globalization, such as the protesters at meetings of the World Trade Organization (WTO), propose other strategies for development that would place greater responsibilities on multinational corporations.

Another kind of unfairness is violating the rules of the marketplace, which is to say engaging in unfair competition and otherwise taking unfair advantage. One example is the ability of multinationals to avoid paying taxes by means of *transfer pricing*. Transfer prices are the values assigned to raw materials and unfinished products that one subsidiary of a company sells to another. Because transfer prices are set by the company and not the market, they can be raised or lowered so that most of the profits are recorded in countries with low tax rates. This use of transfer pricing is facilitated by the fact that multinationals are usually able to avoid disclosure of the relevant financial information. As a result, host countries often have little knowledge of a company's true financial situation.

Tax avoidance through transfer pricing is a critical problem for both developed and developing countries. Approximately a third of world trade or $1.6 trillion takes place within firms, so the possible loss of tax revenues is enormous. Consequently, the major countries of the world are trying to tighten accounting standards to prevent abuses. The OECD Guidelines for Multinational Enterprises requires firms to disclose financial statements on a regular basis and provide relevant information requested by taxing authorities. Furthermore, the Guidelines states that an enterprise should "[r]efrain from making use of the particular facilities available to them, such as transfer pricing which does not conform to arm's length standard, for modifying in ways contrary to national laws the tax base on which members of the group are assessed." This arm's-length standard is feasible when a market exists for the good in question, and for other goods a market price can be approximated by calculating the costs of production.[15]

If avoiding taxes by means of transfer pricing violates no laws, why should multinationals not take full advantage of this opportunity? The same question can be asked of paying bribes, offering kickbacks, and similar practices. One answer, offered by Norman E. Bowie, is that these actions violate the rules that are required for markets to operate. The very possibility of market exchanges depends on the general observance of certain rules of honesty, trust, and fair dealing. Because businesses benefit from the marketplace that these rules make possible, they are taking an unfair advantage, being a freeloader so to speak, by simultaneously violating these rules. Bowie writes:

> If activities that are permitted in other countries violate the morality of the marketplace—for example, undermine contracts or involve freeloading on the rules of the market—they nonetheless are morally prohibited to multinationals that operate

there. Such multinationals are obligated to follow the moral norms of the market. Contrary behavior is ultimately inconsistent and self-defeating.[16]

An obvious difficulty is determining the essential rules for markets to operate. That problem aside, is avoiding taxes by means of transfer pricing really a violation of market morality? One might argue that prices, which play a critical role in markets, should be set in ways that have a clear economic purpose. Setting prices for no other purpose than reducing taxes is not a genuine economic activity but a sham transaction. Indeed, whether a transaction serves a reasonable business purpose is a test that United States courts use to determine whether a tax shelter is legal.

WAGES AND WORKING CONDITIONS

Public concern about multinational corporations has focused in recent years on the footwear and apparel industries and their relations with foreign contractors. Virtually all major American shoe and clothing companies have adopted the strategy pioneered by Nike (Case 15.1) and outsourced the actual assembly of their products to contractors in Southeast Asia and Central America. This development benefits consumers everywhere by lowering the cost of goods, and jobs are created in countries that desperately need them. Overall, the manufacture of goods in countries with low labor costs is advantageous to developed and developing countries alike.

To many critics, however, the benefits must be weighed against a long list of wrongs that include very low wages, substandard working condition, the use of child labor, and association with repressive regimes. Some of the factories operated by multinationals and their foreign contractors are alleged to be "sweatshops" of the kind that operated in developed countries until the passage of protective legislation in the early twentieth century. The critics also question the ability of foreign contracting to advance economic development. Instead of improving the lives of people, they charge, the contracting system leads to greater misery for the bulk of the population and to a wider gap between the rich and the poor. Although the jobs that are exported overseas are a boon to workers in less developed countries, they reduce job opportunities in the developed world.

At the heart of this controversy is the question, how should the standards for wages and working conditions be determined? One answer is that these standards should be set by the market. In developed countries, the determination of wages and working conditions results primarily from the competition among employers for desirable workers, which compels them to offer high wages and good working conditions. On this view, there is nothing unjust about jobs with lower pay and poorer working conditions. As long as workers are willing to accept employment on the terms offered, then any arrangement is justified. On this view, no wage can be too low in a free market. However, using the market to determine the standards for wages and working conditions in developing countries encounters two obstacles.

First, developed countries do not rely solely on the market but set certain minimum conditions by law, such as minimum wage laws, fair labor standards, and health and safety regulations. These conditions reflect, in part, the recognition of certain human rights that ought to be observed in all economic activity. Thus, one rationale for minimum wage laws is that it is unjust to pay workers less than a certain amount. That is, some wages are unjustifiably low, even if enough workers would accept them. Although multinational corporations and their foreign contractors generally pay the legal minimum wage in the countries where they operate, this amount often provides only a basic subsistence for one person, if that. Consequently, critics argue that the standard should be a "living wage" that enables a worker to live with dignity and support a family.

The second obstacle to using the market to set wages and working conditions is the possibility that the conditions for a free market are lacking. In particular, the mass of unemployed, desperately poor people in less developed countries constitutes a pool of laborers willing accept bare subsistence wages. The market for labor in any given country may also be artificially low because of political repression that prevents workers from organizing. Although the role of correcting such market failures generally falls to national governments—by enacting minimum wage laws, for example—this form of protection is often uncertain in less developed countries.

Consequently, we need to consider the extent to which market forces should be allowed to operate in the setting of wages and working conditions and the extent to which principles of human rights ought to be applied. Specifically, should multinationals and foreign contractors be guided only by the market as long as they observe local laws? Or should they observe higher standards than either the law or the market dictates, guided perhaps by certain human rights principles? In some instances, workplace abuses—such as degrading punishment, forced overtime, and confinement to company quarters—have little or no economic justification. Although the wrongness of these abuses is not in question, we still need to ask whether a company such as Nike is responsible for the abusive practices of its foreign contractors and, if it is responsible, what steps should a company take to prevent these abuses.

What Is a Justified Wage?

The wages paid by multinationals and their foreign contractors are usually above the minimum wage and the prevailing market rate. As a result, the jobs in these factories generally pay better than work in local enterprises, and regular employment in the formal economy is vastly superior to work in the informal sector, which includes agriculture, domestic service, and small, unregulated manufacture. Even so, the wages paid for factory work are seldom sufficient to provide what the nation's government calculates as the minimum for a decent standard of living or the minimum physical needs for one person, let alone a family. In many poor countries of Southeast Asia and Central America, the minimum wage set by law is below the official poverty level.

Critics observe that the labor cost for a pair of shoes or a shirt is usually a few cents and that paying a few cents more would add little to the ultimate price. However, economists warn that raising the pay scales in a developing country has serious adverse consequences. Well-intentioned efforts to better the condition of factory workers will ultimately reduce the number of jobs and the level of foreign investment. If a government raises the minimum wage or multinationals and foreign contractors are pressured to pay above-market wages, the result will be a reduced incentive to relocate jobs from higher-wage countries. Because higher-wage countries have more productive workers and a better infrastructure for manufacturing, firms will have little reason to move to a less developed country unless it offers significantly lower labor costs.

This economic argument shows, first, why less developed countries should not raise the minimum wage beyond the market value of its labor. In a developed country, raising the minimum wage generally benefits low-wage workers without much effect on others, but the effect in a less developed country is different. Although a relatively small minority of urban workers, who already make above-average wages, may benefit, the vast majority will suffer for lack of jobs, and the economy will not develop for lack of foreign investment. For this reason, the economist Jagdish Bhagwati contends, "Requiring a minimum wage in an overpopulated, developing country... may actually be morally wicked."[17]

Low labor costs constitute a competitive advantage for a poor country, and attracting investment on this basis provides jobs that can lead to greater development. Indeed, formerly low-wage countries, such as Korea, Taiwan, and Malaysia, have successfully employed this strategy for creating higher-paying jobs. The result is the same if multinational corporations and their foreign contractors are required, say by public pressure, to pay above-market wages. If firms do not take advantage of the low-cost labor in a countries like Indonesia and Vietnam, then they are depriving these countries of the opportunity to use their main competitive edge, namely unemployed workers, to begin the process of development.

This argument about wages also applies to working conditions inasmuch as both are matters of cost. According to a World Bank report, "Reducing hazards in the workplace is costly, and typically the greater the reduction the more it costs. . . . As a result, setting standards too high can actually lower workers' welfare."[18] One reason for this outcome is that investment to improve working conditions may come at the expense of wages. More significantly, if higher standards inhibit foreign investment, then fewer jobs are created and more of those available are in local industries with lower pay and working conditions.

The dispute between those who advocate paying the market rate for labor and those favoring a "living wage" is not over the ultimate end which is to improve the welfare of people in developing countries. The difference lies in the appropriate means. Economists argue that requiring higher wages is counterproductive; it harms the very people we are trying to benefit. The only path to prosperity is economic development, and this requires an attractive climate for foreign investment. Proponents of a "living wage" believe, on the other hand, that payment of wages above a certain level is morally required. To offer less than a "living wage" is to exploit an opportunity for cheap labor.

One difficulty with the economists' position is that it fails to consider the possibility of exploitation. It assumes that whatever wages people are willing to accept is just. However, most of the jobs in question are among the best paying, so the case for exploitation is weak. Moreover, the "living wage" position seems to entail that no job at all is preferable to one below the living-wage standard. Although developing countries are forced to keep wages low in their competition with each other to attract foreign investment, they still seem to prefer all the low-wage jobs they can get, regardless of whether they pay a "living wage."

Toward a Solution

In contrast to the issue of wages, which remains controversial, there is general agreement on standards for working conditions. When the television personality Kathy Lee Gifford was confronted in 1996 with evidence that the line of clothes she endorsed was made in Honduras by young girls 13 and 14 years old, working 20-hour days for 31 cents an hour, she resolved to correct these abuses. The same year, *Life* magazine published a photograph of a 12-year-old boy in Pakistan stitching a Nike soccer ball. Phil Knight replied that "Nike has zero tolerance for under-age labor." Reports on contractors for other American companies described women who were confined to factory compounds, berated and beaten for violating rules or failing to meet quotas, forbidden to use toilets, and fired for protesting or attempting to organize. In some instances, young women have been forced to take contraceptive pills or undergo pregnancy tests, and they have been dismissed for becoming pregnant. Many factories lack adequate ventilation, sanitation facilities, medical supplies, and fire safety provisions, and workers often have little protection from machines and toxic chemicals.

For many companies, the first step has been to adopt codes of conduct for their own operations and those of contractors. In 1992, Nike adopted a "Code of Conduct" and a "Memorandum of Understanding" which are included with all contracts. The Nike code forbids hiring anyone under 18 in shoe manufacture and under 16 for producing clothing (unless higher ages are mandated by law). It stipulates that workers be paid the higher of the legal minimum wage or the prevailing wage, with a clear, written accounting of all hours and deductions. Although forced overtime is permitted, provided employees are informed and fully compensated according to local law, the code requires one day off in seven and no more than 60 hours a week. Nike also developed a comprehensive Management of Environment, Safety and Health (MESH) that provides for safety standards, a safety committee in each factory, and free personal protective equipment for at-risk employees.

Although these actions addressed the major areas of concern, critics charged that Nike had not gone far enough in setting high standards. In 1990, Reebok, a Nike competitor, adopted a far-reaching human rights policy and inserted human rights language in its contracts. Reebok also committed itself to auditing its contractors for compliance with the human rights policy. Nike, too, agreed to auditing by hiring the firm Ernst & Young to conduct site visits of its contractors. In 1996, Nike also hired the civil rights leader Andrew Young to investigate factories in Asia and to report his findings. Both efforts were denigrated by critics, who challenged the competence of these parties to conduct thorough audits and questioned their independence inasmuch as Nike was footing the bill. Nike's commitment was further undermined when an internal Ernst & Young report, leaked to the press in 1997, revealed serious health and safety issues in a Vietnamese factory.

The challenges of managing foreign contractors are beyond the capability of any single company. First, any firm that invests resources when their competitors do not is put at a competitive disadvantage. Second, each firm deals with thousands of contractors, which in turn manufacture for many brands. The only solution, therefore, is an industrywide effort. The first initiative occurred in 1997 with the launch of the Apparel Industry Partnership (AIP), which Nike immediately joined. Convened by the White House, a group of industry, labor, consumer, and human rights leaders committed themselves to develop a strong workplace code of ethics with internal monitoring and an independent, external monitoring system. The AIP was succeeded the next year by the Fair Labor Association (FLA), which has preserved the same goals. In response to the concern of college students, the Worker Rights Consortium (WRC) was organized to address specifically the conditions under which collegiate apparel with a school's logo is manufactured.

The experiences of the FLA and the WRC reveal substantial agreement on principles and standards, although the issue of a "living wage" has been divisive. The main stumbling block has been monitoring. The practical difficulties of monitoring tens of thousands of contractors around the globe—not to mention the cost of $3,000 to $6,000 for each factory visit—is daunting enough, but the participants have sharply disagreed on issues of principle. For starters, who should do the monitoring? Accounting firms often lack expertise in local situations, whereas activist groups may not be wholly objective. What qualifications should monitors have? How should audits be conducted? Should audits be unannounced or scheduled in advance? Should all of a firm's contractors be audited or only a sample? Should the reports be made public? What actions should be taken when violations are discovered?

As these questions suggest, the solution to the problem of sweatshops requires a sustained, committed effort by all concerned parties. Considerable progress has been made on working conditions—but not on wages. A *New York Times* article profiled a woman who was fired in 1995 for protesting conditions at a factory in El Salvador. Six year later, the woman returned to the factory where workers enjoy coffee breaks in a terrace cafeteria and work in clean, breezy surroundings, but she earns only 60 cents an hour, 5 cents more than before.[19]

Child labor presents an especially thorny issue. Although an estimated 150 million children under the age of 14 work worldwide, less than 5 percent of these make goods for export. The vast majority are employed in the informal economy which contains the most dangerous jobs. Virtually every country bans child labor, but enforcement is often ineffective. Although multinationals should abide by the law and refrain from employing children, the main challenge is how to deal with existing factories that employ children. A *New York Times* editorial observes, "American consumers are right to insist that the goods we buy are not made with child labor. But these efforts will backfire if children kicked out of these factories drift to more hazardous occupations."[20]

In response to this problem, the International Labour Organization has worked with governments and businesses to establish special schools for approximately 10,000 children who worked in garment factories and to pay their parents for the lost wages. Ultimately, the solution to the problem of child labor is not merely to prohibit the employment of underage workers but also to provide schooling for children and jobs for parents so that child labor is no longer an economic necessity.

FOREIGN BRIBERY

As the experience of Lockheed in Japan illustrates (see Case 2.1), bribery is one of the most common and controversial issues that multinational corporations face. Bribery is universally condemned, and no government in the world legally permits bribery of its own officials. However, corruption exists to some extent in every country and is endemic to more than a few. The main ethical question is whether companies are justified in making payments when they are necessary for doing business in a corrupt environment. Although the demand for a bribe may be unethical, is it unethical to give in to a demand? The defenders of bribery thus appeal to the slogan, "We don't agree with the Romans, but find it necessary to do things their way."

The issue of bribery is far from simple. First, the term *bribe* is vague. It applies to many different kinds of payments with varying interpretations, ranging from gift giving and influence peddling to kickbacks and extortion. There is need, therefore, to develop a definition and make some distinctions. Second, corruption is a fact of life in global business, and so MNCs and their home governments must decide how to confront it. The United States has legally prohibited certain kinds of payments since the passage of the Foreign Corrupt Practices Act (FCPA) in 1977. For the next two decades, few countries followed this approach, however, and American companies have complained that they had been placed at a competitive disadvantage. Both the overall approach and specific provisions of the FCPA have been questioned, and so these must be examined.

What Is Bribery? The FCPA forbids American corporations to offer or make any payment to a foreign official for the purpose of "influencing any act or decision of such foreign official in his official capacity or of inducing such foreign official to do or omit to do any act in violation of the lawful duty of such official" in order to obtain or retain business.[21] (The act also covers payments to political parties and candidates for office in foreign countries.) This legal prohibition accords with standard philosophical definitions of bribery.[22] The key point is that a bribe is a payment made with an intention to corrupt. More specifically, the payment is made with the intention of causing a person to be dishonest or disloyal or to betray a trust in the performance of official duties. The holder of a government office has a duty to make decisions about such matters as the purchase of aircraft, for example, according to certain criteria. A corrupt official is one whose decision about this purchase is influenced by a payment, and an aircraft manufacturer who offers something of value for this purpose has made a corrupt payment.

It follows from this definition that certain kinds of payments are legally permitted and do not constitute bribes. These include "facilitating payments," which are made to expedite the performance of "routine governmental action." Also called "grease payments," these are small sums paid to lower-level officials to lubricate the rusty machinery that provides government services. Facilitating payments do not induce anyone to violate a duty or a trust. Still, they are generally prohibited by the same governments that create the need for them, although the laws on such matters are rarely enforced. Also excluded from the category of bribes are reasonable expenditures for legitimate expenses, such as entertaining a foreign official in the course of doing business. Finally, any payments that are permitted or required by the written laws of the country in question are legal under the FCPA. Although such payments might still be considered bribes, the drafters of the FCPA did not believe that American corporations should be prosecuted in the United States for abiding by local laws. The stipulation that the laws be written is designed to ensure that the payments are really legal and not merely customary.

Can a line be drawn between bribes and other kinds of payments that are generally considered ethical? In some business cultures, gifts are means for cementing relations and are given (and reciprocated) without the intention of influencing an official's decision. Payments to gain influence are often intended merely to gain access to a decision maker in order to be heard. If demands for bribes are viewed as a kind of extortion, then companies can say that they make payments not for the purpose of obtaining a favorable decision but in order to avoid being harmed illegitimately. The difficulty with all of these arguments is that the distinctions are subjective at best and prone to special pleading. It is instructive that the Japanese, who have traditionally exchanged gifts in business, have curtailed the practice with government officials after a rash of scandals. If American companies were permitted to make extortion payments but not to pay bribes, then foreign officials might well become more threatening in their requests.

What's Wrong with Bribery? The immorality of demanding or accepting bribes is implicit in the definition of bribery: A government official is violating a duty or a trust. In addition, inducing such a violation by offering a bribe is commonly recognized as a wrong. Corrupting others is as wrong as being corrupt oneself. However, these points do not explain why it is wrong to pay a bribe that is demanded or why a government should prohibit its own citizens from bribing the officials of a foreign country. Each country is generally charged with the task of enforcing its own laws.

The FCPA was enacted by Congress in part to protect American interests. The widespread use of slush funds to make payments during the 1970s raised fears that U.S. corporations were engaging in false financial reporting that compromised the integrity of securities markets. When the Securities and Exchange Commission encouraged the voluntary disclosure of foreign payments without fear of prosecution, more than 450 companies admitted to paying a total of $400 million that was not fully accounted for in their books. In addition, Congress was concerned that foreign payments by American corporations were undermining the governments of friendly countries around the world and interfering in the conduct of U.S. foreign policy.

The principal ethical objection to foreign bribery is that systematic and widespread corruption inhibits the development of fair and efficient markets. First, an economy based on bribery does not provide open access to all competitors on equal terms. Not only are some competitors unable to enter certain markets because of existing bribe arrangements, but all businesses are unable to compete on the merits of their products or services. Bribery generally takes place in secret, so that price and quality are less important factors in making a sale than access. Second, bribery-prone economies are, for the same reasons, less efficient. The efficient allocation of resources depends on readily available information and rational decision making. When bribery occurs, prices are invariably higher than they would be in a competitive market. Higher prices also create a disincentive for domestic and foreign firms to invest in the further development of a nation's economy. These costs of bribery are generally reflected in a lower standard of living for the people of the countries where bribery is commonplace. Put bluntly, corruption is a major cause of poverty and underdevelopment in much of the world today.

Although corrupt government officials bear chief responsibility for the economic consequences of bribery, major multinational corporations contribute to the problem when they actively participate in corruption and take no steps to combat it. They also undermine their own long-term interests, which lie in free and open markets worldwide. Corruption in Russia, for example, does not affect the Russian people alone; it also makes Russia a less attractive environment for investment by corporations from the United States, Europe, and other developed countries. As the economies of all countries become more closely linked, the consequences of corruption in one country impact many others. For example, the turmoil in Russia in 1998, which was due in large part to corruption, caused large losses for many major American banks.

What Should Be Done about Bribery? If bribery is morally wrong as well as economically undesirable, then corporations and governments should take reasonable steps to reduce the incidence of bribery around the world. Some question, however, whether the FCPA is a sound approach. Should the U.S. government attempt to prohibit foreign bribery by American firms, and can the law, as written, be effective? Bribery can effectively be addressed only by concerted action among the major developed countries, and so what kind of cooperation should be sought?

The FCPA forms a legal approach to problems of bribery. In addition to the prohibition of corrupt payments, the act contains an accounting provision that requires corporations to maintain records that "accurately and fairly reflect the transactions and disposition of the assets" of the firm. This provision was drafted in response to the discovery that some corporations had failed to record payments in their books or else disguised their true purpose. For violations of the FCPA, corporations may be fined up to $2 million and individuals may be fined up to $100,000 and imprisoned for up to five years (and the fine for an individual may not be paid by the corporation).

An especially problematic part of the FCPA addresses the responsibility of corporations for the conduct of their agents. Much business in foreign countries is conducted through intermediaries, who may pay bribes without the knowledge of the American firm. When American firms enter into joint ventures with foreign companies, they may have a similar lack of control over the conduct of their business partners and the maintenance of their books. The standard adopted by Congress is that a corporation is responsible when it pays an agent or other third party "knowing or having reason to know" that the payments will be used by that person to bribe a foreign official. Deliberately avoiding knowledge of an agent's activities is not an adequate defense, and so companies should take precautions to "know their agents." Among the advised precautions are checking out the reputation of the agent and being sure that the agent has some genuine service to provide; paying the agent only an amount commensurate with the services provided and seeking an accounting of all expenses incurred; avoiding suspicious requests, such as depositing the money is a certain bank account; and obtaining a detailed agreement that includes a pledge not to violate the FCPA and the right to terminate the contract for any violation.

The justification for prohibiting foreign bribery is rather straightforward. A double standard is employed if a country permits its companies to do abroad what they are forbidden to do at home. Bribery imposes harms and violates basic rights. The main arguments to the contrary are that bribery is necessary for doing business in some countries and that a country that forbids its own companies from bribing places them at an unfair competitive advantage. Some have argued that the FCPA is a form of "ethical imperialism" that imposes our values on other countries.

These arguments are not very persuasive. First, if foreign bribery is wrong, then the fact that America's competitors around the globe practice it does not provide a justification. Second, American firms have been able to compete in many instances without bribing, in part by developing better products and services and marketing them aggressively. Anecdotal evidence suggests that bribery is involved in many contracts that U.S. firms fail to attain, but it is difficult to determine who would have obtained the award had there been a level playing field. No academic studies to date have documented substantial loss of business due to the FCPA.[23] Even if some loss of business has occurred, it must be weighed against the other benefits that led Congress to enact the FCPA. That is, is the United States as a nation better off for the passage of this act? Third, the FCPA is scarcely an instance of "ethical imperialism" because it applies only to the conduct of American firms. In addition, the immorality of bribery comes close to being a universal norm. Perhaps the best way to counter all of these arguments is for the United States to persuade other countries to follow America's lead.

Fortunately, world attitudes are changing. The proposed code of conduct developed by the United Nations Commission on Multinationals has long contained a prohibition on bribery, although the code has yet to be adopted by the organization's membership. The United States has fought, largely without success, to gain agreement for a ban on bribery through the World Trade Organization. However, in the 1990s, leading financial organizations, including the International Monetary Fund and the World Bank, have stressed the economic consequences of foreign bribery and placed restrictions on aid recipients to limit the practice. A private organization, Transparency International, compiles an annual list that scores countries on the perceived level of corruption. These latter efforts seek to influence the "demand" side of the equation by applying pressure on countries that receive bribes.

The most significant development has been the adoption in April 1996 of a legally binding treaty by the 29-member Organization for Economic Cooperation and Development (OECD) which commits each member country to change its laws to accord roughly with the FCPA. Specifically, OECD members, which include the world's richest nations, have agreed to prohibit bribery of foreign officials, impose criminal penalties on those found guilty, and allow for the seizure of profits gained by bribery. This initiative concentrates on the "supply" side by changing the conduct of corporations that have been paying bribes. Many observers are optimistic that the OECD treaty will significantly reduce the incidence of foreign bribery and produce a more level playing field for all multinational corporations.

CULTURAL DIFFERENCES

As globalization advances, business practices are becoming increasingly uniform. Business education is much the same everywhere, and managers from diverse countries approach business in similar ways. Ethics, however, remains stubbornly local. Unlike business skills, which are the same worldwide, people's conceptions of ethical conduct remain rooted in particular cultures. The importance of cultural differences is not confined to the problem of different standards that underlies the "When in Rome" question. Although what is considered right and wrong varies from one culture to another, this kind of difference is only one of the many ways in which different cultures pose a challenge for global business.

Our thinking about ethical issues involves the use of concepts, such as duty, rights, equality, welfare, freedom, and trust. Although these concepts have their counterparts in other languages, their meaning is often slightly different, reflecting each country's culture and history. Indeed, the meanings of these concepts have changed within our own culture over the course of time. As a result, we cannot assume that our conception of rights, for example, is shared by people everywhere.

Furthermore, different cultures place different emphases on these concepts. Thus, some foreign critics contend that Americans place to much stress on rights and too little on equality. For example, high executive compensation in the United States is considered excessive by many people abroad. In response to American criticism of human rights in China, some defenders argue not only that human rights are understood differently in Asia but also that they are less important than the goal of improving people's welfare.

Cultural differences like these occasionally result in misunderstandings and accusations of misconduct. For example, Japanese companies are sometimes accused of showing favoritism to other Japanese firms, thereby mistreating potential foreign business partners. The Japanese response is that they are showing loyalty to companies in long-established relationships. They criticize non-Japanese firms for not establishing close relationships and for lacking loyalty once a relationship is established. Because of this emphasis on loyalty, whistle-blowing is viewed negatively in Japan in contrast to the United States where it is often considered a mark of integrity and moral courage.

Conducting business globally, then, requires us to understand the ways in which cultural differences are reflected in people's moral outlook. Some appreciation of how European and Asian managers view ethics is critical for successful interaction with them, as is a recognition of how America is distinctive.

Is America Distinctive?

Broad generalizations about the ethics of any culture must be made cautiously. Nevertheless, observers have noted some distinctive characteristics of Americans that stand in contrast to the ethical outlook of Europeans, Asians, and others. For example, David Vogel finds evidence of an "ethics gap." "By any available measure," he claims, "the level of public, business, and academic interest in issues of business ethics in the United States far exceeds that in any other capitalist country."[24] This heightened level of concern results, in part, from the higher expectations for business that Americans generally hold.

Vogel further observes that Americans tend to regard the individual as the arbiter of right conduct. That is, Americans are more likely than people elsewhere to consult their own values in deciding what is right and wrong. By contrast, Europeans and Asians generally seek guidance from the community and their own business organization. European managers, in particular, are not inclined to navigate by their own "personal moral compass" but usually consult widely with others and consider the impact of any decision on the organization.[25] As a result, ethical decisions in Europe are more commonly made by a group, not an individual.

Americans are also more legalistic and rule-oriented. They tend not only to embody business ethics in laws that are rigorously enforced but also to think of ethics as a set of rules to be observed. Europeans by contrast disdain Americans' reliance on the law and voluntary codes and rely more on informal mechanisms for securing ethical behavior. In the United States, errant managers are more likely than their counterparts abroad to face legal sanctions, including fines and imprisonment. Moreover, rules are regarded by Americans as universal prescriptions that apply impartially to everyone, whereas Asians, in particular, view moral obligations as arising from specific relationships.

Ethics East and West

Although cultural differences affect people's ethical outlook the world over, the greatest contrast exists between East and West. Westerners who do business in Japan, China, and the other countries of Asia are aware of the different approaches of these two parts of the world to matters of ethics. To be sure, ethics East and West still have much in common, especially with regard to fundamental values. The differences reveal themselves in more subtle ways. We should also be wary of broad generalizations about either Eastern or Western ethics. Neither is monolithic, and both have changed over time. Nevertheless, the ethical outlook of Asians is different in ways that are important for business managers to understand.

The Primacy of Relationships. The most striking feature of Asian firms, especially those in Japan, is the central role of long-term relationships. These rest on a high level of trust among all the parties and require careful attention to each party's interests so as to maintain harmony. In Japan, this feature extends to the *keiretsu*, which is a group of companies closely linked with each other. (A similar structure in Korea is the *Chaebol.*)

Such close ties produce some admired features of Japanese firms, such as lifetime employment and strong employee loyalty. Companies in a *keiretsu* come to each other's aid in times of difficulty. The primacy of relationships has some negative consequences, however. Employees are usually unable to leave a firm, and the pressures to conform have resulted in cases of *karoshi*, in which people are said to work themselves to death. Outside firms complain that they are excluded from doing business with *keiretsu* members. In China, an established relationship, known as *guanxi*, is often critical for gaining access.

In a society built on relationships, ethical obligations depend on what each party owes the other in the relationship. The treatment that one can expect results not only from whether one is inside or outside of a group but also from the specific nature of the relationship itself. Accordingly, norms in such a society tend to be relative or situational rather than absolute and universal.[26] Instead of known rules that are applied equally to all, moral decisions are made on a case-by-case basis with attention to specifics. Laws and regulations are often vague, with large grey areas that are interpreted by government officials to fit the case at hand. Foreign companies often find themselves unsure of the rules, which they suspect, in any event, are being manipulated to favor Japanese firms.

A system built on relationships places great emphasis on reciprocity. That is, each party must take care to return all favors received so as to preserve a balance. Moreover, conflicts need to be handled by a mutual accommodation that preserves the harmony of the group. Any individual or company that takes more than a fair share is viewed as disruptive and, for that reason, untrustworthy. In addition, commitments in a relationship are viewed as binding. Americans, by contrast, tend to make sincere but overly optimistic commitments that they may not be able keep.[27] This cultural difference over the the meaning of commitments is a ready source of misunderstanding.

Explaining Japanese Practices. The features of Japanese firms just described help explain some distinctively Japanese business practices. Although the practices themselves do not bear directly on ethics, they depend crucially on these ethical features, thus illustrating the interplay of ethics and management practice.

First, *kaizen* or continuous improvement is a broad-ranging concept that includes just in time delivery, total quality control, and waste elimination. Making continuous improvements is a responsibility of everyone in a firm, including management, work groups, and individuals, and it utilizes such means as detailed statistical measures, quality circles, and structured group activities. By continually improving all aspects of production, Japanese firms have managed to produce improved products at lower costs.

The success of *kaizen* depends on a strong group commitment, whereby each person aims to benefit the whole organization, and on a high level of trust that one's efforts will be recognized and rewarded. Attempts by American firms to emulate *kaizen* have faltered for lack of these enabling conditions. For starters, workers without job security have little incentive to suggest changes that might eliminate their positions. Although commitment to a firm in Japan results, to some extent, from a distinctively Japanese notion of self-realization through group activity, Japanese companies also create it through a variety of techniques that includes hiring workers in cohorts that generally advance together and receive similar pay. Commitment is further developed by a strict system of seniority and respect for people with diverse talents and abilities.[28]

Second, the *keiretsu* is a distinctively Japanese form of organization that represents extensive horizontal and vertical integration of firms. A horizontally integrated *keiretsu*, such as Mitsubishi, consists of companies that manufacture a wide variety of goods, plus key banks and other financial service firms. Toyota, which is an example of a vertical *keiretsu*, ties together its key suppliers and dealers. Such concentration of economic power would generally be prohibited in the United States by antitrust law.

Although the *keiretsu* system developed in response to historical conditions, it also reflects both the emphasis on relationships and a certain conception fairness. Relationships are conceived by the Japanese as a set of concentric circles. The innermost circle is the family, which has the tightest bonds, and successive circles entail looser ties. Whereas Western firms tend to have arm's-length market relations with other companies, *keiretsu* members have closer working relationships with each other and market relations only with firms outside the group. This arrangement reflects a preference for cooperation over competition.

Moreover, fairness within the sphere of cooperation is primarily a matter of reciprocity, which is to say fulfilling the obligations owed to others, whereas the concept of fairness in competition is reserved for activity outside the *keiretsu*. These two concept of fairness—which may be characterized as doing one's fair share and competing fairly—are applicable to Western firms as well. The main difference is their range of application. Western standards for fair treatment of a supplier are primarily those of fair competition, whereas the standards for a Japanese firm in a *keiretsu* are those of doing one's fair share. Thus, coming to the aid of a supplier may be considered just by a Japanese firm because such treatment is owed to a partner as a matter of doing one's fair share. At the same time, it may be regarded as unjust by a Western company for showing favoritism inasmuch as fair competition requires that all suppliers be treated equally.

Third, regulation in Japan is achieved, in part, through a process known as *administrative guidance*. In the United States and Europe, business is regulated by means of detailed, precise rules that are known to all and applied uniformly. These rules are promulgated primarily by legislatures and regulatory bodies (which are established by legislatures), and they are enforced mainly through the courts. In short, regulation is achieved in the West largely through legislation and adjudication. In Japan, by contrast, powerful administrative bodies, such as the Ministry of Finance and the Ministry for International Trade and Industry, regulate by making decisions in a case-by-case process. The rules themselves are often vague, with the result that business leaders must go to these agencies to get clarification.

Such a system, which is based on administration rather than legislation and adjudication, reflects two features of Japanese ethics. One is the view that ethical norms are situational and relative, instead of being absolute and universal. The second feature is the Confusion ideal of an elite group that is selected and trained to rule. Both Japan and China have centuries-old traditions of bureaucracies staffed by wise and dedicated career civil servants, separated from the political leaders. Affairs in Japan are often described in terms of the "iron triangle" of business, the bureaucracy, and politics.

Although administrative guidance can be viewed as an alternative form of regulation, no better or worse than that of the United States and Europe, the need to establish close working relations with the various ministries is a barrier to foreign enterprises and a source of confusion and misunderstanding. To Americans accustomed to clear and precise rules, the need to obtain clarifications of vague rules raises suspicions of a "rigged game."

Is Agreement Possible?

Given the diversity of ethical outlooks in the world, is it possible to agree on a set of standards for business worldwide? Such a goal must be achievable if globalization is to succeed. The theologian Hans Küng has observed that "the very phenomenon of globalization makes it clear that there must also be a globalization of ethics."[29] Of course, full agreement is not necessary or even desirable, but globalization requires a commitment to some core standards or at least a willingness to abide by them.

Substantial agreement is being achieved through a number of codes that have been developed by international organizations involving governments, religious bodies, and private individuals. The foundational document for human rights is the 1948 United Nations Universal Declaration on Human Rights. In 1966, the U.N. adopted two agreements which have subsequently been ratified by the major countries of the world: the International Covenant on Social, Economic, and Cultural Rights and the International Covenant on Civil and Political Rights. Since 1972, the United Nations has been developing a code of conduct for multinational corporations, which has yet to be completed or adopted. However, in 1999, the U.N. Secretary-General Kofi Annan challenged world business leaders to "embrace and enact" the Global Compact, which consists of nine principles covering human rights, labor, and the environment.

The International Labour Organization, which dates from 1919 and is now a specialized agency of the United Nations, sets many international standards, including those of the Tripartite Declaration of Principles concerning Multinational Enterprises and Social Policy (1977). More recently, the Organization for Economic Cooperation and Development (OECD), whose members are the more developed countries of the world, has adopted the OECD Guidelines for Multinational Enterprises. Several interfaith religious bodies have developed codes. The most prominent are the Principles for Global Corporate Responsibility, adopted by the U.S.-based Interfaith Center on Corporate Responsibility and similar organizations in Great Britain, Ireland, and Canada, and the Interfaith Declaration on International Business Ethics, which resulted from a dialog among Christians, Jews, and Muslims.[30] A group of world business leaders, meeting in Caux, Switzerland, developed the Caux Roundtable Principles for Business.

These codes have many guidelines in common and cover the areas of employment practices, consumer protection, environmental preservation, involvement in politics, including bribery, and basic human rights. The main guidelines, which represent substantial international agreement, are summarized in Exhibit 15.1.

Exhibit 15.1
International Codes of Ethics

Although international codes of ethics have diverse sources, they cover many of the same topics and make similar prescriptions. The following is a summary of the most common elements of these codes.

Employment Practices and Policies

Multinational corporations should

- Respect host-country labor standards and upgrade the local work force through training
- Promote favorable working conditions with limited hours, holidays with pay, and protection against unemployment.
- Promote job stability and job security, avoiding arbitrary dismissals and providing severance pay for those dismissed.

- Pay a basic living wage to employees.
- Develop nondiscriminatory employment policies and provide equal pay for equal work.
- Adopt adequate health and safety standards and inform workers of job-related hazards.
- Respect the right of employees to join labor unions and to bargain collectively.

Consumer Protection

Multinational corporations should

- Respect host-country laws and policies regarding the protection of consumers.
- Safeguard the health and safety of consumers by appropriate disclosures, safe packaging, proper labeling, and accurate advertising.

Environmental Protection

Multinational corporations should

- Respect host-country laws and policies concerning the protection of the environment.
- Preserve ecological balance, adopt preventive measures to avoid harm to the environment, and correct any damage done to the environment by operations.
- Promote the development of international environmental standards.

Political Payments and Involvement

Multinational corporations should

- Avoid paying bribes and making other improper payments to public officials.
- Avoid improper or illegal involvement or interference in the internal politics of host countries.
- Avoid improper interference in relations between national governments.

Basic Rights and Freedoms

Multinational Corporations should

- Respect the rights of all persons to life, liberty, and security.
- Respect the rights of all person to equal protection of the law, freedom of thought and expression, freedom of assembly and association, freedom of residence and movement, and freedom of religion.
- Promote a standard of living that supports the health and well-being of workers and their families, with special attention given to mothers and children.

Source: William C. Frederick, "The Moral Authority of Transnational Corporate Codes," *Journal of Business Ethics*, 10 (1991), 166-67.

The emerging order has been described as a "policy regime" to distinguish it from conventional regulation.[31] Regulation within a state consists of precise rules for most foreseeable circumstances that are enforced by law. In the absence of international government, the main alternative is the development of mutually accepted norms that serve to coordinate global business activity. A policy regime provides everyone with a reasonable set of expectations about how others will behave. The various codes that make up a policy regime do not have the force of law but still represent publicly declared commitments that have some binding force.

A policy regime is enforced primarily by the advantages of cooperation and the threat of retaliation for not cooperating. Although compliance is voluntary, countries usually find it in their interest to abide by the major codes. However, these codes also draw strength from their ethical force. That is, the widespread recognition that the guidelines represent universal values further contributes to their observance. Consequently, codes for international business are more likely to be effective when they have considerable moral authority. William C. Frederick finds four sources from which codes derive moral authority. These are (1) *national sovereignty*, which underlies strictures against interfering in the political process of a host country and bribing foreign officials; (2) *equality*, which supports nondiscriminatory treatment on the basis of race, sex, nationality; (3) *market integrity*, which requires property rights, contracts, fair competition, and the other conditions for free markets to operate; and (4) *human rights*.[32] Any guidelines not supported by these sources—to aid in a country's development, for example—do not have strong moral force and hence are less effective.

Of course, agreement on international codes of ethics can mask some substantial cultural differences and people's positions on specific issues. For example, in response to criticism of their human rights record, the leaders of China and some other Asian countries argue that the rights of people to economic well-being take precedence over other rights. Although the legal and political rights that are essential to a fair judicial system and free, democratic government are long-term goals, they must be deferred, these leaders say, until sufficient economic progress permits their realization. This emphasis on economic development before political liberalization has been defended in the name of "Asian values." However, questions can be raised about whether some rights that are being deferred really do require economic progress to be realized. Indeed, some legal and political rights may be instrumental in furthering economic development.[33]

Conclusion

Operating abroad, especially in less developed countries, creates dilemmas that lead to charges of serious ethical failings. Multinational corporations generally recognize a social responsibility and attempt to fulfill their responsibilities everywhere they are located. The major cause of occasional failures to act responsibly is not the lack of effort but the diversity of political and legal systems around the world and differences in economic development. Foreign operations give rise to challenges—and also create opportunities for misconduct—that simply do not exist for purely domestic enterprises.

The main quandary facing all MNCs is deciding which standards to follow. We have seen that neither of the two extreme positions is satisfactory. The familiar adage "When in Rome, do as the Romans do" and the opposite, "When in Rome or anywhere else, do as you would at home," are both inadequate guides. Instead, this chapter offers guidelines for developing special standards for the conduct of international business that can be applied to such matters as so-called sweatshops and to foreign bribery. Ultimately, the solution to many of the ethical problems of international business lies in the development of international agreements and codes of ethics. As the guidelines for multinational corporations become more detailed and comprehensive, the need for special standards of international business may diminish, and business conduct may eventually be the same worldwide.

Case 15.2   H. B. Fuller in Honduras

In 1985, journalists began writing about a new social problem in Honduras that created an acute dilemma for H. B. Fuller Company, based in St. Paul, Minnesota.[34] The news stories described the ravaging effects of glue sniffing among the street children of Tegucigalpa, the capital of Honduras, and other Central American cities. The drug of choice for these addicts was Resistol, a glue produced by a Honduran subsidiary of H. B. Fuller, and the victims of this debilitating habit were known, in Spanish, as *resistoleros*. The negative publicity was sullying the company's stellar reputation for corporate social responsibility, and company executives came under great pressure to address the problem quickly.

Poverty in Honduras had forced many families to send their children into the streets to beg or do odd jobs. The earnings of these children were critical to the support of many families, especially those headed by a single mother. Some children lived in the streets in order to avoid abusive homes; others were abandoned or orphaned. Many children, some as young as five or six, sought relief from their misery by sniffing glue containing volatile solvents which produces a temporary elation and sense of power. These chemicals are addictive and lead to irreversible damage to the brain and liver. The victims of solvent abuse generally stagger as they walk and exhibit tense, aggressive behavior.

Resistol is a brand name for a line of adhesives manufactured by a wholly owned subsidiary of H. B. Fuller and marketed throughout Latin America. The solvent-based adhesives favored by glue sniffers were widely used in shoemaking and shoe repair and were readily available on the street. H. B. Fuller had urged the press not to use the term *resistoleros* because other brands of adhesives were used as well and the problem was with the abuse of Resistol, not the product itself. Nevertheless, the name was commonly used in Honduras to describe the street children addicted to solvents. One of H. B. Fuller's most successful brands had thus become synonymous with a major social problem.

Criticism of H. B. Fuller for the company's involvement in this problem came not only from activists and public health officials in Honduras but also from customers and shareholders in the United States. One shareholder asked, "How can a company like H. B. Fuller claim to have a social conscience and continue to sell Resistol, which is 'literally burning out the brains' of children in Latin America?" The company's mission statement placed its commitment to customers first, followed by its responsibilities to employees and shareholders. And the statement affirms: "H. B. Fuller will conduct business legally and ethically... and be a responsible corporate citizen." When the company acquired its subsidiary in Honduras, the CEO at the time said:

> We were convinced that we had something to offer Latin America that the region did not have locally. In our own small way, we also wanted to be of help to that part of the world. We believed that by producing adhesives in Latin America and by employing only local people, we would create new jobs and help elevate the standard of living. We were convinced that the way to aid world peace was to help Latin America become more prosperous.

Company executives faced the dilemma of whether these expressions of H. B. Fuller's aspirations could be reconciled with the continued production of Resistol in Honduras.

Community activists in Honduras proposed the addition of oil of mustard to all solvent-based adhesives. This chemical, allyl isothiocyanate, produces a reaction that has been compared to getting an overdose of horseradish. Adding it to Resistol would effectively deter anyone attempting to inhale the fumes. However, research revealed that oil of mustard has many side effects, including severe irritation of the eyes, nose, throat, and lungs, and it can even be fatal if inhaled, swallowed, or absorbed through the skin. In addition, adhesives with oil of mustard have a shelf life of only six months. H. B. Fuller executives were convinced that the addition of oil of mustard was not an acceptable solution. However, in 1989, the Honduran legislature passed a law requiring oil of mustard, despite the lobbying efforts of H. B. Fuller.

Another alternative was a community relations effort to alert people about the dangers of glue sniffing and to address the underlying social causes. By working with community groups and the government, the company could spread the responsibility and expand its resources. On the other hand, the community groups in Honduras and elsewhere in the region were not well organized, and the government was unstable and unreliable. In 1982, the Gillette Company had faced a similar problem with its solvent-based typewriter correction fluid, Liquid Paper, which was being abused by youngsters in the United States. Gillette also rejected the possibility of adding oil of mustard, but the company's community relations effort was facilitated by the existing network of private and government-sponsored drug education programs. In Honduras, H. B. Fuller did not have the same base of community and government support. A community relations effort would be much more difficult in a less developed country.

H. B. Fuller executives also considered withdrawing all solvent-based adhesives from the market and perhaps substituting water-based products, but these alternatives were not very attractive from a business point of view. Furthermore, they would have no impact on the critical social problem of glue sniffing by street children. The waste of young lives would continue unless conditions were changed. But what could a modest-sized company located in St. Paul, Minnesota, do to address a problem caused by deep cultural, social, political, and economic forces? A failure to act, however, would seriously damage H. B. Fuller's carefully built reputation for corporate social responsibility.

Case 15.3   Shell Oil in Nigeria

On November 2, 1995, the Nigerian writer and activist Ken Saro-Wiwa was found guilty of ordering the murder of several Ogoni chiefs who were suspected of collaborating with the military government of Nigeria.[35] The Nigerian junta, headed at the time by General Sani Abacha, was criticized worldwide for bringing trumped-up charges against Saro-Wiwa and 14 co-defendants in order to suppress a resistance movement that had criticized the operations of Shell Oil Company in the oil-rich Ogoniland region of Nigeria. A specially created tribunal, widely regarded as a kangaroo court, sentenced Saro-Wiwa and six others to death by hanging (eight of the accused were acquitted). Many world leaders and human rights organizations called upon Shell Oil Company to persuade the Nigerian government not to execute Ken Saro-Wiwa, but Shell executives were reluctant to intervene. According to one company official, "It is not for a commercial organization like Shell to interfere in the legal processes of a sovereign state such as Nigeria." Another executive said, "Our responsibility is very clear. We pay taxes and [abide by] regulation. We don't run the government." However, critics charged that Shell had been actively involved all along in the military suppression of the Ogoni people. A *New York Times* editorial described Shell's position as "untenable." "If the company is determined to stay in Nigeria, it must use its considerable influence there to restrain the government."

Royal Dutch/Shell is the world's largest oil company, earning profits of $6.2 billion in 1994 on $94.9 billion in revenue. With headquarters in both London and The Hague, Netherlands, and with operations in more than one hundred countries, this joint Anglo-Dutch enterprise is truly a transnational corporation.[36] When oil was discovered in the Niger River delta in 1956, Royal Dutch/Shell was the first major company to begin production. Today, Nigeria pumps approximately two million barrels of crude oil, which provides about $10 billion a year for the military government or more than 80 percent of total government revenues. Almost half of this amount is produced by Shell Nigeria, in which the Nigerian government has a 55 percent stake and Royal Dutch/Shell a 30 percent stake (the remaining 15 percent is owned by a

French and an Italian company). The Nigerian government mainly sits back and rakes in the profits; for its efforts in Nigeria, Royal Dutch/Shell receives about $312 million in profit each year. Nigeria is, arguably, the most corrupt country in the world and, for the vast majority of its people, one of the poorest. Most Nigerians live on less than $300 a year, while the country's elite maintain lavish lifestyles. A military government that came to power in 1993 annulled that year's election for president, abolished the major democratic institutions, closed many newspapers, and silenced all opposition. The country's infrastructure continued to deteriorate because funds for public works were being siphoned off by government officials and their henchmen into guarded villas, fleets of luxury cars, and foreign bank accounts. General Abacha was rumored to have stashed more than a billion dollars abroad.

Shell's operations are centered in Ogoniland, a 400-square-mile area at the mouth of the Niger River, where approximately one-half million Ogoni live in crowded, squalid conditions among some of the world's worst pollution. Although much of the pollution is due to overpopulation, oil spills and atmospheric discharges foul the environment. The most serious oil-related harm is due to flaring, which is the burning of the natural gas that results from oil production. Flares from tall vents create an eerie orange glow in the sky and emit greenhouse gases along with heat, soot, and noise. Very little of the wealth that comes from the ground in Ogoniland reaches the local population. Between 1958 and 1994, an estimated $30 billion worth of oil was extracted from the area. Under the formula for sharing revenue with the states, the federal government was obligated to return only 1.5 percent to Ogoniland (the percentage was increased to 3 percent in 1992), but much of this revenue was diverted by corruption. Shell Nigeria also returned some profits to the region, contributing $20 million in 1995 for schools, hospitals, and other services. However, some of this money was also used to build roads to oil installations, and a World Bank study concluded that the impact of the oil company's investment on the quality of life in Ogoniland was "minimal."

Ken Saro-Wiwa was a successful Ogoni businessman turned writer and television producer who developed an interest in political activism in his late forties. He was instrumental in drafting an Ogoni Bill of Rights that demanded a "fair proportion" of oil revenues and greater environmental protection, and he helped form the Movement for the Survival of the Ogoni People (MOSOP). Saro-Wiwa traveled widely in the United States and Europe to build support for the new organization, and for his efforts he was nominated for the Nobel Peace Prize. Although MOSOP sought change from the federal government (and, to dramatize its cause, issued a symbolic demand for $10 billion in compensation from Shell), the organization opposed the use of violence and the idea of secession from Nigeria. Despite the calls for nonviolence, gangs of armed youths staged raids on oil installations, looting and vandalizing the facilities and attacking workers. In January 1993, Shell Nigeria abruptly ceased production in Ogoniland and evacuated its employees from the region. Amid growing civil unrest and military crackdowns, the leadership of MOSOP split. Ken Saro-Wiwa rejected a strategy of cooperating with the federal government to reduce violence in return for concessions. In particular, he called for a boycott of elections scheduled for June 1993, while some other leaders urged participation. On May 21, 1993, several hundred young men attacked a house where a group of dissident Ogoni chiefs (known as the "vultures") was meeting and killed most of them before the police could intervene. Although Saro-Wiwa was far away from the scene of the killing—he had been prevented from attending a MOSOP rally and had returned home—he was arrested the next day and imprisoned for eight months before his trial.

After the first attacks on Shell installations in Ogoniland, the company accepted the protection of the Nigerian police and provided some support for their services. Shell admits that the company provided firearms to the police, but human rights organizations charge that Shell-owned vehicles were used to transport police and soldiers and Shell officials participated in the planning of security operations. In an effort to induce Shell to resume operations in Ogoniland, General Abacha formed a special Internal Security Task Force to suppress opposition to the company's operations in the area. In a memo dated May 12, 1994, the commander of the task force proposed "wasting" operations to undermine the support for MOSOP and advised the government to seek prompt regular inputs from the oil companies. Shell denied making any payments for this purpose.

In May and June of 1994, the task force attacked at least 30 towns and villages, assaulting and killing the inhabitants and looting and destroying homes, fields, and livestock. By the time of Ken Saro-Wiwa's trial, Shell had not returned to Ogoniland, but its operations elsewhere in Nigeria were conducted with round-the-clock military protection. Critics generally accept Shell's arguments that withdrawing from Nigeria would harm the Nigerian people, but a *New York Times* editorial concluded that "Shell can no longer pretend that the political life of Nigeria is none of its business." To the argument that the company is merely a guest in the country, the editorial responds, "Shell, surely, has never hesitated to use its influence on matters of Nigerian tax policy, environmental rules, labor laws and trade policies."

Postscript

Eight days after being sentenced to death, Ken Saro-Wiwa was hanged along with eight others. The United States, Canada, South Africa, and many European countries withdrew their ambassadors in protest, and a consumer boycott was organized by several human rights and environmental organizations. A Shell statement expressed "deep regret" over the executions. Although the company admitted that its top official had appealed privately to General Abacha for clemency on "humanitarian grounds," a spokesperson said, "We can't issue a bold statement about human rights because … it could be considered treasonous by the regime and employees could come under attack. It would only inflame the issues." Within a week of Ken Saro-Wiwa's death, Shell announced plans for a $4 billion liquefied natural gas plant in a partnership with the Nigerian government.

NOTES

1. John H. Cushman, Jr., "Nike to Step Forward on Plant Conditions," *San Diego Union-Tribune*, 13 May 1998, A1.
2. Richard J. Barnet and John Cavanagh, *Global Dreams: Imperial Corporations and the New World Order* (New York: Simon & Schuster, 1994), 326.
3. Adam Schwarz, "Running a Business," *Far Eastern Economic Review*, 20 June 1991, 16.
4. Barnet and Cavanagh, *Global Dreams*, 326.
5. Jeffrey Ballinger, "The New Free Trade Heel," *Harper's Magazine*, August 1992, 64.
6. Norman E. Bowie, "The Moral Obligations of Multinational Corporations," in *Problems of International Justice*, ed. Steven Luper-Foy (Boulder, CO: Westview Press, 1987), 97.
7. Ralph J. Ledogar, *Hungry for Profits: U.S. and Drug Multinationals in Latin American* (New York: IDOC/North America, 1975), 46-47.
8. See Paul Steidlmeier, "Gift Giving, Bribery and Corruption: Management of Business Relationships in China," *Journal of Business Ethics*, 12 (1993), 157-64.
9. Sheila M. Puffer and Daniel J. McCarthy, "Finding Common Ground in Russian and American Business Ethics," *California Management Review*, 37 (Winter 1995), 29-46. The points in this paragraph are drawn from this article.
10. Thomas Donaldson, *The Ethics of International Business* (New York: Oxford University Press, 1989).
11. Ibid., 62.
12. Ibid., 81.
13. Ibid., 124.
14. Manuel Velasquez, "International Business Ethics: The Aluminum Companies in Jamaica," *Business Ethics Quarterly*, 5 (1995), 878.
15. Messaoud Mehafdi, "The Ethics of International Transfer Pricing," *Journal of Business Ethics*, 28 (2000), 367, 374-75.
16. Bowie, "Moral Obligations of Multinational Corporations," 529.
17. Jagdish Bhagwati and Robert E. Hudec, eds., *Fair Trade and Harmonization*, Vol. 1 (Cambridge, MA: MIT Press, 1996), 2.
18. World Bank, *Workers in an Integrating World Economy*, 75.
19. Leslie Kaufman and David Gonzalez, "Labor Standards Clash with Global Reality," *New York Times*, 24 April 2001, A1.
20. "The Invisible Children," *New York Times*, 20 February 2000, sec. 4, p. 12.
21. U.S.C. §78dd-1 (a)(1).
22. Michael Phillips, "Bribery," *Ethics*, 94 (1984), 621-36; Thomas L. Carson, "Bribery, Extortion, and the 'Foreign Corrupt Practices Act,'" *Philosophy and Public Affairs*, 14 (1985) 66-90; John Danley, "Toward a Theory of Bribery," *Business and Professional Ethics Journal*, 2 (1983) 19-39; and Kendall D'Andrade, Jr., "Bribery," *Journal of Business Ethics*, 4 (1985), 239-48.
23. See Kate Gillispie, "Middle East Response to the U.S. Foreign Corrupt Policies Act," *California Management Review*, 29 (1987), 9-30; and Paul J. Beck, Michael W. Maher, and Adrian E Tschoegl, "The Impact of the Foreign Corrupt Policies Act on U.S. Exports," *Managerial and Decision Economics*, 12 (1991), 295-303.
24. David Vogel, "The Globalization of Business Ethics: Why America Remains Distinctive," *California Management Review*, 35 (Fall 1991), 35.
25. Henk van Luijk, "Recent Developments in European Business Ethics," *Journal of Business Ethics*, 9 (1990), 542.

26. Ernest Gundling, "Ethics and Working with the Japanese: The Entrepreneur and the Elite Course," *California Management Review*, 33 (Spring 1991), 27.

27. Ibid., 31.

28. Iwao Taka and Wanda D. Foglia, "Ethical Aspects of 'Japanese Leadership Style,'" *Journal of Business Ethics*, 13 (1994), 135-48.

29. Hans Küng, "A Global Ethics in an Age of Globalization," *Business Ethics Quarterly*, 7 (1997), 18.

30. Simon Webley, "The Interfaith Declaration: Constructing a Code of Ethics for International Business, *Business Ethics: A European Review*, 5 (1996), 52-57.

31. Duane Windsor, "Toward a Transnational Code of Business Conduct," in *Emerging Global Business Ethics*, W. Michael Hoffman et al., eds. (Westport, CT: Quorum Books, 1994), 173.

32. William C. Frederick, "The Moral Authority of Transnational Corporate Codes," *Journal of Business Ethics*, 10 (1991), 165-77.

33. This position is argued in Amartya Sen, *Development as Freedom* (New York: Knopf, 1999). See also, Xiaorong Li, "A Question of Priorities: Human Rights, Development, and 'Asian Values'," *Philosophy and Public Policy*, 18 (1998), 7-12.

34. This case was adapted from "H. B. Fuller in Honduras: Street Children and Substance Abuse," prepared by Norman E. Bowie and Stefanie Ann Lenway. Copyright 1991 by Columbia Graduate School of Business.

35. This case is based on "Shell Oil in Nigeria," by Anne T. Lawrence, in *Case Research Journal*, 17 (Fall–Winter 1997), 1-21; Joshua Hammer, "Nigeria Crude," *Harper's Magazine*, June 1996, 58-68; Paul Lewis, "Rights Groups Say Shell Oil Shares Blame," *New York Times*, 11 November 1995, sec. 1, p. 6; "Shell Game in Nigeria," *New York Times*, 3 December 1995, sec. 4, p. 14; Andy Rowell, "Shell Shocked: Did the Shell Petroleum Company Silence Nigerian Environmentalist Ken Saro-Wiwa?" *The Village Voice*, 21 November 1995, 21; Andy Rowell, "Sleeping with the Enemy: Worldwide Protests Can't Stop Shell Snuggling up to Nigeria's Military," *The Village Voice*, 23 January 1996, 23; Paul Beckett, "Shell Boldly Defends Its Role in Nigeria," *Wall Street Journal*, 27 November 1995, A9.

36. Shell Oil Company U.S. is a wholly owned subsidiary of Royal Dutch/Shell. In 1995, the U.S. company had no direct operations or direct investment in Nigeria.

Index